T0235615

Chain Conditions in Commutative Rings

Ali Benhissi

Chain Conditions in
Commutative Rings

Springer

Ali Benhissi
Department of Mathematics
Faculty of Sciences of Monastir
Monastir, Tunisia

ISBN 978-3-031-10147-2   ISBN 978-3-031-09898-7   (eBook)
https://doi.org/10.1007/978-3-031-09898-7

Mathematics Subject Classification: 13E05, 13E10, 13E15, 13F25, 13F20

© The Editor(s) (if applicable) and The Author(s), under exclusive license to Springer Nature Switzerland AG 2022
This work is subject to copyright. All rights are solely and exclusively licensed by the Publisher, whether the whole or part of the material is concerned, specifically the rights of translation, reprinting, reuse of illustrations, recitation, broadcasting, reproduction on microfilms or in any other physical way, and transmission or information storage and retrieval, electronic adaptation, computer software, or by similar or dissimilar methodology now known or hereafter developed.
The use of general descriptive names, registered names, trademarks, service marks, etc. in this publication does not imply, even in the absence of a specific statement, that such names are exempt from the relevant protective laws and regulations and therefore free for general use.
The publisher, the authors, and the editors are safe to assume that the advice and information in this book are believed to be true and accurate at the date of publication. Neither the publisher nor the authors or the editors give a warranty, expressed or implied, with respect to the material contained herein or for any errors or omissions that may have been made. The publisher remains neutral with regard to jurisdictional claims in published maps and institutional affiliations.

This Springer imprint is published by the registered company Springer Nature Switzerland AG
The registered company address is: Gewerbestrasse 11, 6330 Cham, Switzerland

To Youssef and Lobna Benhissi

Preface

This book is intended to serve as a textbook for an advanced course in chains conditions on commutative rings and as a reference for researchers. It covers recent topics that are not yet exposed in books, and it is based on lectures I gave in the Faculty of Sciences of Monastir, Tunisia, during the last 5 years in the seminar of algebra, master's courses and the doctoral school. I am deeply grateful to my present and ancient students who attend these lectures, especially Dr Walid Maaref, Dr Ahmed Maatallal, Abdel Amir Debbabi, Olfa Zbidi, Sonia Chelligue and Nader Ouni, and many other colleagues and students who attend for curiosity or for their own formation. I have made use, with only silent thanks, of comments and remarks that my listeners have contributed along the way. They have been of tremendous help to me in providing corrections and in some cases simplifications and improvements to my arguments. I also want to thank my student and colleague Dr Mohamed Khalifa who helped me to write a part of his work on isonoetherian rings and my colleague Dr Samir Moulahi whose presence was a benefit for me.

Moreover, the I would like to express my thanks to the referees for their useful comments and suggestions on the first draft of this book. I am also deeply grateful to Robinson Nelson dos Santos of Springer for careful editorial assistance and Ms Saveetha Balasundaram for her precious help to improve the final version of the notes.

The original version of the book is written and exposed in French, the official language used in Tunisian universities in teaching large branches of scientific subjects. The present English version is not only a translation but also contains additional material and some modifications. The book is divided topically in six chapters. Each one of them is followed by a selection of problems with detailed solutions, of varying degrees of difficulty. They should deepen the reader's understanding of the concepts presented. The list of references used in each chapter is given at the end, and the authors of each result are included with the year of publication in order to allow the reader to consult easily the original paper. The reader can find other results that are not included in the book.

I shall now describe the content of this book. The first chapter concerns the new notion of S-Noetherian rings and modules. It begins with the main properties of these structures and then considers the classical constructions, such as polynomial and power series extensions, pullbacks, amalgamation, and idealization.

Chapter 2 deals with the dual notion of the preceding one. It treats the $S-$Artinian rings and modules. Multiplication modules are essential tools. As an application, we include the proof of the generalized principal ideal theorem, which will be frequently used. We finish with a famous example of Nagata, which constructs a Noetherian ring with infinite Krull dimension.

Chapter 3 is devoted to the almost principal polynomial rings and the crucial role played by the ideals of the form $f(X)K[X] \cap A[X]$.

In Chap. 4, we come to the SFT-notion introduced by J.T. Arnold in 1973, to study the Krull dimension of the formal power series ring. We include a new proof due to M. Roitman, 2015, and the very interesting example of Coykendall, 2002, based on the notion of infinite product of formal power series initiated by B.G. Kang and M. Park, 1999. An important place is also given to the construction of the $I-$adic completion of a ring. We add the $t-$SFT generalization.

Chapter 5 discusses the notion of nonnil-Noetherian rings and their $S-$version. Application to the formal power series rings is done.

The last chapter collects results about strongly Hophian, endo-Noetherian, PF and isonoetherian rings. The transfer of these concepts to polynomial and power series extensions is investigated. The injection in zero-dimensional rings plays an important role.

Monastir, Tunisia   Ali Benhissi
January 2022

Contents

Chapter 1
S-Noetherian Modules and Rings

All the rings considered in this chapter are commutative with unity. If no contrary mention is expressed, all the multiplicative sets are supposed containing 1 and not containing 0. The results of the five first sections are due to Anderson and Dumitrescu (2002). Noetherian rings constitute an important tool in commutative algebra. For this reason, many mathematicians lock to generalize this notion in order to generalize known results for Noetherian rings. One possible generalization is S-Noetherian modules and rings, introduced and studied for the first time, by D.D. Anderson and T. Dumitrescu.

1 S-Noetherian Modules

Definition 1.1 Let A be a ring, $S \subset A$ a multiplicative set, and M a $A-$ module. We say that M is S-finite if there is $s \in S$ and a finitely generated sub-module F of M such that $sM \subseteq F$. We say that M is S-Noetherian if all its sub-modules are S-finite.

Examples

(1) A Noetherian module is S-Noetherian.
(2) A sub-module of a S-Noetherian module is S-Noetherian.
(3) Let $S_1 \subseteq S_2$ be two multiplicative sets. A S_1-Noetherian module is S_2-Noetherian. We will see later that the converse is in general false.
(4) For any multiplicative set S of $\mathbb{Z}$, the $\mathbb{Z}$-module $\mathbb{Q}$ is not S-finite. Indeed, suppose there exist $x_1, \ldots, x_n \in \mathbb{Q}$ and $s \in S$ such that $s\mathbb{Q} \subseteq (x_1, \ldots, x_n)$. Then $\mathbb{Q} = (x_1/s, \ldots, x_n/s)$, which is absurd since $\mathbb{Q}$ is not a finitely generated $\mathbb{Z}$-module.
(5) Let $S = \{s_1, \ldots, s_n\}$ be a finite multiplicative set of a ring A and $s = s_1 \ldots s_n$. A $A-$module M is S-Noetherian if and only if for any sub-module N of M, there

© The Author(s), under exclusive license to Springer Nature Switzerland AG 2022
A. Benhissi, *Chain Conditions in Commutative Rings*,
https://doi.org/10.1007/978-3-031-09898-7_1

exist a finitely generated sub-module F of M such that $sN \subseteq F \subseteq N$. Indeed, let N be a sub-module of M. There exist $i \in \{1, \ldots, n\}$ and a finitely generated sub-module F of M such that $s_i N \subseteq F \subseteq N$. Then $sN = s_1 \ldots s_n N \subseteq s_i N \subseteq F \subseteq N$.

Lemma 1.2 *Let A be a ring, $S \subset A$ a multiplicative set, and M a S-finite $A-$module. If N is a sub-module of M maximal among the sub-modules of M which are not S-finite, then $[N : M] = \{a \in A;\ aM \subseteq N\}$ is a prime ideal of A.*

Proof It is clear that $P = [N : M]$ is an ideal of A. If $P = A$, then $1 \in [N : M]$, so $N = M$, which is S-finite and absurd. Suppose that there exist $a, b \in A \setminus P$ such that $ab \in P$. Then $aM \not\subseteq N$, $bM \not\subseteq N$, but $abM \subseteq N$. By maximality of N, $N + aM$ is S-finite. There exist $s \in S$, $n_1, \ldots, n_p \in N$, and $m_1, \ldots, m_p \in M$ such that $s(N + aM) \subseteq (n_1 + am_1, \ldots, np + am_p)$. Since $N \subset [N : a]$ because of bM, then $[N : a]$ is S-finite. There exist $t \in S$ and $q_1, \ldots, q_k \in [N : a]$ such that $t[N : a] \subseteq (q_1, \ldots, q_k)$. Then $aq_1, \ldots, aq_k \in N$. Let $x \in N$. Since $N \subseteq N + aM$, there exist $r_1, \ldots, r_p \in A$ such that $sx = r_1(n_1 + am_1) + \ldots + r_p(n_p + am_p)$. Then $a(r_1 m_1 + \ldots + r_p m_p) = sx - r_1 n_1 - \ldots - r_p n_p \in N$; therefore, the element $y = r_1 m_1 + \ldots + r_p m_p \in [N : a]$. There exist $c_1, \ldots, c_k \in A$ such that $ty = c_1 q_1 + \ldots + c_k q_k$. Then $stx = t(r_1 n_1 + \ldots + r_p n_p) + ta(r_1 m_1 + \ldots + r_p m_p) = t(r_1 n_1 + \ldots + r_p n_p) + tay = t(r_1 n_1 + \ldots + r_p n_p) + a(c_1 q_1 + \ldots + c_k q_k) = tr_1 n_1 + \ldots + tr_p n_p + c_1 a q_1 + \ldots + c_k a q_k$. Then $stN \subseteq (n_1, \ldots, n_p, aq_1, \ldots, aq_k) \subseteq N$; therefore, N is S-finite, which is absurd.   □

A famous theorem of Cohen says that a ring is Noetherian if and only if all its prime ideals are finitely generated. We have an analogue of Cohen's theorem for S-Noetherian modules.

Theorem 1.3 *Let A be a ring, $S \subset A$ a multiplicative set, and M a S-finite $A-$module. Then M is S-Noetherian if and only if for each prime ideal P of A disjoint with S, the sub-module $PM = \{p_1 m_1 + \ldots + p_k m_k;\ k \in \mathbb{N}^*,\ p_1, \ldots, p_k \in P,\ m_1, \ldots, m_k \in M\}$ of M is S-finite.*

Proof Since M is S-finite, there exist $w \in S$ and a finitely generated sub-module F of M such that $wM \subseteq F$. Suppose that M is not S-Noetherian. The set $\mathscr{F}$ of the sub-modules of M which are not S-finite is not empty. It is easy to see that $\mathscr{F}$ is inductive for the inclusion. By Zorn's lemma, $\mathscr{F}$ has a maximal element N. By the preceding lemma, $P = [N : M]$ is a prime ideal of A. We prove that $P \cap S = \emptyset$. If $s \in P \cap S$, then $sM \subseteq N$, so $swN \subseteq swM \subseteq sF \subseteq sM \subseteq N$, and $swN \subseteq sF \subseteq N$ with $sw \in S$ and sF a finitely generated sub module, which contradicts $N \in \mathscr{F}$. We prove that $P = [N : F]$. By definition, $P = [N : M] \subseteq [N : F] \subseteq [N : wM] = P$, since $a \in [N : wM] \Longleftrightarrow (aw)M \subseteq N \Longleftrightarrow aw \in [N : M] = P \Longleftrightarrow a \in P$ because P is a prime ideal of A disjoint with S. Then $P = [N : F]$. Put $F = (f_1, \ldots, f_k)$, then $P = [N : F] = [N : f_1] \cap \ldots \cap [N : f_k]$. Since P is a prime ideal, there exist i such that $P = [N : f_i]$, let $f_i = g$, then $P = [N : g]$. If $g \in N$, then $P = [N : g] = A$, which is absurd. Then $g \notin N$. By maximality of N in $\mathscr{F}$, the sub-module $N + Ag$ is S-finite. There exist $t \in S$, $n_1, \ldots, n_p \in N$, and $a_1, \ldots, a_p \in A$ such that $t(N + Ag) \subseteq (n_1 + a_1 g, \ldots, n_p + a_p g)$. Put $N' = (n_1, \ldots, n_p) \subseteq N$. Let $x \in N$. Since $N \subseteq N + Ag$, there exist $r_1, \ldots, r_p \in A$ such that $tx = r_1(n_1 + a_1 g) + \ldots + r_p(n_p + a_p g) = r_1 n_1 + \ldots + r_p n_p + (r_1 a_1 + \ldots + r_p a_p)g$. Then $(r_1 a_1 + \ldots + r_p a_p)g = tx - r_1 n_1 - \ldots - r_p n_p \in N$ so $r_1 a_1 + \ldots + r_p a_p \in [N : g] = P$. Then $tx = r_1 n_1 + \ldots + r_p n_p + (r_1 a_1 + \ldots + r_p a_p)g \in N' + Pg \subseteq N' + PM$. Then $tN \subseteq N' + PM$. By hypothesis, PM is S-finite, and there exist $v \in S$ and a finitely generated sub-module G such that $v(PM) \subseteq G \subseteq PM \subseteq N$, by definition of P. Then $vtN \subseteq vN' + v(PM) \subseteq vN' + G \subseteq N$. Finally, $vtN \subseteq vN' + G \subseteq N$ with $vt \in S$ and $vN' + G$ a finitely generated sub-module, which contradicts the fact that N is not S-finite.   □

2 S-Noetherian Rings

Definition 2.1 Let A be a ring and $S \subseteq A$ a multiplicative set. An ideal I of A is said to be S-finite if it is S-finite as a $A-$module; i.e., there exist $s \in S$ and a finitely generated ideal J of A such that $sI \subseteq J \subseteq I$. We say that the ring A is S-Noetherian if all its ideals are S-finite; i.e., the $A-$module A is S-Noetherian.

Example 1 A Noetherian ring is S-Noetherian.

Example 2 Let S be a multiplicative set formed by unities of a ring A. Then A is S-Noetherian if and only if it is Noetherian. Indeed, let I be an ideal of A. There exist $s \in S$ and a finitely generated ideal F of A such that $sI \subseteq F \subseteq I$. Since s is a unit, then $sI = I$, and $I = F$.

Example 3 Let S be multiplicative set of a ring A and I an ideal of A. If $I \cap S \neq \emptyset$, then I is S-finite. In particular, an integral domain A is $A \setminus (0)$-Noetherian. Indeed, let $s \in I \cap S$. Then $sI \subseteq sA \subseteq I$.

The following two examples are due to Baeck et al. (2016). They show that S-Noetherian rings have their own benefits. For example, in S-Noetherian rings, by using multiplicative sets, we can control the infinite product of rings and infinitely generated ideals, while Noetherian rings do not contain infinitely generated ideals, and the infinite product of Noetherian rings is never a Noetherian ring.

Example 4 Let p be a prime number and $A = \prod_{n:1}^{\infty} \mathbb{Z}/p^n \mathbb{Z}$, the product ring. The ideal of A formed by the elements of finite supports is not finitely generated, so A is not Noetherian. Let $s = (\bar{1}, \bar{p}, \bar{p}, \bar{0}, \ldots) \in A$. Then $s^2 = (\bar{1}, \bar{0}, \bar{p}^2, \bar{0}, \ldots)$, $s^3 = (\bar{1}, \bar{0}, \bar{0}, \bar{0}, \ldots)$ and for each integer $n \geq 3$, $s^n = s^3$. The set $S = \{1, s, s^2, s^3\}$ of A is multiplicative. Let I be any ideal of A. An element of sI is of the type $(\bar{a}_1, \bar{a}_2, \bar{a}_3, \bar{0}, \ldots)$ with $\bar{a}_i \in \mathbb{Z}/p^i \mathbb{Z}$, $1 \leq i \leq 3$. Then the ideal sI is finite, so finitely generated. It is also contained in I. Then I is S-finite and the ring A is S-Noetherian.

To continue with this example, let's recall that in a Noetherian ring B, if I is a nil ideal, i.e., $I \subseteq Nil(B)$, there exists an integer $n \geq 1$ such that $I^n = (0)$. This property is not true in a S-Noetherian ring. Let J be the ideal of the ring A formed by the elements of finite supports and $I = pJ$. An element of I is of the form $x = (pa_1, \ldots, pa_k, 0, \ldots)$ with $k \in \mathbb{N}^*$ and $a_i \in \mathbb{Z}/p^i\mathbb{Z}$ for $1 \leq i \leq k$.

Then $x^k = (p^k a_1^k, \ldots, p^k a_k^k, 0, \ldots) = 0$. So $I \subseteq Nil(A)$.

For $n \in \mathbb{N}^*$, let $e_{n+1} = (\bar{1}, \ldots, \bar{1}, \bar{0}, \ldots)$, the element of J with the $n+1$ first components, be equal to the unity. Then $(pe_{n+1})^n = (\bar{0}, \ldots, \bar{0}, \bar{p}^n, \bar{0}, \ldots) \neq 0$ because $\bar{p}^n \neq \bar{0}$ in $\mathbb{Z}/p^{n+1}\mathbb{Z}$.

Example 5 Let $(p_n)_{n \in \mathbb{N}}$ be the sequence of the prime numbers. The product ring $A = \prod_{n:0}^{\infty} \mathbb{F}_{p_n}$ is not Noetherian because we have the infinite strictly increasing sequence of ideals: $\mathbb{F}_{p_0} \times 0 \times \ldots \subset \mathbb{F}_{p_0} \times \mathbb{F}_{p_1} \times 0 \times \ldots \subset \mathbb{F}_{p_0} \times \mathbb{F}_{p_1} \times \mathbb{F}_{p_2} \times 0 \ldots \subset \ldots$. Let $A^* = \prod_{n:0}^{\infty} \mathbb{F}_{p_n}^*$, the group of units of A, and fix a nonzero element a of A of finite support. Then $S = \{1\} \cup aA^*$ is a multiplicative set of A. Indeed, it is clear that $0 \notin S$. Let $s_1, s_2 \in S \setminus \{1\}$, $s_1 = ab$ and $s_2 = ac$ with $b, c \in A^*$. Then $s_1 s_2 = a(abc)$ with $bc \in A^*$. Let's define the element a' of A^* by its components as follows: if $a_{p_n} \neq 0$, put $a'_{p_n} = a_{p_n}$ and if $a_{p_n} = 0$, put $a'_{p_n} = 1$. Then $a'bc \in A^*$ and $s_1 s_2 = a(abc) = a(a'bc) \in aA^*$. We prove that the ring A is S-Noetherian. Let I be any ideal of A. The element $a \in S$ and the ideal $F = aI$ of A is contained in I. It is finite, so finitely generated, and then I is S-finite. Indeed, let $p_{k_1}, \ldots, p_{k_n}$ the support of a. Let x be any element of aI. For each prime number $p \notin \{p_{k_1}, \ldots, p_{k_n}\}$, the component x_p of x is zero. For $i \in \{1, \ldots, n\}$, the component of index p_{k_i} of x is in the finite field $\mathbb{F}_{p_i}$. Then F is finite.

To continue with this example, consider the multiplicative set $S' = \{1\}$ of A. Then $S' \subset S$, and the ring A is not S'−Noetherian because it is not Noetherian. But it is S-Noetherian.

In the sequel, we give some corollaries for Theorem 1.3.

Corollary 2.2 *Let A be a ring and $S \subseteq A$ a multiplicative set. Then A is S-Noetherian if and only if each prime ideal of A disjoints with S is S-finite.*

Example (Cohen's Theorem) The ring A is Noetherian if and only if each prime ideal of A is finitely generated. Take $S = \{1\}$.

Corollary 2.3 *Let $A \subseteq B$ be an extension of rings and S a multiplicative set of A. If B is a S-finite A−module and if for each prime ideal P of A disjoint with S, PB is a S-finite ideal of B, then A is a S-Noetherian ring.*

Proof Since A is $A-$sub-module of B, it suffices to prove that B is a $A-$module S-Noetherian. Since B is a S-finite $A-$module, there exist $s \in S$ and $b_1, \ldots, b_n \in B$ such that $sB \subseteq (b_1, \ldots, b_n)A$. Let P be a prime ideal of A disjoint with S. By hypothesis, PB is a S-finite ideal of B. There exist $w \in S$ and $p_1, \ldots, p_m \in P$ such that $w(PB) \subseteq (p_1, \ldots, p_m)B$. Then $swPB \subseteq (p_1, \ldots, p_m)sB \subseteq (p_1, \ldots, p_m)(b_1, \ldots, b_n)A = (p_i b_j; \ 1 \le i \le m, 1 \le j \le n)A$. Then PB is a S-finite $A-$sub-module of B. By Theorem 1.3, B is a S-Noetherian $A-$module. $\square$

Corollary 2.4 *Let $A \subseteq B$ be an extension of rings and S a multiplicative set of A such that B is a S-finite $A-$module. If B is a S-Noetherian ring, then it is the same for A.*

Proof Let P be a prime ideal of A. The ideal PB of the ring B is S-finite. We conclude by the preceding corollary. $\square$

We will prove later that the converse is also true.

Proposition 2.5 *Let $A \subseteq B$ be an extension of rings such that $IB \cap A = I$ for each ideal I of A and let $S \subseteq A$ a multiplicative set. If B is a S-Noetherian ring, then it is the same for A.*

Proof Let I be an ideal of A. Since the ring B is S-Noetherian, there exist $s \in S$ and $g_1, \ldots, g_n \subset I$ such that $sIB \subseteq (g_1, \ldots, g_n)B$. By hypothesis, $sI = sIB \cap A \subseteq (g_1, \ldots, g_n)B \cap A = (g_1, \ldots, g_n)A$. Then I is a S-finite ideal of A. $\square$

Example Let $A \subset B$ be an extension of rings and I an ideal of A. The equality $IB \cap A = I$ is in general false. Take the extension $A = \mathbb{Z} \subset B = \mathbb{Q}$ and $I = 2\mathbb{Z}$. Then $IB = 2\mathbb{Q} = \mathbb{Q}$, so $IB \cap A = \mathbb{Q} \cap \mathbb{Z} = \mathbb{Z} \ne 2\mathbb{Z} = I$.

Proposition 2.6 *A ring A is Noetherian if and only if for each maximal ideal M of A, A is $(A \setminus M)-$Noetherian.*

Proof Let I be an ideal of A. For each maximal ideal M of A, there exist $s_M \in A \setminus M$ and a finitely generated ideal F_M of A such that $s_M I \subseteq F_M \subseteq I$. If the ideal $J = (s_M; \ M \in Max(A)) \ne A$, there exist $M_0 \in Max(A)$ such that $J \subseteq M_0$, then $s_{M_0} \in J \subseteq M_0$, which is absurd. Then $J = A$ and there exist $M_1, \ldots, M_n \in Max(A)$ such that $(s_{M_1}, \ldots, s_{M_n}) = A$. Then $I = AI = (s_{M_1}, \ldots, s_{M_n})I \subseteq F_{M_1} + \ldots + F_{M_n} \subseteq I$ so $I = F_{M_1} + \ldots + F_{M_n}$ is a finitely generated ideal. Then A is Noetherian. $\square$

3 Modules Over *S*-Noetherian Rings

The following theorem characterizes S-Noetherian modules.

Theorem 3.1 *Let A be a ring, $S \subseteq A$ a multiplicative set, M a $A-$module, and N a sub-module of M. The following assertions are equivalent:*

1. *The $A-$modules N and M/N are S-Noetherian.*
2. *The $A-$module M is S-Noetherian.*

Proof $''(2) \implies (1)''$ A sub-module of N is a sub-module of M, so S-finite. A sub-module of M/N is of the form T/N with T a sub module of M containing N. There exist $s \in S$ and $t_1, \ldots, t_n \in T$ such that $sT \subseteq (t_1, \ldots, t_n)$. Then $s(T/N) \subseteq (\bar{t}_1, \ldots, \bar{t}_n)$.

$''(1) \implies (2)''$ Let $\pi : M \longrightarrow M/N$ the canonical map. Let T be a sub-module of M. Then $\pi(T)$ is a sub-module of M/N. There exist $s \in S$ and $t_1, \ldots, t_r \in T$ such that $s\pi(T) \subseteq (\bar{t}_1, \ldots, \bar{t}_r)$. Since $T \cap N$ is a sub-module of N, it is S-finite. There exist $s' \in S$ and $n_1, \ldots, n_k \in T \cap N$ such that $s'(T \cap N) \subseteq (n_1, \ldots, n_k)$. Then $(t_1, \ldots, t_r, n_1, \ldots, n_k) \subseteq T$. Let $x \in T$. Then $\bar{x} \in \pi(T)$. There exist $\alpha_1, \ldots, \alpha_r \in A$ such that $s\bar{x} = \sum_{i:1}^{r} \alpha_i \bar{t}_i$, and then $sx - \sum_{i:1}^{r} \alpha_i t_i \in N \cap T$. There exist $\beta_1, \ldots, \beta_k \in A$ such that $s'(sx - \sum_{i:1}^{r} \alpha_i t_i) = \sum_{i:1}^{k} \beta_i n_i$, then $s'sx = s'(\sum_{i:1}^{r} \alpha_i t_i) + \sum_{i:1}^{k} \beta_i n_i \in (t_1, \ldots, t_r, n_1, \ldots, n_k)$. So $(s's)T \subseteq (t_1, \ldots, t_r, n_1, \ldots, n_k) \subseteq T$. $\square$

Example Let A be a ring, M an $A-$module, and N a sub-module of M. The module M is Noetherian if and only if the $A-$modules N and M/N are Noetherian.

In the following corollary, we give another characterization of S-Noetherian modules with some specific sub-module.

Corollary 3.2 (Kim and Lim 2020) *Let A be a ring, S a multiplicative set of A, M a $A-$module, and $N = \bigcap \{PM; \; P \in Spec(A)\}$.*
The following assertions are equivalent:

1. *The $A-$module M/N is S-Noetherian and N is a S-finite $A-$module.*
2. *The $A-$module M is S-Noetherian.*

Proof $''(2) \implies (1)''$ Apply the preceding theorem.

$''(1) \implies (2)''$ Since N is a S-finite $A-$module, there exist $s \in S$ and $n_1, \ldots, n_l \in N$ such that $sN \subseteq (n_1, \ldots, n_l)$. Since M/N is a S-Noetherian $A-$module, there exist $t_1 \in S$ and $m_1, \ldots, m_k \in M$ such that $t_1(M/N) \subseteq (\bar{m}_1, \ldots, \bar{m}_k)$, and then $t_1 M \subseteq (m_1, \ldots, m_k) + N$. Then $st_1 M \subseteq s(m_1, \ldots, m_k) + sN \subseteq (sm_1, \ldots, sm_k, n_1, \ldots, n_l)$. Then M is a S-finite $A-$module. Let P be a prime ideal of A. By definition of N, PM is a sub-module of M containing N. Then PM/N is a sub-module of M/N. Since M/N is a S-Noetherian $A-$module, there exist $t_2 \in S$ and $x_1, \ldots, x_q \in PM$ such that $t_2(PM/N) \subseteq (\bar{x}_1, \ldots, \bar{x}_q)$, and then $t_2(PM) \subseteq (x_1, \ldots, x_q) + N$ so $st_2(PM) \subseteq s(x_1, \ldots, x_q) + sN \subseteq (sx_1, \ldots, sx_q, n_1, \ldots, n_l) \subseteq PM$. Then PM is a S-finite $A-$module. By Theorem 1.3, the $A-$module M is S-Noetherian. $\square$

Let A be a ring and S a multiplicative set of A. The nil radical of A is $Nil(A) = \cap\{P; P \in Spec(A)\}$. Then $\bar{S} = \{\bar{s}; s \in S\}$ is a multiplicative set of the quotient ring $A/Nil(A)$. Indeed, $\bar{0} \notin \bar{S}$ since $S \cap Nil(A) = \emptyset$. In the following corollary, we give a characterization of the S-Noetherian rings in terms of $Nil(A)$, analogue to that of S-Noetherian modules.

Corollary 3.3 (Kim and Lim 2020) *Let A be a ring and S a multiplicative set of A. The following assertions are equivalent:*

1. *The ring A is S-Noetherian.*
2. *The $A-$module $A/Nil(A)$ is S-Noetherian and the ideal $Nil(A)$ of A is S-finite.*
3. *The ring $A/Nil(A)$ is $\bar{S}-$Noetherian and the ideal $Nil(A)$ of A is S-finite.*

Proof The equivalence $"(1) \Longleftrightarrow (2)"$ comes from the preceding corollary.

$"(1) \Longrightarrow (3)"$ By hypothesis, the ideal $Nil(A)$ is S-finite. An ideal of the ring $A/Nil(A)$ is of the form $I/Nil(A)$ with I an ideal of A containing $Nil(A)$. There exist $s \in S$ and $a_1, \ldots, a_n \in I$ such that $sI \subseteq (a_1, \ldots, a_n)$. Then $\bar{s}(I/Nil(A)) \subseteq (\bar{a}_1, \ldots, \bar{a}_n) \subseteq I/Nil(A)$. It follows that the ring $A/Nil(A)$ is $\bar{S}-$Noetherian.

$"(3) \Longrightarrow (2)"$ By hypothesis, the ring $A/Nil(A)$ is $\bar{S}-$Noetherian. Then it is $\bar{S}-$finite, as an ideal of $A/Nil(A)$. There exist $s \in S$ and $a_1, \ldots, a_n \in A$ such that $\bar{s}(A/Nil(A)) \subseteq (\bar{a}_1, \ldots, \bar{a}_n)(A/Nil(A))$. Then $s(A/Nil(A)) \subseteq A(\bar{a}_1, \ldots, \bar{a}_n)$. So $A/Nil(A)$ is a S-finite $A-$module. Let P be a prime ideal of A. By hypothesis, the ideal $P/Nil(A)$ of the ring $A/Nil(A)$ is S-finite. There exist $t \in S$ and $p_1, \ldots, p_n \in P$ such that $\bar{t}(P/Nil(A)) \subseteq (\bar{p}_1, \ldots, \bar{p}_n)(A/Nil(A))$. Then $tP(A/Nil(A)) \subseteq A(\bar{p}_1, \ldots, \bar{p}_n)$ so the $A-$sub-module $P(A/Nil(A))$ of $A/Nil(A)$ is S-finite. By Theorem 1.3, the $A-$module $A/Nil(A)$ is S-Noetherian. □

Example The following assertions are equivalent for a ring A:

(1) A is Noetherian.
(2) The $A-$module $A/Nil(A)$ is Noetherian and the ideal $Nil(A)$ is finitely generated.
(3) The quotient ring $A/Nil(A)$ is Noetherian and the ideal $Nil(A)$ is finitely generated.

Lemma 3.4 *Let A be a ring, $S \subseteq A$ a multiplicative set, and $n \in \mathbb{N}^*$. If M is a S-Noetherian $A-$module, then it is the same for the product $A-$module M^n.*

Proof By induction on n. If $n = 1$, the $A-$module M is S-Noetherian. Suppose that the $A-$module M^n is S-Noetherian. The map $f : M^{n+1} \longrightarrow M$, given by $f(x_1, \ldots, x_{n+1}) = x_1$, is onto homomorphism of $A-$modules, with kernel $ker\, f = \{0\} \times M^n \simeq M^n$, which is a S-Noetherian $A-$module, by induction hypothesis. By isomorphism theorem $M^{n+1}/ker\, f \simeq M$. By Theorem 3.1, M^{n+1} is a S-Noetherian $A-$module. □

Example Let A be a ring, $S \subseteq A$ a multiplicative set, and $n \in \mathbb{N}^*$. If A is a S-Noetherian ring, the $A-$module A^n is S-Noetherian.

Theorem 3.5 *Let A be a ring and $S \subseteq A$ a multiplicative set such that A is a S-Noetherian ring. If M is a finitely generated $A-$module, then M is S-Noetherian.*

Proof Put $M = (x_1, \ldots, x_n)$. The map $\phi : A^n \longrightarrow M$, defined by $\phi(a_1, \ldots, a_n) = \sum_{i:1}^{n} a_i x_i$, is a homomorphism onto of $A-$modules, so $M \simeq A^n / ker\ \phi$. By the preceding lemma (or Example), the $A-$module A^n is S-Noetherian. By Theorem 3.1, the quotient $A-$ module $A^n / ker\ \phi$ is S-Noetherian. $\square$

Example Let S be a multiplicative set of a ring A and $n \in \mathbb{N}^*$. If A is S-Noetherian, the $A-$module $\mathbb{M}_n(A)$, of $n \times n$ matrix with coefficients in A, is S-Noetherian. In fact, $\mathbb{M}_n(A)$ is generated by the elementary matrix E_{ij}, $1 \leq i,\ j \leq n$. We can replace $\mathbb{M}_n(A)$ by other types of matrix, like upper triangular matrix, diagonal matrix, etc.

The following two corollaries are due to Hamed and Hizem (2015) and Baeck et al. (2016). The hypothesis "M is finitely generated" is weaken to "M is S-finite".

Corollary 3.6 *Let A be a ring and S a multiplicative set of A such that A is S-Noetherian. If M is a S-finite $A-$module, then M is S-Noetherian.*

Proof There exist $s \in S$ and a finitely generated sub-module F of M such that $sM \subseteq F$. By the preceding theorem, F is a S-Noetherian $A-$module. Let N be any sub-module of M. Then $sN \subseteq sM \subseteq F$. Then sN is a sub-module of F, so S-finite. There exist $t \in S$ and a finitely generated sub-module P of M such that $t(sN) \subseteq P \subseteq sN \subseteq N$. Then $(ts)N \subseteq P \subseteq N$ with $ts \in S$, so N is S-finite and M is S-Noetherian. $\square$

Other Proof By Theorem 1.3, it suffices to show that for any prime ideal P of A disjoint with S, the sub-module PM of M is S-finite. But the ring A is S-Noetherian, so the ideal P is S-finite. There exist $s \in S$ and $p_1, \ldots, p_n \in P$ such that $sP \subseteq p_1A + \ldots + p_nA$. By hypothesis, M is S-finite. There exist $s' \in S$ and $x_1, \ldots, x_k \in M$ such that $s'M \subseteq Ax_1 + \ldots + Ax_k$. Then $ss'(PM) = (sP)(s'M) \subseteq (p_1A + \ldots + p_nA)(Ax_1 + \ldots + Ax_k) = (p_i x_j; 1 \leq i \leq n, 1 \leq j \leq k) \subseteq PM$. Then PM is S-finite. $\square$

Example A finitely generated module over a Noetherian ring is Noetherian.

In the following corollary, we prove the converse of Corollary 2.4.

Corollary 3.7 *Let $A \subseteq B$ be an extension of rings and $S \subseteq A$ a multiplicative set. Suppose that B is a S-finite $A-$module. Then the ring A is S-Noetherian if and only if the ring B is S-Noetherian.*

Proof " $\Longleftarrow$ " By Corollary 2.4.

$'' \Longrightarrow ''$ By the preceding corollary, B is a S-Noetherian $A-$module. Let I be an ideal of B. Then I is a $A-$sub-module of B and then S-finite. There exist $s \in S$ and $a_1, \ldots, a_n \in I$ such that $sI \subseteq Aa_1 + \ldots + Aa_n \subseteq I$. Then $sI \subseteq Ba_1 + \ldots + Ba_n \subseteq I$ because I is an ideal of B. Then I is a S-finite ideal of B and the ring B is S-Noetherian. $\qquad\qquad\square$

Example 1 (Eakin-Nagata Theorem) Let $A \subseteq B$ be an extension of rings such that B is a finitely generated $A-$module. Then A is Noetherian if and only if B is Noetherian.

Example 2 The hypothesis "B is a S-finite $A-$module" is necessary in the two ways of the preceding corollary. Indeed, let $A \subseteq B$ be an extension of rings. Then A may be Noetherian, but B is not. For example, for any Noetherian ring A, the polynomial ring $B = A[X_i, i \in \mathbb{N}]$, in infinitely many variables, is not Noetherian. The ring B may be Noetherian, but A is not Noetherian, even when B is an over ring of an integral domain A. For example, $B = \mathbb{Q}[X]$ and $A = \mathbb{Z} + X\mathbb{Q}[X]$. The sequence $\left(\frac{X}{2^n} A \right)_{n \in \mathbb{N}}$ of principal ideals of A is strictly increasing. Indeed, $\frac{X}{2^n} = 2(\frac{X}{2^{n+1}})$. Suppose there exists $n \in \mathbb{N}$ such that $\frac{X}{2^n} A = \frac{X}{2^{n+1}} A$. There exists $f(X) \in A$ such that $\frac{X}{2^{n+1}} = \frac{X}{2^n} f(X)$. Then $f(X) = \frac{1}{2} \notin A$, which is absurd.

Other nontrivial examples, constructed with idealization, will be given in Sect. 8, showing that "B is a S-finite $A-$module" is a necessary condition in the two ways of the equivalence in the preceding corollary.

The remaining of the results in this section are due to Kim and Lim (2020). We summarize many results previously proved in the following corollary.

Corollary 3.8 *Let $A \subseteq B$ be an extension of rings and S a multiplicative set of A. Suppose that B is a S-finite $A-$module. The following assertions are equivalent:*

1. *The ring A is S-Noetherian.*
2. *The ring B is S-Noetherian.*
3. *For any prime ideal P of A, the ideal PB of B is S-finite.*
4. *The $A-$module B is S-Noetherian.*

Proof The equivalence $''(1) \Longleftrightarrow (2)''$ comes from the preceding corollary.

$''(2) \Longrightarrow (3)''$ By definition of a S-Noetherian ring.

$''(3) \Longrightarrow (4)''$ By Theorem 1.3, it suffices to show that if P is a prime ideal of A, the $A-$sub-module PB of B is S-finite. By hypothesis, there exist $s_1 \in S$ and $p_1, \ldots, p_m \in P$ such that $s_1(PB) \subseteq (p_1, \ldots, p_m)B$. Since B is a S-finite $A-$module, there exist $s_2 \in S$ and $b_1, \ldots, b_n \in B$ such that $s_2 B \subseteq (b_1, \ldots, b_n)A$. Then $s_1 s_2(PB) \subseteq (p_1, \ldots, p_m)s_2 B \subseteq (p_1, \ldots, p_m)(b_1, \ldots, b_n)A = (p_i b_j; 1 \le i \le m, 1 \le j \le n)A \subseteq PB$.

$''(4) \Longrightarrow (1)''$ Since A is a sub $A-$module of B, then A is S-Noetherian as $A-$module, so as a ring. $\qquad\qquad\square$

The following proposition generalizes the implication $''(3) \Longrightarrow (4)''$ in the preceding corollary with an analogue proof.

Proposition 3.9 *Let $A \subseteq B$ be an extension of rings, S a multiplicative set of A, and M a B-module. Suppose that M is S-finite as a A-module and for any prime ideal P of A, the ideal PB of B is S-finite. Then M is a S-Noetherian A-module.*

Proof By Theorem 1.3, it suffices to show that if P is a prime ideal of A, the sub-module PM of M is S-finite. By hypothesis, the ideal PB of B is S-finite. There exist $s_1 \in S$ and $p_1, \ldots, p_l \in P$ such that $s_1(PB) \subseteq (p_1, \ldots, p_l)B$. Since M is a S-finite A-module, there exist $s_2 \in S$ and $m_1, \ldots, m_k \in M$ such that $s_2 M \subseteq A(m_1, \ldots, m_k)$. Since $BM = M$, then $PM = P(BM) = (PB)M$. Then $s_2 s_1 (PM) = s_2 s_1 (PB)M \subseteq s_2(p_1, \ldots, p_l)BM = s_2 A(p_1, \ldots, p_l)BM = s_2 A(p_1, \ldots, p_l)M = A(p_1, \ldots, p_l)s_2 M \subseteq A(p_1, \ldots, p_l)A(m_1, \ldots, m_k) = A(p_i m_j; \ 1 \leq i \leq l, 1 \leq j \leq k) \subseteq PM$. □

The following result is classical.

Lemma 3.10 *Let A be a ring, N a A-module, and $k \in \mathbb{N}^*$. Then $Hom_A(A^k, N) \simeq N^k$.*

Proof Let $\Psi : \big(Hom_A(A, N)\big)^k \longrightarrow Hom_A(A^k, N)$ be the map, whose associates to each element $(f_1, \ldots, f_k) \in \big(Hom_A(A, N)\big)^k$ the homomorphism $f : A^k \longrightarrow N$, defined by $f(a_1, \ldots, a_k) = f_1(a_1) + \ldots + f_k(a_k)$ for all $(a_1, \ldots, a_k) \in A^k$. Then Ψ is an isomorphism of A-modules. Indeed, if $f = 0$, for all $i \in \{1, \ldots, n\}$ and $a_i \in A$, $0 = f(0, \ldots, 0, a_i, 0, \ldots, 0) = f_i(a_i)$, then $f_i = 0$. So Ψ is one to one. Let $f \in Hom_A(A^k, N)$. We define $(f_1, \ldots, f_k) \in \big(Hom_A(A, N)\big)^k$ by $f_i(a_i) = f(0, \ldots, 0, a_i, 0, \ldots, 0)$ for $1 \leq i \leq k$ and $a_i \in A$. Then $\Psi(f_1, \ldots, f_k) = f$, so Ψ is onto. So we have the isomorphism $Hom_A(A^k, N) \simeq \big(Hom_A(A, N)\big)^k$. The dual $Hom_A(A, N) \simeq N$, and then $\big(Hom_A(A, N)\big)^k \simeq N^k$. Then $Hom_A(A^k, N) \simeq N^k$. □

Proposition 3.11 *Let S be a multiplicative set of a ring A, M a finitely generated A-module, and N a S-Noetherian A-module. Then $Hom_A(M, N)$ is a S-Noetherian A-module.*

Proof Put $M = (m_1, \ldots, m_k)$. The map $\phi : A^k \longrightarrow M$, defined by $\phi(a_1, \ldots, a_k) = a_1 m_1 + \ldots + a_k m_k$ for each $(a_1, \ldots, a_k) \in A^k$, is a homomorphism onto of A-modules. The map $\Phi : Hom_A(M, N) \longrightarrow Hom_A(A^k, N)$, defined by $\Phi(f) = f \circ \phi$ for each element $f \in Hom_A(M, N)$, is a homomorphism one to one of A-modules. Indeed, if $\Phi(f) = f \circ \phi = 0$, then $\phi(A^k) \subseteq ker\ f$. Since ϕ is onto, $f(M) = (0)$ and $f = 0$. So $Hom_A(M, N) \simeq \Phi\big(Hom_A(M, N)\big) \subseteq Hom_A(A^k, N) \simeq N^k$, by the preceding lemma. By Lemma 3.4, N^k is S-Noetherian. Then $Hom_A(M, N)$ is a S-Noetherian A-module. □

Example The hypothesis "M is finitely generated" is necessary in the preceding proposition. Let p be a prime number; $A = \prod_{n:1}^{\infty} \mathbb{Z}/p^n\mathbb{Z}$, the product ring; and $M =$

$A^{(\mathbb{N}^*)}$, the set of the sequences with finite supports of elements of A. Then M is a free A−module with canonical basis $(e_n)_{n \in \mathbb{N}^*}$, with $e_n = (0, \ldots, 0, 1, 0, \ldots)$, where 1 is the n-th coordinate of e_n. Let $s = (\bar{1}, \bar{p}, \bar{p}, \bar{0}, \ldots) \in A$. We know that $S = \{1, s, s^2, s^3\}$ is a multiplicative set of the ring A.

(1) The A−module M is not S-finite, so it is not finitely generated.
  Indeed, suppose that there exist $t \in S$ and $f_1, \ldots, f_n \in M$ such that $tM \subseteq (f_1, \ldots, f_n)$. Let the element $f = (1, \ldots, 1, 0, \ldots) \in M$ with a finite support but larger than those of $f_1, \ldots, f_n$. Then $tf = (t, \ldots, t, 0, \ldots) \notin (f_1, \ldots, f_n)$, which is absurd.

(2) We know that if the ring A is S-Noetherian, then it is a S-Noetherian A−module.

(3) We have an isomorphism of A−modules $Hom_A(M, A) \simeq A^{\mathbb{N}^*}$.
  Indeed, the map $\phi : Hom_A(M, A) \longrightarrow A^{\mathbb{N}^*}$, defined by $\phi(f) = \big(f(e_n)\big)_{n \in \mathbb{N}^*}$ for each $f \in Hom_A(M, A)$, is an isomorphism of A−modules because M is a free A−module with basis $(e_n)_{n \in \mathbb{N}^*}$.

Finally, the A−module $A^{\mathbb{N}^*}$ is not S-Noetherian because its sub-module $M = A^{(\mathbb{N}^*)}$ is not S-finite.

The hypothesis on M in the preceding proposition may be weaken if all the elements of the multiplicative set S are idempotent; i.e., for each $s \in S, s^2 = s$.

Proposition 3.12 *Let A be a ring, S a multiplicative set of A formed by idempotent elements, M a S-finite A−module, and N a S-Noetherian A−module.*
  Then $Hom_A(M, N)$ is a S-Noetherian A−module.

Proof Since M is S-finite, there exist $s \in S$ and $m_1, \ldots, m_k \in M$ such that $sM \subseteq (m_1, \ldots, m_k)$. Let $\phi : A^k \longrightarrow sM$ the homomorphism defined by $\phi(a_1, \ldots, a_k) = s(a_1 m_1 + \ldots + a_k m_k)$ for each $(a_1, \ldots, a_k) \in A^k$. Then ϕ is onto. Indeed, let $x \in sM, x = sm = s^2 m$ with $m \in M$. Put $sm = a_1 m_1 + \ldots + a_k m_k$ with $a_1, \ldots, a_k \in A$. Then $x = s.sm = s(a_1 m_1 + \ldots + a_k m_k) = \phi(a_1, \ldots, a_k)$.

The map $\Phi : Hom_A(sM, N) \longrightarrow Hom_A(A^k, N)$, defined by $\Phi(f) = f \circ \phi$ for each $f \in Hom_A(sM, N)$, is a homomorphism one to one of A−modules. Indeed, if $\Phi(f) = f \circ \phi = 0$, then $\phi(A^k) \subseteq ker\ f$. But ϕ is onto, and then $f(sM) = (0)$ and $f = 0$. Then $Hom_A(sM, N) \simeq \Phi\big(Hom_A(sM, N)\big) \subseteq Hom_A(A^k, N) \simeq N^k$, by Lemma 3.10. But N^k is S-Noetherian, and then $Hom_A(sM, N)$ is S-Noetherian.

Let $\Psi : Hom_A(sM, N) \longrightarrow s Hom_A(M, N)$, the map which associate to each element $f \in Hom_A(sM, N)$, $\Psi(f) = sg$ with $g : M \longrightarrow N$, the homomorphism defined by $g(m) = f(sm)$ for each $m \in M$. Then Ψ is an isomorphism of A−modules. Indeed, if $\Psi(f) = sg = 0$, then for each $m \in M$, $f(sm) = f(s^2 m) = sf(sm) = sg(m) = 0$. Then $f = 0$ and Ψ is one to one. Let $g \in Hom_A(M, N)$ and $f : sM \longrightarrow N$ be the restriction of g to sM. Then $f \in Hom_A(sM, N)$ and $\psi(f) = sg'$ with $g' : M \longrightarrow N$ the homomorphism defined by $g'(m) = f(sm) = g(sm) = sg(m)$ for each $m \in M$. Then $g' = sg$, so $\psi(f) = sg' = s^2 g = sg$. Then ψ is onto. We conclude that $s Hom_A(M, N)$ is a S-Noetherian A−module. Let L a A−sub-module of $Hom_A(M, N)$. Then sL is a

$A-$sub-module of $sHom_A(M, N)$. There exist $t \in S$ and $f_1, \ldots, f_q \in L$ such that $t(sL) \subseteq (sf_1, \ldots, sf_q) \subseteq L$. Then L is a S-finite $A-$module and $Hom_A(M, N)$ is a S-Noetherian $A-$module.   □

Example The hypothesis "N is S-Noetherian" is necessary in the preceding proposition. Let's consider the ring $A = (\mathbb{Z}/6\mathbb{Z})[X_n; \ n \in \mathbb{N}]$, of polynomials in infinitely countably variables, and its principal ideal $I = (\bar{2}X_0)$. Then $S = \{\bar{1}, \bar{3}\}$ is a multiplicative set of A formed by idempotent elements.

(1) The quotient $A-$module A/I is finitely generated by $\{\bar{1}\}$, so it is S-finite.
(2) The ring A is not S-Noetherian, so it is not a S-Noetherian $A-$module.
    Indeed, suppose that the ideal $J = (X_n; \ n \in \mathbb{N})$ is S-finite. There exist $s \in S = \{\bar{1}, \bar{3}\}$ and $k \in \mathbb{N}$ such that $sJ \subseteq (X_0, \ldots, X_k)$. In particular, $X_{k+1} = f_0X_0 + \ldots + f_kX_k$ with $f_0, \ldots, f_k \in A$, which is absurd.
(3) $Hom_A(A/I, A) \simeq ann_A(I) = (\bar{3})$ is not a S-Noetherian $A-$module.
    Indeed, for any ring A and any ideal I of A, a homomorphism of $A-$modules $f : A/I \longrightarrow A$ is completely determined by $f(\bar{1})$ because for each $\bar{x} \in A/I$, $f(\bar{x}) = f(x\bar{1}) = xf(\bar{1})$. Moreover, $f(A/I) \subseteq ann_A(I)$ because for each $\bar{x} \in A/I$ and each $a \in I$, $af(\bar{x}) = f(a\bar{x}) = f(\overline{ax}) = f(\bar{0}) = 0$. Then the map $\phi : Hom_A(A/I, A) \longrightarrow ann_A(I)$, defined by $\phi(f) = f(\bar{1})$ for each $f \in Hom_A(A/I, A)$, is a homomorphism and one to one of $A-$ modules. It is also onto because if $b \in ann_A(I)$, the map $f : A/I \longrightarrow A$, which associates to each $\bar{x} \in A/I$, $f(\bar{x}) = bx$, is well defined since if $\bar{x} = \bar{y}$, then $x - y \in I$, so if $b(x - y) = 0$, then $bx = by$. It is also a homomorphism of $A-$modules. Moreover, $\phi(f) = f(\bar{1}) = b$. Then $Hom_A(A/I, A) \simeq ann_A(I)$. In our case, $ann_A(I) = (\bar{3})$. Indeed, $\bar{3}I = (\bar{3}.\bar{2}X_0) = (0)$, and then $(\bar{3}) \subseteq ann_A(I)$. Conversely, if $f \in ann_A(I)$, then $\bar{2}X_0f = 0$, so $f \in (\bar{3})$. To conclude that the $A-$module $(\bar{3})$ is not S-Noetherian, suppose that the sub-module $H = (\bar{3}X_n; \ n \in \mathbb{N})$ is S-finite. There exist $s \in S = \{\bar{1}, \bar{3}\}$ and $k \in \mathbb{N}$ such that $sH \subseteq (\bar{3}X_0, \ldots, \bar{3}X_k)$. In particular, $\bar{3}X_{k+1} = \bar{3}X_0 f_0 + \ldots + \bar{3}X_k f_k$ with $f_0, \ldots, f_k \in A$, which is absurd.

Definition 3.13 Let A be a ring and M a $A-$module. We say that M is faithful if the ideal $ann_A(M) = \{a \in A; \ aM = 0\} = (0)$.

Examples

(1) For each ring A, the $A-$modules A and $A[X]$ are faithful.
(2) For each $n \in \mathbb{N}^*$, the $\mathbb{Z}-$module $\mathbb{Z}/n\mathbb{Z}$ is not faithful because $n(\mathbb{Z}/n\mathbb{Z}) = (\bar{0})$.

Lemma 3.14 *Let A be a ring, s a regular element of A, and M a faithful $A-$module. A sub-module of M containing sM is faithful.*

Proof Let N be sub-module of M containing sM. Since $ann_A(N) \subseteq ann_A(sM)$, it suffices to show that $ann_A(sM) = (0)$. Let $a \in ann_A(sM)$. Then $asM = (0)$, so $as \in ann_A(M) = (0)$. Then $as = 0$ and since s is regular, $a = 0$.   □

Example A sub-module of a faithful module is not in general faithful. Let $n \geq 2$ be an integer. The $\mathbb{Z}$-module $M = \mathbb{Z} \times (\mathbb{Z}/n\mathbb{Z})$ is faithful. But its sub-module $N = (0) \times (\mathbb{Z}/n\mathbb{Z})$ is not faithful.

Proposition 3.15 *Let A be a ring, S a multiplicative set of A formed by regular elements, and M a faithful A−module. If M a S-Noetherian A−module, then A is a S-Noetherian ring.*

Proof By hypothesis, M is S-finite. There exist $s \in S$ and $m_1, \ldots, m_k \in M$ such that $sM \subseteq (m_1, \ldots, m_k)$. The map $\phi : A \longrightarrow M^k$, defined by $\phi(a) = (am_1, \ldots, am_k)$ for each $a \in A$, is a homomorphism of A−modules. By the preceding lemma, the sub-module $(m_1, \ldots, m_k)$ of M is faithful. Then $ann_A(m_1, \ldots, m_k) = (0)$. Let $a \in ker\ \phi$. Then $am_1 = \ldots = am_k = 0$, so $a \in ann_A(m_1, \ldots, m_k) = (0)$. Then $ker\ \phi = (0)$ and ϕ is one to one. Then $A \simeq \phi(A) \subseteq M^k$. By Lemma 3.4, M^k is a S-Noetherian A−module. So is A. $\square$

An example constructed with the idealization in Sect. 8 that shows that "S is formed by regular elements of A" is necessary in the preceding proposition.

4 Polynomials Over S-Noetherian Rings

Definition 4.1 Let A be a ring and $S \subseteq A$ a multiplicative set. We say that S is anti-Archimedean if for each $s \in S$, $(\bigcap_{n:1}^{\infty} s^n A) \cap S \neq \emptyset$.

Example 1 If $S \subseteq U(A)$, then S is anti-Archimedean.

Example 2 (Hamed and Hizem 2016) Every finite multiplicative set S of a ring A is anti-Archimedean. Indeed, put $S = \{s_1, \ldots, s_k\}$ and $t = s_1 \ldots s_k \in S$. Let $s \in S$. For each integer $n \geq 1$, $s^n \in S$, and then $s^n = s_{i_n}$ with $i_n \in \{1, \ldots, k\}$. Then $t \in s_{i_n} A = s^n A$, so $t \in (\bigcap_{n:1}^{\infty} s^n A) \cap S$. For example, $S = \{\bar{1}, \bar{2}, \bar{4}, \bar{8}\}$ is an anti-Archimedean multiplicative set of the ring $\mathbb{Z}/12\mathbb{Z}$.

Note that a finite multiplicative set $S = \{s_1, \ldots, s_k\}$ not containing zero divisors is included in $U(A)$. Indeed, since $s_1 \ldots s_k \in S$, there exists $j \in \{1, \ldots, k\}$ such that $s_1 \ldots s_k = s_j$, so $s_1 \ldots s_{j-1}s_{j+1} \ldots s_k = 1$. Then $s_1, \ldots, s_{j-1}, s_{j+1}, \ldots, s_k \in U(A)$. On the other hand, $s_j^2 \in S$, then $s_j^2 = s_i$ with $i \in \{1, \ldots, k\}$. If $i \neq j$, then $s_i \in U(A)$, so $s_j \in U(A)$. If $i = j$, then $s_j^2 = s_j$, so $s_j = 1 \in U(A)$.

Example 3 Let A be an integral domain with quotient field K and complete integral closure equal to K. Then $A \setminus (0)$ is an anti-Archimedean multiplicative set. Indeed, let $0 \neq a \in A$. There exists $0 \neq s \in A$ such that $\frac{s}{a^n} \in A$ for each integer $n \geq 1$, so $s \in a^n A$. Then $s \in \bigcap_{n:1}^{\infty} a^n A$.

Example 4 Let A be a valuation domain which is not a field and without a height one prime ideal. The multiplicative set $A \setminus (0)$ is anti-Archimedean. Indeed, let $0 \neq a \in A$. Suppose that a is in all the nonzero prime ideals of A. The intersection P of all these ideals is prime because they are comparable, and we have $(0) \subset (a) \subseteq P$. Then the height of P is 1, which is absurd. There exists a nonzero prime ideal P_0 of A strictly included in (a). Since A is a valuation domain, for each integer $n \geq 1$, $P_0 \subset a^n A$. Then $(0) \neq P_0 \subseteq \bigcap_{n:1}^{\infty} a^n A$.

Example 5 (Park et al. 2019) We will generalize the preceding example. Let A be a Prüfer domain without a height one prime ideal. Suppose that each principal ideal of A has only finitely many minimal prime ideals. Then $A \setminus (0)$ is an anti-Archimedean multiplicative set. Indeed, let $0 \neq a \in A$ and $P_1, \ldots, P_m$ the minimal prime ideals over (a). Since A has no height one prime ideal, for each i, there is a nonzero prime ideal Q_i of A such that $Q_i \subset P_i$. The minimality of P_i over (a) implies that $a \notin Q_i$. So $a^n \notin Q_i$ for each $n \in \mathbb{N}$. Let $M \in Max(A)$ and $n \in \mathbb{N}^*$.

First case: There is i such that $Q_i \subseteq M$. In this case, $Q_i A_M$ is a prime ideal of A_M. Since $A \cap Q_i A_M = Q_i$, then $a^n \notin Q_i A_M$. Since A_M is a valuation domain, $Q_i A_M \subseteq a^n A_M$. Then $Q_1 \ldots Q_m A_M \subseteq Q_i A_M \subseteq a^n A_M$.

Second case: For each i, $Q_i \not\subseteq M$. In this case, $a \notin M$ because, in the contrary case, M must contain at least one P_i and then it contains Q_i, which is absurd. We have $Q_1 \ldots Q_m \not\subseteq M$. So $Q_1 \ldots Q_m A_M = A_M = a^n A_M$.

We conclude that for each $M \in Max(A)$ and $n \in \mathbb{N}^*$, $Q_1 \ldots Q_m A_M \subseteq a^n A_M$. Thus $(0) \neq Q_1 \ldots Q_m \subseteq \bigcap_{n:1}^{\infty} a^n A$.

Example 6 For each ring A, the mutiplicative set $\{X^n; \ n \in \mathbb{N}\}$ of $A[X]$ and $A[[X]]$ is not anti-Archimedean because $\bigcap_{n:1}^{\infty} (X^n) = (0)$.

Theorem 4.2 *Let A be a ring and S an anti-Archimedean multiplicative set of A. If A is S-Noetherian, then it is the same for the polynomial ring $A[X_1, \ldots, X_n]$.*

Proof Since S is anti-Archimedean in any ring containing A, we may suppose $n = 1$. Put $X_1 = X$. Let I be a nonzero ideal of $A[X]$ and J the ideal of A formed by 0 and the leading coefficients of the nonzero polynomials in I. Since A is S-Noetherian, there exist $s \in S$ and $a_1, \ldots, a_n \in J \setminus (0)$ such that $sJ \subseteq (a_1, \ldots, a_n)A$. Let $f_1, \ldots, f_n \in I$ be such that a_i is the leading coefficient of f_i, $1 \leq i \leq n$. By multiplying the f_i with convenient powers of X, we may suppose that $deg\ f_1 = \ldots = deg\ f_n = d$. Consider the finitely generated A–module $M = A + AX + AX^2 + \ldots + AX^{d-1}$. By Theorem 3.5, M is a S-Noetherian A–module. Its sub-module $M \cap I = T$ is S-finite. There exist $t \in S$ and $g_1, \ldots, g_m \in T$ such that $tT \subseteq (g_1, \ldots, g_m)A$. Let $f \in I$. By repeated subtractions of linear combinations of $f_1, \ldots, f_n$, with coefficients in $A[X]$, we can find $q \in \mathbb{N}^*$ such that $s^q f \in (f_1, \ldots, f_n)A[X] + T$. Indeed, $f = b_0 + b_1 X + \ldots + bX^k, b \in J$, $k \geq d, sb = \alpha_1 a_1 + \ldots + \alpha_n a_n$ with $\alpha_i \in A$. Then $sf - \sum_{i:1}^{n} \alpha_i X^{k-d} f_i \in I$

with degree $k' \leq k - 1$. We restart, $s(sf - \sum_{i:1}^{n} \alpha_i X^{k-d} f_i) - \sum_{i:1}^{n} \beta_i X^{k'-d} f_i \in I$ of degree $k'' \leq k' - 1$ with $\beta_i \in A$. We continue until we arrive at a polynomial of degree $\leq d - 1$. Let $w \in (\bigcap_{j:1}^{\infty} s^j A) \cap S$. Then $w = s^q c$ with $c \in A$. Then $twf = ts^q cf \in (f_1, \ldots, f_n)A[X] + tTA[X] \subseteq (f_1, \ldots, f_n)A[X] + (g_1, \ldots, g_m)A[X] = (f_1, \ldots, f_n, g_1, \ldots, g_m)A[X]$.

Then $(tw)I \subseteq (f_1, \ldots, f_n, g_1, \ldots, g_m)A[X]$ with $f_1, \ldots, f_n, g_1, \ldots, g_m \in I$ et $tw \in S$. $\qquad\square$

Example 1 We retrieve the Hilbert basis theorem. If A is a Noetherian ring, then it is the polynomial ring $A[X_1, \ldots, X_n]$.

Example 2 For any multiplicative set S of a ring A, the ring $B = A[X_n;\ n \in \mathbb{N}]$ is not S-Noetherian. Indeed, suppose that the ideal $I = (X_n;\ n \in \mathbb{N})$ is S-finite. There exist $s \in S$ and $k \in \mathbb{N}$ such that $sI \subseteq (X_0, \ldots, X_k)$. There exist $f_0, \ldots, f_k \in B$ such that $sX_{k+1} = X_0 f_0 + \ldots + X_k f_k$. Putting $X_0 = \ldots = X_k = 0$, we find $sX_{k+1} = 0$, and then $s = 0 \in S$, which is absurd.

We will give, in Remark 1.6 of Chap. 3, an example showing the necessity of the condition S is anti-Archimedean in the preceding theorem.

The following two results are due to Benhissi and Koja (2012).

Lemma 4.3 *Let A be a ring and S and T two multiplicative sets of A with T not containing any zero divisors. If A is S-Noetherian, then A_T is S-Noetherian.*

Proof Since T does not contain any zero divisors, $A \subseteq A_T$. A prime ideal of A_T is of the form P_T with P a prime ideal of A disjoint with T. Since A is S-Noetherian, P is S-finite. There exist $s \in S$ and a finitely generated ideal I of A such that $sP \subseteq I \subseteq P$. Then $sP_T \subseteq I_T \subseteq P_T$ with I_T a finitely generate ideal of A_T. Then P_T is S-finite. By Corollary 2.2, the ring A_T is S-Noetherian. $\qquad\square$

Corollary 4.4 *Let A be a ring and S an anti-Archimedean multiplicative set of A. If A is S-Noetherian, then it is the same for the Laurent polynomial ring $A[X, X^{-1}]$.*

Proof By Theorem 4.2, the ring $A[X]$ is S-Noetherian. On the other hand, $A[X, X^{-1}] = A[X]_T$ with $T = \{X^n;\ n \in \mathbb{N}\}$ a multiplicative set of $A[X]$

not containing zero divisors. By the preceding lemma, the ring $A[X, X^{-1}]$ is S-Noetherian.   □

Example Let K be a field. For $0 \neq f(X) = a_m X^m + a_{m+1} X^{m+1} + \ldots + a_n X^n \in K[X, X^{-1}]$ with $m, n \in \mathbb{Z}, m \leq n, a_i \in K, a_m \neq 0$ and $a_n \neq 0$, let $\sigma(f(X)) = n - m$. We will prove that $K[X, X^{-1}]$ is an Euclidean domain for the sthatm σ.

(1) With the preceding notations, $X^{-m} f(X) = a_m + a_{m+1} X + \ldots + a_n X^{n-m} \in K[X]$ is a polynomial with nonzero constant term a_m.

(2) For each $k \in \mathbb{Z}, \sigma(X^k f(X)) = (n+k) - (m+k) = n - m = \sigma(f(X))$.

(3) For each $0 \neq r(X) \in K[X], \sigma(r(X)) \leq deg\ r(X)$ and if $r(0) \neq 0$, then we have the equality. Indeed, $0 \neq r(X) = a_m X^m + \ldots + a_n X^n \in K[X]$ with $a_m \neq 0, a_n \neq 0$ and $0 \leq m \leq n$. Then $\sigma(r(X)) = n - m \leq n = degr(X)$. If $r(0) \neq 0$, then $m = 0$, so $\sigma(r(X)) = n = deg\ (r(X))$.

(4) Let $f(X)$ and $g(X) \in K[X, X^{-1}] \setminus (0)$. Then $\sigma(f(X)g(X)) \geq \sigma(f(X))$. Indeed, $0 \neq f(X) = a_m X^m + \ldots + a_n X^n \in K[X, X^{-1}]$ with $a_m \neq 0, a_n \neq 0$, $m \leq n$ and $0 \neq g(X) = b_s X^s + \ldots + b_r X^r \in K[X, X^{-1}]$ with $b_s \neq 0, b_r \neq 0$, $s \leq r$. Then $f(X)g(X) = a_m b_s X^{m+s} + \ldots + a_n b_r X^{n+r}$ with $a_m b_s \neq 0$ and $a_n b_r \neq 0$. So $\sigma(f(X)g(X)) = n + r - (m+s) = (n-m) + (r-s) \geq n - m = \sigma(f(X))$.

(5) We will now show that $K[X, X^{-1}]$ is an Euclidean domain for the stathm σ. It is clear that σ takes its values in $\mathbb{N}$. By (4), if $f(X)$ and $h(X)$ are two nonzero elements of $K[X, X^{-1}]$ such that $f(X)$ divides $h(X)$, then $\sigma(f(X)) \leq \sigma(h(X))$.

Let $0 \neq f(X) = a_m X^m + \ldots + a_n X^n \in K[X, X^{-1}]$ with $a_m \neq 0, a_n \neq 0$, $m \leq n$, and $0 \neq g(X) = b_s X^s + \ldots + b_r X^r \in K[X, X^{-1}]$ with $b_s \neq 0$, $b_r \neq 0, s \leq r$. Suppose that $g(X)$ does not divide $f(X)$ in $K[X, X^{-1}]$. By (1), $X^{-m} f(X)$ and $X^{-s} g(X) \in K[X]$ are two polynomials with nonzero constant terms. Suppose that $X^{-s} g(X)$ divides $X^{-m} f(X)$ in $K[X]$. There exists $q(X) \in K[X]$ such that $X^{-m} f(X) = q(X) X^{-s} g(X)$. Then $f(X) = q(X) X^{m-s} g(X)$ with $q(X) X^{m-s} \in K[X, X^{-1}]$. So $g(X)$ divides $f(X)$ in $K[X, X^{-1}]$. But this is absurd. The Euclidean division of $X^{-m} f(X)$ by $X^{-s} g(X)$ in $K[X]$ can be written: $X^{-m} f(X) = X^{-s} g(X)q(X) + r(X)$ with $q(X), r(X) \in K[X], r(X) \neq 0$ and $deg\ r(X) < deg\ X^{-s} g(X)$. Then $f(X) = g(X) X^{m-s} q(X) + X^m r(X)$ with $X^{m-s} q(X), X^m r(X) \in K[X, X^{-1}], X^m r(X) \neq 0$. By the steps (2) and (3), $\sigma(X^m r(X)) = \sigma(r(X)) \leq deg\ r(X) < deg\ X^{-s} g(X) = \sigma(X^{-s} g(X)) = \sigma(g(X))$.

Corollary 4.5 (Baeck et al. 2016) *Let $A \subseteq B$ be an extension of rings and S an anti-Archimedean multiplicative set of A.*

The following assertions are equivalent:

1. *The ring $A + X B[X]$ is S-Noetherian.*
2. *The ring A is S-Noetherian and the $A-$module B is S-finite.*

Proof "(1) $\implies$ (2)" Let I be an ideal of A. Then $I + XB[X]$ is an ideal of the ring $A + XB[X]$. There exist $s \in S$ and $f_1, \ldots, f_n \in I + XB[X]$ such that $s(I + XB[X]) \subseteq f_1(A + XB[X]) + \ldots + f_n(A + XB[X])$. Then $sI \subseteq f_1(0)A + \ldots f_n(0)A \subseteq I$. So the ideal I is S-finite and the ring A is S-Noetherian.

Since $XB[X]$ is an ideal of the ring $A + XB[X]$, there exist $s \in S$ and $g_1, \ldots, g_n \in B[X]$ such that $sXB[X] \subseteq Xg_1(A+XB[X])+\ldots+Xg_n(A+XB[X])$. For each $b \in B$, there exist $h_1, \ldots, h_n \in A+XB[X]$ such that $sXb = Xg_1h_1+\ldots+Xg_nh_n$, and then $sb = g_1h_1+\ldots+g_nh_n$, so $sb = g_1(0)h_1(0)+\ldots+g_n(0)h_n(0)$ with $h_i(0) \in A$. Then $sB \subseteq g_1(0)A + \ldots + g_n(0)A$ with $g_i(0) \in B$, so the A−module B is S-finite.

"(2) $\implies$ (1)" Since B is a A−module S-finite, there exist $s \in S$ and $b_1, \ldots, b_n \in B$ such that $sB \subseteq Ab_1 + \ldots + Ab_n \subseteq B$, and then $sB[X] \subseteq A[X]b_1 + \ldots + A[X]b_n \subseteq B[X]$. Then $B[X]$ is a $A[X]$−module S-finite. On the other hand, by Theorem 4.2, the ring $A[X]$ is S-Noetherian. By Corollary 3.6, the $A[X]$−module $B[X]$ is S-Noetherian.

We will prove that $B[X]$ is a S-Noetherian $A + XB[X]$−module. Let N be a $A + XB[X]$− sub-module of $B[X]$. Then N is a $A[X]$−sub-module of $B[X]$, so it is S-finite. There exist $t \in S$ and $f_1, \ldots, f_m \in N$ such that $tN \subseteq f_1A[X] + \ldots + f_mA[X] \subseteq N$, and then $tN \subseteq f_1(A + XB[X]) + \ldots + f_m(A + XB[X]) \subseteq N$ because N is stable by multiplication by the elements of $A + XB[X]$. Then N is S-finite.

We will show that the ring $A + XB[X]$ is S-Noetherian. Let I be an ideal of $A + XB[X]$. Then I is a $A + XB[X]$−sub-module of $B[X]$. It is S-finite, by the preceding step. There exist $w \in S$ and $g_1, \ldots, g_k \in I$ such that $wI \subseteq g_1(A + XB[X]) + \ldots + g_k(A + XB[X]) \subseteq I$. Then I is a S-finite ideal of the ring $A + XB[X]$. □

Example 1 For any anti-Archimedean multiplicative set S of $\mathbb{Z}$, the $\mathbb{Z}$-module $\mathbb{Q}$ is not S-finite. Then the ring $\mathbb{Z} + X\mathbb{Q}[X]$ is not S-Noetherian.

Example 2 Let $A \subseteq B$ be an extension of rings. Then $A + XB[X]$ is Noetherian if and only if A is Noetherian and B is a finitely generated A−module.

The following results are due to Hamed and Hizem (2015).

Notations Let $\mathscr{A} = (A_n)_{n \in \mathbb{N}}$ be an increasing sequence of rings and $A = \bigcup_{n:0}^{\infty} A_n$, their union. We note $\mathscr{A}[X] = \left\{ f = \sum_{i:0}^{r} a_i X^i \in A[X]; r \in \mathbb{N}, a_i \in A_i, \text{ for each } i \in \mathbb{N} \right\}$, and $\mathscr{A}[[X]] = \left\{ f = \sum_{i:0}^{\infty} a_i X^i \in A[[X]]; a_i \in A_i, \text{ for each } i \in \mathbb{N} \right\}$. Then $\mathscr{A}[X]$ is a sub-ring of $A[X]$ containing $A_0[X]$ and $\mathscr{A}[[X]]$ is a sub-ring of $A[[X]]$ containing $A_0[[X]]$. Note also that $\mathscr{A}[X] \subset \mathscr{A}[[X]]$.

Example With the preceding notations, if $A_1 = A_2 = \ldots$, then $\mathscr{A}[X] = A_0 + XA_1[X]$ and $\mathscr{A}[[X]] = A_0 + XA_1[[X]]$.

Definition 4.6 Let $\mathscr{A} = (A_n)_{n\in\mathbb{N}}$ be an increasing sequence of rings and $S \subseteq A_0$ a multiplicative set.

1. We say that the sequence $(A_n)_{n\in\mathbb{N}}$ is S-stationary if there exist $s \in S$ and $k \in \mathbb{N}$ such that for each integer $n \geq k$, $sA_n \subseteq A_k$.
2. We say that the sequence $(A_n)_{n\in\mathbb{N}}$ is S-Noetherian if it satisfies the following:

    (a) The ring A_0 is S-Noetherian.
    (b) The sequence $(A_n)_{n\in\mathbb{N}}$ is S-stationary.
    (c) For each $n \in \mathbb{N}$, A_n is a S-finite A_0-module.

Example Let $\mathscr{A} = (A_n)_{n\in\mathbb{N}}$ be a S-stationary increasing sequence of rings with $S \subseteq A_0$ a multiplicative set. The sequences $(A_n[X])_{n\in\mathbb{N}}$ and $(A_n[[X]])_{n\in\mathbb{N}}$ are S-stationary. Indeed, there exist $s \in S$ and $k \in \mathbb{N}$ such that for each $n \geq k$, $sA_n \subseteq A_k$, then $sA_n[X] \subseteq A_k[X]$ and $sA_n[[X]] \subseteq A_k[[X]]$.

Theorem 4.7 *Let $\mathscr{A} = (A_n)_{n\in\mathbb{N}}$ be an increasing sequence of rings, $A = \bigcup\limits_{n:0}^{\infty} A_n$ and $S \subseteq A_0$ an anti-Archimedean multiplicative set.*
    The following assertions are equivalent:

1. *The sequence $\mathscr{A} = (A_n)_{n\in\mathbb{N}}$ is S-Noetherian.*
2. *The ring A_0 is S-Noetherian and the A_0-module A is S-finite.*
3. *The ring $\mathscr{A}[X]$ is S-Noetherian.*

Proof $''(1) \implies (2)''$ By hypothesis, the ring A_0 is S-Noetherian. Since the sequence $\mathscr{A} = (A_n)_{n\in\mathbb{N}}$ is S-stationary, there exist $s \in S$ and $k \in \mathbb{N}$ such that for each integer $n \geq k$, $sA_n \subseteq A_k$. But A_k is a S-finite A_0-module, so there exist $s' \in S$ and $x_1, \ldots, x_r \in A_k$ such that $s'A_k \subseteq A_0x_1 + \ldots + A_0x_r$. Then

$$s'sA = s's(\bigcup_{n:0}^{\infty} A_n) \subseteq s'A_k \subseteq A_0x_1 + \ldots + A_0x_r \subseteq A \text{ with } s's \in S. \text{ Then } A \text{ is a}$$

S-finite A_0-module.

$''(2) \implies (3)''$ By Corollary 3.7, the ring A is S-Noetherian. By Theorem 4.2, the ring $A[X]$ is S-Noetherian. On the other hand, A is a S-finite A_0-module. There exist $s \in S$ and $x_1, \ldots, x_r \in A$ such that $sA \subseteq A_0x_1 + \ldots + A_0x_r$. Then $sA[X] \subseteq A_0[X]x_1 + \ldots + A_0[X]x_r \subseteq \mathscr{A}[X]x_1 + \ldots + \mathscr{A}[X]x_r \subseteq A[X]$ because $A_0[X] \subseteq \mathscr{A}[X] \subseteq A[X]$. Then $A[X]$ is a S-finite $\mathscr{A}[X]-$module. By Corollary 3.7, the ring $\mathscr{A}[X]$ is S-Noetherian. $''(3) \implies (1)''$ We will prove that the ring A_0 is S-Noetherian. Let I be an ideal of A_0. Consider the increasing sequence of rings $\mathscr{B} = (A_n)_{n\geq 1}$ and the ring $\mathscr{B}[X] = \{f = \sum\limits_{i:0}^{n} a_i X^i; n \in \mathbb{N}, a_i \in A_{i+1} \text{ for each } i\}$.

Then $I + X\mathscr{B}[X]$ is an ideal of $\mathscr{A}[X]$. Indeed, the stability by addition is clear. Let

$a \in I$, $b \in A_0$, and $f, g \in \mathscr{B}[X]$. Then $(a + Xf)(b + Xg) = ab + X(bf + ag + Xfg) \in I + X\mathscr{B}[X]$. Since the ring $\mathscr{A}[X]$ is S-Noetherian, there exist $s \in S$ and $f_1, \ldots, f_n \in I + X\mathscr{B}[X]$ such that $s(I + X\mathscr{B}[X]) \subseteq f_1\mathscr{A}[X] + \ldots + f_n\mathscr{A}[X]$. Then $sI \subseteq f_1(0)A_0 + \ldots + f_n(0)A_0 \subseteq I$, so I is S-finite.

We will prove that the sequence $\mathscr{A} = (A_n)_{n \in \mathbb{N}}$ is S-stationary. The set I of the elements of $\mathscr{A}[X]$ with zero constant terms is an ideal of $\mathscr{A}[X]$. It is S-finite. There exist $s \in S$ and $f_1, \ldots, f_n \in I \setminus (0)$ such that $sI \subseteq (f_1, \ldots, f_n)\mathscr{A}[X]$. Let $m = max\{deg\ f_i;\ 1 \le i \le n\} \in \mathbb{N}^*$. The number of nonzero monomials in $f_1, \ldots, f_n$ is finite. Denote them by $a_k X^{\alpha_k}$ with $1 \le k \le r$. Then $1 \le \alpha_k \le m$ and $a_k \in A_{\alpha_k}$ for $1 \le k \le r$. Then $a_1, \ldots, a_r \in A_m$ and $sI \subseteq (f_1, \ldots, f_n)\mathscr{A}[X] \subseteq (a_k X^{\alpha_k};\ 1 \le k \le r)\mathscr{A}[X]$.

Fix an integer $i > m$. For each $y \in A_i$, $yX^i \in I$, and then $syX^i \in (a_k X^{\alpha_k};\ 1 \le k \le r)\mathscr{A}[X]$. Put $syX^i = \sum_{k:1}^{r} a_k X^{\alpha_k} g_k$ with $g_k = \sum_j g_{k,j}X^j \in \mathscr{A}[X]$, and then $g_{k,j} \in A_j$. By identification, $sy = \sum_{k:1}^{r} a_k g_{k,i-\alpha_k}$. Since $g_{k,i-\alpha_k} \in A_{i-\alpha_k} \subseteq A_{i-1}$, then $sA_i \subseteq a_1 A_{i-1} + \ldots + a_r A_{i-1} \subseteq A_{i-1}$. So for each integer $i > m$, $sA_i \subseteq A_{i-1}$ and then $s^{i-m}A_i \subseteq A_m$. Since S is anti-Archimedean in A_0, there exists $t \in (\bigcap_{n:1}^{\infty} s^n A_0) \cap S$. For each $n \in \mathbb{N}^*$, there exists $b_n \in A_0$ such that $t = s^n b_n$. For each integer $i > m$, $tA_i = s^{i-m}b_{i-m}A_i \subseteq s^{i-m}A_i \subseteq A_m$ because $b_{i-m} \in A_0 \subseteq A_i$. Then the sequence $\mathscr{A} = (A_n)_{n \in \mathbb{N}}$ is S-stationary.

Let $n \in \mathbb{N}^*$. We will show that the A_0-module A_n is S-finite. The ideal $I = \{f = \sum_i a_i X^i \in \mathscr{A}[X];\ a_0 = \ldots = a_{n-1} = 0\}$ of $\mathscr{A}[X]$ is S-finite. There exist $s \in S$ and $f_1, \ldots, f_m \in I$ such that $sI \subseteq f_1\mathscr{A}[X] + \ldots + f_m\mathscr{A}[X]$. Put $f_i = X^n g_i$ with $g_i \in A_n + XA_{n+1} + \ldots$. Then $sA_n \subseteq g_1(0)A_0 + \ldots + g_m(0)A_0 \subseteq A_n$.

In conclusion, the sequence $\mathscr{A} = (A_n)_{n \in \mathbb{N}}$ is S-Noetherian. $\qquad\square$

Remark 4.8 The preceding theorem helps to retrieve Corollary 4.5. Indeed, let $A \subseteq B$ be an extension of rings and $S \subseteq A$ an anti-Archimedean multiplicative set. Consider the sequence of rings $\mathscr{A} = (A_n)_{n \in \mathbb{N}}$ defined by $A_0 = A$ and for each $n \ge 1$, $A_n = B$. Then $\mathscr{A}[X] = A + XB[X]$. The sequence $(A_n)_{n \in \mathbb{N}}$ is stationary and then S-stationary. Then $\mathscr{A}[X] = A + XB[X]$ is S-Noetherian $\Longleftrightarrow$ the sequence $(A_n)_{n \in \mathbb{N}}$ is S-Noetherian $\Longleftrightarrow$ the ring $A_0 = A$ is S-Noetherian and for each $n \ge 1$, $A_n = B$ is a S-finite $A_0 = A$-module.

Proposition 4.9 *Let $\mathscr{A} = (A_n)_{n \in \mathbb{N}}$ be a S-Noetherian increasing sequence of rings and $S \subseteq A_0$ an anti-Archimedean multiplicative set. For each $n \in \mathbb{N}$, the rings $A_n[X]$ and $A_0 + XA_n[X]$ are S-Noetherian.*

Proof By definition, the ring A_0 is S-Noetherian and the A_0−module A_n is S-finite. By Corollary 3.7, the ring A_n is S-Noetherian. By Theorem 4.2, the ring $A_n[X]$ is S-Noetherian. By Corollary 4.5, the ring $A_0 + XA_n[X]$ is S-Noetherian.   □

5 Formal Power Series Over S-Noetherian Rings

Theorem 5.1 *Let A be a ring and S an anti-Archimedean multiplicative set of A not containing zero divisors. If A is S-Noetherian, it is the same for the formal power series ring $A[[X_1, \ldots, X_n]]$.*

Proof Since S is anti-Archimedean in any ring containing A and it does not contain zero divisors in $A[[X_1, \ldots, X_n]]$, we may suppose $n = 1$. Put $X_1 = X$. By Corollary 2.2, it suffices to show that each prime ideal P of $A[[X]]$ is S-finite. Consider the ideal $P' = \{f(0); \; f \in P\}$ of A. There exist $s \in S$ and $f_1, \ldots, f_n \in P$ such that $sP' \subseteq (f_1(0), \ldots, f_n(0))$.

First case: $X \in P$. Then $P = P' + XA[[X]]$. We will show that $sP \subseteq (f_1, \ldots, f_n, X) \subseteq P$. Put $f_i = f_i(0) + Xh_i$ with $h_i \in A[[X]]$, $1 \leq i \leq n$. Let $f \in P$, $f = f(0) + Xh$ with $h \in A[[X]]$. Then $sf = sf(0) + Xsh =$

$$\sum_{i:1}^{n} a_i f_i(0) + Xsh \text{ with } a_1, \ldots, a_n \in A. \text{ Then } sf = \sum_{i:1}^{n} a_i(f_i - Xh_i) + Xsh =$$

$$\sum_{i:1}^{n} a_i f_i + X(sh - \sum_{i:1}^{n} a_i h_i) \in (f_1, \ldots, f_n, X).$$

Second case: $X \notin P$. Let $t \in (\bigcap_{i:1}^{\infty} s^i A) \cap S$. We will show that $tP \subseteq (f_1, \ldots, f_n) \subseteq P$. Put $a_i = f_i(0)$, $1 \leq i \leq n$. Let $f \in P$. Then $sf(0) =$

$$\sum_{i:1}^{n} b_{i0} a_i \text{ with } b_{i0} \in A, \text{ then } sf - \sum_{i:1}^{n} b_{i0} f_i = Xg_1 \in P. \text{ Then } g_1 \in P \text{ and }$$

$$sf = \sum_{i:1}^{n} b_{i0} f_i + Xg_1. \text{ In the same way, } sg_1 = \sum_{i:1}^{n} b_{i1} f_i + Xg_2 \text{ with } b_{i1} \in A \text{ and }$$

$$g_2 \in P. \text{ Then } s^2 f = \sum_{i:1}^{n} sb_{i0} f_i + Xsg_1 = \sum_{i:1}^{n} sb_{i0} f_i + X(\sum_{i:1}^{n} b_{i1} f_i + Xg_2) =$$

$$\sum_{i:1}^{n} (sb_{i0} + b_{i1}X) f_i + X^2 g_2. \text{ In the same way, } sg_2 = \sum_{i:1}^{n} b_{i2} f_i + Xg_3 \text{ with } b_{i2} \in A$$

and $g_3 \in P$. Then $s^3 f = \sum_{i:1}^{n} (s^2 b_{i0} + sb_{i1}X) f_i + X^2 sg_2 = \sum_{i:1}^{n} (s^2 b_{i0} + sb_{i1}X) f_i +$

$$X^2(\sum_{i:1}^{n} b_{i2} f_i + Xg_3) = \sum_{i:1}^{n} (s^2 b_{i0} + sb_{i1}X + b_{i2}X^2) f_i + X^3 g_3. \text{ By induction, for each}$$

$$k \in \mathbb{N}, s^{k+1} f = \sum_{i:1}^{n} (s^k b_{i0} + s^{k-1} b_{i1} X + \ldots + s b_{ik-1} X^{k-1} + b_{ik} X^k) f_i + X^{k+1} g_{k+1}$$

with $g_{k+1} \in A[[X]]$.

Since S does not contain zero divisor, $A \subseteq S^{-1}A = B$ and in $B[[X]]$,

$$f = \sum_{i:1}^{n} (\frac{b_{i0}}{s} + \frac{b_{i1}}{s^2} X + \ldots + \frac{b_{ik-1}}{s^k} X^{k-1} + \frac{b_{ik}}{s^{k+1}} X^k) f_i + \frac{X^{k+1}}{s^{k+1}} g_{k+1}. \text{ Put}$$

$h_i = \sum_{j:0}^{\infty} \frac{b_{ij}}{s^{j+1}} X^j \in B[[X]]$, for $1 \leq i \leq n$. Put $t = s^j c_j$ with $c_j \in A$ for

$j \in \mathbb{N}$. Then $h_i = \sum_{j:0}^{\infty} \frac{b_{ij} c_{j+1}}{t} X^j$, so $t h_i = \sum_{j:0}^{\infty} b_{ij} c_{j+1} X^j \in A[[X]]$. We will

show that $f = h_1 f_1 + \ldots + h_n f_n$, and then $tf = (th_1) f_1 + \ldots + (th_n) f_n \in$

$(f_1, \ldots, f_n) A[[X]]$. For each $g = \sum_{i:0}^{\infty} x_i X^i \in B[[X]]$ and $k \in \mathbb{N}$, we note

$g^{(k)} = \sum_{i:0}^{k} x_i X^i \in B[X]$ a polynomial of degree $\leq k$ and $g = g^{(k)} + X^{k+1} h$ with

$h \in B[[X]]$. We have $f = f^{(k)} + X^{k+1} u = \sum_{i:1}^{n} h_i^{(k)} (f_i^{(k)} + X^{k+1} r_i) + \frac{X^{k+1}}{s^{k+1}} g_{k+1}$,

with $u, r_i \in B[[X]]$. Then $f^{(k)} - \sum_{i:1}^{n} h_i^{(k)} f_i^{(k)} = X^{k+1} v$, with $v \in B[[X]]$. The

order $\omega (f^{(k)} - \sum_{i:1}^{n} h_i^{(k)} f_i^{(k)}) \geq k + 1$. For each $k \in \mathbb{N}$, the coefficient of X^k in

$f - \sum_{i:1}^{n} h_i f_i$ is equal to that of X^k in $f^{(k)} - \sum_{i:1}^{n} h_i^{(k)} f_i^{(k)}$, which is zero. Then

$f - \sum_{i:1}^{n} h_i f_i = 0$. $\qquad\qquad\qquad\qquad\qquad\qquad\qquad\qquad\qquad\qquad\Box$

Corollary 5.2 (I. Kaplansky) *If A is a Noetherian ring, the formal power series ring $A[[X_1, \ldots, X_n]]$ is also Noetherian.*

Proof Take $S = \{1\}$. $\qquad\qquad\qquad\qquad\qquad\qquad\qquad\qquad\qquad\qquad\qquad\qquad\Box$

Proposition 5.3 (Benhissi and Koja 2012) *Let A be a ring and S an anti-Archimedean multiplicative set of A not containing zero divisors. If A is S-Noetherian, it is the same for the ring $A[[X]][X^{-1}]$.*

Proof By Theorem 5.1, $A[[X]]$ is S-Noetherian. On the other hand, $A[[X]][X^{-1}] = A[[X]]_T$ with $T = \{X^n; n \in \mathbb{N}\}$ a multiplicative set of $A[[X]]$ not containing zero divisors. By Lemma 4.3, the ring $A[[X]][X^{-1}]$ is S-Noetherian. $\qquad\qquad\Box$

Lemma 5.4 *Let S be a multiplicative set of a ring A. If A is S-Noetherian, the ring of fractions $S^{-1}A$ is Noetherian.*

Proof An ideal of $S^{-1}A$ is of the form $S^{-1}I$ with I an ideal of A. There exist $s \in S$ and a finitely generated ideal J of A such that $sI \subseteq J \subseteq I$. Then $S^{-1}I = S^{-1}(sI) \subseteq S^{-1}J \subseteq S^{-1}I$ so $S^{-1}I = S^{-1}J$ is a finitely generated ideal.   □

The converse of the lemma will be studied in Exercise 6.

Example 1 A Noetherian ring is locally Noetherian. Indeed, let P be a prime ideal of a Noetherian ring A and $S = A \setminus P$. Then A est S-Noetherian, so $A_P = S^{-1}A$ is Noetherian.

Example 2 The converse of Example 1 is false. The product ring $A = \prod_{i:1}^{\infty} \mathbb{F}_2$ is not Noetherian since it contains the strictly increasing sequence of ideals $(0) \subset \mathbb{F}_2 \times (0) \times \ldots \subset \mathbb{F}_2 \times \mathbb{F}_2 \times (0) \times \ldots \subset \ldots$. It is clear that the ring A is Boole; i.e., all its elements are idempotent. Each prime ideal P of A is maximal. Indeed, let I be an ideal of A such that $P \subset I \subseteq A$ and $x \in I \setminus P$. Then $x^2 = x$, so $x(x-1) = 0 \in P$, and then $x - 1 \in P \subset I$. So $1 = x - (x-1) \in I$ and $I = A$. For each prime ideal P of A, the ring A_P is a domain. Indeed, if $\frac{a}{s}\frac{b}{t} = 0$ with $a, b \in A$ and $s, t \in A \setminus P$, there exists $r \in A \setminus P$ such that $r(ab) = 0 \in P$, so a or $b \in P$. For example, $a \in P$, and then $a - 1 \notin P$. But $(a-1)a = a^2 - a = 0$, then $\frac{a}{s} = 0$. For each prime ideal P of A, $PA_P = (0)$. Indeed, suppose that this ideal contains a nonzero element $x = \frac{a}{s}$ with $a \in P$ and $s \in A \setminus P$. Since $a^2 = a$, then $a(a-1) = 0$, so $\frac{a}{s}(\frac{a}{s} - \frac{1}{s}) = 0$, and then $x(x - \frac{1}{s}) = 0$. But A_P is a domain, and then $x - \frac{1}{s} = 0$, so $x = \frac{1}{s}$. There exists $t \in A \setminus P$ such that $t(sa - s) = 0$, and then $ts(a-1) = 0 \in P$, so $a - 1 \in P$. Then $1 = a - (a-1) \in P$, which is absurd. We conclude that A_P is a local ring with maximal ideal $PA_P = (0)$. Then A_P is a field, so Noetherian.

Example 3 The converse of the lemma is false. Let $A = \mathbb{Z}_{(2)} + X\mathbb{Q}[X]$ where $\mathbb{Z}_{(2)}$ is the localization of the ring $\mathbb{Z}$ by the prime ideal $2\mathbb{Z}$ and $S = \{2^n; \ n \in \mathbb{N}\}$. Then $S^{-1}A = \mathbb{Q}[X]$ is Noetherian. Indeed, let $f \in \mathbb{Q}[X]$, $f = a + Xg(X)$ with $a \in \mathbb{Q}$ and $g(X) \in \mathbb{Q}[X]$. We have $a = \frac{\alpha}{2^k\beta}$ with $\alpha \in \mathbb{Z}$, $\beta \in \mathbb{N}^*$, and $\Delta(2, \beta) = 1$. Then $f = \frac{1}{2^k}(\frac{\alpha}{\beta} + X2^kg(X)) \in S^{-1}A$. Suppose that the ideal $X\mathbb{Q}[X]$ of A is S-finite. There exist $r \in \mathbb{N}$ and $g_1, \ldots, g_s \in \mathbb{Q}[X]$ such that $2^r X\mathbb{Q}[X] \subseteq Xg_1A + \ldots + Xg_sA \subseteq X\mathbb{Q}[X]$. Then $2^r\mathbb{Q}[X] \subseteq g_1A + \ldots + g_sA \subseteq \mathbb{Q}[X]$, so $\mathbb{Q}[X] \subseteq g_1A + \ldots + g_sA \subseteq \mathbb{Q}[X]$. Then $\mathbb{Q}[X] = g_1A + \ldots + g_sA$, so $\mathbb{Q} = g_1(0)\mathbb{Z}_{(2)} + \ldots + g_s(0)\mathbb{Z}_{(2)}$ with $g_1(0), \ldots, g_s(0) \in \mathbb{Q}$. An element of $\mathbb{Q}$ of the form $\frac{1}{2^t}$ with $t \in \mathbb{N}^*$ sufficiently larger cannot be in $g_1(0)\mathbb{Z}_{(2)} + \ldots + g_s(0)\mathbb{Z}_{(2)}$. We have a contradiction. So the ring A is not S-Noetherian.

Corollary 5.5 *Let A be an integral domain with $S = A \setminus (0)$ an anti-Archimedean multiplicative set of A. Then*

1. *The ring $A[[X_1, \ldots, X_n]]$ is S-Noetherian.*
2. *The ring $A[[X_1, \ldots, X_n]]_{A \setminus (0)}$ is Noetherian.*

Proof

(1) The only prime ideal of A disjoint with S is (0). It is S-finite. By Corollary 2.2, A is S-Noetherian. By Theorem 5.1, $A[[X_1, \ldots, X_n]]$ is S-Noetherian.
(2) By the preceding lemma, the ring $A[[X_1, \ldots, X_n]]_{A \setminus (0)}$ is Noetherian.   □

Example 1 Let A be a valuation domain which is not a field and without a height one prime ideal. By Example 4 to Definition 4.1, the multiplicative set $S = A \setminus (0)$ is anti-Archimedean. Since A is not Noetherian, then it is the same for the ring $A[[X_1, \ldots, X_n]]$. This is an example of S-Noetherian ring that it is not Noetherian.

Example 2 Let A be an integral domain with quotient field K.
The domains $A[[X_1, \ldots, X_n]]_{A \setminus (0)} \subseteq A_{A \setminus (0)}[[X_1, \ldots, X_n]] = K[[X_1, \ldots, X_n]]$ are in general different. The ring $K[[X_1, \ldots, X_n]]$ is Noetherian. Let, for example, $A = \mathbb{Z}$, $K = \mathbb{Q}$, and $(p_k)_{k \in \mathbb{N}}$ the sequence of the prime numbers. Then
$$f = \sum_{k:0}^{\infty} \frac{1}{p_k} X^k \in \mathbb{Q}[[X]].$$ Suppose that $f \in \mathbb{Z}[[X]]_{\mathbb{Z} \setminus (0)}$. There exist $g \in \mathbb{Z}[[X]]$ and $m \in \mathbb{Z} \setminus (0)$ such that $f = \frac{g}{m}$, and then $mf - g \in \mathbb{Z}[[X]]$. So $\frac{m}{p_k} \in \mathbb{Z}$ for each $k \in \mathbb{N}$, then $m = 0$, which is absurd.

Corollary 5.6 (Jiao and Zhang 2009 and Baeck et al. 2016) *Let $A \subseteq B$ be an extension of rings and $S \subseteq A$ an anti-Archimedean multiplicative set not containing zero divisors of A. The following assertions are equivalent:*

1. *The ring $A + XB[[X]]$ is S-Noetherian.*
2. *The ring A is S-Noetherian and the A−module B is S-finite.*

Proof Same proof as Corollary 4.5, by using Theorem 5.1.   □

Example For each anti-Archimedean multiplicative set S of $\mathbb{Z}$, the $\mathbb{Z}$-module $\mathbb{Q}$ is not S-finite. Then the ring $\mathbb{Z} + X\mathbb{Q}[[X]]$ is not S-Noetherian.

Corollary 5.7 (Hizem and Benhissi 2005) *Let $A \subseteq B$ be an extension of rings. The ring $A + XB[[X]]$ is Noetherian if and only if the ring A is Noetherian and the A−module B is finitely generated.*

Proof Take $S = \{1\}$ in the preceding corollary.   □

Theorem 5.8 (Hamed and Hizem 2015) *Let $\mathscr{A} = (A_n)_{n \in \mathbb{N}}$ be an increasing sequence of rings, $A = \bigcup_{n:0}^{\infty} A_n$, and $S \subseteq A_0$ an anti-Archimedean multiplicative set not containing zero divisors of A. The following assertions are equivalent:*

1. *The sequence $\mathscr{A} = (A_n)_{n \in \mathbb{N}}$ is S-Noetherian.*

2. *The ring A_0 is S-Noetherian and the A_0−module A is S-finite.*
3. *The ring $\mathscr{A}[[X]]$ is S-Noetherian.*

Proof $"(1) \implies (2)"$ By Theorem 4.7. $"(2) \implies (3)"$ By Corollary 3.7, the ring A is S-Noetherian. By Theorem 5.1, the ring $A[[X]]$ is S-Noetherian. On the other hand, A is a S-finite A_0−module. There exist $s \in S$ and $x_1, \ldots, x_r \in A$ such that $sA \subseteq A_0 x_1 + \ldots + A_0 x_r$. Then $sA[[X]] \subseteq A_0[[X]]x_1 + \ldots + A_0[[X]]x_r \subseteq \mathscr{A}[[X]]x_1 + \ldots + \mathscr{A}[[X]]x_r \subseteq A[[X]]$ because $A_0[[X]] \subseteq \mathscr{A}[[X]] \subseteq A[[X]]$. Then $A[[X]]$ is a S-finite $\mathscr{A}[[X]]$−module. By Corollary 3.7, the ring $\mathscr{A}[[X]]$ is S-Noetherian.

$"(3) \implies (1)"$ We show that the ring A_0 is S-Noetherian. Let I be an ideal of A_0. Consider the increasing sequence of rings $\mathscr{B} = (A_n)_{n \geq 1}$ and $\mathscr{B}[[X]] = \{f = \sum_{i:0}^{\infty} a_i X^i; a_i \in A_{i+1} \text{ for each } i \in \mathbb{N}\}$. Then $I + X\mathscr{B}[[X]]$ is an ideal of $\mathscr{A}[[X]] = A_0 + X\mathscr{B}[[X]]$. Indeed, the stability by addition is clear. Let $a \in I$, $b \in A_0$, and $f, g \in \mathscr{B}[[X]]$. Then $(a + Xf)(b + Xg) = ab + X(bf + ag + Xfg) \in I + X\mathscr{B}[[X]]$. Since the ring $\mathscr{A}[[X]]$ is S-Noetherian, there exist $s \in S$ and $f_1, \ldots, f_n \in I + X\mathscr{B}[[X]]$ such that $s(I + X\mathscr{B}[[X]]) \subseteq f_1\mathscr{A}[[X]] + \ldots + f_n\mathscr{A}[[X]]$. Then $sI \subseteq f_1(0)A_0 + \ldots + f_n(0)A_0 \subseteq I$, so I is S-finite.

We will show that the sequence $\mathscr{A} = (A_n)_{n \in \mathbb{N}}$ is S-stationary. Since the ring $\mathscr{A}[[X]]$ is S-Noetherian, the ideal $I = (aX^i; i \in \mathbb{N}^*, a \in A_i)\mathscr{A}[[X]]$ is S-finite. There exist $s \in S$ and $a_k X^{\alpha_k}$, $1 \leq k \leq r$ with $\alpha_k \geq 1$ and $a_k \in A_{\alpha_k}$ such that $sI \subseteq (a_k X^{\alpha_k}; 1 \leq k \leq r)\mathscr{A}[[X]]$. Let $m = max\{\alpha_1, \ldots, \alpha_r\} \in \mathbb{N}^*$. Then $a_1, \ldots, a_r \in A_m$.

Fix an integer $i > m$. For each $y \in A_i$, $yX^i \in I$, and then $syX^i \in (a_k X^{\alpha_k}; 1 \leq k \leq r)\mathscr{A}[[X]]$. Put $syX^i = \sum_{k:1}^{r} a_k X^{\alpha_k} g_k$ with $g_k = \sum_{j:0}^{\infty} g_{k,j} X^j \in \mathscr{A}[[X]]$, and then $g_{k,j} \in A_j$. By identification, $sy = \sum_{k:1}^{r} a_k g_{k,i-\alpha_k}$. Since $g_{k,i-\alpha_k} \in A_{i-\alpha_k} \subseteq A_{i-1}$, then $sA_i \subseteq a_1 A_{i-1} + \ldots + a_r A_{i-1} \subseteq A_{i-1}$. So for each integer $i > m$, $sA_i \subseteq A_{i-1}$ and then $s^{i-m}A_i \subseteq A_m$. Since S is anti-Archimedean in A_0, there exist $t \in (\bigcap_{n:1}^{\infty} s^n A_0) \cap S$. For each $n \in \mathbb{N}^*$, there exist $a_n \in A_0$ such that $t = s^n a_n$. For each integer $i > m$, $tA_i = s^{i-m} a_{i-m} A_i \subseteq s^{i-m} A_i \subseteq A_m$ because $a_{i-m} \in A_0 \subseteq A_i$. So the sequence $\mathscr{A} = (A_n)_{n \in \mathbb{N}}$ is S-stationary.

Let $n \in \mathbb{N}^*$. We will show that the A_0−module A_n is S-finite. The ideal $I = \{f = \sum_{i:0}^{\infty} a_i X^i \in \mathscr{A}[[X]]; a_0 = \ldots = a_{n-1} = 0\}$ of $\mathscr{A}[[X]]$ is S-finite. There exist $s \in S$ and $f_1, \ldots, f_m \in I$ such that $sI \subseteq f_1\mathscr{A}[[X]] + \ldots + f_m\mathscr{A}[[X]]$. Put $f_i = X^n g_i$ with $g_i \in A_n + XA_{n+1} + \ldots$. Then $sA_n \subseteq g_1(0)A_0 + \ldots + g_m(0)A_0 \subseteq A_n$.

In conclusion, the sequence $\mathscr{A} = (A_n)_{n \in \mathbb{N}}$ is S-Noetherian.   □

Corollary 5.9 (Hamed and Hizem 2015) *Let $\mathscr{A} = (A_n)_{n\in\mathbb{N}}$ be an increasing sequence of rings and $S \subseteq A_0$ an anti-Archimedean multiplicative set not containing zero divisors of $A = \bigcup_{n:0}^{\infty} A_n$. The following assertions are equivalent:*

1. *The sequence $\mathscr{A} = (A_n)_{n\in\mathbb{N}}$ is S-Noetherian.*
2. *The ring $\mathscr{A}[X]$ is S-Noetherian.*
3. *The ring $\mathscr{A}[[X]]$ is S-Noetherian.*

Example Let $(A_n)_{n\in\mathbb{N}}$ be an increasing sequence of rings. The following assertions are equivalent:

1. The sequence $(A_n)_{n\in\mathbb{N}}$ is Noetherian; i.e., the ring A_0 is Noetherian, the sequence $(A_n)_{n\in\mathbb{N}}$ is stationary, and for each $n \in \mathbb{N}$, A_n is a finitely generated A_0–module.
2. The ring $\mathscr{A}[X]$ is Noetherian.
3. The ring $\mathscr{A}[[X]]$ is Noetherian.

Remark 5.10 Theorem 5.8 (or the preceding corollary) allows us to retrieve some version of Corollary 5.6. Indeed, let $A \subseteq B$ be an extension of rings and $S \subseteq A$ an anti-Archimedean multiplicative set not containing zero divisors of B. Consider the sequence of rings $\mathscr{A} = (A_n)_{n\in\mathbb{N}}$ defined by $A_0 = A$ and for each $n \geq 1$, $A_n = B$. Then $\mathscr{A}[[X]] = A + XB[[X]]$. If the sequence $(A_n)_{n\in\mathbb{N}}$ is stationary, then it is S-stationary. So the ring $\mathscr{A}[[X]] = A + XB[[X]]$ is S-Noetherian $\Longleftrightarrow$ the sequence $(A_n)_{n\in\mathbb{N}}$ is S-Noetherian $\Longleftrightarrow$ the ring $A_0 = A$ is S-Noetherian and for each $n \geq 1$, $A_n = B$ is a S-finite $A_0 = A$–module.

In the remaining of this section, we want to prove that a domain A is Euclidean (resp. PID) if and only if the domain $A[[X]][X^{-1}]$ is Euclidean (resp. PID).

Proposition 5.11 *A DVR of rank one is Euclidean.*

Proof Let A be a DVR of rank one defined by a normalized valuation v. Let $\sigma : A \setminus (0) \longrightarrow \mathbb{N}$ be the map defined by $\sigma(x) = v(x) + 1$. If x divides y, then $v(x) \leq v(y)$, so $v(x) + 1 \leq v(y) + 1$, and then $\sigma(x) \leq \sigma(y)$. If x does not divide y, then $v(x) > v(y)$, so $\frac{x}{y} \in A$. Let $r = y - \frac{x^2}{y} = \frac{y^2-x^2}{y} \in A$. Then $v(r) = v(y^2 - x^2) - v(y)$. But $v(y^2) = 2v(y) < 2v(x) = v(x^2)$, and then $v(y^2 - x^2) = inf\{v(x^2), v(y^2)\} = v(y^2) = 2v(y)$, and $v(r) = 2v(y) - v(y) = v(y)$. So $\sigma(r) = v(r) + 1 = v(y) + 1 < v(x) + 1 = \sigma(x)$ and $y = x\frac{x}{y} + y - \frac{x^2}{y} = x(\frac{x}{y}) + r$. $\qquad\square$

The following corollary relative to the ring of formal power series is analogue to the case of the ring of polynomials.

Corollary 5.12 *The following assertions are equivalent for a ring A:*
(1) $A[[X]]$ *is Euclidean.* (2) $A[[X]]$ *is PID.* (3) A *is a field.*

Proof $''(1) \implies (2)''$ Clear. $''(2) \implies (3)''$ Let $0 \neq a \in A$. Put $(a, X) = f(X)A[[X]]$ with $f(X) \in A[[X]]$. Then $a = f(X)g(X)$ and $X = f(X)h(X)$ with $g(X), h(X) \in A[[X]]$. So $a = f(0)g(0)$, and then $f(0) \neq 0$ and $0 = f(0)h(0)$. So $h(0) = 0$ and $h(X) = Xh_1(X)$ with $h_1(X) \in A[[X]]$. Then $X = Xf(X)h_1(X)$, so $1 = f(X)h_1(X)$ and $f(X)$ is invertible, then $(a, X) = A[[X]]$. Put $1 = aq(X) + Xr(X)$ with $q(X), r(X) \in A[[X]]$. Then $1 = aq(0)$, so a is invertible in A, and then A is a field. $\qquad\qquad\square$

Other proofs

(1) If the domain A is not a field, it must have a nonzero maximal ideal M. Then $(0) \subset M[[X]] \subset M + XA[[X]]$. Then $M[[X]]$ is a nonzero nonmaximal prime ideal of $A[[X]]$. Then $A[[X]]$ is not a PID, which is absurd.

(2) Since $A[[X]]$ is a domain, then A is also a domain. Since $A[[X]]/(X) \simeq A$, then (X) is a nonzero prime ideal of a PID. So (X) is maximal and $A \simeq A[[X]]/(X)$ is a field.

$''(3) \implies (1)''$ Since $A[[X]]$ is a DVR of rank one, we conclude by the preceding proposition. $\qquad\qquad\square$

Notations Let A be a domain and $S = \{X^i; \ i \in \mathbb{N}\}$: a multiplicative set of $A[[X]]$. The quotient ring $S^{-1}A[[X]] = A[[X]][X^{-1}] = A((X))$ is called the domain of Laurent series with coefficients in A. It contains $A[[X]]$. A nonzero element of $A((X))$ can be written as $f(X) = a_n X^n + a_{n+1} X^{n+1} + \ldots$ with $n \in \mathbb{Z}, a_i \in A$, and $a_n \neq 0$. The element a_n is called the initial coefficient of $f(X)$.

For each domain A, the ring $A[[X]]$ is never a field. This is not the case for the Laurent series domain $A((X))$.

Proposition 5.13 *Let A be an integral domain and $0 \neq f(X) \in A((X))$. Then $f(X)$ is invertible in $A((X))$ if and only if its initial coefficient is invertible in A.*

In particular, $A((X))$ is a field if and only if A is a field and in this case, $A((X))$ is the quotient field of $A[[X]]$.

Proof Put $f(X) = a_n X^n + \ldots \in A((X))$ with $a_n \neq 0$.

$'' \implies ''$ There exists $g(X) = b_m X^m + \ldots \in A((X))$ with $b_m \neq 0$ such that $1 = f(X)g(X) = a_n b_m X^{n+m} + \ldots$. Then $a_n b_m = 1$ and a_n is invertible in A.

$'' \impliedby ''$ $f(X) = X^n(a_n + a_{n+1}X + \ldots)$ with X^n invertible in $A((X))$ and $a_n + a_{n+1}X + \ldots$ invertible in $A[[X]]$. Then $f(X)$ is invertible in $A((X))$.

Since $A((X))$ is an over ring of $A[[X]]$, then the two domains have the same quotient field. When A is a field, $A((X))$ is also a field. Then it is the quotient field of $A[[X]]$. $\qquad\qquad\square$

Definition 5.14 Let A be an integral domain. An Euclidean stathm over the domain $A((X))$ is said to be $A-$determined if it depends only on the value of the initial coefficient of the considered series.

The following theorem shows that there is a correspondence between the stathms of the domain A and some remarkable stathms of the domain $A((X))$.

Theorem 5.15 (Dress 1970 and 1971) *Let A be an Euclidean domain. There is a bijection between the set of the Euclidean stathms of A and the set of the A−determined Euclidean stathms of $A((X))$.*

Proof

(1) Let σ be an Euclidean stathm on A. For $F = a_n X^n + \dots \in A((X))$, $n \in \mathbb{Z}$, $a_n \neq 0$, let $\sigma_X(F) = \sigma(a_n)$. Then $\sigma_X(F)$ depends only on the value of the coefficient a_n and the restriction of σ_X to $A \setminus (0)$ coincide with σ. Then the map $\sigma \longrightarrow \sigma_X$ is one to one.

Let F and $G \in A((X)) \setminus (0)$. If G divides F in $A((X))$, the initial coefficient a of G divides the initial coefficient b of F in A. Then $\sigma(a) \leq \sigma(b)$, so $\sigma_X(G) \leq \sigma_X(F)$. Suppose that G does not divide F. Then $\frac{F}{G} = X^s \frac{F_1}{G_1}$ with $s \in \mathbb{Z}$, $F_1, G_1 \in A[[X]]$, $F_1(0) \neq 0$ and $G_1(0) \neq 0$. Let K be the quotient field of A. Then G_1 is invertible in $K[[X]]$ and $\frac{F}{G} = c_s X^s + c_{s+1} X^{s+1} + \dots$ where the $c_i \in K$ and $c_s \neq 0$. This means that $\frac{F}{G} \in K((X))$. Since G does not divide F in $A((X))$, the coefficients c_i are not all in A. Let λ be the smallest index such that $c_\lambda \notin A$. Then $c_s, c_{s+1}, \dots, c_{\lambda-1} \in A$. We have $F - G(c_s X^s + \dots + c_{\lambda-1} X^{\lambda-1}) = (c_\lambda X^\lambda + \dots)G$. The left member is in $A((X))$ and by identification, the initial coefficient of the right member $0 \neq ac_\lambda \in A$, a is always the initial coefficient of G. The coefficient ac_λ is not divisible by a in A since $\frac{ac_\lambda}{a} = c_\lambda \notin A$. Then $ac_\lambda = ac'_\lambda + r$ with $r, c'_\lambda \in A$, $r \neq 0$, and $\sigma(r) < \sigma(a)$. So $c_\lambda = c'_\lambda + \frac{r}{a}$. Let $P = c_s X^s + \dots + c_{\lambda-1} X^{\lambda-1} + c'_\lambda X^\lambda \in A((X))$ and $R = F - GP = (\frac{r}{a} X^\lambda + \dots)G \neq 0$. Then $R = F - GP \in A((X))$ and $F = GP + R$. It suffices to show that $\sigma_X(R) < \sigma_X(G)$. But the initial coefficient of $R = (\frac{r}{a} X^\lambda + \dots)G$ is $\frac{r}{a} a = r$ with $\sigma(r) < \sigma(a)$, and then $\sigma_X(R) < \sigma_X(G)$.

(2) Let σ' be a A−determined Euclidean stathm over $A((X))$ and σ its restriction to A. We will show that σ is a stathm over A. Let $a, b \in A \setminus (0)$. If b divides a in A, it divides it in $A((X))$, so $\sigma'(b) \leq \sigma'(a)$, and then $\sigma(b) \leq \sigma(a)$. Suppose that b does not divide a in A. It does not divide it in $A((X))$, by identification. The division can be written as $a = b(q_s X^s + q_{s+1} X^{s+1} + \dots) + (r_t X^t + r_{t+1} X^{t+1} + \dots)$ with $s, t \in \mathbb{Z}$, $q_i, r_i \in A$, $q_s \neq 0$, $r_t \neq 0$, and $\sigma'(r_t X^t + r_{t+1} X^{t+1} + \dots) < \sigma'(b)$. Since σ' is A−determined, then $\sigma'(r_t) < \sigma'(b)$, so $\sigma(r_t) < \sigma(b)$. (The result is clear in the case where the quotient of a by b in $A((X))$ is zero: $a = b0 + r_0$ with $\sigma(r_0) < \sigma(b)$)

First case: $s \leq -1$: If $s < t$, then $q_s = 0$, which is absurd and if $t < s$, then $r_t = 0$, which is also absurd. Then $t = s$ and $-bq_s = r_t$, and then b divides r_t and $\sigma'(b) \leq \sigma'(r_t)$, which is absurd.

Second case: $s \geq 0$: If $t \leq -1$, then $r_t = 0$, which is absurd. Then $t \geq 0$. So $a = bq_0 + r_0$ with eventually $q_0 = 0$ and/or $r_0 = 0$. Since b does not divide a, then $r_0 \neq 0$. So necessarily $r_0 = r_t =$the initial coefficient of the rest. But in this case, $\sigma(r_0) = \sigma(r_t) = \sigma'(r_t) < \sigma(b)$.

We conclude that σ is a stathm over A and $\sigma_X = \sigma'$ since $\sigma_X(a_n X^n + \ldots) = \sigma(a_n) = \sigma'(a_n) = \sigma'(a_n X^n + \ldots)$ because σ' is A–determined.   □

Example The intersection of two Euclidean subdomains is not necessarily a PID. For example, $\mathbb{Z}((X)) \cap \mathbb{Q}[X] = \mathbb{Z}[X]$ is not a PID. Indeed, let $I = (2, X) = \{2R + XQ; \ R, Q \in \mathbb{Z}[X]\} = \{h(X) \in \mathbb{Z}[X]; \ h(0) \in 2\mathbb{Z}\}$. Suppose that I is principal generated by $P(X)$. Then $P(X)$ divides 2, so $P \in \{-1, 1 - 2, 2\}$. Also, P divides X. Since 2 and -2 do not divide X, then either $P = 1$ or $P = -1$. But these two elements are not in I.

Definition 5.16 Let A be an integral domain. A prestathm over A is a map
$\sigma : A \setminus (0) \longrightarrow \mathbb{N}$ satisfying: For each $a, b \in A \setminus (0)$ with b not dividing a, then $a = bq + r$, where $q, r \in A, r \neq 0$, and $\sigma(r) < \sigma(b)$.

We can endow the set of the prestathms over an integral domain with an order. This allows us to speak about the smallest prestathm and the smallest stathm.

Definition 5.17 Let σ and σ' be two prestathms over an integral domain A. We say that $\sigma \leq \sigma'$ if for each $a \in A \setminus (0), \sigma(a) \leq \sigma'(a)$.

Starting with any prestathm, we can construct a stathm.

Proposition 5.18 *Let A be an integral domain. If A has a prestathm σ, then it has a stathm $\sigma' \leq \sigma$.*

Proof For $x \in A \setminus (0)$, let $\sigma'(x) = \inf\{\sigma(y); \ 0 \neq y \in xA\} \leq \sigma(x)$. If $b|a$, then $aA \subseteq bA$, so $\inf\{\sigma(y); 0 \neq y \in aA\} \geq \inf\{\sigma(y); 0 \neq y \in bA\}$. Then $\sigma'(a) \geq \sigma'(b)$. If b does not divide a, let $\sigma'(b) = \inf\{\sigma(y); \ 0 \neq y \in bA\} = \sigma(y_0)$ with $y_0 = bx_0 \neq 0$. The element y_0 does not divide a, and then $a = y_0 q + r$ where $q, r \in A, r \neq 0$, and $\sigma(r) < \sigma(y_0)$. Then $a = b(x_0 q) + r, r \neq 0, \sigma'(r) \leq \sigma(r) < \sigma(y_0) = \sigma'(b)$.   □

Proposition 5.19 *Let A be an integral domain and (σ_α) a nonempty family of Euclidean stathms over A. The map σ, defined for each $0 \neq a \in A$ by $\sigma(a) = \inf\{\sigma_\alpha(a)\}$, is also an Euclidean stathm over A.*

Proof Let $a, b \in A \setminus (0)$. If b divides a, then for each $\alpha, \sigma_\alpha(b) \leq \sigma_\alpha(a)$, so $\sigma(b) = \inf\{\sigma_\alpha(b)\} \leq \inf\{\sigma_\alpha(a)\} = \sigma(a)$. If b does not divide a, let α_0 be such that $\sigma(b) = \sigma_{\alpha_0}(b)$. Then $a = bq + r$ with $q, r \in A, r \neq 0$, and $\sigma_{\alpha_0}(r) < \sigma_{\alpha_0}(b)$. So $\sigma(r) \leq \sigma_{\alpha_0}(r) < \sigma_{\alpha_0}(b) = \sigma(b)$.   □

Corollary 5.20 *An Euclidean domain A admits a smallest Euclidean stathm, and it is also the smallest Euclidean prestathm over A.*

Proof Each prestathm is bounded from below by a stathm.   □

Lemma 5.21 *Let A be an Euclidean domain, $0 \neq b \in A$, $n \in \mathbb{N}^*$, and σ the smallest Euclidean stathm over A (it is also the smallest Euclidean prestathm).*

Suppose that for each $0 \neq a \in A$ not divisible by b, there exist $r, q \in A$, $r \neq 0$ satisfying $a = bq + r$ and $\sigma(r) < n$. Then $\sigma(b) \leq n$.

Proof Suppose that $\sigma(b) > n$. Define the map $\sigma_1 : A \setminus (0) \longrightarrow \mathbb{N}$ by $\sigma_1(b) = n$ and $\sigma_1(a) = \sigma(a)$ for each $a \in A \setminus \{0, b\}$. We will show that σ_1 is a prestathm over A, which contradicts the minimality of σ.

First case: b plays the role of a divisor: Let $0 \neq a \in A$ be such that b does not divide a. By hypothesis, $a = bq + r$ with $q, r \in A$, $r \neq 0$, and $\sigma(r) < n$. Then $r \neq b$, so $\sigma_1(r) = \sigma(r) < n = \sigma_1(b)$. Then $\sigma_1(r) < \sigma_1(b)$.

Second case: b plays the role of the rest: Let a and $c \in A \setminus \{0, b\}$ be such that c does not divide a. Then $a = cq + r$ with $q, r \in A$, $r \neq 0$, and $\sigma(r) < \sigma(c)$. For $r \neq b$, we immediately have $\sigma_1(r) < \sigma_1(c)$. For $r = b$, $a = cq + b$ with $\sigma_1(b) = n < \sigma(b) < \sigma(c) = \sigma_1(c)$.

Third case: b plays the role of dividend: Let $a \in A \setminus \{0, b\}$ be such that a does not divide b. Then $b = aq + r$ with $q, r \in A$, $r \neq 0$, and $\sigma(r) < \sigma(a)$. For $r \neq b$, we immediately have $\sigma_1(r) < \sigma_1(a)$. For $r = b$, $aq = 0$, then $q = 0$ and $\sigma(b) < \sigma(a)$. So $b = a0 + b$ with $\sigma_1(b) = n < \sigma(b) < \sigma(a) = \sigma_1(a)$.

Conclusion: We construct a prestathm σ_1 over A such that $\sigma_1(b) = n < \sigma(b)$, which is absurd. Then $\sigma(b) \leq n$. $\qquad\qquad\square$

Theorem 5.22 (Dress 1970 and 1971) *Let A be an integral domain. If $A((X))$ is an Euclidean domain, its smallest stathm is A-determined.*

Proof Let σ be the smallest stathm of $A((X))$. We will show that by induction on n that if $f = b_k X^k + \ldots \in A((X))$, $b_k \neq 0$, is a series with image n by σ, then each other series with initial coefficient b_k admits also n as image.

The order 0: Note first that in any Euclidean domain (A, σ), if $\sigma(a) = 0$ with $0 \neq a \in A$, then a is invertible. Indeed, in the contrary case, a does not divide 1, and then $1 = aq + r$ with $q, r \in A$, $r \neq 0$, and $\sigma(r) < \sigma(a) = 0$, which is absurd.

Now since $\sigma(f) = 0$, then f is invertible in $A((X))$, so b_k is invertible in A. Let $g \in A((X))$ with initial coefficient b_k. Then g is invertible in $A((X))$. So f and g divide each other in $A((X))$, then $\sigma(f) \leq \sigma(g)$ and $\sigma(g) \leq \sigma(f)$ so $\sigma(g) = \sigma(f) = 0$.

Induction hypothesis: The property is true until the order $n - 1$.

The order n: Let $f = b_k X^k + \ldots \in A((X))$, $b_k \neq 0$, and $\sigma(f) = n$. Then $f = X^k(b_k + b_{k+1}X \ldots)$. Since X^k invertible in $A((X))$, then $\sigma(b_k + b_{k+1}X \ldots) = \sigma(f) = n$. We may suppose that $f = b_k + b_{k+1}X \ldots$. The goal is to show that for each $G(X) \in A((X))$ with initial coefficient b_k, we have $\sigma(G) = n$.

First Step Each nonzero element a of A may be divided by b_k in the following way: $a = b_k q + r$ with $q, r \in A$, either $r = 0$ or $\sigma(r) < n$.

Indeed, if f divides a in $A((X))$, then b_k divides a in A. If f does not divide a in $A((X))$, then $a = (b_k + b_{k+1}X \ldots)(q_s X^s + q_{s+1}X^{s+1} + \ldots) + (r_t X^t + r_{t+1}X^{t+1} + \ldots)$ with $r_t \neq 0$, $s, t \in \mathbb{Z}$, and $\sigma(r_t X^t + r_{t+1}X^{t+1} + \ldots) < \sigma(f) = n$. Put $r' = r_t X^t + r_{t+1}X^{t+1} + \ldots$. By the induction hypothesis, $\sigma(r_t) = \sigma(r') < n$. We may suppose that $q_s \neq 0$ since if the quotient is zero, $t = 0$ and $r' = r_0 = a$, and then $a = b_k 0 + r_0$ with $\sigma(r_0) < n$.

We distinguish two cases:

First case: $s \leq -1$. Necessarily, $s = t$ and $0 = b_k q_s + r_t$, and then $r_t = -b_k q_s$ and $\sigma(b_k) \leq \sigma(r_t) < n$. By the induction hypothesis, $\sigma(f) = \sigma(b_k + b_{k+1}X \ldots) = \sigma(b_k) < n$. This contradicts $\sigma(f) = n$.

Second case: $s \geq 0$: Necessarily, $t \geq 0$ and $a = b_k q_0 + r_0$ with q_0 and $r_0 \in A$, eventually zero. If $r_0 \neq 0$, then $r_0 = r_t$, so $\sigma(r_0) = \sigma(r_t) = \sigma(r') < n$.

Second Step Let $F(X)$ and $G(X) \in A((X)) \setminus (0)$ with $G(X)$ having an initial term b_k. If $G(X)$ does not divide $F(X)$ in $A((X))$, then $F(X) = G(X)P(X) + R(X)$ with $P(X), R(X) \in A((X))$, $R(X) \neq 0$, and $\sigma(R(X)) < n$.

Indeed, $\frac{F}{G} = X^s \frac{F_1}{G_1}$ with $s \in \mathbb{Z}$, $F_1, G_1 \in A[[X]]$, $F_1(0) \neq 0$, and $G_1(0) \neq 0$. Let K be the quotient field of A. Then G_1 is invertible in $K[[X]]$ and $\frac{F}{G} = c_s X^s + c_{s+1} X^{s+1} + \ldots$ where the $c_i \in K$ and $c_s \neq 0$. This means that $\frac{F}{G} \in K((X))$. Since G does not divide F in $A((X))$, the coefficients c_i are not all in A. Let λ be the smallest index such that $c_\lambda \notin A$. Then $c_s, c_{s+1}, \ldots, c_{\lambda-1} \in A$. So $F - G(c_s X^s + \ldots + c_{\lambda-1} X^{\lambda-1}) = (c_\lambda X^\lambda + \ldots)G$. The member of the left is in $A((X))$ and by identification, the initial coefficient in right member $0 \neq b_k c_\lambda \in A$ and is not divisible by b_k in A since $\frac{b_k c_\lambda}{b_k} = c_\lambda \notin A$. By the first step, $b_k c_\lambda = b_k c'_\lambda + r$ with $r, c'_\lambda \in A$, $r \neq 0$, and $\sigma(r) < n$. Then $c_\lambda = c'_\lambda + \frac{r}{b_k}$. Let $P = c_s X^s + \ldots + c_{\lambda-1} X^{\lambda-1} + c'_\lambda X^\lambda \in A((X))$ and $R = F - GP = (\frac{r}{b_k} X^\lambda + \ldots)G \neq 0$. Then $R = F - GP \in A((X))$ and $F = GP + R$. The initial coefficient of $R = (\frac{r}{b_k} X^\lambda + \ldots)G$ is $\frac{r}{b_k} b_k = r$ with $\sigma(r) < n$. By the induction hypothesis, $\sigma(R(X)) = \sigma(r) < n$.

Third Step Let $G(X) \in A((X)) \setminus (0)$ with initial coefficient b_k. By the second step and the preceding lemma, $\sigma(G(X)) \leq n$. By the induction hypothesis, if $\sigma(G(X)) < n$, then $\sigma(f) = \sigma(b_k + b_{k+1}X + \ldots) = \sigma(G(X)) < n$, which is absurd. Then $\sigma(G(X)) = n$.   □

By Corollary 5.12, if A is an Euclidean domain which is not a field, then the ring $A[[X]]$ is not Euclidean. The following corollary show that $A((X))$ is Euclidean.

Corollary 5.23 *A domain A is Euclidean if and only if $A((X))$ is Euclidean.*

Proof $" \Longrightarrow "$ By Theorem 5.15.

$" \Longleftarrow "$ The smallest stathm over $A((X))$ is A-determined. By the second part of the proof of Theorem 5.15, its restriction to A is a stathm.   □

Proposition 5.24 *A ring A is a PID if and only if $A((X))$ is a PID.*

Proof $" \Longrightarrow "$ Let I be an ideal of A. Then $IA((X)) = aA((X))$ with $a \in I$. So $I = aA$.

$" \Longleftarrow "$ (Dress 1968). Let $\mathscr{I}$ be an ideal of $A((X))$ and I the set of the constant terms of the series in $\mathscr{I}$ of the form $f = \sum_{i:0}^{\infty} a_i X^i$. We will show that I is an ideal of A. Since $0 \in \mathcal{I}$, then $0 \in I$. Let a_0 and $b_0 \in I$ correspond to the series

$$f = \sum_{i:0}^{\infty} a_i X^i \quad \text{and} \quad g = \sum_{i:0}^{\infty} b_i X^i \text{ of } \mathcal{I}. \text{ Then } f - g = \sum_{i:0}^{\infty} (a_i - b_i) X^i \in \mathcal{I},$$

so $a_0 - b_0 \in I$. For each $c \in A$, $cf = \sum_{i:0}^{\infty} ca_i X^i \in \mathcal{I}$, and then $ca_0 \in I$. Put

$I = bA$ as a principal ideal of A. The element b is the constant term of a series
$\beta(X) = b + b_1 X + b_2 X^2 + \ldots \in \mathcal{I}$. We will show that $\mathcal{I} = \beta(X)A((X))$. The
converse inclusion is clear. Since the X^m, $m \in \mathbb{Z}$, are invertible in $A((X))$, for the
direct inclusion, we may just consider the series of $\mathcal{I}$ of the form $f = \sum_{i:0}^{\infty} a_i X^i$.

By definition, $a_0 = bc_0$ with $c_0 \in A$. Suppose by induction that we have n elements

$c_0, c_1, \ldots, c_{n-1} \in A$ such that $\beta(X)(c_0 + c_1 X + \ldots + c_{n-1} X^{n-1}) = \sum_{k:0}^{\infty} \alpha_k X^k$ with

$\alpha_k = a_k$ for $0 \leq k \leq n-1$. Then $f - \beta(X)(c_0 + c_1 X + \ldots + c_{n-1} X^{n-1}) = u_n X^n + u_{n+1} X^{n+1} + \ldots \in \mathcal{I}$. Since X^n is invertible in $A((X))$, $u_n + u_{n+1} X + \ldots \in \mathcal{I}$,
then $u_n \in I$, $u_n = bc_n$ with $c_n \in A$. So $\beta(X)(c_0 + c_1 X + \ldots + c_{n-1} X^{n-1} + c_n X^n) = \beta(X)(c_0 + c_1 X + \ldots + c_{n-1} X^{n-1}) + c_n X^n \beta(X) = f - (u_n X^n + u_{n+1} X^{n+1} + \ldots) + c_n X^n \beta(X) = f - (u_{n+1} X^{n+1} + \ldots) + \sum_{i:1}^{\infty} c_n b_i X^{n+i} = \sum_{k:0}^{\infty} \alpha_k X^k$, with

$\alpha_k = a_k$ for $0 \leq k \leq n$. We obtain a series $g = \sum_{k:0}^{\infty} c_k X^k \in A[[X]]$ such that

$$\beta(X)g(X) = f(X). \qquad \square$$

Remark 5.25 By Corollary 5.12, the ring $A[[X]]$ is a PID if and only if A is a field.

We retrieve a particular case of Corollary 5.2.

Corollary 5.26 *If A is a PID, then $A[[X]]$ is Noetherian.*

Proof Let P be a nonzero prime ideal of $A[[X]]$ and $S = \{X^i; \ i \in \mathbb{N}\}$.

First case: $P \cap S = \emptyset$. Then $S^{-1}P$ is a nonzero prime ideal of $A((X))$. Put
$S^{-1}P = fA((X))$ with $0 \neq f \in P$. Since the X^n, $n \in \mathbb{Z}$, are invertible in $A((X))$,
we may suppose that $f(0) \neq 0$. It is clear that $fA[[X]] \subseteq P$. Conversely, let
$0 \neq g \in P \subseteq fA((X))$. Then $g = fh$ with $h \in A((X))$. Suppose that $h \notin A[[X]]$.
Then $h = \frac{h'}{X^n}$ with $h' \in A[[X]]$, $h'(0) \neq 0$, and $n \geq 1$. So $X^n g = fh'$, then
$f(0)h'(0) = 0$, which is absurd. Then $h \in A[[X]]$ and $P = fA[[X]]$.

Second case: $P \cap S \neq \emptyset$. Since P is prime containing a positive power of X, then
$X \in P$. The set $P' = \{g(0); \ g \in P\}$ is an ideal of A. Put $P' = aA$ with $a = f(0)$,
$f \in P$. Since $f = a + Xf'$ with $f' \in A[[X]]$, then $a \in P$. We have $(a, X) \subseteq P$.
Conversely, let $g \in P$, $g = b + Xg'$ with $b = g(0) = \alpha a$, $\alpha \in A$, and $g' \in A[[X]]$.
So $g \in (a, X)$ and $P = (a, X)$. $\qquad \square$

The preceding results show that many properties of the ring A pass to $A((X))$ and do not pass to $A[[X]]$.

6 S-Noetherian Pull Backs

Pull backs are important tools in commutative algebra. They allow to construct various examples. We retrieve the classical construction $D + M$, which includes $D + (X_1, \ldots, X_n)K[X_1, \ldots, X_n]$ and $D + (X_1, \ldots, X_n)K[[X_1, \ldots, X_n]]$. Other examples of pull backs will be given in the following sections. We admit that a multiplicative set may contain 0.

Lemma 6.1 *Let $f : A \longrightarrow B$ be a homomorphism of rings.*

1. *B may be endowed with a structure of $A-$module for the multiplication defined by $a.b = f(a)b$ for each $a \in A$ and $b \in B$. Moreover, f is a homomorphism of $A-$modules.*
2. *Each ideal of B is a $A-$sub-module of B. If f is onto, the ideals of the ring B are exactly the $A-$sub-modules of B.*
3. *Suppose that f is onto. Let S a multiplicative set of the ring A.*

For each ideal I of the ring B, i.e., a $A-$sub-module of B, the ideal I is $f(S)-$finite in the ring B if and only if I is a S-finite $A-$sub-module of B. In particular, the ring B is $f(S)-$Noetherian if and only if the $A-$module B is S-Noetherian.

Proof

(1) Let $a, a' \in A$ and $b, b' \in B$. Then $1.b = f(1)b = 1b = b$; $(a + a').b = f(a + a')b = f(a)b + f(a')b = a.b + a'.b$; $a.(b + b') = f(a)(b + b') = f(a)b + f(a)b' = a.b + a.b'$; $a.(a'.b) = f(a)(f(a')b) = f(aa')b = (aa').b$.
   Moreover, f is additive by hypothesis and $f(aa') = f(a)f(a') = a.f(a')$.

(2) Let J be an ideal of B. For each $a \in A$ and $j \in J$, $a.j = f(a)j \in J$. Suppose that f is onto. Let J be a $A-$sub-module of B. For each $j \in J$ and $b \in B$, there exists $a \in A$ such that $f(a) = b$. Then $bj = f(a)j = a.j \in J$. So J is an ideal of B.

(3) I is a $f(S)-$finite ideal of the ring $B \Longleftrightarrow$ there exist $s \in S$ and $a_1, \ldots, a_n \in I$ such that $f(s)I \subseteq Ba_1 + \ldots + Ba_n \subseteq I \Longleftrightarrow f(s)I \subseteq f(A)a_1 + \ldots + f(A)a_n \subseteq I \Longleftrightarrow s.I \subseteq A.a_1 + \ldots + A.a_n \subseteq I \Longleftrightarrow I$ is a S-finite $A-$sub-module of B.

$\qquad\qquad\qquad\qquad\qquad\qquad\qquad\qquad\qquad\qquad\qquad\qquad\qquad\qquad\qquad\qquad\qquad\qquad\square$

Example The condition "f is onto" is necessary in (2). Let A be any ring, $B = A[X]$, and $f : A \longrightarrow A[X]$ the canonical injection. For each $n \in \mathbb{N}$, the $A-$sub-module of $A[X]$ generated by $1, X, \ldots, X^n$ is not an ideal of $A[X]$.

Definition 6.2 Let $\alpha : A \longrightarrow C$ and $\beta : B \longrightarrow C$ be two homomorphisms of rings. The sub-ring denoted $D = \alpha \times_C \beta = \{(a, b) \in A \times B; \alpha(a) = \beta(b)\}$, of the product ring $A \times B$, is called the pull back of α and β.

The canonical projections $p_A : D \subseteq A \times B \longrightarrow A$ and $p_B : D \subseteq A \times B \longrightarrow B$ are defined by $p_A(a, b) = a$ and $p_B(a, b) = b$ for each $(a, b) \in D$.

We obtain the following commutative diagram:

$$
\begin{array}{ccc}
D = \alpha \times_C \beta & \xrightarrow{p_A} & A \\
{\scriptstyle p_B}\downarrow & & \downarrow{\scriptstyle \alpha} \\
B & \xrightarrow{\beta} & C
\end{array}
$$

Example 1 Let A be a ring, M a maximal ideal of A, and B a sub-ring of the field A/M. Consider the injection $i : B \longrightarrow A/M$ and the the canonical homomorphism $s : A \longrightarrow A/M$. The set $D = s^{-1}(B) = \{a \in A; s(a) = \bar{a} \in B\}$ is a sub-ring of A. On the other hand, $s \times_{A/M} i = \{(a, b) \in A \times B; s(a) = i(b)\} = \{(a, b) \in A \times B; \bar{a} = b\} = \{(a, \bar{a}); a \in A, \bar{a} \in B\} \simeq D$. Then the pull back is isomorphic to a sub-ring of the the initial ring and we have the following commutative diagram:

$$
\begin{array}{ccc}
D \simeq s \times_{A/M} i & \xrightarrow{p_A} & A \\
{\scriptstyle p_B}\downarrow & & \downarrow{\scriptstyle s} \\
B & \xrightarrow{i} & A/M
\end{array}
$$

Example 2 Let in the preceding example, $A = \mathbb{Q}[X]$ and $M = X\mathbb{Q}[X]$. Then $A/M = \mathbb{Q}[X]/X\mathbb{Q}[X] \simeq \mathbb{Q}$. The sub-ring $\mathbb{Z}$ of $\mathbb{Q}$ is isomorphic to the sub-ring $B = \{n + X\mathbb{Q}[X]; n \in \mathbb{Z}\}$ of $\mathbb{Q}[X]/X\mathbb{Q}[X] = A/M$. The injection $i : \mathbb{Z} \simeq B \longrightarrow \mathbb{Q} \simeq A/M$ is defined by $i(n) = n$ for each $n \in \mathbb{Z}$. The canonical homomorphism $s : A = \mathbb{Q}[X] \longrightarrow A/M = \mathbb{Q}[X]/X\mathbb{Q}[X] \simeq \mathbb{Q}$ is defined by $s(f(X)) = f(0)$ for each $f(X) \in \mathbb{Q}[X]$. The ring $D = s^{-1}(\mathbb{Z}) = \{f(X) \in \mathbb{Q}[X]; s(f(X)) \in \mathbb{Z}\} = \{f(X) \in \mathbb{Q}[X]; f(0) \in \mathbb{Z}\} = \mathbb{Z} + X\mathbb{Q}[X]$. Then $D = \mathbb{Z} + X\mathbb{Q}[X] \simeq s \times_{\mathbb{Q}} i$ and we have the following commutative diagram:

$$
\begin{array}{ccc}
\mathbb{Z} + X\mathbb{Q}[X] & \xrightarrow{p_A} & \mathbb{Q}[X] \\
{\scriptstyle p_B}\downarrow & & \downarrow{\scriptstyle s} \\
\mathbb{Z} & \xrightarrow{i} & \mathbb{Q}
\end{array}
$$

with $p_A(f(X)) = f(X)$ and $p_B(f(X)) = f(0)$ for each $f(X) \in \mathbb{Z} + X\mathbb{Q}[X]$.

The following lemma reformulates Lemma 6.1 in the case of a pull back.

Lemma 6.3 *With the notations of Definition 6.2, we have the following properties:*

1. If the homomorphism β is onto, the projection p_A is onto.

2. *The rings A and B are endowed with structures of D−modules induced by the canonical projections $p_A : D \longrightarrow A$ and $p_B : D \longrightarrow B$, defined for all $a \in A$, $b \in B$, and $d \in D$ by $d.a = p_A(d)a$ and $d.b = p_B(d)b$, and for which p_A and p_B are homomorphisms of D−modules. Moreover, ker β is a D−sub-module of B.*

3. *ker $p_A = (0) \times$ ker β and ker $p_A \simeq$ ker β, a D−modules isomorphism.*

4. *If β is onto, the D−sub-modules of A are the ideals of the ring A.*

5. *Suppose that β is onto. Let S be a multiplicative set of D.*

   The ring A is $p_A(S)$−Noetherian if and only if the D−module A is S-Noetherian.

Proof

(1) Let $a \in A$, $\alpha(a) \in C$. Since β is onto, there exists $b \in B$ such that $\beta(b) = \alpha(a)$. Then $(a, b) \in D$ and $p_A(a, b) = a$. So p_A is onto.

(2) By Lemma 6.1 (2), since *ker* β is an ideal of the ring B, it is a D−sub-module of B.

(3) $ker\ p_A = \{(a, b) \in D;\ p_A(a, b) = 0\} = \{(a, b) \in A \times B;\ \alpha(a) = \beta(b), p_A(a, b) = a = 0\} = \{(0, b);\ b \in B, \beta(b) = 0\} = \{(0, b);\ b \in ker\ \beta\} = (0) \times ker\ \beta$. By (2), the projection p_B induces an isomorphism of D−modules between *ker* p_A and *ker* β.

(4) By (1), the homomorphism p_A is onto. We conclude by Lemma 6.1 (2).

(5) By Lemma 6.1 (3).   □

The following proposition gives necessary and sufficient conditions in order that a pull back is S-Noetherian.

Proposition 6.4 (Lim and Oh 2014) *Let D be a pull back of the ring homomorphisms $\alpha : A \longrightarrow C$ and $\beta : B \longrightarrow C$, with β onto, and let S be a multiplicative set of the ring D. The following assertions are equivalent:*

1. *The ring D is S-Noetherian.*
2. *The ring A is $p_A(S)$−Noetherian and the D−module ker β is S-Noetherian (for the structure of module induced by the canonical projection $p_B : D \longrightarrow B$).*

Proof By the preceding lemma, the canonical projection $p_A : D \longrightarrow A$ is a homomorphism onto of D−modules. So we have the isomorphism $A \simeq D/ker\ p_A$ of D− modules. By Theorem 3.1, the ring D is S-Noetherian $\iff$ the D−module D is S-Noetherian $\iff$ the D−modules A and *ker* p_A are S-Noetherian.

But by the preceding lemma, we have the two following equivalences:

(1) The D−module A is S-Noetherian $\iff$ the ring A is $p_A(S)$−Noetherian.

(2) The D−module *ker* p_A is S-Noetherian $\iff$ the D−module *ker* β is S-Noetherian.

So the ring D is S-Noetherian $\iff$ the ring A is $p_A(S)$−Noetherian and the D−module *ker* β is S-Noetherian.   □

The following corollary deals with the case of Noetherian rings.

Corollary 6.5 (D'anna et al. 2008) *Let D be the pull back of the ring homomorphisms* $\alpha : A \longrightarrow C$ *and* $\beta : B \longrightarrow C$ *with* β *onto.*
The following assertions are equivalent:

1. *The ring D is Noetherian.*
2. *The ring A is Noetherian and the D—module ker* β *is Noetherian.*

Proof If $S = \{(1, 1)\}$, then $p_A(S) = \{1\}$.   □

7 S-Noetherian Amalgamations

In this section, we apply the results proved for pull backs in the case of amalgamation of rings along an ideal with respect to a homomorphism.

Lemma 7.1 (Zhongkui 2007) *Let* $f : A \longrightarrow B$ *be a ring homomorphism onto and S a multiplicative set of A. If the ring A is S-Noetherian, the ring B is* $f(S)$—*Noetherian.*

Proof Let J be an ideal of B. Then $f^{-1}(J)$ is an ideal of A and $J = f(f^{-1}(J))$. There exist $s \in S$ and $a_1, \ldots, a_n \in f^{-1}(J)$ such that $sf^{-1}(J) \subseteq a_1 A + \ldots + a_n A$.
     Then $f(s)f(f^{-1}(J)) \subseteq f(a_1)B + \ldots + f(a_n)B$, so $f(s)J \subseteq f(a_1)B + \ldots + f(a_n)B \subseteq J$ because $f(a_i) \in J$. So J is $f(S)$—finite.   □

Example Let S be a multiplicative set of a ring A. If $A[X]$ (resp. $A[[X]]$) is S-Noetherian, then A is also S-Noetherian. The converse is studied in Sects. 4 and 5 and requires some additional conditions on S. Indeed, the map $\phi : A[X]$ (resp $A[[X]]$) $\longrightarrow A$, which associates to each $f(X)$ its constant term, is a homomorphism onto of rings and $\phi(S) = S$.

Definition 7.2 Let $f : A \longrightarrow B$ be a homomorphism of rings and J an ideal of B. The sub-ring noted $A \bowtie^f J = \{(a, f(a) + j); \ a \in A, j \in J\}$, of the product ring $A \times B$, is called the amalgamation of A with B along J with respect to f.

     Let $\pi : B \longrightarrow B/J$ be the canonical homomorphism and $\hat{f} = \pi \circ f : A \longrightarrow B \longrightarrow B/J$. Then $A \bowtie^f J = \hat{f} \times_{B/J} \pi$ and we have the following commutative diagram:

$$
\begin{array}{ccc}
A\bowtie^f J = \hat{f} \times_{B/J}\pi & \xrightarrow{\ p_A\ } & A \\
{\scriptstyle p_B}\downarrow & & \downarrow{\scriptstyle \hat{f}=\alpha} \\
B & \xrightarrow[\pi=\beta]{} & B/J
\end{array}
$$

     Indeed, $\hat{f} \times_{B/J} \pi = \{(a, b) \in A \times B; \ \hat{f}(a) = \pi(b)\} = \{(a, b) \subset A \times B; \overline{f(a)} = \bar{b} \text{ in } B/J\} = \{(a, b) \in A \times B; \ \exists j \in J, b = f(a) + j\} = \{(a, f(a) + j); \ a \in A, j \in J\} = A \bowtie^f J$.

There exist other ways to describe $A \bowtie^f J$ as a pull back.

The following proposition summarizes the essential properties of the amalgamation.

Proposition 7.3 (D'anna et al. 2008) *Let* $f : A \longrightarrow B$ *be a homomorphism of rings,* J *an ideal of* B, *and* $A \bowtie^f J = \{(a, f(a) + j); \ a \in A, j \in J\}$, *the amalgamation of* A *with* B *along* J *with respect to* f.

1. *The map* $i : A \longrightarrow A \bowtie^f J$, *defined for each* $a \in A$ *by* $i(a) = (a, f(a))$, *is a homomorphism of rings one to one.*
2. *Let* I *be an ideal of* A. *The set noted* $I \bowtie^f J = \{(a, f(a)+j); a \in I, j \in J\}$ *is an ideal of the ring* $A \bowtie^f J$ *and the quotient ring* $(A \bowtie^f J)/(I \bowtie^f J) \simeq A/I$.

    In particular, $I \bowtie^f J$ *is a prime ideal of* $A \bowtie^f J$ *if and only if* I *is a prime ideal of* A.
3. *Let* $p_B : A \bowtie^f J \subseteq A \times B \longrightarrow B$ *be the canonical projection. Then* $p_B(A \bowtie^f J) = f(A) + J$, $\ker p_B = f^{-1}(J) \times (0)$ *and we have the isomorphism* $(A \bowtie^f J)/(f^{-1}(J) \times (0)) \simeq f(A) + J$. *Moreover, if* $f^{-1}(J) = (0)$, *then* $A \bowtie^f J \simeq f(A) + J$.
4. *The canonical projection* $p_A : A \bowtie^f J \subseteq A \times B \longrightarrow A$ *is onto with kernel* $\ker p_A = (0) \times J$ *and we have the isomorphism* $(A \bowtie^f J)/((0) \times J) \simeq A$.
5. *The map* $\gamma : A \bowtie^f J \longrightarrow (f(A) + J)/J$, *defined by* $\gamma(a, f(a) + j) = \overline{f(a)}$ *for each* $a \in A$ *and* $j \in J$, *is a homomorphism onto with kernel* $f^{-1}(J) \times J$ *and we have the isomorphism* $(A \bowtie^f J)/(f^{-1}(J) \times J) \simeq (f(A) + J)/J$.

    Moreover, if f *is onto, then* $(A \bowtie^f J)/(f^{-1}(J) \times J) \simeq B/J$.
6. *Let* Q *be a prime ideal of* B. *The set* $\overline{Q}^f = \{(a, f(a) + j); \ a \in A, j \in J, f(a) + j \in Q\}$ *a prime ideal of the ring* $A \bowtie^f J$.

    Moreover, if $J \subseteq Q$, *then* $\overline{Q}^f = f^{-1}(Q) \bowtie^f J$ *with* $f^{-1}(Q)$ *is a prime ideal of* A.

Proof (1) Clear. (2) The stability of $I \bowtie^f J$ by addition is clear. Let $a \in I, a' \in A$, and $j, j' \in J$. Then $(a, f(a) + j)(a', f(a') + j') = (aa', f(aa') + jf(a') + j'f(a) + jj') \in I \bowtie^f J$. Then $I \bowtie^f J$ is an ideal of the ring $A \bowtie^f J$.

The map $\phi : A \longrightarrow (A \bowtie^f J)/(I \bowtie^f J)$, which associates to $a \in A$ the element $\phi(a) = \overline{i(a)} = \overline{(a, f(a))}$, is an homomorphism onto with kernel I. Indeed, for each $a \in A$ and $j \in J$, $\phi(a) = \overline{(a, f(a))} = \overline{(a, f(a) + j)}$ because $(a, f(a) + j) - (a, f(a)) = (0, j) \in I \bowtie^f J$.

Moreover, $a \in \ker\phi \iff (a, f(a)) \in I \bowtie^f J \iff a \in I$. So we have the isomorphism.

(3) The image $p_B(A \bowtie^f J) = \{f(a) + j; \ a \in A, j \in J\} = f(A) + J$.

The kernel $\ker p_B = \{(a, f(a) + j); \ a \in A, j \in J, f(a) + j = 0\} = \{(a, 0); \ a \in A, f(a) \in J\} = \{(a, 0); \ a \in f^{-1}(J)\} = f^{-1}(J) \times (0)$. So we have $(A \bowtie^f J)/(f^{-1}(J) \times (0)) \simeq f(A) + J$.

(4) For each $a \in A$, $p_A(a, f(a)) = a$, and then p_A is onto. On the other hand, $\ker p_A = \{(a, f(a) + j); \ a = 0, j \in J\} = \{(0, j); \ j \in J\} = (0) \times J$. Then $(A \bowtie^f J)/((0) \times J) \simeq A$.

(5) Note that $\gamma = t \circ p_B$, where $t : f(A) + J \longrightarrow (f(A) + J)/J$ is the canonical homomorphism and $p_B : A \bowtie^f J \longrightarrow f(A) + J$. Then γ is a homomorphism onto, since it is the composite of two homomorphisms onto.

On the other hand, $\ker \gamma = \{(a, f(a) + j); \ a \in A, j \in J, f(a) \in J\} = \{(a, f(a) + j), a \in f^{-1}(J), j \in J\} = f^{-1}(J) \times J$. Then $(A \bowtie^f J)/(f^{-1}(J) \times J) \simeq (f(A) + J)/J$.

If f is onto, $(f(A) + J)/J = (B + J)/J = B/J$.

(6) $\overline{Q}^f = \{(a, f(a) + j); \ a \in A, j \in J, f(a) + j \in Q\} = \{x \in A \bowtie^f J; \ p_B(x) \in Q\} = p_B^{-1}(Q)$ is a prime ideal of $A \bowtie^f J$. If $J \subseteq Q$, for each $a \in A$ and $j \in J$, $f(a) + j \in Q \Longleftrightarrow f(a) \in Q \Longleftrightarrow a \in f^{-1}(Q)$. Then $\overline{Q}^f = \{(a, f(a) + j); \ a \in f^{-1}(Q), j \in J\} = f^{-1}(Q) \bowtie^f J$.

With the notations of the preceding proposition, the sub-ring $f(A) + J$ of the ring B plays an important role in the study of the amalgamation $A \bowtie^f J$.   □

Remark 7.4 (D'anna et al. 2010) One can prove that the only prime ideals of $A \bowtie^f J$ are of two types: $P \bowtie^f J = \{(a, f(a) + j); a \in P, j \in J\}$ with P a prime ideal of A and $\overline{Q}^f = \{(a, f(a) + j); \ a \in A, j \in J, f(a) + j \in Q\}$ with Q a prime ideal of B not containing J.

Example 1 Let $A \subseteq B$ be an extension of rings and $X = \{X_1, \ldots, X_n\}$ a finite set of variables over B. Then $A + XB[X] = \{h \in B[X], h(0) \in A\}$, where 0 is the $n-$uple whose all components are zero and is a sub-ring of $B[X]$. Consider the canonical injection $f : A \longrightarrow B[X]$ and the ideal $J = XB[X]$ of the ring $B[X]$. Then $f^{-1}(J) = (0)$ and $f(A) = A$. By Proposition 7.3 (3), $A \bowtie^f J \simeq f(A) + J = A + XB[X]$.

In the same way, we define $A + XB[[X]]$. If we note $f' : A \longrightarrow B[[X]]$ the canonical injection and the ideal $J' = XB[[X]]$ of $B[[X]]$, then $A \bowtie^{f'} J \simeq f'(A) + J' = A + XB[[X]]$.

Example 2 We generalize the preceding example in the following way. Let $A \subseteq B$ be an extension of rings, I an ideal of B, and $X = \{X_1, \ldots, X_n\}$ a finite set of variables over B. Then $A + XI[X] = \{h \in B[X], h(0) \in A, h(X) - h(0) \in XI[X]\}$ is a bub ring of $B[X]$. Consider the canonical injection $f : A \longrightarrow B[X]$ and the ideal $J = XI[X]$ of $B[X]$. Then $f^{-1}(J) = (0)$ and $f(A) = A$. By Proposition 7.3 (3), we have the isomorphism $A \bowtie^f J \simeq f(A) + J = A + XI[X]$.

In the same way, we define $A + XI[[X]]$. If we note $f' : A \longrightarrow B[[X]]$ the canonical injection and the ideal $J' = XI[[X]]$ of $B[[X]]$, then $A \bowtie^{f'} J \simeq f'(A) + J' = A + XI[[X]]$.

Example 3 Let M be a maximal ideal of an integral domain A and D a sub-ring of A such that $M \cap D = (0)$. The ring $D + M = \{x + m; \ x \in D, m \in M\} \simeq D \bowtie^i M$, where $i : D \longrightarrow A$ is the canonical injection. Indeed, $i(D) = D$ and $i^{-1}(M) = \{x \in D; \ x \in M\} = D \cap M = (0)$. By Proposition 7.3 (3), $D \bowtie^i M \simeq i(D) + M = D + M$.

More generally, let A be an integral domain, D a sub-ring of A, and $\mathscr{F}$ a nonempty set of $Max(A)$ such that $M \cap D = (0)$ for each $M \in \mathscr{F}$. Let $J = \bigcap_{M \in \mathscr{F}} M$. Then $D + J = \{x + j; \ x \in D, j \in J\} \simeq D \bowtie^i J$, where $i : D \longrightarrow A$ is the canonical injection. Indeed, $i^{-1}(J) = \{x \in D, x \in J\} = D \cap J = (0)$ and $i(D) = D$. We conclude by Proposition 7.3 (3). In particular, if $D = K$ is a subfield of A and $J = Jac(A)$, the Jacobson radical of A, then $K + J \simeq K \bowtie^i J$, where $i : K \longrightarrow A$ is the canonical injection.

Example 4 Nagata idealization, 1955.

Let A be a ring and M a A−module. The product set $A \times M$ may be endowed with a structure of ring such that the addition is termwise and the product is defined by $(a, x)(a', x') = (aa', ax' + a'x)$. We note $A(+)M$ this ring and we call it the idealization of M in A. The zero element is $(0, 0)$ and the unit is $(1, 0)$. The map $i_A : A \longrightarrow A(+)M$, defined by $i_A(a) = (a, 0)$ for each $a \in A$, is a homomorphism one to one of rings, since $i_A(aa') = (aa', 0) = (a, 0)(a', 0) = i_A(a)i_A(a')$. This allows us to identify A with the sub-ring $i_A(A) = A \times (0)$ of $A(+)M$. We also say that $A(+)M$ is the trivial extension of A by M. Note that $A(+)M$ is also endowed with a natural structure of A−module, and the map $i_M : M \longrightarrow A(+)M$, defined by $i_M(x) = (0, x)$ for each $x \in M$, is a homomorphism one to one of A−modules. Then we have the isomorphism of A−modules $M \simeq i_M(M) = (0) \times M$. We will show that $i_M(M) = (0) \times M$ is an ideal of the ring $A(+)M$. For each $(0, x) \in i_M(M)$ and $(a', x') \in A(+)M$, $(0, x)(a', x') = (0, a'x) \in i_M(M)$. By identifying M with $i_M(M)$, we can see M as an ideal of $A(+)M$. Let $f = i_A : A \longrightarrow A(+)M$ be the homomorphism of rings defined by $f(a) = i_A(a) = (a, 0)$ for each $a \in A$. Then $A(+)M \simeq A \bowtie^f M$. Indeed, $f^{-1}(M) = \{a \in A; \ f(a) = i_A(a) = (a, 0) \in M\} = \{a \in A; \ (a, 0) \in i_M(M) = (0) \times M\} = (0)$ and $f(A) = i_A(A) = A \times (0)$. So $f(A) + M = A \times (0) + (0) \times M = A(+)M$. By Proposition 7.3 (3), $A(+)M \simeq A \bowtie^f M$.

The following lemma reformulates Lemma 6.1 in the case of the amalgamation.

Lemma 7.5 *Let $f : A \longrightarrow B$ be a homomorphism of rings, J an ideal of B, and $A \bowtie^f J = \{(a, f(a) + j); a \in A, j \in J\}$ the amalgamation of A with B along J with respect to the homomorphism f.*

$$A \bowtie^f J = \hat{f} \times_{B/J} \pi \xrightarrow{\quad p_A \quad} A$$
$$p_B \downarrow \qquad\qquad \downarrow \hat{f}=\alpha$$
$$B \xrightarrow{\quad \pi=\beta \quad} B/J$$

with $\pi : B \longrightarrow B/J$ the canonical homomorphism and $\hat{f} = \pi \circ f : A \longrightarrow B \longrightarrow B/J$.

1. The projection $p_B : A \bowtie^f J = \hat{f} \times_{B/J} \pi \longrightarrow B$ allows us to endow B with a structure of $A \bowtie^f J$−module, defined for each $a \in A$, $j \in J$ and $b \in B$ by $(a, f(a) + j).b = p_B(a, f(a) + j)b = (f(a) + j)b$. Moreover, J is a $A \bowtie^f J$−sub-module of B.
2. The ideals of the ring $f(A) + J$ are exactly the $A \bowtie^f J$−sub-modules of $f(A) + J$.

    In particular, each $A \bowtie^f J$−sub-module J_0 of J is an ideal of $f(A) + J$.
3. Let S be multiplicative set of the ring A. Then $f(S)$ is a multiplicative set of the ring B and $S' = \{(s, f(s)); s \in S\}$ is a multiplicative set of the ring $A \bowtie^f J$.

    Moreover, $p_A(S') = S$ and $p_B(S') = f(S)$.

Proof

(1) The structure of module follows from Lemma 6.1. Since J is an ideal of B, then it is $A \bowtie^f J$−sub-module of B.
(2) By Lemma 7.3 (3), the map $p_B : A \bowtie^f J \longrightarrow f(A) + J$ is a homomorphism onto of rings. By Lemma 6.1 (3), the ideals of the ring $f(A) + J$ are exactly the $A \bowtie^f J$−sub-modules of $f(A) + J$.
(3) Clear. □

In the following theorem, we show the first main result of this section.

Theorem 7.6 (Lim and Oh 2014) Let $f : A \longrightarrow B$ be a homomorphism of rings, J an ideal of B, and S a multiplicative set of A. The following assertions are equivalent:

1. The ring $A \bowtie^f J$ is S'−Noetherian, where $S' = \{(s, f(s)); s \in S\}$.
2. The ring A is S-Noetherian and J is a S'−Noetherian $A \bowtie^f J$−module (for the structure of module induced by the homomorphism $p_B : A \bowtie^f J \longrightarrow B$).
3. The ring A is S-Noetherian and the sub-ring $f(A) + J$ of B is $f(S)$−Noetherian.

Proof "(1) $\Longleftrightarrow$ (2)" By Proposition 6.4, the ring $A \bowtie^f J = \hat{f} \times_{B/J} \pi$ is S'−Noetherian if and only if the ring A is $p_A(S') = S$−Noetherian and the $A \bowtie^f J$−module $\ker \beta = \ker \pi = J$ is S'−Noetherian.

"(1) $\Longrightarrow$ (3)" By Proposition 6.4, the ring A is $p_A(S') = S$−Noetherian.

On the other hand, $p_B(A \bowtie^f J) = f(A) + J$ and $p_B(S') = f(S)$. By Lemma 7.1, the ring $f(A) + J$ is $f(S)$−Noetherian.

$"(3) \implies (2)"$ By hypothesis, the ring A is S-Noetherian. Let J_0 be an $A \bowtie^f$ J—sub-module of J. By the preceding lemma, J_0 is an ideal of the ring $f(A) + J$. There exist $s \in S$ and $j_1, \ldots, j_k \in J_0$ such that $f(s)J_0 \subseteq (f(A) + J)j_1 + \ldots + (f(A) + J)j_k \subseteq J_0$, and then $p_B(s, f(s))J_0 \subseteq p_B(A \bowtie^f J)j_1 + \ldots + p_B(A \bowtie^f J)j_k \subseteq J_0$. So we have the inclusions $(s, f(s)).J_0 \subseteq (A \bowtie^f J).j_1 + \ldots + (A \bowtie^f J).j_k \subseteq J_0$ with $(s, f(s)) \in S'$. Then J_0 is S'—finite.   □

Corollary 7.7 (D'anna et al. 2008) *Let* $f : A \longrightarrow B$ *be a homomorphism of rings and* J *an ideal of* B. *Then* $A \bowtie^f J$ *is Noetherian if and only if the rings* A *and* $f(A) + J$ *are Noetherian.*

Proof Take $S = \{1\}$, then $S' = \{(1, 1)\}$, and $f(S) = \{1\}$. We conclude by the equivalence (1) $\Longleftrightarrow$ (3) of the preceding theorem.   □

In the following theorem is the second main result of this section.

Theorem 7.8 (Lim and Oh 2014) *Let* $f : A \longrightarrow B$ *be a homomorphism of rings,* J *an ideal of* B, $S \subseteq A$ *a multiplicative set, and* $S' = \{(s, f(s)); s \in S\} \subseteq A \bowtie^f J$.

Suppose that A *is* S-Noetherian *and* B *is a* S-finite A—module *for the structure of* A—module *induced by* f. *Then*

1. *The ring* $A \bowtie^f J$ *is* S'—Noetherian.
2. *The ring* $f(A) + J$ *is* $f(S)$—Noetherian.

Proof

(1) It suffices to show that each prime ideal $\mathscr{P}$ of the ring $A \bowtie^f J$ is S'—finite.

First Case $\mathscr{P} = P \bowtie^f J = \{(p, f(p) + j); p \in P, j \in J\}$ with P a prime ideal of A.

Since A is a S-Noetherian ring, there exist $s_1 \in S$ and $p_1, \ldots, p_n \in P$ such that $s_1 P \subseteq Ap_1 + \ldots + Ap_n$. Since B is a S-finite module over the S-Noetherian ring A, then B is a S-Noetherian A—module. But J is a A—sub-module of B, so it is S-finite. There exist $s_2 \in S$ and $J_1, \ldots, j_m \in J$ such that $s_2.J \subseteq A.j_1 + \ldots + A.j_m$, and then $f(s_2)J \subseteq f(A)j_1 + \ldots + f(A)j_m$. Put $s = s_1s_2 \in S$, and then $(s, f(s)) \in S'$. Let $(p, f(p) + j) \in P \bowtie^f J = \mathscr{P}$ any element, with $p \in P$ and $j \in J$. There exist $a_1, \ldots, a_n \in A$ such that $sp = a_1p_1 + \ldots + a_np_n$ and there exist $a'_1, \ldots, a'_m \in A$ such that $f(s)j = f(a'_1)j_1 + \ldots + f(a'_m)j_m$.

Then $(s, f(s))(p, f(p) + j) = (sp, f(s)f(p) + f(s)j) = (sp, f(sp) +$
$$f(s)j) = \left(\sum_{i:1}^{n} a_i p_i, \sum_{i:1}^{n} f(a_i)f(p_i) + \sum_{k:1}^{m} f(a'_k)j_k \right) = \sum_{i:1}^{n} (a_i p_i, f(a_i)f(p_i)) +$$
$$\sum_{k:1}^{m} (0, f(a'_k)j_k)$$
$$= \sum_{i:1}^{n} (a_i, f(a_i))(p_i, f(p_i)) + \sum_{k:1}^{m} (a'_k, f(a'_k))(0, j_k) \in ((p_i, f(p_i)), (0, j_k); 1 \le$$
$i \le n, 1 \le k \le m)(A \bowtie^f J).$

Since $(p_i, f(p_i))$ and $(0, j_k) \in P \bowtie^f J$, then $(s, f(s))(P \bowtie^f J) \subseteq$
$((p_i, f(p_i)), (0, j_k); 1 \leq i \leq n, 1 \leq k \leq m)(A \bowtie^f J) \subseteq P \bowtie^f J = \mathscr{P}.$

Second Case $\mathscr{P} = \overline{Q}^f = \{(a, f(a)+j); \ a \in A, j \in J, f(a)+j \in Q\}$ with Q a prime ideal of B not containing J.

Let $p_B : A \bowtie^f J \longrightarrow B$ be the canonical projection. Then $p_B(\overline{Q}^f)$ is a $A-$sub-module of B for the structure of $A-$module induced by f. Indeed, let $a' \in A$ and $(a, f(a)+j) \in \overline{Q}^f$. Then $a \in A$, $j \in J$ and $f(a)+j \in Q$. Then $a'.p_B(a, f(a)+j) = f(a')(f(a)+j) = f(a'a) + f(a')j = p_B(a'a, f(a'a) + f(a')j)$ with $f(a')j \in J$ and $f(a'a) + f(a')j = f(a')(f(a)+j) \in Q$. Then $(a'a, f(a'a) + f(a')j) \in \overline{Q}^f$.

Since B is a S-Noetherian $A-$module, the $A-$sub-module $p_B(\overline{Q}^f)$ is S-finite. There exist $s_1 \in S$ and $f(a_1) + j_1, \ldots, f(a_n) + j_n \in p_B(\overline{Q}^f)$ with $a_k \in A$ and $j_k \in J$ such that $s_1.p_B(\overline{Q}^f) \subseteq A.(f(a_1)+j_1)+\ldots+A.(f(a_n)+j_n) \subseteq p_B(\overline{Q}^f)$.

Then $f(s_1)p_B(\overline{Q}^f) \subseteq f(A)(f(a_1)+j_1)+\ldots+f(A)(f(a_n)+j_n) \subseteq p_B(\overline{Q}^f)$.
(*)

Note that by definition, the element $(a_k, f(a_k) + j_k) \in \overline{Q}^f$ for $1 \leq k \leq n$.

Since J is an ideal of B, $f^{-1}(J)$ is an ideal of A. But the ring A is S-Noetherian. There exist $s_2 \in S$ and $z_1, \ldots, z_m \in f^{-1}(J)$ such that $s_2 f^{-1}(J) \subseteq z_1 A \mid \ldots + z_m A \subseteq f^{-1}(J)$. (**)

Then $f(z_1), \ldots, f(z_m) \in J$, so $(z_i, 0) = (z_i, f(z_i) - f(z_i)) \in \overline{Q}^f$ for $1 \leq i \leq m$.

Put $s = s_2 s_1 \in S$, and then $(s, f(s)) \in S'$. Let $(a, f(a) + j) \in \overline{Q}^f$ be any element. By (*), there exist $r_1, \ldots, r_n \in A$ such that $f(s_1)(f(a) + j) = \sum_{k:1}^{n} f(r_k)(f(a_k) + j_k)$.

Then $f(s_1 a - \sum_{k:1}^{n} r_k a_k) = \sum_{k:1}^{n} f(r_k)j_k - f(s_1)j \in J$, so $s_1 a - \sum_{k:1}^{n} r_k a_k \in f^{-1}(J)$.

By (**), there exist $r_1', \ldots, r_m' \in A$ such that $s_2(s_1 a - \sum_{k:1}^{n} r_k a_k) = r_1' z_1 + \ldots + r_m' z_m$, and then $sa = s_2 s_1 a = s_2(\sum_{k:1}^{n} r_k a_k) + \sum_{i:1}^{m} r_i' z_i$.

Then $(s, f(s))(a, f(a)+j) = (sa, f(s)(f(a)+j)) = (sa, f(s_2)f(s_1)(f(a)+j))$
$= (s_2(\sum_{k:1}^{n} r_k a_k) + \sum_{i:1}^{m} r_i' z_i, f(s_2)(\sum_{k:1}^{n} f(r_k)(f(a_k) + j_k)))$
$= (\sum_{k:1}^{n} s_2 r_k a_k, \sum_{k:1}^{n} f(s_2 r_k)(f(a_k) + j_k)) + (\sum_{i:1}^{m} r_i' z_i, 0)$

$$= \sum_{k:1}^{n} \left(s_2 r_k a_k, f(s_2 r_k)(f(a_k) + j_k) \right) + \sum_{i:1}^{m} \left(r_i' z_i, 0 \right)$$
$$= \sum_{k:1}^{n} \left(a_k, f(a_k) + j_k \right) \left(s_2 r_k, f(s_2 r_k) \right) + \sum_{i:1}^{m} \left(r_i', f(r_i') \right) (z_i, 0)$$
$$\in \left((a_k, f(a_k) + j_k), (z_i, 0); 1 \leq k \leq n, 1 \leq i \leq m \right) (A \bowtie^f J),$$

with $\left\{ (a_k, f(a_k) + j_k), (z_i, 0); 1 \leq k \leq n, 1 \leq i \leq m \right\} \subseteq \overline{Q}^f$.

Then $(s, f(s)) \overline{Q}^f \subseteq \left((a_k, f(a_k) + j_k), (z_i, 0); 1 \leq k \leq n, 1 \leq i \leq m \right) (A \bowtie^f J) \subseteq \overline{Q}^f$.

(2) Follows from (1) and the equivalence (1) $\Longleftrightarrow$ (3) in Theorem 7.6. $\square$

Corollary 7.9 *Let $A \subseteq B$ be an extension of rings, J an ideal of B, and S a multiplicative set of A. Suppose that A is S-Noetherian and B is a S-finite $A-$module. Then $A + J$ is a S-Noetherian ring.*

Proof Take $f : A \longrightarrow B$ as the canonical injection and apply (2) of the theorem.
$\square$

8 Applications

The interest of Theorem 7.6 is middle because the Noetherian property of $A \bowtie^f J$ is not directly linked to the initial parameters (A, B, f, J), but rather to the ring $f(A) + J$, which is canonically isomorphic to the construction $A \bowtie^f J$ if $f^{-1}(J) = 0$ (see Proposition 7.3 (3)). To obtain practical criteria for Noetherian property of $A \bowtie^f J$, we consider in this section some particular interesting cases.

Corollary 8.1 *Let $A \subseteq B$ be an extension of rings, $\{X_1, \ldots, X_n\}$ a finite set of variables over B, J an ideal of $B[X_1, \ldots, X_n]$, and S a multiplicative anti-Archimedean set of A. Suppose that A is S-Noetherian and B is a S-finite $A-$module. Then $A[X_1, \ldots, X_n] + J$ is a S-Noetherian ring.*

In particular, $A + (X_1, \ldots, X_n)B[X_1, \ldots, X_n]$ is a S-Noetherian ring.

Proof By Theorem 4.2, the ring $A[X_1, \ldots, X_n]$ is S-Noetherian. Since B is a S-finite $A-$module, then $B[X_1, \ldots, X_n]$ is a S-finite $A[X_1, \ldots, X_n]-$module. Indeed, there exist $s \in S$ and $b_1, \ldots, b_k \in B$ such that $sB \subseteq Ab_1 + \ldots + Ab_k \subseteq B$.

Then $sB[X_1, \ldots, X_n] \subseteq A[X_1, \ldots, X_n]b_1 + \ldots + A[X_1, \ldots, X_n]b_k \subseteq B[X_1, \ldots, X_n]$. We apply the preceding corollary. For the particular case, take $J = (X_1, \ldots, X_n)B[X_1, \ldots, X_n]$. $\square$

Corollary 8.2 *Let $A \subseteq B$ be an extension of rings, $\{X_1, \ldots, X_n\}$ a finite set of variables over B, J an ideal of $B[[X_1, \ldots, X_n]]$ and $S \subseteq A$ an anti-Archimedean multiplicative set not containing zero divisors of A. Suppose that A is S-Noetherian and B is a S-finite $A-$module. Then $A[[X_1, \ldots, X_n]] + J$ is a S-Noetherian ring.*

In particular, $A + (X_1, \ldots, X_n)B[[X_1, \ldots, X_n]]$ *is a S-Noetherian ring.*

Proof By Theorem 5.1, the ring $A[[X_1, \ldots, X_n]]$ is S-Noetherian. Since B is a S-finite A−module, $B[[X_1, \ldots, X_n]]$ is a S-finite $A[[X_1, \ldots, X_n]]$−module. Indeed, there exist $s \in S$ and $b_1, \ldots, b_k \in B$ such that $sB \subseteq Ab_1 + \ldots + Ab_k \subseteq B$. Then $sB[[X_1, \ldots, X_n]] \subseteq A[[X_1, \ldots, X_n]]b_1 + \ldots + A[[X_1, \ldots, X_n]]b_k \subseteq B[[X_1, \ldots, X_n]]$. We conclude by Corollary 7.9. For the particular case, $J = (X_1, \ldots, X_n)B[[X_1, \ldots, X_n]]$. $\qquad\square$

Example (Lim and Oh 2015) Let A be a valuation domain without a height one prime ideal, whose quotient field $K \neq A$. By Example 3 of Definition 4.1, $S = A \backslash (0)$ is an anti-Archimedean multiplicative set of A. The ring A is S-Noetherian by Example 1 of Definition 2.1. Let $x_1, \ldots, x_n \in K \setminus A$. The ring $A_n = A[x_1, \ldots, x_n]$ is a S-finite A−module. Indeed, $x_1 = \frac{a_1}{s}, \ldots, x_n = \frac{a_n}{s}$ with $a_1, \ldots, a_n \in A$ and $0 \neq s \in A$. Since S is anti-Archimedean in A, there exists $t \in (\bigcap_{k:1}^{\infty} s^k A) \cap S$. For each $x \in A_n = A[x_1, \ldots, x_n]$, $tx \in A$, and then $tA_n \subseteq A \subseteq A_n$. By the preceding two corollaries, the two rings $A + XA_n[X]$ and $A + XA_n[[X]]$ are S-Noetherian, where X is a finite set of variables over A_n.

Corollary 8.3 (D'anna et al. 2008) *Let $A \subseteq B$ be an extension of rings and $X_1, \ldots, X_n$ a finite number of variables over B. The following assertions are equivalent:*

1. *The ring $A + (X_1, \ldots, X_n)B[X_1, \ldots, X_n]$ is Noetherian.*
2. *The ring $A + (X_1, \ldots, X_n)B[[X_1, \ldots, X_n]]$ is Noetherian.*
3. *The ring A is Noetherian and the A−module B is finitely generated.*

Proof "(3) $\Longrightarrow$ (1) and (2)" Take $S = \{1\}$ in Corollaries 8.1 and 8.2.

"(1) or (2) $\Longrightarrow$ (3)" Note $X = \{X_1, \ldots, X_n\}$, then $(A + XB[X])/XB[X] \simeq A$ (resp. $(A + XB[[X]])/XB[[X]] \simeq A$), and then A is Noetherian.

The ideal of $A + XB[X]$ (resp. $A + XB[[X]]$) generated by $\{bX_k; \ b \in B, 1 \leq k \leq n\}$ is finitely generated by $\{f_1, \ldots, f_m\} \subseteq \{bX_k; \ b \in B, 1 \leq k \leq n\}$. We may suppose that $f_1 = b_1 X_1, \ldots, f_r = b_r X_1, 1 \leq r \leq m$, and the other generators have the form bX_k with $k \geq 2$. We will prove that the A−module B is generated by $\{b_1, \ldots, b_r\}$. Let $b \in B$. There exist $g_1, \ldots, g_m \in A + XB[X]$ (resp. $A + XB[[X]]$) such that $bX_1 = g_1 f_1 + \ldots + g_m f_m$. Put in this equality $X_2 = \ldots = X_n = 0$. By identification of the coefficient of X_1, we find $b = g_1(0)b_1 + \ldots + g_r(0)b_r$ with $g_1(0), \ldots, g_r(0) \in A$. $\qquad\square$

Example 1 Let L/K be an extension of fields. The ring $K + (X_1, \ldots, X_n)L[X_1, \ldots, X_n]$ (resp. $K + (X_1, \ldots, X_n)L[[X_1, \ldots, X_n]]$) is Noetherian if and only if the extension L/K is finite.

Example 2 Let $\overline{\mathbb{Q}}$ be the algebraic closure of $\mathbb{Q}$ in $\mathbb{C}$. Since the extension $\overline{\mathbb{Q}}/\mathbb{Q}$ is infinite, the ring $A = \mathbb{Q} + X\overline{\mathbb{Q}}[X]$ is not Noetherian. The quotient field of A is

$\overline{\mathbb{Q}}(X)$. Since the extension $\overline{\mathbb{Q}}/\mathbb{Q}$ is integral, the integral closure $\overline{A}$ of A contains $\overline{\mathbb{Q}}[X]$. Since the ring $\overline{\mathbb{Q}}[X]$ is integrally closed, then $\overline{A} = \overline{\mathbb{Q}}[X]$ is a PID.

Example 3 (D'anna et al. 2008)

(i) Let $A \subseteq B$ be an extension of rings and X an indeterminate over B. The ideal $J = XB[X]$ of $B[X]$ is never a finitely generated $A-$module for the structure of the $A-$module induced by the canonical injection $f : A \longrightarrow B[X]$. Indeed, for every $g_1, \dots, g_m \in B[X]$, if $n = max\{deg\ g_i;\ 1 \le i \le m\}$, then $X^{n+1} \in J \setminus (Ag_1 + \dots + Ag_m)$.

(ii) Let $f : A \longrightarrow B$ be a homomorphism of rings and J an ideal of B. By Theorem 7.6, if the ring $A \bowtie^f J$ is Noetherian, the $A \bowtie^f J-$module J is Noetherian for the structure of $A \bowtie^f J-$module induced by the canonical projection $p_B : A \bowtie^f J \longrightarrow B$. But we will see in (iii) that J is not necessarily finitely generated for the structure of the $A-$module induced by the homomorphism $f : A \longrightarrow B$.

(iii) By Example 2 after Remark 7.4, $\mathbb{R} \bowtie^f X\mathbb{C}[X] \simeq \mathbb{R} + X\mathbb{C}[X]$ where $f : \mathbb{R} \longrightarrow \mathbb{C}[X]$ is the natural injection. By the preceding corollary, the ring $\mathbb{R} \bowtie^f X\mathbb{C}[X]$ is Noetherian. But by (i), $J = X\mathbb{C}[X]$ is not finitely generated for the structure of the $\mathbb{R}-$module induced by f; i.e., J is not a $\mathbb{R}-$vector space of finite dimension.

Lemma 8.4 (Gilmer 1970) *Let A be a ring and I a finitely generated idempotent ideal ($I^2 = I$) of A. Then I is principal generated by an idempotent element.*

Proof Put $I = (a_1, \dots, a_n)$, and then $I = I^2 = Ia_1 + \dots + Ia_n$. For $1 \le k \le n$, $a_k = \sum_{i:1}^{n} s_{ki}a_i$ with $s_{ki} \in I$, and then $\sum_{i:1}^{n} (\delta_{ki} - s_{ki})a_i = 0$ where δ_{ki} is the Kronecker symbol. The determinant d of the matrix $(\delta_{ki} - s_{ki});\ 1 \le k, i \le n$ is of the form $d = 1 - b$ with $b \in I$. By Cramer rule, $da_i = 0$ for each $1 \le i \le n$, and then $(1 - b)a_i = 0$ so $a_i = ba_i$, which proves that I is principal generated by b. Since $1 - b$ annihilates all the a_i, it annihilates I, and then b. So $(1 - b)b = 0$ then $b^2 = b$.

$\square$

Proposition 8.5 (D'anna et al. 2008) *Let $A \subseteq B$ be an extension of rings, J an ideal of B, and $X = \{X_1, \dots, X_n\}$ a finite number of variables over B. The following conditions are equivalent:*

1. *The ring $A + XJ[X]$ is Noetherian.*
2. *The ring A is Noetherian and J is an idempotent ideal of B and a finitely generated $A-$module.*

Proof $"(1) \Longrightarrow (2)"$ The ring $A \simeq (A + XJ[X])/XJ[X]$, so it is Noetherian.

We will prove that J is an idempotent ideal. Let $a \in J$. The ideal of $A + XJ[X]$ generated by the set $\{aX_1, aX_1^2, \dots\}$ is finitely generated by $\{aX_1, aX_1^2, \dots, aX_1^m\}$ where $m \in \mathbb{N}^*$. Put $aX_1^{m+1} = aX_1 f_1 + \dots + aX_1^m f_m$

with $f_i \in A + XJ[X]$, $f_i = a_i + g_i$ where $g_i \in XJ[X]$. Then $aX_1^{m+1} = (aa_1X_1 + \ldots + aa_mX_1^m) + a(X_1g_1 + \ldots + X_1^mg_m)$. By identification of the coefficient of X_1^{m+1}, we find $a = ab$ with $b \in J$. Then $J = J^2$.

We will prove that J is a finitely generated A−module. The ideal of $A + XJ[X]$ generated by $\{bX_k;\ b \in J, 1 \leq k \leq n\}$ is finitely generated by $\{f_1, \ldots, f_m\} \subseteq \{bX_k;\ b \in J, 1 \leq k \leq n\}$. We may suppose that $f_1 = b_1X_1, \ldots, f_r = b_rX_1$, $1 \leq r \leq m$, and the other generators are of the form bX_k with $k \geq 2$. We will prove that the A−module J is generated by $\{b_1, \ldots, b_r\}$. Let $b \in J$, and there exist $g_1, \ldots, g_m \in A + XJ[X]$ such that $bX_1 = g_1f_1 + \ldots + g_mf_m$. We put in this equality $X_2 = \ldots = X_n = 0$. By identification of the coefficient of X_1, we find $b = g_1(0)b_1 + \ldots + g_r(0)b_r$ with $g_1(0), \ldots, g_r(0) \in A$.

"$(2) \Longrightarrow (1)$" Since J is a finitely generated A−module, then it is a finitely generated ideal of B. By the preceding lemma, $J = eB$ with e an idempotent element of J. Let $\{b_1, \ldots, b_s\}$ be a generating family of the A−module J. We add a new family of variables $Y = \{Y_{ij};\ 1 \leq i \leq n, 1 \leq j \leq s\}$ over B. We define a A−homomorphism of rings $\phi : A[X, Y] \longrightarrow B[X]$ by $\phi(X_i) = eX_i$ and $\phi(Y_{ij}) = b_jX_i$ for $1 \leq i \leq n$ and $1 \leq j \leq s$. It is clear that $\phi(A[X, Y]) \subseteq A + XJ[X]$.

Conversely, let $f \in A + XJ[X]$. Then $f = a + \sum_{i:1}^{n}\left(\sum_{\alpha} b_{\alpha i}X^{\alpha}\right)X_i$ where $a \in A$,

$\alpha = (\alpha_1, \ldots, \alpha_n) \in \mathbb{N}^n$, $b_{\alpha i} \in J$ and $X^{\alpha} = X_1^{\alpha_1} \ldots X_n^{\alpha_n}$. Put $b_{\alpha i} = \sum_{j:1}^{s} a_{\alpha ij}b_j$ with $a_{\alpha ij} \in A$ for $1 \leq i \leq n$ and $1 \leq j \leq s$.

Then $f = a + \sum_{i:1}^{n}\left(\sum_{\alpha}\left(\sum_{j:1}^{s} a_{\alpha ij}b_j\right)X^{\alpha}\right)X_i = a + \sum_{i:1}^{n}\sum_{j:1}^{s}\left(\sum_{\alpha} a_{\alpha ij}X^{\alpha}\right)b_jX_i$.

Since $b_j \in J = eB$ with e an idempotent element, then $b_j = eb_j$ pour $1 \leq j \leq s$.

Then $f = a + \sum_{i:1}^{n}\sum_{j:1}^{s}\left(\sum_{\alpha} a_{\alpha ij}eX^{\alpha}\right)b_jX_i = a + \sum_{i:1}^{n}\sum_{j:1}^{s}\left(\sum_{\alpha} a_{\alpha ij}(eX)^{\alpha}\right)b_jX_i$

where $(eX)^{\alpha} = (eX_1)^{\alpha_1} \ldots (eX_n)^{\alpha_n}$. Let $g = a + \sum_{i:1}^{n}\sum_{j:1}^{s}\left(\sum_{\alpha} a_{\alpha ij}X^{\alpha}\right)Y_{ij} \in A[X, Y]$. Then $\phi(g) = f$. Then $A + XJ[X] \simeq A[X, Y]/ker\phi$, which is a Noetherian ring because $A[X, Y]$ is Noetherian. $\square$

Example (D'anna et al. 2008) Let $A \subseteq B$ be an extension of rings, J an ideal of B, and $X = \{X_1, \ldots, X_n\}$ a finite number of variables over B. The ring $A + XJ[X]$ may be Noetherian, but the ring B is not Noetherian. Take, for example, $A = K$ a field and $B = \prod_{n:1}^{\infty} K$ the product ring. The injection $i : A \longrightarrow B$, defined by $i(a) = (a, a, \ldots)$ for each $a \in A$, allows us to see A as a sub-ring of B. The element $e = (1, 0, \ldots)$ of B is idempotent, and then the ideal $J = eB$ is also idempotent. It is also a cyclic $A-$module generated by e. Indeed, an element of J can be written as $x = (a_1, a_2, \ldots)e = (a_1, 0, \ldots) = a_1 e$. By the preceding proposition, the ring $A + XJ[X]$ is Noetherian. But the ring B is not Noetherian, since it contains the infinite strictly increasing sequence of ideals: $K \times (0) \times \ldots \subset K \times K \times (0) \subset \ldots$.

Remark 8.6 By taking $J = B$ in the preceding proposition, we retrieve the equivalence "The ring $A + (X_1, \ldots, X_n)B[X_1, \ldots, X_n]$ is Noetherian if and only if the ring A is Noetherian and the $A-$module B is finitely generated" in Corollary 8.3.

Corollary 8.7 *Let A be a ring, I an ideal of A, and X a finite set of variables over A. The ring $A + XI[X]$ is Noetherian if and only if A is Noetherian and the ideal I is idempotent.*

Example In the ring $A = \mathbb{Z}/10\mathbb{Z}$, the elements $\bar{5}$ and $\bar{6}$ are idempotent. For $I = (\bar{5})$ or $(\bar{6})$, the ring $A + XI[X]$ is Noetherian.

Corollary 8.8 (Gilmer 1970) *Let A be a ring whose only idempotent elements are 0 and 1. Let X be a finite set of variables over A. If I is a nonzero proper ideal of A, then the ring $A + XI[X]$ is not Noetherian.*

Proof Suppose that the ring $A + XI[X]$ is Noetherian. By the preceding corollary, the ideal I is idempotent and finitely generated. By Lemma 8.4, $I = (0)$ or $I = (1) = A$, which is absurd. □

Examples

(1) Let A be an integral domain. If I is a nonzero proper ideal of A, then the ring $A + XI[X]$ is not Noetherian.
(2) The ring $A = \mathbb{Z}/4\mathbb{Z}$ is not a domain. But its only idempotent elements are $\bar{0}$ and $\bar{1}$. For the ideal $I = (\bar{2})$, the ring $A + XI[X]$ is not Noetherian.

To study the S-Noetherian property in the idealization, we need the following lemma. Note that results about idealization are taken from Anderson and Winders (2009).

Lemma 8.9 *Let A be a ring and M a $A-$module.*

1. *Let I be an ideal of A and N a sub-module of M. The set $I(+)N = \big\{(a, n); a \in I, n \in N\big\}$ is an ideal of the ring $A(+)M$ if and only if $IM \subseteq N$.*
2. *The prime ideals of $A(+)M$ are of the form $P(+)M$ with P a prime ideal of A.*

Proof

(1) The stability of $I(+)N$ by addition is clear.
   The product $\big(A(+)M\big)\big(I(+)N\big) = I(+)(N + IM)$. Then $I(+)N$ is an ideal of $A(+)M \iff I(+)(N+IM) = I(+)N \iff N+IM = N \iff IM \subseteq N$.
(2) Let P be a prime ideal of A, and then $P \neq A$, so $P(+)M \neq A(+)M$. Let (a, m) and $(a', m') \in A(+)M$ such that $(a, m)(a', m') = (aa', am' + a'm) \in P(+)M$.

Then $aa' \in P$, so a or $a' \in P$. Then (a, m) or $(a', m') \in P(+)M$ and $P(+)M$ is a prime ideal of $A(+)M$.

Conversely, let $\mathscr{P}$ be a prime ideal of $A(+)M$. The natural projection p : $A(+)M \longrightarrow A$ is a homomorphism onto. Then $P = p(\mathscr{P}) = \{a \in A;\ \exists m \in M, (a, m) \in \mathscr{P}\}$ is an ideal of A. For each $m \in M$, $(0, m)^2 = (0, 0) \in \mathscr{P}$, and then $(0, m) \in \mathscr{P}$. For each $a \in P$, $(a, 0) \in \mathscr{P}$. Indeed, there exists $m \in M$ such that $(a, m) \in \mathscr{P}$. But $(0, -m) \in \mathscr{P}$, and then $(a, 0) = (a, m) + (0, -m) \in \mathscr{P}$. Now, the inclusion $\mathscr{P} \subseteq P(+)M$ is clear. Conversely, let $(a, m) \in P(+)M$, and then $(a, m) = (a, 0) + (0, m) \in \mathscr{P}$. Then $\mathscr{P} = P(+)M$. The ideal P is prime in A. Indeed, let $a, b \in A$ such that $ab \in P$. Then $(a, 0)(b, 0) = (ab, 0) \in P(+)M = \mathscr{P}$. Then $(a, 0)$ or $(b, 0) \in \mathscr{P}$, so a or $b \in P$. $\square$

Example 1 Consider the ideal $I = 3\mathbb{Z}$ of the ring $A = \mathbb{Z}$ and the sub-module $N = 4\mathbb{Z}$ of the $\mathbb{Z}$-module $M = 2\mathbb{Z}$. By Lemma 8.9 (1), $I(+)N = 3\mathbb{Z}(+)4\mathbb{Z}$ is not an ideal of the ring $A(+)M = \mathbb{Z}(+)2\mathbb{Z}$ because $IM = 3\mathbb{Z}.2\mathbb{Z} = 6\mathbb{Z} \not\subseteq N = 4\mathbb{Z}$.

The ideals of the ring $A(+)M$ are not all of the form $I(+)N$ with I an ideal of A and N a sub-module of M, as can be shown by the following two examples.

Example 2 The principal ideal generated $(2, 2)$ in the ring $\mathbb{Z}(+)2\mathbb{Z}$ is $J = (2, 2)(\mathbb{Z}(+)2\mathbb{Z}) = \{(2, 2)(n, 2m);\ n, m \in \mathbb{Z}\} = \{(2n, 2n + 4m); n, m \in \mathbb{Z}\}$. Let π_1 and π_2 be the canonical projections. Then $\pi_1(J) = \pi_2(J) = 2\mathbb{Z}$. Suppose that $(2, 4) \in J$; there exist $n, m \in \mathbb{Z}$ such that $2n = 2$ and $2n + 4m = 4$, and then $n = 1$ and $m = \frac{1}{2}$, which is absurd. Then J is not of the form $J = I(+)N$ with I an ideal of $\mathbb{Z}$ and N a $\mathbb{Z}$-sub-module of $2\mathbb{Z}$.

Example 3 (Kabbaj and Mahdou 2004) Let A be an integral domain with a nonzero prime ideal P and let $M = A/P$. Let $0 \neq x \in P$. The ideal $J = (x, \bar{1})(A(+)M) = \{(x, \bar{1})(a, \bar{m});\ (a, \bar{m}) \in A(+)M\} = \{(xa, \bar{a} + \overline{xm});\ a \in A, \bar{m} \in M\}$. The images by the canonical projections are $\pi_1(J) = xA$ and $\pi_2(J) = M$. Suppose that $(x, \bar{0}) \in J$. There exist $a \in A$ and $\bar{m} \in M$ such that $(x, \bar{0}) = (xa, \bar{a} + \overline{xm})$. Then $x = ax$ and $\bar{a} + \overline{xm} = \bar{0}$, so $a = 1$ and $1 + xm \in P$, and then $1 \in P$, which is absurd. Then $J \neq xA(+)M$.

Theorem 8.10 (Lim and Oh 2014) *Let A be a ring, M a A-module, and S a multiplicative set of A.*

1. *$S(+)M = \{(s, m);\ s \in S, m \in M\}$ is a multiplicative set of the ring $A(+)M$.*
2. *The ring $A(+)M$ is $S(+)M$-Noetherian if and only if the ring A is S-Noetherian and the A-module M is S-finite.*

Proof (1) Clear. (2) $" \Longrightarrow "$ The projection $f : A(+)M \longrightarrow A$, defined by $f(a, m) = a$ for each $(a, m) \in A(+)M$, is a homomorphism onto of rings with $f(S(+)M) = S$. By Lemma 7.1, the ring A is S-Noetherian. By the preceding lemma, $(0)(+)M$ is an ideal of the ring $A(+)M$. There exist $(s, m) \in S(+)M$ and $m_1, \ldots, m_r \in M$ such that $(s, m)((0)(+)M) \subseteq (0, m_1)(A(+)M) + \ldots +$

$(0, m_r)\big(A(+)M\big)$. Then $(0)(+)sM \subseteq (0)(+)Am_1 + \ldots + (0)(+)Am_r$, so $sM \subseteq Am_1 + \ldots + Am_r \subseteq M$. Then M is a S-finite $A-$module.

Let $\mathscr{P}$ be a prime ideal of $A(+)M$. By the preceding lemma, $\mathscr{P} = P(+)M$ with P a prime ideal of A. Since A is S-Noetherian, there exist $s \in S$ and a finitely generate ideal I of A such that $sP \subseteq I \subseteq P$. Since M is S-finite, there exist $t \in S$ and a finitely generated sub-module F of M such that $tM \subseteq F$. Then $(st, 0) \in S(+)M$ and $(st, 0)\mathscr{P} = (st, 0)(P(+)M) = stP(+)stM \subseteq tI(+)F \subseteq P(+)M = \mathscr{P}$. By the preceding lemma, $tI(+)F$ is an ideal of the ring $A(+)M$ because $tIM \subseteq tM \subseteq F$. Put $I = (a_1, \ldots, a_r)$ and $F = (f_1, \ldots, f_n)$. We will show that $tI(+)F$ is generated by $(ta_i, 0), 1 \leq i \leq r; (0, f_j), 1 \leq j \leq n$. An element of $tI(+)F$ is of the form (tx, f) with $x = \sum_{i:1}^{r} \alpha_i a_i$ and $f = \sum_{j:1}^{n} \beta_j f_j$ where $\alpha_i, \beta_j \in A$. Then

$$(tx, f) = \sum_{i:1}^{r}(\alpha_i ta_i, 0) + \sum_{j:1}^{n}(0, \beta_j f_j) = \sum_{i:1}^{r}(\alpha_i, 0)(ta_i, 0) + \sum_{j:1}^{n}(\beta_j, 0)(0, f_j).$$

Then $\mathscr{P}$ is $S(+)M$-finite.   □

Example 1 For each multiplicative set S of $\mathbb{Z}$, the ring $\mathbb{Z}(+)\mathbb{Q}$ is not $S(+)\mathbb{Q}$-Noetherian because the $\mathbb{Z}-$module $\mathbb{Q}$ is not S-finite.

Example 2 (Kim and Lim 2020) In Corollary 3.7, we proved that if $A \subseteq B$ is an extension of rings and if $S \subseteq A$ is a multiplicative set with B a S-finite $A-$module, then the ring A is S-Noetherian if and only if the ring B is S-Noetherian. We will show by the following examples that "B is a S-finite $A-$module" is a necessary condition in the two ways of the equivalence.

(a) Let $A = \mathbb{Z}(+)\mathbb{Z}$, $B = \mathbb{Z}(+)\mathbb{Z}[X]$, and $S \subseteq \mathbb{Z}$ any multiplicative set. Then $A \subseteq B$ is an extension of rings. By the preceding theorem, A is $S(+)\mathbb{Z}-$Noetherian. By examining the degrees, we can see that B is not a $S(+)\mathbb{Z}-$finite $A-$module. On the other hand, by the preceding theorem, the ring B is not $S(+)\mathbb{Z}[X]-$Noetherian because the $\mathbb{Z}-$module $\mathbb{Z}[X]$ is not S-finite for degrees reasons. Since $S(+)\mathbb{Z} \subseteq S(+)\mathbb{Z}[X]$, then B is not $S(+)\mathbb{Z}-$Noetherian.

(b) Let $A = \mathbb{Z}(+)\mathbb{Z}[X]$, $B = \mathbb{Z}[X](+)\mathbb{Z}[X]$, and $S \subseteq \mathbb{Z}$ any multiplicative set. Then $A \subseteq B$ is an extension of rings. By the preceding theorem, B is $S(+)\mathbb{Z}[X]-$Noetherian. By examining the degrees, we can see that B is not a $S(+)\mathbb{Z}[X]-$finite $A-$module. By the preceding theorem, the ring A is not $S(+)\mathbb{Z}[X]-$Noetherian because the $\mathbb{Z}-$module $\mathbb{Z}[X]$ is not S-finite for degrees reasons.

Example 3 (Kim and Lim 2020) In Proposition 3.15, we proved that if A is a ring, S a multiplicative set of A formed par by regular elements and M is a faithful S-Noetherian $A-$module, and then A is a S-Noetherian ring. We will prove by this example that the condition "M faithful" is necessary in this proposition. Let S' be any multiplicative set of the ring $\mathbb{Z}$. By the preceding theorem, $S = S'(+)\mathbb{Q}$ is a multiplicative set of the ring $A = \mathbb{Z}(+)\mathbb{Q}$. By Lemma 8.9 (1), $M = (0)(+)\mathbb{Z}$ is an ideal of the ring A, so it is a A−module.

(1) By Example 1, the ring A is not S-Noetherian.
(2) The multiplicative set S does not contain zero divisors. Indeed, let $(s, r) \in S$ and $(n, r') \in A$ such that $(0, 0) = (s, r)(n, r') = (sn, nr + sr')$. Then $sn = 0$ and $nr + sr' = 0$. Since $0 \notin S'$, then $s \neq 0$, so $n = 0$ and then $sr' = 0$, so $r' = 0$.
(3) The A−module M is not faithful because $(0, 1)M = \{(0, 1)(0, n); n \in \mathbb{Z}\} = \{(0, 0)\}$.
(4) The A−module M is Noetherian, so it is S-Noetherian. Indeed, let N be a sub-module of M. We will prove that $I = \{k \in \mathbb{Z}; (0, k) \in N\}$ is an ideal of $\mathbb{Z}$. It is clear that $0 \in I$ and that I is stable by addition. Let $k \in I$ and $m \in \mathbb{Z}$. Then $(0, k) \in N$ and $(m, 0) \in A$, and then $(m, 0)(0, k) \in N$ so $(0, mk) \in N$, and then $mk \in I$. There exists $n \in \mathbb{N}$ such that $I = n\mathbb{Z}$. By definition, for each $k \in \mathbb{Z}$, $k \in I$ if and only if $(0, k) \in N$. Then $N = (0)(+)I = (0)(+)n\mathbb{Z}$. We will show that $N = A(0, n)$, a cyclic module generated by the element $(0, n)$. We have $A(0, n) = \{(k, r)(0, n); k \in \mathbb{Z}, r \in \mathbb{Q}\} = \{(0, kn); k \in \mathbb{Z}\} = (0)(+)n\mathbb{Z}$.

Corollary 8.11 *Let A be a ring and M a A−module. The ring $A(+)M$ is Noetherian if and only if A is Noetherian and M is finitely generated.*

Proof Take $S = \{1\}$. An element of $S(+)M$ is of the form $(1, m)$ with $m \in M$, and then it is invertible in $A(+)M$ with inverse $(1, -m)$. □

Example The ring $\mathbb{Z}(+)\mathbb{Q}$ is not Noetherian because $\mathbb{Q}$ is not finitely generated.

9 Polynomials and Formal Power Series with Coefficients in a Module

In this section, we introduce the notions of polynomials, Laurent polynomials, and formal power series with coefficients in a module. By this way, we define new structures of modules. Then, we study its S-Noetherian property. The results are due to Faisol et al. (2019).

Lemma 9.1 *Let A be a ring and M a A−module. Then M is finitely generated if and only if there exist $n \in \mathbb{N}^*$ and a A−sub-module N of A^n such that $M \simeq A^n/N$.*

Proof " $\implies$" Put $M = (x_1, \ldots, x_n)$. The map $f : A^n \longrightarrow M$, defined for each $a = (a_1, \ldots, a_n) \in A^n$ by $f(a) = a_1x_1 + \ldots + a_nx_n$, is a homomorphism onto of A−modules. Then $M \simeq A^n/\ker f$.
" $\impliedby$" Let $\{e_1, \ldots, e_n\}$ be the canonical basis of the A−module A^n.
Then $A^n = (e_1, \ldots, e_n)$, so $M \simeq A^n/N = (\overline{e_1}, \ldots, \overline{e_n})$ is finitely generated. □

Notations Let A be a ring and M a $A-$module. We note:

(1) $M[X] = \left\{ \sum_{i:0}^{k} m_i X^i; \ k \in \mathbb{N}, m_i \in M \right\}$

(2) $M[X, X^{-1}] = \left\{ \sum_{i:-l}^{k} m_i X^i; \ l, k \in \mathbb{N}, m_i \in M \right\}$

(3) $M[[X]] = \left\{ \sum_{i:0}^{\infty} m_i X^i; m_i \in M \right\}.$

Then $M[X]$ is a $A[X]-$module, $M[X, X^{-1}]$ is a $A[X, X^{-1}]-$module, and $M[[X]]$ is a $A[[X]]-$module. For the three structures, the addition is termwise. The external multiplications are defined respectively in:

(1) $M[X]$ by $\left(\sum_{i:0}^{k} a_i X^i \right)\left(\sum_{j:0}^{l} m_j X^j \right) = \sum_{n:0}^{k+l} c_n X^n.$

(2) $M[X, X^{-1}]$ by $\left(\sum_{i:-u}^{k} a_i X^i \right)\left(\sum_{j:-v}^{l} m_j X^j \right) = \sum_{n:-u-v}^{k+l} c_n X^n.$

(3) $M[[X]]$ by $\left(\sum_{i:0}^{\infty} a_i X^i \right)\left(\sum_{j:0}^{\infty} m_j X^j \right) = \sum_{n:0}^{\infty} c_n X^n.$

In the three cases, $c_n = \sum_{i+j=n} a_i m_j \in M.$

The particular sub-modules defined in the following lemma will play an important role in the study of S-Noetherian property for the structures of modules introduced before.

Lemma 9.2 *Let A be a ring, M a $A-$module, and N a sub-module of M. Then:*

1. *$N[X]$ is a $A[X]-$sub-module of $M[X]$.*
2. *$N[X, X^{-1}]$ is a $A[X, X^{-1}]-$sub-module of $M[X, X^{-1}]$.*
3. *$N[[X]]$ is a $A[[X]]-$sub-module of $M[[X]]$.*

Proof Each one of the subsets $N[X]$, $N[X, X^{-1}]$, and $N[[X]]$ of $M[X]$, $M[X, X^{-1}]$, and $M[[X]]$, respectively, are modules for the induced laws. Then they are sub-modules.   □

The modules of polynomials, Laurent polynomials, and formal power series behave like the rings of polynomials, Laurent polynomials, and formal power series, respectively, with respect to the quotients.

Proposition 9.3 *Let A be a ring, M a $A-$module, and N a sub-module of M. We have the following isomorphisms:*

1. $(M/N)[X] \simeq M[X]/N[X]$ of $A[X]-modules.$
2. $(M/N)[X, X^{-1}] \simeq M[X, X^{-1}]/N[X, X^{-1}]$ of $A[X, X^{-1}]-modules.$
3. $(M/N)[[X]] \simeq M[[X]]/N[[X]]$ of $A[[X]]-modules.$

Proof Let $\phi : M[X] \longrightarrow (M/N)[X]$ be the map, defined for $f = m_0 + m_1 X + \ldots + m_k X^k$ by $\phi(f) = \bar{m}_0 + \bar{m}_1 X + \ldots + \bar{m}_k X^k$. Then ϕ is a homomorphism onto of $A[X]-modules$ with kernel $N[X]$. Then $(M/N)[X] \simeq M[X]/N[X]$.

We do the same for the other isomorphisms. □

Proposition 9.4 *Let A be a ring and $n \in \mathbb{N}^*$. We have the following isomorphisms:*

1. $A^n[X] \simeq (A[X])^n$ of $A[X]-modules.$
2. $A^n[X, X^{-1}] \simeq (A[X, X^{-1}])^n$ of $A[X, X^{-1}]-modules.$
3. $A^n[[X]] \simeq (A[[X]])^n$ of $A[[X]]-modules.$

Proof Let $\phi : A^n[X] \longrightarrow (A[X])^n$, the map defined by $\phi(f) = (f_1, \ldots, f_n)$, where $f = \sum_{j:0}^{k} a_j X^j \in A^n[X]$ with $a_j = (\alpha_{1j}, \alpha_{2j}, \ldots, \alpha_{nj}) \in A^n$ and $f_i = \sum_{j:0}^{k} \alpha_{ij} X^j \in A[X]$ for $1 \le i \le n$. Then ϕ is an isomorphism of $A[X]-modules.$

We do the same for the other isomorphisms. □

In the following proposition, we give a sufficient condition for the $A[X]-module$ $M[X]$ (resp. the $A[X, X^{-1}]-module$ $M[X, X^{-1}]$, resp. the $A[[X]]-module$ $M[[X]]$) to be finitely generated.

Proposition 9.5 *Let A be a ring and M a finitely generated $A-module$. Then:*

1. $M[X]$ is a finitely generated $A[X]-module.$
2. $M[X, X^{-1}]$ is a finitely generated $A[X, X^{-1}]-module.$
3. $M[[X]]$ is a finitely generated $A[[X]]-module.$

Proof (1) By Lemma 9.1, we have an isomorphism of $A-modules$ $M \simeq A^n/N$ with $n \in \mathbb{N}^*$ and N a $A-sub-module$ of A^n. By Propositions 9.3 and 9.4, we have the isomorphisms of $A[X]-modules$: $M[X] \simeq (A^n/N)[X] \simeq A^n[X]/N[X] \simeq (A[X])^n/N[X]$, where we identify $N[X]$ and its image by the isomorphism of $A^n[X] \longrightarrow (A[X])^n$. By Lemma 9.1, $M[X]$ is a finitely generated $A[X]-module.$

We do the same for the other properties. □

Proposition 9.6 *Let A be a ring, $S \subseteq A$ a multiplicative set, and M a $A-module$. If one of the following modules is S-Noetherian, the $A[X]-module$ $M[X]$, the $A[X, X^{-1}]$ -module $M[X, X^{-1}]$, or the $A[[X]]-module$ $M[[X]]$, then the $A-module$ M is S-Noetherian.*

Proof Suppose that the $A[X]-module$ $M[X]$ is S-Noetherian. Let N be a $A-sub-module$ of M. By Lemma 9.2, $N[X]$ is a $A[X]-sub-module$ of $M[X]$, so it is S-finite. There exist $s \in S$ and $f_1, \ldots, f_n \in N[X]$ such that $sN[X] \subseteq$ $(f_1, \ldots, f_n)A[X]$.

Let $f_1(0), \ldots, f_n(0)$ be the constant terms of $f_1, \ldots, f_n$, respectively.
Then $sN \subseteq (f_1(0), \ldots, f_n(0))A \subseteq N$, so N is S-finite. $\qquad\square$

Now, we are ready to prove the main result of this section.

Proposition 9.7 *Let A be a ring, $S \subseteq A$ an anti-Archimedean multiplicative set, and M a $A-$module. Suppose that A is S-Noetherian and M is S-finite. Then:*

1. *$M[X]$ is a S-Noetherian $A[X]-$module.*
2. *$M[X, X^{-1}]$ is a S-Noetherian $A[X, X^{-1}]-$module.*
3. *Moreover, if S does not contain zero divisors of A, then $M[[X]]$ is a S-Noetherian $A[[X]]-$module.*

Proof (1) By Theorem 4.2, the ring $A[X]$ is S-Noetherian. Since the $A-$module M is S-finite, there exist $s \in S$ and a finitely generated sub-module N of M such that $sM \subseteq N \subseteq M$. Then $sM[X] \subseteq N[X] \subseteq M[X]$. By Proposition 9.5, the $A[X]-$module $N[X]$ is finitely generated. Then the $A[X]-$module $M[X]$ is S-finite. By Corollary 3.6, the $A[X]-$module $M[X]$ is S-Noetherian.

We do the same for the other properties. $\qquad\square$

Lemma 9.8 *Let A be a ring and S a multiplicative set of A stable by addition. Then:*

1. *$S[X]$ is a multiplicative set of the ring $A[X]$.*
2. *$S[X, X^{-1}]$ is a multiplicative set of the ring $A[X, X^{-1}]$.*
3. *$S[[X]]$ is a multiplicative set of the ring $A[[X]]$.*

Proof (1) Let $f = \sum_{i:0}^{n} s_i X^i$ and $g = \sum_{i:0}^{m} t_i X^i \in S[X]$. Then $fg = \sum_{k:0}^{n+m} c_k X^i$ with $c_k = \sum_{i+j=k} s_i t_j \in S$, and then $fg \in S[X]$.

We prove in the same way the other properties. $\qquad\square$

Example $\mathbb{N}^*$ is a multiplicative set of the ring $\mathbb{Z}$ stable by addition.

Now, we give some sufficient conditions for the $A[X]-$module $M[X]$ (resp. the $A[X, X^{-1}]-$module $M[X, X^{-1}]$, resp. the $A[[X]]-$module $M[[X]]$) to be $S[X]$ (resp. $S[X, X^{-1}]$, resp. $S[[X]]$)-Noetherian.

Proposition 9.9 *Let A be a ring, $S \subseteq A$ an anti-Archimedean multiplicative set stable par addition, and M a S-finite $A-$module. Suppose that A is S-Noetherian. Then:*

1. *$M[X]$ is a $S[X]-$Noetherian $A[X]-$module.*
2. *$M[X, X^{-1}]$ is a $S[X, X^{-1}]-$Noetherian $A[X, X^{-1}]-$module.*

3. *Moreover, if S does not contain zero divisors of A, $M[[X]]$ is a $S[[X]]$*
   —Noetherian $A[[X]]$—module.

Proof (1) By Proposition 9.7, $M[X]$ is a S-Noetherian $A[X]$—module. Since $S \subseteq S[X]$, then $M[X]$ is $S[X]$-Noetherian.

We do the same for the other properties. $\square$

10   S-Noetherian Nagata Rings

In this section, we introduce and study the Nagata ring of a S-Noetherian ring. We also study locally S-Noetherian integral domains. The results are due to Lim (2015).

Definition 10.1

1. A ring A is of finite character if each nonzero element of A belongs to at most a finite number of maximal ideals of A.
2. Let S be a multiplicative set of an integral domain A. We say that A is locally S-Noetherian if for each maximal ideal M of A, the ring A_M is S-Noetherian.

   When $S = \{1\}$, we simply say that A is locally Noetherian.

Example 1 A ring with finite maximal spectrum has finite character.

Example 2 The ring $\mathbb{Z}$ has an infinite maximal spectrum. But it has a finite character. Indeed, let $n \in \mathbb{Z} \setminus \{0, \pm 1\}$, $n = \pm p_1^{\alpha_1} \ldots p_r^{\alpha_r}$, where the p_i are prime numbers and the integers $\alpha_i \geq 1$. The only maximal ideals containing n are the $p_i \mathbb{Z}$, $1 \leq i \leq r$.

More generally, any principal ideal ring A has a finite character. Indeed, let a be a nonzero noninvertible element of A. Then $a = p_1^{\alpha_1} \ldots p_n^{\alpha_n}$, where the p_i are nonassociated irreducible elements of A and the integers $\alpha_i \geq 1$. Then $(p_1), \ldots, (p_n)$ are the only maximal ideals of A containing a. Note that a DVR is principal with only one irreducible element modulo association.

Example 3 The ring $\mathbb{Z}[X]$ has no finite character because for each prime number p, the ideal (p, X) is maximal containing X. Indeed, the map $\phi : \mathbb{Z}[X] \longrightarrow \mathbb{Z}/p\mathbb{Z}$, defined by $\phi(f(X)) = \overline{f(0)}$ for each $f(X) \in \mathbb{Z}[X]$, is a homomorphism onto with kernel (p, X). Then $\mathbb{Z}[X]/(p, X) \simeq \mathbb{Z}/p\mathbb{Z}$ is a field, so (p, X) is maximal.

In the following proposition, we study the transfer of the S-Noetherian property to the rings of fractions.

Proposition 10.2 *Let S be a multiplicative set of an integral domain A such that A is S-Noetherian. For each multiplicative set T of A, the ring A_T is S-Noetherian.*
   In particular, for each prime ideal P of A, the ring A_P is S-Noetherian.

Proof This is the particular case of Lemma 4.3 for integral domains. $\square$

The converse of the preceding proposition is true when the domain A has a finite character.

Proposition 10.3 *Let S be a multiplicative set of an integral domain A.*
If A has a finite character and is locally S-Noetherian, then it is S-Noetherian.

Proof Let I be a nonzero ideal of A and $0 \neq a \in I$. Let $M_1, \dots, M_n$ be the only maximal ideals of A containing a. For $1 \leq i \leq n$, the ring A_{M_i} is S-Noetherian. There exist $s_i \in S$ and a finitely generated ideal $F_i \subseteq I$ of A such that $s_i I A_{M_i} \subseteq F_i A_{M_i}$. Let $s = s_1 \dots s_n \in S$ and $F = (a) + F_1 + \dots + F_n$ be a finitely generated ideal of A contained in I. For each $M \in Max(A) \backslash \{M_1, \dots, M_n\}$, $I A_M = A_M = F A_M$ since the element a of $F \subseteq I$ is invertible in A_M. So for each $M \in Max(A)$, $sI A_M \subseteq F A_M$. Then $sI = \bigcap \{sI A_M; \ M \in Max(A)\} \subseteq \bigcap \{F A_M; \ M \in Max(A)\} = F$. So $sI \subseteq F \subseteq I$.   □

Example Let A be an integral domain with finite character. Then A is locally Noetherian if and only if it is Noetherian.

Notations

(1) Let A be an integral domain with quotient field K and let $f \in K[X]$. The content $c(f)$ of f is the A–sub-module of K generated by all the coefficients of f. If $f \in A[X]$, then $c(f)$ is an ideal of A.

(2) Let A be a ring and $\mathscr{I}$ an ideal of $A[X]$. The set $c(\mathscr{I})$ of the coefficients of all the elements of $\mathscr{I}$ is an ideal of A. Indeed, it is clear that for each $x \in A$, $xc(\mathscr{I}) \subseteq c(\mathscr{I})$. Let $a, b \in c(\mathscr{I})$. There exist $f, g \in \mathscr{I}$ such that a is a coefficient of f and b is a coefficient of g. There exists $m \in \mathbb{N}$ such that $a + b$ is a coefficient of the polynomial $f + X^m g$ or of the polynomial $g + X^m f$. Both of them are in $\mathscr{I}$. Then $a + b \in c(\mathscr{I})$.

Example (Gilmer and Hoffmann 1975) Let A be a ring and $f(X) \in A[X]$. Then $c(f A[X]) = c(f)$. Indeed, the first inclusion is clear. Conversely, put $f(X) = a_0 + a_1 X + \dots + a_n X^n \in A[X]$ and let $x \in c(f)$, $x = b_0 a_0 + \dots + b_n a_n$ with $b_0, \dots, b_n \in A$. Let $g(X) = \left(\sum_{i:0}^{n} b_i X^{n-i} \right) f(X) \in f A[X]$. The coefficient of X^n in $g(X)$ is equal to x. Then $x \in c(f A[X])$.

Lemma 10.4 *Let A be a ring.*

1. *The set $N = \{ f \in A[X]; \ c(f) = A \} = A[X] \backslash \bigcup \{M[X]; \ M \in Max(A)\}$ is multiplicative in the polynomial ring $A[X]$.*
2. *The maximal spectrum $Max(A[X]_N) = \{M[X]_N; \ M \in Max(A)\}$.*

Proof

(1) Let $f \in A[X]$. Then $f \in N \iff c(f) = A \iff \forall M \in Max(A), c(f) \not\subseteq M \iff \forall M \in Max(A), f \notin M[X] \iff f \in A[X] \backslash \bigcup \{M[X]; \ M \in$

$Max(A)\}$. Then $N = A[X] \setminus \bigcup \{M[X]; \ M \in Max(A)\}$. Since $M[X]$ is a
prime ideal of $A[X]$ for each $M \in Max(A)$, then N is a multiplicative set of
the ring $A[X]$.

(2) Let $P \in spec(A[X]_N)$. There exists $Q \in spec(A[X])$ such that $Q \cap N = \emptyset$
and $P = Q_N$. The set of coefficients of all the elements of Q is an ideal I of
A. Suppose that $I = A$. There exists $f \in Q$ admitting 1 as a coefficient. Then
$c(f) = A$ and $f \in N$, which is absurd because $Q \cap N = \emptyset$. Then I is a proper
ideal. It is contained in a maximal ideal M. Then $Q \subseteq I[X] \subseteq M[X]$ and
$P = Q_N \subseteq M[X]_N$. Since $M[X] \cap N = \emptyset$, then $M[X]_N$ is prime. To conclude
that the $M[X]_N$ are exactly the maximal ideals of $A[X]_N$, it suffices to show
that they are incomparable. But if $M[X]_N \subseteq M'[X]_N$ with $M, M' \in Max(A)$,
then $M[X] \subseteq M'[X]$, so $M \subseteq M'$, and then $M = M'$. So $M[X]_N = M'[X]_N$.
$\square$

Example Let A be a ring and $\mathscr{I}$ an ideal of $A[X]$ with $\mathscr{I} \subseteq \bigcup \{M[X]; M \in$
$Max(A)\}$. There exists $M_0 \in Max(A)$ such that $\mathscr{I} \subseteq M_0[X]$. Indeed, $c(\mathscr{I}) \subseteq$
$\bigcup \{M; \ M \in Max(A)\}$. Then $1 \notin c(\mathscr{I})$ and so $c(\mathscr{I})$ is a proper ideal of A. There
exists $M_0 \in Max(A)$ such that $c(\mathscr{I}) \subseteq M_0$, so $\mathscr{I} \subseteq M_0[X]$.

Notation Let A be a ring and $N = \{f \in A[X]; \ c(f) = A\}$: multiplicative set of
$A[X]$. We call Nagata ring and we denote by $A(X) = A\lfloor X \rfloor_N$.

Example Let (A, M) be a local ring. Then $N = A[X] \setminus M[X]$ and so $A(X) =$
$A[X]_{M[X]}$.

Proposition 10.5 *Let A be an integral domain and $S \subseteq A$ an anti-Archimedean
multiplicative set. The following assertions are equivalent:*

1. *The ring A is S-Noetherian.*
2. *The polynomial ring $A[X]$ is S-Noetherian.*
3. *The Nagata ring $A(X)$ is S-Noetherian.*

Proof $"(1) \implies (2)"$ By Theorem 4.2. $"(2) \implies (3)"$ By Proposition 10.2.

$"(3) \implies 1)"$ Let I be an ideal of A. Then $IA[X]_N$ is an ideal of $A(X) = A[X]_N$.
Since $A[X]_N$ is S-Noetherian, there exist $s \in S$ and a finitely generated ideal J of
A contained in I such that $sIA[X]_N \subseteq JA[X]_N$. For each $a \in I$, there exists
$g(X) \in N$ such that $sag(X) \subseteq JA[X]$, so $sa.c(g) \subseteq J$. But $c(g) = A$, then
$sa \in J$. So $sI \subseteq J \subseteq I$ and then I is S-finite. $\square$

Corollary 10.6 (Anderson et al. 1985) *The following assertions are equivalent for
an integral domain A:*

1. *The ring A is Noetherian.*
2. *The polynomial ring $A[X]$ is Noetherian.*
3. *The Nagata ring $A(X)$ is Noetherian.*

Lemma 10.7 *Let* (A, M) *be a local integral domain,* $S \subset A$ *a multiplicative set, and* I *an ideal of* A. *Then* I *is* S-*finite if and only if* $I A[X]_{M[X]}$ *is* S-*finite.*

Proof " $\Longrightarrow$ " There exist $s \in S$ and a finitely generated ideal J of A such that $sI \subseteq J \subseteq I$. Then $sIA[X]_{M[X]} \subseteq JA[X]_{M[X]} \subseteq IA[X]_{M[X]}$ with $JA[X]_{M[X]}$ a finitely generated ideal of the ring $A[X]_{M[X]}$.

" $\Longleftarrow$ " There exist $s \in S$ and a finitely generated ideal J of A such that $sIA[X]_{M[X]} \subseteq JA[X]_{M[X]} \subseteq IA[X]_{M[X]}$, and then $sIA[X]_{M[X]} \cap A \subseteq JA[X]_{M[X]} \cap A \subseteq IA[X]_{M[X]} \cap A$. But for each ideal H of A, $HA[X]_{M[X]} \cap A = H$. Indeed, let $a \in HA[X]_{M[X]} \cap A$, $a = \frac{f(X)}{s(X)}$ with $f(X) \in H[X]$ and $s(X) \in A[X] \setminus M[X]$. A coefficient b of $s(X)$ is not in M, so it is invertible in A. Since $as(X) = f(X)$, then $ab \in H$, so $a \in H$. The two preceding inclusions become $sI \subseteq J \subseteq I$. Then I is S-finite. $\square$

Lemma 10.8 (Arnold 1969) *Let* A *be an integral domain.*

1. *If* P *is a prime ideal of* A *and* S *is a multiplicative set of* A *disjoint with* P, *then* $(A_S)_{PA_S} = A_P$.
2. *For each* $M \in Max(A)$, *we have* $(A[X]_N)_{M[X]_N} = A[X]_{M[X]}$ *where* $N = \{f \in A[X];\ c(f) = A\}$.
3. *For each prime ideal* P *of* A, $(A_P[X])_{(PA_P)[X]} = A[X]_{P[X]}$.

Proof

(1) Let $x \in A_P$, $x = \frac{a}{s}$ with $a \in A$ and $s \in A \setminus P$. Since $PA_S \cap A = P$, then $s \notin PA_S$, so $x = \frac{a}{s} \in (A_S)_{PA_S}$. Conversely, let $z \in (A_S)_{PA_S}$, $z = \frac{x}{y}$ with $x \in A_S$ and $y \in A_S \setminus PA_S$. Then $x = \frac{a}{s}$ and $y = \frac{b}{t}$ with $a, b \in A$ and $s, t \in S$, $b \notin P$ and then $z = \frac{at}{bs}$ with $at, bs \in A$ and $bs \notin P$ because $s \in S$. Then $z = \frac{at}{bs} \in A_P$.

(2) By Lemma 10.4, the multiplicative set $N = A[X] \setminus \bigcup \{M'[X];\ M' \in Max(A)\}$ of $A[X]$ is disjoint with the prime ideal $M[X]$. We conclude by (1).

(3) Let $S = A \setminus P$. Then S is a multiplicative set of $A[X]$ disjoint with the prime ideal $P[X]$, $A_P[X] = (A[X])_S$, and $(PA_P)[X] = P[X].A_P[X] = P[X].A[X]_S$.

By (1), $(A_P[X])_{(PA_P)[X]} = ((A[X])_S)_{P[X]A[X]_S} = A[X]_{P[X]}$. $\square$

Example Let A be an integral domain with quotient field K and P a nonzero prime ideal of $A[X]$ such that $P \cap A = (0)$. Then $A[X]_P$ is a DVR. Indeed, $S = A \setminus (0)$ is a multiplicative set of $A[X]$ disjoint with P. By the preceding lemma, $A[X]_P = (A[X]_S)_{P_S} = (A_S[X])_{P_S} = (K[X])_{P_S}$. Since $K[X]$ is a PID and P_S is a nonzero prime ideal of $K[X]$, then $(K[X])_{P_S}$ is a DVR.

In the following theorem, we characterize the property locally S-Noetherian in terms of Nagata ring.

Theorem 10.9 *Let* A *be an integral domain and* $S \subseteq A$ *an anti-Archimedean multiplicative set. The following assertions are equivalent:*

1. The ring A is locally S-Noetherian.
2. The Nagata ring A(X) is locally S-Noetherian.

Proof "(1) $\implies$ (2)" Let Q be a maximal ideal of $A(X) = A[X]_N$ where $N = \{f \in A[X]; c(f) = A\}$. By Lemma 10.4 (2), $Q = M[X]_N$ with M a maximal ideal of A. By hypothesis, the ring A_M is S-Noetherian. By Theorem 4.2, the ring $A_M[X]$ is S-Noetherian. Since MA_M is a prime ideal of A_M, then $(MA_M)[X]$ is a prime ideal of $A_M[X]$. By Proposition 10.2, the localized $A_M[X]_{(MA_M)[X]}$ is S-Noetherian. But by the preceding lemma, $A_M[X]_{(MA_M)[X]} = A[X]_{M[X]}$ and $(A[X]_N)_Q = (A[X]_N)_{M[X]_N} = A[X]_{M[X]}$. Then $(A[X]_N)_Q$ is S-Noetherian.

"(2) $\implies$ (1)" Let M be a maximal ideal of A. By Lemma 10.4 (2), $M[X]_N$ is a maximal ideal of $A[X]_N$. Since $A[X]_N$ is locally S-Noetherian, $(A[X]_N)_{M[X]_N}$ is S-Noetherian. By the preceding lemma, we have $(A[X]_N)_{M[X]_N} = A[X]_{M[X]} = A_M[X]_{(MA_M)[X]}$. Then $A_M[X]_{(MA_M)[X]}$ is S-Noetherian. Let I be an ideal of A_M. Then $IA_M[X]_{(MA_M)[X]}$ is S-finite. Since A_M is local with maximal ideal MA_M, by Lemma 10.7, the ideal I is S-finite. Then A_M is a S-Noetherian ring and so the ring A is locally S-Noetherian. $\square$

11 S-Noetherian t—Nagata Rings

Notations Let A be an integral domain with quotient field $K \neq A$. A fractional ideal of A is a A—sub-module I of K for which there is $0 \neq a \in A$ such that $aI \subseteq A$. For example, a finitely generated sub A—module of K is a fractional ideal of A. Let $x \in K$ and I the A—sub-module of K generated by $\{x^i; i \in \mathbb{N}\}$. Then I is a fractional ideal of A if and only if x is quasi-integral over A.

To avoid any confusion, an ordinary ideal of A is said to be integral. We denote by $\mathscr{F}(A)$ the set of the nonzero fractional ideals of A. For each $I \in \mathscr{F}(A)$, let $I^{-1} = (A : I) = \{x \in K; xI \subseteq A\}$. It is a fractional ideal of A. Indeed, let $0 \neq d \in A$ be such that $dI \subseteq A$ and $0 \neq a \in I \cap A$. Then $0 \neq da \in A \cap I$, so $da(A : I) \subseteq A$.

For each $I \in \mathscr{F}(A)$, let $I_v = (I^{-1})^{-1}$. It is a fractional ideal of A. We say that I is a v—ideal (or divisorial) if $I_v = I$. The map $I \longrightarrow I_v$ is called the v—operation over A. It satisfies the following properties of star operations:

(1) For each $x \in K^*$ and $J \in \mathscr{F}(A)$, $(x)_v = (x)$ and $(xJ)_v = xJ_v$.
(2) For each $J, L \in \mathscr{F}(A)$, $J \subseteq J_v$, and $(J \subseteq L \implies J_v \subseteq L_v)$.
(3) For each $J \in \mathscr{F}(A)$, $(J_v)_v = J_v$.

Indeed, note first that $(x)^{-1} = (A : x) = \{y \in K; yx \in A\} = \{y \in K; y \in \frac{1}{x}A\} = \frac{1}{x}A = (\frac{1}{x})$. Then $(x)_v = ((x)^{-1})^{-1} = (\frac{1}{x})^{-1} = (x)$.

We also have $(xJ)^{-1} = \{y \in K; yxJ \in A\} = \{y \in K; yx \in J^{-1}\} = \{y \in K; y \in \frac{1}{x}J^{-1}\} = \frac{1}{x}J^{-1}$. Then $(xJ)_v = ((xJ)^{-1})^{-1} = (\frac{1}{x}J^{-1})^{-1} = x(J^{-1})^{-1} = xJ_v$. So (1).

(2) We have $J(A : J) \subseteq A$, and then $J \subseteq A : (A : J) = J_v$. So $J \subseteq J_v$.

If $J \subseteq L$, then $(A : L) \subseteq (A : J)$, so $A : (A : J) \subseteq A : (A : L)$. Then $J_v \subseteq L_v$.

To show (3), we need the following remark: $I_v = \bigcap \{xA; \ x \in K, I \subseteq xA\}$. Indeed, let J be the intersection. Since the principal ideals are v−ideals, then $I_v \subseteq xA$ for each x. Then $I_v \subseteq J$. Suppose that the inclusion is strict. Let $x \in J \setminus I_v$. Then $x \notin A : (A : I)$, so $x(A : I) \nsubseteq A$. There exists $0 \neq y \in (A : I)$ such that $xy \notin A$, and then $x \notin y^{-1}A$. But $yI \subseteq A$, and then $I \subseteq y^{-1}A$. By definition of J, we must have $x \in y^{-1}A$. There is a contradiction. We conclude that $I_v = J$.

(3) By (2), we have $J_v \subseteq (J_v)_v$. Conversely, $J_v = \bigcap \{xA; \ x \in K, J \subseteq xA\}$. We must show that $(J_v)_v$ is included in each member of the intersection. But if $J \subseteq xA$, then by (2) and (1), $J_v \subseteq (xA)_v = xA$. Then $(J_v)_v \subseteq (xA)_v = xA$. So $J_v = (J_v)_v$.

For each $I \in \mathscr{F}(A)$, let $I_t = \bigcup \{J_v; \ 0 \neq J$ a finitely generated fractional subideal of $I\}$. We have $I \subseteq I_t \subseteq I_v$ and if I is finitely generated, $I_t = I_v$. Indeed, for each $0 \neq x \in I$, $(x) = (x)_v \subseteq I_t$, and then $I \subseteq I_t$. For each nonzero finitely generated fractional subideal J of I, we have $J_v \subseteq I_v$. Then $I_t \subseteq I_v$. When I is finitely generated, the inverse inclusion is clear from the definition of I_t. We say that I is a t−ideal if $I_t = I$. If I is a v−ideal, then it is a t−ideal. We will show that I_t is a fractional ideal of A. Let $x, y \in I_t$. There exist two nonzero finitely generated fractional subideals J and L of I such that $x \in J_v$ and $y \in L_v$. Then $x + y \in J_v + L_v \subseteq (J + L)_v \subseteq I_t$. For each $a \in A$, $ax \in aJ_v = (aJ)_v \subseteq I_t$. So I_t is a sub A−module of K. Since $I_t \subseteq I_v$, then I_t is a fractional ideal of A. The map $I \longrightarrow I_t$ is called the t−operation. It satisfies the following properties:

(1) For each $x \in K^*$ and $J \in \mathscr{F}(A)$, $(x)_t = (x)$ and $(xJ)_t = xJ_t$.

(2) For each $J, L \in \mathscr{F}(A)$, $J \subseteq J_t$, and $(J \subseteq L \Longrightarrow J_t \subseteq L_t)$.

(3) For each $J \in \mathscr{F}(A)$, $(J_t)_t = J_t$.

Indeed, $(x) = (x)_v = (x)_t$. We have just seen that $J \subseteq J_t$. It is also clear that if $J \subseteq L$, then $J_t \subseteq L_t$. We want to show that $(xI)_t = xI_t$. Let J be a nonzero finitely generated subideal of xI. Then $J_v = (xx^{-1}J)_v = x(x^{-1}J)_v$ with $x^{-1}J$ a nonzero finitely generated fractional ideal and $x^{-1}J \subseteq x^{-1}xI = I$. By definition of I_t, we have $(x^{-1}J)_v \subseteq I_t$. Then $J_v = x(x^{-1}J)_v \subseteq xI_t$. We conclude that $(xI)_t \subseteq xI_t$. Conversely, let J be a nonzero finitely generated fractional subideal of I. Then $xJ_v = (xJ)_v \subseteq (xI)_t$. So $xI_t \subseteq (xI)_t$, and then we have the equality $(xI)_t = xI_t$.

We have noted before that $I_t \subseteq (I_t)_t$. Conversely, let J be a nonzero finitely generated subideal of I_t, $J = (a_1, \ldots, a_n)$. For each i, there exists a nonzero finitely generated subideal J_i of I such that $a_i \in (J_i)_v$. The subideal $L = J_1 + \ldots + J_n$ of I is nonzero, finitely generated and $J = (a_1, \ldots, a_n) \subseteq (J_1)_v + \ldots + (J_n)_v \subseteq (J_1 + \ldots + J_n)_v = L_v$. Then $J_v \subseteq (L_v)_v = L_v \subseteq I_t$. So $(I_t)_t \subseteq I_t$.

Example 1 Let k be a field, $A = k[[X, Y]]$, and $I = (X, Y)$, the maximal ideal of A. Then $I^{-1} = A$. Indeed, since I is an integral ideal, then $A \subseteq I^{-1}$. Let $0 \neq f \in I^{-1}$, $f = \frac{g}{h}$ with $g, h \in k[[X, Y]] \setminus (0)$. Since $k[[X, Y]]$ is UFD (see example of Corollary 8.6, Chap. 2), we may suppose that g and h are relatively prime. Suppose that $h \notin U(k[[X, Y]])$. Since $f \in I^{-1}$, then $Xf, Yf \in A$. But

$Xfh = Xg$ and $Yfh = Yg$, and then h divides Xg and Yg in A. By Euclide lemma, h divides X and Y. But X and Y are irreducible, and then h is associated with X and with Y. So X and Y are associated, which is absurd. Then $h \in U(k[[X, Y]])$ and $f \in A$. So $I^{-1} = A$, and then $I_v = (I^{-1})^{-1} = A$ and so I is not divisorial. Since I is finitely generated, $I_t = I_v = A$.

Example 2 In a valuation domain A, each nonzero integral ideal I is a $t−$ideal. Indeed, the nonzero finitely generated ideals of A are principal, so they are $v−$ideals. Then $I_t = \bigcup \{F_v; \; F$ is a nonzero finitely generated subideal of $I\} = \bigcup \{F; \; F$ is a nonzero finitely generated subideal of $I\} = I$.

Example 3 Let (A, M) be a valuation domain. Then M is divisorial if and only if it is principal. Indeed, if M is divisorial, then $A \subset M^{-1}$ because if not, $A = M^{-1}$, then $A = A^{-1} = (M^{-1})^{-1} = M_v = M$, which is absurd. Let $x \in M^{-1} \setminus A$, and then $x^{-1} \in A$ and even $x^{-1} \in M$. Then $M = M_v = (M^{-1})^{-1} \subseteq (xA)^{-1} = x^{-1}A \subseteq M$. So $M = x^{-1}A$.

Example 4 Let A be an integral domain and $(I_\alpha)_{\alpha \in \Lambda}$ a nonempty family of divisorial ideals. If $I = \bigcap_{\alpha \in \Lambda} I_\alpha \neq (0)$, then it is a divisorial ideal. Indeed, for each $\alpha \in \Lambda$, $I_v \subseteq (I_\alpha)_v = I_\alpha$. Then $I_v \subseteq \bigcap_{\alpha \in \Lambda} I_\alpha = I$. So $I_v = I$.

Example 5 Let A be an integral domain and I a nonzero fractional ideal. Then I^{-1} is a divisorial ideal. Indeed, $I^{-1} = (A : I) = \bigcap_{0 \neq x \in I} (A : xA) = \bigcap_{0 \neq x \in I} \frac{1}{x}A$, an intersection of divisorial ideals. Then I^{-1} is divisorial by the preceding example.

Proposition 11.1 *Let A be an integral domain, $*$ one of the star operations v or t, and I and J two nonzero fractional ideals of A. Then $(IJ)^* = (IJ^*)^* = (I^*J^*)^*$.*

   In particular, $I^*J^* \subseteq (IJ)^*$.

Proof Since $IJ \subseteq IJ^*$, then $(IJ)^* \subseteq (IJ^*)^*$. For the converse inclusion, it suffices to show that $IJ^* \subseteq (IJ)^*$. Let $x \in IJ^*$. There exist $a_1, \dots, a_n \in I$ such that $x \in a_1 J^* + \dots + a_n J^*$. But $a_1 J^* + \dots + a_n J^* = (a_1 J)^* + \dots + (a_n J)^* \subseteq (a_1 J + \dots + a_n J)^* \subseteq (IJ)^*$, and then $x \in (IJ)^*$. So we have proved the first equality of the proposition. By replacing I with I^*, we obtain $(I^*J^*)^* = (I^*J)^* = (IJ)^*$. For the last assertion, $I^*J^* \subseteq (I^*J^*)^* = (IJ)^*$.   □

Let A be an integral domain. An element maximal in the set of the proper integral $t−$ideals of A is said to be $t−$maximal. We denote by $t − Max(A)$ the set of such ideals.

Proposition 11.2 *Let A be an integral domain. Every proper integral $t−$ideal of A is contained in a $t−$maximal ideal of A.*

Proof The set $\mathscr{F}$ of the proper integral t-ideals of A is not empty. It contains the nonzero proper principal ideals. It is also inductive. Indeed, let $(I_\alpha)_{\alpha \in \Lambda}$ be a totally ordered family of elements of $\mathscr{F}$ and $I = \bigcup_{\alpha \in \Lambda} I_\alpha$. Then I is also a proper integral t-ideal of A which bounds this family in $\mathscr{F}$. To see this, let J be a nonzero finitely generated subideal of I. There exists $\alpha_0 \in \Lambda$ such that $J \subseteq I_{\alpha_0}$, and then $J_v \subseteq (I_{\alpha_0})_t = I_{\alpha_0} \subseteq I$. So $I_t = I$. We conclude by Zorn's lemma. $\quad\square$

Let A be an integral domain which is not a field. As it was said in the preceding proof, the nonzero proper principal integral ideals of A are t-ideals. By the preceding proposition, $t - Max(A)$ is not empty.

Proposition 11.3 *Let A be an integral domain. Every t-maximal ideal of A is prime; i.e., $t - Max(A) \subseteq Spec(A)$.*

Proof Let $P \in t - Max(A)$. Suppose that P is not prime. There exist a and $b \in A \setminus P$ such that $ab \in P$. Let $I = P + aA$ and $J = P + bA$. Then $P \subset I \subseteq I_t \subseteq A$ and $P \subset J \subseteq J_t \subseteq A$. By maximality of P, we have $I_t = J_t = A$. By Proposition 11.1, $(IJ)_t = (I_t J_t)_t = (AA)_t = A$. But $IJ = (P + aA)(P + bA) = P^2 + aP + bP + abA \subseteq P$. Then $A = (IJ)_t \subseteq P_t = P$, so $P = A$, which is absurd. $\quad\square$

Example The inclusion in the preceding proposition may be strict. Recall first from Example 2 at the beginning of this section that in a valuation domain, each nonzero integral ideal is a t-ideal. In particular, its maximal ideal is t-maximal. Let A be a valuation domain of Krull dimension ≥ 2. Every nonzero nonmaximal prime ideal of A is $(t-)$prime but non-t-maximal.

Proposition 11.4 (Kang 1989) *Let A be an integral domain.*

1. *The set $N_v = \{0 \neq f \in A[X]; c(f)_v = c(f)_t = A\} = A[X] \setminus \bigcup \{M[X]; M \in t - Max(A)\}$ is multiplicative in the ring $A[X]$. The ring of fractions $A[X]_{N_v}$ is called t-Nagata ring of A. It is usually noted $A\{X\}$.*
2. *The maximal spectrum $Max(A[X]_{N_v}) = \{M[X]_{N_v}; M \in t - Max(A)\}$.*

Proof

(1) Let $0 \neq f \in A[X]$. By Proposition 11.2, $f \in N_v \iff (c(f))_t = A \iff \forall M \in t - Max(A), (c(f))_t \not\subseteq M \iff \forall M \in t - Max(A), c(f) \not\subseteq M \iff \forall M \in t - Max(A), f \notin M[X] \iff f \in A[X] \setminus \bigcup \{M[X]; M \in t - Max(A)\}$. Then $N_v = A[X] \setminus \bigcup \{M[X]; M \in t - Max(A)\}$. By the preceding proposition, $M[X]$ is a prime ideal of $A[X]$ for each $M \in t - Max(A)$. Then N_v is a multiplicative set of the ring $A[X]$.

(2) Let $0 \neq P \in spec(A[X]_{N_v})$. There exists $0 \neq Q \in spec(A[X])$ such that $Q \cap N_v = \emptyset$ and $P = Q_{N_v}$. The set of the coefficients of all the elements of Q is an ideal I of A. Suppose that $I_t = A$, there exists a nonzero finitely generated ideal $J = (a_1, \ldots, a_k) \subseteq I$ such that $1 \in J_v$, a_i is a coefficient of certain $f_i \in Q$. Put $n_i = deg\, f_i$, $1 \leq i \leq k$, and then $f = f_1 + X^{n_1+1} f_2 + X^{n_1+n_2+2} f_3 +$

$\ldots + X^{n_1+n_2+\ldots+n_{k-1}+k-1} f_k \in Q$ and $a_1, \ldots, a_k$ are coefficients of f. Since $1 \in J_v \subseteq c(f)_v \subseteq A$, then $(c(f))_v = A$, so $f \in N_v$, which is absurd because $Q \cap N_v = \emptyset$. Then I_t is a proper t-ideal. By Proposition 11.2, there exists $M \in t - Max(A)$ such that $I \subseteq I_t \subseteq M$, and then $Q \subseteq I[X] \subseteq M[X]$ and $P = Q_{N_v} \subseteq M[X]_{N_v}$. Since $M[X] \cap N_v = \emptyset$, $M[X]_{N_v}$ is a prime ideal of $A[X]_{N_v}$. To conclude that the $M[X]_{N_v}$ are exactly the maximal ideals of $A[X]_{N_v}$, it suffices to show they are incomparable. But if $M[X]_{N_v} \subseteq M'[X]_{N_v}$ with $M, M' \in t - Max(A)$, then $M[X] \subseteq M'[X]$, so $M \subseteq M'$, and then $M = M'$. So $M[X]_{N_v} = M'[X]_{N_v}$. $\square$

Let A be a ring and $f(X) \in A[X]$ a polynomial. Recall from the preceding section that the content $c(f)$ of $f(X)$ is the ideal of A generated by the coefficients of $f(X)$. For $f(X), g(X) \in A[X]$, the ideal $c(fg)$ has interesting properties. Among them is its comparison with $c(f)c(g)$. If the inclusion $c(fg) \subseteq c(f)c(g)$ is clear, the converse is not true in general. The following theorem gives a formula that links the two ideals in a more setting.

Dedekind-Mertens Theorem 11.5 *Let A be an integral domain with quotient field K and $f, g \in K[X]$ with $m = \deg g$. Then $c(f)^{m+1}c(g) = c(f)^m c(fg)$ where $c(f)$ is the sub A-module of K generated by the coefficients of f.*

Proof Since $c(fg) \subseteq c(f)c(g)$, then $c(f)^m c(fg) \subseteq c(f)^{m+1}c(g)$.

We will prove the converse inclusion by induction on $n = \deg f$ and $m = \deg g$.

The equality is true if f or g is a monomial. Indeed, if $f = a_n X^n$, then $c(f)^{m+1}c(g) = a_n^{m+1}c(g) = (a_n^m)a_n c(g) = c(f)^m c(a_n g) = c(f)^m c(fg)$ and if $g = b_m X^m$, then
$$c(f)^{m+1}c(g) = c(f)^m c(f)(b_m) = c(f)^m c(b_m f) = c(f)^m c(fg).$$
In particular, the equality is true if f or g is a nonzero constant.

Put the induction hypothesis:

(1) If $\deg f = n < r$ and $\deg g = m$, then $c(f)^{m+1}c(g) \subseteq c(f)^m c(fg)$.
(2) If $\deg f = r$ and $\deg g = m < s$, then $c(f)^{m+1}c(g) \subseteq c(f)^m c(fg)$.

Put $f = a_0 + a_1 X + \ldots + a_r X^r$ with $a_r \neq 0$ and $g = b_0 + b_1 X + \ldots + b_s X^s$ with $b_s \neq 0$ and suppose that neither f nor g is a monomial. We will prove that $c(f)^{s+1}c(g) \subseteq c(f)^s c(fg)$. Let $f_1 = f - a_r X^r, g_1 = g - b_s X^s, h = fg, h_1 = f_1 g$, and $h_2 = fg_1$. Then
$$h = c_0 + c_1 X + \ldots + c_{r+s} X^{r+s} \text{ with } c_k = \sum_{i+j=k} a_i b_j, \text{ for } 0 \leq k \leq r+s.$$
$h_1 = c_0^{(1)} + c_1^{(1)} X + \ldots + c_{r+s-1}^{(1)} X^{r+s-1}$ with $c_k^{(1)} = c_k$, for $0 \leq k \leq r - 1$ and $c_k^{(1)} = c_k - b_{k-r}a_r$, for $r \leq k \leq r + s - 1$.

$h_2 = c_0^{(2)} + c_1^{(2)} X + \ldots + c_{r+s-1}^{(2)} X^{r+s-1}$ with $c_k^{(2)} = c_k$, for $0 \leq k \leq s - 1$ and $c_k^{(2)} = c_k - a_{k-s}b_s$, for $s \leq k \leq r + s - 1$. Then

$$c(f_1g) = (c_0^{(1)}, \ldots, c_{r+s-1}^{(1)}) = (c_0, \ldots, c_{r-1}, c_r - b_0 a_r, \ldots, c_{r+s-1} - b_{s-1} a_r) \subseteq (c_0, \ldots, c_{r+s}) + (a_r)(b_0, \ldots, b_{s-1}) = c(fg) + a_r c(g_1)$$ and
$$c(fg_1) = (c_0^{(2)}, \ldots, c_{r+s-1}^{(2)}) = (c_0, \ldots, c_{s-1}, c_s - a_0 b_s,$$
$$\ldots, c_{r+s-1} - a_{r-1} b_s) \subseteq (c_0, \ldots, c_{r+s}) + (b_s)(a_0, \ldots, a_{r-1}) = c(fg) + b_s c(f_1).$$

Since $c(f)^{s+1} c(g)$ is generated by elements of the form $\alpha = a_0^{n_0} a_1^{n_1} \ldots a_r^{n_r} b_i$ with $n_0 + \ldots + n_r = s + 1$ and $0 \leq i \leq s$, it suffices to show that these generators belong to $c(f)^s c(fg)$.

If $n_r \neq 0$ and $i = s$, then $\alpha = a_0^{n_0} a_1^{n_1} \ldots a_r^{n_r-1} a_r b_s = a_0^{n_0} a_1^{n_1} \ldots a_r^{n_r-1} c_{r+s}$, but $c_{r+s} \in c(fg)$, and then $\alpha \in c(f)^s c(fg)$.

If $n_r \neq 0$ and $i < s$, then $\alpha = a_0^{n_0} a_1^{n_1} \ldots a_r^{n_r-1} a_r b_i \in c(f)^s (a_r) c(g_1)$.

Finally, if $n_r = 0$, $\alpha = a_0^{n_0} a_1^{n_1} \ldots a_{r-1}^{n_r-1} b_i \in c(f_1)^{s+1} c(g)$.

Then $c(f)^{s+1} c(g) \subseteq c(f)^s c(fg) + c(f_1)^{s+1} c(g) + c(f)^s (a_r) c(g_1)$.

By (1), $c(f_1)^{s+1} c(g) \subseteq c(f_1)^s c(f_1 g)$, but we always have noted before that $c(f_1 g) \subseteq c(fg) + a_r c(g_1)$, then $c(f_1)^{s+1} c(g) \subseteq c(f_1)^s (c(fg) + a_r c(g_1)) = c(f_1)^s c(fg) + c(f_1)^s (a_r) c(g_1) \subseteq c(f)^s c(fg) + c(f)^s (a_r) c(g_1)$.

Then $c(f)^{s+1} c(g) \subseteq c(f)^s c(fg) + c(f_1)^{s+1} c(g) + c(f)^s (a_r) c(g_1) \subseteq c(f)^s c(fg) + (c(f)^s c(fg) + c(f)^s (a_r) c(g_1)) + c(f)^s (a_r) c(g_1) = c(f)^s c(fg) + c(f)^s (a_r) c(g_1)$.

Denote $\deg g_1 = \lambda < s$. By (2), $c(f)^{\lambda+1} c(g_1) \subseteq c(f)^\lambda c(fg_1)$ and since $\lambda \leq s - 1$, then $c(f)^s c(g_1) \subseteq c(f)^{s-1} c(fg_1)$. By a preceding remark, $c(fg_1) \subseteq c(fg) + b_s c(f_1)$, and then $c(f)^s (a_r) c(g_1) \subseteq c(f)^{s-1} (a_r) c(fg_1) \subseteq c(f)^{s-1} (a_r)(c(fg) + b_s c(f_1)) = c(f)^{s-1} (a_r) c(fg) + c(f)^{s-1} (a_r b_s) c(f_1) \subseteq c(f)^s c(fg) + c(f)^{s-1} (c_{r+s}) c(f_1) = c(f)^s c(fg)$.

Finally, $c(f)^{s+1} c(g) \subseteq c(f)^s c(fg) + c(f)^s (a_r) c(g_1) \subseteq c(f)^s c(fg)$, and this is what we want to prove.   □

Example 1 Let A be a ring and $f(X), g(X) \in A[X]$. The equality $c(fg) = c(f)c(g)$ is in general false. Let Y and Z be two variables over a field K. Consider the ring $A = K[Y, Z]/(Y^2, Z^2)$ and its elements $y = \bar{Y}$ and $z = \bar{Z}$. Let X be a new variable over the ring A, $f(X) = yX + z$ and $g(X) = yX - z \in A[X]$. Then $f(X)g(X) = (yX + z)(yX - z) = y^2 X^2 - z^2 = 0$, so $c(fg) = (0)$. But $c(f)c(g) = (y, z)^2 = (y^2, z^2, yz) = (yz) \neq (0)$.

Example 2 Let A be an integral domain with quotient field K, $f \in A[X]$ and $g \in K[X]$. (i) If $c(f) = A$, then $c(fg) = c(g)$. (ii) If $c(f) = A$ and $c(g) = K$, then $c(fg) = K$.

Example 3 Let a and b be two nonzero elements of an integral domain A. For $f = aX + b$ and $g = aX - b$, $c(fg) = (a^2, b^2)$. By the theorem, $(a, b)^3 = (a, b)(a^2, b^2)$.

Example 4 Let A be a valuation domain (resp. PID) and $f, g \in A[X]$. Then $c(fg) = c(f)c(g)$. In particular, $c(f^n) = c(f)^n$ for each $n \in \mathbb{N}^*$. Indeed, $c(f)$ is a principal ideal. By simplification in the content formula, we have the result.

Proposition 11.6 (Kang 1989) *Le A be an integral domain, I a nonzero fractional ideal of A, and $T \subseteq N_v$ a multiplicative set of $A[X]$. Then:*
1. $(I[X]_T)^{-1} = (I^{-1}[X])_T;$ 2. $(I[X]_T)_v = (I_v[X])_T;$ 3. $(I[X]_T)_t = (I_t[X])_T.$

Proof

(1) Since $I^{-1}[X]_T I[X]_T = (I^{-1}[X] I[X])_T \subseteq (I^{-1}I)[X]_T \subseteq A[X]_T$, then $(I^{-1}[X])_T \subseteq (I[X]_T)^{-1}$. Conversely, let $u \in (I[X]_T)^{-1}$. Then $uI[X]_T \subseteq A[X]_T$. In particular, $uI \subseteq A[X]_T$. Let $0 \neq a \in I$ be a fixed element. Then $ua \in A[X]_T$, so $u \in \frac{1}{a}A[X]_T \subseteq K[X]_T$ where K is the quotient field of A. Put $u = \frac{f}{h}$ with $f \in K[X]$ and $h \in T \subseteq A[X]$. By definition, $f = hu \in (I[X]_T)^{-1}$, and then $fI[X]_T \subseteq A[X]_T$ and in particular, $fI \subseteq A[X]_T$. For each $0 \neq b \in I$, $bf \in A[X]_T$. There exists $g \in T \subseteq N_v$ such that $bfg \in A[X]$, and then $c(bfg) \subseteq A$. By Dedekind-Mertens theorem, $c(g)^m c(bfg) = c(g)^{m+1} c(bf)$, where $m = deg(bf) = deg(f)$, and then $\big(c(g)^m c(bfg)\big)_v = \big(c(g)^{m+1} c(bf)\big)_v$. By Proposition 11.1, $\big((c(g)_v)^m c(bfg)\big)_v = \big((c(g)_v)^{m+1} c(bf)\big)_v$. Since $g \in N_v$, $c(g)_v = A$ and $c(bfg)_v = c(bf)_v$. Then $b.c(f) = c(bf) \subseteq c(bf)_v = c(bfg)_v \subseteq A_v = A$ for each $b \in I$. Then $Ic(f) \subseteq A$, so $c(f) \subseteq I^{-1}$, $f \in I^{-1}[X]$ and $u = \frac{f}{h} \in (I^{-1}[X])_T$.

(2) By (1), $(I[X]_T)_v = ((I[X]_T)^{-1})^{-1} = (I^{-1}[X]_T)^{-1} = (I^{-1})^{-1}[X]_T = (I_v[X])_T$.

(3) Let J be a nonzero finitely generated fractional ideal of $A[X]_T$ contained in $I[X]_T$, $J = (f_1, \ldots, f_n)A[X]_T$ with $f_1, \ldots, f_n \in I[X]$. The fractional ideal $I_0 = c(f_1) + \ldots + c(f_n)$ of A is finitely generated and contained in I. Since $f_1, \ldots, f_n \in I_0[X]$, then $J = (f_1, \ldots, f_n) A[X]_T \subseteq I_0[X]_T$. By (2), $J_v \subseteq (I_0[X]_T)_v = \big((I_0)_v[X]\big)_T \subseteq (I_t[X])_T$. This is for each nonzero finitely generated ideal J of $A[X]_T$ contained in $I[X]_T$. Then $(I[X]_T)_t \subseteq I_t[X]_T$. Conversely, let $u \in I_t[X]_T$, $u = \frac{f}{g}$ with $f = a_0 + a_1 X + \ldots + a_n X^n \in I_t[X]$ and $g \in T$. By definition of I_t, for each i, there exists a nonzero finitely generated fractional ideal J_i of A contained in I such that $a_i \in (J_i)_v$. The fractional ideal $J = J_0 + \ldots + J_n$ of A is finitely generated and contained in I. For each i, $J_i \subseteq J$, and then $(J_i)_v \subseteq J_v$ and $f \in J_v[X]$. By (2), $u = \frac{f}{g} \in J_v[X]_T = (J[X]_T)_v$. Since J is a nonzero finitely generated fractional ideal of A contained in I, then $J[X]_T$ is a nonzero finitely generated fractional ideal of $A[X]_T$ contained in $I[X]_T$. By definition of the t−operation, $(J[X]_T)_v \subseteq (I[X]_T)_t$. Then $u \in (I[X]_T)_t$. $\square$

Corollary 11.7 (Hedstrom 1980 and Kang 1989) *Let A be an integral domain and I a nonzero fractional ideal of A. Then:*

1. $I[X]^{-1} = I^{-1}[X];$ $I[X]_v = I_v[X];$ $I[X]_t = I_t[X]$. In particular, I is a v−ideal (resp. t−ideal) of A if and only if $I[X]$ is a v−ideal (resp. t−ideal) of $A[X]$.

2. $(I[X]_N)^{-1} = I^{-1}[X]_N$; $(I[X]_N)_v = I_v[X]_N$; $(I[X]_N)_t = I_t[X]_N$ with $N = \{f \in A[X]; \ c(f) = A\}$.

3. $(I[X]_{N_v})^{-1} = I^{-1}[X]_{N_v}$; $(I[X]_{N_v})_v = I_v[X]_{N_v}$; $(I[X]_{N_v})_t = I_t[X]_{N_v}$ with $N_v = \{0 \neq f \in A[X]; \ c(f)_v = c(f)_t = A\}$.

Proof Take respectively the multiplicative sets $\{1\}$, N, and N_v of $A[X]$. $\square$

Corollary 11.8 *Let A be an integral domain. The maximal spectrum:*
$$Max(A[X]_{N_v}) = \{M[X]_{N_v}; \ M \in t - Max(A)\} = t - Max(A[X]_{N_v}).$$

Proof By Proposition 11.4, we have the first equality. By the preceding corollary, for each $M \in t - Max(A)$, $(M[X]_{N_v})_t = M_t[X]_{N_v} = M[X]_{N_v}$. Then the $M[X]_{N_v}$ are $t-$ideals and since they are maximal ideals in the ring $A[X]_{N_v}$, then they are $t-$maximal. They are the only $t-$maximal ideals because each other $t-$maximal ideal is contained in one of them. So $t - Max(A[X]_{N_v}) = \{M[X]_{N_v}; \ M \in t - Max(A)\}$.
$\square$

The following lemma is simple and very useful.

Lemma 11.9 (Hamann and Houston 1988) *Let A be an integral domain and $0 \neq g \in A[X]$ such that $c(g)_v = A$; i.e., $g \in N_v$.*
For each $0 \neq a \in A$, $(a, g)_v = A[X]$.

Proof Let $h \in (a, g(X))^{-1}$. Then $ah \in A[X]$ and $gh \in A[X]$. So $c(gh) \subseteq A$ and $h \in \frac{1}{a}A[X] \subseteq K[X]$ where K is the quotient field of A. By the content formula, there exists $n \in \mathbb{N}^*$ such that $c(g)^{n+1}c(h) = c(g)^n c(gh) \subseteq A$. Then $(c(g)^{n+1}c(h))_v \subseteq A$. By Proposition 11.1, $((c(g)_v)^{n+1}c(h))_v \subseteq A$. But $c(g)_v = A$, and then $c(h)_v \subseteq A$, so $c(h) \subseteq A$ and $h \in A[X]$. We conclude that $(a, g(X))^{-1} = A[X]$, so $(a, g)_v = A[X]$. $\square$

Example The hypothesis $a \neq 0$ is necessary in the lemma. Let $g(X) \in A[X]$ a nonconstant polynomial such that $c(g)_v = A$. Then $(g(X)A[X])_v = g(X)A[X] \neq A[X]$.

In the following proposition, we characterize the $t-$maximal ideals of the polynomial ring.

Proposition 11.10 (Houston and Zafrullah 1989) *Let A be an integral domain.*

1. *Let P be a $t-$maximal ideal of $A[X]$ with $P \cap A \neq (0)$. Then $P \cap A$ is a $t-$maximal ideal of A and $P = (P \cap A)[X]$.*
2. *If P is a $t-$maximal ideal of A, then $P[X]$ is a $t-$maximal ideal of $A[X]$.*

Proof

(1) Let $c(P)$ be the ideal of A formed by the coefficients of all the elements of P. First, we will prove that $c(P) \subseteq P$. Since $P \subseteq c(P)[X]$, if $c(P) \nsubseteq P$, then $P \subset c(P)[X] \subseteq (c(P)[X])_t \subseteq A[X]$. By $t-$maximality of P, $(c(P)[X])_t = A[X]$.

By Corollary 11.7, $(c(P))_t[X] = A[X]$, and then $(c(P))_t = A$. There exists a nonzero finitely generated ideal $J \subseteq c(P)$ of A such that $1 \in J_v$. There exist $g_1, \ldots, g_k \in P$ such that $J \subseteq c(g_1) + \ldots + c(g_k)$. Put $r_1 = deg\ g_1, \ldots, r_k = deg\ g_k$ and $g = g_1 + X^{r_1+1}g_2 + X^{r_1+r_2+2}g_3 + \ldots + X^{r_1+\ldots+r_{k-1}+k-1}g_k \in P \subseteq A[X]$. Then $c(g) = c(g_1) + \ldots + c(g_k)$, so $1 \in c(g)_v$, and then $c(g)_v = A$. By hypothesis, there exists $0 \neq a \in P \cap A$. By the preceding lemma, $(a, g)_v = A[X]$. Since $(a, g) \subseteq P$, then $A[X] = (a, g)_v \subseteq P_t = P \subseteq A[X]$. So $P = A[X]$, which is absurd because P is a prime ideal of the ring $A[X]$. We conclude that $c(P) \subseteq P$, and then $c(P)[X] \subseteq P$, so $P = c(P)[X]$. Then $P \cap A = c(P)$ and $P = c(P)[X] = (P \cap A)[X]$.

By Corollary 11.7, $(P \cap A)_t[X] = ((P \cap A)[X])_t = P_t = P = (P \cap A)[X]$, and then $(P \cap A)_t = P \cap A$ so $P \cap A$ is a proper $t-$ideal of A. Let M be a proper $t-$ideal of A containing $P \cap A$. By Corollary 11.7, $M[X] = M_t[X] = (M[X])_t$, and then $M[X]$ is a proper $t-$ideal of $A[X]$ and $P = (P \cap A)[X] \subseteq M[X]$, and then $P = (P \cap A)[X] = M[X]$, so $M = P \cap A$. Then $P \cap A$ is a $t-$maximal ideal of A.

(2) By Corollary 11.7, $(P[X])_t = P_t[X] = P[X]$. Then $P[X]$ is a proper $t-$ideal of $A[X]$. By Proposition 11.2, there exists a $t-$maximal ideal Q of $A[X]$ containing $P[X]$. Then $P = P[X] \cap A \subseteq Q \cap A$, so $Q \cap A \neq (0)$. By (1), $Q \cap A$ is a $t-$maximal ideal of A and $Q = (Q \cap A)[X]$. But $P[X] \subseteq Q = (Q \cap A)[X]$, and then $P \subseteq Q \cap A$. By $t-$maximality of P, we have $P = Q \cap A$, and then $P[X] = (Q \cap A)[X] = Q$, so $P[X]$ is a $t-$maximal ideal of $A[X]$. $\square$

Example 1 Let A be a field. The ideal (X) of $A[X]$ is $t-$maximal because it is maximal and principal. We have $(X) \cap A = (0)$.

Example 2 Let A be an integral domain and P a $t-$maximal ideal of $A[X]$ such that $P \cap A = (0)$. By Example of Lemma 10.8, $A[X]_P$ is a DVR.

Example 3 Let $A \subset B$ be an extension of integral domains and M a $t-$maximal ideal of B such that $M \cap A \neq (0)$. The ideal $M \cap A$ is not necessarily $t-$maximal in A. Take $A = k[X, Y]$ with X and Y two variables over a field k and $I = (X, Y)$. Then $I^{-1} = A$. Indeed, let $0 \neq f \in I^{-1}$, $f = \frac{g}{h}$ with $g, h \in k[X, Y] \setminus (0)$. Since $k[X, Y]$ is UFD, we may suppose that g and h are relatively prime. Suppose that $h \notin U(k(X, Y)) = k^*$. Since $f \in I^{-1}$, then $Xf, Yf \in A$. But $hXf = Xg$ and $hYf = Yg$, and then h divides Xg and Yg in A. By Euclide lemma, h divides X and Y. But X and Y are irreducible, and then h is associated with X and with Y. Then X and Y are associated, which is absurd. Then $h \in k^*$ and $f \in A$. So $I^{-1} = A$. Then $I_v = I_t = A$ and I is not a $t-$ideal of A. There exists an over valuation domain (B, M) of A such that $M \cap A = I$. By Example 2 at the beginning of this section, in a valuation domain, each nonzero integral ideal is a $t-$idéal. Then M is a $t-$maximal ideal of B.

Definition 11.11 Let A be an integral domain and $S \subseteq A$ a multiplicative set. We say that A is t-locally S-Noetherian if for each $P \in t - Max(A)$, the ring A_P is S-Noetherian.

Example An integral domain A is t-locally Noetherian if for each $P \in t - Max(A)$, the ring A_P is Noetherian.

We prove now the main result of this section.

Theorem 11.12 (Lim 2015) *Let A be an integral domain, $S \subseteq A$ an anti-Archimedean multiplicative set, and $N_v = \{0 \neq f \in A[X]; (c(f))_v = A\}$.*
The following assertions are equivalent:

1. *The ring A is t-locally S-Noetherian.*
2. *The polynomial ring $A[X]$ is t-locally S-Noetherian.*
3. *The t-Nagata ring $A[X]_{N_v}$ is locally S-Noetherian.*
4. *The t-Nagata ring $A[X]_{N_v}$ is t-locally S-Noetherian.*

Proof $"(1) \implies (2)"$ Let $M \in t - Max(A[X])$. We will prove that the ring $A[X]_M$ is S-Noetherian.

First case: $M \cap A = (0)$. Let K be the quotient field of A and $T = A[X] \setminus M$, a multiplicative set of $A[X]$ because by Proposition 11.3, the ideal M is prime. Then $A[X]_M = T^{-1}A[X] \subseteq T^{-1}K[X]$. Conversely, let $h \in T^{-1}K[X]$, $h = \frac{f}{g}$ with $f \in K[X]$ and $g \in T = A[X] \setminus M$. Then $f = \frac{f'}{a}$ with $f' \in A[X]$ and $0 \neq a \in A$, and then $h = \frac{f'}{ag}$. Since $a \notin M$ because $M \cap A = (0)$, then $ag \in A[X] \setminus M$. So $h = \frac{f'}{ag} \in A[X]_M$ and then $A[X]_M = T^{-1}K[X]$. Since $K[X]$ is PID, then $T^{-1}K[X] = A[X]_M$ is also PID. So $A[X]_M$ is S-Noetherian.

Second case: $M \cap A \neq (0)$. By Proposition 11.10 (1), $P = M \cap A$ is a t-maximal ideal of A and $M = P[X]$. By hypothesis, A is t-locally S-Noetherian, and then A_P is S-Noetherian. By Theorem 4.2, $A_P[X]$ is S-Noetherian. Since PA_P is a prime ideal of A_P, $(PA_P)[X]$ is a prime ideal of $A_P[X]$. By Proposition 10.2, the localized ring $(A_P[X])_{(PA_P)[X]}$ is S-Noetherian. But by Lemma 10.8 (3), the ring $(A_P[X])_{(PA_P)[X]} = A[X]_{P[X]} = A[X]_M$. Then $A[X]_M$ is S-Noetherian.

$"(2) \implies (3)"$ Let M be a maximal ideal of $A[X]_{N_v}$. We will prove that $(A[X]_{N_v})_M$ is S-Noetherian. By Proposition 11.4 (2), $M = P[X]_{N_v}$ with $P \in t - Max(A)$. By Proposition 11.4, $N_v \cap P[X] = \emptyset$. By Lemma 10.8 (1), $(A[X]_{N_v})_M = (A[X]_{N_v})_{P[X]_{N_v}} = A[X]_{P[X]}$. By Proposition 11.10 (2), $P[X] \in t - Max(A[X])$. But by hypothesis, $A[X]$ is t-locally S-Noetherian, and then the ring $A[X]_{P[X]}$ is S-Noetherian. We conclude that the ring $(A[X]_{N_v})_M$ is S-Noetherian.

$"(3) \implies (1)"$ Let P be a t-maximal ideal of A. We will prove that A_P is S-Noetherian. By Proposition 11.10 (2), $P[X]$ is a t-maximal ideal of $A[X]$ disjoint with N_v. By Proposition 11.4 (2), $P[X]_{N_v}$ is a maximal ideal of $A[X]_{N_v}$. But by hypothesis, $A[X]_{N_v}$ is locally S-Noetherian, and then $(A[X]_{N_v})_{P[X]_{N_v}}$ is S-Noetherian. By Lemma 10.8 (1) and (3), $(A[X]_{N_v})_{P[X]_{N_v}} = A[X]_{P[X]} = (A_P[X])_{(PA_P)[X]}$. Then $A_P[X]_{(PA_P)[X]}$ is S-Noetherian. Let I be an ideal of the ring A_P. Then $IA_P[X]_{(PA_P)[X]}$ is a S-finite ideal. By Lemma 10.7, since A_P is local with maximal ideal PA_P, the ideal I is S-finite. Then A_P is S-Noetherian and A is t−locally S-Noetherian.

"(3) $\Longleftrightarrow$ (4)" This equivalence follows from the equality
$$Max(A[X]_{N_v}) = t - Max(A[X]_{N_v}), \text{ in Corollary 11.8.} \qquad \square$$

Corollary 11.13 (Chang 2005) *The following assertions are equivalent for an integral domain A:*

1. *The ring A is t−locally Noetherian.*
2. *The polynomial ring A[X] is t−locally Noetherian.*
3. *The t−Nagata ring $A[X]_{N_v}$ is locally Noetherian.*
4. *The t−Nagata ring $A[X]_{N_v}$ is t−locally Noetherian.*

Proof Take $S = \{1\}$ in the preceding theorem. $\qquad \square$

12 Formal Power Series Over a Krull Domain

The purpose of this section is to introduce and develop the notion of a Krull domain. Then, we show that formal power series ring with coefficients in Krull domain is again a Krull domain.

Definition 12.1 Let A be an integral domain with quotient field K. We say that A is a Krull domain if there exists a family $(A_\alpha)_{\alpha \in \Lambda}$ of over rings of A satisfying the following:

(a) $A = \bigcap_{\alpha \in \Lambda} A_\alpha$.

(b) For each $\alpha \in \Lambda$, A_α is a DVR of rank one.

(c) Each nonzero element of K is invertible in all the A_α except at most a finite number of them.

(d) For each $\alpha \in \Lambda$, there exists $P_\alpha \in Spec(A)$ such that $A_\alpha = A_{P_\alpha}$.

The set $(A_\alpha)_{\alpha \in \Lambda}$ is called the defining family of the Krull domain A.

Note that the condition (c) is equivalent to each $0 \neq x \in A$ which is invertible in all the A_α but not in at most a finite number of them. The property (a) is called the intersection property. The property (c) is called the finite character property for the collection of rings $(A_\alpha)_{\alpha \in \Lambda}$.

Example Each DVR A of rank one is a Krull domain.

Indeed, if M is the maximal ideal of A, then $U(A) = A \setminus M$. So $A_M = (A \setminus M)^{-1}A = A$. In particular, if K is a field, then $K[[X]]$ is a DVR of rank one, and then it is a Krull domain. More generally, we have the following lemma.

Lemma 12.2 *A UFD is a Krull domain.*

Proof Let A be a UFD with quotient field $K \neq A$. Let Λ be a representative system of the irreducible elements of A. For each $p \in \Lambda$, $pA \in Spec(A)$ and A_{pA} is a DVR of rank one and quotient field K. Let $0 \neq x \in \bigcap_{p \in \Lambda} A_{pA} \subseteq K$, $x = \frac{a}{b}$ with $a, b \in A \setminus (0)$ and $\Delta(a, b) = 1$. Suppose that b is not invertible in A, then b has a divisor $p_0 \in \Lambda$. Since $x \in A_{p_0 A}$, $x = \frac{c}{d}$ with $c \in A$ and $d \in A \setminus p_0 A$, then $x = \frac{a}{b} = \frac{c}{d}$, so $ad = bc$. But p_0/b, and then p_0/ad, but $\Delta(p_0, a) = 1$, and then p_0/d, which is absurd. Then b is invertible in A and $x = \frac{a}{b} \in A$, so $A = \bigcap_{p \in \Lambda} A_{pA}$. Let $0 \neq x \in A$. Then $x = \epsilon p_1 \ldots p_s$ with ϵ an invertible element of A and $p_1, \ldots, p_s \in \Lambda$. For each $p \in \Lambda \setminus \{p_i; 1 \leq i \leq s\}$, $x, \frac{1}{x} \in A_{pA}$. Then A is a Krull domain.   □

Examples

(1) $\mathbb{Z}$ is a Krull domain, defined by the family $\{\mathbb{Z}_{(p)}; \ p \text{ premier}\}$.

   Let K be a field. Then $K[X]$ is a Krull domain, defined by the family $\{K[X]_{(P)}; \ P \in K[X] \text{ monic and irreducible }\}$.

(2) Let A be a PID, and then $A[[X]]$ is UFD. See Example of Corollary 8.6, Chap. 2. Then $A((X)) = S^{-1}A[[X]]$ is also UFD, where $S = \{X^i; \ i \in \mathbb{N}\}$. The domains $A[[X]]$ and $A((X))$ are Krull domains. This is the case if A is a DVR of rank one.

(3) Let K be a field. Since $K[X_i; \ i \in \mathbb{N}]$ is UFD, then it is Krull, but it is not Noetherian.

Lemma 12.3 (Gilmer 1969) *Let $(A_\alpha)_{\alpha \in \Lambda}$ be a family of sub-rings of a field L and $A = \bigcap_{\alpha \in \Lambda} A_\alpha$. Suppose that each nonzero element of A is invertible in all the A_α except at most a finite number of them. If all the A_α are Krull domains, so is A.*

Proof For each $\alpha \in \Lambda$, we consider a family $\left(V_\beta^{(\alpha)}\right)_{\beta \in \Lambda_\alpha}$ of DVR of rank one defining the Krull domain A_α. Let K be the quotient field of A and $\mathcal{F} = \{V_\beta^{(\alpha)} \cap K; \ \alpha \in \Lambda, \beta \in \Lambda_\alpha\}$. Each member of $\mathcal{F}$ is a discrete valuation domain of rank ≤ 1 and their intersection $\bigcap_{\alpha \in \Lambda} \left(\bigcap_{\beta \in \Lambda_\alpha} V_\beta^{(\alpha)} \right) \cap K = \bigcap_{\alpha \in \Lambda} A_\alpha \cap K = A \cap K = A$.

Let x be a nonzero element of A. Then x is invertible in all the A_α except in a finite number $A_{\alpha_1}, \ldots, A_{\alpha_n}$ of them. Since, given i, A_{α_i} is a Krull domain defined by the family $\left(V_\beta^{(\alpha_i)}\right)_{\beta \in \Lambda_{\alpha_i}}$, then x is invertible in all the $V_\beta^{(\alpha_i)}$, $\beta \in \Lambda_{\alpha_i}$, except in a finite number of them. For all $\alpha \in \Lambda$ and $\beta \in \Lambda_\alpha$, there exists $P_\beta^{(\alpha)} \in Spec(A_\alpha)$ such that $V_\beta^{(\alpha)} = (A_\alpha)_{P_\beta^{(\alpha)}}$. Put $Q_\beta^{(\alpha)} = P_\beta^{(\alpha)} \cap A \in Spec(A)$, and then $A_{Q_\beta^{(\alpha)}} = V_\beta^{(\alpha)} \cap K$, if it is not a field (Exercise). Then A is a Krull domain defined by a subfamily of $\mathcal{F}$.   □

Example A finite intersection of Krull domains is a Krull domain.

Theorem 12.4 (Gilmer 1969) *If A is a Krull domain, so is $A[[X]]$. More generally, $A[[X_1, \ldots, X_n]]$ is a Krull domain.*

Proof Let K be the quotient field of A and $(A_\alpha)_{\alpha \in \Lambda}$ a defining family of DVR of rank one for A. Then $A[[X]] = K[[X]] \cap (\bigcap_{\alpha \in \Lambda} A_\alpha((X)))$. Indeed, let $f \in K[[X]] \cap$

$(\bigcap_{\alpha \in \Lambda} A_\alpha((X)))$; $f = \sum_{i=0}^{\infty} a_i X^i$ with $a_i \in A_\alpha$, for each $\alpha \in \Lambda$; then $a_i \in \bigcap_{\alpha \in \Lambda} A_\alpha = A$. Then $f \in A[[X]]$. By the preceding examples, $K[[X]]$ and the $A_\alpha((X))$ are Krull domains. Let $0 \neq f \in A[[X]]$, $f = X^n g$, where $n \geq 0$ and $g \in A[[X]]$ with $0 \neq g(0) \in A$. By hypothesis, there exists a finite set $F \subset \Lambda$ such that for each $\alpha \in \Lambda \setminus F$, $g(0)$ is invertible in A_α, then g is invertible in $A_\alpha[[X]]$, and $f = X^n g$ is invertible in $A_\alpha((X))$. By the preceding lemma, $A[[X]]$ is a Krull domain. $\square$

Lemma 12.5 *Let A be a Krull domain defined by a family $(A_\alpha)_{\alpha \in \Gamma}$, P a nonzero prime ideal of A, $S = A \setminus P$, and $\Gamma' = \{\alpha \in \Gamma; \ S^{-1}A_\alpha = A_\alpha\}$. Then $\Gamma' \neq \emptyset$ and $A_P = \bigcap_{\alpha \in \Gamma'} A_\alpha$.*

Proof First, we will show that $A_P = \bigcap_{\alpha \in \Gamma} S^{-1}A_\alpha$. Since $A \subseteq A_\alpha$, then $A_P = S^{-1}A \subseteq S^{-1}A_\alpha$. We have the direct inclusion. Conversely, let $0 \neq x \in \bigcap_{\alpha \in \Gamma} S^{-1}A_\alpha$, and x is invertible in all the A_α (in particular, $x \in A_\alpha$) except for $\alpha_1, \ldots, \alpha_n$. Put $x = \frac{a_1}{s} = \ldots = \frac{a_n}{s}$ with $a_i \in A_{\alpha_i}$, $1 \leq i \leq n$ and $s \in S$. Then $sx \in A_{\alpha_i}$ for $1 \leq i \leq n$. So $sx \in \bigcap_{\alpha \in \Gamma} A_\alpha = A$, and then $x = \frac{sx}{s} \in A_P$. Since $P \neq (0)$, $A_P \neq frac(A)$ because if $0 \neq p \in P$, $\frac{1}{p} \notin A_P$. For each $\alpha \in \Gamma$, A_α is a DVR of rank one with prime spectrum $(0) \subset M_\alpha$ and $S^{-1}A_\alpha$ is an over ring of A_α, and then $S^{-1}A_\alpha \in \{(A_\alpha)_{M_\alpha} = A_\alpha; \ (A_\alpha)_{(0)} = qf(A)\}$. So $A_P = \bigcap \{S^{-1}A_\alpha; \ \alpha \in \Gamma, S^{-1}A_\alpha = A_\alpha\} = \bigcap_{\alpha \in \Gamma'} A_\alpha$. In particular, $\Gamma' \neq \emptyset$. $\square$

Proposition 12.6 *Let A be a Krull domain defined by a family $(A_\alpha)_{\alpha \in \Gamma}$. For each prime ideal P of A of height one, A_P is a DVR of rank one. More precisely, there exists $\alpha \in \Gamma$ such that $A_P = A_\alpha$.*

Proof With the notations of the preceding lemma, $\Gamma' \neq \emptyset$. We will show that for each $\alpha \in \Gamma'$, $A_P = A_\alpha$. Indeed, the maximal ideal M_α of A_α is nonzero, $A \subseteq A_P = S^{-1}A \subseteq S^{-1}A_\alpha = A_\alpha \subseteq K = qf(A)$. If $M_\alpha \cap A_P = (0)$, all the nonzero elements of A_P will be invertible in A_α, and then $K = qf(A_P) \subseteq A_\alpha$, so $A_\alpha = K$ is a field, which is absurd. Then $M_\alpha \cap A_P \neq (0)$. Since $ht_A(P) = 1$, $Spec(A_P) = \{(0), PA_P\}$ so $M_\alpha \cap A_P = PA_P$ for each $\alpha \in \Gamma'$. On the other hand, there exists $P_\alpha \in$

$Spec(A)$ such that $A_\alpha = A_{P_\alpha}$, and then $M_\alpha = P_\alpha A_{P_\alpha}$, so $P_\alpha A_{P_\alpha} \cap A_P = P A_P$. By intersecting with A, we find $P_\alpha A_{P_\alpha} \cap A = P A_P \cap A$, and then $P = P_\alpha$ and $A_P = A_{P_\alpha} = A_\alpha$ for each $\alpha \in \Gamma'$. $\qquad\square$

Notation Let A be an integral domain. We denote $Spec^1(A) = \{P \in Spec(A);\ (0) \subset P$ are consecutive $\}$, the set of height one prime ideals of A.

Thanks to the preceding proposition, we have following equivalent definition.

Definition 12.7 An integral domain A is Krull if:

(a) $A = \bigcap \{A_P;\ P \in Spec^1(A)\}$.
(b) For each $P \in Spec^1(A)$, A_P is a DVR of rank one.
(c) Each element $0 \neq x \in qf(A)$ is invertible in all the A_P, $P \in Spec^1(A)$, except in a finite number of them.

Notation Let V be a valuation domain defined by a valuation v and C a nonempty subset of V. We denote $min\ C = min\ v(C) = min\{v(x);\ x \in C\}$ as the smallest element of $v(C)$ if it exists.

Lemma 12.8 *Let I and J be two ideals of V such that $min\ I$ and $min\ J$ exist. Then $min(IJ)$ exists and is equal to $min\ I + min\ J$.*

Proof First, note that $IJ = \{xy;\ x \in I, y \in J\}$. Indeed, an element of IJ is of the form $z = x_1 y_1 + \ldots + x_n y_n$ with $x_i \in I$ and $y_i \in J$. By changing the indexation, we may suppose that $\frac{x_i}{x_1} \in V$ for $1 \leq i \leq n$. Then $z = x_1(y_1 + \frac{x_2}{x_1}y_2 + \ldots + \frac{x_n}{x_1}y_n)$ with $y_1 + \frac{x_2}{x_1}y_2 + \ldots + \frac{x_n}{x_1}y_n \in J$. Put $min\ I = v(x_0)$ and $min\ J = v(y_0)$ with $x_0 \in I$ and $y_0 \in J$. For each $x \in I$ and $y \in J$, $v(xy) = v(x) + v(y) \geq v(x_0) + v(y_0) = v(x_0 y_0)$ with $x_0 y_0 \in IJ$. Then $min(IJ) = v(x_0 y_0) = v(x_0) + v(y_0) = min\ I + min\ J$. $\qquad\square$

Lemma 12.9 *Let A be an integral domain, V an over valuation domain of A defined by a valuation v, $f = \sum_{i:0}^{\infty} a_i X^i \in A[[X]]$, and $C(f)$ the content of f in A. Suppose that the set $\{v(a_i);\ i \in \mathbb{N}\}$ admits a smallest element denoted $v^*(f)$. Then $min(C(f)V) = v^*(f)$.*

Proof By hypothesis, $\{a_i;\ i \in \mathbb{N}\} \subseteq C(f)V$ and $v^*(f) = min\{v(a_i);\ i \in \mathbb{N}\}$ exists. Let $x \in C(f)V$. There exist $\alpha_0, \ldots, \alpha_n \in V$ such that $x = \alpha_0 a_0 + \ldots + \alpha_n a_n$. Then $v(x) \geq min\{v(\alpha_i) + v(a_i);\ 0 \leq i \leq n\} \geq min\{v(a_i);\ 0 \leq i \leq n\} \geq v^*(f)$. So $min(C(f)V) = v^*(f)$. $\qquad\square$

Notation Let A be an integral domain and V an over valuation domain of A defined by a valuation v. We say that v satisfies the property $(*)$ with respect to A if for each sequence $(a_i)_{i \in \mathbb{N}}$ of elements of A, the set $\{v(a_i);\ i \in \mathbb{N}\}$ admits a smallest element. The following lemma shows that there are two important cases in which this condition is satisfied.

Lemma 12.10

1. *If v is a discrete valuation of rank one, the property $(*)$ is satisfied for each sub-ring A of the valuation domain of v.*
2. *If A is a Noetherian domain, the property $(*)$ is satisfied for each over valuation domain V of A.*

Proof

1. Follows from the inclusion $\{v(a_i); \ i \in \mathbb{N}\} \subseteq \mathbb{N}$.
2. Since the ideal $I = (a_i; \ i \in \mathbb{N})$ of A is finitely generated, there exists $n \in \mathbb{N}$ such that $I = (a_0, \ldots, a_n)$. Let $v(a_k) = min\{v(a_i), 0 \leq i \leq n\}$. For each $l \in \mathbb{N}$, there exist $\alpha_0, \ldots, \alpha_n \in A$ such that $a_l = \alpha_0 a_0 + \ldots + \alpha_n a_n$, and then $v(a_l) \geq min\{v(\alpha_i) + v(a_i); \ 0 \leq i \leq n\} \geq min\{v(a_i); \ 0 \leq i \leq n\} = v(a_k)$. So $min\{v(a_i); \ i \in \mathbb{N}\} = v(a_k)$. $\qquad\square$

Theorem 12.11 (Gilmer and Heinzer 1968) *Let A be an integral domain and V an over valuation domain of A defined by a valuation v. Suppose that v satisfies the property $(*)$ with respect to the domain A. For $f = \sum_{i:0}^{\infty} a_i X^i \in A[[X]]$, let $v^*(f) = min\{v(a_i); \ i \in \mathbb{N}\}$. The function v^* extends in a unique way to a valuation over $qf(A[[X]])$ that we denote also by v^*. If P is the trace over A of the maximal ideal of v, then $P[[X]]$ is the trace over $A[[X]]$ of the maximal ideal of v^*.*

Proof

(1) It suffices to show that v^* is a valuation over $A[[X]]$. Let $f, g \in A[[X]]$. Then $C(fg) \subseteq C(f)C(g)$, so $C(fg)V \subseteq (C(f)C(g))V = C(f)V.C(g)V$. By the preceding two lemmas, $v^*(fg) = min(C(fg)V) \geq min(C(f)V.C(g)V)$

$$= min(C(f)V) + min(C(g)V) = v^*(f) + v^*(g). \text{ Put } f = \sum_{i:0}^{\infty} a_i X^i,$$

$g = \sum_{i:0}^{\infty} b_i X^i$, $v^*(f) = v(a_{i_0})$ and $v^*(g) = v(b_{j_0})$ where i_0 ad j_0 are the

smallest indexes with this property. Then $\displaystyle\sum_{i+j=i_0+j_0} a_i b_j = \sum_{\substack{i+j=i_0+j_0 \\ i<i_0}} a_i b_j +$

$a_{i_0} b_{j_0} + \displaystyle\sum_{\substack{i+j=i_0+j_0 \\ j<j_0}} a_i b_j$. For $i < i_0$, $v(a_i b_j) = v(a_i) + v(b_j) > v(a_{i_0}) + v(b_{j_0}) =$

$v(a_{i_0} b_{j_0})$, then $v\left(\displaystyle\sum_{\substack{i+j=i_0+j_0 \\ i<i_0}} a_i b_j\right) > v(a_{i_0} b_{j_0})$. Also, $v\left(\displaystyle\sum_{\substack{i+j=i_0+j_0 \\ j<j_0}} a_i b_j\right) >$

$v(a_{i_0} b_{j_0})$. Then $v\left(\displaystyle\sum_{i+j=i_0+j_0} a_i b_j\right) = v(a_{i_0} b_{j_0}) = v(a_{i_0}) + v(b_{j_0}) = v^*(f) +$

$v^*(g)$. So we have $v^*(fg) \leq v\left(\sum_{i+j=i_0+j_0} a_i b_j \right) \leq v^*(f) + v^*(g)$. Finally,

$v^*(fg) = v^*(f) + v^*(g)$. On the other hand, $C(f + g) \subseteq C(f) + C(g)$,
and then $C(f + g)V \subseteq C(f)V + C(g)V$, which is equal either to $C(f)V$ or
$C(g)V$. Then either $v^*(f + g) = min\big(C(f + g)V\big) \geq min\big(C(f)V\big) = v^*(f)$
or $v^*(f + g) = min\big(C(f + g)V\big) \geq min\big(C(g)V\big) = v^*(g)$. Then we have
$v^*(f + g) \geq min\{v^*(f); v^*(g)\}$. So v^* is a valuation over $A[[X]]$.

(2) Let $f = \sum_{i:0}^{\infty} a_i X^i \in A[[X]]$. Then f is in the trace over $A[[X]]$ of the ideal of

$v^* \Longleftrightarrow v^*(f) = min\{v(a_i); \ i \in \mathbb{N}\} > 0 \Longleftrightarrow \forall \, i \in \mathbb{N}, \ v(a_i) > 0 \Longleftrightarrow i \in \mathbb{N},$
$a_i \in P \Longleftrightarrow f \in P[[X]]$.   □

Approximation Lemma 12.12 *Let A be a Krull domain with quotient field K
defined by a family $\mathscr{F}$ of discrete valuation domains. If $w \in \mathscr{F}$ and $m \in \mathbb{Z}$, there
exists $x \in K$ such that $w(x) = m$ and for each $v \in \mathscr{F} \setminus \{w\}$, $v(x) \geq 0$.*

Proof Since $w(K) = \mathbb{Z}$, there exists $a \in K$ such that $w(a) = m$. By the third axiom
of Krull domains, there exists a finite number of valuations $v_1, \ldots, v_k \in \mathscr{F} \setminus \{w\}$
such that $v_i(a) = -m_i < 0$ for $1 \leq i \leq k$. The other valuations satisfy $v(a) \geq 0$.
Let $P(w) = \{x \in A; \ w(x) > 0\}$ and $P_i = \{x \in A; \ v_i(x) > 0\}$ the height one
prime ideals of A. The valuation domains of w and of v_i, $1 \leq i \leq k$, are $A_{P(w)}$ and
A_{P_i}, and their maximal ideals are respectively $P(w)A_{P(w)}$ and $P_i A_{P_i}$, $1 \leq i \leq k$.
These ideals are principal, and put $P_i A_{P_i} = p_i A_{P_i}$ with $p_i \in P_i$, $1 \leq i \leq k$.
Since the domains $A_{P(w)}$ and A_{P_i} are different, the ideals $P(w)$ and P_i, $1 \leq i \leq k$,
are also different and then incomparable since their height is one. Then the ideal
$I = P_1^{m_1} P_2^{m_2} \ldots P_k^{m_k} \nsubseteq P(w)$. There exists $y \in I \subset A$ such that $y \notin P(w)$. Then
$w(y) = 0$ but $y \in P_i^{m_i} A_{P_i} = p_i^{m_i} A_{P_i}$, so $v_i(y) \geq m_i$, $1 \leq i \leq k$. Take $x = ay \in K$,
and then $w(x) = w(a) + w(y) = m$, $v_i(x) = v_i(a) + v_i(y) \geq -m_i + m_i = 0$ for
$1 \leq i \leq k$. For $v \in \mathscr{F} \setminus \{w, v_1, \ldots, v_k\}$, $v(a) \geq 0$ and $v(y) \geq 0$ since $y \in A$. Then
$v(x) = v(a) + v(y) \geq 0$.   □

Theorem 12.13 (Gilmer and Heinzer 1968) *Let P be a height one prime ideal of
a Krull domain A. Then $A[[X]]_{P[[X]]}$ is a DVR of rank one. In particular, $P[[X]]$ is
a height one prime ideal of $A[[X]]$.*

Proof Let (P_α) be the family of the height one prime ideals of A other than P.
Let v be the valuation over the quotient field K of A, A_P its ring and for each α,
v_α the valuation over K with ring A_{P_α}. Since v and the v_α are discrete of rank
one, they admit extensions v^* and v_α^* to $qf(A[[X]])$, defined in Theorem 12.11. We
will show that the ring of v^* is $V = A[[X]]_{P[[X]]}$. By Theorem 12.11, $P[[X]]$ is
the trace of the maximal ideal of v^* over $A[[X]]$. Let $F \in A[[X]]_{P[[X]]}$, $F = \frac{f}{g}$
with $f \in A[[X]]$ and $g \in A[[X]] \setminus P[[X]]$. Then $v^*(f) \geq 0$ and $v^*(g) = 0$, so
$v^*(F) = v^*(f) - v^*(g) = v^*(f) \geq 0$, and then $F \in V$. Conversely, let $0 \neq F \in
V \subseteq qf(A[[X]])$, $F = \frac{f}{g}$ with $f, g \in A[[X]]$ and $0 \leq v^*(F) = v^*(f) - v^*(g)$.

Put $v^*(f) = n$ and $v^*(g) = m$, natural integers satisfying $n \geq m$. By the preceding lemma, there exists $a \in K$ such that $v(a) = -m$ and $v_\alpha(a) \geq 0$ for each α. Then $v^*(af) = v(a) + v^*(f) = n - m \geq 0$ and $v_\alpha^*(af) = v_\alpha(a) + v_\alpha^*(f) \geq 0$. Then $af \in (A_P \cap \bigcap_\alpha A_{P_\alpha})[[X]] = A[[X]]$. Also, $v^*(ag) = v(a) + v^*(g) = -m + m = 0$ and $v_\alpha^*(ag) = v_\alpha(a) + v_\alpha^*(g) \geq 0$. Then $ag \in (A_P \cap \bigcap_\alpha A_{P_\alpha})[[X]] = A[[X]]$ and $ag \notin P[[X]]$. Then $F = \frac{f}{g} = \frac{af}{ag} \in A[[X]]_{P[[X]]}$. Finally, $ht\, P[[X]] = dim\, A[[X]]_{P[[X]]} = 1$. $\qquad\square$

13 Polynomials Over a Krull Domain

Lemma 13.1 *Let v be a valuation over a field K. We can extend v to a valuation v^* on the field of rational fractions $K(X)$ by $v^*(f) = min\{v(a_i); 0 \leq i \leq n\}$ for $f = a_0 + a_1 X + \ldots + a_n X^n \in K[X]$ and $v^*(\frac{f}{g}) = v^*(f) - v^*(g)$ for $f, g \in K[X] \setminus (0)$.*

Proof It suffices to show that v^* is a valuation on $K[X]$. Let $f = \sum a_i X^i$ and $g = \sum b_i X^i \in K[X]$. Then we have $f + g = \sum(a_i + b_i)X^i$ with $v(a_i + b_i) \geq min\{v(a_i), v(b_i)\} \geq min\{v^*(f), v^*(g)\}$. Then $v^*(f + g) = min\{v(a_i + b_i)\} \geq min\{v^*(f), v^*(g)\}$. Put $fg = \sum c_k X^k$ with $c_k = \sum_{i+j=k} a_i b_j$. Since $v(a_i b_j) = v(a_i) + v(b_j) \geq v^*(f) + v^*(g)$, $v(c_k) \geq min\{v(a_i b_j); i + j = k\} \geq v^*(f) + v^*(g)$. Then $v^*(fg) = min\{v(c_k)\} \geq v^*(f) + v^*(g)$. To show inverse inequality, let $i_0 = inf\{i; v^*(f) = v(a_i)\}$, $j_0 = inf\{j; v^*(g) = v(b_j)\}$ and $k_0 = i_0 + j_0$. Then

$$c_{k_0} = \sum_{i+j=k_0} a_i b_j = \sum_{\substack{i+j=k_0 \\ i<i_0}} a_i b_j + a_{i_0} b_{j_0} + \sum_{\substack{i+j=k_0 \\ j<j_0}} a_i b_j.$$ In the first sum, $v(a_i b_j) =$

$v(a_i) + v(b_j) > v^*(f) + v^*(g)$, and then $v\left(\sum_{\substack{i+j=k_0 \\ i<i_0}} a_i b_j \right) \geq min\{v(a_i b_j); i + j = $

$k_0, i < i_0\} > v^*(f) + v^*(g)$. By symmetry, $\left(\sum_{\substack{i+j=k_0 \\ j<j_0}} a_i b_j \right) > v^*(f) + v^*(g)$. On the other hand, $v(a_{i_0} b_{j_0}) = v(a_{i_0}) + v(b_{j_0}) = v^*(f) + v^*(g)$. Then $v(c_{k_0}) = v(a_{i_0} b_{j_0}) = v^*(f) + v^*(g)$. Since $v^*(fg) \leq v(c_k) = v^*(f) + v^*(g)$, then $v^*(fg) = v^*(f) + v^*(g)$. $\qquad\square$

Remark 13.2 With the notations of the preceding lemma, the groups of v and v^* are identical.

Lemma 13.3 *Let (A, M) be a valuation domain with quotient field K defined by a valuation v. The ring of the valuation v^*, defined in the preceding lemma, is $V = A[X]_{M[X]}$.*

Proof " $\supseteq$ " Let $F = \frac{f}{g} \in A[X]_{M[X]}$ with $f \in A[X]$ and $g \in A[X] \setminus M[X]$. Then $v^*(f) \geq 0$ and $v^*(g) = 0$ since one of the coefficients of g belongs to $A \setminus M = U(A)$, so its valuation is zero. Then $v^*(F) = v^*(f) \geq 0$, so $F \in V$.

" $\subseteq$ " Let $0 \neq F \in V \subseteq K(X)$, $F = \frac{f}{g}$ with $f, g \in K[X] \setminus (0)$. Then $v^*(F) \geq 0$, so $v^*(f) \geq v^*(g)$. Put $f = \sum a_i X^i$, $g = \sum b_i X^i$, $v^*(f) = v(a_{i_0})$ and $v^*(g) = v(b_{j_0})$. Then $v(a_{i_0}) \geq v(b_{j_0})$. For each i, $v(a_i) \geq v(a_{i_0}) \geq v(b_{j_0})$. Then $f' = \frac{f}{b_{j_0}} \in A[X]$ and for each j, $v(b_j) \geq v(b_{j_0})$, so $g' = \frac{g}{b_{j_0}} \in A[X]$. Since the coefficient of X^{j_0} in g' is equal to 1, then $g' \notin M[X]$. So $F = \frac{f'}{g'} \in A[X]_{M[X]}$. $\square$

Lemma 13.4 *Let (V, M) be a DVR of rank one with quotient field K. Then $V[X]_{M[X]} \cap K[X] = V[X]$.*

Proof The inverse inclusion is clear. Let $F(X) \in V[X]_{M[X]} \cap K[X]$, $F(X) = \frac{f(X)}{g(X)} = \frac{h(X)}{a}$ with $f(X), h(X) \in V[X]$, $g(X) \in V[X] \setminus M[X]$ and $0 \neq a \in V$. Since $V[X]$ is UFD, we may suppose that $h(X)$ and a are relatively prime. The equality $af(X) = g(X)h(X)$ implies that a divides $g(X)$ in $V[X]$. If $a \in M$, then $g(X) \in M[X]$, which is absurd. Then $a \in U(V)$ and $F(X) = \frac{h(X)}{a} \in V[X]$. $\square$

Lemma 13.5 *Let A be a Krull domain with quotient field K and $P \in Spec^1(A)$. Then $K[X] \cap A[X]_{P[X]} = A_P[X]$.*

Proof Since A_P is a DVR of rank one and maximal ideal PA_P, by applying Lemma 10.8 (3) and then the preceding lemma, we find $K[X] \cap A[X]_{P[X]} = K[X] \cap (A_P[X])_{PA_P[X]} = A_P[X]$. $\square$

Lemma 13.6 *Let A be a Krull domain with quotient field K. Then $A[X] = K[X] \cap \bigcap \{A[X]_{P[X]}; \ P \in Spec^1(A)\}$, an intersection with finite character.*

Proof Since A is a Krull domain, by Definition 12.7, $A = \cap \{A_P; \ P \in Spec^1(A)\}$. By the preceding lemma, $A[X] = \cap \{A_P[X]; \ P \in Spec^1(A)\} = \cap \{K[X] \cap A[X]_{P[X]}; \ P \in Spec^1(A)\} = K[X] \cap \bigcap \{A[X]_{P[X]}; \ P \in Spec^1(A)\}$.

Let $0 \neq f(X) = a_0 + a_1 X + \ldots + a_n X^n \in A[X]$, $a_n \neq 0$. There exist $P_1, \ldots, P_k \in Spec^1(A)$ such that for each $P \in Spec^1(A) \setminus \{P_1, \ldots, P_k\}$, $a_n \notin P$, then $f(X) \notin P[X]$, and then $f(X)$ is invertible in $A[X]_{P[X]}$. The intersection has a finite character. $\square$

Theorem 13.7 *If A is a Krull domain, then so is the domain $A[X_1, \ldots, X_n]$.*

Proof By induction, it suffices to show that $A[X]$ is a Krull domain. By the preceding lemma, $A[X] = K[X] \cap \{A[X]_{P[X]}; \ P \in Spec^1(A)\}$, an intersection with finite character. Since $K[X]$ is a PID, and then Krull, it suffices to show that $A[X]_{P[X]}$ is a DVR of rank one, for each $P \in Spec^1(A)$ in order to apply Lemma 12.3. By Lemma 10.8, $A[X]_{P[X]} = \left(A_P[X]\right)_{PA_P[X]}$. By hypothesis, A_P is a DVR of rank one. By Lemma 13.3 and Remark 13.2, $\left(A_P[X]\right)_{PA_P[X]}$ is a DVR of rank one.

$\square$

14 Formal Power Series Over a Generalized Krull Domain

In this section, we introduce a class of rings that generalize Krull domains, called generalized Krull domains. Then, we study the transfer of this property to the formal power series extension. The results are due to Paran and Temkin (2010).

Definition 14.1 Let A be an integral domain with quotient field K. We say that A is a generalized Krull domain if there exists a family $\mathscr{F}$ of valuations of rank one, not necessarily discrete, over K satisfying the following:

1. For each $v \in \mathscr{F}$, the domain A_v of v is the localized of A by the prime ideal $M_v = \{a \in A; \ v(a) > 0\}$.
2. $A = \bigcap_{v \in \mathscr{F}} A_v$.
3. For each $0 \neq a \subset \Lambda$, $v(a) = 0$ for each $v \in \mathscr{F}$, except for a finite number of them.

Examples

(1) Each Krull domain is a generalized Krull domain.
(2) A nondiscrete valuation domain of rank one is a generalized Krull domain that is not a Krull domain.

Lemma 14.2 *Let K be a field and $(\alpha_n)_{n \in \mathbb{N}}$ a sequence of elements of K. For each $n \in \mathbb{N}$, let $f_n(X) = \prod_{i:0}^{n} (1 - \alpha_i X^{2^i})$.*

1. The sequence $(f_n)_{n \in \mathbb{N}}$ converges, with respect to the (X)–adic topology, to a formal power series $f(X) = \sum_{i:0}^{\infty} \lambda_i X^i \in K[[X]]$ that we denote by $f(X) = \prod_{i:0}^{\infty} (1 - \alpha_i X^{2^i})$.

2. For each index $i < 2^n$, with $n \in \mathbb{N}^*$, the coefficient λ_i is either zero or of the form $\lambda_i = \pm \prod_{j \in I} \alpha_j$, where $I \subseteq \{0, 1, \ldots, n-1\}$.

3. *For each $n \in \mathbb{N}$, let $g_n(X) = \frac{f(X)}{1 - \alpha_n X^{2^n}} \in K[[X]]$. A coefficient of $g_n(X)$ is either zero or of the form $\pm \prod_{j \in I} \alpha_j$, for a finite set $I \subseteq \mathbb{N}$.*

Proof

1. The domain $K[[X]]$ is complete for the (X)−adic topology. See Theorem 6.4, Chap. 4. It suffices to show that the sequence (f_n) is Cauchy. We have $f_n - f_{n+1} = f_n - f_n(1 - \alpha_{n+1} X^{2^{n+1}}) = \alpha_{n+1} X^{2^{n+1}} f_n$. Let v be the natural valuation of $K[[X]]$. Then $v(f_n - f_{n+1}) \geq 2^{n+1} \longrightarrow +\infty$ when $\longrightarrow +\infty$. We conclude by Remark 6.1 (1), Chap. 4.

2. For $i < 2^n$, the coefficient λ_i of $f(X)$ is the same as that of X^i in the finite product $\prod_{j:0}^{n-1} (1 - \alpha_j X^{2^j}) = f_{n-1}(X)$, which is either zero or of the form $\prod_{j \in I} (-\alpha_j)$ with $I \subseteq \{0, 1, \ldots, n-1\}$. There is only one term by the uniqueness of the expression of i in the binary basis.

3. Since $g_n(X) = \frac{f(X)}{1 - \alpha_n X^{2^n}} = \prod_{\substack{i:0 \\ i \neq n}}^{\infty} (1 - \alpha_i X^{2^i})$ is of the form $f(X)$ with $\alpha_n = 0$, the coefficients of $g_n(X)$ are either zero or of the form $\pm \prod_{j \in I} \alpha_j$, with I a finite subset of $\mathbb{N}$. $\square$

Lemma 14.3 *Let A be a generalized Krull domain defined by a family $\mathscr{F}$ of real valuations. If $w, v_1, \ldots, v_n \in \mathscr{F}$ are nonassociated valuations and $M_1, \ldots, M_n$ are real numbers, there exists $a \in A$ such that $w(a) = 0$ and $v_i(a) > M_i$ for $1 \leq i \leq n$.*

Proof It suffices to find $a \in A$ such that $w(a) = 0$ and $v_i(a) > 0$ for $1 \leq i \leq n$. Then, we replace a by a convenient power of a if necessary. Consider the prime ideals $P = \{b \in A; \ w(b) > 0\}$ and $P_i = \{b \in A; \ v_i(b) > 0\}, 1 \leq i \leq n$, of A. The localized domains A_P and $A_{P_i}, 1 \leq i \leq n$, are different valuation domains of rank one. Then $ht \ P = dim \ A_P = 1$ and $ht \ P_i = dim \ A_{P_i} = 1, 1 \leq i \leq n$. Then the ideals $P, P_1, \ldots, P_n$ are different and incomparable. Take $a_i \in P_i \setminus P, 1 \leq i \leq n$. Then $a = a_1 \ldots a_n \in P_i$, for $1 \leq i \leq n$, but $a \notin P$. Then $w(a) = 0$ and $v_i(a) > 0$, $1 \leq i \leq n$. $\square$

Proposition 14.4 *Let A be a generalized Krull domain defined by a family $\mathscr{F}$ of real valuations. Suppose that $w \in \mathscr{F}$ is nondiscrete. There exist two sequences $(b_i)_{i \in \mathbb{N}}$ and $(c_i)_{i \in \mathbb{N}^*}$ of nonzero elements of A such that:*

1. $w(b_0) > 0$ and $0 = w(c_i) < w(b_i) < \frac{w(b_{i-1})}{i}$, *for each $i \in \mathbb{N}^*$.*

2. $f(X) = b_0 \prod_{i:1}^{\infty} \left(1 - (\frac{c_i}{b_i} X)^{2^{i-1}}\right) \in A[[X]]$.

3. *For each $i \in \mathbb{N}^*$, $b_i^{2^{i-1}} f(X)$ is divisible by $b_i - c_i X$ in $A[[X]]$.*

Proof We will prove (1) and (2). Let K be the quotient field of A, $P(w) = \{x \in A; \; w(x) > 0\}$ and $A_w = A_{P(w)} = \{x \in K; \; w(x) \geq 0\}$, the ring of w. It is clear that $w(A) \subseteq w(A_w)$. Conversely, let $x \in A_w$, $x = \frac{a}{b}$ with $a \in A$ and $b \in A \setminus P(w)$. Then $w(x) = w(a) - w(b) = w(a) \in w(A)$. So $w(A) = w(A_w)$. Since w is nondiscrete, for each real number $\epsilon > 0$, there exists $a \in A$ such that $0 < w(a) < \epsilon$. Fix an element $b_0 \in A$ such that $0 < w(b_0) < 1$, and then an element $b_1 \in A$ such that $0 < w(b_1) < w(b_0)$. By the third axiom of generalized Krull domains, the family $\mathscr{F}_1 = \{v \in \mathscr{F}; \; v(\frac{b_0}{b_1}) < 0\}$ is finite. Since $w(\frac{b_0}{b_1}) = w(b_0) - w(b_1) > 0$, then $w \notin \mathscr{F}_1$. By the preceding lemma, there exists $c_1 \in A$ such that $w(c_1) = 0$ and $v(c_1) > -v(\frac{b_0}{b_1})$ for each $v \in \mathscr{F}_1$. Then for each $v \in \mathscr{F}$, $v(\frac{b_0 c_1}{b_1}) \geq 0$ since for $v \in \mathscr{F} \setminus \mathscr{F}_1$, $v(\frac{b_0}{b_1}) \geq 0$. Then for each $v \in \mathscr{F}$, $\frac{b_0 c_1}{b_1} \in A_v$, so $\frac{b_0 c_1}{b_1} \in \bigcap_{v \in \mathscr{F}} A_v = A$.

Then we always find elements $b_1, c_1 \in A$ such that $\frac{b_0 c_1}{b_1} \in A$ and $0 = w(c_1) < w(b_1) < w(b_0)$.

Suppose by induction that we have constructed nonzero elements $b_1, \ldots, b_n$ and $c_1, \ldots, c_n \in A$ satisfying the following:

(i) For each subset $I \subseteq \{1, \ldots, n\}$, $b_0 \prod_{i \in I} (\frac{c_i}{b_i})^{2^{i-1}} \in A$.

(ii) For $1 \leq i \leq n$, $0 = w(c_i) < w(b_i) < \frac{w(b_{i-1})}{i}$.

By considering $\epsilon = \frac{w(b_n)}{n+1}$, we find an element $b_{n+1} \in A$ such that $0 < w(b_{n+1}) < \frac{w(b_n)}{n+1}$. The set $C = \{\frac{b_0}{b_{n+1}^{2^n}} \prod_{i \in I} (\frac{c_i}{b_i})^{2^{i-1}}; \; I \subseteq \{1, \ldots, n\}\}$ of K is finite since $\{1, \ldots, n\}$ is finite. By the third axiom of generalized Krull domains, there exists a finite number of valuations in $\mathscr{F}$ taking negative values each on at least an element of C. Let $\mathscr{F}'$ be the set of such valuations except w (if necessary). Put $\mathscr{F}' = \{v_1, \ldots, v_k\}$ and $M_1 = \sup\{-v_1(a); \; a \in C\}, \ldots, M_k = \sup\{-v_k(a); \; a \in C\} \in \mathbb{R}$. By the preceding lemma, there exists $c_{n+1} \in A$ such that $w(c_{n+1}) = 0$ and for each $v \in \mathscr{F}'$, $v(c_{n+1}) > M_i$, $1 \leq i \leq k$. Let now $v \in \mathscr{F}$ and $a \in C$, and then $v(c_{n+1}^{2^n} a) = 2^n v(c_{n+1}) + v(a) = (2^n - 1)v(c_{n+1}) + v(c_{n+1}) + v(a) \geq v(c_{n+1}) + v(a)$ since $c_{n+1} \in A \subseteq A_v$, so $v(c_{n+1}) \geq 0$. If $v \notin \mathscr{F}'$, since $v(a) \geq 0$, then $v(c_{n+1}) + v(a) \geq 0$. If $v \in \mathscr{F}'$, then $v(c_{n+1}) > -v(a)$, so $v(c_{n+1}) + v(a) > 0$. We conclude that for each $v \in \mathscr{F}'$ and each $a \in C$, $v(c_{n+1}^{2^n} a) \geq 0$. Then for each subset $I \subseteq \{1, \ldots, n+1\}$ and for each $v \in \mathscr{F}$, $v(b_0 \prod_{i \in I} (\frac{c_i}{b_i})^{2^{i-1}}) \geq 0$ so $b_0 \prod_{i \in I} (\frac{c_i}{b_i})^{2^{i-1}} \in \bigcap_{v \in \mathscr{F}} A_v = A$. The induction is then done. By Lemma 14.2,

$$f(X) = b_0 \prod_{i:1}^{\infty} \left(1 - (\frac{c_i}{b_i} X)^{2^{i-1}}\right) \in K[[X]]$$ and its coefficients are either zero or of

the form $\pm b_0 \prod_{i \in I} (\frac{c_i}{b_i})^{2^{i-1}} \in A$, for a finite set $I \subseteq \mathbb{N}$. Then $f(X) \in A[[X]]$. We have always proved (1) and (2).

For (3), fix an index $i \in \mathbb{N}^*$. By Lemma 14.2 (3), the coefficients of $\frac{f(X)}{1-(\frac{c_i}{b_i}X)^{2^{i-1}}}$

are either zero or of the form $\pm b_0 \prod_{j \in I} (\frac{c_j}{b_j})^{2^{j-1}} \in A$, for a finite set $I \subseteq \mathbb{N}$. Then

$g(X) = \frac{f(X)}{1-(\frac{c_i}{b_i}X)^{2^{i-1}}} \in A[[X]]$ and $b_i^{2^{i-1}} f(X) = (b_i^{2^{i-1}} - c_i^{2^{i-1}} X^{2^{i-1}})g(X)$ is

divisible by $b_i - c_i X$ in $A[[X]]$.   □

Lemma 14.5 *Let A be a generalized Krull domain defined by a family $\mathscr{F}$ of valuations. Let $w \in \mathscr{F}$ and $b, c \in A$ be such that $0 = w(c) < w(b)$. Suppose that $A[[X]]$ is a generalized Krull domain defined by a family $\mathscr{F}^*$ of valuations. Then there exists a valuation $v \in \mathscr{F}^*$ such that $v(b - cX) > v(b) = 0$.*

Proof Suppose that does not exist any valuation $v \in \mathscr{F}^*$ satisfying $v(b - cX) > v(b) = 0$. By the third axiom of generalized Krull domains, there exists a finite number of valuations $v_1, \ldots, v_n \in \mathscr{F}^*$ that do not annihilate on $b - cX$, and then $v_i(b - cX) > 0$ so $v_i(b) > 0$ for $1 \leq i \leq n$. For $k \in \mathbb{N}$ sufficiently big, we have $v_i(b^k) > v_i(b - cX)$ for $1 \leq i \leq n$. Then for each $v \in \mathscr{F}^*$, $v(\frac{b^k}{b-cX}) \geq 0$ since for $v \in \mathscr{F}^* \setminus \{v_1, \ldots, v_n\}$, $v(b - cX) = 0$. Then $\frac{b^k}{b-cX} \in \bigcap_{v \in \mathscr{F}^*} A[[X]]_v = A[[X]]$ so

$$\frac{b^k}{b-cX} = b^{k-1} \sum_{j:0}^{\infty} (\frac{c}{b}X)^j \in A[[X]].$$ In particular, the coefficient $b^{k-1}(\frac{c}{b})^k = \frac{c^k}{b} \in$

A, and then $0 \leq w(\frac{c^k}{b}) = kw(c) - w(b) = -w(b)$, so $w(b) \leq 0$, which is absurd.
                                                                                             □

Theorem 14.6 *Let A be a generalized Krull domain. If A is not Krull, then $A[[X]]$ is not a generalized Krull domain.*

Proof Let $\mathscr{F}$ be a family of valuations that define A. Since A is not Krull, there exists a valuation $w \in \mathscr{F}$ that is not discrete. Suppose that $A[[X]]$ is a generalized Krull domain defined by a family $\mathscr{F}^*$ of real valuations. By Proposition 14.4, there exist two sequences $(b_i)_{i \in \mathbb{N}}$ and $(c_i)_{i \in \mathbb{N}^*}$ of nonzero elements of A satisfying the following:

(a) $w(b_0) > 0$ and $0 = w(c_i) < w(b_i) < \frac{w(b_{i-1})}{i}$, for each $i \in \mathbb{N}^*$.

(b) $f(X) = b_0 \prod_{i:1}^{\infty} (1 - (\frac{c_i}{b_i}X)^{2^{i-1}}) \in A[[X]]$.

(c) For each $i \in \mathbb{N}^*$, $b_i^{2^{i-1}} f(X)$ is divisible by $b_i - c_i X$ in $A[[X]]$.

For each $i \in \mathbb{N}^*$, let $\mathscr{F}_i^* = \{v \in \mathscr{F}^*; \ v(b_i - c_i X) > v(b_i) = 0\}$. By the

preceding lemma, $\mathscr{F}_i^* \neq \emptyset$. Let $\mathscr{F}_f^* = \bigcup_{i:1}^{\infty} \mathscr{F}_i^*$. We will show that $\mathscr{F}_f^*$ is infinite.

In the contrary case, there exist two indexes $i < j$ such that $\mathscr{F}_i^* = \mathscr{F}_j^*$. For each

$v \in \mathscr{F}_i^*$, $v(b_i - c_i X) > v(b_i) = 0$, and then $v(1 - \frac{c_i}{b_i} X) > 0$ and $v(c_i X) = 0$ (so $v(c_i) = v(X) = 0$) since for any valuation v, if $v(a) \neq v(b)$, then $v(a + b) = min\{v(a), v(b)\}$. For the same reasons, since $v \in \mathscr{F}_i^* = \mathscr{F}_j^*$, $v(1 - \frac{c_j}{b_j} X) > 0$. Then $v\big((\frac{c_i}{b_i} - \frac{c_j}{b_j})X\big) = v\big(\frac{c_i}{b_i} X - 1 + 1 - \frac{c_j}{b_j} X\big) \geq min\big\{v(\frac{c_i}{b_i} X - 1); \; v(1 - \frac{c_j}{b_j} X)\big\} > 0$. Then $0 < v\big((\frac{c_i}{b_i} - \frac{c_j}{b_j})X\big) = v\big(\frac{c_i}{b_i} - \frac{c_j}{b_j}\big)$. On the other hand, by (a), $w(\frac{c_i}{b_i}) = -w(b_i) < -w(b_j) = w(\frac{c_j}{b_j})$, and then $\frac{c_i}{b_i} \neq \frac{c_j}{b_j}$, so $\frac{c_i}{b_i} - \frac{c_j}{b_j} \in K^*$ with K the quotient field of A. We conclude that v is nonzero over K and then nonzero over A. Then the valuations of $\mathscr{F}_i^*$ are nonzero over A.

Let now $v \in \mathscr{F}^*$ be any valuation such that $v(b_i - c_i X) \neq 0$. Then $v(b_i - c_i X) > 0$. If $v(b_i) > 0$, then v is nonzero over A. If $v(b_i) = 0$, then $v(b_i - c_i X) > v(b_i) = 0$, so $v \in \mathscr{F}_i^*$ then v is nonzero over A.

By the third axiom of generalized Krull domains, there exists a finite number of valuations $v_1, \ldots, v_n \in \mathscr{F}^*$ nonzero on $b_i - c_i X$. By the preceding step, they are nonzero over A. We can choose elements $a_1, \ldots, a_n \in A$ such that $v_k(a_k) > v_k(b_i - c_i X)$ for $1 \leq k \leq n$. Then $a = a_1 \ldots a_n \in A$ and $v_k(a) > v_k(b_i - c_i X)$ for $1 \leq k \leq n$.

For each $v \in \mathscr{F}^* \setminus \{v_1, \ldots, v_n\}$, $v(a) \geq 0 = v(b_i - c_i X)$. Then $\frac{a}{b_i} \sum_{l:0}^{\infty} (\frac{c_i}{b_i} X)^l = \frac{a}{b_i - c_i X} \in \bigcap_{v \in \mathscr{F}^*} A[[X]]_v - A[[X]]$. In particular, $w(\frac{ac_i^l}{b_i^{l+1}}) \geq 0$, and then $w(ac_i^l) \geq w(b_i^{l+1})$ for each $l \in \mathbb{N}$. But $w(b_i) > w(c_i) = 0$, so for $l \in \mathbb{N}^*$ sufficiently big, $w(b_i^{l+1}) > w(a) = w(a) + w(c_i^l) = w(ac_i^l)$ and then $w(b_i^{l+1}) > w(ac_i^l)$. There is a contradiction. We conclude that $\mathscr{F}_f^*$ is an infinite set.

Let $v \in \mathscr{F}_f^* = \bigcup_{i:1}^{\infty} \mathscr{F}_i^*$. There exists $i \in \mathbb{N}^*$ such that $v \in \mathscr{F}_i^*$, and then $v(b_i - c_i X) > v(b_i) = 0$. By (c), $b_i^{2^{i-1}} f(X) = (b_i - c_i X)g(X)$ with $g(X) \in A[[X]]$, then $v(b_i^{2^{i-1}}) + v(f(X)) = v(b_i - c_i X) + v(g(X))$, so $v(f(X)) = v(b_i - c_i X) + v(g(X)) \geq v(b_i - c_i X) > 0$. Then $v(f(X)) > 0$. Since $\mathscr{F}_f^*$ is an infinite subset of $\mathscr{F}^*$, there exist an infinity of valuations in $\mathscr{F}^*$ that do not annihilate on $f(X)$. This contradicts the third axiom of generalized Krull domains. □

Problems

The three following exercises are due to Baeck et al. (2016).

Exercise 1 Let A be a ring and $S \subseteq A$ a multiplicative set. Show that a finite sum of S-finite ideals of A is S-finite.

Exercise 2 Let A be a ring and $S_1 \subseteq S_2$ two multiplicative sets. Suppose that for each $s \in S_2$, there exists $r \in S_2$ such that $rs \in S_1$. Show that A is S_1-Noetherian if and only if it is S_2-Noetherian.

Exercise 3 Let $S_1, \ldots, S_n$ be multiplicative sets respectively of the rings $A_1, \ldots, A_n$ and $S = S_1 \times \ldots \times S_n$.

(1) Show that the product ring $A = A_1 \times \ldots \times A_n$ is S-Noetherian if and only if A_i is S_i-Noetherian for each $i \in \{1, \ldots, n\}$.
(2) Show by an example that the result is not true for an infinite product.

Exercise 4 (Hamed and Hizem 2016) Let A be a ring, $S \subseteq A$ a multiplicative set, and M a A-module. An increasing sequence $(N_n)_{n \in \mathbb{N}}$ of sub-modules of M is said to be S-stationary if there exist $s \in S$ and $k \in \mathbb{N}$ such that for each integer $n \geq k$, $sN_n \subseteq N_k$.

Let $\mathscr{F}$ be a nonempty family of sub-modules of M. An element N of $\mathscr{F}$ is said to be S-maximal if there exists $s \in S$ such that for each $L \in \mathscr{F}$, when $N \subseteq L$ and then $sL \subseteq N$.

Suppose that $S = \{s_1, \ldots, s_n\}$ is finite and let $s = s_1 \ldots s_n$.

(1) Show that an increasing sequence $(N_n)_{n \in \mathbb{N}}$ of sub-modules of M is S-stationary if there exists $k \in \mathbb{N}$ such that for each integer $n \geq k$, $sN_n \subseteq N_k$.
(2) Show that an element N of a family $\mathscr{F}$, of sub-modules of M, is S-maximal if for each $L \in \mathscr{F}$ such that $N \subseteq L$, we have $sL \subseteq N$.

Exercise 5 (Hamed and Hizem 2016) Let A be a ring, $S \subseteq A$ a multiplicative set, and M a A-module. A sub-module N of M is said to be extended if it is of the form $N = IM$ with I an ideal of A.

Suppose that M is S-finite and consider the following properties:

(1) Each nonempty family of extended sub-modules of M has a S-maximal element.
(2) Each extended sub-module of M is S-finite.
(3) Each sub-module of M of the form PM, with P a prime ideal of A disjoint with S, is S-finite.
(4) The module M is S-Noetherian.
(5) Each increasing sequence of extended sub-modules of M is S-stationary.

Show the following implications (1) $\implies$ (2) $\implies$ (3) $\iff$ (4) $\implies$ (5). Moreover, if S is finite or S countable and if the ideals of A are comparable, then (5) $\implies$ (1).

Exercise 6 Let A be a ring and S a multiplicative set of A not containing zero divisors. Show that A is S-Noetherian if and only if the ring $S^{-1}A$ is Noetherian and for each finitely generated ideal I of A, there exists $s \in S$ such that $(S^{-1}I) \cap A = (I : s)_A$.

Exercise 7 Let S be a multiplicative set of a ring A and $0 \longrightarrow M' \longrightarrow M \longrightarrow M'' \longrightarrow 0$ a short exact sequence of A-modules. Show that M is S-Noetherian if and only if M' and M'' are S-Noetherian.

Exercise 8 (Cahen and Haouat 1987) Let $(A_n)_{n \in \mathbb{N}}$ be an increasing sequence of rings, $A = \bigcup_{n:0}^{\infty} A_n$ and $\mathscr{A}[X] = \{f = \sum_{i:0}^{n} a_i X^i \in A[X]; \ n \in \mathbb{N}, a_i \in A_i$ for each $i \in \mathbb{N}\}$.

(1) Show that for each $n \in \mathbb{N}$, $X^n A_n[X] \subseteq \mathscr{A}[X]$.
(2) Show that $S = \{X^n; \ n \in \mathbb{N}\}$ is a multiplicative set of $\mathscr{A}[X]$ formed by regular elements and $\mathscr{A}[X] \subseteq A[X] \subseteq S^{-1}\mathscr{A}[X]$.
(3) Show that the prime ideals of the ring $\mathscr{A}[X]$ containing X have the form $J_P = \{f \in \mathscr{A}[X]; \ f(0) \in P\}$ where P is a prime ideal of A_0.
(4) Show that the prime ideals of $\mathscr{A}[X]$ not containing X have the form $\mathscr{P} \cap \mathscr{A}[X]$ with $\mathscr{P}$ is a prime ideal of $A[X]$ not containing X.

The following five exercises are due to Haouat (1988) and Hizem (2012).

Exercise 9 Let $\mathscr{A} = (A_n)_{n \in \mathbb{N}}$ be an increasing sequence of rings, $A = \bigcup_{n \in \mathbb{N}} A_n$ and $\mathscr{A}[[X]] = \{f = \sum_{n:0}^{\infty} a_n X^n \in A[[X]]; \ a_n \in A_n$ for each $n \in \mathbb{N}\}$.

(1) Show that $\mathscr{A}[[X]]$ is a sub-ring of $A[[X]]$.
(2) Show that $f = \sum_{n:0}^{\infty} a_n X^n$ is invertible in $\mathscr{A}[[X]]$ if and only if a_0 is invertible in A_0.
(3) Show that $\mathscr{A}[[X]]$ is an integral domain $\Longleftrightarrow A$ is an integral domain $\Longleftrightarrow \forall n \in \mathbb{N}, A_n$ is an integral domain.
(4) Let $f = \sum_{n:0}^{\infty} a_n X^n \in \mathscr{A}[[X]]$ be a nilpotent element. Show that for each $n \in \mathbb{N}$, a_n is nilpotent. Deduce that $\mathscr{A}[[X]]$ is reduced $\Longleftrightarrow A$ is reduced $\Longleftrightarrow \forall n \in \mathbb{N}, A_n$ is reduced.
(5) We consider the sequence of rings $\mathscr{B} = (A_n)_{n \geq 1}$. Let I be an ideal of A_0.

   (a) Show that $I + X\mathscr{B}[[X]]$ is an ideal of $\mathscr{A}[[X]]$.
   (b) Show that $I + X\mathscr{B}[[X]]$ is a prime (resp. maximal, resp. radical) ideal of $\mathscr{A}[[X]]$ if and only if I is a prime (resp. maximal, resp. radical) ideal of A_0.
   (c) Show that $\sqrt{I + X\mathscr{B}[[X]]} \subseteq \sqrt{I} + X\mathscr{B}[[X]]$.

(6) Show that $Max(\mathscr{A}[[X]]) = \{\mathscr{M} + X\mathscr{B}[[X]]; \ \mathscr{M} \in Max(A_0)\}$.
(7) Let P be a prime ideal of $\mathscr{A}[[X]]$ containing X. Show that there exists a prime ideal I of A_0 such that $P = I + X\mathscr{B}[[X]]$.

(8) Show that if all the rings A_n are fields, then $spec(\mathscr{A}[[X]]) = \{0; \ X\mathscr{B}[[X]]\}$.
(9) Let $\mathscr{P} = (P_n)_{n\in\mathbb{N}}$ with P_n be a prime ideal of A_n and $P_n = A_n \cap P_{n+1}$ for each $n \in \mathbb{N}$.

    (a) Show that $P = \bigcup_{n\in\mathbb{N}} P_n$ is a prime ideal of $A = \bigcup_{n\in\mathbb{N}} A_n$.

    (b) Show that for $0 \le n \le m$, $P_n = A_n \cap P_m$. Deduce that $P_n = A_n \cap P$.

    (c) Show that $\mathscr{P}[[X]] = P[[X]] \cap \mathscr{A}[[X]]$ and $\mathscr{P}[[X]]$ is a prime ideal of $\mathscr{A}[[X]]$ not containing X.

(10) Let P be a prime ideal of A, $P_n = A_n \cap P$ for $n \in \mathbb{N}$ and $\mathscr{P} = (P_n)_{n\in\mathbb{N}}$. Show that $\mathscr{P}[[X]] = P[[X]] \cap \mathscr{A}[[X]]$.

(11) Let I be an ideal of A, $I_n = A_n \cap I$ for $n \in \mathbb{N}$ and $\mathscr{I} = (I_n)_{n\in\mathbb{N}}$. Show that $\sqrt{\mathscr{I}[[X]]} \subseteq \sqrt{I}[[X]] \cap \mathscr{A}[[X]]$.

(12) Suppose that A is an integral domain.

    (a) Show that if the sequence $\mathscr{A} = (A_i)_{i\in\mathbb{N}}$ is stationary at the rank n, the quotient field $qf(\mathscr{A}[[X]]) = qf(A_n[[X]])$.

    (b) Show that $\mathscr{A}[[X]]$ is integrally closed if and only if the following conditions are satisfied:

        (i) For each integer $n \ge 1$, $A_n = A_1$.
        (ii) The domain A_0 is integrally closed in A_1.
        (iii) The domain $A_1[[X]]$ is integrally closed.

    (c) Show that the domain $\mathscr{A}[[X]]$ is completely integrally closed if and only if the sequence $\mathscr{A} = (A_n)_{n\in\mathbb{N}}$ is constant and A_0 is completely integrally closed.

Exercise 10 Let $\mathscr{A} = (A_n)_{n\in\mathbb{N}}$ be an increasing sequence of integral domains. Show that:

1. The element X is irreducible in $\mathscr{A}[[X]]$ if and only if $U(A_0) = U(A_1) \cap A_0$.
2. The ideal $X\mathscr{A}[[X]]$ is prime if and only if the sequence $(A_n)_{n\in\mathbb{N}}$ is constant.

Exercise 11 Let B be an integral domain and $x \in B$. We say that x is primal if when x divides a product $a_1 a_2$ with $a_1, a_2 \in B$, there exist $x_1, x_2 \in B$ such that $x = x_1 x_2$ with x_1 dividing a_1 and x_2 dividing a_2.

    Let $\mathscr{A} = (A_n)_{n\in\mathbb{N}}$ be an increasing sequence of integral domains. Show that if the element X is primal in $\mathscr{A}[[X]]$, then for each integer $n \ge 1$, $A_n = S^{-1}A_0$ where $S = U(A_1) \cap A_0$. Deduce that in this case $\mathscr{A}[[X]] = A_0 + X(S^{-1}A_0)[[X]]$.

Exercise 12 Let $\mathscr{A} = (A_n)_{n\in\mathbb{N}}$ be an increasing sequence of integral domains.

(1) Show that $\mathscr{A}[[X]]$ satisfies the acc for principal ideals if and only if for each $n \in \mathbb{N}$, each increasing sequence $b_1 A_0 \subseteq b_2 A_0 \subseteq \ldots$, with $0 \neq b_k \in A_n$, $k \in \mathbb{N}^*$, is stationary.
(2) Deduce that if A_0 is a field, $\mathscr{A}[[X]]$ satisfies the acc for the principal ideals.

(3) Suppose that the quotient field $qf(A_0) \subseteq A_1$. Show that $\mathscr{A}[[X]]$ satisfies the acc for the principal ideals if and only if A_0 is a field.

Exercise 13 An integral domain B is said to be Archimedean if for each $b \in B \setminus U(B)$, $\bigcap_{i:0}^{\infty} b^i B = (0)$. Let $\mathscr{A} = (A_n)_{n \in \mathbb{N}}$ be an increasing sequence of integral domains.

(1) Show that $\mathscr{A}[[X]]$ is Archimedean if and only if for each $a \in A_0 \setminus U(A_0)$ and for each $k \in \mathbb{N}$, $\bigcap_{i:0}^{\infty} a^i A_k = (0)$.

(2) Suppose that A_0 is Archimedean and for each $k \in \mathbb{N}$, there exists $0 \neq d_k \in A_0$ such that $d_k A_k \subseteq A_0$. Show that $\mathscr{A}[[X]]$ is Archimedean.

(3) Suppose that for each $k \in \mathbb{N}$, A_k is Archimedean and $U(A_k) \cap A_0 = U(A_0)$.

Show that $\mathscr{A}[[X]]$ is Archimedean.

Exercise 14 (D'anna et al. 2008) Let $f : A \longrightarrow B$ be a homomorphism of rings, J an ideal of B, $\pi : B \longrightarrow B/J$ the canonical homomorphism, and $\hat{f} = \pi \circ f : A \longrightarrow B \longrightarrow B/J$. Suppose that the ring B is Noetherian and that B/J is a finitely generated A−module for the structure of module induced by $\hat{f}$.

Show that the ring $A \bowtie^f J$ is Noetherian if and only if A is Noetherian.

Exercise 15 (D'anna et al. 2008) Let $f : A \longrightarrow B$ be a homomorphism of rings, J a nonzero ideal of B, and $A \bowtie^f J$ the amalgamation of A with B along J with respect to f.

(1) Show that the $A \bowtie^f J$ is an integral domain if and only if $f(A) + J$ is an integral domain and $f^{-1}(J) = (0)$.

Deduce that if B is an integral domain and if $f^{-1}(J) = (0)$, then $A \bowtie^f J$ is integral.

(2) Show that $A \bowtie^f J$ is reduced if and only if A is reduced and $Nil(B) \cap J = (0)$.

Deduce that if the rings A and B are reduced, then $A \bowtie^f J$ is reduced, and if J is a radical ideal of B and $A \bowtie^f J$ is reduced, then the rings A and B are reduced.

Exercise 16 Let A be a ring and M a A−module.

(1) Show that the group of unities $U(A(+)M) = U(A)(+)M$.

(2) Give the form of the maximal ideals of the ring $A(+)M$. Deduce that A is local if and only if $A(+)M$ is local and in this case, the residual fields are isomorphic.

(3) Show that $\mathbb{Z}/2\mathbb{Z}$ can be endowed with the structure of $\mathbb{Z}/4\mathbb{Z}$−module.

(4) Show that the ring $(\mathbb{Z}/4\mathbb{Z})(+)(\mathbb{Z}/2\mathbb{Z})$ is local and give its maximal ideal $\mathscr{M}$.

(5) Show that the principal ideal $\mathscr{I}$ of the ring $(\mathbb{Z}/4\mathbb{Z})(+)(\mathbb{Z}/2\mathbb{Z})$ generated by $(\bar{2}, \bar{1})$ is not of the form $I(+)N$ with I an ideal of the ring $\mathbb{Z}/4\mathbb{Z}$ and N a sub-module of $\mathbb{Z}/2\mathbb{Z}$.

(6) Show that $\sqrt{\mathscr{I}} = \mathscr{M}$.

Exercise 17 Let $A \subseteq B$ be an extension of rings and X a variable over B. Show that $A(+)B \simeq (A + XB[X])/X^2B[X]$.

Exercise 18 Let A be a ring, M a A−module, and $\mathscr{M}$ the set of matrices of the form $\left(\begin{smallmatrix} a & 0 \\ x & a \end{smallmatrix}\right)$ with $a \in A$ and $x \in M$. Show that $\mathscr{M}$ is a commutative ring with unity for the natural addition and multiplication of matrices and $\mathscr{M} \simeq A(+)M$.

Exercise 19 Let A be an integral domain and $(A_\lambda)_{\lambda \in \Lambda}$ a family of finite character of sub-rings of A such that $A = \bigcap_{\lambda \in \Lambda} A_\lambda$. Show that if S is a multiplicative set of A, then $S^{-1}A = \bigcap_{\lambda \in \Lambda} S^{-1}A_\lambda$ and the family $(S^{-1}A_\lambda)_{\lambda \in \Lambda}$ has a finite character.

Exercise 20 Let A be a ring and $f(X, Y), g(X, Y) \in A[X, Y]\backslash(0)$. Show that there exist $f^*(X), g^*(X) \in A[X] \backslash (0)$ such that $c_f = c_{f^*}$, $c_g = c_{g^*}$, and $c_{fg} = c_{f^*g^*}$, where c_h is the set of the coefficients of h for each $h \in A[X, Y]$.

The four following exercises are due to Ribenboim (1986a, b).

Exercise 21 An ideal I of a ring A is said to be real if the following condition is satisfied: if $n \in \mathbb{N}^*$ and $a_1, \ldots, a_n \in A$ are such that $a_1^2 + \ldots + a_n^2 \in I$, then $a_1, \ldots, a_n \in I$. The ring A is said to be real if its zero ideal is real in A.

(1) Give the real ideals of the following rings $\mathbb{F}_2, \mathbb{C}, \mathbb{R}$, and $\mathbb{Z}$.
(2) Show that a real ideal I is radical. What is the converse? Deduce that a real ring is reduced.
(3) Show that an integral domain A is real if and only if its quotient field K is orderable.
(4) Show that the ideal I is real if and only if either $I = A$ or the ring A/I is real.
(5) Let (A, M) be a valuation domain with quotient field K. Show that:

  (a) M is a real ideal if and only if A/M is an orderable field.
  (b) (0) is a real ideal of A if and only if K is an orderable field.
  (c) Each real ideal of A is either prime or equal to A.
  (d) If $P \neq A$ is a real ideal, each prime ideal Q of A contained in P is real.
  (e) The existence of a non-Noetherian ring having a finite number of real ideals.

Exercise 22 Let A be a ring.

(1) Show that if $(I_\lambda)_{\lambda \in \Lambda}$ is a totally ordered family of real ideals of A, then $\bigcup_{\lambda \in \Lambda} I_\lambda$ is a real ideal of A.

(2) Show that if $(I_\lambda)_{\lambda \in \Lambda}$ is a family of real ideals of A, then $\bigcap_{\lambda \in \Lambda} I_\lambda$ is a real ideal of A.

(3) (a) Show that if S is a nonempty set of A, there exists a smallest real ideal of A containing S, noted $R(S)$.

    (b) Show that if $I = \sum AS$ is the ideal of A generated by the set S, then $R(I) = R(S)$.

    (c) Show that if $I \subseteq J$ are two ideals of A, then $R(I) \subseteq R(J)$.

    (d) Give an example of a ring A and an ideal $I \neq A$ with $R(I) = A$.

(4) Show that if I is an ideal of A, then $\sqrt{I} \subseteq R(I)$.

(5) Let I and J be two ideals of A. Show that $R(I + R(J)) = R(I + J)$.

(6) Show that if I is a real ideal and J any ideal, then $I : J$ is a real ideal.

(7) Let I be an ideal of A. Show that $R(I) = \{a \in A; \exists m \in \mathbb{N}^*, \exists n \in \mathbb{N}, \exists a_1, \ldots, a_n \in A, a^{2m} + a_1^2 + \ldots + a_n^2 \in I\} = \bigcap P$, where P runs the set of real prime ideals of A containing I, A is included.

(8) Let I be an ideal of A. We define the sequence $(R_k(I))_{k \in \mathbb{N}^*}$, by $R_1(I) = \{a \in A; \exists n \in \mathbb{N}, \exists a_1, \ldots, a_n \in A, a^2 + a_1^2 + \ldots + a_n^2 \in I\}$ and for each integer $k \geq 1$, $R_{k+1}(I) = R_1(R_k(I))$.

    (a) Show that $R_1(I)$ is an ideal of A containing I.

    (b) Show that $(R_k(I))_{k \in \mathbb{N}^*}$ is an increasing sequence of ideals of A and
$$R(I) = \bigcup_{k \in \mathbb{N}^*} R_k(I).$$

(9) Let I and J be two ideals of A. Show the relations $IR(J) \subseteq R(I)R(J) \subseteq R(IJ) = R(I \cap J) = R(I) \cap R(J)$.

(10) Let $\phi : A \longrightarrow B$ be an homomorphism between rings.

    (a) If J is an ideal of A, show that $\phi(R(J)) \subseteq R(\phi(J))$.

    (b) Show that if I is a real ideal of B, then $\phi^{-1}(I)$ is a real ideal of A.

    (c) If ϕ is onto, then I is a real ideal of B and if $\phi^{-1}(I) = R(\Sigma AS)$, show that $I = R(\Sigma B\phi(S))$.

(11) A real ideal I of A is said to be really finitely generated if there exists a finite set $S \subseteq A$ such that $I = R(\Sigma AS) = R(S)$.

    Show that if I is really finitely generated by a finite set $S = \{x_1, \ldots, x_n\}$, then I is really generated by $\{x_1^2 + \ldots + x_n^2\}$ or also by $\{x_1^2 + \ldots + x_k^2, x_{k+1}, \ldots, x_n\}$, $1 \leq k \leq n - 1$.

(12) Let $I = R(S)$ be a real ideal of A, really generated by a set S of A. Show that if I is really finitely generated, there exists a finite subset S' of S such that $I = R(S')$.

Exercise 23 We say that the ring A satisfies the ascending chain condition (acc) for real ideals if the following property is satisfied: when $(I_n)_{n \in \mathbb{N}}$ is an increasing sequence of real ideals of A, there exists $n_0 \in \mathbb{N}$ such that for each $n \geq n_0$, $I_n = I_{n_0}$.

(1) Show that A satisfies the acc for real deals if and only if each real ideal of A is really finitely generated.

(2) Let $\phi : A \longrightarrow B$ be a homomorphism onto between rings. Show that if A satisfies the acc for real ideals, then the same is true for B.

(3) Show that A satisfies the acc for real ideals if and only if each real prime ideal of A is really finitely generated.

(4) Show that if I is a real ideal of A, then $I[X]$ is a real ideal of $A[X]$.

(5) Show that if the real ideal I of A is really finitely generated, then the real ideal $I[X]$ is also really finitely generated.

(6) Show that A satisfies the acc for the real ideals if and only if $A[X]$ satisfies this property.

(7) If A satisfies the acc for real ideals, show that each $A-$algebra of finite type satisfies the acc for the real ideals.

Exercise 24

(1) Let A be a ring. Show that if $A[[X]]$ satisfies the acc for real ideals, then so is A.

In the sequel, A is a valuation domain defined by a valuation v, with value group $\mathbb{R}$, maximal ideal M, and residual field $A/M \simeq \mathbb{R}$.

(2) What are the real ideals of A? Deduce that A satisfies the acc for the real ideals.

(3) Give an example of a valuation domain satisfying the preceding hypotheses.

(4) Show that $XM[[X]]$ is a real ideal of $A[[X]]$.

(5) Let $\delta > 0$ be a real number, $k \geq 1$, $h \geq 0$ natural integers and $c_1, \ldots, c_k$, $b_1, \ldots, b_h \in A$ be such that $v(b_i) \geq \delta$, for $1 \leq i \leq h$ and $v\left(c_1^2 + \ldots + c_k^2 + b_1 + \ldots + b_h\right) \geq \delta$.

Show that $v(c_j) \geq \frac{\delta}{2}$, and pour $1 \leq j \leq k$.

(6) If Y is a nonempty subset of $A[[X]]$ and $n \in \mathbb{N}$, we denote by $\gamma_n(Y) = \inf\{v(s_i);\ S = \sum_{j:0}^{\infty} s_j X^j \in Y,\ 0 \leq i \leq n\}$. Let J be an ideal of $A[[X]]$ such that $J \subseteq XM[[X]]$ and $\gamma_n(J) > 0$, for each $n \in \mathbb{N}$. Show that for each $n \geq 1$ and each $k \geq 1$, $\gamma_n(R_k(J)) \geq \frac{1}{2^{(2^k-1)n}}\gamma_{2^k n}(J)$.

(7) Let $T = \sum_{i:0}^{\infty} t_i X^i \in XM[[X]]$. Show that for each $n \in \mathbb{N}$, $\gamma_n(TA[[X]]) > 0$.

(8) Show that the ideal $XM[[X]]$ of $A[[X]]$ is not really finitely generated. Deduce that $A[[X]]$ does not satisfy the acc for the real ideals.

Exercise 25 Let A be an integral domain and I a nonzero integral ideal of A. Show the $A-$modules I^{-1} and $Hom_A(I, A)$ are isomorphic.

Exercise 26 Let A be a fractional domain and I an integral divisorial ideal of A containing a power of its radical. Show that the ideal $\sqrt{I}$ is also divisorial.

Exercise 27 Let A be an integral domain with quotient field K and $\mathscr{F}(A)$ the set of the nonzero fractional ideals of A. An integral ideal J of A is said to be Glaz-Vasconcelos (or GV-ideal) if it is finitely generated and satisfies $J^{-1} = A$. We denote by $GV(A)$ the set of the GV-ideals of the domain A.

(1) Show that if $J_1, J_2 \in GV(A)$, then the product $J_1 J_2 \in GV(A)$.

(2) Let $I \in \mathscr{F}(A)$. Show that $I_w = \bigcup\limits_{J \in GV(A)} (I : J) \in \mathscr{F}(A)$.

(3) Show that the map $w : \mathscr{F}(A) \longrightarrow \mathscr{F}(A)$, defined by $w(I) = I_w$ for each $I \in \mathscr{F}(A)$, is a star operation of finite character; i.e., for each $I \in \mathscr{F}(A)$, $I_w = \cup J_w$, where J runs over the finitely generated nonzero fractional subideals of I.

(4) Show that every proper integral $w-$ideal of A is contained in a $w-$maximal ideal of A, i.e., maximal in the set of the proper integral $w-$ideals of A.

(5) (a) Let $I_1, I_2 \in \mathscr{F}(A)$. Show that $(I_1 \cap I_2)_w = (I_1)_w \cap (I_2)_w$. Deduce that the intersection of a finite number of fractional $w-$ideals of A is a $w-$ideal.

    (b) More generally, let $(I_\alpha)_{\alpha \in \Lambda}$ be a nonempty family of fractional $w-$ideals of A. Show that if $I = \bigcap\limits_{\alpha \in \Lambda} I_\alpha \neq (0)$, then it is a $w-$ideal.

    (c) Deduce that for each nonzero fractional ideal I, then I^{-1} is a $w-$ideal.

(6) Let I be a fractional $w-$ideal of A. Show that $I = \bigcap \{IA_M;\ M \in w - Max(A)\}$.

    Deduce that $A = \bigcap \{A_M;\ M \in w - Max(A)\}$.

(7) Let $I, J \in \mathscr{F}(A)$. Show that $(IJ)_w = (IJ_w)_w = (I_w J_w)_w$. Deduce that $I_w J_w \subseteq (IJ)_w$.

(8) Let $I \in \mathscr{F}(A)$.

    (a) Show that $(I^{-1})_w = I^{-1} = (I_w)^{-1}$.

    (b) Show that $I_v = (I_w)_v$. Deduce that $I_w \subseteq I_v$ and each $v-$ideal is a $w-$ideal.

    (c) Deduce that $I \subseteq I_w \subseteq I_t$ and each fractional $t-$ideal is a $w-$ideal.

(9) Let $I \in \mathscr{F}(A)$ be $w-$finite; i.e., $I_w = F_w$ with $F \in \mathscr{F}(A)$ a finitely generated ideal. Show that we can choose $F \subseteq I$.

Exercise 28 (Fanggui and McCasland 1997) Let A be an integral domain.

(1) Let I an integral $w-$ideal of A. Show that $\sqrt{I}$ is also an integral $w-$ideal.

(2) Let I an integral ideal of A. Show that $(\sqrt{I})_w = \sqrt{I_w}$.

Solutions

Exercise 1 Let $I_1, \ldots, I_n$ be S-finite ideals of A and $I = I_1 + \ldots + I_n$. There exist $s_1, \ldots, s_n \in S$ and finitely generated ideals $F_1, \ldots, F_n$ of A such that $s_i I_i \subseteq F_i \subseteq$

I_i, for $1 \leq i \leq n$. Then $s = s_1 \ldots s_n \in S$, the ideal $F = F_1 + \ldots + F_n$ is finitely generated, and $sI \subseteq F \subseteq I$.

Exercise 2 $" \Longrightarrow "$ Clear. $" \Longleftarrow "$ Let I be an ideal of A. There exist $s \in S_2$ and a finitely generated ideal J of A such that $sI \subseteq J \subseteq I$. By hypothesis, there exists $r \in S_2$ such that $sr \in S_1$. Then $(sr)I \subseteq sI \subseteq J \subseteq I$. Then A is S_1-Noetherian.

Exercise 3

(1) $" \Longrightarrow "$ Fix an index $i \in \{1, \ldots, n\}$. Let I_i be an ideal of A_i. Consider the ideal $I = 0 \times \ldots \times 0 \times I_i \times 0 \ldots \times 0$ of A. There exist $s = (s_1, \ldots, s_n) \in S$ and a finitely generated ideal F of A such that $sI \subseteq F \subseteq I$. We know $F = F_1 \times \ldots \times F_n$ with F_j an ideal of A_j, $1 \leq j \leq n$ (see Lemma 2.2, Chap. 2). Each F_j is necessarily zero for $j \neq i$ and F_i is finitely generated with $s_i I_i \subseteq F_i \subseteq I_i$. Then I_i is S_i-finite in A_i.

$" \Longleftarrow "$ An ideal of A is of the form $I = I_1 \times \ldots \times I_n$ with I_i an ideal of A_i (see Lemma 2.2, Chap. 2). For each i, there exist $s_i \in S_i$ and a finitely generated ideal F_i of A_i such that $s_i I_i \subseteq F_i \subseteq I_i$. Then $s = (s_1, \ldots, s_n) \in S$, $F = F_1 \times \ldots \times F_n$ is a finitely generated ideal of A and we have $sI \subseteq F \subseteq I$.

(2) The ring $\mathbb{Z}$ is $\{1\}$-Noetherian. Let $A = \prod_{i:1}^{\infty} \mathbb{Z}$ and $S = \prod_{i:1}^{\infty} \{1\} = \{s\}$ with $s = (1, 1, \ldots)$. The ideal I of the elements of A with finite supports is not finitely generated. Suppose that I is S-finite. There exists a finitely generated ideal F of A such that $sI \subseteq F \subseteq I$. Since $sI = I$, then $I = F$ is finitely generated, which is absurd.

Exercise 4

(1) There exist $i \in \{1, \ldots, n\}$ and $k \in \mathbb{N}$ such that for each integer $l \geq k$, $s_i N_l \subseteq N_k$. Then $sN_l = s_1 \ldots s_n N_l \subseteq s_i N_l \subseteq N_k$.
(2) The same.

Exercise 5 $"(1) \Longrightarrow (2)"$ Let IM be an extended sub-module of M with I an ideal of A. Let $\mathscr{F}$ be the set of the S-finite extended sub-modules of M contained in IM. Since $(0) \in \mathscr{F}$, $\mathscr{F} \neq \emptyset$. By hypothesis, $\mathscr{F}$ has a S-maximal element LM with L an ideal of A. There exists $s' \in S$ such that if $JM \in \mathscr{F}$, with J an ideal of A, and $LM \subseteq JM$, then $s'(JM) \subseteq LM$. Since LM is S-finite, there exist $s \in S$ and $x_1, \ldots, x_m \in LM \subseteq IM$ such that $s(LM) \subseteq F = Ax_1 + \ldots + Ax_m \subseteq IM$. We will prove that $ss'(IM) \subseteq F$. Let $\alpha \in IM$. If $\alpha \in LM$, then $ss'\alpha \in F$. If $\alpha \notin LM$, $\alpha = y_1 m_1 + \ldots + y_r m_r$ with $y_i \in I$ and $m_i \in M$. Let $Q = LM + y_1 M + \ldots + y_r M$. Then $Q \subseteq IM$ and Q is S-finite. Indeed, since M is S-finite, there exist $t \in S$ and a finitely generated sub-module T of M such that $tM \subseteq T$. Then $t(y_1 M + \ldots + y_r M) = y_1 tM + \ldots + y_r tM \subseteq y_1 T + \ldots + y_r T \subseteq y_1 M + \ldots + y_r M$ with $y_1 T + \ldots + y_r T$ a finitely generated sub-module of M. Then the sub-module $y_1 M + \ldots + y_r M$ is S-finite and Q is S-finite. Since $Q = (L + y_1 A + \ldots + y_r A)M$, then Q is extended. So $Q \in \mathscr{F}$ and $LM \subseteq Q$. By S-maximality of Q in $\mathscr{F}$, $s'Q \subseteq LM$. Then $ss'Q \subseteq s(LM) \subseteq F$. In particular, $ss'\alpha \in F$. Then $ss'(IM) \subseteq F \subseteq IM$ and so IM is S-finite.

$"(2) \Longrightarrow (3)"$ Clear. $"(3) \Longleftrightarrow (4)"$ Theorem 1.3.

$"(4) \Longrightarrow (5)"$ Let $(I_n M)_{n \in \mathbb{N}}$ be an increasing sequence of extended sub-modules of M. Consider the sub-module $N = \bigcup_{n:0}^{\infty} I_n M$ of M and the ideal $J = \sum_{n:0}^{\infty} I_n$ of A. Then $N = JM$. Since M is S-Noetherian, N is S-finite. There exist $s \in S$ and $a_1, \ldots, a_p \in N$ such that $sN \subseteq F = Aa_1 + \ldots + Aa_p$. For $1 \leq i \leq p$, there exists $n_i \in \mathbb{N}$ such that $a_i \in I_{n_i} M$. Let $k = max\{n_i; \ 1 \leq i \leq p\}$. Then $a_1, \ldots, a_p \in I_k M$. For each integer $n \geq k$, $sI_n M \subseteq sN \subseteq F \subseteq I_k M$. Then the sequence $(I_n M)_{n \in \mathbb{N}}$ is S-stationary.

$"(5) \Longrightarrow (1)"$ First case: $S = \{s_1, \ldots, s_n\}$ is finite. Let $s = s_1 \ldots s_n \in S$. Let $\mathscr{F}$ be a nonempty family of extended sub-modules of M. Suppose that $\mathscr{F}$ does not have a S-maximal element. Let $I_0 M \in \mathscr{F}$. It is not S-maximal. There exists $I_1 M \in \mathscr{F}$ such that $I_0 M \subseteq I_1 M$ and $sI_1 M \nsubseteq I_0 M$. By induction, we construct an increasing sequence $(I_k M)_{k \in \mathbb{N}}$ of extended sub-modules of M which is not S-stationary.

Second case: $S = (s_n)_{n \in \mathbb{N}}$ is countable and the ring A is chained. Let $\mathscr{F}$ be a nonempty family of extended sub-modules of M. Suppose that $\mathscr{F}$ does not have a S-maximal element. Let $I_0 M \in \mathscr{F}$. It is not S-maximal. There exists $I_1 M \in \mathscr{F}$ such that $I_0 M \subseteq I_1 M$ and $s_1 I_1 M \nsubseteq I_0 M$. Also, $I_1 M$ is not S-maximal. There exist $J_1 M$ and $J_2 M \in \mathscr{F}$ such that $I_1 M \subseteq J_1 M$ and $s_1 J_1 M \nsubseteq I_1 M$ and $I_1 M \subseteq J_2 M$ and $s_2 J_2 M \nsubseteq I_1 M$. The ideals J_1 and J_2 of A are comparable. Let I_2 be the biggest one. Then $I_1 M \subseteq I_2 M$ and $s_1 J_1 M \subseteq s_1 I_2 M$, $s_2 J_2 M \subseteq s_2 I_2 M$, so $s_1 I_2 M \nsubseteq I_1 M$ and $s_2 I_2 M \nsubseteq I_1 M$. Suppose constructed elements $I_0 M \subseteq I_1 M \subseteq \ldots \subseteq I_n M$ of $\mathscr{F}$ such that for $1 \leq k \leq n$ and $1 \leq i \leq k$, $s_i I_k M \nsubseteq I_{k-1} M$. Since $I_n M$ is not S-maximal, for each s_i, $1 \leq i \leq n+1$, there exists $J_i M \in \mathscr{F}$ such that $I_n M \subseteq J_i M$ and $s_i J_i M \nsubseteq I_n M$. The ideals $J_1, \ldots, J_{n+1}$ of A are comparable. Let I_{n+1} be the biggest one. Then $I_n M \subseteq I_{n+1} M$ and $s_i J_i M \subseteq s_i I_{n+1} M$, so $s_i I_{n+1} M \nsubseteq I_n M$ for $1 \leq i \leq n + 1$. We will prove that the sequence $(I_n M)_{n \in \mathbb{N}}$ is not S-stationary. Suppose that there exist $s = s_m \in S$, $m \in \mathbb{N}^*$, and $k \in \mathbb{N}$ such that for each integer $n \geq k$, $sI_n M \subseteq I_k M$. For $n > max\{k, m\}$, $s_m I_n M \subseteq I_k M \subseteq I_{n-1} M$, which is absurd.

Exercise 6 $" \Longrightarrow "$ By Lemma 5.4, the ring $S^{-1}A$ is Noetherian. Since S does not contain zero divisors, $A \subseteq S^{-1}A$. Let I be any ideal of A. Then $(S^{-1}I) \cap A$ is an ideal of A, so S-finite. There exist $s_1 \in S$ and a finitely generated ideal J of A such that $s_1\big((S^{-1}I) \cap A\big) \subseteq J \subseteq (S^{-1}I) \cap A$. Since J is finitely generated, there exists $s_2 \in S$ such that $s_2 J \subseteq I$. Let $s = s_1 s_2 \in S$. Then $s\big((S^{-1}I) \cap A\big) \subseteq s_2 J \subseteq I$, so $(S^{-1}I) \cap A \subseteq (I : s)_A$. Conversely, let $x \in (I : s)_A$. Then $sx \in I$, so $x = \frac{sx}{s} \in (S^{-1}I) \cap A$. Then $(S^{-1}I) \cap A = (I : s)_A$.

$" \Longleftarrow "$ Let I be any ideal of A. Since $S^{-1}A$ is Noetherian, the ideal $S^{-1}I$ is finitely generated. Then $S^{-1}I = S^{-1}J$ with J a finitely generated ideal of A

contained in I. Then $I \subseteq (S^{-1}I) \cap A = (S^{-1}J) \cap A = (J : s)_A$ with $s \in S$. Then $sI \subseteq J \subseteq I$ and so I is S-finite.

Exercise 7 Let $\phi : M' \longrightarrow M$ be the homomorphism one to one and $\psi : M \longrightarrow M''$ the homomorphism onto of the exact sequence. We conclude by Theorem 3.1 as follows:

$" \Longrightarrow "$ The module $M' \simeq \phi(M') \subseteq M$, so M' is S-Noetherian. Also, $M'' \simeq M/\ker \psi$, which is S-Noetherian.

$" \Longleftarrow "$ Since the A−modules $\ker \psi = \phi(M') \simeq M'$ and $M/\ker \psi \simeq M''$ are S-Noetherian, then M is S-Noetherian.

Exercise 8

(1) Let $f = a_0 + a_1 X + \ldots + a_k X^k \in A_n[X]$. Then $a_i \in A_n$ for $0 \le i \le k$. Then $X^n f = a_0 X^n + a_1 X^{n+1} + \ldots + a_k X^{n+k}$ with $a_i \in A_n \subseteq A_{n+i}$ for $0 \le i \le k$. Then $X^n f \in \mathscr{A}[X]$. So $X^n A_n[X] \subseteq \mathscr{A}[X]$.

(2) It is clear that S is a multiplicative set of $\mathscr{A}[X]$. Let $n \in \mathbb{N}$ and $f \in \mathscr{A}[X]$ such that $X^n f = 0$. Then $f = 0$, so X^n is regular. Then $\mathscr{A}[X] \subseteq S^{-1}\mathscr{A}[X]$. Let $f \in A[X]$. There exists $n \in \mathbb{N}$ such that $f \in A_n[X]$. By (1), $X^n f \in \mathscr{A}[X]$, and then $f \in S^{-1}\mathscr{A}[X]$ and so $A[X] \subseteq S^{-1}\mathscr{A}[X]$. Then $\mathscr{A}[X] \subseteq A[X] \subseteq S^{-1}\mathscr{A}[X]$.

(3) Let P be a prime ideal of A_0 and $J_P = \{f \in \mathscr{A}[X]; \ f(0) \in P\}$. We will prove that J_P is a prime ideal of $\mathscr{A}[X]$ containing X. The map $\phi : \mathscr{A}[X] \longrightarrow A_0$, defined by $\phi(f) = f(0)$ for each $f \in \mathscr{A}[X]$, is a homomorphism onto of rings and $\phi^{-1}(P) = J_P$. Then J_P is a prime ideal of $\mathscr{A}[X]$. Moreover, $X \in J_P$ because its constant term is zero.

Conversely, let $\mathscr{P}$ be a prime ideal of $\mathscr{A}[X]$ containing X and $P = \mathscr{P} \cap A_0 \in spec(A_0)$. We will show that $\mathscr{P} = J_P$. Let $f \in \mathscr{P}$, $f = a_0 + Xh$ with $a_0 \in A_0$ and $Xh \in \mathscr{A}[X]$. There exists $n \in \mathbb{N}$ such that $h \in A_n[X]$, and then $X^n h^{n+1} \in X^n A_n[X] \subseteq \mathscr{A}[X]$ and $X(X^n h^{n+1}) \in \mathscr{P}$ so $(Xh)^{n+1}) \in \mathscr{P}$ with $Xh \in \mathscr{A}[X]$. Since $\mathscr{P}$ is a prime ideal of $\mathscr{A}[X]$, then $Xh \in \mathscr{P}$. So $a_0 = f - Xh \in \mathscr{P}$, and then $a_0 \in \mathscr{P} \cap A_0 = P$ and $f \in J_P$. Then $\mathscr{P} \subseteq J_P$. Let now $f \in J_P \subseteq \mathscr{A}[X]$, $f = a_0 + Xh$ with $a_0 \in P \subseteq \mathscr{P}$ and $Xh \in \mathscr{A}[X]$. There exists $n \in \mathbb{N}$ such that $h \in A_n[X]$, and then $X^n h^{n+1} \in X^n A_n[X] \subseteq \mathscr{A}[X]$. So $X(X^n h^{n+1}) \in \mathscr{P}$, and then $(Xh)^{n+1}) \in \mathscr{P}$ with $Xh \in \mathscr{A}[X]$. Since $\mathscr{P}$ is a prime ideal of $\mathscr{A}[X]$, then $Xh \in \mathscr{P}$. So $f = a_0 + Xh \in \mathscr{P}$, then $J_P \subseteq \mathscr{P}$.

(4) Let $\mathscr{P}$ be a prime ideal of $A[X]$ not containing X. Then $\mathscr{P} \cap \mathscr{A}[X]$ is a prime ideal of $\mathscr{A}[X]$ not containing X. Conversely, let P be a prime ideal of $\mathscr{A}[X]$ not containing X. Then $P \cap S = \emptyset$ with $S = \{X^n; \ n \in \mathbb{N}\}$ a multiplicative set of $\mathscr{A}[X]$. Then $S^{-1}P$ is a prime ideal of $S^{-1}\mathscr{A}[X]$. Then $(S^{-1}P) \cap A[X]$ is a prime ideal of $A[X]$ not containing X because if not, $X = \frac{f}{X^n}$ with $n \in \mathbb{N}$ and $f \in P \subseteq \mathscr{A}[X]$, and then $X^{n+1} = f \in P$, which is absurd. We will prove now that $((S^{-1}P) \cap A[X]) \cap \mathscr{A}[X] = P$ or $(S^{-1}P) \cap \mathscr{A}[X] = P$. An element of the intersection can be written as $f = \frac{g}{X^n}$ with $f \in \mathscr{A}[X]$, $g \in P$, and $n \in \mathbb{N}$.

Then $X^n f = g \in P$ with $X^n \notin P$, and then $f \in P$. The converse inclusion is clear.

Exercise 9

(1) Let $f = \sum_{i:0}^{\infty} a_i X^i$ and $g = \sum_{i:0}^{\infty} b_i X^i \in \mathscr{A}[[X]]$. Then $fg = \sum_{n:0}^{\infty} c_n X^n$ with

$c_n = \sum_{i+j=n} a_i b_j$. Since $a_i \in A_i \subseteq A_n$ and $b_j \in A_j \subseteq A_n$, then $c_n \in A_n$ and $fg \in \mathscr{A}[[X]]$.

(2) $" \Longrightarrow "$ Let $g = \sum_{i:0}^{\infty} b_i X^i \in \mathscr{A}[[X]]$ such that $fg = 1$. Then $a_0 b_0 = 1$ and $a_0 \in U(A_0)$.

$" \Longleftarrow "$ We construct $g = \sum_{i:0}^{\infty} b_i X^i \in \mathscr{A}[[X]]$ in such a way that $fg = 1$.

We have $b_0 = a_0^{-1} \in A_0$ and $b_n = -a_0^{-1}(a_n b_0 + a_{n-1} b_1 + \ldots + a_1 b_{n-1}) \in A_n$.

(3) The second equivalence is clear because $A = \bigcup_{n:0}^{\infty} A_n$.

$" \Longrightarrow "$ Let $a, b \in A$ such that $ab = 0$. There exist $n, m \in \mathbb{N}$ such that $a \in A_n$ and $h \in A_m$. Let $f = aX^n$ and $g = bX^m \in \mathscr{A}[[X]]$. Then $fg = abX^{n+m} = 0$. So $f = 0$ or $g = 0$ and then $a = 0$ or $b = 0$.

$" \Longleftarrow "$ Since A is an integral domain, $A[[X]]$ is integral. Since $\mathscr{A}[[X]] \subseteq A[[X]]$, then $\mathscr{A}[[X]]$ is integral.

(4) Let $f = \sum_{i:0}^{\infty} a_i X^i \in Nil(\mathscr{A}[[X]]) \subseteq Nil(A[[X]]) \subseteq Nil(A)[[X]])$. Then for each $i \in \mathbb{N}$, $a_i \in Nil(A) \cap A_i = Nil(A_i)$.

The second equivalence is clear.

$" \Longrightarrow "$ Let $a \in Nil(A)$, there exists $i \in \mathbb{N}$ such that $a \in A_i$, and then $aX^i \in Nil(\mathscr{A}[[X]])$. So $aX^i = 0$ and then $a = 0$.

$" \Longleftarrow "$ If A is reduced, $A[[X]]$ is also reduced, and then $\mathscr{A}[[X]]$ is reduced.

(5) (a) The stability by addition is clear. Let $a \in I$, $b \in A_0$, and $f, g \in \mathscr{B}[[X]]$. Then $(a + Xf)(b + Xg) = ab + X(bf + ag + Xfg) \in I + X\mathscr{B}[[X]]$.

(b) Let $\phi : \mathscr{A}[[X]] \longrightarrow A_0/I$, the map defined for $f = \sum_{i:0}^{\infty} a_i X^i$ by $\phi(f) = \bar{a}_0$. Then ϕ is a homomorphism onto with kernel $I + X\mathscr{B}[[X]]$. Then $\mathscr{A}[[X]]/I + X\mathscr{B}[[X]] \simeq A_0/I$. So $I + X\mathscr{B}[[X]]$ is prime (resp. maximal, resp. radical) if and only if I is prime (resp. maximal, resp. radical).

(c) $\sqrt{I} + X\mathscr{B}[[X]]$ is a radical ideal of $\mathscr{A}[[X]]$ containing $I + X\mathscr{B}[[X]]$. Then $\sqrt{I + X\mathscr{B}[[X]]} \subseteq \sqrt{I} + X\mathscr{B}[[X]]$.

(6) For each $\mathscr{M} \in Max(A_0)$, $\mathscr{M} + X\mathscr{B}[[X]] \in Max(\mathscr{A}[[X]])$.

Conversely, let $M \in Max(\mathscr{A}[[X]])$ and $\mathscr{M} = \{f(0); \ f \in M\}$. Then $\mathscr{M}$ is a proper ideal of A_0, by the characterization of the unities of $\mathscr{A}[[X]]$ and

$M \subseteq \mathcal{M} + X\mathcal{B}[[X]]$. Since M is maximal, we have the equality. Since M is maximal, $\mathcal{M}$ is also maximal.

(7) Let $I = \{f(0);\ f \in P\}$: ideal of A_0. Then $P \subseteq I + X\mathcal{B}[[X]]$. Conversely, if $f \in \mathcal{B}[[X]]$, then $(Xf)^2 = X(Xf^2) \in P$ because $X \in P$ and $Xf^2 \in \mathcal{A}[[X]]$. Then $Xf \in P$ so $X\mathcal{B}[[X]] \subseteq P$. Let $a \in I$. There exists $g \in P$ such that $g(0) = a$. Then $g = a + Xf$ with $f \in \mathcal{B}[[X]]$. So $a = g - Xf \in P$ because $Xf \in X\mathcal{B}[[X]] \subseteq P$. Then $I + X\mathcal{B}[[X]] \subseteq P$ and we have the equality. Since P is prime, I is also prime in A_0.

(8) Let P be a nonzero prime ideal of $\mathcal{A}[[X]]$ and $0 \neq f = a_n X^n + \ldots \in P$ with $a_n \neq 0$. Then $f = X^n g$ with $g \in A_n + XA_{n+1} + \ldots$ and invertible by (2) because A_n is a field. Then $X^n g^{-1} \in \mathcal{A}[[X]]$ and $X^{2n} = X^{2n} g g^{-1} = f(X^n g^{-1}) \in P$, and then $X \in P$. By (7), $P = I + X\mathcal{B}[[X]]$ with I a prime ideal of the field A_0. Then $I = (0)$ and $P = X\mathcal{B}[[X]]$.

(9) (a) Let $x, y \in P$. There exists $i \in \mathbb{N}$ such that $x, y \in P_i$. Then $x - y \in P_i \subseteq P$. Let $a \in A$, and there exist $i_1 \geq i$ such that $a \in A_{i_1}$. Since $x \in P_i \subseteq P_{i_1}$, then $ax \in P_{i_1} \subseteq P$. So P is an ideal of A.

Let $x, y \in A$ such that $xy \in P$ and $x \notin P$. For each i, $x \notin P_i$. There exists i_0 such that $xy \in P_{i_0}$. Let $i_1 \geq i_0$ such that $x, y \in A_{i_1}$. Then $xy \in P_{i_0} \subseteq P_{i_1}$: a prime ideal of A_{i_1} with $x \notin P_{i_1}$, and then $y \in P_{i_1} \subseteq P$. Then P is a prime ideal of A.

(b) By hypothesis, $P_n = A_n \cap P_{n+1}$. Put the induction hypothesis $P_n = A_n \cap P_{n+k}$. Then $P_n = A_n \cap P_{n+k} = A_n \cap (A_{n+k} \cap P_{n+k+1}) = (A_n \cap P_{n+k+1}) \cap A_{n+k} = A_n \cap P_{n+k+1}$. Then we have the result. The sequence $(P_m)_{m \in \mathbb{N}}$ is increasing. Then $A_n \cap P = A_n \cap (\bigcup_{m:0}^{\infty} P_m) = A_n \cap (\bigcup_{m:n}^{\infty} P_m) = \bigcup_{m:n}^{\infty} (A_n \cap$

$$P_m) = \bigcup_{m:n}^{\infty} P_n = P_n.$$

(c) $P[[X]] \cap \mathcal{A}[[X]] = \{f = \sum_{n:0}^{\infty} a_n X^n;\ a_n \in P \cap A_n\} = \{f =$

$\sum_{n:0}^{\infty} a_n X^n;\ a_n \in P_n\} = \mathcal{P}[[X]]$. Since $P \in spec(A)$, then $P[[X]] \in spec(A[[X]])$, so $\mathcal{P}[[X]] = P[[X]] \cap \mathcal{A}[[X]]$ is a prime ideal of $\mathcal{A}[[X]]$. Since $1 \notin P$, then $X \notin P[[X]]$ so $X \notin \mathcal{P}[[X]]$.

(10) $P_n = A_n \cap P = (A_n \cap A_{n+1}) \cap P = A_n \cap (A_{n+1} \cap P) = A_n \cap P_{n+1}$ and $P = P \cap A = P \cap (\bigcup_{n:0}^{\infty} A_n) = \bigcup_{n:0}^{\infty} (P \cap A_n) = \bigcup_{n:0}^{\infty} P_n$. We conclude by the preceding question.

(11) Let P be a prime ideal of A containing I, $P_n = A_n \cap P \supseteq A_n \cap I = I_n$, $\mathcal{I} = (I_n)_{n \in \mathbb{N}}$ and $\mathcal{P} = (P_n)_{n \in \mathbb{N}}$. By the preceding question, $\mathcal{P}[[X]] = P[[X]] \cap \mathcal{A}[[X]]$ is a prime ideal of $\mathcal{A}[[X]]$ and $\mathcal{I}[[X]] \subseteq \mathcal{P}[[X]]$. Then $\sqrt{\mathcal{I}[[X]]} \subseteq \mathcal{P}[[X]] \subseteq P[[X]]$. This is for each $P \in spec(A)$. Then $\sqrt{\mathcal{I}[[X]]} \subseteq \bigcap\{P[[X]];\ I \subseteq P \in spec(A)\} = \sqrt{I}[[X]]$.

So $\sqrt{\mathscr{I}[[X]]} \subseteq \sqrt{I}[[X]] \cap \mathscr{A}[[X]]$.

(12) (a) Since $\mathscr{A}[[X]] \subseteq A_n[[X]]$, then $qf(\mathscr{A}[[X]]) \subseteq qf(A_n[[X]])$. Conversely,
let $f = \sum_{i:0}^{\infty} a_i X^i \in A_n[[X]]$. Then $f = \sum_{i:0}^{n-1} a_i X^i + \sum_{i:n}^{\infty} a_i X^i$ with

$\sum_{i:n}^{\infty} a_i X^i \in \mathscr{A}[[X]]$ and $\sum_{i:0}^{n-1} a_i X^i = \sum_{i:0}^{n-1} \frac{a_i X^n}{X^{n-i}} \in qf(\mathscr{A}[[X]])$. Then
$A_n[[X]] \subseteq qf(\mathscr{A}[[X]])$ so $qf(A_n[[X]]) \subseteq qf(\mathscr{A}[[X]])$.

(b) $'' \Longrightarrow ''$

(i) Let $n \geq 1$ be an integer and $a \in A_n$. Then $aX = \frac{aX^n}{X^{n-1}} \in qf(\mathscr{A}[[X]])$ and $(aX)^n = a^n X^n \in \mathscr{A}[[X]]$. Then aX is integral over $\mathscr{A}[[X]]$, so $aX \in \mathscr{A}[[X]]$ and $a \in A_1$. Then $A_n = A_1$.

(ii) Let $a \in A_1$ be an integral element over A_0. Then a is integral over $\mathscr{A}[[X]]$ with $a \in A_1 \subset A_1[[X]] \subseteq qf(\mathscr{A}[[X]])$, by (a). Then $a \in \mathscr{A}[[X]]$ so $a \in A_0$ because a is a constant series.

(iii) Let $f \in qf(A_1[[X]]) = qf(\mathscr{A}[[X]]) \subseteq K((X))$ be an integral element over $A_1[[X]] \subseteq K[[X]]$, where K is the quotient field of A. Since $K[[X]]$ is integrally closed, then $f \in K[[X]]$ and the order $\omega(f) \geq 0$. There exist $g_0, \ldots, g_{n-1} \in A_1[[X]]$ such that $f^n + g_{n-1} f^{n-1} + \ldots + g_1 f + g_0 = 0$. We multiply by X^n, and we obtain $(Xf)^n + X g_{n-1}(Xf)^{n-1} + \ldots + X^{n-1} g_1(Xf) + X^n g_0 = 0$ with $Xf \in qf(\mathscr{A}[[X]])$ and $X^i g_j \in \mathscr{A}[[X]]$, $1 \leq i \leq n$. Then Xf is integral over $\mathscr{A}[[X]]$, so $Xf \in \mathscr{A}[[X]] \subseteq A_1[[X]]$, and then $f \in A_1[[X]]$ because $\omega(f) \geq 0$. Then $A_1[[X]]$ is integrally closed.

$'' \Longleftarrow ''$ Let $f \in qf(\mathscr{A}[[X]]) = qf(A_1[[X]])$ an integral element over $\mathscr{A}[[X]] \subseteq A_1[[X]]$ be integrally closed. Then $f \in A_1[[X]]$ and $f(0) \in A_1$. But f is integral over $\mathscr{A}[[X]]$, then $f(0)$ is integral over A_0 so $f(0) \in A_0$ and $f \in \mathscr{A}[[X]]$.

(c) $'' \Longrightarrow ''$ Let $n \in \mathbb{N}$ and $a \in A_n$. Then $a = \frac{aX^n}{X^n} \in qf(\mathscr{A}[[X]])$. For each $k \in \mathbb{N}$, $a^k X^n \in \mathscr{A}[[X]]$ with $0 \neq X^n \in \mathscr{A}[[X]]$. Since $\mathscr{A}[[X]]$ is completely integrally closed, $a \in \mathscr{A}[[X]]$ so $a \in A_0$. Then $A_n = A_0$.

$'' \Longleftarrow ''$ For each $n \in \mathbb{N}$, $A_n = A_0$ and then $\mathscr{A}[[X]] = A_0[[X]]$: completely integrally closed because A_0 is completely integrally closed. See Example 2 (1) to Corollary 14.2, Chap. 4.

Exercise 10

(1) $'' \Longrightarrow ''$ Let $b \in U(A_1) \cap A_0$. Then $X = b(\frac{1}{b}X)$ with b and $\frac{1}{b}X \in \mathscr{A}[[X]]$. Since X is irreducible in $\mathscr{A}[[X]]$, then $b \in U(\mathscr{A}[[X]]) = U(A_0)$.

$'' \Longleftarrow ''$ Let $f, g \in \mathscr{A}[[X]]$ such that $X = fg$. For example, $f = a_0 + a_1 X + \ldots$ and $g = b_1 X + \ldots$ with $a_0 \neq 0$ and $b_1 \neq 0$. Then $a_0 b_1 = 1$, so $a_0 \in U(A_1) \cap A_0 = U(A_0)$ and $f \in U(\mathscr{A}[[X]])$.

(2) $"\Longrightarrow"$ Suppose that the sequence $(A_n)_{n\in\mathbb{N}}$ is not constant. Let $n \geq 1$ be the least integer such that $A_n \neq A_0$. There exists $b \in A_n \setminus A_0$. Then X divides $(bX^n)(bX^n) = X(b^2X^{2n-1})$ in $\mathscr{A}[[X]]$ because $2n - 1 \geq n$. But X does not divide bX^n in $\mathscr{A}[[X]]$ because $bX^{n-1} \notin \mathscr{A}[[X]]$ since $b \notin A_{n-1}$.

$"\Longleftarrow"$ We have $\mathscr{A}[[X]] = A_0[[X]]$ with $A_0[[X]]/(X) \simeq A_0$: an integral domain.

Exercise 11 Let $n \geq 1$ be an integer and $b \in A_n$. Then X divides $(bX^n)(bX^n) = X(b^2X^{2n-1})$ in $\mathscr{A}[[X]]$ because $2n - 1 \geq n$. There exist $f, g \in \mathscr{A}[[X]]$ such that $X = fg$ with both f and g dividing bX^n. Put $bX^n = fu = gv$ with $u, v \in \mathscr{A}[[X]]$. We may suppose that $f(0) \neq 0$. Then $f = a_0 + a_1X + \ldots, g = b_1X + \ldots$ and $v = \beta_{n-1}X^{n-1} + \ldots$. Since $X = fg$, then $a_0b_1 = 1$ so $a_0 \in U(A_1) \cap A_0 = S$. Since $bX^n = gv$, then $b = b_1\beta_{n-1} = \frac{1}{a_0}\beta_{n-1} \in S^{-1}A_{n-1}$. So $A_n \subseteq S^{-1}A_{n-1}$ and then $A_n \subseteq S^{-1}A_0$. The inverse inclusion is clear because the elements of S are invertible in $A_1 \subseteq A_n$. Then for each integer $n \geq 1$, $A_n = S^{-1}A_0$ and $\mathscr{A}[[X]] = A_0 + (S^{-1}A_0)X + (S^{-1}A_0)X^2 + \ldots = A_0 + (S^{-1}A_0)[[X]]$.

Exercise 12

(1) $"\Longrightarrow"$ Fix an integer n. Let $b_1, b_2, \ldots \in A_n \setminus (0)$ such that $b_1A_0 \subseteq b_2A_0 \subseteq \ldots$. Since $b_1X^n, b_2X^n, \ldots \in \mathscr{A}[[X]]$, we have the increasing sequence $b_1X^n\mathscr{A}[[X]] \subseteq b_2X^n\mathscr{A}[[X]] \subseteq \ldots$ of principal ideals of $\mathscr{A}[[X]]$. There exist $k_0 \in \mathbb{N}$ such that for each integer $k \geq k_0$, $b_kX^n\mathscr{A}[[X]] = b_{k+1}X^n\mathscr{A}[[X]]$. Then $b_{k+1}X^n \in b_kX^n\mathscr{A}[[X]]$ so $b_{k+1} \in b_k\mathscr{A}[[X]]$. There exists $a \in A_0$ such that $b_{k+1} = b_ka \in b_kA_0$. The sequence $b_1A_0 \subseteq b_2A_0 \subseteq \ldots$ is stationary.

$"\Longleftarrow"$ Let $f_1\mathscr{A}[[X]] \subseteq f_2\mathscr{A}[[X]] \subseteq \ldots$ be an increasing sequence of principal ideals of $\mathscr{A}[[X]]$. Then $ord(f_{k+1}) \leq ord(f_k)$. By omitting a finite number of terms of the sequence, we may suppose that all the f_k have the same order. Let b_k be the initial coefficient of f_k. Then all the $b_k, k \in \mathbb{N}^*$, belong to the same ring A_n. Since $f_k = f_{k+1}g_k$ with $g_k \in \mathscr{A}[[X]]$ and $ord(f_{k+1}) = ord(f_k)$, the initial coefficient of g_k is necessarily in $A_0 \setminus (0)$ and $b_k \in b_{k+1}A_0$. Then we have the following increasing sequence $b_1A_0 \subseteq b_2A_0 \subseteq \ldots$ with $b_1, b_2, \ldots \in A_n$. By hypothesis, there exists $k_0 \in \mathbb{N}^*$ such that for each $k \geq k_0$, $b_kA_0 = b_{k+1}A_0$, and then $\frac{b_k}{b_{k+1}} \in U(A_0)$. So $g_k(0) \in U(A_0)$ and $g_k \in U(\mathscr{A}[[X]])$. Then $f_k\mathscr{A}[[X]] = f_{k+1}\mathscr{A}[[X]]$.

(2) If A_0 is a field and if $(0) \subset b_1A_0 \subseteq b_2A_0 \subseteq \ldots$ with $0 \neq b_k \in A_n$, then $b_k = b_{k+1}a$ with $0 \neq a \in A_0$. Then $b_{k+1} = \frac{1}{a}b_k \in b_kA_0$.

(3) $"\Longleftarrow"$ By (2). $"\Longrightarrow"$ Let $0 \neq a \in A_0$. Then $\frac{1}{a} \in A_1$ and $\frac{1}{a}A_0 \subseteq \frac{1}{a^2}A_0 \subseteq \ldots$. By (1), there exists k such that $\frac{1}{a^k}A_0 = \frac{1}{a^{k+1}}A_0$. Then $A_0 = \frac{1}{a}A_0$, so $\frac{1}{a} \in A_0$.

Exercise 13

(1) $" \implies "$ Let $a \in A_0 \setminus U(A_0)$ and $k \in \mathbb{N}$. Then $\bigcap\limits_{i:0}^{\infty} a^i \mathscr{A}[[X]] = (0)$. Let $b \in$

$\bigcap\limits_{i:0}^{\infty} a^i A_k$. For each $i \geq 0$, $b \in a^i A_k$, and then $b = a^i c_k$ with $c_k \in A_k$, so

$bX^k = a^i c_k X^k \in a^i \mathscr{A}[[X]]$, and then $bX^k \in \bigcap\limits_{i:0}^{\infty} a^i \mathscr{A}[[X]] = (0)$. So $bX^k = 0$

and $b = 0$. Then $\bigcap\limits_{i:0}^{\infty} a^i A_k = (0)$.

$" \impliedby "$ Let $f \in \mathscr{A}[[X]] \setminus U(\mathscr{A}[[X]])$. If $ord(f) \geq 1$, let $A = \bigcup\limits_{n:0}^{\infty} A_n$.

In this case, $\bigcap\limits_{i:0}^{\infty} f^i \mathscr{A}[[X]] \subseteq \bigcap\limits_{i:0}^{\infty} f^i A[[X]] = (0)$. If $ord(f) = 0$, let $f(0) =$

$a \in A_0 \setminus U(A_0)$, $a \neq 0$. Let $g \in \bigcap\limits_{i:0}^{\infty} f^i \mathscr{A}[[X]]$. If $g \neq 0$, $g = b_k X^k + \ldots$

with $b_k \neq 0$. For each $n \in \mathbb{N}$, there exists $g_n \in \mathscr{A}[[X]]$ such that $g = f^n g_n$,

$g_n = \sum\limits_{p \geq k} b_{p,n} X^p$ with $0 \neq b_{k,n} \in A_k$, and then $b_k = a^n b_{k,n} \in a^n A_k$. So

$b_k \in \bigcap\limits_{n:0}^{\infty} a^n A_k = (0)$, by hypothesis. Then $b_k = 0$, which is absurd.

(2) Let $a \in A_0 \setminus U(A_0)$ and $k \in \mathbb{N}$. There exists $0 \neq d_k \in A_0$ such that $d_k A_k \subseteq A_0$.

Then $d_k \left(\bigcap\limits_{i:0}^{\infty} a^i A_k \right) = \bigcap\limits_{i:0}^{\infty} (a^i d_k A_k) \subseteq \bigcap\limits_{i:0}^{\infty} a^i A_0 = (0)$, so $\bigcap\limits_{i:0}^{\infty} a^i A_k = (0)$. By

(1), $\mathscr{A}[[X]]$ is Archimedean.

(3) Let $a \in A_0 \setminus U(A_0)$ and $k \in \mathbb{N}$. Then $a \in A_k \setminus U(A_k)$, so $\bigcap\limits_{i:0}^{\infty} a^i A_k = (0)$

because A_k is Archimedean. We conclude by (1).

Exercise 14 $" \implies "$ By Corollary 7.7. $" \impliedby "$ By Corollary 7.7, it suffices that the sub-ring $f(A) + J$ of B is Noetherian. By Eakin-Nagata theorem, it suffices to show that B is a finitely generated $f(A) + J$−module. By hypothesis, $B/J = A.(\bar{b}_1, \ldots, \bar{b}_n)$, with $b_1, \ldots, b_n \in B$. For each $b \in B$, there exist $a_1, \ldots, a_n \in A$ such that $\bar{b} = a_1.\bar{b}_1 + \ldots + a_n.\bar{b}_n = \overline{f(a_1)b_1 + \ldots + f(a_n)b_n}$, so $b = f(a_1)b_1 + \ldots + f(a_n)b_n + j$ with $j \in J$. Then $B = (b_1, \ldots, b_n, 1)(f(A) + J)$.

Exercise 15

(1) $" \Longleftarrow "$ By Proposition 7.3 (3), since $f^{-1}(J) = (0)$, $A \bowtie^f J \simeq f(A) + J$ is an integral domain.

$" \Longrightarrow "$ Suppose that there exists $0 \neq a \in f^{-1}(J)$. Then $f(a) \in J$ and the element $(a, 0) = (a, f(a) - f(a)) \in (A \bowtie^f J) \setminus \{(0, 0)\}$. Let $0 \neq j \in J$. Then $(0, j) \in (A \bowtie^f J) \setminus \{(0, 0)\}$ and $(a, 0)(0, j) = (0, 0)$. This contradicts the fact that $A \bowtie^f J$ is an integral domain. Then $f^{-1}(J) = (0)$. In this case, $f(A) + J \simeq A \bowtie^f J$ is an integral domain.

Deduction: if B is an integral domain, its sub-ring $f(A) + J$ is also an integral domain. We conclude by the preceding equivalence.

(2) $" \Longrightarrow "$ By Proposition 7.3 (1), the map $i : A \longrightarrow A \bowtie^f J$, defined by $i(a) = ((a, f(a))$ for each $a \in A$, is one to one. This allows us to conclude that A is reduced.

Let $j \in Nil(B) \cap J$. There exists $n \in \mathbb{N}^*$ such that $j^n = 0$. Then $(0, j)^n = 0$ with $(0, j) \in A \bowtie^f J$. So $(0, j) \in Nil(A \bowtie^f J) = (0)$, and then $j = 0$. So $Nil(B) \cap J = (0)$.

$" \Longleftarrow "$ Let $(a, f(a) + j) \in Nil(A \bowtie^f J)$ with $a \in A$ and $j \in J$. There exists $n \in \mathbb{N}^*$ such that $(a, f(a) + j)^n = 0$. Then $a^n = 0$ and $(f(a) + j)^n = 0$. Then $a \in Nil(A) = (0)$ and $j^n = 0$, so $j \in Nil(B) \cap J = (0)$. Then $(a, f(a) + j) = 0$. So $A \bowtie^f J$ is reduced.

Deduction: if the rings A and B are reduced, then $Nil(B) \cap J = (0)$. By the preceding equivalence, the ring $A \bowtie^f J$ is reduced.

Suppose that J is a radical ideal of B and the ring $A \bowtie^f J$ is reduced. The natural homomorphism $A \longrightarrow A \bowtie^f J$ is one to one and allows us to say that the ring A is reduced. By the preceding equivalence, $Nil(B) \cap J = (0)$. Let $b \in Nil(B)$. There exists $n \in \mathbb{N}^*$ such that $b^n = 0 \in J$. Since J is a radical ideal, $b \in J$. Then $b \in Nil(B) \cap J = (0)$.

Exercise 16

(1) Let $(a, m) \in U(A(+)M)$. There exists $(b, n) \in A(+)M$ such that $(1, 0) = (a, m)(b, n) = (ab, bm + an)$, then $ab = 1$. So $a \in U(A)$ and $(a, m) \in U(A)(+)M$. Conversely, let $(a, m) \in U(A)(+)M$. There exists $b \in A$ such that $ab = 1$. We want an element $n \in M$ such that $(1, 0) = (a, m)(b, n) = (ab, bm + an)$, and then $bm + an = 0$. We multiply by b, and we find $b^2 m + abn = 0$, so $n = -b^2 m$. Now, we calculate $(a, m)(b, -b^2 m) = (ab, -ab^2 m + bm) = (1, -bm + bm) = (1, 0)$. Then $(a, m) \in U(A(+)M)$.

(2) By Lemma 8.9, the prime ideals of the ring $A(+)M$ are of the form $P(+)M$ with P a prime ideal of A. Then the maximal ideals of $A(+)M$ are of the form $\mathcal{M}(+)M$ with $\mathcal{M}$ a maximal ideal of A. So A is local with maximal ideal $\mathcal{M}$ if and only if $A(+)M$ is local with maximal ideal $\mathcal{M}(+)M$. In this case, the map $\phi : A(+)M \longrightarrow A/\mathcal{M}$, defined by $\phi(a, m) = \bar{a}$ for each $(a, m) \in A(+)M$, is a homomorphism onto with kernel $\mathcal{M}(+)M$. Then $A(+)M / \mathcal{M}(+)M \simeq A/\mathcal{M}$.

(3) For $\bar{a} \in \mathbb{Z}/4\mathbb{Z}$ and $\bar{x} \in \mathbb{Z}/2\mathbb{Z}$, let $\bar{a}\bar{x} = \overline{ax}$ in $\mathbb{Z}/2\mathbb{Z}$. This multiplication is well defined. Indeed, if $\bar{a} = \bar{b}$ in $\mathbb{Z}/4\mathbb{Z}$ and $\bar{x} = \bar{y}$ in $\mathbb{Z}/2\mathbb{Z}$, then $4/a - b$ and $2/x - y$. So $ax - by = a(x - y) + y(a - b)$ is divisible by 2 and so $\overline{ax} = \overline{by}$ in $\mathbb{Z}/2\mathbb{Z}$. The modules axioms are easy to see from the definition.

(4) The only prime ideal of $\mathbb{Z}$ containing $4\mathbb{Z}$ is $2\mathbb{Z}$. Then the only prime ideal of $\mathbb{Z}/4\mathbb{Z}$ is $2\mathbb{Z}/4\mathbb{Z} = (\bar{2}) = \{\bar{0}, \bar{2}\}$. So $\mathbb{Z}/4\mathbb{Z}$ is local with maximal ideal $(\bar{2})$. By (2), $\mathbb{Z}/4\mathbb{Z}(+)\mathbb{Z}/2\mathbb{Z}$ is local with maximal ideal $\mathscr{M} = (\bar{2})(+)\mathbb{Z}/2\mathbb{Z} = \{(\bar{0}, \bar{0}); (\bar{0}, \bar{1}); (\bar{2}, \bar{0}); (\bar{2}, \bar{1})\}$.

(5) To establish the list of the elements of the principal ideal $\mathscr{I}$, we must find all the multiples of the element $(\bar{2}, \bar{1})$ in the ring $\mathbb{Z}/4\mathbb{Z}(+)\mathbb{Z}/2\mathbb{Z}$:

$(\bar{0}, \bar{0})(\bar{2}, \bar{1}) = (\bar{0}, \bar{0}); (\bar{0}, \bar{1})(\bar{2}, \bar{1}) = (\bar{0}, \bar{0}); (\bar{1}, \bar{0})(\bar{2}, \bar{1}) = (\bar{2}, \bar{1}); (\bar{1}, \bar{1})(\bar{2}, \bar{1}) = (\bar{2}, \bar{1});$

$(\bar{2}, \bar{0})(\bar{2}, \bar{1}) = (\bar{0}, \bar{0}); (\bar{2}, \bar{1})(\bar{2}, \bar{1}) = (\bar{0}, \bar{0}); (\bar{3}, \bar{0})(\bar{2}, \bar{1}) = (\bar{2}, \bar{1}); (\bar{3}, \bar{1})(\bar{2}, \bar{1}) = (\bar{2}, \bar{1}).$

Then $\mathscr{I} = \{(\bar{0}, \bar{0}); (\bar{2}, \bar{1})\}$. Suppose that $\mathscr{I} = I(+)N$ with I an ideal of $\mathbb{Z}/4\mathbb{Z}$ and N a sub-module of $\mathbb{Z}/2\mathbb{Z}$. Let $\pi_1 : \mathbb{Z}/4\mathbb{Z}(+)\mathbb{Z}/2\mathbb{Z} \longrightarrow \mathbb{Z}/4\mathbb{Z}$ and $\pi_2 : \mathbb{Z}/4\mathbb{Z}(+)\mathbb{Z}/2\mathbb{Z} \longrightarrow \mathbb{Z}/2\mathbb{Z}$ the canonical projections. Then $\pi_1(\mathscr{I}) = I = \{\bar{0}, \bar{2}\} = (\bar{2})$ and $\pi_2(\mathscr{I}) = N = \{\bar{0}, \bar{1}\} = \mathbb{Z}/2\mathbb{Z}$. So $\mathscr{I} = I(+)N = (\bar{2})(+)\mathbb{Z}/2\mathbb{Z} = \{(\bar{0}, \bar{0}); (\bar{0}, \bar{1}); (\bar{2}, \bar{0}); (\bar{2}, \bar{1})\}$, which is absurd because $\mathscr{I} = \{(\bar{0}, \bar{0}); (\bar{2}, \bar{1})\}$.

(6) Since $\mathscr{I} \subseteq \mathscr{M}$. a maximal ideal, then $\sqrt{\mathscr{I}} \subseteq \mathscr{M}$. Conversely, the elements $(\bar{0}, \bar{0})$ and $(\bar{2}, \bar{1})$ of $\mathscr{M}$ belong to $\mathscr{I} \subseteq \sqrt{\mathscr{I}}$. On the other hand, $(\bar{0}, \bar{1})^2 = (\bar{0}, \bar{1})(\bar{0}, \bar{1}) = (\bar{0}, \bar{0}) \in \mathscr{I}$, and then $(\bar{0}, \bar{1}) \in \sqrt{\mathscr{I}}$ and $(\bar{2}, \bar{0})^2 = (\bar{2}, \bar{0})(\bar{2}, \bar{0}) = (\bar{0}, \bar{0}) \in \mathscr{I}$, so $(\bar{2}, \bar{0}) \in \sqrt{\mathscr{I}}$. Then $\sqrt{\mathscr{I}} = \mathscr{M}$.

Exercise 17 Let $\phi : A + XB[X] \longrightarrow A(+)B$ be the map, defined by $\phi(f) = (a_0, a_1)$ for each $f = a_0 + a_1 X + \ldots \in A + XB[X]$. Then ϕ is a homomorphism onto of rings with kernel $X^2 B[X]$. Indeed, $\phi(1) = (1, 0)$, the unity of $A(+)B$. Let $g = b_0 + b_1 X + \ldots \in A + XB[X]$. It is clear that $\phi(f + g) = \phi(f) + \phi(g)$. Since $fg = a_0 b_0 + (a_0 b_1 + a_1 b_0)X + \ldots$, then $\phi(fg) = (a_0 b_0, a_0 b_1 + a_1 b_0) = (a_0, a_1)(b_0, b_1) = \phi(f)\phi(g)$. For each $(a_0, a_1) \in A(+)B$, $\phi(a_0 + a_1 X) = (a_0, a_1)$. Finally, $f = a_0 + a_1 X + \ldots \in \ker \phi$ if and only if $a_0 = a_1 = 0$, and then $f \in X^2 B[X]$. So we have the result.

Exercise 18 $\begin{pmatrix} a & 0 \\ x & a \end{pmatrix} + \begin{pmatrix} b & 0 \\ y & b \end{pmatrix} = \begin{pmatrix} a+b & 0 \\ x+y & a+b \end{pmatrix}$ and $\begin{pmatrix} a & 0 \\ x & a \end{pmatrix} \begin{pmatrix} b & 0 \\ y & b \end{pmatrix} = \begin{pmatrix} ab & 0 \\ bx+ay & ab \end{pmatrix}$.

Zero element: $\begin{pmatrix} 0 & 0 \\ 0 & 0 \end{pmatrix}$. Unity element: $\begin{pmatrix} 1 & 0 \\ 0 & 1 \end{pmatrix}$. The map $\phi : A(+)M \longrightarrow \mathscr{M}$, defined by $\phi(a, x) = \begin{pmatrix} a & 0 \\ x & a \end{pmatrix}$, is bijective and additive. Moreover, $\phi(1, 0) = \begin{pmatrix} 1 & 0 \\ 0 & 1 \end{pmatrix}$ and $\phi((a, x)(b, y)) = (ab, ay + bx) = \begin{pmatrix} ab & 0 \\ bx+ay & ab \end{pmatrix} = \begin{pmatrix} a & 0 \\ x & a \end{pmatrix} \begin{pmatrix} b & 0 \\ y & b \end{pmatrix} = \phi(a, x)\phi(b, y)$.

Exercise 19 The inclusion $S^{-1}A \subseteq \bigcap_{\lambda \in \Lambda} S^{-1}A_\lambda$ is clear. Let $0 \neq x \in \bigcap_{\lambda \in \Lambda} S^{-1}A_\lambda$. There exist $\lambda_1, \ldots, \lambda_n \in \Lambda$ such that x is not invertible only in $A_{\lambda_1}, \ldots, A_{\lambda_n}$. Since

$x \in S^{-1}A_{\lambda_i}$, $1 \le i \le n$, there exists $s \in S$ such that $sx \in A$ for $1 \le i \le n$. Then $sx \in \bigcap_{\lambda \in \Lambda} A_\lambda = A$, so $x = \frac{sx}{s} \in S^{-1}A$. It is clear that the family $\left(S^{-1}A_\lambda\right)_{\lambda \in \Lambda}$ has a finite character.

Exercise 20 For $0 \ne t = t(X, Y) \in A[X, Y]$, $t = \sum_{i:0}^{k} t_i Y^i$ with $t_i \in A[X]$, let

$\partial t = max\{deg_X t_i,\ 0 \le i \le k\}$. For each integer $m > \partial t$, the coefficients of the polynomial $t^* = t(X, X^m) = \sum_{i:0}^{k} t_i X^{im} \in A[X]$ are the same as the coefficients of t. Indeed, for $0 \le i \le k$, $t_i \ne 0$, $deg_X t_i < m$. For each monomial ξ in $t_i X^{im}$, $mi \le deg_X \xi < m + im = m(i + 1)$. Then the nonzero monomials in $t_i X^{im}$ are different from the ones in $t_j X^{jm}$ for $i \ne j$. Then the coefficients of t and t^* are the same, so $c_t = c_{t^*}$. The map $\phi : A[X, Y] \longrightarrow A[X]$, defined by $\phi(t) = t^* = t(X, X^m)$ for each $t = t(X, Y) \in A[X, Y]$, is a homomorphism onto of rings. Then $\phi(fg) = \phi(f)\phi(g)$, so $(fg)^* = f^*g^*$. Then $c_f = c_{f^*}$, $c_g = c_{g^*}$ and $c_{fg} = c_{(fg)^*} = c_{f^*g^*}$.

Exercise 21

(1) The only real ideal of $\mathbb{F}_2$ is $\mathbb{F}_2$ since $1^2 + 1^2 = 0$. The only real ideal of $\mathbb{C}$ is $\mathbb{C}$ since $1^2 + i^2 = 0$. In particular, the rings $\mathbb{F}_2$ and $\mathbb{C}$ are not real.

The ideals (0) and $\mathbb{R}$ of $\mathbb{R}$ are real since if $x^2 + y^2 = 0$, then $x = y = 0$.

In $\mathbb{Z}$, the only real ideals are (0) and $\mathbb{Z}$. Indeed, by Lagrange theorem, if $n \in \mathbb{N} \setminus \{0, 1\}$, there exist integers not all zero $n_1, n_2, n_3, n_4 \in \mathbb{N}$ such that $n = n_1^2 + n_2^2 + n_3^2 + n_4^2$. For example, if $n_1 \ne 0$, then $n_1 < n$ so $n_1 \notin n\mathbb{Z}$.

(2) (i) Let $a \in A$ and $n \in \mathbb{N}^*$ be such that $a^n \in I$. Either $n = 2m$ and in this case $a^n = (a^m)^2 \in I$, so $a^m \in I$, or $n = 2m + 1$, and then $a^{n+1} = aa^n \in I$, so $(a^{m+1})^2 \in I$ and $a^{m+1} \in I$, ... By induction, we show that $a \in I$.

   (ii) In $\mathbb{Z}$, if p is a prime number, the ideal $p\mathbb{Z}$ is prime, and then radical, but it is not real.

   (iii) Let A be a real ring. The ideal (0) is real, and then radical, so $\sqrt{(0)} = (0)$. Then A is reduced.

(3) The ring A is real $\iff$ the field K is real $\iff$ K est orderable.

(4) $"\implies"$ Suppose that $I \ne A$, and then $A/I \ne (0)$. Let $a_1, \ldots, a_n \in A$ be such that $\bar{a}_1^2 + \ldots + \bar{a}_n^2 = 0$. Then $a_1^2 + \ldots + a_n^2 \in I$, so $a_1, \ldots, a_n \in I$, and then $\bar{a}_1 = \ldots = \bar{a}_n = \bar{0}$.

   $"\impliedby"$ Let $a_1, \ldots, a_n \in A$ be such that $a_1^2 + \ldots + a_n^2 \in I$. Then $\bar{a}_1^2 + \ldots + \bar{a}_n^2 = \bar{0}$, so $\bar{a}_1 = \ldots = \bar{a}_n = \bar{0}$, and then $a_1, \ldots, a_n \in I$.

(5) (a) and (b) follow from the questions (4) and (3) and Artin-Schreier theorem.

   (c) Let $I \ne A$ be a real ideal. Let a and $b \in A$ be such that $ab \in I$. Either $\frac{a}{b} \in A$ or $\frac{b}{a} \in A$. Suppose, for example, that $\frac{b}{a} = t \in A$. Then $b = at$ and $b^2 = (at)^2 = a(at)t = abt \in I$, so $b \in I$.

(d) Let $z = a_1^2 + \ldots + a_n^2 \in Q \subseteq P$ with $a_1, \ldots, a_n \in A$. We may suppose that $\frac{a_i}{a_1} \in A$ for $2 \le i \le n$. We have $z = a_1^2\left(1 + (\frac{a_2}{a_1})^2 + \ldots + (\frac{a_n}{a_1})^2\right) = a_1^2 t$ with $t = 1 + (\frac{a_2}{a_1})^2 + \ldots + (\frac{a_n}{a_1})^2 \in A$. We cannot have $t \in P$ because in the contrary case, since P is real, we will have $1 \in P$, which is absurd. Since $Q \subseteq P$, then $t \notin Q$. Since $z = a_1^2 t \in Q$ and Q is prime, then $a_1 \in Q$. By induction, we show that $a_1, \ldots, a_n \in Q$ so Q is real.

(e) It suffices to take a nondiscrete valuation domain of rank one or a finite rank ≥ 2. By (c), since its prime spectrum is finite, it has a finite number of real ideals.

Exercise 22 The questions (1), (2), and (3-a-b-c) are easy.

(d) $A = \mathbb{Z}$, $I = n\mathbb{Z}$ with $n \in \mathbb{N} \setminus \{0, 1\}$, and then $R(n\mathbb{Z}) = \mathbb{Z}$.

(4) By definition, $I \subseteq R(I)$. Since a real ideal is radical, by the question (2) of the preceding exer, then $\sqrt{I} \subseteq R(I)$.

(5) Since $J \subseteq I+J \subseteq R(I+J)$, then $R(J) \subseteq R(I+J)$. We have also $I \subseteq R(I+J)$, and then $I + R(J) \subseteq R(I+J)$ so $R(I + R(J)) \subseteq R(I+J)$. The inverse inclusion is clear.

(6) Let $a_1, \ldots, a_n \in A$ with $a_1^2 + \ldots + a_n^2 \in (I : J)$. Then $(a_1^2 + \ldots + a_n^2)J \subseteq I$. For each $x \in J$, $(a_1 x)^2 + \ldots + (a_n x)^2 = (a_1^2 + \ldots + a_n^2)x^2 \in I$. Since I is real, $a_1 x, \ldots, a_n x \in I$. Then $a_1, \ldots, a_n \in (I : J)$.

(7) For each ideal I of A, let $\tilde{I} = \{a \in A; \; \exists m \in \mathbb{N}^*, \; \exists n \in \mathbb{N}, \; \exists a_1, \ldots, a_n \in A, \; a^{2m} + a_1^2 + \ldots + a_n^2 \in I\}$. Then $I \subseteq \tilde{I}$ and if I is real, then $\tilde{I} = I$. Indeed, if $a \in \tilde{I}$, there exist $m \in \mathbb{N}^*$, $n \in \mathbb{N}$, and $a_1, \ldots, a_n \in A$ such that $a^{2m} + a_1^2 + \ldots + a_n^2 \in I$. Since I is real, $a^m \in I$, but each real ideal is radical, then $a \in I$. Note also that if $I \subseteq J$ are two ideals of A, then $\tilde{I} \subseteq \tilde{J}$. We start by showing that $\tilde{I} = \bigcap P$, where P runs the set of real prime ideals of A containing I. For each P, $I \subseteq \tilde{I} \subseteq \tilde{P} = P$, and then $\tilde{I} \subseteq \bigcap P$. Conversely, if $a \notin \tilde{I}$, we will show that there exists a real prime ideal P containing I such that $a \notin P$. Let $S_a = \{a^{2m} + a_1^2 + \ldots + a_n^2; \; m \in \mathbb{N}^*, \; n \in \mathbb{N}, \; a_1, \ldots, a_n \in A\}$. Then S_a is a multiplicative set. Indeed, let $\sigma = a_1^2 + \ldots + a_n^2$ and $\sigma' = a_1'^2 + \ldots + a_{n'}'^2 \in \Sigma A^2$, $m, m' \in \mathbb{N}^*$. Then $(a^{2m} + \sigma)(a^{2m'} + \sigma') = a^{2(m+m')} + (\Sigma \text{ squares}) \in S_a$. Since $a \notin \tilde{I}$, then $I \cap S_a = \emptyset$. Let $\mathscr{F}$ be the set of ideals J of A satisfying $J \cap S_a = \emptyset$. Then $I \in \mathscr{F}$ and $\mathscr{F}$ is inductive for the inclusion. By Zorn's lemma, there exists P maximal in $\mathscr{F}$ satisfying $I \subseteq P$. Suppose that P is not a prime ideal. There exist $x, y \in A \setminus P$ such that $xy \in P$. Then $P \subset P+xA$ and $P \subset P+yA$. By maximality of P in $\mathscr{F}$, $(P+xA) \cap S_a \ne \emptyset$ and $(P + yA) \cap S_a \ne \emptyset$. Let s and t be two elements respectively in these two sets. Put $s = p + xc$ and $t = p' + yd$ with $p, p' \in P$ and $c, d \in A$. Then $st = p(p' + yd) + (xy)cd + xcp' \in P$ and since S_a is stable by multiplication, $st \in S_a$, then $P \cap S_a \ne \emptyset$, which is absurd. Then P is prime. Suppose that P is not real, there exist $b_1, \ldots, b_k \in A$, $b_1 \notin P$ such that $b_1^2 + \ldots + b_k^2 \in P$. Since $P \subset P+b_1 A$, by maximality of P in $\mathscr{F}$, $S_a \cap (P + b_1 A) \ne \emptyset$. Let s be an element of this set, $s = p + b_1 x$ with $p \in P$ and $x \in A$. Since $s^2 = p^2 + 2pb_1 x + b_1^2 x^2$, by adding the quantity $x^2(b_2^2 + \ldots + b_k^2)$ to both members of the equality, we find

$s^2 + (xb_2)^2 + \ldots + (xb_k)^2 = p(p + 2b_1 x) + x^2(b_1^2 + b_2^2 + \ldots + b_k^2) \in P$. On the other hand, S_a is stable by multiplication, then $s^2 \in S_a$, put $s^2 = a^{2m} + a_1^2 + \ldots + a_n^2$ with $m \in \mathbb{N}^*$ and $a_1, \ldots, a_n \in A$, then $s^2 + (xb_2)^2 + \ldots + (xb_k)^2 = a^{2m} + a_1^2 + \ldots + a_n^2 + (xb_2)^2 + \ldots + (xb_k)^2 \in S_a$, and then $P \cap S_a \neq \emptyset$, which is absurd. Then P is real. Since $a^2 \in S_a$, $a^2 \notin P$, then $a \notin P$. We always showed that $\tilde{I} = \bigcap P$, where P runs the set of real prime ideals of A containing I. In particular, $\tilde{I}$ is a real ideal containing I, and then $R(I) \subseteq \tilde{I}$. Conversely, let $a \in \tilde{I}$. There exist $m \in \mathbb{N}^*$ and $a_1, \ldots, a_n \in A$ such that $a^{2m} + a_1^2 + \ldots + a_n^2 \in I \subseteq R(I)$. Since $R(I)$ is a real ideal, $a^m \in R(I)$, but each real ideal is radical, then $a \in R(I)$. So $\tilde{I} = R(I)$.

(8) (a) For each $a \in I$, $a^2 \in I$, then $a \in R_1(I)$ so $I \subseteq R_1(I)$. Let $a \in R_1(I)$ and $c \in A$. There exist $n \in \mathbb{N}$ and $a_1, \ldots, a_n \in A$ such that $a^2 + a_1^2 + \ldots + a_n^2 \in I$, then $(ca)^2 + (ca_1)^2 + \ldots + (ca_n)^2 = c^2(a^2 + a_1^2 + \ldots + a_n^2) \in I$ so $ca \in R_1(I)$. If b is another element of $R_1(I)$, there exist $m \in \mathbb{N}$ and $b_1, \ldots, b_m \in A$ such that $b^2 + b_1^2 + \ldots + b_m^2 \in I$. Then $2(a^2 + a_1^2 + \ldots + a_n^2) + 2(b^2 + b_1^2 + \ldots + b_m^2) \in I$ so $(a + b)^2 + (a - b)^2 + 2a_1^2 + \ldots + 2a_n^2 + 2b_1^2 + \ldots + 2b_m^2 \in I$, and then $a + b, a - b \in R_1(I)$.

(b) " $\supseteq$ " Let $a \in R_1(I)$. There exist $n \in \mathbb{N}$ and $a_1, \ldots, a_n \in A$ such that $a^2 + a_1^2 + \ldots + a_n^2 \in I$. By (7), $a \in R(I)$. Suppose by induction that $R_k(I) \subseteq R(I)$. Let $a \in R_{k+1}(I)$. There exist $n \in \mathbb{N}$ and $a_1, \ldots, a_n \in A$ such that $a^2 + a_1^2 + \ldots + a_n^2 \in R_k(I) \subseteq R(I)$. Since $R(I)$ is a real ideal, $a, a_1, \ldots, a_n \in R(I)$. Then $a \in R(I)$ and $\bigcup_{k \in \mathbb{N}^*} R_k(I) \subseteq R(I)$.

" $\subseteq$ " Since $I \subseteq R_1(I) \subseteq \bigcup_{k:1}^{\infty} R_k(I)$, to conclude that $R(I) \subseteq \bigcup_{k:1}^{\infty} R_k(I)$, it suffices to show that $\bigcup_{k:1}^{\infty} R_k(I)$ is a real ideal. Let $a_1, \ldots, a_n \in A$ be such that $a_1^2 + \ldots + a_n^2 \in \bigcup_{k:1}^{\infty} R_k(I)$. There exists $k \in \mathbb{N}^*$ such that $a_1^2 + \ldots + a_n^2 \in R_k(I)$. For each $j \in \{1, \ldots, n\}$, $a_j^2 + \sum_{i \neq j} a_i^2 \in R_k(I)$, then $a_j \in R_1(R_k(I)) = R_{k+1}(I)$. So for each $j \in \{1, \ldots, n\}$, $a_j \in \bigcup_{k:1}^{\infty} R_k(I)$.

(9) (i) $I \subseteq R(I) \implies IR(J) \subseteq R(I)R(J)$.

(ii) Let $a \in R(I)$ and $a' \in R(J)$. By (7), there exist $m, m' \in \mathbb{N}^*$, $n, n' \in \mathbb{N}$, $a_1, \ldots, a_n, a_1', \ldots, a_{n'}' \in A$ such that $a^{2m} + a_1^2 + \ldots + a_n^2 \in I$ and $a'^{2m'} + a_1'^2 + \ldots + a_{n'}'^2 \in J$. We can suppose that $m' \geq m$. We have $a^{2(m'-m)}(a^{2m} + a_1^2 + \ldots + a_n^2) \in I$, and then $a^{2m'} + (a_1 a^{m'-m})^2 + \ldots + (a_n a^{m'-m})^2 \in I$. By multiplication, we find $(aa')^{2m'} + (\sum \text{squares}) \in IJ$. By (7), $aa' \in R(IJ)$. Then $R(I)R(J) \subseteq R(IJ)$.

(iii) $IJ \subseteq I \cap J \implies R(IJ) \subseteq R(I \cap J)$.

(iv) $I \cap J \subseteq I \implies R(I \cap J) \subseteq R(I)$. We also have $R(I \cap J) \subseteq R(J)$, and then $R(I \cap J) \subseteq R(I) \cap R(J)$. Let $a \in R(I) \cap R(J)$. By (7), there exist $m, m' \in \mathbb{N}^*$,

$n, n' \in \mathbb{N}, a_1, \ldots, a_n, a'_1, \ldots, a'_{n'} \in A$ such that $a^{2m} + a_1^2 + \ldots + a_n^2 \in I$ and $a^{2m'} + a_1'^2 + \ldots + a_{n'}'^2 \in J$. By multiplication, we find $a^{2(m+m')} + (\sum \text{squares}) \in IJ$. By (7), $a \in R(IJ)$. So we have the inclusion $R(I) \cap R(J) \subseteq R(IJ)$ and then the desired equalities.

(10) (a) Let $a \in R(J)$. By (7), there exist $m \in \mathbb{N}^*, n \in \mathbb{N}$, and $a_1, \ldots, a_n \in A$ such that $a^{2m} + a_1^2 + \ldots + a_n^2 \in J$, and then $\phi(a)^{2m} + \phi(a_1)^2 + \ldots + \phi(a_n)^2 \in \phi(J)$. By (7), $\phi(a) \in R(\phi(J))$. Then $\phi(R(J)) \subseteq R(\phi(J))$.

(b) Let $x_1, \ldots, x_n \in A$ be such that $x_1^2 + \ldots + x_n^2 \in \phi^{-1}(I)$. Then $\phi(x_1)^2 + \ldots + \phi(x_n)^2 = \phi(x_1^2 + \ldots + x_n^2) \in I$. Since I is real, $\phi(x_1), \ldots, \phi(x_n) \in I$, then $x_1, \ldots, x_n \in \phi^{-1}(I)$.

(c) Since ϕ is onto, $I = \phi(\phi^{-1}(I)) = \phi(R(\Sigma AS)) \subseteq R(\phi(\Sigma AS)) = R(\Sigma B\phi(S))$, then $I \subseteq R(\Sigma B\phi(S))$. Conversely, by hypothesis, $S \subseteq \phi^{-1}(I)$, and then $\phi(S) \subseteq I$, so $\Sigma B\phi(S) \subseteq I$ and since I is real, $R(\Sigma B\phi(S)) \subseteq I$. Then we have the desired equality.

(11) For $1 \le k \le n, S = \{x_1, \ldots, x_n\} \subseteq R(\{x_1^2 + \ldots + x_k^2, x_{k+1}, \ldots, x_n\})$ since the last ideal is real. Then $I = R(S) \subseteq R(\{x_1^2 + \ldots + x_k^2, x_{k+1}, \ldots, x_n\})$. On the other hand, $\{x_1^2 + \ldots + x_k^2, x_{k+1}, \ldots, x_n\} \subseteq R(S)$, then $R(\{x_1^2 + \ldots + x_k^2, x_{k+1}, \ldots, x_n\}) \subseteq R(S)$. So we have the desired equality.

(12) Let $I = R(T)$ with T a finite subset of I. By (7), for each $t \in T \subseteq I = R(S) = R(\Sigma AS)$, there exist $m_t \in \mathbb{N}^*, n_t \in \mathbb{N}$, and $a_{1,t}, \ldots, a_{n_t,t} \in A$ such that $z_t = t^{2m_t} + a_{1,t}^2 + \ldots + a_{n_t,t}^2 \in \Sigma AS$. There exists a finite set $S_t \subseteq S$ such that $z_t \in \Sigma AS_t$. Consider the finite subset $S' = \bigcup_{t \in T} S_t$ of S. For each $t \in T, z_t = t^{2m_t} + a_{1,t}^2 + \ldots + a_{n_t,t}^2 \in \Sigma AS_t \subseteq \Sigma AS'$. By (7), for each $t \in T, t \in R(\Sigma AS') = R(S')$, and then $T \subseteq R(S')$, which is a real ideal. Then $I = R(T) \subseteq R(S') \subseteq R(S) = I$. So $I = R(S')$.

Exercise 23

(1) " $\Longrightarrow$ " Suppose that there exists a real ideal I of A that is not really finitely generated. Let $a_1 \in I$, and then $R(a_1 A) \subset I$; let $a_2 \in I \setminus R(a_1 A)$, and then $R(R(a_1 A) + a_2 A) = R(a_1 A + a_2 A) \subset I$ and $R(a_1 A) \subset R(a_1 A + a_2 A), \ldots$ etc. By iteration of this process, we construct a strictly increasing sequence of real ideals of A, which is absurd.

"$\Longleftarrow$" Let $(I_n)_{n \in \mathbb{N}}$ be an increasing sequence of real ideals of A. Since $\bigcup_{n \in \mathbb{N}} I_n$ is a real ideal, it is really finitely generated. Let $a_1, \ldots, a_m \in A$ such that $\bigcup_{n \in \mathbb{N}} I_n = R(\{a_1, \ldots, a_m\})$. There exists $n_0 \in \mathbb{N}$ such that $a_1, \ldots, a_m \in I_{n_0}$. For each $n \ge n_0, I_n = I_{n_0}$.

(2) By (1), it suffices to show that each real ideal I of B is really finitely generated. By the question (9-b) of the preceding exer, $\phi^{-1}(I)$ is a real ideal of A. Since A satisfies the acc for real ideals, there exists a finite set $S \subseteq A$ such that

$\phi^{-1}(I) = R(S) = R(\Sigma AS)$. Since ϕ is onto, by the question (10-c) of the preceding exer, $I = R(\Sigma B\phi(S)) = R(\phi(S))$ with $\phi(S)$ is finite.

(3) Let $\mathscr{F}$ be the family of real ideals of A that are not really finitely generated. Suppose that $\mathscr{F} \neq \emptyset$, then $(\mathscr{F}, \subset)$ is inductive. Indeed, if $(I_\lambda)_{\lambda \in \Lambda}$ is a chain of real ideals that are not really finitely generated, then it is the same for the ideal $\bigcup_{\lambda \in \Lambda} I_\lambda$. By Zorn's lemma, there exists a real ideal P which is not really finitely generated and is maximal for this property. We will show that P is prime, which is absurd, and then $\mathscr{F} = \emptyset$. Suppose that P is not prime; there exist $a, b \in A \setminus P$ such that $ab \in P$. Let $I = R(P + aA)$. Then $P \subset R(P + aA) = I$, so I is really finitely generated by the maximality of P in $\mathscr{F}$. By the question (12) of the preceding exer, there exist $x_1, \ldots, x_n \in P$ such that $I = R(P + aA) = R(\{x_1, \ldots, x_n, a\})$. Let $f = x_1^2 + \ldots + x_n^2 \in P$, and by the question (11) of the preceding exer, $I = R(fA + aA$. The inclusion $P \subset P : aA$ is strict because of b. By the question (6) of the preceding exer, since P is real, $P : aA$ is also real. By maximality of P in $\mathscr{F}$, the ideal $P : aA$ is really finitely generated. Put $P : aA = R(gA)$ with $g \in A$. Let $x \in P \subset I = R(fA + aA)$. By the question (7) of the preceding exer, there exist $m \in \mathbb{N}^*, n \in \mathbb{N}, b_1, \ldots, b_n \in A$ and $s, t \in A$ such that $x^{2m} + b_1^2 + \ldots + b_n^2 = tf + sa$. On the other hand, since $x \in P \subset P : aA = R(gA)$, there exist $m' \in \mathbb{N}^*, n' \in \mathbb{N}, b_1', \ldots, b_{n'}' \in A$ and $t' \in A$, such that $x^{2m'} + b_1'^2 + \ldots + b_{n'}'^2 = t'g$. By multiplication member by member of the two preceding equalities, we obtain $x^{2(m+m')} + (\Sigma \text{ squares}) = t''f + s''ag$ with $t'', s'' \in A$. By the question (7) of the preceding exer, $x \in R(fA + agA)$. Then $P \subseteq R(fA + agA)$. On the other hand, $P : aA = R(gA)$, and then $ag \in P$ so $R(fA + agA) \subseteq P$, and then $R(fA + agA) = P$. Then P is really finitely generated, which is absurd. Then P is prime.

(4) Let $f(X) = (f_1(X))^2 + \ldots + (f_n(X))^2 \in I[X]$ with $f_i(X) = \displaystyle\sum_{j:0}^{\infty} a_{ij}X^j \in A[X]$.

The constant term of $f(X)$ is $a_{10}^2 + \ldots + a_{n0}^2 \in I$, and then $a_{10}, \ldots, a_{n0} \in I$ since I is real. Put the following hypothesis of induction: For each $i \in \{1, \ldots, n\}$ and each integer $j < m, a_{ij} \in I$. The coefficient of X^{2m} in $f(X)$ is $a = \displaystyle\sum_{i:1}^{n} \Big(\sum_{j+k=2m} a_{ij}a_{ik} \Big) \in I$. If $j < m$, by the induction hypothesis, $a_{ij} \in I$, then $a_{ij}a_{ik} \in I$. If $j > m$, then $k < m$, so $a_{ik} \in I$ and $a_{ij}a_{ik} \in I$. If $j = m$, then $k = m$. Then a is of the form $a = b + a_{1m}^2 + \ldots + a_{nm}^2$ with $b \in I$, and then $a_{1m}^2 + \ldots + a_{nm}^2 \in I$ so $a_{1m}, \ldots, a_{nm} \in I$.

(5) Put $I = R_A(S) = R_A(\Sigma AS)$ with S a finite subset of A. We will show that $I[X] = R_{A[X]}(S) = R_{A[X]}(\Sigma A[X]S)$. By the preceding question, the ideal $I[X]$ is real and since $S \subseteq I \subset I[X]$, then $R_{A[X]}(S) \subseteq I[X]$. Conversely, let $f(X) = b_0 + b_1X + \ldots + b_rX^r \in I[X]$. For each $i \in \{0, \ldots, r\}, b_i \in I = R_A(S)$. By the question (7) of the preceding exer, there exist $m \in \mathbb{N}^*, n \in \mathbb{N}, a_1, \ldots, a_n \in A$ (all depending on i) such that $b_i^{2m} + a_1^2 + \ldots + a_n^2 \in \Sigma AS \subset$

$\Sigma A[X]S$. Then $b_i \in R_{A[X]}(\Sigma A[X]S)$, so $f(X) = b_0 + b_1 X + \ldots + b_r X^r \in R_{A[X]}(\Sigma A[X]S)$.

(6) $" \Longleftarrow "$ Let $\phi : A[X] \longrightarrow A$ be the map defined by $\phi(a_0 + a_1 X + \ldots + a_n X^n) = a_0$. Then ϕ is a homomorphism onto. We conclude by (2).

$" \Longrightarrow "$ Suppose that $A[X]$ does not satisfy the acc for the real ideals. By (3), the set $\mathscr{F}$, of the real ideals of $A[X]$ which are not really finitely generated, is nonempty. It is inductive for the inclusion. There exists a real ideal M of $A[X]$ which is not really finitely generated and is maximal for this property. Then $M \cap A$ is a real ideal of A, so really finitely generated. By the question (11) of Exercise 22, it is really generated by an element s of A. Let $N = (M \cap A)[X] \subseteq M$. By (5), N is really generated by the element s, and then $N \subset M$. Let $f(X) \in M \setminus N$ be a polynomial with minimum degree. Put $f(X) = a_0 + a_1 X + \ldots + a_{e-1} X^{e-1} + a X^e$ with $a, a_0, \ldots, a_{e-1} \in A, a \neq 0, e \in \mathbb{N}^*$. Then $a \notin M$. Indeed, if not, $a X^e \in M$, then $a_0 + a_1 X + \ldots + a_{e-1} X^{e-1} = f(X) - a X^e \in M$ with $deg(a_0 + a_1 X + \ldots + a_{e-1} X^{e-1}) < deg\ f(X)$, and then $a_0 + a_1 X + \ldots + a_{e-1} X^{e-1} \in N$ and since $a \in M \cap A$, then $a X^e \in N$, which implies that $f(X) = (a_0 + a_1 X + \ldots + a_{e-1} X^{e-1}) + a X^e \in N$, which contradicts the choice of $f(X)$. We conclude that $M \subset R(M \cup \{a\})$. By maximality of M in $\mathscr{F}$, the real ideal $R(M \cup \{a\})$ is really finitely generated. By the question (11) of the preceding exer, there exists $t(X) \in M$ such that $R(M \cup \{a\}) = R(A[X]t(X) + A[X]a)$. Now, we will prove that for each $g(X) \in M, ag(X) \in R(\{f(X)\} \cup N)$. Indeed, if $g(X) \in N$, the affirmation is clear and if $g(X) \in M \setminus N$, let $m = deg\ g(X) \geq deg\ f(X) = e \geq 1$. The Euclidean division of $a^m g(X)$ by $f(X)$ can be written as $a^m g(X) = f(X)q(X) + r(X)$ with $q(X)$ and $r(X) \in A[X]$, either $r(X) = 0$ or $deg\ r(X) < e$. Then $r(X) = a^m g(X) - f(X)q(X) \in M$ since $g(X), f(X) \in M$. By minimality of $deg\ f(X)$ in $M \setminus N, r(X) \in N$. Then $(ag(X))^m = a^m g(X)(g(X))^{m-1} = (f(X)q(X) + r(X))(g(X))^{m-1} = f(X)q(X)(g(X))^{m-1} + r(X)(g(X))^{m-1} \in A[X]f(X) + N \subseteq R(A[X]f(X) + N)$. Since each real ideal is radical, $ag(X) \in R(A[X]f(X)+N) = R(A[X]f(X)+R(sA[X])) = R(A[X]f(X)+ sA[X])$. Then $aM \subseteq R(A[X]f(X) + sA[X])$. For each $h(X) \in M$, we have $(h(X))^2 \in M^2 \subseteq M.R(A[X]t(X) + A[X]a) \subseteq R(M(A[X]t(X) + A[X]a)) \subseteq R(A[X]t(X) + aM) \subseteq R(A[X]t(X) + R(A[X]f(X) + A[X]s)) = R(A[X]t(X) + A[X]f(X) + A[X]s) = R(t(X), f(X), s)$. Since each real ideal is radical, we have $h(X) \in R(t(X), f(X), s)$. Since $t(X), f(X)$, and $s \in M$, then $M \subseteq R(t(X), f(X), s) \subseteq M$, so $M = R(t(X), f(X), s)$, which is absurd.

(7) Let $B = A[x_1, \ldots, x_n]$ be an A−algebra of finite type. The map $\phi : A[X_1, \ldots, X_n] \longrightarrow B$, defined by $\phi(f) = f(x_1, \ldots, x_n)$, is a homomorphism onto. By (2), since A satisfies the acc for the real ideals, then so is B.

Exercise 24

(1) The map $\phi : A[[X]] \longrightarrow A$, defined by $\phi(f) = f(0)$, is a homomorphism onto. By question (2) of the preceding exer, A satisfies the acc for the real ideals.

(2) By the question (5) of Exercise 21, since $A/M \simeq \mathbb{R}$ is an orderable field, then M is a real ideal. The proper real ideals of A are prime and all the prime ideals of A are real. Then the real ideals of A are (0), M, and A. So A satisfies the acc for the real ideals.

(3) The ring of generalized power series with coefficients in $\mathbb{R}$ and well-ordered supports in $\mathbb{R}+$, with the natural valuation.

(4) In the sequel, the elements of $A[[X]]$ will be denoted $S = \sum_{i:0}^{\infty} s_i X^i$. Let $S_1, \ldots, S_r \in A[[X]]$ be such that $S_1^2 + \ldots + S_r^2 \in XM[[X]]$. Then $s_{1,0}^2 + \ldots + s_{r,0}^2 = 0$. Since (0) is a real ideal of A, then $s_{1,0} = \ldots = s_{r,0} = 0$. Consider the coefficient of X^2 in $S_1^2 + \ldots + S_n^2$. Then $s_{1,1}^2 + \ldots + s_{r,1}^2 \in M$, so $s_{1,1}, \ldots, s_{r,1} \in M$ since M is a real ideal in A. Suppose by induction that $s_{i,l} \in M$, for $1 \leq i \leq r$ and $1 \leq l \leq n-1$. The coefficient of X^{2n} in $S_1^2 + \ldots + S_n^2$ is $a = \sum_{i:1}^{r} \sum_{j+k=2n} s_{i,j} s_{i,k} \in M$. If $j < n$, $s_{i,j} \in M$, then $s_{i,j} s_{i,k} \in M$. If $j > n$, then $k < n$, so $s_{i,k} \in M$, and then $s_{i,j} s_{i,k} \in M$. If $j = n$, then $k = n$. So $a = b + s_{1,n}^2 + \ldots + s_{r,n}^2 \in M$ with $b \in M$, so $s_{1,n}^2 + \ldots + s_{r,n}^2 \in M$. Since M is a real ideal, then $s_{1,n}, \ldots, s_{r,n} \in M$. We conclude that $S_1, \ldots, S_r \in XM[[X]]$ so $XM[[X]]$ is a real ideal in $A[[X]]$.

(5) Suppose that $min\{v(c_j); \ 1 \leq j \leq k\} = v(c_{j_0}) < \frac{\delta}{2}$ with $j_0 \in \{1, \ldots, k\}$. Put $c_{j_0} = x$. The inequalities $v(c_1^2 + \ldots + c_k^2 + b_1 + \ldots + b_h) \geq \delta$ and $v(x) < \frac{\delta}{2}$ give $v(x^{-2}) + v(c_1^2 + \ldots + c_k^2 + b_1 + \ldots + b_h) > 0 \Longrightarrow v((x^{-1}c_1)^2 + \ldots + (x^{-1}c_k)^2 + x^{-2}b_1 + \ldots + x^{-2}b_h) > 0 \Longrightarrow v\left(1 + \sum_{j \neq j_0} (x^{-1}c_j)^2 + x^{-2}b_1 + \ldots + x^{-2}b_h\right) > 0$

$\Longrightarrow 1 + \sum_{j \neq j_0} (x^{-1}c_j)^2 + x^{-2}b_1 + \ldots + x^{-2}b_h \in M$. Since $v(x^{-2}b_i) = -2v(x) + v(b_i) > 0$, then $x^{-2}b_i \in M$, for $1 \leq i \leq h$. So $1 + \sum_{j \neq j_0} (x^{-1}c_j)^2 \in M$. Since M is a real ideal and $x^{-1}c_j \in A$ for $1 \leq j \leq k$, then $1 \in M$, which is absurd.

(6) The proof will be done by double induction on n and k. Take $k = 1$ and $n = 1$. Let $U \in R_1(J)$. There exist $h \in \mathbb{N}$ and $U_1, \ldots, U_h \in A[[X]]$ such that $U' = U^2 + U_1^2 + \ldots + U_h^2 \in J$. Since $J \subseteq XM[[X]]$ and $XM[[X]]$ is a real ideal of $A[[X]]$, then $U, U_1, \ldots, U_n \in XM[[X]]$. So $u_0 = 0$ and $u_{i,0} = 0$, for $1 \leq i \leq h$. The coefficient of X^2 in U' is $u_1^2 + u_{1,1}^2 + \ldots + u_{h,1}^2$. By definition, $v(u_1^2 + u_{1,1}^2 + \ldots + u_{h,1}^2) \geq \gamma_2(J)$. By the preceding question, $v(u_1) \geq \frac{1}{2}\gamma_2(J)$ and $v(u_{i,1}) \geq \frac{1}{2}\gamma_2(J)$, for $1 \leq i \leq h$. In particular, $\gamma_1(R_1(J)) \geq \frac{1}{2}\gamma_2(J)$. Suppose by induction that $\gamma_{n-1}(R_1(J)) \geq \frac{1}{2^{n-1}}\gamma_{2(n-1)}(J)$ and $v(u_j) \geq \frac{1}{2^j}\gamma_{2j}(J)$ and

$v(u_{i,j}) \geq \frac{1}{2^j}\gamma_{2j}(J)$, for $1 \leq i \leq h$ and $1 \leq j \leq n-1$. Consider the coefficient of X^{2n} in U', by definition, $v\big(2u_1u_{2n-1} + 2u_2u_{2n-2} + \ldots + 2u_{n-1}u_{n+1} + u_n^2 +$

$$2\sum_{i:1}^{h} u_{i,1}u_{i,2n-1} + 2\sum_{i:1}^{h} u_{i,2}u_{i,2n-2} + \ldots + 2\sum_{i:1}^{h} u_{i,n-1}u_{i,n+1} + \sum_{i:1}^{h} u_{i,n}^2\big) \geq$$

$\gamma_{2n}(J) > \frac{1}{2^{n-1}}\gamma_{2n}(J)$. On the other hand, by the induction hypothesis, for each $j \in \{1, \ldots, n-1\}$ and $i \in \{1, \ldots, h\}$, we have $v(u_ju_{2n-j}) \geq v(u_j) \geq \frac{1}{2^j}\gamma_{2j}(J) \geq \frac{1}{2^{n-1}}\gamma_{2n}(J) > 0$ and $v(2u_{i,j}u_{i,2n-j}) \geq v(u_{i,j}) \geq \frac{1}{2^j}\gamma_{2j}(J) \geq \frac{1}{2^{n-1}}\gamma_{2n}(J) > 0$. By the preceding question, $v(u_n) \geq \frac{1}{2^n}\gamma_{2n}(J))$ and $v(u_{i,n}) \geq \frac{1}{2^n}\gamma_{2n}(J)$. Then $\gamma_n(R_1(J)) \geq \frac{1}{2^n}\gamma_{2n}(J)$, for each $n \geq 1$.

Suppose now that $k \geq 1$ and $\gamma_n(R_k(J)) \geq \frac{1}{2^{(2^k-1)n}}\gamma_{2^kn}(J) > 0$ for each $n \geq 1$. Since $R_k(J) \subseteq R(J) \subseteq R(XM[[X]]) = MX[[X]]$, then we have the inequalities $\gamma_n(R_{k+1}(J)) = \gamma_n(R_1(R_k(J))) \geq \frac{1}{2^n}\gamma_{2n}(R_k(J)) \geq \frac{1}{2^n}\frac{1}{2^{(2^k-1)2n}}\gamma_{2^{k+1}n}(J) = \frac{1}{2^{(2^{k+1}-1)n}}\gamma_{2^{k+1}n}(J)$. Then we obtain the result.

(7) For $S = \sum_{i:0}^{\infty} s_i X^i \in A[[X]]$, $ST = \sum_{i:0}^{\infty}\big(\sum_{j+k=i} t_js_k\big)X^i$, with $min\{v\big(\sum_{j+k=i} t_js_k\big);$ $0 \leq i \leq n\} \geq min\{v(t_j); \ 0 \leq i \leq n, 0 \leq j \leq i\} = min\{v(t_j); \ 0 \leq i \leq n\} = \gamma_n(T) > 0$.

(8) Suppose that $XM[[X]]$ is really finitely generated. Then it is of the form $XM[[X]] = R(TA[[X]])$ with $T \in XM[[X]]$. By (6), for each $n \in \mathbb{N}^*$, we have $\gamma_n(R_n(TA[[X]])) \geq \frac{1}{2^{(2^n-1)n}}\gamma_{2^nn}(TA[[X]])$. By the preceding question, $\gamma_{2^nn}(TA[[X]]) > 0$, and then we have $\big(\frac{1}{2^{(2^n-1)n}}\gamma_{2^nn}(TA[[X]])\big)_{n\in\mathbb{N}^*}$ a decreasing sequence of strictly positive numbers. We can construct a strictly decreasing sequence $(\alpha_n)_{n\in\mathbb{N}^*}$ of strictly positive real numbers, such that for each $n \in \mathbb{N}^*$, $0 < \alpha_n < \frac{1}{2^{(2^n-1)n}}\gamma_{2^nn}(TA[[X]])$. For each $n \in \mathbb{N}^*$, there exists $u_n \in A$ such that $v(u_n) = \alpha_n$. Let $U = \sum_{i:1}^{\infty} u_i X^i \in XM[[X]] = R(TA[[X]])$. Let $n \in \mathbb{N}$ be such that $U \in R_n(TA[[X]])$. Then we have $v(u_n) = \alpha_n \geq \gamma_n(R_n(TA[[X]])) \geq \frac{1}{2^{(2^n-1)n}}\gamma_{2^nn}(TA[[X]]) > \alpha_n$, which is absurd.

Exercise 25 Let $\phi : I^{-1} \longrightarrow Hom_A(I, A)$ be the map defined by $\phi(q) : I \longrightarrow A$ with $\phi(q)(x) = qx$, for each $q \in I^{-1}$ and $x \in I$. The map $\phi(q)$ takes its value in A since $qx \in I^{-1}I \subseteq A$. It is clear that $\phi(q)$ is A−linear, and then $\phi(q) \in Hom_A(I, A)$. Let $q, q' \in I^{-1}$ be such that $\phi(q) = \phi(q')$. Then $qx = q'x$ for each $x \in I$. For $0 \neq x \in I$, we obtain $q = q'$. Then ϕ is injective. Let $f \in Hom_A(I, A)$. We want to find $q \in I^{-1}$ such that $\phi(q) = f$; i.e., for each $x \in I$, $qx = f(x)$. It suffices to show that $\frac{f(x)}{x}$ is constant in I^{-1}, independent of $x \in I \setminus (0)$. Let $x, y \in I \setminus (0)$. Since $I \subseteq A$ and $f : I \longrightarrow A$ is a A−linear map, then $xf(y) = f(xy) = yf(x)$, so $\frac{f(x)}{x} = \frac{f(y)}{y}$. Moreover, $y\frac{f(x)}{x} = f(y) \in A$ for each $y \in I$, and

then $\frac{f(x)}{x} \in I^{-1}$. Let $q = \frac{f(x)}{x}$ for each $x \in I \setminus (0)$, and then $f(x) = qx = \phi(q)(x)$.
So $f = \phi(q)$ and ϕ is surjective.

Exercise 26 There is $n \in \mathbb{N}^*$ such that $(\sqrt{I})^n \subseteq I$. By Proposition 11.1, we have
$((\sqrt{I})_v)^n \subseteq (((\sqrt{I})_v)^n)_v = ((\sqrt{I})^n)_v \subseteq I_v = I$. Then $(\sqrt{I})_v \subseteq \sqrt{I}$, so
$(\sqrt{I})_v = \sqrt{I}$.

Exercise 27

(1) It is clear that $J_1 J_2$ is finitely generated. Let $x \in K$. Then $x \in (J_1 J_2)^{-1} = (A : J_1 J_2) \iff x(J_1 J_2) \subseteq A \iff x J_1 \subseteq (A : J_2) = J_2^{-1} = A \iff x \in (A : J_1) = J_1^{-1} = A$. Then $(J_1 J_2)^{-1} = A$. So $J_1 J_2 \in GV(A)$.

(2) For each $J \in GV(A)$, since $IJ \subseteq IA = I$, then $I \subseteq (I : J) \subseteq I_w$. In particular $I_w \neq \emptyset$.

Let $x_1, x_2 \in I_w$. There exist $J_1, J_2 \in GV(A)$ such that $x_1 \in (I : J_1)$ and $x_2 \in (I : J_2)$. Then $x_1 J_1 \subseteq I$ and $x_2 J_2 \subseteq I$. So $(x_1 + x_2)(J_1 J_2) \subseteq x_1 J_1 J_2 + x_2 J_1 J_2 \subseteq I$ with $J_1 J_2 \in GV(A)$. Then $x_1 + x_2 \in I_w$. For each $a \in A$, since $(I : J_1)$ is a fractional ideal of A, then $ax_1 \in (I : J_1) \subseteq I_w$. We conclude that I_w is a A−sub-module of K. Since $I \in \mathcal{F}(A)$, there is $0 \neq d \in A$ such that $dI \subseteq A$. Then $dI_w = \bigcup_{J \in GV(A)} d(I : J) = \bigcup_{J \in GV(A)} (dI : J) \subseteq \bigcup_{J \in GV(A)} (A : J) = \bigcup_{J \in GV(A)} J^{-1} = A$. We conclude that $I_w \in \mathcal{F}(A)$.

(3)

   (i) Let $a \in K^*$. For each $J \in GV(A)$, $(aA : J) = a(A : J) = aJ^{-1} = aA$. So $(aA)_w = \bigcup_{J \in GV(A)} (aA : J) = aA$.

   (ii) Let $I \subseteq H$ be two element of $\mathcal{F}(A)$. Then $I_w = \bigcup_{J \in GV(A)} (I : J) \subseteq \bigcup_{J \in GV(A)} (H : J) = H_w$.

   (iii) Let $I \in \mathcal{F}(A)$. We have always noted that $I_w \subseteq (I_w)_w$. Conversely, let $x \in (I_w)_w$. There exist $J = (a_1, \dots, a_n) \in GV(A)$ such that $x \in (I_w : J)$. Then $xa_1, \dots, xa_n \in I_w$. There are $J_1, \dots, J_n \in GV(A)$ such that $xa_i \in (I : J_i)$, $1 \leq i \leq n$. So $xa_i J_i \subseteq I$, $1 \leq i \leq n$. By (1), we have $J J_1 \dots J_n \in GV(A)$ with $x J J_1 \dots J_n = x(a_1 A + \dots + a_n A) J_1 \dots J_n = xa_1 J_1 \dots J_n + \dots + xa_n J_1 \dots J_n \subseteq I$. Then $x \in (I : J J_1 \dots J_n) \subseteq I_w$. So $I_w = (I_w)_w$.

   (iv) Let $I \in \mathcal{F}(A)$ and $u \in K^*$. We have $(uI)_w = \bigcup_{J \in GV(A)} (uI : J) = \bigcup_{J \in GV(A)} u(I : J) = u \bigcup_{J \in GV(A)} (I : J) = uI_w$. We conclude that w is a star operation on A.

Let $I \in \mathscr{F}(A)$ and $0 \neq x \in I_w$. There exists $J \in V(A)$ such that $x \in (I : J)$. Then xJ is a finitely generated nonzero subideal of I. Since $x \in (xJ : J)$, then $x \in (xJ)_w$. We conclude that w has a finite character.

(4) The set $\mathscr{F}$ of the proper integral w—ideals of A is not empty. It contains the nonzero proper principal ideals. It is also inductive. Indeed, let $(I_\alpha)_{\alpha \in \Lambda}$ be a totally ordered family of elements of $\mathscr{F}$ and $I = \bigcup_{\alpha \in \Lambda} I_\alpha$. Then I is also a proper integral w—ideal of A which bounds this family in $\mathscr{F}$. To see this, let J be a nonzero finitely generated subideal of I. There exists $\alpha_0 \in \Lambda$ such that $J \subseteq I_{\alpha_0}$, and then $J_w \subseteq (I_{\alpha_0})_w = I_{\alpha_0} \subseteq I$. So $I_w = I$. We conclude by Zorn's lemma.

(5)

(a) The inclusion $(I_1 \cap I_2)_w \subseteq (I_1)_w \cap (I_2)_w$ is clear. Conversely, let $x \in (I_1)_w \cap (I_2)_w$. There exist $J_1, J_2 \in GV(A)$ such that $x \in (I_1 : J_1)$ and $x \in (I_2 : J_2)$. Since $xJ_1 \subseteq I_1$ and $xJ_2 \subseteq I_2$, then $xJ_1J_2 \subseteq I_1 \cap I_2$. So $x \in (I_1 \cap I_2 : J_1J_2)$ with $J_1J_2 \in GV(A)$, by (1). Then $x \in (I_1 \cap I_2)_w$. The finite case follows by induction.

(b) For each $\alpha \in \Lambda$, $I_w \subseteq (I_\alpha)_w = I_\alpha$. Then $I_w \subseteq \bigcap_{\alpha \in \Lambda} I_\alpha = I$. So $I_w = I$.

(c) We have $I^{-1} = (A : I) = \bigcap_{0 \neq x \in I} (A : xA) = \bigcap_{0 \neq x \in I} \frac{1}{x} A$, an intersection of w—ideals. Then I^{-1} is a w—ideal by (b).

(6) Let $0 \neq x \in \bigcap\{IA_M;\ M \in w - Max(A)\}$. For each $M \in w - Max(A)$, there exists $s_M \in A \setminus M$ such that $s_M x \in I$. We have $(\frac{1}{x}I)_w = \frac{1}{x}I_w = \frac{1}{x}I$. By (5), $J = A \cap \frac{1}{x}I$ is an integral w—ideal, as an intersection of two w—ideals, and $s_M \in J$ for each $M \in w - Max(A)$. Suppose that $J \neq A$. By (4), there exists $M_0 \in w - Max(A)$ such that $J \subseteq M_0$, and then $s_{M_0} \in J \subseteq M_0$, which is absurd. Then $J = A$ and $A \subseteq \frac{1}{x}I$, so $xA \subseteq I$, and then $x \in I$.

(7) Since $IJ \subseteq IJ_w$, then $(IJ)_w \subseteq (IJ_w)_w$. For the converse inclusion, it suffices to show that $IJ_w \subseteq (IJ)_w$. Let $x \in IJ_w$; there exist $a_1, \ldots, a_n \in I$ such that $x \in a_1J_w + \ldots + a_nJ_w$. But $a_1J_w + \ldots + a_nJ_w = (a_1J)_w + \ldots + (a_nJ)_w \subseteq (a_1J + \ldots + a_nJ)_w \subseteq (IJ)_w$, and then $x \in (IJ)_w$. So we have proved the first required equality. By replacing I by I_w, we obtain $(I_wJ_w)_w = (I_wJ)_w = (IJ)_w$. For the last assertion, $I_wJ_w \subseteq (I_wJ_w)_w = (IJ)_w$.

(8) (a) By (5-c), I^{-1} is a w—ideal. Then we have the first equality. Let $x \in K^*$. Then $x \in I^{-1} = (A : I) \iff xI \subseteq A \iff (xI)_w \subseteq A \iff xI_w \subseteq A \iff x \in (A : I_w) = (I_w)^{-1}$. Then $I^{-1} = (I_w)^{-1}$.

(b) We apply the second equality of (a) to I^{-1}. We obtain $(I^{-1})^{-1} = ((I^{-1})_w)^{-1}$. By the first equality of (a), we have $I_v = ((I_w)^{-1})^{-1} = (I_w)_v$. The last assertion is clear.

(c) By (3), we have $I_w = \cup J_w$, where J runs over the finitely generated nonzero fractional subideals of I. By (b), $J_w \subseteq J_v$. Then $I_w \subseteq I_t$.

(9) Let $F = (a_1, \ldots, a_n)$. By (3), w has a finite character. Then $a_i \in (F_i)_w$, where F_i is a finitely generated subideal of I. Let $F' = F_1 + \ldots + F_n$ be a finitely generated subideal of I. We have $F \subseteq (F_1)_w + \ldots + (F_n)_w \subseteq (F')_w$. Then $I_w = F_w \subseteq ((F')_w)_w = (F')_w \subseteq I_w$. So $I_w = (F')_w$.

Exercise 28

(1) Let $x \in (\sqrt{I})_w$. There exists an ideal $J = (a_1, \ldots, a_n) \in GV(A)$ such that $x \in (\sqrt{I} : J)$. Since $xJ \subseteq \sqrt{I}$, there exists $m \in \mathbb{N}^*$ such that $(xa_i)^m \in I$, $1 \le i \le n$. Let $k = mn$. Then $x^k J^k \subseteq I$. By (1) of the preceding exer, $J^k \in GV(A)$. Then $x^k \in (I : J^k) \subseteq I_w = I$ and $x \in \sqrt{I}$. So $(\sqrt{I})_w = \sqrt{I}$.

(2) Since $I \subseteq I_w \subseteq A$, then $\sqrt{I} \subseteq \sqrt{I_w}$. By (1), $\sqrt{I_w}$ is a w-ideal. So $(\sqrt{I})_w \subseteq \sqrt{I_w}$. Conversely, let $x \in \sqrt{I_w}$. There exists $n \in \mathbb{N}^*$ such that $x^n \in I_w$. There exists an ideal $J = (a_1, \ldots, a_r) \in GV(A)$ such that $x^n \in (I : J)$. Since $x^n J \subseteq I$, then $x^n J^n \subseteq I$. In particular, $(xa_i)^n \in I$, so $xa_i \in \sqrt{I}$ for $1 \le i \le r$. Then $xJ \subseteq \sqrt{I}$ and $x \in (\sqrt{I} : J) \subseteq (\sqrt{I})_w$. So $\sqrt{I_w} = (\sqrt{I})_w$.

References

D.D. Anderson, T. Dumitrescu, S-Noetherian rings. Comm. Algebra **30**(9), 4407–4416 (2002)

D.D. Anderson, M. Winders, Idealization of a module. J. Commutative Algebra **1**(1), 3–56 (2009)

D.D. Anderson, D.F. Anderson, R. Markanda, The ring $R(X)$ and $R < X >$. J. Algebra **95**, 96–115 (1985)

J.T. Arnold, On the ideal theory of the Kronecker function ring and the domain D(X). Canad. J. Math. **21**, 558–563 (1969)

J. Baeck, G. Lee, J.W. Lim, S-Noetherian rings and their extension. Taiwanese J. Math. **20**(6), 1231–1250 (2016)

A. Benhissi, F. Koja, Basic properties of Hurwitz series rings. Ricerche Mat. **61**, 255–273 (2012)

P.-J. Cahen, Y. Haouat, Spectre d'anneaux de polynômes sur une suite croissante d'anneaux. Arch. Math. **49**, 281–285 (1987)

G.W. Chang, Strong Mori domains and the ring $D[X]_{N_v}$. J. Pure Appl. Algebra **197**, 293–304 (2005)

M. D'anna, C.A. Finocchiaro, M. Fontana, Amalgamated algebra along an ideal, in ed. by M. Fontana et al., *Commutative Algebra and Its Applications: Proceedings of the Fifth International Fez Conference on Commutative Algebra and Its Applications*, Fez, Marocco (W. de Gruyter Publisher, Berlin, 2008), pp. 155–172

M. D'anna, C.A. Finocchiaro, M. Fontana, Properties of chains of prime ideals in a amalgamated algebra along an ideal. J. Pure Appl. Algebra **214**, 1633–1641 (2010)

F. Dress, Familles de séries formelles et ensembles de nombres algébriques. Ann. Scient. Ec. Norm. Sup.,Série 4 **1**, 1–44 (1968)

F. Dress, Stathmes euclidiens et séries formelles. Séminaire Delange -Pisot -Poitou. Théorie des Nombres t. **12**, Exp. 2, 1–7 (1970–1971)

F. Dress, Stathmes euclidiens et séries formelles. Acta Arithmetica **19**, 261–265 (1971)

A. Faisol, B. Surodjo, S. Wahymi, The sufficient conditions for $R[X]$-module $M[X]$ to be $S[X]$-Noetherian. Euro. J. Math. Sci. **5**(1), 1–13 (2019)

W. Fanggui, R.L. McCasland, On ω-modules over strong Mori domains. Comm. Algebra **25**(4), 1285–1306 (1997)

R. Gilmer, Power series rings over Krull domains. Pacific J. Math. **29**(3), 543–549 (1969)

R. Gilmer, An existence theorem for non-Noetherian rings. Amer. Math. Month. **77**(6), 621–623 (1970)

R. Gilmer, W. Heinzer, Rings of formal power series over a Krull domain. Math. Z. **106**, 379–387 (1968)

R. Gilmer, J.F. Hoffmann, A characterization of Prufer domains in terms of polynomials. Pacific J. Math. **60**(1), 81–85 (1975)

E. Hamann, E. Houston, J.L. Johnson, Properties of uppers to zero in $R[X]$. Pacific J. Math. **135**(1), 65–79 (1988)

A. Hamed, S. Hizem, S-Noetherian rings of the form $\mathscr{A}[X]$ and $\mathscr{A}[[X]]$. Comm. Algebra **43**, 3848–3856 (2015)

A. Hamed, S. Hizem, Modules satisfying the S-Noetherian property and S-ACCR. Comm. Algebra **44**, 941–951 (2016)

Y. Haouat, Anneaux de polynômes; Thèse d'Etat (1988); Faculté des Sciences de Tunis

J.R. Hedstrom, E.G. Houston, Some remarks on star-operations. J. Pure Appl. Algebra **18**, 37–44 (1980)

S. Hizem, Power series over an ascending chain of rings. Comm. Algebra **40**, 4263–4275 (2012)

S. Hizem, A. Benhissi, When is $A + XB[[X]]$ Noetherian. C.R. Acad. Sci. Paris, Ser. I **340**, 5–7 (2005)

E.G. Houston, M. Zafrullah, On t−invertibility II. Comm. Algebra **17**(8), 1955–1969 (1989)

Y.-J. Jiao, S.-G. Zhang, S-Noetherianess of $A + XB[[X]]$. J. Shandong Univ. (Nat. Sci.) **44**(8), 58–61 (2009)

S. Kabbaj, N. Mahdou, Trivial extensions defined by coherent like conditions. Comm. Algebra **22**(10), 3937–3953 (2004)

B.G. Kang, Prufer v-multiplication domains and the ring $R[X]_{N_v}$. J. Algebra **123**(1), 151–170 (1989)

D.K. Kim, J.W. Lim, The Cohen type theorem and the Eakin-Nagata type theorem for S-Noetherian rings revisited. Rocky Mountain J. Math. **50**(2), 619–630 (2020)

J.W. Lim, A note on S-Noetherian domains. Kyungpook Math. J. **55**, 507–514 (2015)

J.W. Lim, D.Y. Oh, S-Noetherian properties on amalgamated algebras along an ideal. J. Pure Appl. Algebra **218**, 1075–1080 (2014)

J.W. Lim, D.Y. Oh, S-Noetherian properties of composite ring extensions. Comm. Algebra **43**, 2820–2829 (2015)

E. Paran, M. Temkin, Power series over generalized Krull domains. J. Algebra **323**, 546–550 (2010)

M.H. Park, A. Hamed, W. Maaref, Anti-Archimedean property and the formal power series rings. Comm. Algebra **47**(8), 3190–3197 (2019)

P. Ribenboim, The ascending chain condition for real ideals. C. R. Math. Rep. Acad. Sci. Canada **VIII**(1), 65–68 (1986a)

P. Ribenboim, The ascending chain condition for real ideals. Manuscripta Math. **57**, 109–124 (1986b)

L. Zhongkui, On S−Noetherian rings. Archivum Math. (Brno) **43**, 55–60 (2007)

Chapter 2
S-Artinian Rings and Modules

In this chapter, all the rings considered are commutative with unity. A multiplicative set contains 1 and does not contain 0.

1 Saturated Multiplicative Sets of a Ring

The saturated multiplicative sets of rings will play an important role in this chapter. For this reason, it deserves a full section.

Definition 1.1 Let S be a multiplicative set of a ring A. We say that S is saturated if for each $a \in A$ and $b \in S$ such that a divides b, and then $a \in S$.

Example 1 The multiplicative set $S = \{1\}$ of $\mathbb{Z}$ is not saturated since -1 divides 1 and does not belong to S.

Example 2 The group of the unities of any ring is a saturated multiplicative set.

Example 3 The set S of the regular elements of any ring A is a saturated multiplicative set. Indeed, let $a \in A$ and $b \in S$ such that a divides b. There exists $c \in A$ such that $b = ca$. Let $x \in A$ such that $ax = 0$. Then $bx = cax = 0$, so $x = 0$ and $a \in S$.

Example 4 The multiplicative set $S = \mathbb{Z} \times \mathbb{Q}^*$ of the product ring $A = \mathbb{Z} \times \mathbb{Q}$ is saturated. Indeed, let $(n, a) \in A$ and $(m, b) \in S$ such that (n, a) divides (m, b). Since $b \neq 0$, then $a \neq 0$, so $(n, a) \in S$.

Example 5 Let A be a ring and M a nonzero A-module. The set denoted $U_M(A) = \{a \in A; \ aM = M\}$ is a saturated multiplicative set of A containing $U(A)$. Indeed, let $a, b \in A$ be such that a divides b and $bM = M$. Then $b = ac$ with $c \in A$. So $M = acM \subseteq aM \subseteq M$, and then $aM = M$ and $a \in U_M(A)$.

© The Author(s), under exclusive license to Springer Nature Switzerland AG 2022
A. Benhissi, *Chain Conditions in Commutative Rings*,
https://doi.org/10.1007/978-3-031-09898-7_2

Example 6 Let A be an integral domain. The set $S = \left\{ a \in A; \ \bigcap\limits_{i:0}^{\infty} a^i A \neq 0 \right\}$ is a saturated multiplicative set of A. Indeed, it is clear that $0 \notin S$ and $1 \in S$. Let $a, a' \in S$. There exist $0 \neq x \in \bigcap\limits_{i:0}^{\infty} a^i A$ and $0 \neq y \in \bigcap\limits_{i:0}^{\infty} a'^i A$. For each $i \in \mathbb{N}$, there exist $c_i, c_i' \in A$ such that $x = a^i c_i$ and $y = a'^i c_i'$, and then $0 \neq xy = (aa')^i c_i c_i'$. So $\bigcap\limits_{i:0}^{\infty} (aa')^i A \neq (0)$ and $aa' \in S$.

Let a and a' be two elements of A such that $aa' \in S$. Then $(0) \neq \bigcap\limits_{i:0}^{\infty} (aa')^i A \subseteq \bigcap\limits_{i:0}^{\infty} a^i A$ and $(0) \neq \bigcap\limits_{i:0}^{\infty} (aa')^i A \subseteq \bigcap\limits_{i:0}^{\infty} a'^i A$, so $a, a' \in S$.

Krull has always shown that if S is a multiplicative set of a ring A, each ideal of A disjoint with S may be enlarged to an ideal P maximal to avoid S and such an ideal is necessarily prime. We use this result in the following proposition to characterize the saturated multiplicative sets of the ring A.

Proposition 1.2 *Let S be a nonempty set of a ring A. Then S is a saturated multiplicative set if and only if S is the complementary of the union of a family of prime ideals.*

Proof " $\Longleftarrow$ " Let $(P_\alpha)_{\alpha \in \Lambda}$ be a family of prime ideals of A such that $S = A \setminus \bigcup\limits_{\alpha \in \Lambda} P_\alpha = \bigcap\limits_{\alpha \in \Lambda} (A \setminus P_\alpha)$: a multiplicative set. Let $a \in A$ and $b \in S$ such that a divides b. Then $b = ac$ with $c \in A$. Suppose that $a \notin S$. There exists $\alpha_0 \in \Lambda$ such that $a \notin (A \setminus P_{\alpha_0})$, and then $a \in P_{\alpha_0}$. So $b = ac \in P_{\alpha_0}$ and then $b \notin S$, which is absurd.

" $\Longrightarrow$ " For each $x \in A \setminus S$, the principal ideal $(x) = \{ ax; \ a \in A \}$ is disjoint with S because S is saturated. There exists a prime ideal P_x of A containing (x) and disjoint with S. Then $A \setminus S \subseteq \bigcup\limits_x (x) \subseteq \bigcup\limits_x P_x$, so $A \setminus \bigcup\limits_x P_x \subseteq S$. Conversely, for each $x \in A \setminus S$, $S \cap P_x = \emptyset$, and then $S \subseteq A \setminus P_x$. So $S \subseteq \bigcap\limits_x (A \setminus P_x) = A \setminus \bigcup\limits_x P_x$ and then $S = A \setminus \bigcup\limits_x P_x$. $\qquad\qquad\square$

Example 1 For any ring A, $U(A) = A \setminus \bigcup \{ M; \ M \in Max(A) \} = A \setminus \bigcup \{ P; \ P \in Spec(A) \}$. In particular, we do not have uniqueness in the preceding proposition.

Example 2 The set $Z(A)$ of the zero divisors of a ring A is the union of prime ideals. Indeed, $Z(A)$ is the complementary of the set of the regular elements of A, which is a saturated multiplicative set.

Example 3 If P is a prime ideal of a ring A, then $A \setminus P$ is a saturated multiplicative set of A.

Example 4 Let A be any ring. By Lemma 10.4 of Chap. 1, the set $N = \{f \in A[X]; \ c(f) = A\} = A[X] \setminus \bigcup \{M[X]; \ M \in Max(A)\}$ is a saturated multiplicative set of the polynomial ring $A[X]$.

Example 5 Let A be an integral domain. By Proposition 11.4 of Chap. 1, the set $N_v = \{0 \neq f \in A[X]; \ c(f)_v = c(f)_t = A\} = A[X] \setminus \bigcup \{M[X]; \ M \in t - Max(A)\}$ is a saturated multiplicative set of the polynomial ring $A[X]$, where v and t are respectively the v-operation and the t-operation over the domain A.

We have analogue examples for power series. For a ring A and $f \in A[[X]]$, the content of f is the ideal $c(f)$ generated by its coefficients.

Example 6 Let A be a ring. The set $N = \{f \in A[[X]]; \ c(f) = A\} = A[[X]] \setminus \bigcup \{M[[X]]; \ M \in Max(A)\}$ is multiplicative and saturated in the power series ring $A[[X]]$.

Indeed, let $f \in A[[X]]$. Then $f \in N \iff c(f) = A \iff \forall M \in Max(A), c(f) \not\subseteq M \iff \forall M \in Max(A), f \notin M[[X]] \iff f \in A[[X]] \setminus \bigcup \{M[X]; \ M \in Max(A)\}$. Then $N = A[[X]] \setminus \bigcup \{M[[X]]; \ M \in Max(A)\}$. Since $M[[X]]$ is a prime ideal of $A[[X]]$ for each $M \in Max(A)$, then N is a saturated multiplicative set of the ring $A[[X]]$.

Example 7 Let A be an integral domain. The set $N_t = \{0 \neq f \in A[[X]]; \ c(f)_t = A\} = A[[X]] \setminus \bigcup \{M[[X]]; M \in t - Max(A)\}$ is multiplicative and saturated in the ring $A[[X]]$. Indeed, let $0 \neq f \in A[X]$. By Proposition 11.2 of Chap. 1, $f \in N_t \iff c(f)_t = A \iff \forall M \in t - Max(A), c(f)_t \not\subseteq M \iff \forall M \in t - Max(A), c(f) \not\subseteq M \iff \forall M \in t - Max(A), f \notin M[[X]] \iff f \in A[[X]] \setminus \bigcup \{M[[X]]; \ M \in t - Max(A)\}$. Then $N_t = A[[X]] \setminus \bigcup \{M[[X]]; M \in t - Max(A)\}$. By Proposition 11.3 of Chap. 1, $M[[X]]$ is a prime ideal of $A[[X]]$ for each $M \in t - Max(A)$. Then N_t is a saturated multiplicative set of the ring $A[[X]]$.

Proposition 1.3 *Let S be a multiplicative set of a ring A.*

1. *The set $S^* = \{a \in A; \ \frac{a}{1} \in U(S^{-1}A)\} = \{a \in A; \ \exists b \in A, ab \in S\}$, the set of the divisors of the elements of S, is the smallest saturated multiplicative set of A containing S. We say that S^* is the saturation of S in A.*
2. *Let $(P_\lambda)_{\lambda \in \Lambda}$ be the family of the prime ideals of A disjoints with S. Then $S^* = A \setminus \bigcup_{\lambda \in \Lambda} P_\lambda$. More precisely, let $(P_\lambda)_{\lambda \in \Lambda'}$ be the subfamily of the prime ideals of A maximal to avoid S. Then $S^* = A \setminus \bigcup_{\lambda \in \Lambda'} P_\lambda$.*
3. *The multiplicative set S is saturated if and only if $S = S^*$.*

Proof

(1)   (i) Let's start by proving the equality of the two sets. Let $a \in A$ be such that $\frac{a}{1} \in U(S^{-1}A)$. There exist $b \in A$ and $s \in S$ such that $\frac{a}{1}\frac{b}{s} = \frac{1}{1}$. There exists $t \in S$ such that $t(ab - s) = 0$. Then $a(tb) = st \in S$. Conversely, let $a, b \in A$ be such that the element $s = ab \in S$. Then $\frac{b}{s} \in S^{-1}A$ and $\frac{a}{1}\frac{b}{s} = \frac{1}{1}$ because $1(ab - s) = ab - ab = 0$. So $\frac{a}{1} \in U(S^{-1}A)$.

(ii) It is clear that S^* is a multiplicative set of A. Let $s \in S$. Then $\frac{s}{1} \in U(S^{-1}A)$, so $s \in S^*$ and $S \subseteq S^*$.

(iii) Let $a \in S^*$ and $b \in A$ such that b divides a. Then $\frac{b}{1}$ divides $\frac{a}{1}$ in the ring $S^{-1}A$. But $\frac{a}{1}$ is invertible, and then $\frac{b}{1}$ is also invertible. So $b \in S^*$ and then S^* is saturated.

(iv) Let $\overline{S}$ be a saturated multiplicative set of A containing S. Let $a \in S^*$, and then $\frac{a}{1} \in U(S^{-1}A)$. There exist $b \in A$ and $s \in S$ such that $\frac{a}{1}\frac{b}{s} = \frac{1}{1}$. There exists $t \in S$ such that $tab = ts \in S \subseteq \overline{S}$. But $\overline{S}$ is saturated, and then $a \in \overline{S}$ and then $S^* \subseteq \overline{S}$.

(2)   (i) " $\subseteq$" By the preceding proposition, $S' = A \setminus \bigcup_{\lambda \in \Lambda} P_\lambda$ is a saturated multiplicative set of A. By definition, for each $\lambda \in \Lambda$, $P_\lambda \cap S = \emptyset$, and then $S \subseteq A \setminus P_\lambda$. So $S \subseteq \bigcap_{\lambda \in \Lambda}(A \setminus P_\lambda) = A \setminus \bigcup_{\lambda \in \Lambda} P_\lambda = S'$. By (1), $S^* \subseteq S'$.

" $\supseteq$" By the preceding proposition, since S^* is saturated, there exists a family $(Q_\alpha)_{\alpha \in \Gamma}$ of prime ideals of A such that $S^* = A \setminus \bigcup_{\alpha \in \Gamma} Q_\alpha$. For each $\alpha \in \Gamma$, Q_α is a prime ideal of A disjoint with S^*, so it is disjoint with S because $S \subseteq S^*$. Then $(Q_\alpha)_{\alpha \in \Gamma}$ is a subfamily of $(P_\lambda)_{\lambda \in \Lambda}$. Then $\bigcup_{\alpha \in \Gamma} Q_\alpha \subseteq \bigcup_{\lambda \in \Lambda} P_\lambda$, so $A \setminus \bigcup_{\lambda \in \Lambda} P_\lambda \subseteq A \setminus \bigcup_{\alpha \in \Gamma} Q_\alpha$, and then $S' \subseteq S^*$. We conclude that $S' = S^*$.

(ii) Since $\Lambda' \subseteq \Lambda$, then $S^* = A \setminus \bigcup_{\lambda \in \Lambda} P_\lambda \subseteq A \setminus \bigcup_{\lambda \in \Lambda'} P_\lambda$. Conversely, by Zorn's lemma, for each $\lambda \in \Lambda$, there exists $\lambda' \in \Lambda'$ such that $P_\lambda \subseteq P_{\lambda'}$. Then $\bigcup_{\lambda \in \Lambda} P_\lambda \subseteq \bigcup_{\lambda \in \Lambda'} P_\lambda$. So $A \setminus \bigcup_{\lambda \in \Lambda'} P_\lambda \subseteq A \setminus \bigcup_{\lambda \in \Lambda} P_\lambda = S^*$.

(3) Clear by (1).   □

Example 1 Let $A \subseteq B$ be an extension of rings. The saturation of the multiplicative set $S = U(A)$ in B is $S^* = U(B)$.

Indeed, $S^* = \{x \in B; \; x \in U(S^{-1}B)\} = \{x \in B; \; x \in U(B)\} = U(B)$.

Example 2 Let S be a multiplicative set of an integral domain A and S^* its saturation. Then $S^{-1}A = S^{*-1}A$. Indeed, since $S \subseteq S^*$, then $S^{-1}A \subseteq S^{*-1}A$. Conversely, let $x \in S^{*-1}A$, $x = \frac{a}{s'}$ with $a \in A$ and $s' \in S^*$. By the preceding proposition, there exists $t \in A$ such that $s't \in S$, and then $x = \frac{a}{s'} = \frac{at}{s't} \in S^{-1}A$. So $S^{-1}A = S^{*-1}A$.

Example 3 Let I be a proper ideal of a ring A. Then $S = 1 + I$ is a multiplicative set of A with saturation $S^* = A \setminus \bigcup \{M \in Max(A); \ I \subseteq M\}$.

Indeed, note first that if P is a prime ideal of A, then $P \cap S \neq \emptyset \iff \exists p \in P, \ \exists a \in I$ such that $p = 1 + a \iff \exists p \in P, \ \exists a \in I$ such that $1 = p - a \iff P + I = A$.

Then $P \cap S = \emptyset \iff P + I \neq A \iff \exists M \in Max(A)$ such that $P + I \subseteq M$.

Let $(P_\lambda)_{\lambda \in \Lambda}$ be the family of the prime ideals of A which are disjoint with S. By Proposition 1.3 (2), $S^* = A \setminus \bigcup_{\lambda \in \Lambda} P_\lambda$. It suffices to show that $\bigcup \{M \in Max(A); \ I \subseteq M\} = \bigcup_{\lambda \in \Lambda} P_\lambda$. For the first inclusion, let $M \in Max(A)$ with $I \subseteq M$. Then M is a prime ideal of A and $M + I \subseteq M$. By the preceding observation, $M \cap S = \emptyset$. Then M is some P_λ. Conversely, let $\lambda \in \Lambda$. Then $P_\lambda \cap S = \emptyset$. By the preceding observation, there exists $M_\lambda \in Max(A)$ such that $P_\lambda + I \subseteq M_\lambda$. Then $M_\lambda \in Max(A)$ with $I \subseteq M_\lambda$ and $P_\lambda \subseteq M_\lambda$.

Proposition 1.4 (Gilmer and Hoffmann 1975) *Let $A \subseteq B$ be an integral extension of rings. We consider the multiplicative sets:*

1. $N = \{f \in A[X]; \ c(f) = A\} = A[X] \setminus \bigcup \{M[X]; \ M \in Max(A)\}$ *of the ring* $A[X]$.
2. $N' = \{f \in B[X]; \ c(f) = B\} = B[X] \setminus \bigcup \{M[X]; \ M \in Max(B)\}$ *of the ring* $B[X]$.

Then N' is the saturation of N in the ring $B[X]$. Moreover, if B is an integral domain, then $B[X]_N = B[X]_{N'} = B(X)$, the Nagata ring associated with the domain B.

Proof Let N^* be the saturation of N in $B[X]$. By the preceding proposition, since N' is saturated in $B[X]$ and $N \subseteq N'$, then $N^* \subseteq N'$. Let $\mathscr{P}$ be the set of the prime ideals of $B[X]$ that are maximal to avoid N. By the preceding proposition, $N^* = B[X] \setminus \bigcup \{P; \ P \in \mathscr{P}\}$. To prove the inclusion $N' \subseteq N^*$, it suffices to show that $\mathscr{P} \subseteq \{M[X]; \ M \in Max(B)\}$. Let $P' \in \mathscr{P}$ and $P = P' \cap A[X]$. Since $P' \cap N = \emptyset$, then $P \cap N = \emptyset$, so $P \subseteq \bigcup \{M[X]; \ M \in Max(A)\}$. By the example of Lemma 10.4, Chap. 1, there exists $M_0 \in Max(A)$ such that $P \subseteq M_0[X]$. Since the extension $A[X] \subseteq B[X]$ is integral, by the GU property, there exists $Q' \in spec(B[X])$ such that $P' \subseteq Q'$ and $Q' \cap A[X] = M_0[X]$. Then $(Q' \cap B) \cap A = (Q' \cap A[X]) \cap A = M_0[X] \cap A = M_0$. Since the extension $A \subseteq B$ is integral and $M_0 \in Max(A)$, the ideal $M' = Q' \cap B \in Max(B)$. We have $M' \cap A = M_0$ and $M'[X] \subseteq Q'$ because $M' \subset Q'$. Since the extension $A[X] \subseteq B[X]$ is integral and $Q' \cap A[X] = M_0[X] = (M' \cap A)[X] = M'[X] \cap A[X]$, by the INC property, $M'[X] = Q'$. Then we have the inclusion $P' \subseteq Q' = M'[X]$ with $M'[X] \cap N = $

$M'[X] \cap N \cap A[X] = (M' \cap A)[X] \cap N = M_0[X] \cap N = \emptyset$. By the maximality of P' to avoid N, we have $P' = M'[X]$.

The last assertion follows from Example 2 of the preceding proposition. $\square$

2 S-Artinian Rings

The results of the three following sections are due to Sevim et al. (2020). Let's start by recalling the definition of an Artinian ring.

Definition 2.1 A ring A is said to be Artinian if each decreasing sequence of ideals of A is stationary.

Examples

(1) The domain $\mathbb{Z}$ is Noetherian, but not Artinian because for each prime number p, the sequence $p\mathbb{Z} \supset p^2\mathbb{Z} \supset p^3\mathbb{Z} \supset \ldots$ is strictly decreasing.
(2) Any finite ring is Artinian and Noetherian. This is the case for $\mathbb{Z}/n\mathbb{Z}$ with $n \in \mathbb{N}^*$.
(3) Let K be a field. The polynomial ring $K[X_1, X_2, \ldots]$, in countable variables over K, is neither Noetherian nor Artinian because $(X_1) \supset (X_1^2) \supset (X_1^3) \supset \ldots$.
(4) A sub-ring of an Artinian ring is not necessarily Artinian. For example, $\mathbb{Z} \subset \mathbb{Q}$.
(5) For any ring A, the polynomial ring $A[X]$ and the formal power series ring $A[[X]]$ are neither Artinian because $(X) \supset (X^2) \supset \ldots$.

The following classical lemma is very useful in the sequel.

Lemma 2.2 *Let $A_1, \ldots, A_n$ be rings. The ideals of the product ring $A_1 \times \ldots \times A_n$ are of the form $I_1 \times \ldots \times I_n$ with I_i an ideal of A_i for $1 \le i \le n$.*

Proof By induction, it suffices to show the result for $n = 2$. The canonical projections $\pi_1 : A_1 \times A_2 \longrightarrow A_1$ and $\pi_2 : A_1 \times A_2 \longrightarrow A_2$ are onto. If I is an ideal of $A_1 \times A_2$, then $I_1 = \pi_1(I)$ and $I_2 = \pi_2(I)$ are respectively ideals of A_1 and A_2. We will show that $I = I_1 \times I_2$. The first inclusion is clear. Conversely, let $(x, y) \in I_1 \times I_2$. There exist $x_1 \in A_1$ and $y_2 \in A_2$ such that $(x, y_2) \in I$ and $(x_1, y) \in I$. Since I is an ideal of $A_1 \times A_2$, then $(x, 0) = (x, y_2)(1, 0) \in I$ and $(0, y) = (x_1, y)(0, 1) \in I$. Then $(x, y) = (x, 0) + (0, y) \in I$. So $I_1 \times I_2 \subseteq I$. $\square$

Example The result is not true for groups. The subgroup $H = \{(n, n); n \in \mathbb{Z}\}$ of $\mathbb{Z} \times \mathbb{Z}$ is not of the form $H_1 \times H_2$ with H_1 and H_2 two subgroups of $\mathbb{Z}$. Indeed, $\pi_1(H) = \pi_2(H) = \mathbb{Z}$. But $H \ne \mathbb{Z} \times \mathbb{Z}$.

We generalize the classical definition of Artinian rings using multiplicative sets.

Definition 2.3 Let S be a multiplicative set of a ring A. We say that A is S-Artinian if each decreasing sequence $I_1 \supseteq I_2 \supseteq \ldots$ of ideals of A is S-stationary; i.e., there exist $s \in S$ and $k \in \mathbb{N}^*$ such that for each integer $n \ge k$, $sI_k \subseteq I_n$.

With the hypotheses and the notations of the definition, $s I_k \subseteq \bigcap_{n:k}^{\infty} I_n = \bigcap_{n:1}^{\infty} I_n$.

Example 1 An Artinian ring is *S*-Artinian with $S = \{1\}$.

Example 2 An Artinian ring is *S*-Artinian for each multiplicative set *S* of *A*. The converse is true if $S \subseteq U(A)$, the group of unities of the ring *A*.

Example 3 Let *D* be an integral domain that is not a field and *K* be a field. The product ring $A = D \times K$ is not Artinian. Indeed, let $0 \neq a \in D$ which is not a unit. We have $aD \times (0) \supset a^2 D \times (0) \supset \ldots$. Consider the multiplicative set $S = D \times K^*$ of *A*. By the preceding lemma, a decreasing sequence of ideals of *A* is of the form: $I_1^{(1)} \times I_2^{(1)} \supseteq I_1^{(2)} \times I_2^{(2)} \supseteq \ldots$. Then $I_1^{(1)} \supseteq I_1^{(2)} \supseteq \ldots$ and $I_2^{(1)} \supseteq I_2^{(2)} \supseteq \ldots$ are two decreasing sequences of ideals, respectively, of *D* and of *K*. Since *K* is a field, the second sequence is stationary at certain rank *k*. Let $s = (0, 1) \in S$. For each integer $n \geq k$, $s\left(I_1^{(k)} \times I_2^{(k)}\right) = (0) \times I_2^{(k)} \subseteq I_1^{(n)} \times I_2^{(n)}$. The initial sequence of *A* is *S*-stationary. Then *A* is *S*-Artinian.

Example 4 (W. Maaref, A. Hamed, A. Benhissi) Let *A* be a domain and *K* its field of fractions. An element *x* in *K* is said to be almost integral over *A* if there is a $0 \neq d \in A$ such that $dx^n \in D$ for all $n \geq 0$. The set A_0 of elements of *K* almost integral over *A* is called the complete integral closure of *A*, and if $A = A_0$, we say that *A* is completely integrally closed. Now, let *A* be a completely integrally closed domain which is not a field. For an arbitrary multiplicative subset *S* of *A*, we claim that *A* is not an *S*-Artinian domain. Indeed, choose $0 \neq a \in A$ such that *a* is not a unit. Then $(a^n A)_{n \in \mathbb{N}}$ is a strictly descending chain of ideals of *A*. Suppose there exists a multiplicative subset *S* of *A* such that *A* is an *S*-Artinian domain. Then there exist $s \in S$ and $n_0 \in \mathbb{N}$ such that $sa^{n_0} \in a^n A$ for all integers $n \geq n_0$. Thus, $s(\frac{1}{a})^n \in A$ for all $n \in \mathbb{N}$ and $\frac{1}{a}$ is almost integral over *A*. Since *A* is a completely integrally closed domain, we have $\frac{1}{a} \in A$; therefore, *a* is a unit, a contradiction.

The following lemma is very useful.

Lemma 2.4 *Let A be a ring.*

1. *Let $S_1 \subseteq S_2$ be two multiplicative sets of A. If the ring A is S_1-Artinian, then it is S_2-Artinian.*
2. *Let S be a multiplicative set of A and S^* its saturation. Then A is S-Artinian if and only if it is S^*-Artinian.*

Proof (1) Clear. (2) By (1), if *A* is *S*-Artinian, then it is S^*-Artinian. Conversely, let $I_1 \supseteq I_2 \supseteq \ldots$ be a decreasing sequence of ideals of *A*. By hypothesis, there exist $r \in S^*$ and $k \in \mathbb{N}^*$ such that for each integer $n \geq k$, $r I_k \subseteq I_n$. By definition of S^*, the element $\frac{r}{1} \in U(S^{-1}A)$. There exist $a \in A$ and $s \in S$ such that $\frac{a}{s} \frac{r}{1} = \frac{1}{1}$. There exists $u \in S$ such that $uar = us$. The element $s' = us \in S$ and for each integer $n \geq k$, $s'I_k = usI_k = uarI_k \subseteq rI_k \subseteq I_n$. The sequence $I_1 \supseteq I_2 \supseteq \ldots$ is S-stationary. Then A is S-Artinian.   □

Example The converse of (1) is in general false. Let D be an integral domain that is not a field and K be a field. Let $A = D \times K$, the product ring, and $S_1 = \{(1, 1)\} \subset S_2 = D \times K^*$. We have just seen that the ring A is S_2-Artinian but it is not Artinian. So A is not S_1-Artinian.

Proposition 2.5 *Let S be a multiplicative set of a ring A and S^* its saturation. If A is S-Artinian, then S^* contains the regular elements of A.*

   In particular, if S is a saturated multiplicative set of A and if A is S-Artinian, then the regular elements of A belong to S.

Proof By the preceding lemma, A is S^*-Artinian. Let a be a regular element of A. The decreasing sequence $(a) \supseteq (a^2) \supseteq \ldots$ is S^*-stationary. There exist $s \in S^*$ and $n \in \mathbb{N}^*$ such that $s(a^n) \subseteq (a^{n+1})$, so $sa^n = ba^{n+1}$ with $b \in A$. Since a is regular, $ba = s \in S^*$. But S^* is saturated, and then $a \in S^*$.   □

Example With the notations and the hypotheses of the preceding proposition, all the regular elements of A belong to S^*. But they are not all in S. Indeed, let K be a field and $S = \{1\}$. Since a field is Artinian, K is S-Artinian. The set of the regular elements of K is $K^* \nsubseteq S$. By Proposition 1.3 (2), the saturation $S^* = K \setminus (0) = K^*$.

   As a consequence of the preceding proposition, we have the following result.

Corollary 2.6

1. *Let S be a saturated multiplicative set of an integral domain A.*
   If A is S-Artinian, then $S = A \setminus (0)$.
2. *Let A be an integral domain which is not a field and P a nonzero prime ideal of A. For $S = A \setminus P$, the ring A is not S-Artinian.*

Proof

(1) The set of the regular elements of the integral domain A is $A \setminus (0)$. By the preceding proposition, $A \setminus (0) \subseteq S^* = S \subseteq A \setminus (0)$. Then $S = A \setminus (0)$.
(2) Since $P \neq (0)$, then $S \neq A \setminus (0)$. By (1), the ring A is not S-Artinian.   □

Example 1 Let A be an integral domain. The ideal (0) is prime in A. By Proposition 1.2, $S = A \setminus (0)$ is a saturated multiplicative set. But the ring A is not necessarily S-Artinian. Indeed, let $A = \mathbb{Z}$ and let p be a prime number. We have the decreasing sequence $p\mathbb{Z} \supset p^2\mathbb{Z} \supset \ldots$ of ideals of $\mathbb{Z}$. Suppose that there exist $s \in \mathbb{Z} \setminus (0)$ and $k \in \mathbb{N}^*$ such that for each integer $n \geq k$, $sp^k\mathbb{Z} \subseteq p^n\mathbb{Z}$. Then p^n divides sp^k for each integer $n \geq k$, which is absurd.

Example 2 The hypothesis of integrity for the ring A is necessary in (1). Indeed, the multiplicative set $S = \mathbb{Z} \times \mathbb{Q}^*$ of the product ring $A = \mathbb{Z} \times \mathbb{Q}$ is saturated. The ring A is S-Artinian and it is not integral since $(1, 0)(0, 1) = (0, 0)$. But $S \neq \mathbb{Z} \times \mathbb{Q} \setminus \{(0, 0)\}$.

Example 3 An integral domain is Artinian if and only if it is a field.

Indeed, a field is clearly Artinian. Conversely, let A be an Artinian integral domain. The multiplicative set $S = U(A)$ is saturated and A is S-Artinian. By (1), $U(A) = S = A \setminus (0)$. All the nonzero elements of A are invertible. This means that A is a field.

Lemma 2.7 *Let A be a ring, S a multiplicative set of A, and $I_1, \ldots, I_n$ ideals of A. Then $S^{-1}(I_1 \cap \ldots \cap I_n) = (S^{-1}I_1) \cap \ldots \cap (S^{-1}I_n)$.*

Proof By induction, it suffices to show the property for $n = 2$. The direct inclusion is clear. Let $x \in (S^{-1}I_1) \cap (S^{-1}I_2)$, $x = \frac{a_1}{s} = \frac{a_2}{s}$ with $a_1 \in I_1$, $a_2 \in I_2$ and $s \in S$. There exists $t \in S$ such that $tsa_1 = tsa_2 \in I_1 \cap I_2$. Then $x = \frac{tsa_1}{ts^2} \in S^{-1}(I_1 \cap I_2)$. $\qquad \square$

Example The result is false for an infinite intersection. Let $A = \mathbb{Z}$ and $S = \mathbb{Z} \setminus (0)$. Then $S^{-1}A = \mathbb{Q}$. The intersection $\bigcap \{p\mathbb{Z}, \ p \text{ a prime number}\} = (0)$. Then $S^{-1}(\bigcap_p p\mathbb{Z}) = S^{-1}(0) = (0)$. But for each prime number p, the ideal $S^{-1}(p\mathbb{Z})$ contains invertible elements of the ring $S^{-1}A$, and then $S^{-1}(p\mathbb{Z}) = S^{-1}A = \mathbb{Q}$. So $\bigcap \{S^{-1}(p\mathbb{Z}), \ p \text{ a prime number}\} = \mathbb{Q} \neq (0)$.

Proposition 2.8 *Let S be a multiplicative set of a ring A. If A is S-Artinian, the ring of fractions $S^{-1}A$ is Artinian.*

Proof Let $I_1 \supseteq I_2 \supseteq \ldots$ be a decreasing sequence of ideals of the ring $S^{-1}A$. For each $n \in \mathbb{N}^*$, there exists an ideal J_n of the ring A such that $I_n = S^{-1}J_n$. The decreasing sequence $J_1 \supseteq J_1 \cap J_2 \supseteq J_1 \cap J_2 \cap J_3 \supseteq \ldots$ is S-stationary. There exist $k \in \mathbb{N}^*$ and $s \in S$ such that for each integer $n \geq k$, $s(J_1 \cap J_2 \cap \ldots \cap J_k) \subseteq J_1 \cap J_2 \cap \ldots \cap J_n$, and then $S^{-1}(J_1 \cap J_2 \cap \ldots \cap J_k) \subseteq S^{-1}(J_1 \cap J_2 \cap \ldots \cap J_n)$. By the preceding lemma, for each integer $n \geq k$, $(S^{-1}J_1) \cap (S^{-1}J_2) \cap \ldots \cap (S^{-1}J_k) \subseteq (S^{-1}J_1) \cap (S^{-1}J_2) \cap \ldots \cap (S^{-1}J_n)$ and then $I_1 \cap I_2 \cap \ldots \cap I_k \subseteq I_1 \cap I_2 \cap \ldots \cap I_n$, so $I_k \subseteq I_n$. Then for each integer $n \geq k$, $I_n = I_k$. $\qquad \square$

Example 1 The converse of the preceding proposition is false. Consider the multiplicative set $S = \mathbb{Z} \setminus (0)$ of the ring $\mathbb{Z}$. The ring of fractions $S^{-1}\mathbb{Z} = \mathbb{Q}$ is a field, and then it is Artinian. Suppose that the ring $\mathbb{Z}$ is S-Artinian. Let p be a prime number and consider the decreasing sequence $p\mathbb{Z} \supset p^2\mathbb{Z} \supset p^3\mathbb{Z} \ldots$ of ideals of $\mathbb{Z}$. There exist $k \in \mathbb{N}^*$ and $s \in S$ such that for each integer $n \geq k$, $sp^k\mathbb{Z} \subseteq p^n\mathbb{Z}$. Put $s = p^m t$ with $m \in \mathbb{N}$ and t a natural number nondivisible by p. For each integer $n \geq k$, p^n divides $p^{m+k}t$, which is absurd.

The following example shows that the converse of the preceding proposition is true when the multiplicative set S is finite.

Example 2 (Ozen et al. 2021) Let S be a finite multiplicative set of a ring A. Then A is S-Artinian if and only if the ring of fractions $S^{-1}A$ is Artinian. Indeed, let $I_1 \supseteq I_2 \supseteq \ldots$ be a decreasing sequence of ideals of the ring A. Then $S^{-1}I_1 \supseteq S^{-1}I_2 \supseteq \ldots$ is a decreasing sequence of ideals of the ring $S^{-1}A$. By hypothesis, there exists $k \in \mathbb{N}^*$ such that for each integer $n \geq k$, $S^{-1}I_k = S^{-1}I_n$. Put $S = \{s_1, \ldots, s_r\}$ and $s = s_1 \ldots s_r \in S$. Let $x \in I_k$, $\frac{x}{1} \in S^{-1}I_k = S^{-1}I_n$ for each integer $n \geq k$. There exists $s_i \in S$ that depends on n such that $s_i x \in I_n$, and then $sx \in I_n$. So $sI_k \subseteq I_n$ for each integer $n \geq k$. Then A is S-Artinian.

Let $f : A \longrightarrow B$ be a homomorphism of rings and S a multiplicative set of A. It is clear that $f(S)$ is a subset of B stable by multiplication and containing 1. Moreover, if $0 \notin f(S)$, then $f(S)$ is a multiplicative set of B.

Proposition 2.9 *Let* $f : A \longrightarrow B$ *be a homomorphism onto of rings and* S *a multiplicative set of* A *such that* $0 \notin f(S)$. *If* A *is* S-Artinian, then B is $f(S)$-*Artinian.*

Proof Let $J_1 \supseteq J_2 \supseteq \ldots$ be a decreasing sequence of ideals of B. Then $f^{-1}(J_1) \supseteq f^{-1}(J_2) \supseteq \ldots$ is a decreasing sequence of ideals of A. There exist $k \in \mathbb{N}^*$ and $s \in S$ such that for each integer $n \geq k$, $sf^{-1}(J_k) \subseteq f^{-1}(J_n)$, and then $f(s)f(f^{-1}(J_k)) \subseteq f(f^{-1}(J_n))$. Since f is onto, $f(s)J_k \subseteq J_n$ for each $n \geq k$. Then B is $f(S)$-Artinian. □

Example Let A be a ring, M a A-module, and S a multiplicative set of A. For each sub-module N of M, $S(+)N = \{(s, n); \ s \in S, n \in N\}$ is a multiplicative set of the ring $A(+)M$. If $A(+)M$ is $S(+)N$-Artinian, then A is S-Artinian.

Corollary 2.10 *Let* A *be an* S-Artinian ring for some multiplicative set S of A, I *an ideal of* A *disjoint with* S, *and* $\pi : A \longrightarrow A/I$ *the canonical homomorphism.*
  The quotient ring A/I *is* $\pi(S)$-*Artinian.*
  In particular, the quotient of an Artinian ring is Artinian.

Proof Since $I \cap S = \emptyset$, then $\bar{0} \notin \pi(S))$. We conclude by the preceding proposition. □

Example The converse of the corollary is false. Take $A = \mathbb{Z}$, $S = \{1\}$ and $I = n\mathbb{Z}$ with $n \geq 2$ an integer. The quotient ring $A/I = \mathbb{Z}/n\mathbb{Z}$ is finite, so it is Artinian. But the ring $A = \mathbb{Z}$ is not Artinian.

We characterize the Artinian rings in terms of S-Artinian rings.

Theorem 2.11 *The following assertions are equivalent for a ring* A:

1. *The ring* A *is Artinian.*
2. *For each* $P \in Spec(A)$, *the ring* A *is* S-Artinian where $S = A \setminus P$.

3. *For each $M \in Max(A)$, the ring A is S-Artinian where $S = A \setminus M$.*

Proof The implications $"(1) \implies (2) \implies (3)"$ are clear.

$"(3) \implies (1)"$ Let $I_1 \supseteq I_2 \supseteq \ldots$ be a decreasing sequence of ideals of A. For each $M \in Max(A)$, there exist $s_M \in A \setminus M$ and $k_M \in \mathbb{N}^*$ such that for each integer $n \geq k_M$, $s_M I_{k_M} \subseteq I_n$. Consider the ideal $J = (s_M; \ M \in Max(A))$ of A. If $J \neq A$, there exists $M_0 \in Max(A)$ such that $J \subseteq M_0$, and then $s_{M_0} \in M_0$, which is absurd. Then $J = A$. There exist $M_1, \ldots, M_r \in Max(A)$ such that $(s_{M_1}, \ldots, s_{M_r}) = A$. Let $k = max\{k_{M_1}, \ldots, k_{M_r}\}$. Then $I_k \subseteq I_{k_{M_i}}$ for each $i \in \{1, \ldots, r\}$. For each integer $n \geq k$, $I_k = AI_k = (s_{M_1}, \ldots, s_{M_r})I_k = s_{M_1} I_k + \ldots + s_{M_r} I_k \subseteq s_{M_1} I_{k_{M_1}} + \ldots + s_{M_r} I_{k_{M_r}} \subseteq I_n$, and then $I_k = I_n$ so A is Artinian. $\qquad\square$

Lemma 2.12 (Sheldon 1971) *Let A be an integral domain and S a multiplicative subset of A. The following assertions are equivalent:*

1. *The domain $(S^{-1}A)[[X]]$ is an over ring of $A[[X]]$.*

2. *We have $(S^{-1}A)[[X]] \subseteq (A \setminus 0)^{-1} A[[X]]$.*

3. *For each sequence $(s_i)_{i \in \mathbb{N}}$ of elements of S, the intersection $\bigcap_{i=0}^{\infty} s_i A \neq (0)$.*

Proof (Gilmer 1967) $"(1) \implies (3)"$ Let $(s_i)_{i \in \mathbb{N}}$ be a sequence of elements of S. For each $k \in \mathbb{N}$, let $c_k = \frac{1}{s_0 s_1 \ldots s_k}$. By hypothesis, the series $f = \sum_{k=0}^{\infty} c_k X^k \in (S^{-1}A)[[X]] \subseteq qf(A[[X]])$. There exist $g = \sum_{i=0}^{\infty} a_i X^i$ and $0 \neq h = \sum_{i=0}^{\infty} b_i X^i \in A[[X]]$ such that $f = \frac{g}{h}$, so $fh = g$. Canceling by a convenient power of X, we may suppose that $b_0 \neq 0$. For each $n \in \mathbb{N}$, we have $a_n = \sum_{i=0}^{n} b_i c_{n-i} = \frac{b_0}{s_0 s_1 \ldots s_n} + \frac{b_1}{s_0 s_1 \ldots s_{n-1}} + \ldots + \frac{b_n}{s_0}$

$\implies a_n s_0 s_1 \ldots s_n = b_0 + b_1 s_n + b_2 s_n s_{n-1} + \ldots + b_n s_n s_{n-1} \ldots s_1$

$\implies b_0 = s_n(a_n s_0 s_1 \ldots s_{n-1} - b_1 - b_2 s_{n-1} - \ldots - b_n s_{n-1} \ldots s_1) \in s_n A$, and then $0 \neq b_0 \in \bigcap_{n=0}^{\infty} s_n A$.

$"(3) \implies (2)"$ Let $f \in (S^{-1}A)[[X]]$. Then $f = \sum_{n=0}^{\infty} \frac{a_n}{s_n} X^n$ with $a_n \in A$ and $s_n \in S$. Let $0 \neq b \in \bigcap_{n=0}^{\infty} s_n A$. Then $b = s_n c_n$ with $c_n \in A$ for each $n \in \mathbb{N}$. So

$$f = \sum_{n=0}^{\infty} \frac{a_n c_n}{b} X^n = \frac{1}{b} \sum_{n=0}^{\infty} a_n c_n X^n \in (A \setminus 0)^{-1} A[[X]].$$

"(2) $\Longrightarrow$ (1)" By hypothesis, $A[[X]] \subseteq (S^{-1}A)[[X]] \subseteq (A \setminus 0)^{-1}A[[X]] \subseteq qf(A[[X]])$. $\qquad\qquad\qquad\qquad\qquad\qquad\qquad\qquad\qquad\qquad\qquad\qquad\qquad\qquad$ □

Example The preceding lemma works for a multiplicative set of the form $S = A \setminus P$ where P is a prime ideal of an integral domain A. For example, if p is a prime number, then $qf(\mathbb{Z}_{(p)}[[X]]) \neq qf(\mathbb{Z}[[X]])$. Indeed, let $q \neq p$ be a prime number. Then $\bigcap_{i=0}^{\infty} q^i \mathbb{Z} = (0)$. In particular, $qf(\mathbb{Z}[[X]]) \neq \mathbb{Q}((X))$.

By taking $S = A \setminus 0$ in the preceding lemma, we obtain the following proposition.

Proposition 2.13 (Gilmer 1967) *Let A be an integral domain with quotient field K. The following assertions are equivalent:*

1. $qf(A[[X]]) = K((X))$.
2. $K[[X]] = (A \setminus 0)^{-1}A[[X]]$.
3. *For each sequence $(s_i)_{i \in \mathbb{N}}$ of nonzero elements of A, we have $\bigcap_{i=0}^{\infty} s_i A \neq (0)$.*

Example Let A be a valuation domain of rank one, defined by a valuation v with values in $\mathbb{R}$, and let K be its quotient field. Then $qf(A[[X]]) \neq K((X))$. Indeed, let (a_n) be a sequence of elements in A such that $\lim_{n \to +\infty} v(a_n) = +\infty$, for example, $a_n = a^n$ where $a \in A$ with $v(a) > 0$. If $x \in \bigcap_{n=1}^{\infty} A a_n$, then for each $n \in \mathbb{N}^*$, there exists $b_n \in A$ such that $x = b_n a_n$, so $v(x) = v(b_n) + v(a_n) \geq v(a_n) \longrightarrow +\infty$. Then $v(x) = +\infty$ and $x = 0$.

We now define a special type of anti-Archimedean multiplicative subsets.

Definition 2.14 A multiplicative subset S of a ring A is called strongly anti-Archimedean if $\left(\bigcap_{i \geq 1} s_i A\right) \cap S \neq \emptyset$, for every subset $\{s_i, \ i \geq 1\}$ of S.

A domain A is called strongly anti-Archimedean if $A \setminus \{0\}$ is a strongly anti-Archimedean multiplicative subset.

Note that every strongly anti-Archimedean multiplicative subset is anti-Archimedean. The converse is not true.

The remaining results of this section are due to W. Maaref, A. Hamed, and A. Benhissi. The next theorem determines which integral domains are S-Artinian and for which choice of the multiplicative subset S of A.

Theorem 2.15 *Let A be an integral domain with quotient field K, S a multiplicative set of A with saturation S^*, and $T = A \setminus (0)$. The following statements are equivalent:*

1. *A is an S-Artinian domain.*

2. *The multiplicative subset T is strongly anti-Archimedean and $S^* = T$.*
3. $S^* = T$ *and* $K[[X]] = A[[X]]_T$.
4. $S^* = T$ *and* $qf(A[[X]]) = qf(K[[X]])$.
5. $S^* = T$ *and* $qf(A[[X]]) = K((X))$.

Proof $''(1) \implies (2)''$ By Lemma 2.4, the domain A is S^*-Artinian. By Corollary 2.6, since A is an integral domain, then $S^* = T$. Let $(a_i)_{i \in \mathbb{N}}$ be a sequence of nonzero elements of A. Consider the sequence $(b_i)_{i \in \mathbb{N}}$, where $b_i = a_0 a_1 \ldots a_i \neq 0$. Hence, $b_1 A \supseteq b_2 A \supseteq \ldots$ is a descending chain of principal ideals of A. As A is an S-Artinian domain, there exist $s \in T$ and $n_0 \in \mathbb{N}$ such that $0 \neq s b_{n_0} \in \bigcap_{i:0}^{\infty} b_i A \subseteq$

$\bigcap_{i:0}^{\infty} a_i A$.

$''(2) \implies (1)''$ By Lemma 2.4, A being an S-Artinian domain is equivalent to A being T-Artinian domain, since $T = S^*$. Consider any descending sequence $(I_k)_{k \in \mathbb{N}}$ of nonzero ideals of A. From each I_k, choose a nonzero element b_k. As T is a strongly anti-Archimedean multiplicative subset of A, $\bigcap_{k:0}^{\infty} b_k A \neq 0$. Let s be a nonzero element in this intersection Thus, $s I_0 \subseteq I_n$ for all $n \geq 0$ and A is S-Artinian.

The equivalences $''(2) \iff (3) \iff (4) \iff (5)''$ follow from Proposition 2.13. $\square$

We are now able to give a nontrivial example of a S-Artinian domain.

Example Let E be the ring of entire functions; N the multiplicative subset of E consisting of those functions f such that $\mathscr{Z}(f)$, the set of zeros of f, is finite; and K the field of fractions of E. Let A be the ring E_N. Then $A \neq K$. Consider any sequence $(g_i)_{i \geq 0}$ of nonzero elements of A, and then $g_i = \frac{f_i}{h_i}$, where $f_i \in E$ and $h_i \in N$. Let $a_1, \ldots, a_{n_0}$ be the zeros of f_0 that belong to the closed ball $B(0, 1)$ and m_0 the maximum multiplicity of these zeros. Let $a_{n_0+1}, \ldots, a_{n_1}$ be the zeros of $f_0 f_1$ that belong to the left-open, right-closed annulus $(0, 1, 2) = \{z \in \mathbb{C}, \ 1 <| z |\leq 2\}$ and m_1 the maximum multiplicity of these zeros. More generally, let $a_{n_k+1}, \ldots a_{n_{k+1}}$ be the zeros of $\prod_{j \leq k+1} f_j$ in the left-open, right-closed annulus $(0, k+1, k+2) = \{z \in \mathbb{C}, \ k <| z |\leq k+1\}$ and m_{k+1} the maximum multiplicity of these zeros. By the Weierstrass theorem, there exists $f \in E$ such that $\mathscr{Z}(f) = \{a_1, \ldots, a_{n_0}, a_{n_0+1}, \ldots, a_{n_1}, a_{n_1+1}, \ldots\}$ and m_{k+1} is the multiplicity of $a_{n_k+1}, \ldots a_{n_{k+1}}$. Clearly, $\frac{f}{f_0} \in E$ and $\frac{f}{f_1}$ is holomorphic on $\mathbb{C}$ except maybe at the zeros of f_1 that belong to the closed ball $B(0, 1)$. The set of zeros of f_1 that belong to the closed ball $B(0, 1)$ is finite; thus, $\frac{f}{f_1} \in A$. Similarly, $\frac{f}{f_k}$ is holomorphic on $\mathbb{C}$ except maybe at the zeros of f_k that belong to the closed ball $B(0, k)$. The set of zeros of f_k that belong to the closed ball $B(0, k)$ is finite; thus, $\frac{f}{f_k} \in A$. Hence,

$f = g_k(\frac{f}{f_k}h_k) \in g_k A$ for all $k \in \mathbb{N}$, and $f \in \bigcap_{k \geq 0} g_k A$. Thus, the multiplicative closed set $S = A \setminus (0)$ is strongly anti-Archimedean and the domain A is S-Artinian. Consequently, $qf(A[\![X]\!]) = K(\!(X)\!)$.

From the implication $"(1) \implies (2)"$ of the preceding theorem, we deduce the following.

Corollary 2.16 *Let A be an integral domain and S a multiplicative subset of A. If A is S-Artinian, then A is anti-Archimedean.*

Even though an anti-Archimedean domain need not be an S-Artinian domain, the anti-Archimedean property helps put restrictions on the class of S-Artinian domains.

Corollary 2.17 *Let A be an integral domain which is not a field. Suppose that A satisfies one of the following conditions:*

1. *A satisfies the ascending chain condition on principal ideals.*
2. *A is completely integrally closed.*

Then A is not S-Artinian for any multiplicative subset S of A.

Proof In both cases, we will show that A is not anti-Archimedean.

(1) Let $0 \neq a \in A$ be a noninvertible element. Suppose that there exists $0 \neq b \in \bigcap_{n:1}^{\infty} a^n A$. For each $n \in \mathbb{N}$, there exists $0 \neq b_n \in A$ such that $b = a^n b_n$. Then $a^n b_n = b = a^{n+1} b_{n+1}$, so $b_n = a b_{n+1}$. The sequence $(b_n A)_{n \in \mathbb{N}}$ is increasing, so stationary. Let $n \in \mathbb{N}$ be such that $b_n A = b_{n+1} A$. Then b_n and b_{n+1} are associated, which is absurd since the coefficient of proportionality a is not invertible.

(2) Let $0 \neq a \in A$ be a noninvertible element. Suppose that there exists $0 \neq b \in \bigcap_{n:1}^{\infty} a^n A$. For each $n \in \mathbb{N}$, there exists $b_n \in A$ such that $b = a^n b_n$, and then $b\frac{1}{a^n} = b_n \in A$. So $\frac{1}{a}$ is quasi-integral over A, which is absurd. $\square$

Corollary 2.18 *Let A be an integral domain which is not a field and S a multiplicative subset of A. If A is S-Artinian, then any over ring of A is also S-Artinian.*

Proof Let K be the quotient field of A and B an over ring of A. Then $A[\![X]\!] \subseteq B[\![X]\!] \subseteq K(\!(X)\!)$. By Theorem 2.15, since A is an S-Artinian domain, then $qf(A[\![X]\!]) = K(\!(X)\!)$. Thus, $qf(B[\![X]\!]) = K(\!(X)\!)$. Let S_A^* and S_B^* be the saturations of the multiplicative set S respectively in the domains A and B. By Theorem 2.15, $S_A^* = A \setminus (0)$. By Proposition 1.3 (1), this means that each element of $A \setminus (0)$ is a divisor of an element of S. Since B is an over ring of A, then each element of $B \setminus (0)$ is a divisor of an element of $A \setminus (0)$. It follows that each element of $B \setminus (0)$ is a divisor of an element of S. By Proposition 1.3 (1), $S_B^* = B \setminus (0)$. Again, by Theorem 2.15, B is S-Artinian. □

By Example 3 of Corollary 2.6, an Artinian integral domain is a field. It is also known and easy to see that a finite integral domain is also a field. The following corollary treats countable integral S-Artinian domains for a multiplicative set S of A.

Corollary 2.19 *Let A be a countable integral domain and S a multiplicative subset of A. If A is S-Artinian, then it is a field.*

Proof Since A is countable, then $A = \{0\} \cup \{a_i; \ i \in \mathbb{N}^*\}$ with $a_i \neq 0$. By Theorem 2.15, the domain A is strongly anti-Archimedean. Then $\bigcap_{i:1}^{\infty} a_i A \neq (0)$. Let c be a nonzero element in this intersection. Since $c^2 = a_{i_0}$ for some $i_0 \in \mathbb{N}^*$, then $c \in c^2 A$ and c is invertible in A. Now, all the a_i are divisors of c, so they are invertible. □

Following Definition 2.1 of Chap. 1, we say that an ideal I of a ring A is S-finite, where $S \subseteq A$ is a multiplicative subset, if $sI \subseteq J \subseteq I$ for some finitely generated ideal J of A and some $s \in S$. We say that A is S-Noetherian if each ideal of A is S-finite. It is well known, and will be proved in Theorem 7.5, that an Artinian ring is Noetherian. We next give a S-variant of this result for integral domains.

Corollary 2.20 *Let A be an integral domain and S a multiplicative subset of A. If A is S-Artinian, then it is S-Noetherian.*

Proof By Theorem 2.15, the saturation of S is $S^* = A \setminus (0)$. By Proposition 1.3 (2), for every nonzero prime ideal P of A, we have $P \cap S \neq \emptyset$. By Example 3 of Definition 2.1, Chap. 1, the ideal P is S-finite. By Corollary 2.2 of Chap. 1, the domain A is S-Noetherian. □

An element a of a domain A is said to be bounded if $\bigcap_{n:1}^{\infty} a^n A \neq (0)$.

Lemma 2.21 (M.H. Park, A. Hamed, W. Maaref; 2019) *Let A be an integral domain and $0 \neq f(X) \in A[[X]]$. The following assertions are equivalent:*

1. *The element $f(X)$ is bounded in $A[[X]]$.*
2. *The element $f(0)$ is bounded in A.*

Moreover, in this case, $f(X)$ is a unit in element of $A[[X]]_{A \setminus (0)}$.

Proof "$\Longrightarrow$" We have $f(0) \neq 0$ since in the contrary case, $\bigcap_{n:1}^{\infty} f^n A[[X]] \subseteq \bigcap_{n:1}^{\infty} X^n A[[X]] = (0)$, which is absurd. Choose $0 \neq g(X) \in \bigcap_{n:1}^{\infty} f(X)^n A[[X]]$, $g(X) = X^m h(X)$ where $m \geq 0$ and $h(X) \in A[[X]]$ with $h(0) \neq 0$. For each $n \in \mathbb{N}$,

$X^m h(X) = f(X)^n h_n(X)$ with $h_n(X) \in A[[X]]$. Since $f(0) \neq 0$ and $h(0) \neq 0$, then X^m divides $h_n(X)$ and $h(X) \in f(X)^n A[[X]]$. So $h(X) \in \bigcap_{n:1}^{\infty} f(X)^n A[[X]]$.

Then $h(0) \in \bigcap_{n:1}^{\infty} f(0)^n A$ and $f(0)$ is bounded.

"$\Longleftarrow$" Put $a = f(0) \neq 0$. Choose $0 \neq c \in \bigcap_{n:1}^{\infty} a^n A$. For each $n \in \mathbb{N}^*$, write $f(X)^n = a^n - g_n(X)$ avec $g_n(X) \in XA[[X]]$. Then $\left(a^n - g_n(X)\right)^{-1} = a^{-n}\left(1 - \frac{g_n(X)}{a^n}\right)^{-1}$

$= a^{-n}\left(1 + \frac{g_n(X)}{a^n} + \frac{g_n(X)^2}{a^{2n}} + \frac{g_n(X)^3}{a^{3n}} + \frac{g_n(X)^4}{a^{4n}} + \dots\right) = \frac{1}{a^n} + \frac{g_n(X)}{a^{2n}} + \frac{g_n(X)^2}{a^{3n}} + \frac{g_n(X)^3}{a^{4n}} + \dots \in K[[X]]$ where K is the quotient field of A. Note that for each $n \in \mathbb{N}^*$, $c\left(a^n - g_n(X)\right)^{-1} \in A[[X]]$. Then $c = f(X)^n c\left(a^n - g_n(X)\right)^{-1} \in f(X)^n A[[X]]$. So $0 \neq c \in \bigcap_{n:1}^{\infty} f(X)^n A[[X]]$ and hence $f(X)$ is a bounded element in $A[[X]]$.

Now, put $c = f(X)g(X)$ with $g(X) \in A[[X]]$. Then $\frac{1}{f(X)} = \frac{g(X)}{c} \in A[[X]]_{A\setminus(0)}$. $\qquad\square$

Let A be a ring. By Example 6 of Proposition 1.2, the set $N = \{f \in A[[X]]; \ c(f) = A\}$ is multiplicative in the power series ring $A[[X]]$. We denote by $A((X)) = A[[X]]_N$ the quotient ring.

Proposition 2.22 *Let A be an integral domain and S a multiplicative set of A. If A is S-Artinian, then so is $A((X))$.*

Proof Let S_A^* and $S_{A((X))}^*$ be the saturations of S respectively in the domains A and $A((X))$. By Theorem 2.15, $S_A^* = A \setminus (0)$ and A is strongly anti-Archimedean. We must show that $S_{A((X))}^* = A((X)) \setminus (0)$ and $A((X))$ is strongly anti-Archimedean. For the first point, by Proposition 1.3 (1), we must show that each nonzero element of $A((X))$ is a divisor of an element of S. But such an element is of the form $f(X) = \frac{X^m g(X)}{h(X)}$ where $m \in \mathbb{N}$, $g(X) \in A[[X]]$ with $g(0) \neq 0$ and $h(X) \in N$. Since X^m and $h(X)$ are units in $A((X))$, we may suppose that $f(X) \in A[[X]]$ with $f(0) \neq 0$. Since A is strongly anti-Archimedean, the element $f(0)$ is bounded in A. By the preceding lemma, $f(X)$ is a unit in element of $A[[X]]_{A\setminus(0)}$. There exist $l(X) \in A[[X]]$ and $0 \neq a \in A$ such that $f(X)\frac{l(X)}{a} = 1$. By hypothesis, there exist $b \in A$ and $s \in S$ such that $s = ab$. Then $bf(X)l(X) = ab = s$. So $f(X)$ divides s. For the second point, let $\left(f_n(X)\right)_{n \in \mathbb{N}^*}$ be a sequence of nonzero elements of $A((X))$. We must show that $\bigcap_{n:1}^{\infty} f_n(X)A[[X]] \neq (0)$. As before, we may suppose that $f_n(X) \in A[[X]]$ with $f_n(0) \neq 0$ for each $n \in \mathbb{N}^*$. As before, for each $n \in \mathbb{N}^*$, there exits

$0 \neq c_n \in f_n(X)A[[X]] \cap A$. Since A is strongly anti-Archimedean, $(0) \neq \bigcap_{n:1}^{\infty} c_n A \subseteq$
$\bigcap_{n:1}^{\infty} c_n A[[X]] \subseteq \bigcap_{n:1}^{\infty} f_n(X)A[[X]]$. $\qquad\qquad\qquad\qquad\qquad\qquad$ □

Let A be an integral domain. By Example 7 of Proposition 1.2, the set $N_t = \{0 \neq f \in A[[X]]; \ c(f)_t = A\}$ is multiplicative in the power series domain $A[[X]]$. We denote by $A\{\{X\}\} = A[[X]]_{N_t}$ the quotient domain.

Corollary 2.23 *Let A be an integral domain and S a multiplicative set of A. If A is S-Artinian, then so is $A\{\{X\}\}$.*

Proof Since $N \subseteq N_t$, then $A((X)) \subseteq A\{\{X\}\}$ are over rings of $A[[X]]$. So we conclude by Corollary 2.18s. $\qquad\qquad\qquad\qquad\qquad\qquad\qquad\qquad\qquad\qquad$ □

3 S-Artinian Modules

We define the notion of S-Artinian modules.

Definitions 3.1 Let A be a ring and M a A-module.

1. We say that M is Artinian if each decreasing sequence of sub-modules of M is stationary.
2. Let S be a multiplicative set of A. We say that M is S-Artinian if each decreasing sequence $N_1 \supseteq N_2 \supseteq \ldots$ of sub-modules of M is S-stationary; i.e., there exist $s \in S$ and $k \in \mathbb{N}^*$ such that for each integer $n \geq k$, $s N_k \subseteq N_n$.

With the hypotheses and the notations of the definition, $s N_k \subseteq \bigcap_{n:k}^{\infty} N_n = \bigcap_{n:1}^{\infty} N_n$.

Example 1 Let (A, σ) be an Euclidian domain and $0 \neq d \in A$. The A-module quotient A/dA is Artinian. Indeed, a sub-module of A/dA is of the form xA/dA with x a divisor of d in A. Then $0 \leq \sigma(x) \leq \sigma(d)$ are integers. Let $x_1 A/dA \supseteq x_2 A/dA \supseteq \ldots$ a decreasing sequence of sub A-modules of A/dA. Then $x_1 A \supseteq x_2 A \ldots \supseteq dA$. So $0 \leq \sigma(x_1) \leq \sigma(x_2) \leq \ldots \leq \sigma(d)$ are integers. There exists $k \in \mathbb{N}^*$ such that for each integer $n \geq k$, $\sigma(x_k) = \sigma(x_n)$ and since $x_n A \subseteq x_k A$, then x_k divides x_n. We will prove that for each integer $n \geq k$, the elements x_k and x_n are associated with A and then $x_n A/dA = x_k A/dA$. Suppose that x_n does not divide x_k. Then $x_k = x_n q + r$ with $q, r \in A$, $r \neq 0$, and $\sigma(r) < \sigma(x_n)$. Since x_k divides x_n, then $x_n = c x_k$ with $0 \neq c \in A$. So $x_k = x_n q + r = c x_k q + r$, and then $r = x_k(1 - cq)$. So $\sigma(x_k) \leq \sigma(r) < \sigma(x_n)$, which contradicts $\sigma(x_k) = \sigma(x_n)$.

Example 2 The following assertions are equivalent for a vector space E: (i) E is Noetherian $\Longleftrightarrow$ (ii) $dim \, E < \infty \Longleftrightarrow$ (iii) E is Artinian. Indeed,

$''(i)$ or $(iii) \implies (ii)''$ If $dim\ E = \infty$, then E contains a free family $(e_i)_{i \in \mathbb{N}^*}$ of vectors. Then we have a strictly increasing sequence $(e_1) \subset (e_1, e_2) \subset \ldots$ and a strictly decreasing $(e_1, e_2, \ldots) \supset (e_2, e_3, \ldots) \supset \ldots$ of sub-vector spaces of E, which is absurd.

$''(ii) \implies (i)''$ Let $E_1 \subseteq E_2 \subseteq \ldots$ be an increasing sequence of sub-vector spaces of E. Then $dim\ E_1 \le dim\ E_2 \le \ldots \le dim\ E < \infty$. The sequence of the dimensions is stationary. Then the sequence of the sub-vector spaces is stationary.

$''(ii) \implies (iii)''$ Let $E_1 \supseteq E_2 \supseteq \ldots$ be a decreasing sequence of sub-vector spaces of E. Then $dim\ E \ge dim\ E_1 \ge dim\ E_2 \ge \ldots$ with $dim\ E < \infty$. The sequence of dimensions is stationary. It is the same for the sequence of the sub-vector spaces.

Example 3 An Artinian module is S-Artinian for each multiplicative set S.

Example 4 Let A be a ring and S a multiplicative set of A. The A-module A is S-Artinian if and only if the ring A is S-Artinian. Let's see again Example 3 of Definition 2.3. The ring $A = \mathbb{Z} \times \mathbb{Q}$ is not Artinian. But it is S-Artinian for the multiplicative set $S = \mathbb{Z} \times \mathbb{Q}^*$. In particular, the A-module A is S-Artinian but it is not Artinian.

Example 5 A sub-module of an S-Artinian module is S-Artinian.

Example 6 (Ozen et al. 2021) Let A be a ring, S a multiplicative set of A, and M a A-module. If $S \cap ann_A(M) \ne \emptyset$, then M is S-Artinian. Indeed, let $s \in S \cap ann_A(M)$ and $N_1 \supseteq N_2 \supseteq \ldots$ a decreasing sequence of sub-modules of M. Then $sN_1 = (0) \subseteq N_i$ for each $i \in \mathbb{N}^*$.

For example, the $\mathbb{Z}$-module $\mathbb{F}_2[[X]]$ is S-Artinian for the multiplicative set $S = \mathbb{Z} \setminus (0)$ since $2 \in S \cap ann_{\mathbb{Z}}(\mathbb{F}_2[[X]])$. But it is not Artinian because we have the strictly decreasing sequence of sub-modules $X\mathbb{F}_2[[X]] \supset X^2\mathbb{F}_2[[X]] \supset \ldots \supset X^n\mathbb{F}_2[[X]] \supset \ldots$.

Example 7 (Ozen et al. 2021) Let A_1 be an integral domain that is not a field and A_2 an Artinian ring. By Example 3 of Corollary 2.6, the ring A_1 is not Artinian. Consider the product ring $A = A_1 \times A_2$. The sub-modules of the A-module A are the ideals of A. Since the ring A_1 is not Artinian, the A-module A is also not Artinian. Consider the multiplicative set $S = A_1 \times U(A_2)$ of the ring A. By Lemma 2.2, a decreasing sequence of ideals of A is of the form $I_1^{(1)} \times I_1^{(2)} \supseteq I_2^{(1)} \times I_2^{(2)} \supseteq I_3^{(1)} \times I_3^{(2)} \supseteq \ldots$ with $I_1^{(1)} \supseteq I_2^{(1)} \supseteq \ldots$ (1) and $I_1^{(2)} \supseteq I_2^{(2)} \supseteq \ldots$ (2) two decreasing sequences of ideals of A_1 and A_2, respectively. Since the ring A_2 is Artinian, the sequence (2) is stationary at a rank k. Let $s = (0, 1) \in S$. For each integer $n \ge k$, $s(I_k^{(1)} \times I_k^{(2)}) = (0) \times I_k^{(2)} = (0) \times I_n^{(2)} \subseteq I_n^{(1)} \times I_n^{(2)}$. Then the A-module A is S-Artinian.

Proposition 3.2 *Let S be a multiplicative set of a ring A and I an ideal of A disjoint with S. The following assertions are equivalent:*

1. *The ring A is S-Artinian.*
2. *The ring A/I is $\pi(S)$-Artinian and the A-module I is S-Artinian, where π : $A \longrightarrow A/I$ is the canonical homomorphism.*

Proof "(1) $\implies$ (2)" The ring A/I is $\pi(S)$-Artinian by Corollary 2.10. The A-submodules of I are the ideals of A contained in I. Then I is a S-Artinian A-module.

"(2) $\implies$ (1)" Let $J_1 \supseteq J_2 \supseteq \ldots$ be a decreasing sequence of ideals of ring A. Then $(J_1+I)/I \supseteq (J_2+I)/I \supseteq \ldots$ is a decreasing sequence of ideals of ring A/I. But A/I is $\pi(S)$-Artinian. There exist $s \in S$ and $k_1 \in \mathbb{N}^*$ such that for each integer $n \geq k_1, \bar{s}\big((J_{k_1}+I)/I\big) \subseteq (J_n+I)/I$, and then $sJ_{k_1} \subseteq J_n+I$. We have also $J_1 \cap I \supseteq J_2 \cap I \supseteq \ldots$ a decreasing sequence of A-sub-modules of I. There exist $s' \in S$ and $k_2 \in \mathbb{N}^*$ such that for each integer $n \geq k_2$, $s'(J_{k_2} \cap I) \subseteq J_n \cap I$. Let $t = ss' \in S$ and $k = max(k_1, k_2) \in \mathbb{N}^*$. For each integer $n \geq k$, $tJ_k \subseteq sJ_k \subseteq sJ_{k_1} \subseteq J_n + I$ and $t(J_k \cap I) \subseteq s'(J_k \cap I) \subseteq s'(J_{k_2} \cap I) \subseteq J_n \cap I$. Let $n \geq k$ be an integer. We will prove that $t^2 J_k \subseteq J_n$. Let $a_k \in J_k$. Then $ta_k \in tJ_k \subseteq J_n + I$, so $ta_k = a_n + x$ with $a_n \in J_n \subseteq J_k$ and $x \in I$. Then $x = ta_k - a_n \in J_k$, so $x \in J_k \cap I$, and then $tx \in J_n \cap I$. So $t^2 a_k = ta_n + tx \in J_n + (J_n \cap I) = J_n$. Then $t^2 J_k \subseteq J_n$. $\square$

Corollary 3.3 *Let I be a proper ideal of a ring A. Then A is Artinian if and only if the quotient ring A/I is Artinian and the A-module I is Artinian.*

Proof Take $S = \{1\}$ in the preceding proposition. $\square$

We characterize the Artinian modules in terms of S-Artinian modules.

Theorem 3.4 *Let A be a ring and M a A-module.*
The following assertions are equivalent:

1. *The module M is Artinian.*
2. *For each $P \in Spec(A)$, the module M is S-Artinian where $S = A \setminus P$.*
3. *For each $P \in Max(A)$, the module M is S-Artinian where $S = A \setminus P$.*

Proof The implications "(1) $\implies$ (2) $\implies$ (3)" are clear.

"(3) $\implies$ (1)" Let $N_1 \supseteq N_2 \supseteq \ldots$ be a decreasing sequence of sub-modules of M. For each $P \in Max(A)$, there exist $s_P \in A \setminus P$ and $k_P \in \mathbb{N}^*$ such that for each integer $n \geq k_P$, $s_P N_{k_P} \subseteq N_n$. Consider the ideal $J = \big(s_P; \ P \in Max(A)\big)$ of A. If $J \neq A$, there exists $P_0 \in Max(A)$ such that $J \subseteq P_0$, and then $s_{P_0} \in P_0$, which is absurd. Then $J = A$. There exist $P_1, \ldots, P_r \in Max(A)$ such that $\big(s_{P_1}, \ldots, s_{P_r}\big) = A$. Let $k = max\{k_{P_1}, \ldots, k_{P_r}\}$. Then $N_k \subseteq N_{k_{P_i}}$ for each $i \in \{1, \ldots, r\}$. For each integer $n \geq k$, $N_k = AN_k = \big(s_{P_1}, \ldots, s_{P_r}\big)N_k \subseteq s_{P_1}N_k + \ldots + s_{P_r}N_k \subseteq s_{P_1}N_{k_{P_1}} + \ldots + s_{P_r}N_{k_{P_r}} \subseteq N_n$, and then $N_k = N_n$ so M is Artinian. $\square$

Let $A_1, \ldots, A_n$ be rings and S_i a multiplicative set of A_i, $1 \leq i \leq n$. Then $S_1 \times \ldots \times S_n$ is a multiplicative set of the product ring $A_1 \times \ldots \times A_n$.

Proposition 3.5 *Let S_i be a multiplicative set of a ring A_i, $1 \leq i \leq n$.*
Put $A = A_1 \times \ldots \times A_n$ and $S = S_1 \times \ldots \times S_n$. The following assertions are
equivalent:

1. *The ring A is S-Artinian.*
2. *The ring A_i is S_i-Artinian for $1 \leq i \leq n$.*

Proof $"(1) \implies (2)"$ The canonical projections $\pi_i : A \longrightarrow A_i$, $1 \leq i \leq n$, are
onto. By Proposition 2.9, $\pi_i(A) = A_i$ is $\pi_i(S) = S_i$-Artinian.

$"(2) \implies (1)"$ Let $J_1 \supseteq J_2 \ldots$ be a decreasing sequence of ideals of the ring
A. By Lemma 2.2, for each $k \in \mathbb{N}^*$, $J_k = I_1^{(k)} \times \ldots \times I_n^{(k)}$ where $I_i^{(k)}$ is an ideal
of the ring A_i. For $1 \leq i \leq n$, $I_i^{(1)} \supseteq I_i^{(2)} \supseteq \ldots$ is a decreasing sequence of
ideals of the ring A_i. But A_i is S_i-Artinian. There exist $s_i \in S_i$ and $k_i \in \mathbb{N}^*$ such
that $s_i I_i^{(k_i)} \subseteq I_i^{(m)}$ for each integer $m \geq k_i$. Put $k = max\{k_1, \ldots, k_n\} \in \mathbb{N}^*$ and
$s = (s_1, \ldots, s_n) \in S$. For each integer $m \geq k$, $s J_k = (s_1, \ldots, s_n)(I_1^{(k)} \times \ldots \times$
$I_n^{(k)}) = s_1 I_1^{(k)} \times \ldots \times s_n I_n^{(k)} \subseteq s_1 I_1^{(k_1)} \times \ldots \times s_n I_n^{(k_n)} \subseteq I_1^{(m)} \times \ldots \times I_n^{(m)} = J_m$.
Then A is S-Artinian. $\square$

Example A finite product of Artinian rings is Artinian. The result is not true for
infinite product. Let $(K_i)_{i \in \mathbb{N}^*}$ be a family of fields, so Artinian. But $A = \prod_{i:1}^{\infty} K_i$
is not Artinian since we have the sequence of ideals: $\prod_{i:1}^{\infty} K_i \supset (0) \times \prod_{i:2}^{\infty} K_i \supset$
$(0) \times (0) \times \prod_{i:3}^{\infty} K_i \supset \ldots$.

Proposition 3.6 *Let A be an S-Artinian ring where S is a multiplicative set of A.*
All the prime ideals of A disjoint with S are maximal to avoid S and are minimal
prime ideals of A.

Proof Let P be a prime ideal of A disjoint with S.

(1) There exists a prime ideal P' of A maximal to avoid S such that $P \subseteq P'$.
    Suppose that $P \neq P'$. Let $x \in P' \setminus P$. The decreasing sequence $P + Ax \supseteq$
    $P + Ax^2 \supseteq \ldots$ of ideals of A is S-stationary. There exist $s \in S$ and $k \in \mathbb{N}^*$
    such that $s(P + Ax^k) \subseteq P + Ax^{k+1}$, and then $sx^k \in P + Ax^{k+1}$. There exists
    $a \in A$ such that $sx^k - ax^{k+1} \in P$, then $x^k(s - ax) \in P$. But $x^k \notin P$, and then
    $s - ax \in P$, so $s \in P + Ax \subseteq P'$, which is absurd since $s \in S$ and $S \cap P' = \emptyset$.
    Then $P = P'$ is maximal to avoid S.
(2) Let $Q \subseteq P$ be a prime ideal of A. Then Q is disjoint with S. By (1), Q is
    maximal to avoid S, and then $P = Q$. We conclude that P is a minimal prime
    of A. $\square$

Corollary 3.7 *Let P be a prime ideal of a ring A and $S = A \setminus P$. If A is S-Artinian, then P is a minimal prime of A.*

Corollary 3.8 *An Artinian ring is Krull zero dimensional.*

Proof Let A be an Artinian ring and P a prime ideal of A. Then A is S-Artinian where $S = A \setminus P$. By the preceding corollary, P is minimal. Then $ht\ P = 0$ and $dim\ A = 0$. ☐

Example 1 The converse of the preceding corollary is false. The ring $A = \displaystyle\prod_{i:1}^{\infty} \mathbb{F}_2$ is zero dimensional. Indeed, in Example 2 of Lemma 5.4, Chap. 1, we have always shown that each ideal of A is maximal. The ring A is not Artinian since we have the sequence of ideals $\displaystyle\prod_{i:1}^{\infty} \mathbb{F}_2 \supset (0) \times \prod_{i:2}^{\infty} \mathbb{F}_2 \supset (0) \times (0) \times \prod_{i:3}^{\infty} \mathbb{F}_2 \supset \ldots$.

Example 2 Let P be prime ideal of any ring A. Then $P[X] \subset P + XA[X]$ and $P[[X]] \subset P + XA[[X]]$ are two chains of prime ideals in $A[X]$ and in $A[[X]]$, respectively. Then $dim\ A[X] \geq 1$ and $dim\ A[[X]] \geq 1$. Then $A[X]$ and $A[[X]]$ are neither Artinian.

Theorem 3.9 *Let A be an S-Artinian ring with S a multiplicative set of A. Then A has a finite number of prime ideals disjoint with S.*

Proof Suppose A has an infinitely many different prime ideals $P_1, P_2, \ldots$ disjoint with S. The decreasing sequence $P_1 \supseteq P_1 P_2 \supseteq \ldots$ of ideals of A is S-stationary. There exist $s \in S$ and $k \in \mathbb{N}^*$ such that $s(P_1 P_2 \ldots P_k) \subseteq P_1 P_2 \ldots P_k P_{k+1} \subseteq P_{k+1}$. Since $s \notin P_{k+1}$, there exists $i \leq k$ such that $P_i \subseteq P_{k+1}$. By Proposition 3.6, P_{k+1} is minimal, so $P_i = P_{k+1}$, which is absurd. ☐

Corollary 3.10 *An Artinian ring has a finite prime spectrum.*

Proof Take $S = \{1\}$. All the prime ideals are disjoint with S. ☐

In the first section of this chapter, we have always shown that the set $Z(A)$ of the zero divisors of a ring A is union of prime ideals of A. In the following lemma, we show that $Z(A)$ contains all the minimal prime ideals of A.

Lemma 3.11 (Henriksen and Jerison 1965) *Let A be a ring.*

1. *Let I be a proper ideal of A and P a prime ideal of A minimal over I. Then $\sqrt{IA_P} = PA_P$ and for each $x \in P$, there exist $n \in \mathbb{N}^*$ and $s \in A \setminus P$ such that $sx^n \in I$.*
2. *A minimal prime ideal of a ring A is formed by zero divisors of A.*
3. *Moreover, if A is reduced, the set $Z(A)$ of the zero divisors of A is equal to the union of the minimal prime ideals of A.*

Proof

(1) The only prime ideal of A_P containing IA_P is PA_P. Indeed, such ideal is of
    the form QA_P with $Q \in spec(A)$ and $Q \subseteq P$. Moreover, $I \subseteq Q$ since if $x \in I$,
    then $\frac{x}{1} \in IA_P \subseteq QA_P$, so $\frac{x}{1} = \frac{q}{s}$ with $q \in Q$ and $s \in A \setminus P$. There exists
    $r \in A \setminus P$ such that $r(sx - q) = 0$, and then $rsx = rq \in Q$. Since $r, s \notin Q$,
    then $x \in Q$. So $I \subseteq Q \subseteq P$, and then $Q = P$. So $\sqrt{IA_P} = PA_P$. Let
    $x \in P$, and then $\frac{x}{1} \in PA_P = \sqrt{IA_P}$. There exists $n \in \mathbb{N}^*$ such that $\frac{x^n}{1} \in IA_P$.
    Then $\frac{x^n}{1} = \frac{a}{s}$ with $a \in I$ and $s \in A \setminus P$. There exists $r \in A \setminus P$ such that
    $r(sx^n - a) = 0$. Then $rsx^n = ra \in I$ with $rs \in A \setminus P$.
(2) Let $P \in Min(A)$. Then P is minimal over (0). Let $a \in P$. By (1), there exist
    $n \geq 1$ and $s \in A \setminus P$ such that $sa^n = 0$. Choose s and n in such a way that
    n has the smallest value. Then $sa^{n-1} \neq 0$ and $asa^{n-1} = 0$. Then a is a zero
    divisor. So $P \subseteq Z(A)$.
(3) Let $a \in Z(A)$. There exists $0 \neq b \in A$ such that $ab = 0$. Since $(0) = Nil(A) =$
    $\cap\{P;\ P \in Min(A)\}$, there exists $P_0 \in Min(A)$ such that $b \notin P_0$. But $ab =$
    $0 \in P_0$, and then $a \in P_0$.   □

Example In general, $Z(A) \neq \cup\{P;\ P \in Min(A)\}$. Let K be a field and
consider the ring $A = K[X, Y]/(X^2, XY)$. Since $\overline{XY} = \bar{0}$ and $\overline{X} \neq \bar{0}$ because
$X \notin (X^2, XY)$, then $\overline{Y}$ is a zero divisor in A. We will prove that the only minimal
prime ideal of A is $(\overline{X})$ and it does not contain $\overline{Y}$. Let $\mathscr{P}$ be such an ideal of A.
Then $\mathscr{P} = P/(X^2, XY)$ with P a prime ideal of $K[X, Y]$ containing (X^2, XY).
So $X \in P$, and then $(X) \subseteq P$. But (X) is a prime ideal of $K[X, Y]$ containing
(X^2, XY) since $K[X, Y]/(X) \simeq K[Y]$. Then $(X)/(X^2, XY) \subseteq P/(X^2, XY) = \mathscr{P}$,
so $\mathscr{P} = (\overline{X})$. If $\overline{Y} \in (\overline{X})$, there exists $f(X, Y) \in K[X, Y]$ such that
$Y - Xf(X, Y) \in (X^2, XY)$, which is absurd.

Corollary 3.12 *Let S be a multiplicative set formed by regular elements of a ring
A. If A is S-Artinian, then A has a finite number of minimal prime ideals.*

Proof Let P be a minimal prime ideal of A. By the preceding lemma, $P \subseteq Z(A)$,
the set of the zero divisors of A. Since S is formed by regular elements, $P \cap S = \emptyset$.
By Theorem 3.9, the set of the prime ideals of A disjoints with S is finite. Then
$Min(A)$ is finite.   □

The remaining results of this section are due to W. Maaref, A. Hamed, and A.
Benhissi.

Proposition 3.13 *Let A be a ring, $S \subseteq A$ a multiplicative subset, and $f :
M \longrightarrow M'$ a surjective A-module homomorphism. If M is S-Artinian, then M'
is S-Artinian.*

Proof Let $N'_1 \supseteq N'_2 \supseteq \ldots$ be a descending chain of sub-modules of M'. Then
$f^{-1}(N'_1) \supseteq f^{-1}(N'_2) \supseteq \ldots$ is a descending chain of sub-modules of M. As M
is S-Artinian, there exist $s \in S$ and $n_0 \in \mathbb{N}$ such that $sf^{-1}(N'_{n_0}) \subseteq f^{-1}(N'_n)$ for

all $n \geq n_0$. Since f is an epimorphism, $f(sf^{-1}(N'_n)) = s(f(f^{-1}(N'_n))) = sN'_n$. Hence, $sN'_{n_0} \subseteq N'_n$ for all $n \geq n_0$. Therefore, M' is S-Artinian. $\qquad\square$

Example Let S be a multiplicative subset of a ring A, M a S-Artinian A-module, and N a sub-module of M. Then the quotient module M/N is also an S-Artinian.

Theorem 3.14 *Let A be a ring, $S \subseteq A$ a multiplicative subset, M an A-module, and N a sub-module of M. The following statements are equivalent:*

1. *The module M is S-Artinian.*
2. *The modules M/N and N are S-Artinian.*

Proof "(1) $\implies$ (2)" Follows from the preceding example.

"(2) $\implies$ (1)" Let $N_1 \supseteq N_2 \supseteq \ldots$ be a descending chain of sub-modules of M. Then $N_1 + N \supseteq N_2 + N \supseteq \ldots$ and $N_1 \cap N \supseteq N_2 \cap N \supseteq \ldots$ are two descending chains of sub-modules of M. The members of the first chain contain N and the members of the second one are contained in N. Hence, $(N_1 + N)/N \supseteq (N_2 + N)/N \supseteq \ldots$ is a descending chain of sub-modules of M/N. As M/N is S-Artinian, there exist $s \in S$ and $n_1 \in \mathbb{N}$ such that $s((N_{n_1} + N)/N) \subseteq (N_n + N)/N$ for all $n \geq n_1$. Hence, $sN_{n_1} \subseteq N_n + N$ for all $n \geq n_1$. Also, since N is S-Artinian, there exist $s' \in S$ and $n_2 \in \mathbb{N}$ such that $s'(N_{n_2} \cap N) \subseteq N_n \cap N$ for all $n \geq n_2$. Put $n_0 = max\{n_1, n_2\}$ and $t = ss' \in S$. For all $n \geq n_0$, we have $tN_{n_0} \subseteq sN_{n_1} \subseteq N_n + N$ and $t(N_{n_0} \cap N) \subseteq s'(N_{n_2} \cap N) \subseteq N_n \cap N$. We claim that $t^2 N_{n_0} \subseteq N_n$ for all $n \geq n_0$. Indeed, let $a_{n_0} \in N_{n_0}$. Then $ta_{n_0} \in N_n + N$. Thus, there exist $a_n \in N_n$ and $x \in N$ such that $ta_{n_0} = a_n + x$. As $N_n \subseteq N_{n_0}$, $a_n \in N_{n_0}$ and $x = ta_{n_0} - a_n \in N_{n_0}$. Thus, $x \in N_{n_0} \cap N$. Hence $tx \in N_n \cap N$. Therefore $t^2 a_{n_0} = ta_n + tx \in N_n$ and so $t^2 N_{n_0} \subseteq N_n$. Hence, M is S-Artinian. $\qquad\square$

Proposition 3.15 *Let A be a ring, S a multiplicative subset of A, and $n \in \mathbb{N}^*$. If A is an S-Artinian ring, then A^n is an S-Artinian module.*

Proof We will show this via induction. Let $P(n)$ be the property that A^n is an S-Artinian module whenever A is an S-Artinian ring. For $n = 1$, A is an S-Artinian module if and only if A is an S-Artinian ring. Suppose that the property holds for $k \leq n$. We prove $P(n + 1)$. The module A^n is isomorphic to the sub-module $N = A^n \times \{0\}$ of A^{n+1}. Hence, by the induction hypothesis, N is S-Artinian. Clearly, $A^{n+1}/N \simeq A$ is S-Artinian by the case $n = 1$. Thus, by Theorem 3.14, A^{n+1} is S-Artinian. $\qquad\square$

Note that the product ring A^n is S^n-Artinian by Proposition 3.5.

Proposition 3.16 *Let A be a ring, S a mutiplicative subset of A, and M a finitely generated A-module. If A is an S-Artinian ring, then M is an S-Artinian A-module.*

Proof There exist $n \in \mathbb{N}^*$ and a sub-module N of A^n such that $M \simeq A^n/N$. By the preceding proposition, A^n is an S-Artinian module and by Theorem 3.14, the module M is S-Artinian. $\qquad\square$

Let A be a ring and $S \subseteq A$ a multiplicative set. Following Definition 1.1 of Chap. 1, we say that a A-module M is S-finite if $sM \subseteq F$ for some $s \in S$ and some finitely generated sub-module F of M.

Proposition 3.17 *Let A be a ring, $S \subseteq A$ a multiplicative subset, and M a S-finite A-module. If A is an S-Artinian ring, then M is an S-Artinian A-module.*

Proof Since M is S-finite, there exist $s \in S$ and a finitely generated submodule F of M such that $sM \subseteq F$. By the preceding proposition, F is S-Artinian. Let $N_1 \supseteq N_2 \supseteq \ldots$ be a descending chain of sub-modules of M. Then $sN_1 \supseteq sN_2 \supseteq \ldots$ is a descending chain of sub-modules of F. Hence, there exist $s' \in S$ and $n_0 \in \mathbb{N}$ such that $s's N_{n_0} \subseteq sN_n \subseteq N_n$ for all $n \geq n_0$. Therefore, M is S-Artinian. $\square$

4   S-Cofinite Rings and Modules

Definitions 4.1 Let A be a ring, S a multiplicative set of A, $\mathscr{F}$ a nonempty family of ideals of A, and $I \in \mathscr{F}$. We say that I is a S-minimal element of $\mathscr{F}$ if there exists $s \in S$ such that for each element $J \in \mathscr{F}$ satisfying $J \subseteq I$, we have the inclusion $sI \subseteq J$.

We say that the ring A has the property $(S - Min)$ if each nonempty family of ideals of A has a S-minimal element.

Examples Let A be a ring and $S \subseteq U(A)$ a multiplicative set of A.

(1) Let $\mathscr{F}$ be a nonempty family of ideals of a ring A. A S-minimal element of $\mathscr{F}$ is a minimal element of $(\mathscr{F}, \subseteq)$.
(2) The ring A has the property $(S - Min)$ if each nonempty family of ideals of A has a minimal element for the inclusion.

Definition 4.2 Let A be a ring, S a multiplicative set of A, and $\mathscr{F}$ a nonempty family of ideals of A. We say that $\mathscr{F}$ is S-cosaturated if for each ideal J of A, if there exist $s \in S$ and $I \in \mathscr{F}$ such that $sI \subseteq J$, then $J \in \mathscr{F}$.

Definitions 4.3

1. We say that a ring A is cofinite if for each nonempty family $(I_i)_{i \in \Delta}$ of ideals of A satisfying $\bigcap_{i \in \Delta} I_i = (0)$, there exists a nonempty finite subset $\Delta' \subseteq \Delta$ such that
$$\bigcap_{i \in \Delta'} I_i = (0).$$

2. Let S be a multiplicative set of a ring A. We say that A is S-cofinite if for each nonempty family $(I_i)_{i \in \Delta}$ of ideals of A satisfying $\bigcap_{i \in \Delta} I_i = (0)$, there exist $s \in S$ and a nonempty finite subset $\Delta' \subseteq \Delta$ such that $s\left(\bigcap_{i \in \Delta'} I_i\right) = (0)$.

Example 1 The ring $\mathbb{Z}$ is not cofinite.

Indeed, $\bigcap_{n:1}^{\infty} n\mathbb{Z} = (0)$. But for each $n_1, \ldots, n_k \in \mathbb{N}^*, 0 \neq n_1 \ldots n_k \in \bigcap_{i:1}^{k} n_i \mathbb{Z}$.

Example 2 A cofinite ring A is S-cofinite for each multiplicative set S of A.

Conversely, if S is a multiplicative set formed by regular elements of A (e.g., $S \subseteq U(A)$) and if A is S-cofinite, then it is cofinite.

Example 3 The product ring $A = \mathbb{Z} \times \mathbb{Q}$ is not cofinite.

Indeed, $\bigcap_{n:1}^{\infty} (n\mathbb{Z} \times (0)) = \left(\bigcap_{n:1}^{\infty} n\mathbb{Z}\right) \times (0) = (0) \times (0)$. But for each $n_1, \ldots, n_k \in \mathbb{N}^*, (0,0) \neq (n_1 \ldots n_k, 0) \in \bigcap_{i:1}^{k} (n_i \mathbb{Z} \times (0))$.

Consider the multiplicative set $S = \mathbb{Z} \times \mathbb{Q}^*$ of A. Let $\left(I_1^{(i)} \times I_2^{(i)}\right)_{i \in \Delta}$ be a family of ideals of A satisfying $\bigcap_{i \in \Delta} \left(I_1^{(i)} \times I_2^{(i)}\right) = (0) \times (0)$ with $\left(I_1^{(i)}\right)_{i \in \Delta}$ a family of ideals of $\mathbb{Z}$ and $\left(I_2^{(i)}\right)_{i \in \Delta}$ a family of ideals of $\mathbb{Q}$. The only ideals of $\mathbb{Q}$ are (0) and $\mathbb{Q}$. There exists $i \in \Delta$ such that $I_2^{(i)} = (0)$. Let $s = (0, 1) \in S$. Then $s\left(I_1^{(i)} \times I_2^{(i)}\right) = (0) \times (0)$. So A is S-cofinite.

We will give Sect. 7 a nontrivial example of a cofinite ring.

The notions defined before have a link with the S-Artinian property.

Theorem 4.4 *Let A be a ring and S a multiplicative set of A.*
Consider the following properties:

1. *Every nonempty S-cosaturated family of ideals of A has a minimal element for the inclusion.*
2. *The ring A satisfies the property $(S - Min)$.*
3. *For each ideal I of A such that $I \cap S = \emptyset$, the quotient ring A/I is $\pi(S)$-cofinite, where $\pi : A \longrightarrow A/I$ is the canonical homomorphism.*
4. *The ring A is S-Artinian.*

We have the following implications: $(1) \Longrightarrow (2) \Longrightarrow (3) \Longrightarrow (4)$.
Proof "$(1) \Longrightarrow (2)$" Let $\mathscr{F}$ be a nonempty family of ideals of A and $\mathscr{F}^* = \{J : \text{ideal of } A, \exists s \in S, \exists I \in \mathscr{F}, sI \subseteq J\}$. Then $\mathscr{F} \subseteq \mathscr{F}^*$ because $1 \in S$, so $\mathscr{F}^* \neq \emptyset$.

We will show that $\mathscr{F}^*$ is S-cosaturated. Let J be an ideal of A, $s \in S$ and $I \in \mathscr{F}^*$ such that $sI \subseteq J$. We must show that $J \in \mathscr{F}^*$. By the definition of $\mathscr{F}^*$, there exist

$s' \in S$ and $I' \in \mathscr{F}$ such that $s'I' \subseteq I$. Then $ss'I' \subseteq sI \subseteq J$, so $ss'I' \subseteq J$ with $ss' \in S$ and $I' \in \mathscr{F}$. Then $J \in \mathscr{F}^*$ so $\mathscr{F}^*$ is S-cosaturated. By hypothesis, $(\mathscr{F}^*, \subseteq)$ has a minimal element $M \in \mathscr{F}^*$. By the definition of $\mathscr{F}^*$, there exist $s \in S$ and $I \in \mathscr{F}$ such that $sI \subseteq M$.

We will show that I is a S-minimal element of $\mathscr{F}$. Let $J \in \mathscr{F}$ such that $J \subseteq I$. Then $sJ \subseteq sI \subseteq M$, so $sJ \subseteq M \cap J$. By the definition of $\mathscr{F}^*$, $M \cap J \in \mathscr{F}^*$. Since $M \cap J \subseteq M$ with M minimal in $(\mathscr{F}^*, \subseteq)$, then $M \cap J = M$, so $M \subseteq J$. Then $sI \subseteq M \subseteq J$, so $sI \subseteq J$. Then I is a S-minimal element of $\mathscr{F}$.

"(2) $\Longrightarrow$ (3)" Let I be an ideal of A such that $S \cap I = \emptyset$. Then $\bar{0} \notin \pi(S)$. Let $(\mathscr{I}_i)_{i \in \Delta}$ be a nonempty family of ideals of A/I satisfying $\bigcap_{i \in \Delta} \mathscr{I}_i = (\bar{0})$. Put $\mathscr{I}_i = I_i/I$ with I_i an ideal of A containing I. Then $\bigcap_{i \in \Delta} I_i = I$. Let $\mathscr{F}$ be the family of ideals of A of the form $\bigcap_{i \in \Delta'} I_i$ with Δ' a nonempty finite subset of Δ. By the $(S - Min)$ property of A, $\mathscr{F}$ has a S-minimal element $I' = \bigcap_{i \in \Delta_0} I_i$ with Δ_0 a nonempty finite subset of Δ. There exists $s \in S$ such that for each $J \in \mathscr{F}$ satisfying $J \subseteq I'$, $sI' \subseteq J$. For each $k \in \Delta \setminus \Delta_0$, $I_k \cap I' \in \mathscr{F}$ and $I_k \cap I' \subseteq I'$, and then $sI' \subseteq I_k \cap I'$. So $sI' \subseteq \bigcap_{i \in \Delta} I_i = I$ then $\bar{s}(I'/I) = (\bar{0})$. So $\bar{s}\left(\bigcap_{i \in \Delta_0} \mathscr{I}_i\right) = (\bar{0})$ and the quotient ring A/I is $\pi(S)$-cofinite.

"(3) $\Longrightarrow$ (4)" Let $I_1 \supseteq I_2 \supseteq \ldots$ be a decreasing sequence of ideals of the ring A and $I = \bigcap_{k:1}^{\infty} I_k$. If $I \cap S \neq \emptyset$, let $s \in I \cap S$. For each integer $n \geq 1$, $s \in I_n$, and then $sI_1 \subseteq I_n$. The sequence is S-stationary from the rank 1. If $I \cap S = \emptyset$, by hypothesis, the ring A/I is $\pi(S)$-cofinite, where $\pi : A \longrightarrow A/I$ is the canonical homomorphism. Since $\bigcap_{n:1}^{\infty}(I_n/I) = \left(\bigcap_{n:1}^{\infty} I_n\right)/I = I/I = (\bar{0})$, there exist $s \in S$ and some integers $1 \leq n_1 < \ldots < n_r$ such that $\bar{s}\left(\bigcap_{i:1}^{r}(I_{n_i}/I)\right) = (\bar{0})$, and then $\bar{s}(I_{n_r}/I) = (\bar{0})$, so $sI_{n_r} \subseteq I = \bigcap_{n:1}^{\infty} I_n$. For each integer $n \geq 1$, $sI_{n_r} \subseteq I_n$. $\qquad\square$

We retrieve the following classical result.

Corollary 4.5 *The following assertions are equivalent for a ring A:*

1. *Every nonempty family of ideals of A has a minimal element for the inclusion.*
2. *For every proper ideal I of A, the quotient ring A/I is cofinite.*
3. *The ring A is Artinian.*

Proof The implications "(1) $\Longrightarrow$ (2) $\Longrightarrow$ (3)" follow from the preceding theorem with the multiplicative set $S = \{1\}$.

"(3) $\Longrightarrow$ (1)" Let $\mathscr{F}$ be a nonempty family of ideals of the ring A. If $(\mathscr{F}, \subseteq)$ does not have a minimal element, we construct by induction a strictly decreasing sequence of ideals of A.   □

Proposition 4.6 *Let A be a ring.*

1. *Let $S_1 \subseteq S_2$ be two multiplicative sets of A. If A is S_1-cofinite, then it is S_2-cofinite.*
2. *Let S be a multiplicative set of A and S^* its saturation. Then A is S-cofinite if and only if it is S^*-cofinite.*

Proof (1) Clear. (2) " $\Longrightarrow$ " By (1). " $\Longleftarrow$ " Let $(I_i)_{i \in \Delta}$ be a nonempty family of ideals of A satisfying $\bigcap_{i \in \Delta} I_i = (0)$. By hypothesis, there exist $s \in S^*$ and a nonempty finite subset Δ' of Δ such that $s\left(\bigcap_{i \in \Delta'} I_i \right) = (0)$. By definition of S^*, there exist $a \in A$ and $u \in S$ such that $\frac{s}{1}\frac{a}{u} = \frac{1}{1}$. There exists $t \in S$ such that $tsa = tu$. Let $t' = tu \in S$. Then $t'\left(\bigcap_{i \in \Delta'} I_i \right) = tsa\left(\bigcap_{i \in \Delta'} I_i \right) \subseteq s\left(\bigcap_{i \in \Delta'} I_i \right) = (0)$. So A is S-cofinite.   □

We characterize now the S-cofinite integral domains.

Proposition 4.7 *Let A be an integral domain and S a multiplicative set of A. Then A is S-cofinite if and only if it is a field.*

Proof " $\Longleftarrow$ " The only ideals of a field A are (0) and A. A nonempty family of ideals of A having (0) as intersection contains (0).

" $\Longrightarrow$ " Let $(I_\lambda)_{\lambda \in \Delta}$ be the family of all the nonzero ideals of A and I their intersection. Suppose that $I = (0)$. Since A is S-cofinite, there exist $s \in S$ and a nonempty finite subset Δ' of Δ such that $s\left(\bigcap_{\lambda \in \Delta'} I_\lambda \right) = (0)$. Since A is an integral domain and $s \neq 0$, $\bigcap_{\lambda \in \Delta'} I_\lambda = (0)$. Since (0) is a prime ideal and Δ' is finite, there exists $\lambda_0 \in \Delta'$ such that $I_{\lambda_0} = (0)$, which is absurd. Then $I \neq (0)$. Let $0 \neq a \in I$. Then $aA \subseteq I$. But the ideal aA is a member of the family $(I_\lambda)_{\lambda \in \Delta}$, and then $I \subseteq aA$, so $I = aA$. Let x be any nonzero element of A. We must show that x is invertible in A. Since $0 \neq xa \in I$, then $I = xaA$. So $aA = xaA$. Since A is an integral domain, $A = xA$, then x is invertible.   □

Example An integral domain is cofinite if and only if it is a field.

Proposition 4.8 *Let S_i be a multiplicative set of a ring A_i, $1 \leq i \leq n$. Put $A = A_1 \times \ldots \times A_n$ and $S = S_1 \times \ldots \times S_n$. The ring A is S-cofinite if and only if the rings A_i are S_i-cofinite, $1 \leq i \leq n$.*

Proof $''\Longrightarrow''$ Fix an index $1 \leq i \leq n$. Let $\left(J_k^{(i)}\right)_{k \in \Delta}$ be a nonempty family of ideals of A_i satisfying $\bigcap_{k \in \Delta} J_k^{(i)} = (0)$. For each $k \in \Delta$, we define the ideal $I_k = (0) \times \ldots \times J_k^{(i)} \times \ldots \times (0)$ of A. Then $\bigcap_{k \in \Delta} I_k = (0)$. By hypothesis, there exist $s = (s_1, \ldots, s_n) \in S$ and a nonempty finite subset Δ' of Δ such that $s\left(\bigcap_{k \in \Delta'} I_k\right) = (0)$. Then $s_i\left(\bigcap_{k \in \Delta'} J_k^{(i)}\right) = (0)$. So A_i is S_i-cofinite.

$''\Longleftarrow''$ Let $\left(I_k\right)_{k \in \Delta}$ be a nonempty family of ideals of A satisfying $\bigcap_{k \in \Delta} I_k = (0)$. By Lemma 2.2, for each $k \in \Delta$, $I_k = J_k^{(1)} \times \ldots \times J_k^{(n)}$ with $J_k^{(i)}$ an ideal of A_i for $1 \leq i \leq n$. Since $(0) = \bigcap_{k \in \Delta} I_k = \left(\bigcap_{k \in \Delta} J_k^{(1)}\right) \times \ldots \times \left(\bigcap_{k \in \Delta} J_k^{(n)}\right)$, then $\bigcap_{k \in \Delta} J_k^{(i)} = (0)$ for $1 \leq i \leq n$. Since the ring A_i is S_i-cofinite, there exist $s_i \in S_i$ and a nonempty finite subset Δ_i of Δ such that $s_i\left(\bigcap_{k \in \Delta_i} J_k^{(i)}\right) = (0)$. Put $s = (s_1, \ldots, s_n) \in S$ and $\Delta' = \Delta_1 \cup \ldots \cup \Delta_n \subseteq \Delta$. Then $s\left(\bigcap_{k \in \Delta'} I_k\right) = s_1\left(\bigcap_{k \in \Delta'} J_k^{(1)}\right) \times \ldots \times s_n\left(\bigcap_{k \in \Delta'} J_k^{(n)}\right) \subseteq s_1\left(\bigcap_{k \in \Delta_1} J_k^{(1)}\right) \times \ldots \times s_n\left(\bigcap_{k \in \Delta_n} J_k^{(n)}\right) = (0)$. Then A is S-cofinite. $\qquad\square$

Example Let $A_1, \ldots, A_n$ be a finite number of rings. Then $A_1 \times \ldots \times A_n$ is cofinite if and only if $A_1, \ldots, A_n$ are cofinite.

We characterize now the cofinite property in terms of S-cofinite.

Theorem 4.9 *The following assertions are equivalent for a ring A:*

1. *The ring A is cofinite.*
2. *For each $P \in Spec(A)$, the ring A is S-cofinite where $S = A \setminus P$.*
3. *For each $M \in Max(A)$, the ring A is S-cofinite where $S = A \setminus M$.*

Proof The implications $''(1) \Longrightarrow (2) \Longrightarrow (3)''$ are clear.

$''(3) \Longrightarrow (1)''$ Let $(I_k)_{k \in \Delta}$ be a nonempty family of ideals of A satisfying $\bigcap_{k \in \Delta} I_k = (0)$. By hypothesis, for each $M \in Max(A)$, there exist $s_M \in A \setminus M$ and a nonempty finite subset Δ_M of Δ such that $s_M\left(\bigcap_{k \in \Delta_M} I_k\right) = (0)$. Consider the ideal $J = \left(s_M; \ M \in Max(A)\right)$ of A. If $J \neq A$, there exists $M_0 \in Max(A)$ such that $J \subseteq M_0$, and then $s_{M_0} \in M_0$, which is absurd. Then $J = A$. There exist $M_1, \ldots, M_r \in Max(A)$ such that $\left(s_{M_1}, \ldots, s_{M_r}\right) = A$. Let $\Delta' = \Delta_{M_1} \cup \ldots \cup \Delta_{M_r} \subseteq \Delta$. Then $\bigcap_{k \in \Delta'} I_k = A\left(\bigcap_{k \in \Delta'} I_k\right) = \left(s_{M_1}, \ldots, s_{M_r}\right)\left(\bigcap_{k \in \Delta'} I_k\right) =$

$$s_{M_1}\Big(\bigcap_{k\in\Delta'} I_k\Big)+\ldots+s_{M_r}\Big(\bigcap_{k\in\Delta'} I_k\Big)\subseteq s_{M_1}\Big(\bigcap_{k\in\Delta_{M_1}} I_k\Big)+\ldots+s_{M_r}\Big(\bigcap_{k\in\Delta_{M_r}} I_k\Big)=(0).$$

Then A is cofinite. $\square$

The remaining of the results in this section are due to Ozen et al. 2021.

Definitions 4.10 Let A be a ring, S a multiplicative set of A, M a A-module, $\mathscr{F}$ a nonempty family of sub-modules of M, and $X\in\mathscr{F}$. We say that X is a S-minimal element of $\mathscr{F}$ if there exists $s\in S$ such that for each element $Y\in\mathscr{F}$ satisfying $Y\subseteq X$, we have the inclusion $sX\subseteq Y$.

We say that the module M has the property $(S-Min)$ if each nonempty family of sub-modules of M admits a S-minimal element.

Definition 4.11 Let A be a ring, S a multiplicative set of A, M a A-module, and $\mathscr{F}$ a nonempty family of sub-modules of M. We say that $\mathscr{F}$ is S-cosaturated if for each sub-module Y of M, if there exist $s\in S$ and $X\in\mathscr{F}$ such that $sX\subseteq Y$, then $Y\in\mathscr{F}$.

The notion of finitely generated modules has a dual in the following definition that generalizes the notion of cofinite rings.

Definitions 4.12 Let A be a ring and M a A-module.

1. We say that M is cofinite if for each nonempty family $(X_i)_{i\in\Delta}$ of sub-modules of M satisfying $\bigcap_{i\in\Delta} X_i=(0)$, there exists a nonempty finite set $\Delta'\subseteq\Delta$ such that $\bigcap_{i\in\Delta'} X_i=(0)$.

2. Let S be a multiplicative set of A. We say that M is S-cofinite if for each nonempty family $(X_i)_{i\in\Delta}$ of sub-modules of M satisfying $\bigcap_{i\in\Delta} X_i=(0)$, there exist $s\in S$ and a finite set $\Delta'\subseteq\Delta$ such that $s\Big(\bigcap_{i\in\Delta'} X_i\Big)=(0)$.

Example 1 By Example 1 of Definition 4.3, the $\mathbb{Z}$-module $\mathbb{Z}$ is not cofinite.

Example 2 Let $p_1=2$, $p_2=3,\ldots$ be the sequence of the prime numbers. The product $\mathbb{Z}$-module $M=\prod_{i:1}^{\infty}\mathbb{Z}/p_i\mathbb{Z}$ is not cofinite. Indeed, for each $i\in\mathbb{N}^*$, let the sub-module $M_i=\mathbb{Z}/p_1\mathbb{Z}\times\ldots\times\mathbb{Z}/p_{i-1}\mathbb{Z}\times(0)\times\mathbb{Z}/p_{i+1}\mathbb{Z}\times\ldots$ of M. Then $\bigcap_{i:1}^{\infty} M_i=(0)$. But for each nonempty finite set $\Delta\subset\mathbb{N}^*$, $\bigcap_{i\in\Delta} M_i\neq(0)$.

Example 3 Let A be a ring and M a A-module. If M is cofinite, then M is S-cofinite for each multiplicative set S of A.

Example 4 Let A be a ring, M a A-module, and $S_1 \subseteq S_2$ two multiplicative sets of A. If M is S_1-cofinite, then M est S_2-cofinite.

Theorem 4.13 *Let A be a ring, S a multiplicative set of A, and M a A-module. We consider the following properties:*

1. *Each nonempty S-cosaturated family of sub-modules of M admits a minimal element for the inclusion.*
2. *The module M satisfies the property $(S - Min)$.*
3. *For each sub-module X of M, the quotient A-module M/X is S-cofinite.*
4. *The module M is S-Artinian.*

We have the following implications: $(1) \implies (2) \implies (3) \implies (4)$.

Proof $"(1) \implies (2)"$ Let $\mathscr{F}$ be a nonempty family of sub-modules of M. Consider $\mathscr{F}^* = \{Y : \text{sub-module of } M, \exists s \in S, \exists X \in \mathscr{F}, sX \subseteq Y\}$. Then $\mathscr{F} \subseteq \mathscr{F}^*$ since $1 \in S$. So $\mathscr{F}^* \neq \emptyset$. We will show that the family $\mathscr{F}^*$ is S-cosaturated. Let Y be a sub-module of M, $s \in S$ and $X \in \mathscr{F}^*$ such that $sX \subseteq Y$. We must show that $Y \in \mathscr{F}^*$. By definition of $\mathscr{F}^*$, there exist $s' \in S$ and $X' \in \mathscr{F}$ such that $s'X' \subseteq X$. Then $ss'X' \subseteq sX \subseteq Y$, so $ss'X' \subseteq Y$ with $ss' \in S$ and $X' \in \mathscr{F}$. Then $Y \in \mathscr{F}^*$, so $\mathscr{F}^*$ is S-cosaturated. By hypothesis, $(\mathscr{F}^*, \subseteq)$ admits a minimal element $\mathscr{M} \in \mathscr{F}^*$. By definition of $\mathscr{F}^*$, there exist $s \in S$ and $X \in \mathscr{F}$ such that $sX \subseteq \mathscr{M}$. We will show that X is a S-minimal element of $\mathscr{F}$. Let $Y \in \mathscr{F}$ be such that $Y \subseteq X$. Then $sY \subseteq sX \subseteq \mathscr{M}$, so $sY \subseteq \mathscr{M} \cap Y$. By definition of $\mathscr{F}^*$, $\mathscr{M} \cap Y \in \mathscr{F}^*$. Since $\mathscr{M} \cap Y \subseteq \mathscr{M}$ with $\mathscr{M}$ minimal in $(\mathscr{F}^*, \subseteq)$, then $\mathscr{M} \cap Y = \mathscr{M}$, so $\mathscr{M} \subseteq Y$. Then $sX \subseteq \mathscr{M} \subseteq Y$, so $sX \subseteq Y$. Then X is a S-minimal element of $\mathscr{F}$.

$"(2) \implies (3)"$ Let X be a sub-module of M and $(\mathscr{X}_i)_{i \in \Delta}$ a nonempty family of sub-modules of M/X satisfying $\bigcap_{i \in \Delta} \mathscr{X}_i = (\bar{0})$. Put $\mathscr{X}_i = X_i/X$ with X_i a sub-module of M containing X. Then $\bigcap_{i \in \Delta} X_i = X$. Let $\mathscr{F}$ be the family of the sub-modules of M of the form $\bigcap_{i \in \Delta'} X_i$ with Δ' a nonempty finite subset of Δ. By the $(S - Min)$ property of M, $\mathscr{F}$ admits a S-minimal element $X' = \bigcap_{i \in \Delta_0} X_i$ with Δ_0 a nonempty finite subset of Δ. There exists $s \in S$ such that for each $Y \in \mathscr{F}$ satisfying $Y \subseteq X', sX' \subseteq Y$. For each $k \in \Delta \setminus \Delta_0$, $X_k \cap X' \in \mathscr{F}$ and $X_k \cap X' \subseteq X'$, and then $sX' \subseteq X_k \cap X'$. So $sX' \subseteq \bigcap_{i \in \Delta} X_i = X$, and then $s(X'/X) = (\bar{0})$. So $s\left(\bigcap_{i \in \Delta_0} \mathscr{X}_i\right) = (\bar{0})$ and the A-module M/X is S-cofinite.

$"(3) \implies (4)"$ Let $X_1 \supseteq X_2 \supseteq \ldots$ be a decreasing sequence of sub-modules of M and $X = \bigcap_{k:1}^{\infty} X_k$. Then $\bigcap_{n:1}^{\infty} (X_n/X) = (\bigcap_{n:1}^{\infty} X_n)/X = X/X = (\bar{0})$. By hypothesis, the A-module A/X is S-cofinite. There exist $s \in S$ and integers $1 \leq n_1 < \ldots < n_r$

such that $s\left(\bigcap_{i:1}^{r}(X_{n_i}/X)\right) = (\bar{0})$. Then $s(X_{n_r}/X) = (\bar{0})$, so $sX_{n_r} \subseteq X = \bigcap_{n:1}^{\infty} X_n$.

For each integer $n \geq 1$, $sX_{n_r} \subseteq X_n$. $\square$

We retrieve the following classical result.

Corollary 4.14 *Let A be a ring and M a A-module. The following assertions are equivalent:*

1. *Each nonempty family of sub-modules of M admits a minimal element for the inclusion.*
2. *For each sub-module X of M, the quotient A-module M/X is cofinite.*
3. *The module M is Artinian.*

Proof The implications $"(1) \implies (2) \implies (3)"$ follow from the preceding theorem with the multiplicative set $S = \{1\}$.

$"(3) \implies (1)"$ Let $\mathscr{F}$ be a nonempty family of sub-modules of M. If $(\mathscr{F}, \subseteq)$ does not have a minimal element, we construct by induction a strictly decreasing sequence of sub-modules of M. $\square$

Now, we characterize the cofinite modules in terms of S-cofinite.

Proposition 4.15 *Let A be a ring and M a A-module. The following assertions are equivalent:*

1. *The module M is cofinite.*
2. *For each $P \in Spec(A)$, the module M is S-cofinite where $S = A \setminus P$.*
3. *For each $P \in Max(A)$, the module M is S-cofinite where $S = A \setminus P$.*

Proof The implications $"(1) \implies (2) \implies (3)"$ are clear.

$"(3) \implies (1)"$ Let $(X_k)_{k \in \Delta}$ be a nonempty family of sub-modules of M satisfying $\bigcap_{k \in \Delta} X_k = (0)$. By hypothesis, for each $P \in Max(A)$, there exist $s_P \in A \setminus P$ and a nonempty finite subset Δ_P of Δ such that $s_P\left(\bigcap_{k \in \Delta_P} X_k\right) = (0)$.

Consider the ideal $J = (s_P; P \in Max(A))$ of A. If $J \neq A$, there exists $P_0 \in Max(A)$ such that $J \subseteq P_0$, and then $s_{P_0} \in P_0$, which is absurd. So $J = A$. Let $P_1, \ldots, P_r \in Max(A)$ be such that $(s_{P_1}, \ldots, s_{P_r}) = A$ and $\Delta' = \Delta_{P_1} \cup \ldots \cup \Delta_{P_r} \subseteq \Delta$. Then $\bigcap_{k \in \Delta'} X_k = A\left(\bigcap_{k \in \Delta'} X_k\right) = (s_{P_1}, \ldots, s_{P_r})\left(\bigcap_{k \in \Delta'} X_k\right) = $

$$s_{P_1}\left(\bigcap_{k \in \Delta'} X_k\right) + \ldots + s_{P_r}\left(\bigcap_{k \in \Delta'} X_k\right) \subseteq s_{P_1}\left(\bigcap_{k \in \Delta_{P_1}} X_k\right) + \ldots + s_{P_r}\left(\bigcap_{k \in \Delta_{P_r}} X_k\right) = (0).$$

Then M is cofinite. $\square$

5 Multiplication Modules

The notion of multiplication modules is introduced by Krull in the case of rings. We define it here in the general setting of modules.

Definitions 5.1 Let A be a ring and M a A-module.

1. We say that M is a multiplication module if for each sub-module N of M, $N = (N :_A M)M$, where $(N :_A M) = ann_A(M/N) = \{a \in A;\ aM \subseteq N\}$ an ideal of A.
2. Let P be a maximal ideal of A. We say that M is P-cyclic if there exist $p \in P$ and $x \in M$ such that $(1 - p)M \subseteq Ax$.

In general, for each sub-module N of M, we have the inclusion $(N :_A M)M \subseteq N$. Note that M is a multiplication module if and only if for each sub-module N of M, there exists an ideal I of A such that $N = IM$. Indeed, if M is a multiplication module and N is a sub-module of M, we can take $I = (N :_A M)$. Conversely, let N be a sub-module of M and I an ideal of A satisfying $N = IM$. Then $I \subseteq (N :_A M)$, so $N = IM \subseteq (N :_A M)M \subseteq N$, then $N = (N :_A M)M$.

To give examples of multiplication modules, we recall from Chap. 1 that a A-module M is faithful if the ideal $ann_A(M) = \{a \in A;\ aM = 0\}$ of A is zero. An element x of M is is said of torsion if there exists $0 \neq a \in A$ such that $ax = 0$. The module M is said of torsion if all its elements are of torsion. It is torsion-free if 0 is the only element of torsion. A torsion-free module is faithful. The converse is false. The $\mathbb{Z}$-module $\mathbb{Q}/\mathbb{Z}$ is with torsion and faithful.

Example 1 Let A be a ring. A cyclic A-module M is a multiplication module.

Indeed, $M = Am$ with $m \in M$. Let N be a sub-module of M. Every element x of N can be written as $x = am$ with $a \in A$. Since $aM = aAm = Aam = Ax \subseteq N$, then $a \in (N : M)$, so $x = am \in (N : M)M$. Then $N \subseteq (N : M)M$, so $N = (N : M)M$. For example, the $\mathbb{Z}$-module $\mathbb{Z}/n\mathbb{Z} = (\bar{1})$, $n \geq 2$ an integer, is a multiplication module but not faithful.

Example 2 The $\mathbb{Z}$-module $\mathbb{Q}$ is faithful but not a multiplication module.

Indeed, $(\mathbb{Z} :_{\mathbb{Z}} \mathbb{Q})\mathbb{Q} = (0)\mathbb{Q} = (0) \neq \mathbb{Z}$.

Example 3 Let K be a field and M a K-vector space of dimension 2 and with basis $\{u, v\}$. Then M is faithful but not a multiplication module.

Indeed, let $N = (u)$, and then $(N : M)M = (0)M = (0) \neq N$.

The following example concerns multiplication ideals. Let A be a ring and I an ideal of A. Then I is a multiplication ideal (as a A-module) if each ideal of A contained in I is a multiple of I (by an ideal of A).

Example 4 (Smith 1988) Consider the ring $A = \mathbb{Z}[\sqrt{5}]$ and its ideals $N = (-1 + \sqrt{5})A \subseteq M = 2A + (-1 + \sqrt{5})A$. Then $(N : M) = ((-1 + \sqrt{5})A : 2A + (-1 + \sqrt{5})A) = ((-1 + \sqrt{5})A : 2A) = M$. Indeed, it is clear that $-1 + \sqrt{5} \in$

$((-1+\sqrt{5})A : 2A)$ and since $2.2. = 4 = (-1+\sqrt{5})(1+\sqrt{5}) \in (-1+\sqrt{5})A$, then $M = 2A + (-1+\sqrt{5})A \subseteq ((-1+\sqrt{5})A : 2A)$. Conversely, let $x = a + b\sqrt{5} \in ((-1+\sqrt{5})A : 2A)$ with $a, b \in \mathbb{Z}$. Then $2x \in (-1+\sqrt{5})A$, so $2x = (-1+\sqrt{5})(c+d\sqrt{5})$ wih $c, d \in \mathbb{Z}$. Then $2(a+b\sqrt{5}) = -c+5d+(c-d)\sqrt{5}$, so $2a = -c + 5d$ and $2b = c - d$. By addition term by term and simplification, we find $a + b = 2d$, and then $a = 2d - b$. So $x = a + b\sqrt{5} = 2d - b + b\sqrt{5} = 2d + b(-1+\sqrt{5}) \in M$. We conclude that $(N : M) = M$, and then $(N : M)M = M^2 = (2A+(-1+\sqrt{5})A)^2 = 4A+2(-1+\sqrt{5})A+(6-2\sqrt{5})A = 4A+2(-1+\sqrt{5})A$ because $6-2\sqrt{5} = 4-2(-1+\sqrt{5})$. So $(N : M)M = 2(2A+(-1+\sqrt{5})A) = 2M$. Suppose that the element $-1+\sqrt{5}$ of N belongs to $(N : M)M = 2M$ and put $-1+\sqrt{5} = 4(a+b\sqrt{5})+2(-1+\sqrt{5})(c+d\sqrt{5}) = 4a-2c+10d+(4b+2c-2d)\sqrt{5}$ with $a, b, c, d \in \mathbb{Z}$. Then $1 = 4b+2c-2d = 2(2b+c-d)$, which is absurd. Then $(N : M)M \neq N$, so the A-module M is not multiplication. But it is faithful since A is an integral domain.

Notation Let A be a ring, M a A-module, and P a maximal ideal of A.

Let $T_P(M) = \{x \in M; \exists p \in P, (1 - p)x = 0\}$: it is a sub-module of M.

Indeed, it is clear that $0 \in T_P(M)$. The stability of $T_P(M)$ by multiplication by the elements of A is clear. Let $x, y \in T_P(M)$. There exist $p, q \in P$ such that $(1 - p)x = (1 - q)y = 0$. Then $(1 - p)(1 - q)(x + y) = 0$ and $(1 - p)(1 - q) = 1 - (p + q - pq)$ with $p + q - pq \in P$. Then $x + y \in T_P(M)$.

When $M = T_P(M)$, for each $x \in M$, there exists a nonzero element of A of the form $1 - p$ with $p \in P$ such that $(1 - p)x = 0$. In this case, we say that M is P-torsion.

The following proposition is the principal result of this section. It characterizes multiplication modules by means of maximal ideals of their basic rings.

Proposition 5.2 (Abd El-Bast and Smith 1988) *Let A be a ring and M a A-module. Then M is a multiplication module if and only if for each maximal ideal P of A, either $M = T_P(M)$ or M is P-cyclic.*

Proof $" \Longrightarrow"$ Let P be a maximal ideal of the ring A.

First case: $PM = M$. Let $x \in M$. Since M is a multiplication A-module, there exists an ideal I of A such that $Ax = IM$. Then $Ax = I(PM) = P(IM) = PAx = Px$. There exists $p \in P$ such that $x = px$, and then $(1 - p)x = 0$, so $x \in T_P(M)$. Then $M = T_P(M)$.

Second case: $PM \subset M$. Let $x \in M \setminus PM$. Since M is a multiplication module, there exists an ideal J of A such that $Ax = JM$. If $J \subseteq P$, and then $x \in JM \subseteq PM$, which is absurd. Then $J \nsubseteq P$. Let $q \in J \setminus P$, and then $P + qA = A$ since P is maximal. Put $1 = p + qa$ with $p \in P$ and $a \in A$, then $1 - p = qa \in J$ and $(1 - p)M \subseteq JM = Ax$. So M is P-cyclic.

$" \Longleftarrow"$ Let N be sub-module of M and $I = (N :_A M)$. We will show that $IM = N$. The inclusion $IM \subseteq N$ is clear. Conversely, let $x \in N$. Consider the ideal $K = (IM :_A x) = \{a \in A; ax \in IM\}$ of the ring A. Suppose that $K \neq A$.

There exists a maximal ideal P of A containing K. By hypothesis, we have the following two cases:

First case: $M = T_P(M)$. In particular, $x \in T_P(M)$. There exists $p \in P$ such that $(1 - p)x = 0 \in IM$. Then $1 - p \in K \subseteq P$, which is absurd.

Second case: The module M is P-cyclic. There exist $p \in P$ and $y \in M$ such that $(1-p)M \subseteq Ay$. In particular, $(1-p)N \subseteq Ay$. Consider the ideal $J = \big((1-p)N :_A y\big) = \big\{a \in A; \ ay \in (1-p)N\big\}$ of the ring A. Then $(1 - p)N = Jy$. Indeed, the inverse inclusion follows from the definition of J. Let $z \in (1 - p)N \subseteq Ay$. Then $z = ay$ with $a \in A$. So $a \in J$ by definition of J and $z = ay \in Jy$. We have always shown the equality $(1 - p)N = Jy$. Then $(1 - p)JM = J(1 - p)M \subseteq JAy = Jy = (1 - p)N \subseteq N$ so $(1 - p)J \subseteq (N :_A M) = I$, by definition of I. Then $(1 - p)^2x \in (1 - p)^2N = (1 - p)(1 - p)N = (1 - p)Jy \subseteq IM$. So $(1 - p)^2 \in (IM :_A x) = K \subseteq P$, and then $1 - p \in P$, which is absurd.

We conclude that $K = A$, and then $1 \in K$ so $x \in IM$. We have the equality $IM = N$. $\square$

Example 1

(1) (Bernard 1981). Let (A, P) be a local ring and M a A-module. Then M is a multiplication module if and only if it is cyclic. In particular, the multiplication ideals of a local ring are its principal ideals.

   Indeed, by Example 1 of Definition 5.1, each cyclic module is a multiplication module. Conversely, if $M = T_P(M)$, for each $x \in M$, there exists $p \in P$ such that $(1 - p)x = 0$. But $1 - p$ is invertible in the local ring (A, P), and then $x = 0$. So $M = (0)$. If M is P-cyclic, there exist $p \in P$ and $x \in M$ such that $(1 - p)M \subseteq Ax$, and then $M \subseteq Ax$. So $M = Ax$.

(2) Let $K \subset L$ be an extension of fields with infinite algebraic degree. The domain $A = K + XL[[X]]$ is local with maximal ideal $P = XL[[X]]$. Then P is not a multiplication ideal. Indeed, in the contrary case, by (1), there exists $l(X) \in L[[X]]$ such that $P = XL[[X]] = Xl(X)A$, and then $L[[X]] = l(X)A = l(X)\big(K + XL[[X]]\big)$. So $L = l(0)K$ with $0 \neq l(0) \in L$, the constant term of $l(X)$. Then $[L : K] = 1$, which is absurd.

Example 2 (Anderson 1980) Let I be a multiplication ideal of a ring A. Then I is locally principal; i.e., for each maximal ideal P of A, the ideal IA_P is principal. Indeed, let P be a maximal ideal of A. We have the following two cases:

First case: $I = T_P(I)$. Let $x \in IA_P$, $x = \frac{a}{s}$ with $a \in I$ and $s \in A \setminus P$. There exists $p \in P$ such that $(1 - p)a = 0$. Then $x = \frac{a}{s} = \frac{(1-p)a}{(1-p)s} = 0$. So $IA_P = (0)$ is a principal ideal.

Second case: The ideal I is P-cyclic. There exist $p \in P$ and $a \in I$ such that $(1 - p)I \subseteq aA$. Let $x \in IA_P$, $x = \frac{b}{s}$ with $b \in I$ and $s \in A \setminus P$. There exists $c \in A$ such that $(1 - p)b = ac$. So $x = \frac{(1-p)b}{(1-p)s} = \frac{ac}{(1-p)s} \in aA_P$ since $(1-p)s \in A \setminus P$. Then $IA_P = aA_P$ is a principal ideal.

We have always seen in preceding examples that the two notions, multiplication and faithful, are independent for a module. In the following proposition, we characterize the faithful multiplication modules using the ideals of the basic ring.

Proposition 5.3 (Abd El-Bast and Smith 1988) *Let A be a ring and M a faithful A-module. Then M is a multiplication module if and only if:*

1. *For each nonempty family $(I_\lambda)_{\lambda \in \Lambda}$ of ideals of A, $\left(\bigcap_{\lambda \in \Lambda} I_\lambda \right) M = \bigcap_{\lambda \in \Lambda} (I_\lambda M)$.*
2. *For each sub-module N of M and for each ideal I of A satisfying $N \subset IM$, there exists an ideal $J \subset I$ of A such that $N \subseteq JM$.*

Proof "$\Longrightarrow$" 1. Let $(I_\lambda)_{\lambda \in \Lambda}$ be a nonempty family of ideals of A and $I = \bigcap_{\lambda \in \Lambda} I_\lambda$.

Then $IM \subseteq \bigcap_{\lambda \in \Lambda} (I_\lambda M)$. Conversely, let $x \in \bigcap_{\lambda \in \Lambda} (I_\lambda M)$ and $J = (IM :_A x)$ ideal of A. Suppose that $J \neq A$. There exists a maximal ideal P of A containing J. By the preceding proposition, either $M = T_P(M)$ or M is P-cyclic. We will arrive at a contradiction in each of the two cases.

First case: $M = T_P(M)$. Then $x \in T_P(M)$. There exists $p \in P$ such that $(1 - p)x = 0 \in IM$, and then $1 - p \in (IM : x) = J \subseteq P$ so $1 = (1 - p) + p \in P$, which is absurd.

Second case: M is P-cyclic. There exist $p \in P$ and $m \in M$ such that $(1 - p)M \subseteq Am$.

For each $\lambda \in \Lambda$, $(1 - p)I_\lambda M \subseteq I_\lambda m$, so $(1 - p)\left(\bigcap_{\lambda \in \Lambda} (I_\lambda M) \right) \subseteq \bigcap_{\lambda \in \Lambda} (1 - p)(I_\lambda M) \subseteq \bigcap_{\lambda \in \Lambda} (I_\lambda m)$. Since $x \in \bigcap_{\lambda \in \Lambda} (I_\lambda M)$, then $(1 - p)x = a_\lambda m$ with $a_\lambda \in I_\lambda$ for each $\lambda \in \Lambda$.

Fix $\lambda_0 \in \Lambda$. For each $\lambda \in \Lambda$, $a_{\lambda_0} m = a_\lambda m$, and then $(a_{\lambda_0} - a_\lambda)m = 0$. By multiplying the inclusion $(1 - p)M \subseteq Am$ by $a_{\lambda_0} - a_\lambda$, we obtain $(1 - p)(a_{\lambda_0} - a_\lambda)M = 0$. Since M is faithful, $(1 - p)(a_{\lambda_0} - a_\lambda) = 0$, and then $(1 - p)a_{\lambda_0} = (1 - p)a_\lambda \in I_\lambda$ for each $\lambda \in \Lambda$.

Then $(1 - p)a_{\lambda_0} \in \bigcap_{\lambda \in \Lambda} I_\lambda = I$. So $(1 - p)^2 x = (1 - p)(1 - p)x = (1 - p)a_{\lambda_0} m \in IM$, and then $(1 - p)^2 \in (IM :_A x) = J \subseteq P$. But P is a maximal ideal of A, and then $1 - p \in P$, so $1 = (1 - p) + p \in P$, which is absurd.

Conclusion: $J = A$, then $1 \in J = (IM :_A x)$ so $x \in IM$.

2. Let N be a sub-module of M and I an ideal of A satisfying $N \subset IM$. Since M is a multiplication module, there exists an ideal L of A such that $N = LM$. Let the ideal $J = I \cap L \subseteq I$. By (1), $N = (LM) \cap (IM) = (L \cap I)M = JM$. Necessarily, $J \subset I$.

"$\Longleftarrow$" Let N be a sub-module of M. Let $\mathscr{S}$ be the set of the ideals I of A such that $N \subseteq IM$. Then $A \in \mathscr{S}$ and $(\mathscr{S}, \supseteq)$ is inductive. Indeed, let $(I_\lambda)_{\lambda \in \Lambda}$ be a nonempty family of elements of $\mathscr{S}$. By (1), $N \subseteq \bigcap_{\lambda \in \Lambda} (I_\lambda M) = \left(\bigcap_{\lambda \in \Lambda} I_\lambda \right) M$.

Then $\bigcap_{\lambda \in \Lambda} I_\lambda \in \mathscr{S}$ and bounds the family. By Zorn's lemma, $(\mathscr{S}, \supseteq)$ has a maximal element I_0. Then I_0 is an ideal of A satisfying $N \subseteq I_0 M$. Suppose that $N \neq I_0 M$. By (2), there exists an ideal $J \subset I_0$ of A such that $N \subseteq JM$. Then $J \in \mathscr{S}$ and $I_0 \supset J$, which contradicts the maximality of I_0 in $(\mathscr{S}, \supseteq)$. Then $N = I_0 M$. So M is a multiplication module.   □

Example The equality in (1) of the proposition is in general false even for a finite number of ideals of a ring A and M a faithful A-module.

Let K be a field, $A = K[X^2, X^3]$, $M = K[X]$ and $I_1 = X^2 K[X^2, X^3]$, $I_2 = X^3 K[X^2, X^3]$. Then $I_1 M \cap I_2 M = X^2 K[X] \cap X^3 K[X] = X^3 K[X]$.

On the other hand, $(I_1 \cap I_2)M = (X^2 K[X^2, X^3] \cap X^3 K[X^2, X^3])K[X] = X^5 K[X]$.

Indeed, $X^5 = X^2 X^3 \in X^2 K[X^2, X^3] \cap X^3 K[X^2, X^3] \subseteq (X^2 K[X^2, X^3] \cap X^3 K[X^2, X^3])K[X]$.

Conversely, it suffices to show that $X^2 K[X^2, X^3] \cap X^3 K[X^2, X^3] \subseteq X^5 K[X]$.

A nonzero monomial of an element of $X^2 K[X^2, X^3] \cap X^3 K[X^2, X^3]$ is of the form $f = a X^2 X^{2\alpha} X^{3\beta} = b X^3 X^{2\alpha'} X^{3\beta'}$ with $a, b \in K^*$ and $\alpha, \beta, \alpha', \beta' \in \mathbb{N}$. Then $f = a X^{2+2\alpha+3\beta} = b X^{3+2\alpha'+3\beta'}$. It suffices to show that the integer $2 + 2\alpha + 3\beta = 3 + 2\alpha' + 3\beta' \geq 5$. To do this, it suffices that $(\beta \geq 1)$ or $(\beta = 0$ and $\alpha \geq 2)$ or $(\alpha' \geq 1)$ or $(\alpha' = 0$ and $\beta' \geq 1)$. In the contrary case, $\beta = 0$ and $0 \leq \alpha \leq 1$ and $\alpha' = 0$ and $\beta' = 0$. In this case, the second member of the equality $2 + 2\alpha + 3\beta = 3 + 2\alpha' + 3\beta'$ is 3 and the first member is 2 or 4, which is absurd. Then $I_1 M \cap I_2 M \neq (I_1 \cap I_2)M$.

Other proof for the inclusion: $X^2 K[X^2, X^3] \cap X^3 K[X^2, X^3] \subseteq X^5 K[X]$. Indeed, the set $\{2\alpha + 3\beta; \alpha, \beta \in \mathbb{N}\} = \mathbb{N} \setminus \{1\}$. Then an element of $K[X^2, X^3]$ is of the form $a_0 + a_2 X^2 + a_3 X^3 + \ldots$ where the $a_i \in K$. An element of $X^2 K[X^2, X^3]$ can be written as $a_0 X^2 + a_2 X^4 + a_3 X^5 + \ldots$, and an element of $X^3 K[X^2, X^3]$ can be written as $b_0 X^3 + b_2 X^5 + b_3 X^6 + \ldots$ where the $b_i \in K$. By the uniqueness of the coefficients, an element f of $X^2 K[X^2, X^3] \cap X^3 K[X^2, X^3]$ has a constant term zero but also the coefficients of X, X^2, X^3 and X^4. Then $f \in X^5 K[X]$.

In the following proposition, we are concerned with relationships between the ideals of a ring A and the sub-modules of a multiplication module M.

Proposition 5.4 (Smith 1988) *Let A be a ring, M a multiplication A-module with annihilator $L = \mathrm{ann}_A(M)$, and I and J two ideals of A. Then $IM \subseteq JM$ if and only if $I \subseteq J + L$ or $M = ((J + L) : I)M$.*

Proof "$\Longleftarrow$" If $I \subseteq J + L$, then $IM \subseteq (J + L)M = JM + LM = JM$. If $M = ((J + L) : I)M$, then $IM = I((J + L) : I)M \subseteq (J + L)M = JM$.

"$\Longrightarrow$" First case: $L = \mathrm{ann}_A(M) = (0)$, i.e., the A-module M is faithful.

In this case, we must show that either $I \subseteq J$ or $M = (J : I)M$. Suppose that $I \not\subseteq J$ and show that $M = (J : I)M$. Note that $(J : I) = \bigcap_{a \in \Lambda}(J : aA)$

with $\Lambda = I \setminus J$. By the preceding proposition, $(J : I)M = \left(\bigcap_{a \in \Lambda} (J : aA) \right) M = \bigcap_{a \in \Lambda} (J : aA)M$. It suffices to show that for each $a \in \Lambda$, $(J : aA)M = M$. Fix an element $a \in \Lambda$. Note that the ideal $C = (J : aA) = \{x \in A; \; xa \in J\}$ of A is proper since $a \notin J$. We will show that $CM = M$. The set $\mathscr{P}$ of prime ideals of A containing C is nonempty and $S = \sqrt{C} = \bigcap_{P \in \mathscr{P}} P$. Suppose that there exists $P \in \mathscr{P}$ such that $PM \neq M$. Let $x \in M \setminus PM$. Since M is a multiplication module, there exists an ideal D of A such that $Ax = DM$. Then $D \not\subseteq P$. Let $c \in D \setminus P$, and then $cM \subseteq DM = Ax$. By hypothesis, $IM \subseteq JM$, and then $caM \subseteq cIM \subseteq cJM = J(cM) \subseteq JAx = Jx$. There exists $b \in J$ such that $cax = bx$, and then $(ca - b)x = 0$. Since $cM \subseteq Ax$, then $c(ca - b)M = (0)$. But M is faithful, and then $c(ca - b) = 0$, so $c^2 a = cb \in J$. Then $c^2 \in (J : aA) = C \subseteq P$, so $c \in P$, which is absurd. We conclude that $PM = M$ for each $P \in \mathscr{P}$. By the preceding proposition, $M = \bigcap_{P \in \mathscr{P}} (PM) = \left(\bigcap_{P \in \mathscr{P}} P \right) M = SM$. Then $M = SM$.

Let now $m \in M$. Since M is a multiplication module, there exists an ideal I' of A such that $Am = I'M$. Then $Am = I'(SM) = S(I'M) = SAm = Sm$. There exists $s \in S$ such that $m = sm = s^2 m = \ldots$. Since $s \in S = \sqrt{C}$, there exists $k \in \mathbb{N}^*$ such that $s^k \in C$. Then $m = s^k m \in CM$. So $M = CM$, which completes the proof in this case.

General case: Consider the quotient ring $\bar{A} = A/ann_A(M) = A/L$. Then M is a faithful $\bar{A}$-module. The ideals $\bar{I} = (I + L)/L$ and $\bar{J} = (J + L)/L$ satisfy $\bar{I}M = IM$ and $\bar{J}M = JM$. Then $\bar{I}M \subseteq \bar{J}M$. By the preceding particular case, either $\bar{I} \subseteq \bar{J}$ or $M = (\bar{J} : \bar{I})M$.

If $\bar{I} \subseteq \bar{J}$, then $I + L \subseteq J + L$, so $I \subseteq J + L$.

If $M = (\bar{J} : \bar{I})M$, for each $m \in M$, there exist $a_1, \ldots, a_k \in A$ and $m_1, \ldots, m_k \in A$ such that $\bar{a}_1, \ldots, \bar{a}_k \in (\bar{J} : \bar{I})$ and $m = \bar{a}_1 m_1 + \ldots + \bar{a}_k m_k = a_1 m_1 + \ldots + a_k m_k$. Since $a_i(I + L) \subseteq J + L$, then $a_i I \subseteq J + L$. So $a_i \in ((J + L) : I)$, $1 \leq i \leq k$. Then $M = ((J + L) : I)M$. $\qquad\square$

Corollary 5.5 (Smith 1988) *Let A be a ring, M a finitely generated multiplication A-module, and I and J two ideals of A. Then $IM \subseteq JM$ if and only if $I \subseteq J + ann_A(M)$.*

Proof "$\Longleftarrow$" We have $IM \subseteq (J + ann_A(M))M = JM + ann_A(M)M = JM$.

"$\Longrightarrow$" Put $L = ann_A(M)$. By the preceding proposition, it suffices to show that if $M = ((J + L) : I)M$, then $I \subseteq J + L$. Put $K = ((J + L) : I)$ and $M = (x_1, \ldots, x_n)$. Then $M = KM$ and for $1 \leq i \leq n$, $x_i = \sum_{j:1}^{n} \lambda_{ji} x_j$ with $\lambda_{ji} \in K$. So $\sum_{j:1}^{n} (\delta_{ji} - \lambda_{ji}) x_j = 0$. The determinant of this system is of the form

$1 - a$ with $a \in K$ and $(1 - a)x_1 = \ldots = (1 - a)x_n = 0$. Then $(1 - a)M = (0)$, i.e., $1 - a \in ann_A(M) = L$. Since $1 = a + (1 - a)$, then $A = K + L$, so $I = IA = I(K + L) = IK + IL \subseteq J + L + IL = J + L$. Then we have the result. $\square$

Proposition 5.6 (Ozen et al. 2021) *Let A be a ring, S a multiplicative set of A, M a multiplication A-module, and $\pi : A \longrightarrow A/ann_A(M)$ the canonical homomorphism.*

1. *If the quotient ring $A/ann_A(M)$ is $\pi(S)$-Artinian, the module M is S-Artinian.*
2. *If M is finitely generated and S-Artinian, the ring $A/ann_A(M)$ is $\pi(S)$-Artinian.*

Proof

(1) Let $M_1 \supseteq M_2 \supseteq \ldots$ be a decreasing sequence of sub-modules of M. Then $(M_1 : M)/ann_A(M) \supseteq (M_2 : M)/ann_A(M) \supseteq \ldots$ is a decreasing sequence of ideals of $A/ann_A(M)$. There exist $s \in S$ and $k \in \mathbb{N}^*$ such that for each integer $n \geq k$, $\pi(s)(M_k : M)/ann_A(M) \subseteq (M_n : M)/ann_A(M)$. Let $n \geq k$ be an integer. Then $s(M_k : M) \subseteq (M_n : M)$, so $s(M_k : M)M \subseteq (M_n : M)M$. Since M is a multiplication module, $sM_k \subseteq M_n$. Then the sequence $M_1 \supseteq M_2 \supseteq \ldots$ is S-stationary and the module M is S-Artinian.

(2) Let $J_1 \supseteq J_2 \supseteq \ldots$ be a decreasing sequence of ideals of $A/ann_A(M)$. For each $i \in \mathbb{N}^*$, there exists an ideal I_i of A containing $ann_A(M)$ such that $J_i = I_i/ann_A(M)$. Then $I_1 \supseteq I_2 \supseteq \ldots$ is a decreasing sequence of ideals of A, and $I_1 M \supseteq I_2 M \supseteq \ldots$ is a decreasing sequence of sub-modules of M. Since M is S-Artinian, there exist $s \in S$ and $k \in \mathbb{N}^*$ such that for each integer $n \geq k$, $sI_k M \subseteq I_n M$. By the preceding corollary, $sI_k \subseteq I_n + ann_A(M) = I_n$, and then $\pi(s)J_k \subseteq J_n$ for each integer $n \geq k$. So the ring $A/ann_A(M)$ is $\pi(S)$-Artinian. $\square$

We deduce the following corollary

Corollary 5.7 (Ozen et al. 2021) *Let A be a ring, S a multiplicative set of A, M a finitely generated multiplication A-module, and $\pi : A \longrightarrow A/ann_A(M)$ the canonical homomorphism. The module M is S-Artinian if and only if the quotient ring $A/ann_A(M)$ is $\pi(S)$-Artinian.*

Definition 5.8 Let A be a ring, M a A-module, and N a sub-module of M. We say that N is strongly prime if for each nonempty family $(N_i)_{i \in \Delta}$ of sub-modules of M satisfying $\bigcap_{i \in \Delta} N_i \subseteq N$, there exists $i \in \Delta$ such that $N_i \subseteq N$.

In the following proposition, we characterize certain S-cofinite modules in terms of strongly prime sub-modules.

Proposition 5.9 (Ozen et al. 2021) *Let A be a ring, M a multiplication A-module, and S a multiplicative set of A such that $ann_A(M)$ is a prime ideal of A and $ann_A(M) \cap S = \emptyset$. The following assertions are equivalent:*

1. The module M is S-cofinite.
2. The zero sub-module of M is strongly prime.

Proof *"(1) $\Longrightarrow$ (2)"* Let $(N_i)_{i \in \Delta}$ be a nonempty family of sub-modules of M satisfying $\bigcap_{i \in \Delta} N_i = (0)$. Since M is S-cofinite, there exist $s \in S$ and $i_1, \ldots, i_n \in \Delta$ such that $s(N_{i_1} \cap \ldots \cap N_{i_n}) = (0)$. Since M is a multiplication module, the last equality becomes $s(M : N_{i_1} \cap \ldots \cap N_{i_n})M = (0)$, and then $s(M : N_{i_1} \cap \ldots \cap N_{i_n}) \subseteq ann_A(M)$. But $(M : N_{i_1} \cap \ldots \cap N_{i_n}) = (M : N_{i_1}) \cap \ldots \cap (M : N_{i_n})$. Since $ann_A(M)$ is a prime ideal of A and $s \notin ann_A(M)$, there exists k such that $(M : N_{i_k}) \subseteq ann_A(M)$, and then $(M : N_{i_k})M = (0)$. Since M is a multiplication module, $N_{i_k} = (0)$.

"(2) $\Longrightarrow$ (1)" Let $(N_i)_{i \in \Delta}$ be a nonempty family of sub-modules of M satisfying $\bigcap_{i \in \Delta} N_i = (0)$. Since (0) is a strongly prime sub-module of M, there exists $i \in \Delta$ such that $N_i = (0)$. Then M is cofinite and so S-cofinite.   □

6   S-Artinian and S-Cofinite Idealization

Recall from Lemma 8.9 of Chap. 1 that if A is a ring, M a A-module, I an ideal of A, and N a sub-module of M, the set $I(+)N = \{(a, n); a \in I, n \in N\}$ is an ideal of $A(+)M$ if and only if $IM \subseteq N$. Note that results about idealization are taken from Anderson and Winders (2009).

Definition 6.1 Let A be a ring and M a A-module. An ideal of $A(+)M$ of the form $I(+)N$, with I an ideal of A and N a sub-module of M satisfying $IM \subseteq N$, is said to be homogenous.

Examples

(1) Let A be a ring, I an ideal of A, and M a A-module. Then $I(+)IM$ is a homogenous ideal of $A(+)M$.
(2) By Example 2 of Lemma 8.9, Chap. 1, the principal ideal generated by the element (2, 2) in the ring $\mathbb{Z}(+)2\mathbb{Z}$ is not homogenous.
(3) The principal ideal J generated by (2, 1) in the ring $\mathbb{Z}(+)\mathbb{Z}$ is not homogenous. Indeed, $J = (2, 1)(\mathbb{Z}(+)\mathbb{Z}) = \{(2, 1)(a, m); a, m \in \mathbb{Z}\} = \{(2a, a + 2m); a, m \in \mathbb{Z}\}$. Let π_1 and $\pi_2 : \mathbb{Z}(+)\mathbb{Z} \longrightarrow \mathbb{Z}$ be the canonical projections. Then $\pi_1(J) = 2\mathbb{Z}$ and $\pi_2(J) = \mathbb{Z}$. Suppose that J is homogenous. Then $J = 2\mathbb{Z}(+)\mathbb{Z}$, so $(2, 2) \in J$. There exist $a, m \in \mathbb{Z}$ such that $2a = 2$ and

$a + 2m = 2$, and then $a = 1$ and $2m = 1$, which is absurd. So J is not homogenous.

Let A be a ring and M a A-module. By Lemma 8.9, Chap. 1, the prime ideals of the ring $A(+)M$ are homogenous. The preceding example shows that the other ideals are not necessarily homogenous. In the following lemma, we characterize another class of homogenous ideals of the idealization.

Lemma 6.2 *Let A be a ring and M a A-module.*

1. *An ideal of $A(+)M$ containing $(0)(+)M$ is of the form $I(+)M$ with I an ideal of the ring A.*
2. *An ideal of $A(+)M$ contained in $(0)(+)M$ is of the form $(0)(+)N$ with N a sub-module of M.*
3. *All the ideals of the ring $A(+)M$ are comparable to $(0)(+)M$ if and only if the only ideals of $A(+)M$ are of the form $I(+)M$ or $(0)(+)N$ with I an ideal of A and N a sub-module of M. In this case, all the ideals of $A(+)M$ are homogenous.*

Proof

(1) Let $\mathscr{I}$ be an ideal of $A(+)M$ containing $(0)(+)M$ and $I = \{a \in A; \ \exists m \in M, (a, m) \in \mathscr{I}\}$. Then I is an ideal of A. Indeed, the stability of I by addition is clear. Let $a \in I$ and $b \in A$. There exists $m \in M$ such that $(a, m) \in \mathscr{I}$, and then $(b, 0)(a, m) = (ba, bm) \in \mathscr{I}$, so $ba \in I$. It is clear that $\mathscr{I} \subseteq I(+)M$. Conversely, let $(a, m) \in I(+)M$. Then $a \in I$ and $m \in M$. There exists $m' \in M$ such that $(a, m') \in \mathscr{I}$. Then $(a, m) = (a, m') + (0, m - m') \in \mathscr{I} + (0)(+)M = \mathscr{I}$. So $\mathscr{I} = I(+)M$.

(2) Let $\mathscr{I}$ be an ideal of $A(+)M$ contained in $(0)(+)M$ and $N = \{m \in M; \ (0, m) \in \mathscr{I}\}$. Then N is a sub-module of M. Indeed, the stability of N by addition is clear. Let $a \in A$ and $m \in N$. Then $(0, m) \in \mathscr{I}$, so $(a, 0)(0, m) = (0, am) \in \mathscr{I}$, and then $am \in N$. On the other hand, for each $m \in M$, by definition of N, $(0, m) \in (0)(+)N \Longleftrightarrow m \in N \Longleftrightarrow (0, m) \in \mathscr{I}$. Then $(0)(+)N = \mathscr{I}$.

(3) $"\Longrightarrow"$ Follows from (1) and (2).

$"\Longleftarrow"$ We have $(0)(+)M \subseteq I(+)M$ and $(0)(+)N \subseteq (0)(+)M$.   □

Example Let K be a field and V a K-vector space. The proper ideals of the ring $K(+)V$ are of the form $(0)(+)N$ with N a sub-vector space of V.

Indeed, it is clear that for each sub-vector space N of V, $(0)(+)N$ is a proper ideal of $K(+)V$. Conversely, let $\mathscr{I}$ be a proper ideal of $K(+)V$. Suppose that $\mathscr{I}$ contains an element (k, v) with $k \neq 0$. Since the unity $(1, 0) = (\frac{1}{k}, -\frac{1}{k^2}v)(k, v) \in \mathscr{I}$, then $\mathscr{I} = K(+)V$, which is absurd. Then $\mathscr{I} \subseteq (0)(+)V$. By the preceding lemma, $\mathscr{I} = (0)(+)N$ with N a sub-vector space of V.

Lemma 6.3 *Let G be a commutative group noted additively and A, B, C three subgroups of G such that $C \subseteq B$. Then $B \cap (C + A) = C + (B \cap A)$.*

Proof Since $C \subseteq B$ and $B \cap A \subseteq B$, then $C + (B \cap A) \subseteq B \cap (C + A)$. Conversely, let $b \in B \cap (C + A)$, $b = c + a$ with $c \in C$ and $a \in A$. Then $a = b - c \in B + C = B$, so $a \in B \cap A$ and then $b = c + a \in C + (B \cap A)$.   □

Example 1 Let I, J, and L three ideals of a ring A. Then $I \cap J + I \cap L \subseteq I \cap (J + L)$. But in general, we do not have the distributivity of the intersection over the addition. Indeed, let K be a field. Consider the ideals $I = (X + Y)$, $J = (X)$, and $L = (Y)$ of the ring $K[X, Y]$. Then $I \cap (J + L) = (X + Y) \cap (X, Y) = (X + Y)$ and $I \cap J + I \cap L = (X + Y) \cap (X) + (X + Y) \cap (Y) = \big((X + Y)X\big) + \big((X + Y)Y\big)$. Suppose that $X + Y \in I \cap J + I \cap L$. There exist $f, g \in K[X, Y]$ such that $X + Y = (X + Y)Xf + (X + Y)Yg$, and then $1 = Xf + Yg$. Put $X = Y = 0$, then $1 = 0$, which is absurd.

Example 2 (Smith 1988) Let A be a ring and I and J be two ideals of A such that J is generated by idempotent elements. For each ideal K of A, $K \cap (I + J) = K \cap I + K \cap J$. Indeed, let $x \in K \cap (I + J)$, $x = y + z$ with $y \in I$ and $z \in J$. By hypothesis, there exist idempotent elements $e_1, \ldots, e_n \in J$ and elements $a_1, \ldots, a_n \in A$ such that $z = a_1 e_1 + \ldots + a_n e_n$. The element $e = 1 - (1 - e_1) \ldots (1 - e_n) \in J$ and satisfies $(1 - e)z = (1 - e_1) \ldots (1 - e_n)(a_1 e_1 + \ldots + a_n e_n) = 0$. But $0 = (1 - e)z = (1 - e)(x - y)$, and then $(1 - e)x = (1 - e)y \in K \cap I$. So $x = (1 - e)x + ex \in K \cap I + K \cap J$. Then we have the result.

Proposition 6.4 (Ali 2008) *Let A be a ring and M a A-module.*
  Then $A(+)M$ is Artinian if and only if the ring A and the module M are Artinian.

Proof " $\Longrightarrow$ " 1) Let $I_1 \supseteq I_2 \supseteq \ldots$ be a decreasing sequence of ideals of A. Then $I_1(+)I_1 M \supseteq I_2(+)I_2 M \supseteq \ldots$ is a decreasing sequence of ideals of the ring $A(+)M$. There exists $k \in \mathbb{N}^*$ such that for each integer $n \geq k$, $I_k(+)I_k M = I_n(+)I_n M$, and then $I_k = I_n$. So A is Artinian. We can also remark that $A \simeq A(+)M / (0)(+)M$ is the quotient of an Artinian ring.
   2) Let $M_1 \supseteq M_2 \supseteq \ldots$ be a decreasing sequence of sub-modules of M. Then $(0)(+)M_1 \supseteq (0)(+)M_2 \supseteq \ldots$ is a decreasing sequence of ideals of the ring $A(+)M$. There exists $k \in \mathbb{N}^*$ such that for each integer $n \geq k$, $(0)(+)M_k = (0)(+)M_n$, and then $M_k = M_n$. So M is Artinian.
   " $\Longleftarrow$ " Let $\mathscr{I}_1 \supseteq \mathscr{I}_2 \supseteq \ldots$ be a decreasing sequence of ideals of $A(+)M$. Then $\mathscr{I}_1 + (0)(+)M \supseteq \mathscr{I}_2 + (0)(+)M \supseteq \ldots$ and $\mathscr{I}_1 \cap (0)(+)M \supseteq \mathscr{I}_2 \cap (0)(+)M \supseteq \ldots$ are two decreasing sequences of ideals of $A(+)M$. By Lemma 6.2, $\mathscr{I}_n + (0)(+)M = I_n(+)M$ and $\mathscr{I}_n \cap (0)(+)M = (0)(+)M_n$ with I_n an ideal of A and M_n a sub-module of M, for each $n \in \mathbb{N}^*$. So we have a decreasing sequence $I_1 \supseteq I_2 \supseteq \ldots$ of ideals of A and a decreasing sequence $M_1 \supseteq M_2 \supseteq \ldots$ of sub-modules of M. By hypothesis, they are stationary at a rank k. It is the same for the two sequences $\mathscr{I}_1 + (0)(+)M \supseteq \mathscr{I}_2 + (0)(+)M \supseteq \ldots$ and $\mathscr{I}_1 \cap (0)(+)M \supseteq \mathscr{I}_2 \cap (0)(+)M \supseteq \ldots$. By the modular law, for each integer $n \geq k$, $\mathscr{I}_k = \big(\mathscr{I}_k + (0)(+)M\big) \cap \mathscr{I}_k = \big(\mathscr{I}_n + (0)(+)M\big) \cap \mathscr{I}_k = \mathscr{I}_n + \mathscr{I}_k \cap (0)(+)M = \mathscr{I}_n + \mathscr{I}_n \cap (0)(+)M = \mathscr{I}_n$. Then the ring $A(+)M$ is Artinian.   □

To generalize the preceding proposition using a multiplicative set, we need the following lemma.

Lemma 6.5 *Let S be a multiplicative set of a ring A and $0 \longrightarrow M' \longrightarrow M \longrightarrow M'' \longrightarrow 0$ a short exact sequence of A-modules. Then M is S-Artinian if and only if M' and M'' are S-Artinian.*

Proof Let $\phi : M' \longrightarrow M$ the injective homomorphism and $\psi : M \longrightarrow M''$ the surjective homomorphism in the short exact sequence. We conclude by Theorem 3.14 as follows.

$"\Longrightarrow"$ The module $M' \simeq \phi(M') \subseteq M$, so M' is S-Artinian. Also, $M'' \simeq M/ker\,\psi$, which is S-Artinian.

$"\Longleftarrow"$ Since the A-modules $ker\,\psi = \phi(M') \simeq M'$ and $M/ker\,\psi \simeq M''$ are S-Artinian, then M is S-Artinian. $\square$

Let A be a ring, $S \subseteq A$ a multiplicative set, and M a A-module. Then $S(+)M$ is a multiplicative set of the ring $A(+)M$. The next theorem generalizes Proposition 6.4.

Theorem 6.6 (W. Maaref, A. Hamed, A. Benhissi) *Let A be a ring, $S \subseteq A$ a multiplicative set, and M a A-module. Then $A(+)M$ is an $S(+)M$-Artinian ring if and only if A is an S-Artinian ring and M is an S-Artinian A-module.*

Proof $"\Longrightarrow"$ (1) Let $I_1 \supseteq I_2 \supseteq \ldots$ be a descending chain of ideals of the ring A. Then $I_1(+)I_1M \supseteq I_2(+)I_2M \supseteq \ldots$ is a descending chain of ideals of the ring $A(+)M$. As $A(+)M$ is $S(+)M$-Artinian, then there exist $(s, m) \in S(+)M$ and $n_0 \in \mathbb{N}$ such that $(s, m)(I_{n_0}(+)I_{n_0}M) \subseteq I_n(+)I_nM$ for all $n \geq n_0$. Hence, for all $n \geq n_0, sI_{n_0} \subseteq I_n$. Therefore, A is S-Artinian.

(2) Let $N_1 \supseteq N_2 \supseteq \ldots$ be a descending chain of sub-modules of M. Then $0(+)N_1 \supseteq 0(+)N_2 \supseteq \ldots$ is a descending chain of ideals of $A(+)M$. Hence, there exist $(s, m) \in S(+)M$ and $n_0 \in \mathbb{N}$ such that $(s, m)(0(+)N_{n_0}) \subseteq 0(+)N_n$ for all $n \geq n_0$. Thus, $sN_{n_0} \subseteq N_n$ for all $n \geq n_0$. Therefore, M is an S-Artinian A-module.

$"\Longleftarrow"$ We consider the short exact sequence of A-modules $(0) \longrightarrow M \longrightarrow A(+)M \longrightarrow A \longrightarrow (0)$ where $i : M \longrightarrow A(+)M$ is the injection and $p : A(+)M \longrightarrow A$ is the projection. By Lemma 6.5, the A-module $A(+)M$ is S-Artinian. Since $S \subseteq S(+)M$, via the inclusion $g : A \longrightarrow A(+)M$, then the A-module $A(+)M$ is $S(+)M$-Artinian. Since each ideal of the ring $A(+)M$ is a A-sub-module of $A(+)M$, via the injection g, then each descending chain of ideals of $A(+)M$ is $S(+)M$-stationary. Thus, the ring $A(+)M$ is $S(+)M$-Artinian. $\square$

Corollary 6.7 (W. Maaref, A. Hamed, A. Benhissi) *Let A be a ring, $S \subseteq A$ a multiplicative set, and M a S-finite A-module. Then $A(+)M$ is an $S(+)M$-Artinian ring if and only if A is an S-Artinian ring.*

Proof $"\Longleftarrow"$ By Proposition 3.17, since A is an S-Artinian ring and M a S-finite A-module, then M is S-Artinian. We conclude by the preceding theorem. $\square$

Recall from Definition 4.3 that the ring A is said S-cofinite if it is S-cofinite as A-module; i.e., for each nonempty family $(I_i)_{i \in \Delta}$ of ideals of A satisfying $\bigcap_{i \in \Delta} I_i = (0)$, there exist $s \in S$ and a nonempty finite subset $\Delta' \subseteq \Delta$ such that $s \left(\bigcap_{i \in \Delta'} I_i \right) = (0)$.

Proposition 6.8 (Sevim et al. 2020) *Let A be a ring, S a multiplicative set of A, and M a A-module.*

1. *Suppose that the ring $A(+)M$ is $S(+)M$-cofinite. Then M is a S-cofinite module. Moreover, if M is a faithful multiplication A-module, then A is S-cofinite.*
2. *Suppose that all the ideals of the ring $A(+)M$ are homogenous. If the ring A and the A-module M are S-cofinite, then the ring $A(+)M$ is $S(+)M$-cofinite.*
3. *Suppose that M is a faithful multiplication A-module and all the ideals of the ring $A(+)M$ are homogenous. Then the ring $A(+)M$ is $S(+)M$-cofinite if and only if the ring A and the A-module M are S-cofinite.*

Proof (1) Let $(M_i)_{i \in \Delta}$ be a nonempty family of sub-modules of M satisfying $\bigcap_{i \in \Delta} M_i = (0)$. Then $\bigcap_{i \in \Delta} ((0)(+)M_i) = (0)(+)(0)$. Since $A(+)M$ is $S(+)M$-cofinite, there exist $(s, m) \in S(+)M$ and a finite set $\emptyset \neq \Delta' \subseteq \Delta$ satisfying $(s, m) \bigcap_{i \in \Delta'} ((0)(+)M_i) = (0)(+)(0)$. Then $(0)(+)s \left(\bigcap_{i \in \Delta'} M_i \right) = (0)(+)(0)$, so $s \left(\bigcap_{i \in \Delta'} M_i \right) = (0)$. Then M is a S-cofinite module.

Suppose now that M is a faithful multiplication A-module. Let $(I_i)_{i \in \Delta}$ be a nonempty family of ideals of A satisfying $\bigcap_{i \in \Delta} I_i = (0)$. By Proposition 5.3,

$$\bigcap_{i \in \Delta} (I_i(+)(I_iM)) = \left(\bigcap_{i \in \Delta} I_i \right)(+)\left(\bigcap_{i \in \Delta} (I_iM) \right) = \left(\bigcap_{i \in \Delta} I_i \right)(+)\left(\bigcap_{i \in \Delta} I_i \right)M = $$

$(0)(+)(0)$. But the ring $A(+)M$ is $S(+)M$-cofinite. There exist $(s, m) \in S(+)M$ and a nonempty finite subset $\Delta' \subseteq \Delta$ satisfying $(s, m) \bigcap_{i \in \Delta'} (I_i(+)I_iM) = (0)(+)(0)$. In particular, $s \left(\bigcap_{i \in \Delta'} I_i \right) = (0)$. Then A is S-cofinite.

(2) By hypothesis, the ideals of $A(+)M$ are all of the form $I(+)N$ with I an ideal of A and N a sub-module of M satisfying $IM \subseteq N$. Let $(I_i)_{i \in \Delta}$ be a nonempty family of ideals of A and $(M_i)_{i \in \Delta}$ a family of sub-modules of M such that $\bigcap_{i \in \Delta} (I_i(+)M_i) = (0)(+)(0)$. Then $\bigcap_{i \in \Delta} I_i = (0)$ and $\bigcap_{i \in \Delta} M_i = (0)$. There exist $s_1, s_2 \in S$ and two nonempty finite subsets Δ_1 and Δ_2 of Δ such that $s_1 \left(\bigcap_{i \in \Delta_1} I_i \right) = (0)$ and $s_2 \left(\bigcap_{i \in \Delta_2} M_i \right) = (0)$. For $s = s_1 s_2 \in S$, $(s, 0) \in S(+)M$ and $(s, 0) \left(\bigcap_{i \in \Delta_1 \cup \Delta_2} (I_i(+)M_i) \right) = (s, 0) \left(\left(\bigcap_{i \in \Delta_1 \cup \Delta_2} I_i \right)(+)\left(\bigcap_{i \in \Delta_1 \cup \Delta_2} M_i \right) \right) = $

$$s\Big(\bigcap_{i\in\Delta_1\cup\Delta_2} I_i\Big)(+)s\Big(\bigcap_{i\in\Delta_1\cup\Delta_2} M_i\Big) \subseteq s_1\Big(\bigcap_{i\in\Delta_1} I_i\Big)(+)s_2\Big(\bigcap_{i\in\Delta_2} M_i\Big) = (0)(+)(0).$$ Then $A(+)M$ is a $S(+)M$-cofinite ring.

(3) Consequence of (1) and (2). $\square$

We will show by Example 2 of Theorem 7.5 that the condition (2) in the preceding proposition is not necessary for the ring $A(+)M$ to be cofinite.

7 Artinian Rings and Modules

Lemma 7.1 *A proper ideal in a Noetherian ring has a finite number of minimal prime ideals.*

In particular, a Noetherian ring has a finite number of minimal prime ideals.

Proof Let A be a Noetherian ring and $\mathscr{F}$ the set of the proper ideals of A having infinitely many minimal prime ideals. Suppose that $\mathscr{F}$ is not empty. We will show that $(\mathscr{F}, \subseteq)$ has a maximal element. Let $I_0 \in \mathscr{F}$. If I_0 is not maximal, there exists $I_1 \in \mathscr{F}$ such that $I_0 \subset I_1$. If I_1 is not maximal, there exists $I_2 \in \mathscr{F}$ such that $I_0 \subset I_1 \subset I_2$. Since the ring A is Noetherian, the process must stop. Let I be a maximal element of $(\mathscr{F}, \subseteq)$. It is not prime, since in the contrary case, it will be the only minimal prime over itself. There exist $a, b \in A \setminus I$ such that $ab \in I$. Since $I \subset I + aA$ and $I \subset I + bA$, then each of the ideals $I + aA$ and $I + bA$ has a finite number of minimal prime ideals. But $(I + aA)(I + bA) = I^2 + aI + bI + abA \subseteq I$, and then each minimal prime ideal over I contains either $I + aA$ or $I + bA$ and then it is a minimal prime over one of them. So one of them has infinitely many minimal prime ideals, which is absurd. Then $\mathscr{F} = \emptyset$ and each proper ideal of A has a finite number of minimal prime ideals. $\square$

Example By Example 1 of Corollary 3.8, the product ring $A = \prod_{i:0}^{\infty} \mathbb{F}_2$ is zero dimensional. Then all its prime ideals are minimal. For each integer $n \in \mathbb{N}$, the set $P_n = \{(x_i)_{i\in\mathbb{N}} \in A; \ x_n = 0\}$ is a prime ideal of A. Indeed, it is clear that P_n is a proper ideal of A since $1 \notin P_n$. Let $x = (x_i)_{i\in\mathbb{N}}$ and $y = (y_i)_{i\in\mathbb{N}} \in A$ such that $xy = 0$. Then $x_n y_n = 0$ in the field $\mathbb{F}_2$, and then either $x_n = 0$ or $y_n = 0$, so $x \in P_n$ or $y \in P_n$. We conclude that A has infinitely many minimal prime ideals. The ring A is not Noetherian since $\mathbb{F}_2 \times (0) \times \ldots \subset \mathbb{F}_2 \times \mathbb{F}_2 \times (0) \times \ldots \subset \ldots$.

An ideal I of a ring A is said to be nilpotent if there exists $n \in \mathbb{N}^*$ such that $I^n = (0)$.

Proposition 7.2 *The nilradical of an Artinian ring is a nilpotent ideal.*

Proof Let A be an Artinian ring and $I = Nil(A)$. The sequence $I \supseteq I^2 \supseteq \ldots$ is stationary. There exists an integer $i \geq 1$ such that $I^i = I^{i+1} = \ldots = J$. Suppose that $J \neq (0)$. The set $E = \{ J' \text{ ideal of } A \text{ such that } J'J \neq (0) \}$ is nonempty. It contains A and even J. The set E has a minimal element $\hat{J}$ for the inclusion because in the contrary case, A will have a strictly decreasing sequence of ideals. Since $\hat{J}J \neq (0)$, there exists $a \in \hat{J}$ such that $aJ \neq (0)$. Then $(a) \subseteq \hat{J}$ and $(a)J = aJ \neq (0)$, so $(a) \in E$. By the minimality of $\hat{J}$, we must have $(a) = \hat{J}$. On the other hand, $aJ \subseteq (a) = \hat{J}$ and $(aJ)J = aJ^2 = aJ \neq (0)$ since $J^2 = J$. Then $aJ \in E$. By the minimality of $\hat{J}$, we must have $aJ = \hat{J}$. Then $(a) = aJ$. So there exists $b \in J$ such that $a = ab$. Since $J \subseteq I = Nil(A)$, there exists $n \in \mathbb{N}^*$ such that $b^n = 0$. Then $a = ab = ab^2 = \ldots = ab^n = 0$, which contradicts $aJ \neq (0)$. Then $J = (0)$ and I is a nilpotent ideal.   □

Example 1 Let (A, M) be an Artinian local ring. By Corollary 3.8, $dim\, A = 0$, and then M is the only prime ideal of A. By the preceding proposition, there exists $n \in \mathbb{N}^*$ such that $M^n = (0)$. An element of A is either a unity or nilpotent. This is, for example, the case of $A = \mathbb{Z}/(p^n)$, where p is a prime number and $n \in \mathbb{N}^*$.

The following two examples show that the nilradical of a ring is not in general nilpotent.

Example 2 Let p be a prime number. The ring $A = \prod_{i:1}^{\infty} \mathbb{Z}/p^i\mathbb{Z}$ is not Artinian since we have the sequence $\prod_{i:1}^{\infty} \mathbb{Z}/p^i\mathbb{Z} \supset (0) \times \prod_{i:2}^{\infty} \mathbb{Z}/p^i\mathbb{Z} \supset (0) \times (0) \times \prod_{i:3}^{\infty} \mathbb{Z}/p^i\mathbb{Z} \supset$ $\ldots$. Let $I = Nil(A)$. Suppose that there exists $n \in \mathbb{N}^*$ such that $I^n = (0)$. The element $x = (\bar{0}, \ldots, \bar{0}, \bar{p}, \bar{0}, \ldots)$, where $\bar{p}$ is at the $n + 1$th position, is nilpotent in A since $x^{n+1} = 0$. But $x^n = (\bar{0}, \ldots, \bar{0}, \bar{p}^n, \bar{0}, \ldots) \neq 0$, which is absurd. Then $Nil(A)$ is not nilpotent.

Example 3 Let K be a field and $A = K[X_1, X_2, X_3, \ldots]/(X_1, X_2^2, X_3^3, \ldots)$. The only prime ideal of A is $P = Nil(A) = (\bar{X}_1, \bar{X}_2, \ldots)$. For each $n \in \mathbb{N}^*$, $\bar{X}_{n+1}^n \neq \bar{0}$, and then $P^n \neq (0)$.

The Artinian part in the following proposition follows from Theorem 3.14. But the proof here becomes very easy when $S = \{1\}$. The Noetherian part is proved in Theorem 3.1, Chap. 1. We give here a new proof.

Proposition 7.3 *Let A be a ring, M a A-module, and N a sub-module of M.*

The module M is Artinian (resp. Noetherian) if and only if the A-modules N and M/N are Artinian (resp. Noetherian).

Proof " $\Longrightarrow$ " (1) The sub-modules of N are sub-modules of M, so they satisfy the dcc (resp. acc). Then N is Artinian (resp. Noetherian).

(2) The sub-modules of M/N are in an increasing bijection with the sub-modules of M containing N. Then they satisfy the dcc (resp. acc). So M/N is Artinian (resp. Noetherian).

" $\Longleftarrow$ " Let $\pi : M \longrightarrow M/N$ be the canonical homomorphism and $(M_i)_{i \in \mathbb{N}}$ a decreasing (resp. increasing) sequence of sub-modules of M. Then $\left(\pi(M_i)\right)_{i \in \mathbb{N}}$ is a decreasing (resp. increasing) sequence of sub-modules of M/N, and $(M_i \cap N)_{i \in \mathbb{N}}$ is a decreasing (resp. increasing) sequence of sub-modules of N. By hypothesis, there exists $n \in \mathbb{N}$ such that for each integer $i \geq n$, $\pi(M_i) = \pi(M_n)$ and $M_i \cap N = M_n \cap N$. Let $i \geq n$ and $x \in M_n$ (resp. M_i). Then $\bar{x} = \pi(x) \in \pi(M_n) = \pi(M_i)$. There exists $x' \in M_i$ (resp. M_n) such that $\bar{x} = \bar{x}'$, and then $x - x' \in M_n \cap N = M_i \cap N$. So $x = (x - x') + x' \in M_i$ (resp. M_n). Then $M_n \subseteq M_i$ (resp. $M_i \subseteq M_n$), so $M_n = M_i$.   □

Lemma 7.4 *Let A be a ring having some maximal ideals not necessarily different $M_1, \ldots, M_n$ with a product $M_1 M_2 \ldots M_n = (0)$.*

Then the ring A is Artinian if and only if it is Noetherian.

Proof For $1 \leq i \leq n$, put $I_i = M_1 M_2 \ldots M_i$. Then $(0) = I_n \subseteq I_{n-1} \subseteq \ldots \subseteq I_2 \subseteq I_1 \subseteq I_0 = A$. We apply the preceding proposition many times: The ring A is Artinian (resp. Noetherian) $\Longleftrightarrow$ the A-module A is Artinian (resp. Noetherian) $\Longleftrightarrow$ the A-modules I_1 and I_0/I_1 are Artinian (resp. Noetherian) $\Longleftrightarrow$ the A-modules I_2, I_1/I_2, and I_0/I_1 are Artinian (resp. Noetherian) $\ldots \Longleftrightarrow$ the A-modules I_{n-1}/I_n, I_{n-2}/I_{n-1}, $\ldots$, I_1/I_2, I_0/I_1 are Artinian (resp. Noetherian).

It suffices to show that the A-modules I_{i-1}/I_i, $1 \leq i \leq n$, are Artinian if and only if they are Noetherian.

To do this, note that I_{i-1}/I_i is a $K_i = A/M_i$-vector space for the multiplication: $\bar{a}.\hat{x} = \widehat{ax}$, for each $a \in A$ and $x \in I_{i-1}$. Indeed, this operation is well defined since if $\bar{a} = \bar{b}$ in A/M_i and $\hat{x} = \hat{y}$ in I_{i-1}/I_i, and then $a - b \in M_i$ and $x - y \in I_i$, so $ax - by = (a - b)x + b(x - y) \in M_i I_{i-1} + I_i = I_i + I_i = I_i$. Then $\widehat{ax} = \widehat{by}$ in I_{i-1}/I_i. The axioms of the vector spaces are easy to check. Note also that a subset of I_{i-1}/I_i is a A-sub-module if and only if it is a K_i-sub-vector space. In fact, the stability of this subset of I_{i-1}/I_i by multiplication by the elements of A is equivalent to its stability by multiplication by the elements of $K_i = A/M_i$.

Finally, By Example 2 of Definition 3.1, on the characterization of Noetherian and Artinian vector spaces, the A-module I_{i-1}/I_i is Noetherian $\Longleftrightarrow$ the $K_i = A/M_i$-vector space I_{i-1}/I_i is Noetherian $\Longleftrightarrow$ $\dim_{K_i}\left(I_{i-1}/I_i\right) < \infty \Longleftrightarrow$ the $K_i = A/M_i$-vector space I_{i-1}/I_i is Artinian $\Longleftrightarrow$ the A-module I_{i-1}/I_i is Artinian.   □

Examples

(1) A Noetherian ring is not necessarily Artinian. This is the case of $\mathbb{Z}$.
(2) The maximal ideals of $\mathbb{Z}/12\mathbb{Z}$ are $M_1 = (\bar{2})$ and $M_2 = (\bar{3})$. They satisfy $M_1^2 M_2 = (\bar{0})$.

From Corollary 2.20, we can deduce that an Artinian integral domain is Noetherian. We will generalize this result for any ring.

Theorem 7.5 *A ring is Artinian if and only if it is Noetherian and zero dimensional.*

Proof " $\Longrightarrow$ " By Corollary 3.8, an Artinian ring A is zero dimensional. By Corollary 3.10, $Spec(A) = Max(A) = \{M_1, \ldots, M_k\}$ is finite. By Proposition 7.2, the ideal $I = Nil(A) = M_1 \cap \ldots \cap M_k$ is nilpotent. There exists $n \in \mathbb{N}^*$ such that $I^n = (0)$. Since $M_1 \ldots M_k \subseteq M_1 \cap \ldots \cap M_k = I$, then $(M_1 \ldots M_k)^n = (0)$. By the preceding lemma, the ring A is Noetherian.

" $\Longleftarrow$ " Let A be a zero-dimensional Noetherian ring. By Lemma 7.1, we have $Spec(A) = Max(A) = Min(A) = \{M_1, \ldots, M_k\}$ which is finite. Then $Nil(A) = M_1 \cap \ldots \cap M_k$ and the product $I = M_1 \ldots M_k \subseteq Nil(A)$. Since A is Noetherian, the ideal $I = (x_1, \ldots, x_s)$ is finitely generated. There exists $m \in \mathbb{N}^*$ such that $x_1^m = \ldots = x_s^m = 0$. Let $n = ms$, and then $I^n = (0)$. Indeed, the generators of I^n are the monomials of the form $x_1^{\alpha_1} \ldots x_s^{\alpha_s}$ with $\alpha_1 + \ldots + \alpha_s = n$. If $\alpha_1 < m, \ldots, \alpha_s < m$, then $n = \alpha_1 + \ldots + \alpha_s < sm = n$, which is absurd. There exists an index i such that $\alpha_i \geq m$, and then $x_1^{\alpha_1} \ldots x_s^{\alpha_s} = 0$. We conclude that $I^n = M_1^n \ldots M_k^n = (0)$. By the preceding lemma, A is Artinian. $\square$

Recall that the ring $\mathbb{Z}$ is Noetherian but it is not Artinian, and the product ring $\prod_{i:1}^{\infty} \mathbb{F}_2$ is zero dimensional but it is not Artinian. In the following example, we give a way to construct Artinian rings.

Example 1 Let B be any ring, M a finitely generated maximal ideal of B, $k \in \mathbb{N}^*$, and I an ideal of B containing M^k. The ring $A = B/I$ is Artinian.

Indeed, a prime ideal of A is of the form P/I with P a prime ideal of B satisfying $M^k \subseteq I \subseteq P$. Then $M \subseteq P$, so $M = P$. The only prime ideal of A is M/P. It is finitely generated. By the preceding theorem, the ring A is Artinian.

By the preceding theorem, an Artinian ring is Noetherian. In contrast to the case of rings, an Artinian module is not necessarily Noetherian. In the following two examples, we give a $\mathbb{Z}$-module (a commutative group) which is Artinian but it is not Noetherian.

Example 2 (1) Let p be a prime number. The set $A_{p^\infty} = \{z \in \mathbb{C}; \exists n \in \mathbb{N}, z^{p^n} = 1\}$ is a multiplicative group, called the p-group of Prufer. For each $n \in \mathbb{N}$, let $\xi_n = exp \frac{2i\pi}{p^n}$, a primitive p^nth root of the unity. Since $\xi_{n+1}^p = (exp \frac{2i\pi}{p^{n+1}})^p = exp \frac{2i\pi}{p^n} = \xi_n$, then $(\xi_n) \subset (\xi_{n+1})$. So the $\mathbb{Z}$-module A_{p^∞} is not Noetherian.

We will show that $A_{p^\infty} = \bigcup_{n:0}^{\infty} (\xi_n)$. Let $z \in A_{p^\infty}$. There exists $n \in \mathbb{N}$ such that $z^{p^n} = 1$. There exists an integer k, $0 \leq k \leq p^n - 1$ such that $z = exp \frac{2ik\pi}{p^n} = \xi_n^k$, and then $z \in (\xi_n)$.

The only proper subgroups of A_{p^∞} are the (ξ_n), $n \in \mathbb{N}$. Indeed, let B such subgroup. We have $\xi_0 = 1 \in B$ and B does not contain all the ξ_n. Let n be the smallest natural integer such that $\xi_{n+1} \notin B$. We will show that $B = (\xi_n)$. By hypothesis, $\xi_n \in B$, and then $(\xi_n) \subseteq B$. Conversely, let $z \in B$. There exist $m \in \mathbb{N}$ and $0 \leq k \leq p^m - 1$ such that $z = \xi_m^k$. Since $\xi_m^p = \xi_{m-1}$, we may suppose that p does not divide k. By the Bézout identity, there exist $r, s \in \mathbb{Z}$ such that

$rk + sp^m = 1$. Then $\xi_m = \xi_m^{rk+sp^m} = (\xi_m^k)^r(\xi_m^{p^m})^s = z^r \in B$, so $m < n+1$.
Then $z = \xi_m^k = exp\frac{2ik\pi}{p^m} = exp\frac{2ikp^{n-m}\pi}{p^n} = (exp\frac{2i\pi}{p^n})^{kp^{n-m}} = (\xi_n)^{kp^{n-m}} \in (\xi_n)$.
Then $B = (\xi_n)$. We conclude that the $\mathbb{Z}$-module A_{p^∞} does not contain any strictly
decreasing sequence of sub-modules (subgroups). It is an Artinian module.

(2) By Proposition 6.4, since the ring $\mathbb{Z}$ is not Artinian, the ring $\mathbb{Z}(+)A_{p^\infty}$ is also
not Artinian. We will prove that $\mathbb{Z}(+)A_{p^\infty}$ is an example of nontrivial (i.e., is not a
field) cofinite ring and it is not in the context of Proposition 6.6 (2) since the ring $\mathbb{Z}$
is not cofinite (see Example 1 of Definition 4.3).

Recall that A_{p^∞} is a multiplicative group with zero element 1 and the operations
in the ring $\mathbb{Z}(+)A_{p^\infty}$ are defined by $(a, z) + (a', z') = (a + a', zz')$ and
$(a, z).(a', z') = (aa', z^{a'}z'^a)$. The zero element of the idealization is $(0, 1)$ and the
unit element is $(1, 1)$. Let $\mathscr{I}$ be a nonzero ideal of $\mathbb{Z}(+)A_{p^\infty}$, i.e., $\mathscr{I} \neq \{(0, 1)\}$.
We will show that the ideal $\mathscr{I} \cap ((0)(+)A_{p^\infty})$ is nonzero. Let $(a, z) \in \mathscr{I} \setminus \{(0, 1)\}$.
If $a = 0$, then $(0, 1) \neq (0, z) \in \mathscr{I} \cap ((0)(+)A_{p^\infty})$. If $a \neq 0$, there exists $z' \in A_{p^\infty}$
such that $z'^a \neq 1$ (take a p^n-primitive root of the unity of order $> |a|$). Then
$(0, 1) \neq (0, z'^a) = (a, z)(0, z') \in \mathscr{I} \cap ((0)(+)A_{p^\infty})$. By Lemma 6.2 (2), the
ideal $\mathscr{I} \cap ((0)(+)A_{p^\infty})$ is of the form $(0)(+)H$ with H a sub module (subgroup)
of A_{p^∞} different from $\{1\}$. But the subgroups of A_{p^∞} different from $\{1\}$ have a
smallest element (ξ_1). Then $\{(0, 1)\} \neq (0)(+)(\xi_1) \subseteq (0)(+)H \subseteq \mathscr{I}$. Let $(\mathscr{I}_i)_{i\in\Delta}$
be a nonempty family of ideals of $\mathbb{Z}(+)A_{p^\infty}$ such that $\bigcap_{i\in\Delta} \mathscr{I}_i = \{(0, 1)\}$. By the
preceding remark, one of the $\mathscr{I}_i$ is equal to $\{(0, 1)\}$. Then the ring $\mathbb{Z}(+)A_{p^\infty}$ is
cofinite.

Example 3 Let p be a prime number. We will show that the quotient $\mathbb{Z}$-module
$\mathbb{Z}[\frac{1}{p}]/\mathbb{Z}$ is Artinian but not Noetherian. By definition, the ring $\mathbb{Z}[\frac{1}{p}] = \{k_0 + \frac{k_1}{p} +$
$\ldots + \frac{k_n}{p^n}; \ n \in \mathbb{N}, \ k_i \in \mathbb{Z}\} = \{\frac{k}{p^n}; \ k \in \mathbb{Z}, \ n \in \mathbb{N}\}$; this is also the $\mathbb{Z}$-module
generated by the set $\{\frac{1}{p^n}; \ n \in \mathbb{N}\}$. A nonzero proper subgroup of $\mathbb{Z}[\frac{1}{p}]/\mathbb{Z}$ is of
the form $N/\mathbb{Z}$ with N a proper sub group of $\mathbb{Z}[\frac{1}{p}]$ strictly containing $\mathbb{Z}$. There
exists $n \in \mathbb{N}$ such that $\frac{1}{p^{n+1}} \notin N$. For each integer $m \geq n+1$, $\frac{1}{p^m} \notin N$. Let
x be any element of $N \setminus \mathbb{Z}$, $x = \frac{k}{p^m}$ with $k \in \mathbb{Z} \setminus (0)$ and $m \in \mathbb{N}^*$. We may
suppose that p does not divide k, and then p^m and k are relatively prime. There
exist $a, b \in \mathbb{Z}$ such that $1 = ap^m + bk$, and then $\frac{1}{p^m} = a + bx \in N$. So $m \leq n$
and $x = \frac{k}{p^m} = \frac{kp^{n-m}}{p^n}$ and then $N \subseteq \frac{1}{p^n}\mathbb{Z}$, so $N/\mathbb{Z} \subseteq (\frac{1}{p^n}\mathbb{Z})/\mathbb{Z}$. But $(\frac{1}{p^n}\mathbb{Z})/\mathbb{Z} =$
$\{\frac{r}{p^n} + \mathbb{Z}; \ 0 \leq r \leq p^n - 1\}$ is a finite group, and then $N/\mathbb{Z}$ is also finite. Since the
proper subgroups of $\mathbb{Z}[\frac{1}{p}]/\mathbb{Z}$ are finite, each decreasing sequence of such subgroups
is stationary. Then the quotient $\mathbb{Z}$-module $\mathbb{Z}[\frac{1}{p}]/\mathbb{Z}$ is Artinian. Since we have the
strictly increasing sequence $\mathbb{Z} \subset \frac{1}{p}\mathbb{Z} \subset \frac{1}{p^2}\mathbb{Z} \subset \frac{1}{p^3}\mathbb{Z} \ldots$ of subgroups of $\mathbb{Z}[\frac{1}{p}]$,
then we have the corresponding strictly increasing sequence $(\bar{0}) \subset (\frac{1}{p}\mathbb{Z})/\mathbb{Z} \subset$

$(\frac{1}{p^2}\mathbb{Z})/\mathbb{Z} \subset (\frac{1}{p^3}\mathbb{Z})/\mathbb{Z}\ldots$ of subgroups of $\mathbb{Z}[\frac{1}{p}]/\mathbb{Z}$. So the quotient $\mathbb{Z}$-module $\mathbb{Z}[\frac{1}{p}]/\mathbb{Z}$ is not Noetherian.

Let A be a ring and M a A-module. In Proposition 6.4, we proved that the ring $A(+)M$ is Artinian if and only if the ring A and the module M are both Artinian. In the following corollary, we give another characterization of the Artinian idealization.

Corollary 7.6 *Let A be a ring and M a A-module. Then $A(+)M$ is Artinian if and only if the ring A is Artinian and the module M is finitely generated.*

Proof " $\Longrightarrow$" By Corollary 2.10, the quotient ring $A(+)M/(0)(+)M \simeq A$ is Artinian. By the preceding theorem, the ring $A(+)M$ is Noetherian. By Corollary 8.11, Chap. 1, the module M is finitely generated.

" $\Longleftarrow$" By the preceding theorem, the ring A is Noetherian and zero dimensional. By Corollary 8.11, Chap. 1, the ring $A(+)M$ is Noetherian. By Lemma 8.9 (2), Chap. 1, $Spec(A(+)M) = \{P(+)M;\ P \in Spec(A)\}$. Then $dim(A(+)M) = dim\ A = 0$. By the preceding theorem, $A(+)M$ is Artinian. □

Chinese Remainder Theorem (CRT) 7.7 *Let A be a ring and $I_1, \ldots, I_n$ ($n \geq 2$) comaximal ideals of A, i.e., $I_i + I_j = A$ for $i \neq j$.*

1. *The map $\phi : A \longrightarrow A/I_1 \times, \ldots \times A/I_n$, defined by $\phi(x) = (x+I_1, \ldots, x+I_n)$ for each $x \in A$, is a homomorphism onto of rings, with kernel $ker\ \phi = I_1 \cap \ldots \cap I_n$.*
2. *The product $I_1 \ldots I_n = I_1 \cap \ldots \cap I_n$.*
3. *For each $k \in \mathbb{N}^*$, the ideals $I_1^k, \ldots, I_n^k$ are comaximal.*

Proof

(1) It is clear that ϕ is a homomorphism of kernel $ker\ \phi = I_1 \cap \ldots \cap I_n$. To show that ϕ is onto, it suffices to show that for each $x_1, \ldots, x_n \in A$, there exists $x \in A$ such that $x \equiv x_i (mod\ I_i)$, $1 \leq i \leq n$. By induction, if $n = 2$, there exist $a_1 \in I_1$ and $a_2 \in I_2$ such that $a_1 + a_2 = 1$, we take $x = x_2 a_1 + x_1 a_2$, and then $x - x_1 = x_2 a_1 + x_1 (a_2 - 1) = x_2 a_1 - x_1 a_1 = (x_2 - x_1)a_1 \in I_1$. We have also $x - x_2 \in I_2$.

Suppose that the property is true for $n - 1$ ideals. For each $i \geq 2$, we can find $a_i \in I_1$ and $b_i \in I_i$ such that $a_i + b_i = 1$. The product $(a_2 + b_2)(a_3 + b_3) \ldots (a_n + b_n) = 1$ and belongs to $I_1 + I_2 I_3 \ldots I_n$, and then $I_1 + I_2 I_3 \ldots I_n = A$. By the case $n = 2$, we can find $y_1 \in A$ such that $y_1 \equiv 1(mod\ I_1)$ and $y_1 \equiv 0(mod\ I_2 I_3 \ldots I_n)$. By the same way, we find elements $y_2, \ldots, y_n \in A$ such that $y_j \equiv 1(mod\ I_j)$ and $y_j \equiv 0(mod\ I_i)$ for each $i \neq j$. The element $x = x_1 y_1 + \ldots + x_n y_n$ satisfies the required property since $x - x_j = x_1 y_1 + \ldots + x_{j-1} y_{j-1} + x_j(y_j - 1) + x_{j+1} y_{j+1} + \ldots + x_n y_n \in I_j$.

(2) By induction. For $n = 2$, the inclusion $I_1 I_2 \subseteq I_1 \cap I_2$ is clear and there exist $x \in I_1$ and $y \in I_2$ such that $1 = x + y$. For each $z \in I_1 \cap I_2$, $z = zx + zy \in I_1 I_2$. Then $I_1 I_2 = I_1 \cap I_2$. Suppose that $I_1 \ldots I_{n-1} = I_1 \cap \ldots \cap I_{n-1}$. For $1 \leq i \leq n - 1$, $I_i + I_n = A$, there exist $x_i \in I_i$ and $y_i \in I_n$ such that

$x_i + y_i = 1$. Then $x_1 \ldots x_{n-1} = (1 - y_1) \ldots (1 - y_{n-1}) = 1 + x$ with $x \in I_n$, so $x_1 \ldots x_{n-1} - x = 1$ and then $I_1 \ldots I_{n-1} + I_n = A$. By the case $n = 2$, $(I_1 \ldots I_{n-1})I_n = (I_1 \ldots I_{n-1}) \cap I_n = I_1 \cap \ldots \cap I_{n-1} \cap I_n$.

(3) For $i \neq j$, $I_i + I_j = A$. There exist $x \in I_i$ and $y \in I_j$ such that $1 = x + y$.

Then $1 = (x + y)^{2k} = \sum_{l:0}^{2k} C_{2k}^l x^l y^{2k-l}$. For $0 \leq l \leq k - 1$, $2k - l \geq k$, and then $C_{2k}^l x^l y^{2k-l} \in I_j^k$ and for $k \leq l \leq 2k$, $C_{2k}^l x^l y^{2k-l} \in I_i^k$. Then $I_i^k + I_j^k = A$. □

Example 1 Since $60 = 4.3.5$ and each couple of the integers $4, 3, 5$ is relatively prime, then $\mathbb{Z}/60\mathbb{Z} \simeq \mathbb{Z}/4\mathbb{Z} \times \mathbb{Z}/3\mathbb{Z} \times \mathbb{Z}/5\mathbb{Z}$.

Example 2 Let A be a PID and $m, n \in A$ two relatively prime elements. There exist $u, v \in A$ such that $um + vn = 1$, and then $mA + nA = A$. So $mA \cap nA = mnA$ and we have the isomorphism $A/(m) \times A/(n) \simeq A/(mn)$.

Other Proof for the CRT in the Ring $\mathbb{Z}$: Recall first that Euler function ϕ : $\mathbb{N}^* \longrightarrow \mathbb{N}$ is defined by $\phi(n) = card\{k; \ 1 \leq k \leq n, \ \Delta(k, n) = 1\}$ for each $n \in \mathbb{N}^*$. It is multiplicative; i.e., $\phi(mn) = \phi(m)\phi(n)$ if $\Delta(m, n) = 1$ and that $\phi(n) = n \prod_{p/n} (1 - \frac{1}{p})$. In particular, $\phi(p^k) = p^k - p^{k-1}$ for each prime number p and each $k \in \mathbb{N}^*$. We have also Euler theorem: If $n \in \mathbb{N}^*$ and $k \in \mathbb{Z}$ are such that $\Delta(n, k) = 1$, then $k^{\phi(n)} \equiv 1 (mod \ n)$.

Let $n_1, \ldots, n_r$ be natural integers with each couple relatively prime and $a_1, \ldots, a_r \in \mathbb{Z}$. We will show that the linear system $x \equiv a_i \ (mod \ n_i), 1 \leq i \leq r$ has a unique solution in $\mathbb{Z}$ modulo $n_1 n_2 \ldots n_r$. Let $N_i = \prod_{j:1, j \neq i}^{r} n_j$ for $1 \leq i \leq r$. Then $N_i \equiv 0 \ (mod \ n_j)$ for $1 \leq i \neq j \leq r$. The integer $x = a_1 N_1^{\phi(n_1)} + \ldots + a_r N_r^{\phi(n_r)}$ is a solution. Indeed, $x \equiv a_i N_i^{\phi(n_i)} (mod \ n_i)$ for $1 \leq i \leq r$. But by Euler theorem, since $\Delta(N_i, n_i) = 1$, $N_i^{\phi(n_i)} \equiv 1 (mod \ n_i)$, then $x \equiv a_i \ (mod \ n_i)$ for $1 \leq i \leq r$. Let x and x' be two solutions of the system. Then $x \equiv x' (mod \ n_i)$ for $1 \leq i \leq r$, so n_i divides $x - x'$, so $n_1 n_2 \ldots n_r$ divides $x - x'$ and then $x \equiv x' (mod \ n_1 n_2 \ldots n_r)$. □

Example 1 Solve in $\mathbb{Z}$ the system $x \equiv 1 \ (mod \ 3), x \equiv 2 \ (mod \ 5), x \equiv 3 \ (mod \ 7)$. With the preceding notations, $a_1 = 1, a_2 = 2, a_3 = 3$ and $n_1 = 3, n_2 = 5, n_3 = 7$ and $N_1 = n_2 n_3 = 5.7 = 35, N_2 = n_1 n_3 = 3.7 = 21, N_3 = n_1 n_2 = 3.5 = 15$. Then $\phi(n_1) = \phi(3) = 3 - 1 = 2, \phi(n_2) = \phi(5) = 5 - 1 = 4, \phi(n_3) = \phi(7) = 7 - 1 = 6$ and $x = a_1 N_1^{\phi(n_1)} + a_2 N_2^{\phi(n_2)} + a_3 N_3^{\phi(n_3)} = 1.35^2 + 2.21^4 + 3.15^6$. The other solutions are deduced from x modulo $n_1 n_2 n_3 = 3.5.7 = 105$.

Example 2 Let $n \geq 2$ be an integer, $n = p_1^{\alpha_1} \ldots p_r^{\alpha_r}$, where the p_i are different prime numbers and the integers $\alpha_i \geq 1$. Then $\mathbb{Z}/n\mathbb{Z} \simeq \mathbb{Z}/p_1^{\alpha_1}\mathbb{Z} \times \ldots \times \mathbb{Z}/p_r^{\alpha_r}\mathbb{Z}$.

Theorem 7.8 *A ring is Artinian if and only if it is isomorphic to a finite product of local Artinian rings.*

Proof By Proposition 3.5, a finite product of (local) Artinian rings is Artinian. Conversely, let A be an Artinian ring. By Corollaries 3.8 and 3.10, $Spec(A) = Max(A) = \{M_1, \ldots, M_n\}$ is finite. The maximal ideals of A are comaximal: $M_i + M_j = A$ for $i \neq j$. By CRT, $M_1 \cap \ldots \cap M_n = M_1 \ldots M_n$. By Proposition 7.2, the ideal $I = Nil(A) = M_1 \cap \ldots \cap M_n = M_1 \ldots M_n$ is nilpotent. There exists $k \in \mathbb{N}^*$ such that $(0) = I^k = M_1^k \ldots M_n^k$. By CRT, the ideals $M_1^k, \ldots, M_n^k$ are also comaximal: $M_i^k + M_j^k = A$ for $i \neq j$. Then $M_1^k \cap \ldots \cap M_n^k = M_1^k \ldots M_n^k = (0)$. The map $\phi : A \longrightarrow A/M_1^k \times \ldots \times A/M_n^k$ in the CRT is a homomorphism onto with kernel $ker\ \phi = M_1^k \cap \ldots \cap M_n^k = (0)$. Then ϕ is an isomorphism. By Corollary 2.10, the quotient rings A/M_i^k, $1 \leq i \leq n$, are Artinian. They are local with maximal ideals M_i/M_i^k, respectively. Indeed, a prime ideal of A/M_i^k is of the form P/M_i^k with P a prime ideal of A containing M_i^k. Then $M_i \subseteq P$, so $P = M_i$. □

Example We have the isomorphism $\mathbb{Z}/60\mathbb{Z} \simeq \mathbb{Z}/4\mathbb{Z} \times \mathbb{Z}/3\mathbb{Z} \times \mathbb{Z}/5\mathbb{Z}$. The finite ring $\mathbb{Z}/60\mathbb{Z}$ is Artinian but it is not local. The rings $\mathbb{Z}/4\mathbb{Z}$, $\mathbb{Z}/3\mathbb{Z}$, and $\mathbb{Z}/5\mathbb{Z}$ are local. The two last ones are fields.

Proposition 7.9 *A reduced Artinian ring is isomorphic to a finite product of fields.*

Proof With the notations of the proof of the preceding theorem, $Spec(A) = Max(A) = \{M_1, \ldots, M_n\}$ and $(0) = Nil(A) = M_1 \cap \ldots \cap M_n$. The map $\phi : A \longrightarrow A/M_1 \times \ldots \times A/M_n$, defined by $\phi(x) = (x + M_1, \ldots, x + M_n)$ for each $x \in A$, is an isomorphism. □

8 Application to the Theorem of Generalized Principal Ideal

The Nakayama lemma is a key stool in commutative algebra. It represents a crucial step in many proofs.

Nakayama Lemma 8.1 *Let A be a ring, E a finitely generated A-module, and $J = \bigcap\{M, M \in Max(A)\}$ the Jacobson radical of A. If $JE = E$, then $E = (0)$.*

Proof Let $\{x_1, \ldots, x_n\}$ be a minimal finite family of generators of E. Since $x_n \in E = JE$, there exist $a_1, \ldots, a_n \in J$ such that $x_n = a_1 x_1 + \ldots + a_n x_n$. Suppose that $n \geq 2$. By definition of J, for each $M \in Max(A)$, $a_n \in M$, and then $1 - a_n \notin M$. So $1 - a_n \in U(A)$ and $x_n = \frac{a_1}{1-a_n} x_1 + \ldots + \frac{a_{n-1}}{1-a_n} x_{n-1}$, which contradicts the minimality of $\{x_1, \ldots, x_n\}$ as a family of generators of E. Then $n = 1$ and $x_1 = a_1 x_1$, so $(1 - a_1)x_1 = 0$ with $1 - a_1 \in U(A)$. Then $x_1 = 0$ and $E = (x_1) = (0)$. □

Example Let P be a principal prime ideal of a Noetherian ring A. Then $ht(P) \le 1$. Indeed, put $P = pA$ and let $Q \in Spec(A)$ (if it exists) such that $Q \subset P = pA$. We will show that $Q = pQ$. For each $q \in Q$, there exists $a \in A$ such that $q = pa$. But $p \notin Q$, and then $a \in Q$ so $Q = pQ$. The Jacobson radical $J(A_P) = PA_P$, the ideal QA_P of A_P is finitely generated since A is Noetherian, and $PA_P QA_P = PQA_P = pQA_P = QA_P$. By Nakayama lemma, $QA_P = (0)$. By the increasing bijection between the prime ideals of A contained in P and $Spec(A_P)$, Q cannot contain strictly a prime ideal of A. Then Q is a minimal prime ideal, so $ht(P) \le 1$.

Lemma 8.2 *Let S be a multiplicative set of a ring A and $\psi : A \longrightarrow S^{-1}A$ the natural homomorphism, defined by $\psi(a) = \frac{a}{1}$ for each $a \in A$. For any ideal $\mathscr{I}$ of $S^{-1}A$, $S^{-1}\psi^{-1}(\mathscr{I}) = \mathscr{I}$.*

Proof Let $x \in S^{-1}\psi^{-1}(\mathscr{I})$, $x = \frac{a}{s}$ with $s \in S$ and $a \in \psi^{-1}(\mathscr{I}) \subseteq A$. Then $\frac{a}{1} = \psi(a) \in \mathscr{I}$. So $x = \frac{a}{s} = \frac{1}{s}\frac{a}{1} \in \mathscr{I}$. Then we have the inclusion $S^{-1}\psi^{-1}(\mathscr{I}) \subseteq \mathscr{I}$. Conversely, let $x \in \mathscr{I} \subseteq S^{-1}A$, $x = \frac{a}{s}$ with $s \in S$ and $a \in A$. Then $x = \frac{1}{s}\frac{a}{1}$ with $\frac{1}{s} \in U(S^{-1}A)$, so $\frac{a}{1} \in \mathscr{I}$. Since $\psi(a) = \frac{a}{1} \in \mathscr{I}$, then $a \in \psi^{-1}(\mathscr{I})$ and $x = \frac{a}{s} \in S^{-1}\psi^{-1}(\mathscr{I})$. Then $\mathscr{I} \subseteq S^{-1}\psi^{-1}(\mathscr{I})$. $\square$

In the following theorem, we generalize the example of Lemma 8.1 to prime ideals (nonnecessarily principal) but minimal over principal ideals. This is the first version of the principal ideal theorem.

Krull Principal Ideal Theorem 8.3 *Let A be a Noetherian ring and P a prime ideal of A. If P is minimal over a principal ideal aA, then $ht(P) \le 1$.*

Proof If a is nilpotent, for each prime ideal Q of A contained in P, $a \in Nil(A) \subseteq Q \subseteq P$. By the minimality of P over the ideal aA, $Q = P$. Then $ht(P) = 0$. We suppose that the element a is not nilpotent. Then $S = \{a^n, n \in \mathbb{N}\}$ is a multiplicative set of A.

Particular case: The ring A is local with maximal ideal P.

In this case, the ring $A/(a)$ has only one prime ideal $P/(a)$, and then $dim(A/(a)) = 0$. By Theorem 7.5, the ring $A/(a)$ is Artinian. We want to show that the ring $S^{-1}A$ is also Artinian. Let $\pi : A \longrightarrow A/(a)$ be the canonical homomorphism and $\psi : A \longrightarrow S^{-1}A$ the natural homomorphism. We will first prove that if $J \subseteq I$ are two ideals of $S^{-1}A$ such that $\pi(\psi^{-1}(I)) = \pi(\psi^{-1}(J))$, then $I = J$. Let $x \in \psi^{-1}(I)$. Then $\pi(x) \in \pi(\psi^{-1}(I)) = \pi(\psi^{-1}(J))$. There exists $y \in \psi^{-1}(J)$ such that $\pi(x) = \pi(y)$, and then $\bar{x} = \bar{y}$ in $A/(a)$, so $x - y \in (a)$. Put $x - y = az$ with $z \in A$. Then $az = x - y \in \psi^{-1}(I)$, so $\psi(az) \in I$, and then $\psi(a)\psi(z) \in I$. But $\psi(a) = \frac{a}{1}$ is invertible in the ring $S^{-1}A$, and then $\psi(z) \in I$, so $z \in \psi^{-1}(I)$. Then $x = y + az \in \psi^{-1}(J) + a\psi^{-1}(I)$. We conclude that $\psi^{-1}(I) \subseteq \psi^{-1}(J) + a\psi^{-1}(I)$. Since the ring A is Noetherian, its ideal $\psi^{-1}(I)$ is finitely generated. The quotient A-module $M = \psi^{-1}(I)/\psi^{-1}(J)$ is then finitely generated. On the other hand, the inclusion $\psi^{-1}(I) \subseteq \psi^{-1}(J) + a\psi^{-1}(I)$ implies that $\psi^{-1}(I)/\psi^{-1}(J) \subseteq a\left(\psi^{-1}(I)/\psi^{-1}(J)\right)$, and then $M \subseteq aM \subseteq PM \subseteq M$, so

$PM = M$. Since A is local with maximal ideal P, the Jacobson radical $J(A) = P$. By Nakayama lemma, $M = (0)$, then $\psi^{-1}(I) = \psi^{-1}(J)$, so $S^{-1}\psi^{-1}(I) = S^{-1}\psi^{-1}(J)$. By the preceding lemma, $I = J$.

Let, now, (I_n) be a decreasing sequence of ideals of $S^{-1}A$. Then $\left(\pi(\psi^{-1}(I_n))\right)$ is a decreasing sequence of ideals of the Artinian ring $A/(a)$, and then it is stationary. By the preceding step, the sequence (I_n) is also stationary. We conclude that the ring $S^{-1}A$ is Artinian. By Theorem 7.5, $dim(S^{-1}A) = 0$. Let $Q_1 \subseteq Q_2 \subset P$ be prime ideals of A. By minimality of P over the ideal aA, $a \notin Q_2$, and then $S \cap Q_2 = \emptyset$. So $S^{-1}Q_1 \subseteq S^{-1}Q_2$ are prime ideals of the ring $S^{-1}A$. But $dim(S^{-1}A) = 0$, and then $S^{-1}Q_1 = S^{-1}Q_2$, so $Q_1 = Q_2$. Then $ht(P) \leq 1$.

General case: P is a prime ideal of A minimal over the principal ideal aA.

The ring A_P is local with maximal ideal PA_P and PA_P is the only prime ideal of A_P containing $\frac{a}{1}$, so it is minimal over the ideal $\frac{a}{1}A_P$. Indeed, such ideal is of the form QA_P with Q a prime ideal of A contained in P. Since $\frac{a}{1} \in QA_P$, $\frac{a}{1} = \frac{q}{s}$ with $q \in Q$ and $s \in A \setminus P$. There exists $r \in A \setminus P$ such that $r(sa - q) = 0$, and then $rsa = rq \in Q$. But $r, s \notin Q$, and then $a \in Q$. So $aA \subseteq Q \subseteq P$. By minimality of P over the ideal aA, $Q = P$.

By the particular case, $ht(PA_P) \leq 1$. Then $ht(P) = dim\, A_P = ht(PA_P) \leq 1$. □

Example 1 Let A be a Noetherian ring and x a regular noninvertible element of A. If P is a prime ideal of A minimal over xA, then $ht(P) = 1$.

Indeed, by the preceding theorem, $ht(P) \leq 1$. Suppose that $ht(P) = 0$, and then P is a minimal prime ideal of A. By Lemma 3.11 (2), the elements of P are zero divisors of A, which contradicts $x \in P$. Then $ht(P) = 1$.

Example 2 The Noetherian hypothesis for A is necessary in the theorem. Let K be a field and $A = K + XK[X, Y] \subset K[X, Y]$. Then $M = XK[X, Y]$ is a maximal ideal of A since $A/M \simeq K$. Moreover, $M = (X, XY, XY^2, \ldots)$ since an element of M is of the form $X\left(\sum a_{ij}X^iY^j\right) = \sum a_{ij}X^i(XY^j)$ with $a_{ij} \in K$. Since the converse inclusion is clear, we have the equality. Suppose that M is finitely generated, and then $M = (X, XY, \ldots, XY^n)$ for some $n \in \mathbb{N}^*$. There exist $h_0, \ldots, h_n \in A$ such that $XY^{n+1} = h_0X + h_1XY + \ldots + h_nXY^n$, and then $Y^{n+1} = h_0 + h_1Y + \ldots + h_nY^n$ (*). Pour $0 \leq i \leq n$, $h_i = a_i + Xf_i(X, Y)$ with $a_i \in K$ and $f_i(X, Y) \in K[X, Y]$. Put $X = 0$ in (*), and we find $Y^{n+1} = a_0 + a_1Y + \ldots + a_nY^n$, which is absurd for degrees reasons. In particular, the ring A is not Noetherian. We can also remark that $A = K + X(K[Y])[X]$ with $dim_K K[Y] = \infty$, so $K[Y]$ is not a finitely generated K-module. By Corollary 4.5, Chap. 1, the ring A is not Noetherian. The only prime ideal P of A containing X is M. Indeed, for each $i \in \mathbb{N}$, $(XY^i)^2 = X(XY^{2i}) \in P$, and then$XY^i \in P$ so $M \subseteq P$. Since M is maximal, $P = M$. In particular, M is a minimal prime over the principal ideal $I = XA$. The ideal $Q = (XY, XY^2, \ldots) \subset M$ since $X \notin Q$. Indeed, in the contrary case, $X = h_1XY + h_2XY^2 + \ldots + h_nXY^n$ with $h_1, \ldots, h_n \in A$. By simplification, we find $1 = h_1Y + h_2Y^2 + \ldots + h_nY^n$. Put $Y = 0$, and we find $1 = 0$, which is absurd. The ideal Q is prime since $A/Q \simeq K[X]$. Indeed, the map $\phi : A = K + XK[X, Y] \longrightarrow K[X]$, defined by

$\phi(f(X, Y)) = f(X, 0)$, is a homomorphism onto since $K[X] \subset A$ and for each $g(X) \in K[X]$, $\phi(g(X)) = g(X)$. Let $f(X, Y) = a + X \sum a_{ij} X^i Y^j \in A$ such that $0 = \phi(f(X, Y)) = f(X, 0) = a + X \sum a_{i0} X^i$. Then $a = 0$ and $a_{i0} = 0$ for each i. Then $f(X, Y) = X \sum_{j \geq 1} a_{ij} X^i Y^j = \sum_{j \geq 1} a_{ij} X^i (X Y^j) \in Q$. The inclusion $Q \subseteq ker\ \phi$ is clear. So $ker\ \phi = Q$ and $A/Q \simeq K[X]$. We obtain the chain $(0) \subset Q \subset M$ of prime ideals of A. Then $ht(M) \geq 2$.

The following technical result is very useful.

Avoidance Lemma 8.4 *Let A be a ring, I an ideal of A, and $P_1, \ldots, P_n$ prime ideals of A. If $I \subseteq P_1 \cup \ldots \cup P_n$, there exists at least one index $1 \leq i \leq n$ such that $I \subseteq P_i$.*

Proof By omitting some P_i, we may suppose that they are incomparable. Suppose that I is not contained in any P_i. For each $i \in \{1, \ldots, n\}$, $I \cap \left(\bigcap_{j:1, j \neq i}^n P_j \right) \not\subseteq P_i$ since no one of the ideals $I, P_1, \ldots, P_{i-1}, P_{i+1}, \ldots, P_n$ is contained in P_i. There exists $a_i \in I \cap \left(\bigcap_{j:1, j \neq i}^n P_j \right) \setminus P_i$. Put $a = a_1 + \ldots + a_n \in I$. Since $I \subseteq P_1 \cup \ldots \cup P_n$, there exists $k \in \{1, \ldots, n\}$ such that $a \in P_k$. Then $a_k = a - \left(\sum_{i:1, i \neq k}^n a_i \right) \in P_k$, which is absurd. $\square$

The avoidance lemma consists to an exchange of quantifiers: If for each $x \in I$, there exists $i \in \{1, \ldots, n\}$ such that $x \in P_i$, then there exists i such that for each $x \in I, x \in P_i$. The name avoidance attributed to this lemma is clear in the converse: If I is not contained in any P_i, there exists $x \in I$ which is not in any P_i, i.e., who avoids all the P_i.

Example 1 The lemma is in general false for infinitely many prime ideals.

Indeed, let K be a field and $A = K[X_1, X_2, \ldots]$ the polynomial ring in countably infinite variables. For each $k \in \mathbb{N}^*$, the ideal $P_k = (X_1, \ldots, X_k)$ is prime. Let $I = (X_1, X_2, \ldots)$ be the maximal ideal of A formed by the polynomials having zero constant terms. Then $I = \bigcup_{k:1}^{\infty} P_k$. But for each $k \in \mathbb{N}^*$, $I \not\subseteq P_k$.

Example 2 The lemma may be true for an infinity of prime ideals.

Let A be any ring and I an ideal of A contained in $\bigcup\{M;\ M \in Max(A)\}$. Then $1 \notin I$, so $I \neq A$. There exists $M_0 \in Max(A)$ such that $I \subseteq M_0$.

The Krull principal ideal theorem has many important applications. We give now a corollary in the integral domain case.

Corollary 8.5 *Let A be a Noetherian integral domain of dimension* ≥ 2. *Then A has an infinity of prime ideals of height one. Their intersection is zero.*

Proof Let a be a nonzero noninvertible element of A. By Zorn's lemma, there exists a minimal prime ideal P over aA. By the Krull principal ideal theorem, $ht(P) = 1$. Then each maximal ideal of A is contained in a union of height one prime ideals. By the avoidance lemma, if A has a finite number of height one prime ideals, then each maximal ideal of A will be contained and then equal to a height one prime ideal. Then $dim(A) = 1$, which is absurd. We conclude that A has an infinity $(P_\alpha)_{\alpha \in \Lambda}$ of height one prime ideals. Suppose that there exists $0 \neq x \in \bigcap_{\alpha \in \Lambda} P_\alpha$. For each $\alpha \in \Lambda$, $P_\alpha/(x)$ is a minimal prime ideal of the ring $A/(x)$. Indeed, a prime ideal of $A/(x)$ contained in $P_\alpha/(x)$ is of the form $P/(x)$ with P a prime ideal of A such that $(0) \neq (x) \subseteq P \subseteq P_\alpha$. But $ht(P_\alpha) = 1$, and then $P = P_\alpha$. So the Noetherian ring $A/(x)$ has an infinity of minimal prime ideals, which contradicts Lemma 7.1. Then $\bigcap_{\alpha \in \Lambda} P_\alpha = (0)$. $\qquad\qquad\square$

Example The converse of the preceding corollary is false. Indeed, the domain $\mathbb{Z}$ is Noetherian of dimension 1. For each prime number p, the ideal $p\mathbb{Z}$ is prime of height 1. The intersection $\bigcap_p p\mathbb{Z} = (0)$.

We give another consequence of Theorem 8.3 in the integral domain case.

Corollary 8.6 *Let A be a Noetherian integral domain. Then A is UFD if and only if each height one prime ideal of A is principal.*

Proof Let P be a height one prime ideal of A. There exists $0 \neq a \in P$. Since A is a UFD, $a = p_1 \ldots p_n$ with $p_1, \ldots, p_n$ irreducible elements of A. There exists i such that $p_i \in P$. Since A is a UFD, $p_i A$ is a prime ideal of A. But $(0) \subset p_i A \subseteq P$ with $ht(P) = 1$, and then $P = p_i A$ is a principal ideal.

" $\Longleftarrow$ " (1) Let a be a nonzero noninvertible element of A. There exists a minimal prime ideal P over aA. By Theorem 8.3, $ht(P) = 1$. By hypothesis, $P = p_1 A$ with $p_1 \in A$ a prime element (i.e., generates a prime ideal of A). There exists $0 \neq a_1 \in A$ such that $a = p_1 a_1$. If a_1 is not a unit, with the same argument, we show that $a_1 = p_2 a_2$ where $p_2 \in A$ is a prime element and $a_2 \in A$. Then we have a strictly increasing sequence $(a) \subset (a_1) \subset (a_2) \subset \ldots$ of ideals of A. Since A is Noetherian, there exists n such that $a_n \in U(A)$. Then $a = a_n p_1 \ldots p_n$ with $p_1, \ldots, p_n$ prime elements and a_n a unit of A.

(2) Let p be an irreducible element of A. We must show that it is prime. There exists a prime ideal P of A minimal over pA. By Theorem 8.3, $ht(P) = 1$. By hypothesis, P is principal, and then $P = qA$ with q a prime element of A. Since $p \in P$, then q divides p. So these two elements are associated. Then $pA = qA = P$ is a prime ideal. $\qquad\qquad\square$

Example (Buchsbaum 1961 and Samuel 1961) If A is a PID, then the formal power series ring in one variable $A[[X]]$ is UFD.

Indeed, by Corollary 5.2, Chap. 1, $A[[X]]$ is a Noetherian integral domain. Let P be a height one prime ideal of $A[[X]]$. If $X \in P$, then $(0) \subset (X) \subseteq P$, so $P = (X)$ is principal. If $X \notin P$, the ideal $P' = \{f(0); \ f(X) \in P\}$ of A is principal generated by an element a. Let $f(X) \in P$ such that $f(0) = a$. By the proof of Theorem 5.1, Chap. 1, $P = f(X)A[[X]]$ is a principal ideal. By the preceding corollary, $A[[X]]$ is UFD.

We give now a second version of the Krull principal ideal theorem. It passes from principal ideals to finitely generated ideals. Its proof uses Theorem 8.3. This result is also called the theorem of the small dimension.

Generalized Krull Principal Ideal Theorem 8.7 *Let A be a Noetherian ring and P a prime ideal of A minimal over a finitely generated ideal $(a_1, \ldots, a_n)$ with n generators. Then $ht(P) \leq n$.*

Proof Particular case: The ring A is local with maximal ideal P.

By induction on n. If $n = 0$, P is a minimal prime ideal, then $ht(P) = 0$. If $n = 1$, $ht(P) \leq 1$, by Krull principal ideal theorem. Suppose that $n \geq 2$ and the property is true for the order $n - 1$. By the ascending chain condition, we can find a prime ideal Q of A such that $Q \subset P$ are adjacent. It suffices to show that for each $Q \in Spec(A)$ with this property, $ht(Q) \leq n - 1$. By the minimality of P, $(a_1, \ldots, a_n) \nsubseteq Q$. By changing the indexation, we may suppose that $a_1 \notin Q$. Then P is minimal over $(a_1) + Q$ because if $S \in Spec(A)$ is such that $(a_1) + Q \subseteq S \subseteq P$, then $a_1 \in S$ and $Q \subseteq S \subseteq P$. Since $Q \subset P$ are adjacent and the equality $Q = S$ is impossible because of a_1, then $S = P$. Since A is local with maximal ideal P, the unique prime ideal containing $(a_1) + Q$ is P, and then $\sqrt{(a_1) + Q} = P$. There exists $k \in \mathbb{N}^*$ such that $a_2^k, \ldots, a_n^k \in (a_1) + Q$. Put $a_i^k = b_i + x_i a_1$ with $b_i \in Q$ and $x_i \in A$ for $2 \leq i \leq n$. Since $(a_1, a_2^k, \ldots, a_n^k) = (a_1, b_2, \ldots, b_n)$, P is minimal over $(a_1, b_2, \ldots, b_n)$ and then $P/(b_2, \ldots, b_n)$ is minimal over the ideal $(a_1, b_2, \ldots, b_n)/(b_2, \ldots, b_n) = (\overline{a_1})$ of the ring $A/(b_2, \ldots, b_n)$. By Krull principal ideal theorem, $ht(P/(b_2, \ldots, b_n)) \leq 1$. Since $Q/(b_2, \ldots, b_n) \subset P/(b_2, \ldots, b_n)$, then $ht(Q/(b_2, \ldots, b_n)) = 0$. So Q is a minimal prime over the ideal $(b_2, \ldots, b_n)$. By the induction hypothesis, $ht(Q) \leq n - 1$, and then $ht(P) \leq n$.

General case: P is a minimal prime ideal over the ideal $(a_1, \ldots, a_n)$. Then PA_P is a prime ideal of A_P minimal over $(\frac{a_1}{1}, \ldots, \frac{a_n}{1})$. Since A_P is local with maximal ideal PA_P, by the particular case, $ht(PA_P) \leq n$. Then $ht(P) = dim(A_P) = ht(PA_P) \leq n$.   □

Example Let k be a field. In the ring $k[X_1, \ldots, X_n]$, $ht(X_1, \ldots, X_n) = n$. Indeed, since we have the chain $(0) \subset (X_1) \subset (X_1, X_2) \subset \ldots \subset (X_1, \ldots, X_n)$ of length n, $ht(X_1, \ldots, X_n) \geq n$. By the preceding theorem, $ht(X_1, \ldots, X_n) \leq n$. So we have the equality.

Corollary 8.8 *Let A be a Noetherian ring and P a prime ideal of A. If P is generated by n elements, then $ht(P) \leq n$.*

In particular, if A is a local Noetherian ring with maximal ideal M, then $dim(A) = ht(M)$ is finite and bounded by the number of generators of M.

Example Two maximal ideals of different heights in an integral Noetherian domain of dimension 2. Let k be a field and $A = k[X]_{(X)}[t]$. Let $\phi : A = k[X]_{(X)}[t] \longrightarrow k(X)$, the map defined by $\phi(f(t)) = f(\frac{1}{X})$. Then ϕ is a homomorphism onto. Indeed, an element of $k(X)$ is of the form $\frac{g(X)}{X^n h(X)}$ with $n \in \mathbb{N}$ and $g(X), h(X) \in K[X], h(0) \neq 0$, and then $\frac{g(X)}{h(X)} \in K[X]_{(X)}$, $\frac{g(X)}{h(X)}t^n \in A$ and $\phi\left(\frac{g(X)}{h(X)}t^n\right) = \frac{g(X)}{X^n h(X)}$.

We will prove that $ker\ \phi = (tX - 1)$. The converse inclusion is clear. Let $f(t) \in ker\ \phi \subseteq A = k[X]_{(X)}[t] \subseteq k(X)[t]$. Then $\phi(f(t))) = f(\frac{1}{X}) = 0$. So $f(t) = (t - \frac{1}{X})g(t)$ with $g(t) \in k(X)[t]$ and then $f(t) = (Xt-1)\frac{1}{X}g(t) = (Xt-1)h(t)$ with $h(t) \in k(X)[t]$. It suffices to prove that $h(t) \in A = k(X)_{(X)}[t]$. Put $f(t) = a_0 + a_1 t + \ldots$ with $a_i \in k[X]_{(X)} \subseteq k(X)$ and $h(t) = b_0 + b_1 t + \ldots$ with $b_i \in k(X)$. Then $f(t) = -h(t) + Xth(t)$, so $a_0 = -b_0$, and then $b_0 \in k[X]_{(X)}$; $a_1 = -b_1 + Xb_0$, so $b_1 = -a_1 + Xb_0 \in k[X]_{(X)}$; $a_2 = -b_2 + Xb_1$, and then $b_2 = -a_2 + Xb_1 \in k[X]_{(X)}$, $\ldots$. By induction, $h(t) \in k[X]_{(X)}[t] = A$. Then $f(t) = (Xt - 1)h(t) \in (Xt - 1)A$. So $A/(Xt - 1) \simeq k(X)$ and then $(Xt - 1)$ is a maximal ideal of A. By the preceding corollary, $ht(Xt - 1) \leq 1$, and since we have the chain $(0) \subset (Xt - 1)$, then $ht(Xt - 1) = 1$.

The nonzero ideal (t) of A is prime but nonmaximal since $A/(t) = k[X]_{(X)}[t]/(t) \simeq k[X]_{(X)}$ is an integral domain which is not a field (in fact, it is a DVR of rank 1). The map $\psi : A = k[X]_{(X)}[t] \longrightarrow k$, defined by $\psi(f(X, t)) = f(0, 0)$ for each $f(X, t) \in A$, is well defined and a homomorphism onto since for each $a \in k$, $\psi(a) = a$. It is clear that the ideal $(X, t) \subseteq ker\ \psi$. Conversely, let $f(X, t) = g_0(X) + g_1(X)t + \ldots + g_n(X)t^n \in ker\ \psi$ a polynomial with $g_0(X), g_1(X), \ldots, g_n(X) \in k[X]_{(X)}$. Its constant term $g_0(X) = \frac{h(X)}{l(X)}$ with $h(X), l(X) \in k[X]$ and $l(0) \neq 0$. Since $0 = \psi(f(X, t)) = g_0(0) = \frac{h(0)}{l(0)}$, then $h(0) = 0$, so $h(X) = Xr(X)$ with $r(X) \in k[X]$. Then $f(X, t) = X\frac{r(X)}{l(X)} + t\left(g_1(X) + g_2(X)t + \ldots + g_n(X)t^{n-1}\right) \in (X, t)A$. So $ker\ \psi = (X, t)A$, and then $A/(X, t) \simeq k$, so (X, t) is a maximal ideal of A and we have the chain of prime ideals $(0) \subset (t) \subset (X, t)$, and then $ht(X, t) \geq 2$. But by the preceding corollary, $ht(X, t) \leq 2$. Then $ht(X, t) = 2$ and $dim\ A \geq 2$. In fact, $dim\ A = 2$. Indeed, $k[X]_{(X)}$ is a DVR with rank 1. Then $dim\ A = dim\ k[X]_{(X)}[t] = 2$.

A Noetherian ring is not necessarily Artinian, but satisfies the decreasing chain condition for some type of ideals.

Corollary 8.9 *A Noetherian ring satisfies the decreasing chain condition for prime ideals.*

Proof The length of a decreasing chain of prime ideals starting at a prime ideal P is bounded by the minimal number of generators of P.   □

Theorem 8.10 (The Converse of the Generalized Krull Principal Ideal Theorem) *Let A be a Noetherian ring and P a prime ideal of height n in A, $n \in \mathbb{N}$.*

There exist n elements $a_1, \ldots, a_n \in A$ such that P is minimal over the ideal $(a_1, \ldots, a_n)$.

Proof By induction on n. If $n = 0$, P is minimal, we take an empty set of elements of A. Suppose that $n \geq 1$ and the property is true for any prime ideal of height $\leq n - 1$. By Lemma 7.1, since A is Noetherian, it has a finite number of minimal prime ideals $Q_1, \ldots, Q_k$. Since $ht(P) = n \geq 1$ and $ht(Q_i) = 0$, then $P \not\subseteq Q_i$, $1 \leq i \leq k$. By the avoidance lemma, $P \not\subseteq Q_1 \cup \ldots \cup Q_k$. Let $a_1 \in P \setminus Q_1 \cup \ldots \cup Q_k$. In the Noetherian quotient ring $A/(a_1)$, $ht(P/(a_1)) \leq n - 1$. Indeed, in the contrary case, there exist a chain of length n of prime ideals of $A/(a_1)$ arriving at $P/(a_1)$. It is of the form $P_0/(a_1) \subset P_1/(a_1) \subset \ldots \subset P_n/(a_1) = P/(a_1)$ with $P_0 \subset P_1 \subset \ldots \subset P_n = P$ a chain of length n of prime ideals of A containing (a_1). Since a_1 does not belong to any minimal prime ideal of A, P_0 is not minimal. There exists $i \in \{1, \ldots, k\}$ such that $Q_i \subset P_0 \subset P_1 \subset \ldots \subset P_n = P$. We obtain a chain of length $n + 1$ of prime ideals of A arriving at P, which contradicts $ht(P) = n$. By the induction hypothesis, there exist $a_2, \ldots, a_n \in P$ such that $P/(a_1)$ is minimal over the ideal $(\overline{a_2}, \ldots, \overline{a_n})$ in the ring $A/(a_1)$. We will show that P is minimal over $(a_1, a_2, \ldots, a_n)$ in A. Let Q be a prime ideal of A such that $(a_1, a_2, \ldots, a_n) \subseteq Q \subseteq P$. In the ring $A/(a_1)$, $(\overline{a_2}, \ldots, \overline{a_n}) \subseteq Q/(a_1) \subseteq P/(a_1)$. Then $Q/(a_1) = P/(a_1)$, so $P = Q$.   □

Example If K is a field, then $\dim K[X_1, \ldots, X_n] = n$. We will show the result by induction on n. If $n = 1$, $K[X_1]$ is a PID. All its nonzero prime ideals are maximal. Then $\dim(K[X_1]) = 1$. Suppose the property true for the order $n - 1$. Since $ht(X_1, \ldots, X_n) = n$, then $\dim K[X_1, \ldots, X_n] \geq n$. Let M be a maximal ideal of $K[X_1, \ldots, X_n]$ and $S = K[X_1] \setminus (0)$: a multiplicative set.

If $S \cap M = \emptyset$, then $S^{-1}M$ is a prime ideal of the ring $S^{-1}K[X_1, \ldots, X_n]$ and $ht(M) = ht(S^{-1}M) \leq \dim(S^{-1}K[X_1, \ldots, X_n])$. By the induction hypothesis, $S^{-1}K[X_1, \ldots, X_n] = S^{-1}K[X_1][X_2, \ldots, X_n] = K(X_1)[X_2, \ldots, X_n]$ has a dimension $n - 1$. Then $ht(M) \leq n - 1$.

If $S \cap M \neq \emptyset$, let $g(X_1) \in S \cap M$. It is not constant. Then $g(X_1)$ is a finite product of irreducible elements in $K[X_1]$. One factor $f(X_1) \in M$ and $K[X_1, \ldots, X_n]/(f(X_1)) = K[X_1][X_2, \ldots, X_n]/(f(X_1)) \simeq (K[X_1]/(f(X_1)))[X_2, \ldots, X_n]$. But $K[X_1]/(f(X_1))$ is a field. By the induction hypothesis, we have $\dim(K[X_1]/(f(X_1)))[X_2, \ldots, X_n] = n-1$. Then $\dim(K[X_1, \ldots, X_n]/(f(X_1))) = n - 1$ so $ht(M/(f(X_1)) \leq n - 1$.

Since $K[X_1, \ldots, X_n]/(f(X_1))$ is Noetherian, by the preceding theorem, $M/(f(X_1))$ is minimal over an ideal generated by $n - 1$ elements $\overline{g_1}, \ldots, \overline{g_{n-1}}$. Then M is minimal over the ideal $(f_1(X), g_1, \ldots, g_{n-1})$. By the generalized Krull principal ideal theorem, $ht(M) \leq n$. Then $\dim(K[X_1, \ldots, X_n]) \leq n$.

Corollary 8.11 *Let A be a Noetherian local ring with maximal ideal M.*

Then $dim(A) = ht(M) = inf\{n \in \mathbb{N}; \exists a_1, \ldots, a_n \in M, M = \sqrt{(a_1, \ldots, a_n)}\}$.

Proof Since A is Noetherian, $M = (b_1, \ldots, b_s)$ is a finitely generated ideal. Then $M = \sqrt{(b_1, \ldots, b_s)}$. If $a_1, \ldots, a_n$ are elements of M such that $M = \sqrt{(a_1, \ldots, a_n)}$, then M is minimal over the ideal $(a_1, \ldots, a_n)$. Indeed, let P be a prime ideal of A such that $(a_1, \ldots, a_n) \subseteq P \subseteq M$. Then $M = \sqrt{(a_1, \ldots, a_n)} \subseteq P \subseteq M$, so $P = M$. By the generalized Krull principal ideal theorem, $ht(M) \leq n$. Then $ht(M) \leq n_0 = inf\{n \in \mathbb{N}; \exists a_1, \ldots, a_n \in M, M = \sqrt{(a_1, \ldots, a_n)}\}$. Suppose that $ht(M) = k < n_0$. By the preceding theorem, there exist $a_1, \ldots, a_k \in M$ such that M is minimal over the ideal $(a_1, \ldots, a_k)$. But the only prime ideal P containing the ideal $(a_1, \ldots, a_k)$ is M. Indeed, $(a_1, \ldots, a_k) \subseteq P \subseteq M$ since (A, M) is local. Then $P = M$ so $\sqrt{(a_1, \ldots, a_k)} = M$, which contradicts the definition of n_0. □

9 Nagata's Example

The goal of this example is to construct a Noetherian integral domain of infinite dimension. We start by reformulating the example of Proposition 10.3, Chap. 1.

Proposition 9.1 (Fujita 1975) *Let A be an integral domain satisfying the following two properties:*

1. *For each maximal ideal M of A, A_M is Noetherian; i.e., A is locally Noetherian.*
2. *Each nonzero element of A belongs to at most a finite number of maximal ideals of A; i.e., the ring A has a finite character.*

Then the ring A is Noetherian.

Example 1 Let A be an integral domain with finite numbers $M_1, \ldots, M_n$ of maximal ideals. Then A is Noetherian if and only if $A_{M_1}, \ldots, A_{M_n}$ are Noetherian.

Example 2 An integral domain may be Noetherian without satisfying the condition (2) of the proposition. Indeed, the domain $\mathbb{Z}[X]$ is Noetherinn. But for each prime number p, the ideal (p, X) is maximal containing X. In fact, $\mathbb{Z}[X]/(p, X) \simeq \mathbb{Z}/p\mathbb{Z}$. See Example 3 of Definition 10.1, Chap. 1.

The following lemma is a particular case of Lemma 10.8 (2), Chap. 1.

Lemma 9.2 *Let R be an integral domain, $(\mathscr{P}_i)_{i \in \mathbb{N}^*}$ a sequence of prime ideals of R, $S = R \setminus \bigcup_{i:1}^{\infty} \mathscr{P}_i$: a multiplicative set of R, $A = S^{-1}R$ and $M_i = S^{-1}\mathscr{P}_i \in Spec(A)$, $i \in \mathbb{N}^*$. Then $A_{M_i} = R_{\mathscr{P}_i}$ for each $i \in \mathbb{N}^*$.*

Lemma 9.3 *Let K be a field, $R = K[X_1, X_2, \ldots]$ the polynomial ring in countable variables over K, and $P = (X_1, \ldots, X_n) \in Spec(R)$, with $n \in \mathbb{N}^*$. Then R_P is Noetherian.*

Proof Let $B = K(X_{n+1}, X_{n+2}, \ldots)[X_1, \ldots, X_n]$. Then $R \subset B$ is an extension of rings.

We will show that PB is a maximal ideal of B, $PB \cap R = P$ and $R_P = B_{PB}$.

(1) Since $PB = (X_1, \ldots, X_n)B = (X_1, \ldots, X_n)K(X_{n+1}, X_{n+2}, \ldots)[X_1, \ldots, X_n]$, then $B/PB \simeq K(X_{n+1}, X_{n+2}, \ldots)$ is a field. So PB is a maximal ideal of the ring B.

(2) Let $f \in PB \cap R$. Then $f = f_1 X_1 + \ldots + f_n X_n \in R = K[X_1, X_2, \ldots]$ where the polynomials $f_1, \ldots, f_n \in K(X_{n+1}, X_{n+2}, \ldots)[X_1, \ldots, X_n]$. Put $X_1 = \ldots = X_n = 0$ in this equality, and we find $f(0, \ldots, 0, X_{n+1}, \ldots) = 0$, and then $f \in (X_1, \ldots, X_n)K[X_1, X_2, \ldots] = P$. The converse inclusion is clear. Then we have the equality $PB \cap R = P$.

(3) Let $x \in R_P$, $x = \frac{a}{s}$ with $a \in R \subseteq B$ and $s \in R \setminus P$. Suppose that $s \in PB$, and then $s \in PB \cap R = P$, which is absurd. Then $x = \frac{a}{s} \in B_{PB}$ and we have the inclusion $R_P \subseteq B_{PB}$.

For the converse inclusion, a nonzero element of $K[X_{n+1}, X_{n+2}, \ldots]$ does not belong to $P = (X_1, \ldots, X_n)R$, and then it is invertible in R_P. So we have $K(X_{n+1}, X_{n+2}, \ldots) \subseteq R_P$. Since $X_1, \ldots, X_n \in R \subseteq R_P$, then $B = K(X_{n+1}, X_{n+2}, \ldots)[X_1, \ldots, X_n] \subseteq R_P$. It suffices then to prove that for each element $f \in B \setminus PB$, its inverse $\frac{1}{f} \in R_P$.

Put $f = \sum f_{i_1 \ldots i_n} X_1^{i_1} \ldots X_n^{i_n}$, a finite sum, with $0 \neq f_{i_1 \ldots i_n} \in K(X_{n+1}, X_{n+2}, \ldots)$. Since $f \notin PB = (X_1, \ldots, X_n)K(X_{n+1}, X_{n+2}, \ldots)[X_1, \ldots, X_n]$, the constant term $f_{0, \ldots, 0} \neq 0$. Put $f_{i_1 \ldots i_n} = \frac{g_{i_1 \ldots i_n}}{g}$ with $g_{i_1 \ldots i_n}, g \in K[X_{n+1}, X_{n+2}, \ldots] \setminus (0)$ and $g_{0, \ldots, 0} \neq 0$.

Then $f = \frac{1}{g} \sum g_{i_1 \ldots i_n} X_1^{i_1} \ldots X_n^{i_n}$, so $\frac{1}{f} = \frac{g}{\sum g_{i_1 \ldots i_n} X_1^{i_1} \ldots X_n^{i_n}}$ with g and $\sum g_{i_1 \ldots i_n} X_1^{i_1} \ldots X_n^{i_n} \in R$. Since $g_{0, \ldots, 0} \neq 0$, then $\sum g_{i_1 \ldots i_n} X_1^{i_1} \ldots X_n^{i_n} \notin P$. So $\frac{1}{f} \in R_P$.

(4) Since $K(X_{n+1}, X_{n+2}, \ldots)$ is a field, the ring $B = K(X_{n+1}, X_{n+2}, \ldots)[X_1, \ldots, X_n]$ is Noetherian, and then B_{PB} is also Noetherian so R_P is Noetherian. $\square$

Remark 9.4 With the notations of the preceding lemma, the ring $R = K[X_1, X_2, \ldots]$ itself is not Noetherian.

Example 1 Consider the extension of rings $A = \mathbb{Z} \subset B = \mathbb{Z}[\sqrt{2}]$. The ideal $P = 2\mathbb{Z}$ of A is prime. But the ideal $PB = 2\mathbb{Z}[\sqrt{2}]$ of B is not prime because $\sqrt{2}.\sqrt{2} = 2 \in PB$ with $\sqrt{2} \in B$ but $\sqrt{2} \notin 2\mathbb{Z}[\sqrt{2}]$.

Example 2 Let $A \subseteq B$ be an integral extension of rings and P a prime ideal of A. Then $(PB) \cap A = P$. Indeed, there exists a prime ideal Q of B such that $Q \cap A = P$. Then $P \subseteq (PB) \cap A \subseteq Q \cap A = P$, so $(PB) \cap A = P$.

But the equality $(PB) \cap A = P$ is in general false. Let the extension $A = \mathbb{Z} \subset B = \mathbb{Q}$ and the prime ideal $P = 2\mathbb{Z}$ of A. Then $PB = 2\mathbb{Q} = \mathbb{Q}$. So $(PB) \cap A = \mathbb{Q} \cap \mathbb{Z} = \mathbb{Z} \neq 2\mathbb{Z}$.

Example 3 Let A be an integral domain, S a multiplicative set of A, and P a prime ideal of A disjoint with S. Then $A \subseteq S^{-1}A$, $S^{-1}P \in Spec(S^{-1}A)$, $S^{-1}P = P(S^{-1}A)$ and $(S^{-1}P) \cap A = P$. Indeed, let $a \in (S^{-1}P) \cap A$, $a = \frac{p}{s}$ with $p \in P$ and $s \in S$. So $as = p \in P$, and then $a \in P$ since $s \notin P$.

Example 4 Let A be a ring and P a prime ideal of A. Consider the extension $A \subset A[X]$. Then $PA[X] = P[X]$ is a prime ideal of $A[X]$ and $(PA[X]) \cap A = P[X] \cap A = P$.

Nagata's example Let K be a field, $R = K[X_1, X_2, \ldots]$ the polynomial ring in a countably variables over K, $\mathscr{P}_1 = (X_1)$, $\mathscr{P}_2 = (X_2, X_3)$, $\mathscr{P}_3 = (X_4, X_5, X_6)$, $\ldots \in Spec(R)$. Then $S = R \setminus \bigcup_{i:1}^{\infty} \mathscr{P}_i$ is a multiplicative set of R. Let $A = S^{-1}R$ and $M_i = S^{-1}\mathscr{P}_i \in Spec(A)$, $i \in \mathbb{N}^*$. We keep all these notations in the sequel.

Proposition 9.5 *The only maximal ideals of A are the M_i, $i \in \mathbb{N}^*$.*

Proof The $\mathscr{P}_i$ are incomparable prime ideals of R. Then $M_i = S^{-1}\mathscr{P}_i$, $i \in \mathbb{N}^*$ and are incomparable prime ideals of A. A prime ideal of A is of the form $S^{-1}P$ with P a prime ideal of R contained in $\bigcup_{i:1}^{\infty} \mathscr{P}_i$. It suffices to show that each nonzero ideal I of R contained in $\bigcup_{i:1}^{\infty} \mathscr{P}_i$ is contained in some $\mathscr{P}_i$. Suppose the contrary. For each $n \in \mathbb{N}^*$, $I \subseteq \mathscr{P}_1 \cup \ldots \cup \mathscr{P}_n \cup \left(\sum_{i>n} \mathscr{P}_i \right)$ since the sum of ideals is generated by their union. Since $\sum_{i>n} \mathscr{P}_i = (X_i; i \in N)$, with N an infinite subset of $\mathbb{N}^*$, then $\sum_{i>n} \mathscr{P}_i$ is a prime ideal of $R = K[X_1, X_2, \ldots]$. By the avoidance lemma, $I \subseteq \sum_{i>n} \mathscr{P}_i$. On the other hand, $\sum_{i>n} \mathscr{P}_i = (X_{k_n}, X_{k_n+1}, \ldots)$, where the sequence k_n, $n \in \mathbb{N}^*$, is strictly increasing. Then $\bigcap_{n:1}^{\infty} (X_{k_n}, X_{k_n+1}, \ldots) = (0)$ so $I = (0)$, which is absurd. $\qquad \square$

Proposition 9.6 $dim(A) = \infty$.

Proof For each $i \in \mathbb{N}^*$, the ideal $\mathscr{P}_i = (X_m, X_{m+1}, \ldots, X_{m+i-1})$ is generated by i variables. So we have the chain $(X_m) \subset (X_m, X_{m+1}) \subset \cdots \subset (X_m, X_{m+1}, \ldots, X_{m+i-1}) = \mathscr{P}_i$ of prime ideals of R, and then $ht(\mathscr{P}_i) \geq i$. By Lemma 9.2, for each $i \in \mathbb{N}^*$, $ht(M_i) = dim(A_{M_i}) = dim(R_{\mathscr{P}_i}) = ht(\mathscr{P}_i) \geq i$.

By the preceding proposition, $dim(A) = sup\{ht(M_i); \ i \in \mathbb{N}^*\} = \infty$. $\square$

Corollary 9.7 *The domain A is Noetherian.*

Proof

(1) Let $0 \neq f = f(X_1, \ldots, X_n) \in R = K[X_1, X_2, \ldots]$. Then f does not belong to $\mathscr{P}_i$ for i sufficiently larger. A nonzero element g of $A = S^{-1}R$ belongs to at most a finite number of $M_i = S^{-1}\mathscr{P}_i$, $i \in \mathbb{N}^*$. Indeed, $g = \frac{f}{s}$ with $f \in R$ and $s \in S$. Then $g \in M_i$ if and only if $f \in M_i$ since $\frac{1}{s}$ is a unit of the ring $A = S^{-1}R$. Then $f \in M_i \cap R$. But by Example 3, $M_i \cap R = S^{-1}\mathscr{P}_i \cap R = \mathscr{P}_i$. We conclude by the preceding step. Since the M_i, $i \in \mathbb{N}^*$, are the only maximal ideals of A, then A has a finite character.

(2) By Lemma 9.2 for each $i \in \mathbb{N}^*$, $A_{M_i} = R_{\mathscr{P}_i}$. By Lemma 9.3, the ring $R_{\mathscr{P}_i}$ is Noetherian. Then the ring A is locally Noetherian.

(3) By Proposition 9.1, the ring A is Noetherian. $\square$

Problems

Exercise 1 (Samuel 1957) Let A be a sub-ring of a field K and $S = K \setminus A$.

Show that S is multiplicatively stable if and only if A is a valuation domain with quotient field K.

Exercise 2 Let A be an integral domain, S a saturated multiplicative set of A, and a, b two nonzero elements of A such that $\frac{a}{b} \in S^{-1}A$. Is $b \in S$?

Exercise 3 (Visweswaran 2013) Let I be a proper ideal of a ring A.

(1) Show that the multiplicative set $S = 1 + I$ of A is saturated if and only if the group of unities $U(A/I) = \{\bar{1}\}$.

(2) Suppose that A is a Boole ring; i.e., for each $a \in A$, $a^2 = a$. Show that $S = 1+I$ is saturated.

Exercise 4 (Visweswaran 2013) Let A be a ring and S a multiplicative set of A; i.e., $1 \in S$, $0 \notin S$ and for each $s_1, s_2 \in S$, $s_1 s_2 \in S$. We say that S satisfies the property (*) if for each $a, b \in A$ such that $ab \in S$ and if one of the elements a or b belongs to S, then the two elements $a, b \in S$.

(1) Let I be an ideal of A. Show that the multiplicative set $S = 1 + I$ satisfies (*).

(2) Let a be a noninvertible regular element of A. Show that the multiplicative set $S = \{a^n;\ n \in \mathbb{N}\}$ satisfies (*). Deduce that the multiplicative set $S = \{2^n;\ n \in \mathbb{N}\}$ of $\mathbb{Z}$ satisfies (*) and it is not of the form $S = 1 + I$ with I an ideal of $\mathbb{Z}$.

(3) Suppose that A is an integral domain containing at least three elements. Let S be the set of polynomials of $A[X]$ whose leading coefficients are units. Show that S is a multiplicative set that satisfies (*), which is not of the form $S = 1 + \mathscr{I}$ with $\mathscr{I}$ is an ideal of $A[X]$.

(4) Suppose that A is an integral domain that is not a field. Let I be a nonzero proper ideal of A. Show that $S = \{1\} \cup (I \setminus (0))$ is a multiplicative set of A that does not satisfy (*).

(5) Let S be a multiplicative set of A. Show that S satisfies (*) if and only if $A \setminus S = \bigcup_{a \in A \setminus S} aS$.

Exercise 5 Let A be a ring and Ω the family of the multiplicative sets of A. Show that Ω has maximal elements for the inclusion and $S \in \Omega$ is maximal if and only if $A \setminus S$ is a minimal prime ideal of A.

Exercise 6 Let A be a ring and $S_0 \subset A$ a multiplicative set formed by regular elements.

(1) Show that S_0 is the biggest multiplicative set S of A for which the natural homomorphism $A \longrightarrow S^{-1}A$ is one to one.

(2) Show that each element of $S_0^{-1}A$ is a unit or a zero divisor.

(3) Show that if each element of A is a unit or a zero divisor, then the natural homomorphism $\phi : A \longrightarrow S_0^{-1}A$ is an isomorphism.

Exercise 7 (Smith 1988) Let A be a ring and M a A-module. Show that M is a multiplication module if and only if for each maximal ideal P of A, there exist a multiplication sub-module N of M and $p \in P$ such that $(1 - p)M \subseteq N$.

Exercise 8 (Abd El-Bast and Smith 1988) Let A be a ring and M a A-module.

(1) Show that M is a multiplication module if and only if for each $x \in M$, there exists an ideal I of A such that $Ax = IM$.

(2) Suppose that M is generated by a family $(x_\lambda)_{\lambda \in \Lambda}$. Show that M is a multiplication module if and only if for each $\lambda \in \Lambda$, there exists an ideal I_λ of A such that $Ax_\lambda = I_\lambda M$.

(3) Deduce that an ideal I of A generated by idempotent elements is a multiplication A-module.

Exercise 9 (Bernard 1981) Let A be a ring and M a multiplication A-module. Suppose that there exists an ideal $I \subseteq Jac(A)$ such that $M = IM$. Show that $M = (0)$.

Exercise 10 (Smith 1988) Let A be a ring, P a maximal ideal of A, and M a multiplication A-module. Show that $PM = M$ if and only if $T_P(M) = M$.

Exercise 11 (Ozen et al. 2021) Let A be a ring, M a A-module, S a multiplicative set of A, and S^* its saturation.

Show that M is S-Artinian if and only if it is S^*-Artinian.

Exercise 12 (Ozen et al. 2021) Let A be a ring, S a multiplicative set of A, and M a A-module.

(1) Let $M_1, \ldots, M_n$ be sub-module of M. Show that $S^{-1}(M_1 \cap \ldots \cap M_n) = (S^{-1}M_1) \cap \ldots \cap (S^{-1}M_n)$.
(2) Show that the result in (1) is false for an infinite intersection.
(3) Show that if the A-module M is S-Artinian, then the $S^{-1}A$-module of fractions $S^{-1}M$ is Artinian.
(4) Show that the converse of (3) is in general false.
(5) Suppose that S is finite. Show that the A-module M is S-Artinian if and only if the $S^{-1}A$-module of fractions $S^{-1}M$ is Artinian.

Exercise 13 Let A be a ring and I a proper ideal of A.

(1) Consider the multiplicative set $S = \{X^i; \ i \in \mathbb{N}\}$ of $A[X]$. Show that $A[X]$ is a sub-ring of $S^{-1}A[X]$.
(2) Show that the map $\phi : A[X] \longrightarrow S^{-1}A[X]$, defined by $\phi(f(X)) = f(\frac{1}{X})$ for each $f(X) \in A[X]$, is a homomorphism of rings.
(3) Let $a \in \bigcap_{M \in Max(A[X]), \ I \subseteq M} M \cap A$.

   (a) Show that $I[X] + (aX - 1)A[X] = A[X]$.
   (b) Show that there exist $k \in \mathbb{N}^*$, $h(X) \in I[X]$ and $g(X) \in A[X]$ such that $X^k = h(X) + (a - X)g(X)$.

(4) Show that $\sqrt{I} = \bigcap_{M \in Max(A[X]), \ I \subseteq M} M \cap A$.

Exercise 14 Let A be a ring, I a proper ideal of A, $Min(I)$ the set of the minimal prime ideals over I in A, and $Min(A)$ the minimal prime spectrum of the ring A.

(1) Show that for each prime ideal P of A, $P[X]$ is a prime ideal of $A[X]$.
(2) Show that $Min(I[X]) = \{P[X]; \ P \in Min(I)\}$.
(3) Deduce that $Min(A[X]) = \{P[X]; \ P \in Min(A)\}$.

Exercise 15 Let K be a field.

(1) Show that the two principal ideals $(X^2 - 1)$ and $(X^2 - 2)$ are comaximal in $K[X]$.
(2) Let $r_1(X)$ and $r_2(X) \in K[X]$ be two fixed polynomials. Give a particular solution for the system of equations

$$\begin{cases} g(X) \equiv r_1(X) \ (mod \ X^2 - 1) \\ g(X) \equiv r_2(X) \ (mod \ X^2 - 2) \end{cases}$$

Hint: Think of the proof of RCT.

(3) Suppose that the degrees of the polynomials $r_1(X)$ and $r_2(X)$ are ≤ 1. Let $f(X) \in K[X]$ be a monic polynomial of degree 4 such that the remains of its Euclidean divisions by $X^2 - 1$ and $X^2 - 2$ are respectively $r_1(X)$ and $r_2(X)$. Express $f(X)$ in terms of $r_1(X)$ and $r_2(X)$.

Exercise 16 (Anderson 1994) Part 1. Let A be a ring, I a proper ideal of A, and $Min(I)$ the set of the minimal prime ideals over I in A. Suppose that each element of $Min(I)$ is a finitely generated ideal. We purpose to show that the set $Min(I)$ is finite. Denote by $\pi(I) = \{P_1 P_2 \ldots P_n; \ n \in \mathbb{N}^*, \ P_i \in Min(I)\}$.

(1) Suppose that I does not contain any element of $\pi(I)$.

   (a) Show that the set $E = \{J \text{ ideal of } A; \ I \subseteq J, L \not\subseteq J \text{ for each } L \in \pi(I)\}$ is nonempty and is inductive for the inclusion.

   (b) Show that each maximal element in $(E, \subseteq)$ is a prime ideal of A.

   (c) Find a contradiction.

(2) Show that there exist $P_1, \ldots, P_n \in Min(I)$ such that $P_1 \ldots P_n \subseteq I$.

Deduce that $Min(I) = \{P_1, \ldots, P_n\}$.

Part 2. Let $B = \prod_{i:0}^{\infty} \mathbb{F}_2 = \{(f_i)_{\in \mathbb{N}}; \ f_i \in \mathbb{F}_2\}$ be the infinite product ring.

(1) Show that B is a Boole ring; i.e., all its elements are idempotent.

(2) Let A be the subset of B formed by the stationary sequences. Show that A is a zero-dimensional sub-ring of B.

(3) Show that for each $n \in \mathbb{N}$, the set $M_n = \{(f_i)_{i \in \mathbb{N}} \in A; \ f_n = 0\}$ is a principal prime ideal of A.

(4) Let $f = (f_i)_{i \in \mathbb{N}} \in A$. We denote by $S(f) = \{i \in \mathbb{N}; \ f_i = 1\}$ the support of f.

   (a) Show that $S(f) = S(g)$ if and only if $f = g$ and $fA \subseteq gA$ if and only if $S(f) \subseteq S(g)$.

   (b) Let $f, g, h \in A$. Show that $fA + gA = hA$ if and only if $S(f) \cup S(g) = S(h)$.

Deduce that every finitely generated ideal of A is principal.

(5) Show that the set $M_\infty = \{(f_i)_{i \in \mathbb{N}} \in A; \ f_i = 0 \text{ from some rank }\}$ is a prime ideal of A that is not finitely generated.

(6) Explain the case of the ring A regarding the first part.

Exercise 17 Let $A \subseteq B$ be an extension of rings and S a multiplicative subset of A. Show that if B is an S-Artinian A-module, then the ring B is S-Artinian.

Solutions

Exercise 1 $"\Longrightarrow"$ For each $0 \neq x \in K$, $xx^{-1} = 1 \notin S$, then either $x \in A$ or $x^{-1} \in A$. Then $qf(A) = K$ and A is a valuation domain.
$"\Longleftarrow"$ Le M be the maximal ideal of A. Let $x, y \in S = K \setminus A$. Since A is a valuation domain with quotient field K, then $x^{-1}, y^{-1} \in M$. So $(xy)^{-1} = x^{-1}y^{-1} \in M$. Then $xy \in K \setminus A = S$.

Exercise 2 Let $A = \mathbb{Z}$ and $S = \{ \pm 2^n; \ n \in \mathbb{N} \}$. Then $\frac{15}{12} = \frac{5}{4} \in S^{-1}A$. But $12 \notin S$.

Exercise 3 (1) $"\Longrightarrow"$ Let $\bar{a} \in U(A/I)$, $a \in A$. There exists $b \in A$ such that $\bar{a}\bar{b} = \bar{1}$ in A/I, and then $ab \in 1 + I = S$. Since S is saturated, $a \in S = 1 + I$, then $\bar{a} = \bar{1}$.
$"\Longleftarrow"$ Let $x, y \in A$ be such that $xy \in S = 1 + I$. Then $\bar{x}\bar{y} = \bar{1}$ in A/I, so $\bar{x} \in U(A/I) = \{\bar{1}\}$. Then $x \in 1 + I = S$.
(2) It is clear that the quotient of a Boole ring is a Boole ring. It suffices to show that if B is a Boole ring, then $U(B) = \{1\}$. Let $x \in U(B)$. Then $x(x - 1) = 0$. Since a unit cannot be a zero divisor, then $x - 1 = 0$, so $x = 1$.

Exercise 4

(1) Let $a, b \in A$ be such that $a \in S$ and $ab \in S$. There exist $x, y \in I$ such that $a = 1 + x$ and $ab = 1 + y$. The equality $(1 + x)b = 1 + y$ gives $b = 1 - bx + y \in 1 + I = S$.

(2) Let $x, y \in A$ be such that $x \in S$ and $xy \in S$. Put $x = a^n$ and $xy = a^m$ with $n, m \in \mathbb{N}$. Then $a^m = a^n y$. Suppose that $n > m$. Then $a^m(1 - a^{n-m}y) = 0$. Since a is regular, $1 - a^{n-m}y = 0$, then $a^{n-m}y = 1$ so a is invertible, which is absurd. Then $n \leq m$ and $a^n(y - a^{m-n}) = 0$. Since a is regular, $y = a^{m-n} \in S$. Since 2 is a regular noninvertible element of the ring $\mathbb{Z}$, the multiplicative set $S = \{2^n; \ n \in \mathbb{N}\}$ satisfies (*). Suppose that there exists an ideal I of $\mathbb{Z}$ such that $S = 1 + I$. Then $I = \{ -1 + 2^n; \ n \in \mathbb{N} \} \subseteq \mathbb{N}$ is not an ideal of $\mathbb{Z}$, which is absurd.

(3) Since the product of two polynomials, whose leading coefficients are units, has the same property, then S is a multiplicative set of $A[X]$. Let $f(X), g(X) \in A[X]$ be such that $f(X), f(X)g(X) \in S$. Put $f(X) = b_0 + b_1 X + \ldots + b_{n-1}X^{n-1} + X^n$, $g(X) = c_0 + c_1 X + \ldots + c_r X^r$ and $f(X)g(X) = a_0 + a_1 X + \ldots + a_{m-1}X^{m-1} + X^m$. By identification, $c_r = 1$. Then $g(X)$ has a unit as leading coefficient and S satisfies (*). Suppose that there exists an ideal $\mathscr{I}$ of $A[X]$ such that $S = 1 + \mathscr{I}$, and then $\mathscr{I} = -1 + S$. So $-1 + X \in \mathscr{I}$. Let $a \in A \setminus \{0, 1\}$. Since $a(-1 + X) = -a + aX = -1 + (1 - a + aX) \notin -1 + S$, then $\mathscr{I}$ is not an ideal, which is absurd.

(4) Let $0 \neq a \in I$ and $b \in A \setminus I$ a noninvertible element. Then $ab \in S$, $a \in S$ but $b \notin S$.

(5) $"\Longrightarrow"$ $"\subseteq"$ Let $a \in A \setminus S$. Since $1 \in S$, then $a = a.1 \in aS$.

$"\supseteq"$ Let $a \in A \setminus S$ and $x \in aS$, $x = as$ with $s \in S$. Suppose that $x \in S$, and then $as \in S$. But S satisfies (*), then $a \in S$, which is absurd. Then $x \in A \setminus S$.

$"\Longleftarrow"$ Let $b \in S$ and $c \in A$ be such that $bc \in S$. Suppose that $c \notin S$, and then $c \in A \setminus S$. By hypothesis, there exist $a \in A \setminus S$ such that $c \in aS$, and then $c = as$ with $s \in S$. So $bc = (bs)a \in aS$ since $bs \in S$. But by hypothesis, $aS \subseteq A \setminus S$, $bc \in A \setminus S$, which is absurd. Then $c \in S$.

Exercise 5 We have $\{1\} \in \Omega$. Every chain of Ω is bounded by the union, that is, a multiplicative set of A. We apply Zorn's lemma.

$"\Longrightarrow"$ Let $S \in \Omega$ be a maximal element. There exists a prime ideal P of A disjoint with S. Then $S \subseteq A \setminus P$. Since $A \setminus P$ is a multiplicative set of A, by maximality of S in Ω, we must have $S = A \setminus P$. Let P' a prime ideal of A contained in P. Then $S = A \setminus P \subseteq A \setminus P'$. By maximality of S, we have $S = A \setminus P'$, and then $P = P'$. So P is a minimal prime.

$"\Longleftarrow"$ By hypothesis, $P = A \setminus S$ is a minimal prime ideal of A. By Zorn's lemma, there exists $S' \in \Omega$ a maximal element such that $S \subseteq S'$. By the first way, $P' = A \setminus S'$ is a (minimal) prime ideal of A. Since $P' \subseteq P$, then $P = P'$ and $S = S'$ is a maximal element.

Exercise 6

(1) Let S be a multiplicative set of A and $\phi : A \longrightarrow S^{-1}A$ the natural homomorphism. Then ϕ is one to one $\Longleftrightarrow \frac{a}{1} = \frac{0}{1}$ implies $a = 0 \Longleftrightarrow as = 0$ for certain $s \in S$ implies $a = 0 \Longleftrightarrow S \subseteq S_0$.

(2) Let $x = \frac{a}{s} \in S_0^{-1}A$ with $a \in A$ and $s \in S_0$. If $a \in S_0$, then x is invertible and $x^{-1} = \frac{s}{a}$. If $a \notin S_0$, there exists $0 \neq b \in A$ such that $ab = 0$, and then $x\frac{b}{1} = \frac{a}{s}\frac{b}{1} = 0$ with $\frac{b}{1} = \phi(b) \neq \frac{0}{1}$ since the natural homomorphism $\phi : A \longrightarrow S_0^{-1}A$ is one to one, by (1). Then x is a zero divisor.

(3) The homomorphism ϕ is one to one. By hypothesis, the elements of S_0 are units of A. Let $y = \frac{a}{s} \in S_0^{-1}A$ with $a \in A$ and $s \in S_0$. Then $y = as^{-1} = \frac{as^{-1}}{1} = \phi(as^{-1})$. So ϕ is onto.

Exercise 7 The first way follows directly from Proposition 4.2. Conversely, let P be a maximal ideal of A. Suppose that M is not P-torsion. By hypothesis, there exist a multiplication sub-module N of M and $p_1 \in P$ such that $(1 - p_1)M \subseteq N$. Then N is not also P-torsion. Indeed, in the contrary case, there exists $p_2 \in P$ such that $(1 - p_2)N = (0)$. Then $(1 - p_2)(1 - p_1)M \subseteq (1 - p_2)N = (0)$ with $(1-p_2)(1-p_1) = 1-(p_1+p_2-p_1p_2)$, which is absurd. By Proposition 4.2, since N is multiplication module, there exist $p_3 \in P$ and $x \in N$ such that $(1 - p_3)N \subseteq Ax$. Then $(1 - p_3)(1 - p_1)M \subseteq (1 - p_3)N \subseteq Ax$ with $(1 - p_3)(1 - p_1) = 1 - (p_1 + p_3 - p_1p_3)$. By Proposition 4.2, M is multiplication module.

Exercise 8

(1) The first way follows directly from the definition. Conversely, let N be a sub-module of M. For each $x \in N$, there exists an ideal I_x of A such that $Ax = I_x M$. Then $I = \sum_{x \in N} I_x$ is an ideal of A and $N = IM$.

(2) The first way is clear. Conversely, let P be a maximal ideal of A.
First case: There exists $\lambda_0 \in \Lambda$ such that $I_{\lambda_0} \not\subseteq P$. Then $I_{\lambda_0} + P = A$, so $1 = a + p$ with $a \in I_{\lambda_0}$ and $p \in P$. So $(1 - p)M = aM \subseteq I_{\lambda_0}M = Ax_{\lambda_0}$. Then M is P-cyclic.
Second case: For each $\lambda \in \Lambda$, $I_\lambda \subseteq P$. By hypothesis, for each $\lambda \in \Lambda$, $Ax_\lambda = I_\lambda M \subseteq PM$. But $M = (x_\lambda)_{\lambda \in \Lambda}$, and then $M \subseteq PM$. So $M = PM$. We apply Proposition 4.2.

(3) If e is an idempotent generator of I, then $Ae = Ae^2 = eI$.

Exercise 9 Let $x \in M$. Since M is a multiplication module, there exists an ideal J of A such that $Ax = JM = J(IM) = I(JM) = IAx = Ix$. There exists $a \in I$ such that $x = ax$, and then $(1 - a)x = 0$. Since $1 - a$ is invertible in A, then $x = 0$. So $M = (0)$.

Exercise 10 $" \Longrightarrow "$ This way is done in the proof of Proposition 5.2.
$" \Longleftarrow "$ Let $x \in M = T_P(M)$. There exists $p \in P$ such that $(1 - p)x = 0$. Then $x = px \in PM$.

Exercise 11 It is clear that if M is S-Artinian, then it is S^*-Artinian. Conversely, let $M_1 \supseteq M_2 \supseteq \ldots$ be a decreasing sequence of sub-modules of M. By hypothesis, there exist $r \in S^*$ and $k \in \mathbb{N}^*$ such that for each integer $n \geq k$, $rM_k \subseteq M_n$. By definition of S^*, the element $\frac{r}{1} \in U(S^{-1}A)$. There exist $a \in A$ and $s \in S$ such that $\frac{a}{s}\frac{r}{1} = \frac{1}{1}$. There exists $u \in S$ such that $uar = us$. Then $s' = us \in S$ and for each integer $n \geq k$, $s'M_k = usM_k = uarM_k \subseteq rM_k \subseteq M_n$. The sequence $M_1 \supseteq M_2 \supseteq \ldots$ is then S-stationary. So the module M is S-Artinian.

Exercise 12

(1) By induction, it suffices to show the property for $n = 2$. The direct inclusion is clear. Let $x \in (S^{-1}M_1) \cap (S^{-1}M_2)$, $x = \frac{a_1}{s} = \frac{a_2}{s}$ with $a_1 \in M_1$, $a_2 \in M_2$, and $s \in S$. There exists $t \in S$ such that $tsa_1 = tsa_2 \in M_1 \cap M_2$. Then $x = \frac{a_1}{s} = \frac{tsa_1}{ts^2} \in S^{-1}(M_1 \cap M_2)$.

(2) Take $A = M = \mathbb{Z}$ and $S = \mathbb{Z} \setminus (0)$. Then $S^{-1}A = S^{-1}M = \mathbb{Q}$. The intersection $\bigcap \{p\mathbb{Z};\ p \text{ a prime number}\} = (0)$. Then $S^{-1}(\bigcap_p p\mathbb{Z}) = S^{-1}(0) = (0)$. But for each prime number p, the ideal $S^{-1}(p\mathbb{Z})$ contains invertible elements of the ring $S^{-1}A$, and then $S^{-1}(p\mathbb{Z}) = S^{-1}A = \mathbb{Q}$. So $\bigcap \{S^{-1}(p\mathbb{Z});\ p \text{ a prime number}\} = \mathbb{Q} \neq (0)$.

(3) Let $M_1 \supseteq M_2 \supseteq \ldots$ be a decreasing sequence of sub-modules of the $S^{-1}A$-module $S^{-1}M$. For each $n \in \mathbb{N}^*$, there exists a A-sub-module N_n of M such that $M_n = S^{-1}N_n$. The decreasing sequence $N_1 \supseteq N_1 \cap N_2 \supseteq N_1 \cap N_2 \cap N_3 \supseteq \ldots$ is S-stationary. There exist $k \in \mathbb{N}^*$ and $s \in S$ such that for each integer $n \geq k$, $s(N_1 \cap N_2 \cap \ldots \cap N_k) \subseteq N_1 \cap N_2 \cap \ldots \cap N_n$, and then $S^{-1}(N_1 \cap N_2 \cap \ldots \cap N_k) \subseteq S^{-1}(N_1 \cap N_2 \cap \ldots \cap N_n)$. By (1), for each integer $n \geq k$, $(S^{-1}N_1) \cap (S^{-1}N_2) \cap \ldots \cap (S^{-1}N_k) \subseteq (S^{-1}N_1) \cap (S^{-1}N_2) \cap \ldots \cap (S^{-1}N_n)$ then $M_1 \cap M_2 \cap \ldots \cap M_k \subseteq M_1 \cap M_2 \cap \ldots \cap M_n$, so $M_k \subseteq M_n$. Then for each integer $n \geq k$, $M_n = M_k$.

(4) Take $A = M = \mathbb{Z}$ and $S = \mathbb{Z} \setminus (0)$. Then $S^{-1}A = S^{-1}M = S^{-1}\mathbb{Z} = \mathbb{Q}$ is a field, and then it is also Artinian. Suppose that the module $M = \mathbb{Z}$ is S-Artinian. Let p be a prime number and consider the decreasing sequence $p\mathbb{Z} \supset p^2\mathbb{Z} \supset p^3\mathbb{Z} \ldots$ of sub-modules of $\mathbb{Z}$. There exist $k \in \mathbb{N}^*$ and $s \in S$ such that for each integer $n \geq k$, $sp^k\mathbb{Z} \subseteq p^n\mathbb{Z}$. Put $s = p^m t$ with $m \in \mathbb{N}$ and t an integer nondivisible by p. For each integer $n \geq k$, p^n divides $p^{m+k}t$, which is absurd.

(5) Let $M_1 \supseteq M_2 \supseteq \ldots$ be a decreasing sequence of sub-modules of M. Then $S^{-1}M_1 \supseteq S^{-1}M_2 \supseteq \ldots$ is a decreasing sequence of sub-modules of $S^{-1}M$. By hypothesis, there exists $k \in \mathbb{N}^*$ such that for each integer $n \geq k$, $S^{-1}M_k = S^{-1}M_n$. Put $S = \{s_1, \ldots, s_r\}$ and $s = s_1 \ldots s_r \in S$. Let $x \in M_k$, $\frac{x}{1} \in S^{-1}M_k = S^{-1}M_n$ for each integer $n \geq k$. There exists $s_i \in S$ depending of n such that $s_i x \in M_n$, and then $sx \in M_n$. So $sM_k \subseteq M_n$ for each integer $n \geq k$. Then M is S-Artinian.

Exercise 13

(1) The elements of S are regular.

(2) Clear.

(3) (a) Suppose that $I[X] + (aX - 1)A[X] \neq A[X]$. There exists $M \in Max(A[X])$ such that $I[X] + (aX - 1)A[X] \subseteq M$. Since $a \in M$ and M is an ideal of $A[X]$, then $aX \in M$. So $1 = aX - (aX - 1) \in M$, which is absurd.

(b) Let $1 = h'(X) + (aX - 1)g'(X)$ with $h'(X) \in I[X]$ and $g'(X) \in A[X]$. Apply ϕ to this equality. We obtain $1 = h'(\frac{1}{X}) + (\frac{a}{X} - 1)g'(\frac{1}{X})$. Let $k \geq max\{deg\ h'(X),\ deg\ g'(X) + 1\}$ be an integer. Then $X^k = h(X) + (a - X)g(X)$ with $h(X) \in I[X]$ and $g(X) \in A[X]$.

(4) The direct inclusion is clear. Conversely, let a be an element of the intersection. Then $X^k = h(X) + (a - X)g(X)$ with $k \in \mathbb{N}^*$, $h(X) \in I[X]$ and $g(X) \in A[X]$. Put $X = a$. Then $a^k = h(a) \in I$, so $a \in \sqrt{I}$.

Exercise 14

(1) Follows from the isomorphism $A[X]/P[X] \simeq (A/P)[X]$.

(2) " $\subseteq$ " Let $\mathscr{P} \in Min(I[X])$. Since $I[X] \subseteq \mathscr{P}$, then $I = A \cap I[X] \subseteq A \cap \mathscr{P} = P \in Spec(A)$. Then $I[X] \subseteq P[X] \subseteq \mathscr{P}$. So $\mathscr{P} = P[X]$. It remains to show that $P \in Min(I)$. Let $Q \in Spec(A)$ be such that $I \subseteq Q \subseteq P$. Then

$I[X] \subseteq Q[X] \subseteq P[X] = \mathscr{P}$. So $Q[X] = P[X] = \mathscr{P}$. Then $P = Q$ and
$P \in Min(I)$.

" $\supseteq$ " Let $P \in Min(I)$. Since $I \subseteq P$, then $I[X] \subseteq P[X] \in Spec(A[X])$. Let
$\mathscr{P} \in Spec(A[X])$ be such that $I[X] \subseteq \mathscr{P} \subseteq P[X]$. Then $I = A \cap I[X] \subseteq$
$A \cap \mathscr{P} \subseteq A \cap P[X] = P$. So $I \subseteq A \cap \mathscr{P} \subseteq P$ with $A \cap \mathscr{P} \in Spec(A)$. Then
$A \cap \mathscr{P} = P$. So $P[X] \subseteq \mathscr{P}$ and $P[X] \in Min(I[X])$.

(3) Take $I = (0)$.

Exercise 15

(1) $X^2 - 1 - (X^2 - 2) = 1$. Then $(X^2 - 1)K[X] + (X^2 - 2)K[X] = K[X]$.

(2) By the proof of RCT, a particular solution is $g_0(X) = r_2(X)(X^2 - 1) - r_1(X)(X^2 - 2)$.

(3)

$$\begin{cases} f(X) = q_1(X)(X^2 - 1) + r_1(X) \\ f(X) = q_2(X)(X^2 - 2) + r_2(X) \end{cases}$$

with $q_1(X)$ and $q_2(X) \in K[X]$. Then

$$\begin{cases} f(X) \equiv r_1(X) \equiv g_0(X) \ (mod \ X^2 - 1) \\ f(X) \equiv r_2(X) \equiv g_0(X) \ (mod \ X^2 - 2) \end{cases}$$

This means that the polynomials $X^2 - 1$ and $X^2 - 2$ divide $f(X) - g_0(X)$. Since
by (1) $GCD(X^2 - 1, X^2 - 2) = 1$, then $(X^2 - 1)(X^2 - 2)$ divides $f(X) - g_0(X)$.
There exists $h(X) \in K[X]$ such that $f(X) - g_0(X) = (X^2 - 1)(X^2 - 2)h(X)$.
Then $f(X) = r_2(X)(X^2 - 1) - r_1(X)(X^2 - 2) + (X^2 - 1)(X^2 - 2)h(X)$. Since
the degree of $r_2(X)(X^2 - 1) - r_1(X)(X^2 - 2)$ is ≤ 3 and $deg \ f(X) = 4$, then
$h(X) = 1$. So $f(X) = r_2(X)(X^2 - 1) - r_1(X)(X^2 - 2) + (X^2 - 1)(X^2 - 2)$.

Exercise 16 Part 1.

(1) (a) We have $I \in E$. Let $(J_\lambda)_{\lambda \in \Lambda}$ be a totally ordered family of elements of E
and $J = \bigcup_{\lambda \in \Lambda} J_\lambda$. Then J is an ideal of A, $I \subseteq J$. Suppose that there exists
$L \in \pi(I)$ such that $L \subseteq J$. Since L is finitely generated, as a product of
finitely generated ideals, there exists $\lambda_0 \in \Lambda$ such that $L \subseteq J_{\lambda_0}$, which is
absurd. So $J \in E$ and bounds the family $(J_\lambda)_{\lambda \in \Lambda}$.

   (b) Let Q be a maximal element in $(E, \subseteq)$. Then Q is an ideal of A, $I \subseteq Q$
and for each $L \in \pi(I)$, $L \nsubseteq Q$. Suppose that Q is not prime. There exist
$x, y \in A \setminus Q$ such that $xy \in Q$. Then $Q \subset Q + xA$ and $Q \subset Q + yA$.
So $Q + xA \notin E$ and $Q + yA \notin E$. There are $L_1, L_2 \in \pi(I)$ such that
$L_1 \subseteq Q + xA$ and $L_2 \subseteq Q + yA$. We have $L_1 L_2 \in \pi(I)$ and $L_1 L_2 \subseteq$
$(Q + xA)(Q + yA) = xQ + yQ + xyQ \subseteq Q$, which is absurd. Then Q is
prime.

(c) Since $(E, \subseteq)$ is inductive, by Zorn's lemma, it admits a maximal element
Q. By (b), Q is prime ideal of A containing I. There exists a minimal prime
ideal P over I contained in Q. Since $P \in \pi(I)$, we have a contradiction
with $Q \in E$.

(2) There exists $L_0 = P_1 \dots P_n \in \pi(I)$, the $P_i \in Min(I)$, such that $L_0 \subseteq I$. Let
$P \in Min(I)$. Then $P_1 \dots P_n \subseteq P$. There exists i such that $P_i \subseteq P$. So $P_i = P$.
We conclude that $Min(I) = \{P_1, \dots, P_n\}$.

Part 2.

(1) Clear.

(2) That A is a sub-ring of B is clear. Suppose that there are two prime ideals
$P \subset Q$ in A. Let $f \in Q \setminus P$. Since $f^2 = f$, then $f(f - 1) = 0 \in P$, so
$f - 1 \in P$. Then $1 = f - (f - 1) \in Q$, which is absurd. So $dim\, A = 0$.

(3) It is easy to see that M_n is a principal ideal generated by the element $f = (f_i)_{i \in \mathbb{N}} \in A$ with $f_n = 0$ and $f_i = 1$ for each integer $i \neq n$. It is also prime.
Indeed, let $f = (f_i)_{i \in \mathbb{N}}$ and $g = (g_i)_{i \in \mathbb{N}} \in A$ be such that $fg \in M_n$. Then
$f_n g_n = 0$ in $\mathbb{F}_2$. So either $f_n = 0$ and $f \in M_n$ or $g_n = 0$ and $g \in M_n$.

(4) (a) The first equivalence follows from the fact that $\mathbb{F}_2 = \{0, 1\}$. If $f = gh$
with $h \in A$, then $S(f) = S(gh) = S(g) \cap S(h) \subseteq S(g)$. Conversely, if
$S(f) \subseteq S(g)$, then $fg = f^2 = f$. So $f \in gA$.

(b) " $\Longrightarrow$ " By (4-a), since $f, g \in hA$, then $S(f) \cup S(g) \subseteq S(h)$. Conversely,
there are $k, l \in A$ such that $h = kf + lg$. Then $S(h) = S(kf + lg) \subseteq$
$S(kf) \cup S(lg) \subseteq S(f) \cup S(g)$. So $S(f) \cup S(g) = S(h)$.
" $\Longleftarrow$ " We define $k = (k_i)$ and $l = (l_i)$ by

$$k_i = \begin{cases} h_i & \text{if } f_i = 1 \\ 0 & \text{if } f_i = 0 \end{cases} \quad \text{and } l_i = \begin{cases} 0 & \text{if } f_i = 1 \\ h_i & \text{if } f_i = 0 \text{ and } g_i = 1 \\ 0 & \text{if } f_i = g_i = 0 \end{cases}.$$

The sequences $(k_i)_i$ and $(l_i)_i$ are stationary, so $k, l \in A$. Also $h = kf + lg \in$
$fA + gA$. So $hA \subseteq fA + gA$. Conversely, since $S(f) \subseteq S(h)$ and $S(g) \subseteq S(h)$,
then $f, g \in hA$. So $fA + gA \subseteq hA$.

(5) It is clear that M_∞ is a proper ideal of A. Let $f = (f_i)$ and $g = (g_i) \in A$ be
such that $fg \in M_\infty$. The sequence $fg = (f_i g_i)$ becomes zero from some rank.
The sequences (f_i) and (g_i) are stationary, but they cannot be both equal to 1
from some rank. Then one of them belongs to M_∞. So M_∞ is prime. Suppose
that this ideal is finitely generated. By (4-b), it will be principal, generated by
an element $f = (f_i)$. There exists $n \in \mathbb{N}$ such that $f_i = 0$ for each $i \geq n$.
Let $g = (g_i) \in A$ be the element defined by $g_{n+1} = 1$ and $g_i = 0$ for each
$i \neq n + 1$. Then $g \in M_\infty \setminus fA$, which is absurd.

(6) $Min(0)$ is infinite since one of its elements M_∞ is not finitely generated.

Exercise 17 Let $I_1 \supseteq I_2 \supseteq \dots$ be a decreasing sequence of ideals of the ring
B. Since each ideal of the ring B is also a A-sub-module of the A-module B, this
sequence is S-stationary. There exists $n_0 \in \mathbb{N}^*$ and $s \in S$ such that for each integer
$n \geq n_0$, we have $sI_{n_0} \subseteq I_n$.

References

Z. Abd El-Bast, P. Smith, Multiplication modules. Comm. Algebra **16**(4), 755–779 (1988)

M.M. Ali, Multiplication modules and homogeneous idealization III. Beiträge Algebra Geom. **49**(2), 449–479 (2008)

D.D. Anderson, Some remarks on multiplication ideals. Math. Japon. **25**(4), 463–469 (1980)

D.D. Anderson, A note on minimal prime ideals. Proc. AMS **122**(1), 13–14 (1994)

D.D. Anderson, M. Winders, Idealization of a module. J. Commutat. Algebra **1**(1), 3–56 (2009)

A. Bernard, Multiplication modules. J. Algebra **71**, 174–178 (1981)

D.A. Buchsbaum, Some remarks in power series rings. J. Math. Mech. **10**(5), 749–753 (1961)

K. Fujita, Some counterexamples related to prime chains in integral domains. Hiroshima Math. J. **5**, 473–485 (1975)

R. Gilmer, A note on the quotient field of of the domain $D[[X]]$. Proc. AMS **18**, 1138–1140 (1967)

R. Gilmer, J.F. Hoffmann, A characterization of Prufer domains in terms of polynomials. Pacific J. Math. **60**(1), 81–85 (1975)

M. Henriksen, M. Jerison, The space of minimal prime ideals of a commutative ring. Trans. Amer. Math. Soc. **115**, 110–130 (1965)

W. Maaref, A. Hamed, A. Benhissi, Generalization of Artinian rings and the formal power series rings. Rend. Cir. Mat. Palermo. Accepted

M. Ozen, O.A. Naji, U. Tekir, K.P. Shum, Characterization theorems of S-Artinian modules. C. R. Acad. Bulg. Sci. **74**(4), 496–505 (2021)

P. Samuel, La notion de place dans un anneau. Bull. Soc. Math. France **85**, 123–133 (1957)

P. Samuel, On unique factorization domains. Illinois J. Math. **5**, 1–17 (1961)

E.S. Sevim, U. Tekir, S. Koc, S-Artinian rings and finitely S-cogenerated rings. J. Algebra Its Appl. **19**(3), 16 (2020)

P.B. Sheldon, How changing $D[[X]]$ changes its quotient field. Trans. AMS **159**, 223–244 (1971)

P.F. Smith, Some remarks on multiplication modules. Arch. Math. **50**, 223–235 (1988)

S. Visweswaran, Some remarks on multiplicatively closed sets. Arab. J. Math. **2**, 409–425 (2013)

Chapter 3
Almost Principal Polynomial Rings

Let A be an integral domain. In this chapter, we define a notion of almost principal for the domain $A[X]$. Then we characterize those A with this property. All the rings considered in this chapter are commutative with identity.

1 Generalities

Definition 1.1 Let A be a ring and S a multiplicative set of A.

1. An ideal I of A is said $S-$principal if there exist $s \in S$ and $a \in I$ such that $sI \subseteq aA$.
2. The ring A is said $S-$principal if all its ideals are $S-$principal.

Examples Let A be a ring and S a multiplicative set of A.

(1) A principal ideal of A is $S-$principal.
(2) Let I be an ideal of A. If $S \cap I \neq \emptyset$, then I is $S-$principal.
    Indeed, let $s \in S \cap I$. Then $sI \subseteq sA \subseteq I$. In particular, if A is an integral domain and $S = A \setminus (0)$, then A is $S-$principal.
(3) Let $S \subseteq T$ be two multiplicative sets of A. If A is $S-$principal, then it is $T-$principal.
(4) If S is formed by units of A, then A is $S-$principal if and only if it is principal (not necessarily an integral domain).
(5) If A is $S-$principal, then $S^{-1}A$ is principal.

Indeed, an ideal of $S^{-1}A$ is of the form $S^{-1}I$ with I an ideal of A. There exist $s \in S$ and a principal ideal J of A such that $sI \subseteq J \subseteq I$. Then $S^{-1}I \subseteq S^{-1}J \subseteq S^{-1}I$, so $S^{-1}I = S^{-1}J$ is a principal ideal of $S^{-1}A$.

The results in the following two sections are due to Anderson et al. (1995).

© The Author(s), under exclusive license to Springer Nature Switzerland AG 2022   183
A. Benhissi, *Chain Conditions in Commutative Rings*,
https://doi.org/10.1007/978-3-031-09898-7_3

Definition 1.2 Let A be an integral domain with quotient field K.

1. A nonzero ideal I of $A[X]$ is said almost finitely generated if there exist $0 \neq s \in A$ and nonconstant polynomials $f_1, \ldots, f_n \in I$ such that $sI \subseteq (f_1, \ldots, f_n)$.
2. We say that the ring $A[X]$ is almost Noetherian if each nonzero ideal I of $A[X]$, satisfying $IK[X] \neq K[X]$, is almost finitely generated.
3. A nonzero ideal I of $A[X]$ is said almost principal if there exist $0 \neq s \in A$ and nonconstant polynomial $f \in I$ such that $sI \subseteq (f)$.
4. We say that the ring $A[X]$ is almost principal if each nonzero ideal I of $A[X]$, such that $IK[X] \neq K[X]$, is almost principal.

Example Let A be an integral domain, I a nonzero ideal of $A[X]$, and $S = A \setminus (0)$. If I is almost principal, then it is S−principal. The converse is false. Let a be a nonzero element of A. The ideal $aA[X]$ is principal, and then it is S−principal. Suppose that it is almost principal. There exist $0 \neq s \in A$ and a nonconstant polynomial $f(X) \in aA[X]$ such that $sI \subseteq f(X)A[X]$. Then $sa = f(X)g(X)$ with $g(X) \in A[X]$. This is absurd for degree reasons.

Remark 1.3 Let A be an integral domain with quotient field K and I a nonzero ideal of $A[X]$. If I is almost principal, then $IK[X] \neq K[X]$.

Indeed, since the domain $K[X]$ is principal, there exists $0 \neq h(X) \in I$ such that $IK[X] = h(X)K[X]$. Suppose that $IK[X] = K[X]$, then $h(X) = a \in K^* \cap I$. Since I is almost principal, there exist $0 \neq s \in A$ and a nonconstant polynomial $f(X) \in I$ such that $sI \subseteq fA[X]$. There exists $0 \neq g(X) \in A[X]$ such that $sa = f(X)g(X)$ a nonconstant polynomial. This is absurd.

Proposition 1.4 *Let A be an integral domain with quotient field K and I a nonzero ideal of $A[X]$ such that $IK[X] \neq K[X]$. Then I is almost finitely generated if and only if it is almost principal. In particular, $A[X]$ is almost Noetherian if and only if it is almost principal.*

Proof "$\Longleftarrow$" Clear. "$\Longrightarrow$" There exist $0 \neq s \in A$ and nonconstants polynomials $f_1, \ldots, f_n \in I$ such that $sI \subseteq (f_1, \ldots, f_n)A[X]$. Since $K[X]$ is a principal ideal domain, then $IK[X] = (f_1, \ldots, f_n)K[X] = fK[X]$ with $0 \neq f \in (f_1, \ldots, f_n)A[X] \subseteq I$. Since $IK[X] \neq K[X]$, then f is not constant. Put $f_i = g_i f$ with $g_i \in K[X]$, $1 \leq i \leq n$. There exists $0 \neq t \in A$ such that $tg_1, \ldots, tg_n \in A[X]$. Since $tf_i = (tg_i)f$, then $(ts)I \subseteq (tf_1, \ldots, tf_n)A[X] \subseteq fA[X]$. So I is almost principal.   □

Proposition 1.5 (Houston 1994) *Let A be an integral domain with quotient field K and t an element of K that it is not quasi-integral over A but for which there exist $0 \neq a \in A$ and infinitely many natural integers i such that $at^i \in A$.*
The homomorphism $\phi : A[X] \longrightarrow K$, defined by $\phi(f(X)) = f(t)$ for each $f(X) \in A[X]$, has a kernel $P = (X - t)K[X] \cap A[X]$, a nonzero prime ideal of $A[X]$ of zero trace over A, and it is not almost principal.

Proof It is clear that $ker\phi = (X - t)K[X] \cap A[X] = P \neq (0)$. Then $A[X]/P \simeq \phi(A[X]) \subseteq K$, a field. So P is prime. By definition of ϕ, $P \cap A = (0)$. Suppose that P is almost principal in $A[X]$. There exist $0 \neq b \in A$ and a nonconstant polynomial $f(X) \in P$ such that $bP \subseteq f(X)A[X]$. Put $f(X) = (X - t)g(X)$ with $g(X) \in K[X]$. There exists $0 \neq c \in A$ such that $cg(X) \in A[X]$ and then $cbP \subseteq cf(X)A[X] = (X - t)cg(X)A[X] \subseteq (X - t)A[X]$. So $dP \subseteq (X - t)A[X]$ with $0 \neq d = cb \in A$.

Let $i \in \mathbb{N}^*$ be such that $at^i \in A$ and $h(X) = a(X^{i-1} + tX^{i-2} + \ldots + t^{i-2}X + t^{i-1}) \in K[X]$. Then $(X - t)h(X) = a(X - t)(X^{i-1} + tX^{i-2} + \ldots + t^{i-2}X + t^{i-1}) = a(X^i - t^i) = aX^i - at^i \in A[X] \cap (X - t)K[X] = P$. So $d(X - t)h(X) \in dP \subseteq (X - t)A[X]$. By simplification, $dh(X) \in A[X]$, and then $daX^{i-1} + datX^{i-2} + \ldots + dat^{i-2}X + dat^{i-1} \in A[X]$. So $dat^j \in A$ for $0 \leq j \leq i - 1$ with $0 \neq da \in A$ an element independent of i. Since i may be so big as we want, then $dat^j \in A$ for each $j \in \mathbb{N}$. Then t is quasi-integral over A, which is absurd. We conclude that P is not almost principal. $\square$

Example (Anderson and Zafrullah 2007) Let A be a Noetherian integral domain with quotient field K and $x \in K$. Then x is integral over A if and only if there exist $0 \neq a \in A$ and an infinity of natural integers n such that $ax^n \in A$.

Indeed, let $0 \neq a \in A$ and $0 < n_1 < n_2 < \ldots$ an infinite strictly increasing sequence of integers such that $ax^{n_i} \in A$ for each $i \in \mathbb{N}^*$. The ideal $I = (ax^{n_1}, ax^{n_2}, \ldots)$ of A is finitely generated. There exists $r \in \mathbb{N}^*$ such that $I = (ax^{n_1}, ax^{n_2}, \ldots, ax^{n_r})$. There exist $b_1, \ldots, b_r \in A$ such that $ax^{n_{r+1}} = b_1 ax^{n_1} + b_2 ax^{n_2} + \ldots + b_r ax^{n_r}$. By simplification, we find $x^{n_{r+1}} - b_r x^{n_r} - \ldots - b_1 x^{n_1} = 0$. So x is integral over A. The converse is clear.

Remark 1.6 With the notations and the hypotheses of the preceding proposition, the ideal $P = (X - t)K[X] \cap A[X]$ is not S-finite, where $S = A \setminus (0)$. Indeed, in the contrary case, there exist $s \in S$ and $f_1, \ldots, f_n \in P \setminus (0)$ such that $sP \subseteq (f_1, \ldots, f_n)A[X]$. Since $P \cap A = (0)$, the polynomials $f_1, \ldots, f_n$ are not constant, and then P is almost finitely generated. By Proposition 1.4, P is almost principal, which is absurd. By Example 2 of Definition 1.1, since A is an integral domain, then it is S-Noetherian (and even S-principal). But the ring $A[X]$ is not S-Noetherian.

Example (J.T. Arnold) A concrete example of a domain A with the hypotheses of Proposition 1.5. Let s and t be the two indeterminates over a field F and $A = F\left[s; \{st^{2^n}, n \in \mathbb{N}\}\right] \subset F[s, t]$. Since $t = \frac{st}{s}$, the quotient field of A is equal to $F(s, t)$. Note that a monomial $s^i t^j$ of $F[s, t]$ belongs to A if and only if $i \geq \phi(j)$, where $\phi(j)$ is the number of 1 in the development in binary basis of the integer j. It follows that any $0 \neq f(s, t) \in A$ does not exist such that $f(s, t)t^k \in A$ for each $k \in \mathbb{N}$. Then t is not quasi-integral over A. Note that the multiplicative set $S = A \setminus (0)$ is not anti-Archimedian since $\bigcap_{n:1}^{\infty} s^n A \subseteq \bigcap_{n:1}^{\infty} s^n F[s, t] = (0)$.

Proposition 1.7 (Hamann et al. 1988) *Let A be an integral domain with quotient field K and I a nonzero ideal of $A[X]$ generated by a family (eventually infinite) of polynomials of bounded degrees such that $IK[X] \neq K[X]$. Then I is almost principal.*

Proof Since $K[X]$ is a PID, then $IK[X] = f(X)K[X]$ where $f(X) = X^l(a_n X^n + \ldots + a_1 X + a_0) \in I$ with $a_0 \neq 0$, $a_n \neq 0$, and $deg\ f(X) = l + n \geq 1$ because $IK[X] \neq K[X]$. Then $f(X) \notin A$. We suppose that the degrees of the generators of I are bounded by $l + n + m$ (they are all multiple of $f(X)$ in $K[X]$) with $m \in \mathbb{N}$. Put $s = a_0^{m+1} \in A \setminus (0)$. We will show that $sI \subseteq f(X)A[X]$. To do this, let $h(X)$ be a generator of I, $h(X) = f(X)g(X)$ with $g(X) = b_m X^m + \ldots + b_1 X + b_0 \in K[X]$. Since $h(X) = f(X)g(X) \in I \subseteq A[X]$, the product $(a_n X^n + \ldots + a_1 X + a_0)(b_m X^m + \ldots + b_1 X + b_0) \in A[X]$. For $l \leq m$, $a_0^{l+1} b_l \in A$. Indeed, by induction, $a_0 b_0 \in A$. Suppose that $a_0^{k+1} b_k \in A$ for each

$k < l$. The coefficient of X^l in the preceding product is $c_l = \sum_{i:0}^{l} a_i b_{l-i} \in A$ and

then $a_0^l c_l = \sum_{i:0}^{l} a_i(a_0^l b_{l-i}) \in A$. By the induction hypothesis, for $1 \leq i \leq l$, $a_0^l b_{l-i} \in A$. Then $a_0^{l+1} b_l \in A$. The induction property is then shown. Let now, $l = m$. Then $a_0^{m+1} g(X) = a_0^{m+1}(b_m X^m + \ldots + b_1 X + b_0) \in A[X]$. So $a_0^{m+1} h(X) = f(X)a_0^{m+1} g(X) \in f(X)A[X]$ and then $sI \subseteq f(X)A[X]$. ☐

The class of Noetherian integral domains is an interesting particular case of the preceding proposition.

Corollary 1.8 (Johnson 1982) *Let A be a Noetherian integral domain. The polynomial ring $A[X]$ is almost principal.*

Proof Let I be a nonzero ideal of $A[X]$ such that $IK[X] \neq K[X]$, where K is the quotient field of A. Since $A[X]$ is Noetherian, I is generated by a finite number of polynomials and then of bounded degrees. By the preceding proposition, I is almost principal. ☐

Example 1 Let A be a Noetherian integral domain which is not a field. Then $A[X]$ is almost principal but not principal.

Example 2 Let B be an over ring of an integral domain A. If the domain $A[X]$ is almost principal, the domain $B[X]$ is not necessarily almost principal.

Indeed, let s and t be the two indeterminates over a field F and $B = F[s; \{st^{2^n}, n \in \mathbb{N}\}]$. The ring $B[X]$ is not almost principal. We have the inclusions $A = F[s, st] \subset B \subset F[s, t]$. The quotient field of the integral domain A is $F(s, t)$ since $t = \frac{st}{t}$. Also A is Noetherian. Indeed, the map $\phi : F[s, t] \longrightarrow A = F[s, st]$, defined by $\phi(f(s, t)) = f(s, st)$, is a homomorphism onto. Then

$A \simeq F[s, t]/ker\ \phi$ is Noetherian. By the preceding corollary, the ring $A[X]$ is almost principal.

Example 3 Let s and t be the two indeterminates over a field F and $A = F[s; \{st^{2^n}, n \in \mathbb{N}\}]$. The ring $A[X]$ is not almost principal. For each $n \in \mathbb{N}$, the integral domain $A_n = F[s; \{st^{2^i}, 0 \leq i \leq n\}] \subset F[s, t]$ is Noetherian. Indeed, let $\phi_n : F[X_0, X_1, \ldots, X_{n+1}] \longrightarrow A_n$ be the map defined by $\phi(f(X_0, X_1, \ldots, X_{n+1})) = f(s, st, st^2, \ldots, st^{2^n})$. It is a homomorphism onto. Then $A_n \simeq F[X_0, X_1, \ldots, X_{n+1}]/ker\ \phi_n$ is Noetherian. By the preceding corollary, $A_n[X]$ is almost principal. Note that the sequence $(A_n)_{n \in \mathbb{N}}$ is increasing and $A = \bigcup_{n:0}^{\infty} A_n$, so $A[X] = \bigcup_{n:0}^{\infty} A_n[X]$.

We can improve Proposition 1.7 in the following corollary.

Corollary 1.9 *Let A be an integral domain with quotient field K and J a nonzero ideal of $A[X]$ such that $JK[X] \neq K[X]$. Suppose that J is of the form $J = (f_\alpha(X);\ \alpha \in \Lambda)_v$ where the family $\{f_\alpha(X);\ \alpha \in \Lambda\}$ of polynomials of $A[X]$ is of bounded degrees and v the v-operation on $A[X]$. Then J is almost principal.*

In particular, a v-finite ideal J of $A[X]$ such that $JK[X] \neq K[X]$ is almost principal.

Proof Let $I = (f_\alpha(X);\ \alpha \in \Lambda)$. Then $IK[X] \neq K[X]$ since $IK[X] \subseteq I_v k[X] = JI[X] \neq K[X]$. By Proposition 1.7, I is almost principal. There exist $0 \neq s \in A$ and $f(X) \in I \subseteq J$ a nonconstant polynomial such that $sI \subseteq f(X)A[X]$. Then $sJ = sI_v = (sI)_v \subseteq (f(X)A[X])_v = f(X)A[X]$. So J is almost principal. $\square$

In the following theorem, we give many equivalent conditions to the property $A[X]$ almost principal.

Theorem 1.10 *Let A be an integral domain with quotient field K.*

The following assertions are equivalent:

1. *The ring A is agreeable; i.e, for each fractional ideal I of $A[X]$ with $I \subseteq K[X]$, there exists $0 \neq s \in A$ such that $sI \subseteq A[X]$.*
2. *If I is an integral ideal of $A[X]$, then $I = \frac{u(X)}{r} J$ where $u(X) \in A[X], 0 \neq r \in A$ and J is an integral ideal of $A[X]$ with $J \cap A \neq (0)$.*
3. *If I is a fractional ideal of $A[X]$, then $I = \frac{u(X)}{v(X)} J$ where $u(X), v(X) \in A[X]$ and J is an integral ideal of $A[X]$ with $J \cap A \neq (0)$.*
4. *For each $0 \neq f(X) \in K[X]$ (or even just $0 \neq f(X) \in A[X]$), there exists an integral ideal J of $A[X]$ (with $J \cap A \neq (0)$) and $0 \neq r \in A$ such that $f(X)K[X] \cap A[X] = f(X)\frac{1}{r} J$.*
5. *The ring A satisfies the property $(*)$: For each $0 \neq f(X) \in K[X]$ (or even just $0 \neq f(X) \in A[X]$), there exists $0 \neq s \in A$ such that for each $g(X) \in K[X]$ satisfying $f(X)g(X) \in A[X]$, $sg(X) \in A[X]$.*
6. *For each $0 \neq f(X) \in K[X]$ (or even just $0 \neq f(X) \in A[X]$), there exists $0 \neq s \in A$ such that $f(X)A[X] \cap sdA[X] \subseteq df(X)A[X]$ for each $d \in A$.*

7. *The ring $A[X]$ is almost principal; i.e, for each nonzero integral ideal I of $A[X]$*
   such that $I K[X] \neq K[X]$, there exists a nonconstant polynomial $f(X) \in I$ and
   $0 \neq s \in A$ such that $sI \subseteq f(X)A[X]$.

Proof $''(1) \Longrightarrow (2)''$ Let I be an integral ideal of $A[X]$. If $I \cap A \neq (0)$ or $I = (0)$,
the result is clear. Suppose that $I \cap A = (0)$ and $I \neq (0)$. Since $K[X]$ is a PID,
then $I K[X] = f(X)K[X]$ with $0 \neq f(X) \in I$. Let $E = [I : f(X)]_{K[X]}$.
Then E is stable by addition and by multiplication by the elements of $A[X]$ and
$f(X)E \subseteq I \subseteq A[X]$. Then E is a fractional ideal of $A[X]$ contained in $K[X]$. In
fact, $I = f(X)E$ since if $g(X) \in I \subseteq I K[X] = f(X)K[X]$, there exists $h(X) \in$
$K[X]$ such that $g(X) = f(X)h(X)$ and then $h(X) \in [I : f(X)]_{K[X]} = E$. By
(1), there exists $0 \neq r \in A$ such that $J = rE \subseteq A[X]$. Then $I = f(X)E =$
$\frac{f(X)}{r}rE = \frac{f(X)}{r}J$. Since $f(X)K[X] = I K[X] = \frac{f(X)}{r}JK[X] = f(X)JK[X]$,
by simplification by $f(X)$, we find $K[X] = JK[X]$. Then $J \cap K \neq (0)$. But
$J \subseteq A[X]$, so $J \cap A \neq (0)$.

$''(2) \Longrightarrow (3)''$ Let I be a fractional ideal of $A[X]$. There exists $0 \neq v(X) \in A[X]$
such that $v(X)I = L \subseteq A[X]$ and then $I = \frac{1}{v(X)}L$. If $L \cap A \neq (0)$, we have the
result. If not, by (2), there exist $0 \neq r \in A$, $u(X) \in A[X]$ and an integral ideal J
of $A[X]$ with $J \cap A \neq (0)$ such that $L = \frac{u(X)}{r}J$. Then $I = \frac{1}{v(X)}L = \frac{u(X)}{rv(X)}J$.

$''(3) \Longrightarrow (4)''$ Let $0 \neq f(X) \in A[X]$. We apply (3) to the ideal $f(X)K[X] \cap A[X]$
of $A[X]$. There exist $u(X), v(X) \in A[X]$ and an integral ideal J' of $A[X]$ with
$J' \cap A \neq (0)$ such that $f(X)K[X] \cap A[X] = \frac{u(X)}{v(X)}J'$. Then $\big(f(X)K[X] \cap$
$A[X]\big)K[X] = \big(\frac{u(X)}{v(X)}K[X]\big)\big(J'K[X]\big)$, so $f(X)K[X] = \frac{u(X)}{v(X)}K[X]$. Then
$\frac{u(X)}{v(X)} \in K[X]$, and it is associated with $f(X)$. There exist $a, r \in A \setminus (0)$ such that
$\frac{u(X)}{v(X)} = f(X)\frac{a}{r}$. The integral ideal $J = aJ'$ of $A[X]$ satisfies $J \cap A \neq (0)$ since
$J' \cap A \neq (0)$ and $f(X)K[X] \cap A[X] = \frac{u(X)}{v(X)}J' = f(X)\frac{a}{r}J' = f(X)\frac{1}{r}J$.
For $0 \neq f(X) \in K[X]$, there exists $0 \neq a \in A$ such that $af(X) \in A[X]$. By the
preceding particular case, $af(X)K[X] \cap A[X] = af(X)\frac{1}{r}J$ with $0 \neq r \in A$ and
J an integral ideal of $A[X]$ such that $J \cap A \neq (0)$. Then $f(X)K[X] \cap A[X] =$
$f(X)\frac{1}{r}(aJ)$, the ideal aJ has the same properties as J.

$''(4) \Longrightarrow (5)''$ Let $0 \neq f(X) \in A[X]$. By (4), there exist an integral ideal J of $A[X]$
and $0 \neq s \in A$ such that $f(X)K[X] \cap A[X] = f(X)\frac{1}{s}J$. Let $g(X) \in K[X]$ such
that $f(X)g(X) \in A[X]$. Then $f(X)g(X) \in f(X)K[X] \cap A[X] = f(X)\frac{1}{s}J$. By
simplification, $g(X) \in \frac{1}{s}J$ and then $sg(X) \in J \subseteq A[X]$.
For $0 \neq f(X) \in K[X]$, there exists $0 \neq a \in A$ such that $af(X) \in A[X]$. By the
preceding particular case, there exists $0 \neq s \in A$ such that for each $g(X) \in K[X]$
satisfying $af(X)g(X) \in A[X]$ and $sg(X) \in A[X]$. Let $g(X) \in K[X]$ satisfying
$f(X)g(X) \in A[X]$. Then $af(X)g(X) \in A[X]$, so $sg(X) \in A[X]$.

$''(5) \Longrightarrow (6)''$ Let $0 \neq f(X) \in A[X]$ and $0 \neq s \in A$ be as in (5). Let $0 \neq d \in A$
and $h(X) \in f(X)A[X] \cap sdA[X]$. Then $h(X) = f(X)g(X) = sdr(X)$ with
$g(X), r(X) \in A[X]$. So $f(X)\frac{g(X)}{sd} = r(X) \in A[X]$ with $\frac{g(X)}{sd} \in K[X]$. By
definition of s, $s\big(\frac{g(X)}{sd}\big) \in A[X]$, so $\frac{g(X)}{d} \in A[X]$. So $h(X) = f(X)g(X) =$
$df(X)\frac{g(X)}{d} \in df(X)A[X]$. Then $f(X)A[X] \cap sdA[X] \subseteq df(X)A[X]$.

For $0 \neq f(X) \in K[X]$, there exists $0 \neq a \in A$ such that $af(X) \in A[X]$. By the preceding particular case, there exists $0 \neq s \in A$ such that $af(X)A[X] \cap sdA[X] \subseteq daf(X)A[X]$ for each $d \in A$. But $af(X)A[X] \cap sdaA[X] \subseteq af(X)A[X] \cap sdA[X] \subseteq daf(X)A[X]$. After the simplification by a, we find $f(X)A[X] \cap sdA[X] \subseteq df(X)A[X]$ for each $d \in A$.

"(6) $\Longrightarrow$ (7)" Let I be a nonzero ideal of $A[X]$ such that $IK[X] \neq K[X]$. Then $I \cap A = (0)$. Since $K[X]$ is a PID, $IK[X] = f(X)K[X]$ with $f(X) \in I$ a nonconstant polynomial. Let $0 \neq s \in A$ be as in (6). Then $f(X)A[X] \cap sdA[X] \subseteq df(X)A[X]$ for each $d \in A$. Let $g(X) \in I \subseteq f(X)K[X]$ any element. Then $g(X) = f(X)\frac{h(X)}{d}$ with $h(X) \in A[X]$ and $0 \neq d \in A$. Then $sdg(X) = sf(X)h(X) \in f(X)A[X] \cap sdA[X] \subseteq df(X)A[X]$, so $sg(X) \in f(X)A[X]$. Then $sI \subseteq f(X)A[X]$.

"(7) $\Longrightarrow$ (1)" Let F be a fractional ideal of $A[X]$ contained in $K[X]$. Then $F = \frac{1}{f(X)}I$ with I an integral ideal of $A[X]$ and $0 \neq f(X) \in A[X]$. If $f(X)$ is a constant or $I = (0)$, the result is clear. Suppose that $f(X)$ is not constant and $I \neq (0)$. Then $I = f(X)F \subseteq f(X)K[X] \neq K[X]$. Let $I' = f(X)K[X] \cap A[X]$. Then $I \subseteq I'$. Since $I'K[X] \subseteq f(X)K[X]$, then $I'K[X] \neq K[X]$. We apply (7) to the ideal I'. There exist $0 \neq g(X) \in I'$ and $0 \neq s \in A$ such that $sI' \subseteq g(X)A[X]$. Then $I'K[X] = sI'K[X] \subseteq g(X)K[X] \subseteq I'K[X] \subseteq f(X)K[X] \subseteq I'K[X]$, so $g(X)K[X] = f(X)K[X]$. Then $g(X)$ and $f(X)$ are associated with $K[X]$. Put $g(X) = \frac{a}{b}f(X)$ with $a, b \in A \setminus (0)$. Since $I \subseteq I'$, $sI \subseteq sI' \subseteq g(X)A[X] = \frac{a}{b}f(X)A[X]$. Then $bs\frac{1}{f(X)}I \subseteq aA[X]$, so $bsF \subseteq aA[X] \subseteq A[X]$. □

2 Applications of Anderson-Kwak-Zafrullah Theorem

The application of Anderson-Kwak-Zafrullah theorem allows us to find sufficient conditions under which the ring $A[X]$ is almost principal.

Lemma 2.1 *Let $A \subseteq B$ be an extension of rings. The set $(A : B) = \{x \in B; \ xB \subseteq A\}$ is the biggest common ideal of A and B, called the conductor of A in B.*

Proof It is clear that $(A : B)$ contains 0 and is stable by addition. For each $x \in (A : B)$, $x = x.1 \in xB \subseteq A$ and then $(A : B) \subseteq A$. Let $x \in (A : B)$ and $y \in B$. Then $(xy)B = x(yB) \subseteq xB \subseteq A$, so $xy \in (A : B)$. Then $(A : B)$ is an ideal of B, so an ideal of A. Let I be a common ideal of A and B. For each $x \in I$, $xB \subseteq I \subseteq A$ and then $x \in (A : B)$. So $I \subseteq (A : B)$ and then $(A : B)$ is the biggest common ideal of A and B. □

Example The ring $A = \mathbb{Q} + X\mathbb{C}[[X]]$ has the same conductor $C = X\mathbb{C}[[X]]$ in both rings $B_1 = \mathbb{R} + X\mathbb{C}[[X]]$ and $B_2 = \mathbb{C}[[X]]$. Indeed, $X\mathbb{C}[[X]]$ is a maximal ideal in the three rings since $A/X\mathbb{C}[[X]] \simeq \mathbb{Q}$, $B_1/X\mathbb{C}[[X]] \simeq \mathbb{R}$, and $B_2/X\mathbb{C}[[X]] \simeq \mathbb{C}$.

Proposition 2.2 *Let $A \subseteq B$ be an extension of integral domains such that the conductor $(A : B) \neq (0)$. Then $A[X]$ is almost principal if and only if $B[X]$ is almost principal.*

Proof Since $(A : B) \neq (0)$, B is an over ring of A. Let K be their common quotient field and $0 \neq t \in (A : B)$. We use condition (5) of Theorem 1.10.

"$\implies$" Let $0 \neq f(X) \in K[X]$. There exists $0 \neq s \in A$ such that for each $g(X) \in K[X]$, satisfying $f(X)g(X) \in A[X]$, $sg(X) \in A[X]$.
   Let $g(X) \in K[X]$ satisfying $f(X)g(X) \in B[X]$. Then $f(X)\big(tg(X)\big) = t(f(X)g(X)) \in A[X]$, so $s(tg(X)) \in A[X] \subseteq B[X]$, then $(st)g(X) \in B[X]$ with $0 \neq st \in B$.

"$\impliedby$" Let $0 \neq f(X) \in K[X]$. There exists $0 \neq s \in B$ such that for each $g(X) \in K[X]$, satisfying $f(X)g(X) \in B[X]$ and $sg(X) \in B[X]$.
   Let $g(X) \in K[X]$ satisfying $f(X)g(X) \in A[X] \subseteq B[X]$. Then $sg(X) \in B[X]$. But by definition, $0 \neq ts \in A$ and $(ts)g(X) = t\big(sg(X)\big) \in tB[X] \subseteq A[X]$. $\square$

Example 1 Let A be an integral domain and B an over ring of A which is a finitely generated $A-$module. Then the ring $A[X]$ is almost principal if and only if $B[X]$ is almost principal.

Example 2 The hypothesis on the conductor is necessary in the proposition. Let s and t be the two indeterminates over a field F, $A = F[s, st]$, and $B = F\big[s; \{st^{2^n}, n \in \mathbb{N}\}\big]$. Then $A \subset B \subset F[s, t]$. By Example 2 of Corollary 1.8, $A[X]$ is almost principal. But $B[X]$ is not almost principal, by Example of Remark 1.6. In fact, there does not exist $0 \neq f(s, st) \in F[s, st]$ such that $f(s, st)t^{2^n} \in F[s, st]$ for each $n \in \mathbb{N}$.

Recall that a Mori ring is an integral domain A that satisfies the acc for integral divisorial ideals. This is equivalent to the fact that for each nonzero integral ideal I of A there is a nonzero finitely generated ideal $J \subseteq I$ such that $I^{-1} = J^{-1}$.

Indeed, suppose that A is a Mori domain, and let Ω be the set of the nonzero finitely generated subideals of I. Since A is a Mori domain, the set $\Omega' = \{L_v; L \in \Omega\}$, formed by integral divisorial ideals, admits a maximal element J_v with $J \in \Omega$. Note that J^{-1} is a minimal element of the set $\Omega'' = \{L^{-1}; L \in \Omega\}$. Suppose that $I \nsubseteq J_v$ and let $b \in I \setminus J_v$. The ideal $J' = J + bA \subseteq I$ and is finitely generated. Then $J' \in \Omega$ and $J'^{-1} \in \Omega''$. We have $J'^{-1} = (J + bA)^{-1} = J^{-1} \bigcap \frac{1}{b}A \subseteq J^{-1}$. By the minimality of J^{-1} in Ω'', we have $J'^{-1} = J^{-1}$. Then $J^{-1} \subseteq \frac{1}{b}A$, so $bJ^{-1} \subseteq A$. This means that $b \in (J^{-1})^{-1} = J_v$, which is absurd. We conclude that $I \subseteq J_v$. Then $(J_v)^{-1} \subseteq I^{-1}$. By Example 4 at the beginning of Chap. 1, the ideal J^{-1} is divisorial. So $(J_v)^{-1} = (J^{-1})_v = J^{-1}$. Then $J^{-1} \subseteq I^{-1}$. On the other hand, since $J \subseteq I$, then $I^{-1} \subseteq J^{-1}$. So $I^{-1} = J^{-1}$.

Conversely, let $(I_i)_{i \in \mathbb{N}}$ be an increasing sequence of integral divisorial ideals of A and $I = \bigcup_{i=0}^{\infty} I_i$. By hypothesis, there exists a finitely generated ideal $J =$

$b_1 A + \ldots + b_k A \subseteq I$ such that $J^{-1} = I^{-1}$. Let $m \in \mathbb{N}$ be such that $b_1, \ldots, b_k \in I_m$. For each integer $n \geq m$, we have $J \subseteq I_n$ aand then $I_n^{-1} \subseteq J^{-1} = I^{-1}$, but $I_n \subseteq I$, so $I^{-1} \subseteq I_n^{-1}$, and then $I_n^{-1} = I^{-1}$. Since I_n is divisorial, then $I_n = (I^{-1})^{-1}$. So $I_m = I_{m+1} = \cdots$.

Proposition 2.3 *Let A be an integral domain. If the polynomial ring $A[X]$ is a Mori domain, then it is almost principal.*

Proof We use condition (5) of Theorem 1.10. Let K be the quotient field of A and $0 \neq f(X) \in A[X]$. We will show first that $I = f(X)K[X] \cap A[X]$ is an integral t−ideal of $A[X]$. Let J be a nonzero finitely generated subideal of I. Then $J_v \subseteq I$. Indeed, put $J = \big(f(X)g_1(X), \ldots, f(X)g_n(X)\big)$ with $g_1(X), \ldots, g_n(X) \in K[X]$. Let $0 \neq d \in A$ such that $dg_1(X), \ldots, dg_n(X) \in A[X]$. Then $dJ = \big(f(X)dg_1(X), \ldots, f(X)dg_n(X)\big) \subseteq f(X)A[X]$. By applying v, we find $dJ_v \subseteq f(X)A[X]$. Then $J_v \subseteq J_v K[X] \cap A[X] = (dJ_v K[X]) \cap A[X] \subseteq f(X)K[X] \cap A[X] = I$. We conclude that I is an integral t−ideal of $A[X]$. Since $A[X]$ is a Mori ring, there exist a nonzero finitely generated subideal H of I such that $I^{-1} = H^{-1}$, then $I_v = H_v$. But t and v coincide over a Mori domain, then $I = H_v$. Put $H = \big(f(X)h_1(X), \ldots, f(X)h_m(X)\big)$ with $h_1(X), \ldots, h_m(X) \in K[X] \setminus (0)$. There exists $0 \neq s \in A$ such that $sh_1(X), \ldots, sh_m(X) \in A[X]$. Let now $g(X) \in K[X]$ such that $f(X)g(X) \in A[X]$. Then $f(X)g(X) \in f(X)K[X] \cap A[X] = I = H_v = \big(f(X)h_1(X), \ldots, f(X)h_m(X)\big)_v = f(X)\big(h_1(X), \ldots, h_m(X)\big)_v$. After simplification, $g(X) \in \big(h_1(X), \ldots, h_m(X)\big)_v$ and $sg(X) \in \big(sh_1(X), \ldots, sh_m(X)\big)_v \subseteq A[X]$. □

Example Let A be a Krull domain. Since $A[X]$ is a Krull domain, then it is a Mori ring and so it is almost principal.

Lemma 2.4 *Let $A \subseteq K$ be an extension of rings with K a field and $\overline{A}$ the integral closure of A in K. If $g(X), h(X) \in K[X]$, with $h(X)$ a monic polynomial, are such that $g(X)h(X) \in A[X]$ and then $g(X) \in \overline{A}[X]$.*

Proof Put $g(X)h(X) = f(X) = \sum_{i:0}^{n} a_i X^i \in A[X]$ a polynomial of degree n. If $h(X)$ is constant, $h(X) = 1$, and then $g(X) = f(X) \in A[X]$. If the degree of $h(X)$ is 1, $h(X) = X - s$ with $s \in K$. We will prove, by induction on n, that the coefficients of $g(X)$ belong to $\overline{A}$. Put $g(X) = \sum_{i:0}^{n-1} b_i X^i \in K[X]$. If $n = 1$, $g(X) = b_0$, and the equality $g(X)h(X) = f(X)$ becomes $b_0(X - s) = a_0 + a_1 X$, then $b_0 = a_1 \in A$. Suppose that the lemma is true for $n = k$, i.e., $deg\ g(X) = k - 1$. We prove the property for $n = k+1$. In this case, $f(X) = (X - s)\big(\sum_{i:0}^{k} b_i X^i\big) = \sum_{i:0}^{k+1} a_i X^i$, and then $\displaystyle\sum_{i:0}^{k} b_i X^{i+1} - \sum_{i:0}^{k} sb_i X^i = \sum_{i:0}^{k+1} a_i X^i$, so $\displaystyle\sum_{i:1}^{k+1} b_{i-1} X^i - \sum_{i:0}^{k} sb_i X^i = \sum_{i:0}^{k+1} a_i X^i$.

For $1 \le i \le k$, $b_{i-1} - sb_i = a_i$, $-sb_0 = a_0$, and $b_k = a_{k+1}$. Since the leading coefficient of $f(X) \in A[X]$ is a_{k+1}, $s \in K$, and $f(s) = 0$, then sa_{k+1} is integral over A. Then $sb_k = sa_{k+1}$ is also integral over A. So $b_{k-1} = sb_k + a_k$ is integral over A. Then the extension $A \subseteq A[b_{k-1}]$ is integral. On the other hand, $(X - $

$\displaystyle s)\left(\sum_{i:0}^{k-1} b_i X^i\right) = \sum_{i:0}^{k-1} b_i X^{i+1} - \sum_{i:0}^{k-1} sb_i X^i = \sum_{i:1}^{k} b_{i-1} X^i - \sum_{i:0}^{k-1} sb_i X^i = \sum_{i:1}^{k-1} (b_{i-1} -$

$\displaystyle sb_i)X^i - sb_0 + b_{k-1}X^k = \sum_{i:0}^{k-1} a_i X^i + b_{k-1}X^k \in A[b_{k-1}][X] \subseteq K[X].$

By the induction hypothesis, the coefficients $b_0, b_1, \ldots, b_{k-1}$ are integral over the

ring $A[b_{k-1}]$, so over A. Since $b_k = a_{k+1} \in A$, then $g(X) = \displaystyle\sum_{i:0}^{k} b_i X^i \in \overline{A}[X]$.

We conclude that $g(X) \in \overline{A}[X]$ when the degree of $h(X)$ is 1. Suppose that the property is true for $deg\ h(X) = k$. Consider the case when $f(X) = g_1(X)h_1(X) \in A[X]$ with $g_1(X), h_1(X) \in K[X]$, and $h_1(X)$ is monic of degree $k+1$. Let t be a root of $h_1(X)$ in an algebraic closure $\overline{K}$ of K. Then $h_1(X) = (X-t)h_2(X)$ with $h_2(X) \in \overline{K}[X]$ is a monic polynomial of degree k. Since $f(X) = \big((X - t)g_1(X)\big)h_2(X)$, by induction hypothesis, the coefficients of $(X - t)g_1(X)$ are integral over A. Let A_1 be the sub-ring of $\overline{K}$ obtained by adjunct the coefficients of $(X - t)g_1(X)$ to A. By construction, the extension $A \subseteq A_1$ is integral. By the case when the monic factor is linear, the coefficients of $g_1(X)$ are integral over A_1 and then over A. Since $g_1(X) \in K[X]$, then $g_1(X) \in \overline{A}[X]$. $\square$

Example 1 Let A be an integral domain with quotient field K and integral closure $\overline{A}$. If $f(X)$ and $g(X) \in K[X]$ are two monic polynomials such that $f(X)g(X) \in A[X]$, then $f(X)$ and $g(X) \in \overline{A}[X]$.

Example 2 The hypothesis that $f(X)$ and $g(X)$ are monic is necessary in the preceding example. Indeed, the domain $\mathbb{Z}$ is integrally closed with quotient field $\mathbb{Q}$. The polynomials $f(X) = 3X + \frac{9}{2}$ and $g(X) = 4X^2 - \frac{2}{3}X - \frac{4}{3} \in \mathbb{Q}[X] \setminus \mathbb{Z}[X]$. But $f(X)g(X) = 12X^3 + 16X^2 - 13X \in \mathbb{Z}[X]$.

Example 3 With the same notations and hypotheses of Example 1, the polynomials $f(X)$ and $g(X)$ do not belong necessarily to $A[X]$. Indeed, the quotient field of the integral domain $A = \mathbb{Z}[\sqrt{5}] = \{a + b\sqrt{5};\ a, b \in \mathbb{Z}\}$ is $K = \mathbb{Q}(\sqrt{5})$. The golden number $\theta = \frac{1+\sqrt{5}}{2} \in K \setminus A$ and satisfies $\theta^2 - \theta - 1 = 0$. Then θ is integral over A. Note that $(X-\theta)(X+\theta-1) = X^2 - X - 1 \in A[X]$. But $X - \theta$ and $X + \theta - 1 \notin A[X]$.

Proposition 2.5 (Johnson 1982) *Let A be an integrally closed domain. The ring $A[X]$ is almost principal.*

Proof We use condition (5) of Theorem 1.10. Let $0 \neq f(X) \in A[X]$ and $0 \neq a \in A$ its leading coefficient. Let $g(X) \in K[X]$ with $f(X)g(X) \in A[X]$, where K is the quotient field of A. Then $(\frac{1}{a}f(X))(ag(X)) \in A[X]$ where $\frac{1}{a}f(X), ag(X) \in K[X]$ with $\frac{1}{a}f(X)$ is monic. By the preceding lemma, $ag(X) \in \overline{A}[X] = A[X]$, where $\overline{A}$ is the integral closure of A.   □

Example 1 Let $\phi : A \longrightarrow B$ be a homomorphism onto of integral domains. If the ring $A[X]$ is almost principal, the ring $B[X]$ is not necessarily almost principal.

Let s and t be the two indeterminates over a field F and $B = F\big[s; \{st^{2^n}, n \in \mathbb{N}\}\big]$. The ring $B[X]$ is not almost principal. Let $\{X_n;\ n \in \mathbb{N}\}$ be a sequence of indeterminates over F. The ring $A = F[X_n;\ n \in \mathbb{N}]$ is UFD and then integrally closed. By the preceding proposition, $A[X]$ is almost principal. The map $\phi : A \longrightarrow B$, defined by $\phi\big(f(X_0, X_1, \ldots, X_n)\big) = f(s, st, st^2, \ldots, st^{2^{n-1}})$, is a homomorphism onto.

Example 2 (Houston 1994) Let s and t be the two indeterminates over a field F and $A = F\big[s; \{st^{2^n}, n \in \mathbb{N}\}\big]$. By the example of Remark 1.6, the ring $A[X]$ is not almost principal. We will show that the integral closure of A is $B = F[st^n; n \in \mathbb{N}] \neq A$ since $st^n \notin A$ for $\phi(n) > 1$, where $\phi(n)$ is the number of 1 in the development in binary basis of the integer n. The quotient field of A is $K = F(s, t)$ and $A \subseteq B \subset F[s, t] \subset K$ with $F[s, t]$ integrally closed. For each $n \in \mathbb{N}$, choose $k \in \mathbb{N}$ such that $2^k > n$ and then $(st^n)^{2^k} = s^{2^k}t^{n2^k} = s^{2^k-n}(st^{2^k})^n \in A$. Then the extension $A \subseteq B$ is integral. It suffices to show that B is integrally closed in $F[s, t]$. Let $f(s, t) \in F[s, t]$ be an integral element over B. Then $f(s, t)^k + a_{k-1}(s, t)f(s, t)^{k-1} + \ldots + a_1(s, t)f(s, t) + a_0(s, t) = 0$ with $a_i(s, t) \in B$, so $a_i(0, t) \in F$. Suppose that $f(s, t) \notin B$ and then $f(0, t) \in F[t] \setminus F$ and satisfies $f(0, t)^k + a_{k-1}(0, t)f(0, t)^{k-1} + \ldots + a_1(0, t)f(0, t) + a_0(0, t) = 0$. This is absurd for degree reasons. By the preceding proposition, the ring $B[X]$ is almost principal.

Corollary 2.6 *Let A be an integral domain with integral closure $\overline{A}$. If the conductor $(A : \overline{A}) \neq (0)$, the ring $A[X]$ is almost principal.*

Proof By the preceding proposition, $\overline{A}[X]$ is almost principal. By Proposition 2.2, since $(A : \overline{A}) \neq (0)$, the ring $A[X]$ is almost principal.   □

Example Let s and t be the two indeterminates over a field F and $A = F\big[s; \{st^{2^n}, n \in \mathbb{N}\}\big]$. The integral closure of A is $\overline{A} = F[st^n; n \in \mathbb{N}]$. The conductor $(A : \overline{A}) = (0)$. Indeed, there does not exist $0 \neq f(s, t) \in A$ such that $f(s, t)st^n \in A$ for each $n \in \mathbb{N}$.

Proposition 2.7 *Let A be an integral domain and S a multiplicative set of A. If the ring $A[X]$ is almost principal, then the ring $(S^{-1}A)[X]$ is also almost principal.*

Proof We use condition (5) of Theorem 1.10. We have $A \subseteq S^{-1}A \subseteq K$, the quotient field of A. Let $0 \neq f(X) \in K[X]$. There exists $0 \neq s \in A \subseteq S^{-1}A$ such that $sg(X) \in A[X]$ for each $g(X) \in K[X]$ satisfying $f(X)g(X) \in A[X]$. Let $g(X) \in K[X]$ such that $f(X)g(X) \in S^{-1}A[X]$. There exists $0 \neq t \in S$ such that $f(X)(tg(X)) \in A[X]$ and then $stg(X) \in A[X]$, so $sg(X) \in \frac{1}{t}A[X] \subseteq S^{-1}A[X]$.
                                                                                                                                                                       □

Corollary 2.8 *Let $A \subseteq B$ be an extension of integral domains with $B = \bigcap_{\alpha \in \Gamma} A_{P_\alpha}$, where $(P_\alpha)_{\alpha \in \Gamma}$ is a family of prime ideals of A. If the ring $A[X]$ is almost principal, then the ring $B[X]$ is also almost principal.*

Proof Let $0 \neq f(X) \in K[X]$, where K is the common quotient field of A and B. By hypothesis, there exists $0 \neq s \in A$ such that $sg(X) \in A[X]$ for each $g(X) \in K[X]$ satisfying $f(X)g(X) \in A[X]$. Let $g(X) \in K[X]$ satisfying $f(X)g(X) \in B[X] \subseteq A_{P_\alpha}[X]$ for each $\alpha \in \Gamma$. By the proof of the preceding proposition, $sg(X) \in A_{P_\alpha}[X]$ pour tout $\alpha \in \Gamma$. Then $sg(X) \in \bigcap_{\alpha \in \Gamma}(A_{P_\alpha}[X]) = (\bigcap_{\alpha \in \Gamma} A_{P_\alpha})[X] = B[X]$.
                                                                                                                                                                       □

Corollary 2.9 *Let A be a Noetherian integral domain and $(P_\alpha)_{\alpha \in \Gamma}$ a family of prime ideals of A. For $B = \bigcap_{\alpha \in \Gamma} A_{P_\alpha}$, the ring $B[X]$ is almost principal.*

Proof By Corollary 1.8, the ring $A[X]$ is almost principal. By the preceding corollary, the ring $B[X]$ is almost principal.   □

Note that in the preceding corollary, the domain B is not necessarily Noetherian. In the following lemma, we generalize the notion of ring of fractions.

Lemma 2.10 *Let A be an integral domain with quotient field K and $\mathscr{S}$ a nonempty family of nonzero integral ideals of A stable by multiplication. The set $A_\mathscr{S} = \bigcup_{I \in \mathscr{S}} I^{-1} = \{x \in K; \exists I \in \mathscr{S}, xI \subseteq A\}$ is an over ring of A, called the $\mathscr{S}$-transform of A.*

Proof For all $x \in A$ and $I \in \mathscr{S}$, $xI \subseteq A$, then $A \subseteq A_\mathscr{S} \subseteq K$. Let $x, y \in A_\mathscr{S}$, there exist $I, J \in \mathscr{S}$ such that $xI \subseteq A$ and $yJ \subseteq A$. By hypothesis, $IJ \in \mathscr{S}$, and we have $(x+y)(IJ) = (xI)J + I(yJ) \subseteq A$ and $(xy)(IJ) = (xI)(yJ) \subseteq A$. Then $x + y, xy \in A_\mathscr{S}$, so $A_\mathscr{S}$ is a sub-ring of K.   □

Example 1 Let I be a nonzero integral ideal of an integral domain A. The set $\mathscr{S} = \{I^n; \ n \in \mathbb{N}^*\}$ is stable by multiplication. The domain $A_\mathscr{S} = \bigcup_{n \in \mathbb{N}^*} I^{-n}$, with $I^{-n} =$

$(A : I^n)$, is called Nagata transform of the ring A with respect to the ideal I and it is noted $T_A(I)$.

Example 2 Let S be a multiplicative set of an integral domain A. The set of principal ideals $\mathscr{S} = \{sA; \ s \in S\}$ is stable by multiplication, and we have $A_{\mathscr{S}} = S^{-1}A$. Indeed, let $x \in A_{\mathscr{S}}$. There exists $s \in S$ such that $sAx \subseteq A$ and then $sx \in A$. So $x \in \frac{1}{s}A \subseteq S^{-1}A$. Conversely, let $x \in S^{-1}A$, $x = \frac{a}{s}$ with $a \in A$ and $s \in S$. Then $sx \in A$, so $sAx \subseteq A$. Then $x \in A_{\mathscr{S}}$.

Example 3 Let A be an integral domain with quotient field K, $0 \neq a \in A$ and $I = aA$. The Nagata transform $T_A(I) = \{x \in K; \ \exists n \in \mathbb{N}^*, xa^n A \subseteq A\} = \{x \in K; \ \exists n \in \mathbb{N}^*, x \in \frac{1}{a^n}A\} = A[\frac{1}{a}] = S^{-1}A$, with $S = \{a^n; n \in \mathbb{N}\}$.

Example 4 Let A be an integral domain and $\mathscr{S}$ a family of nonzero integral ideals of A stable by multiplication. The set $\overline{\mathscr{S}} = \{J$ integral ideal of $A; \ \exists I \in \mathscr{S}, I \subseteq J\}$ is also stable by multiplication and contains $\mathscr{S}$. Moreover, $A_{\overline{\mathscr{S}}} = A_{\mathscr{S}}$. Indeed, since $\mathscr{S} \subseteq \overline{\mathscr{S}}$, then $A_{\mathscr{S}} \subseteq A_{\overline{\mathscr{S}}}$. Conversely, let $x \in A_{\overline{\mathscr{S}}}$. There exists $J \in \overline{\mathscr{S}}$ such that $xJ \subseteq A$. But there exists $I \in \mathscr{S}$ such that $I \subseteq J$ and then $xI \subseteq A$ so $x \in A_{\mathscr{S}}$. We say that $\overline{\mathscr{S}}$ is the saturation of $\mathscr{S}$, and if $\overline{\mathscr{S}} = \mathscr{S}$ we say that $\mathscr{S}$ is saturated.

Example 5 (Arnold and Brewer 1971) Let B be an over ring of an integral domain A. Suppose that B is a flat $A-$module; i.e., for each $x \in B$, the ideal $(A :_A x)B = B$. Then there exists a nonempty family $\mathscr{S}$ of nonzero integral ideals of A stable by multiplication such that $B = A_{\mathscr{S}}$ and $IB = B$ for each $I \in \mathscr{S}$. In this case, if Q is a prime ideal of B and $P = Q \cap A$, then $Q = P_{\mathscr{S}}$; i.e, $Q = \{x \in B; \ \exists I \in \mathscr{S}, xI \subseteq P\}$ and we have $B_Q = A_P$.

Indeed, let $\mathscr{P}$ be the set of the prime ideals P of A such that $PB \neq B$ and $\mathscr{S}$ the set of the ideals I of A such that $I \not\subseteq P$ for each $P \in \mathscr{P}$. The two sets are both nonempty. For example, $0 \in \mathscr{P}$ and $A \in \mathscr{S}$. We will prove that $\mathscr{S}$ is stable by multiplication. Let $I, J \in \mathscr{S}$. Suppose that there exists $P \in \mathscr{P}$ such that $IJ \subseteq P$ and then either $I \subseteq P$ or $J \subseteq P$, which is absurd. Then $IJ \in \mathscr{S}$. Suppose that there exists $I \in \mathscr{S}$ such that $IB \neq B$. There exists $P' \in spec(B)$ such that $IB \subseteq P'$. Then $I \subseteq P' \cap A$ with $P' \cap A \in spec(A)$. But $(P' \cap A)B \subseteq P' \neq B$ and then $P' \cap A \in \mathscr{P}$, which contradicts $I \in \mathscr{S}$. Then $IB = B$ for each $I \in \mathscr{S}$. We will show that $B = A_{\mathscr{S}}$. Let $x \in A_{\mathscr{S}}$. There exists $I \in \mathscr{S}$ such that $xI \subseteq A$ and then $xIB \subseteq B$. But $IB = B$ and then $xB \subseteq B$, so $x \in B$. Conversely, let $x \in B$. Since B is a flat $A-$module, $(A :_A x)B = B$. The ideal $(A :_A x) \not\subseteq P$ for each $P \in \mathscr{P}$ because if not, $PB = B$, which is absurd. Then $(A :_A x) \in \mathscr{S}$, and since $x(A :_A x) \subseteq A$, then $x \in A_{\mathscr{S}}$. So $B = A_{\mathscr{S}}$. Let Q be a prime ideal of B and $P = Q \cap A$. We will show that $Q = P_{\mathscr{S}}$. Let $x \in Q \subseteq B$. There exists $I \in \mathscr{S}$ such that $xI \subseteq A$ and then $xI \subseteq A \cap Q = P$, so $x \in P_{\mathscr{S}}$. Conversely, let $x \in P_{\mathscr{S}}$. There exists $I \in \mathscr{S}$ such that $xI \subseteq P$. Then $xIB \subseteq PB \subseteq Q$. But $IB = B$ and then $xB \subseteq Q$, so $x \in Q$. Then $Q = P_{\mathscr{S}}$. We will show that $B_Q = A_P$. Let $x \in A_P$, $x = \frac{a}{s}$ with $a \in A$ and $s \in A \setminus P$. Then $s \notin Q$ because if not, $s \in Q \cap A = P$, which is absurd. Then $x = \frac{a}{s} \in B_Q$ so $A_P \subseteq B_Q$. Conversely, let $x \in B_Q$, $x = \frac{b}{s}$ with $b \in B$ and $s \in B \setminus Q$. Since $B = A_\mathscr{S}$, there exist I_1 and $I_2 \in \mathscr{S}$ such that $bI_1 \subseteq A$ and $sI_2 \subseteq A$. Since $\mathscr{S}$ is stable by multiplication, $I_1I_2 \in \mathscr{S}$ and then $(I_1I_2)B = B$. So $I_1I_2 \nsubseteq P$ because if not, $I_1I_2 \subseteq P$ and then $B = (I_1I_2)B \subseteq PB \subseteq Q$, so $Q = B$, which is absurd. Let $t \in I_1I_2 \setminus P$. Then $bt \in (bI_1)I_2 \subseteq AI_2 = I_2 \subseteq A$ and $st \in sI_1I_2 \subseteq sI_2 \subseteq A$. Since $t \in A \setminus P$, then $t \notin Q$, so $st \notin Q$ and then $st \notin P$. So $st \in A \setminus P$ and $x = \frac{b}{s} = \frac{bt}{st} \in A_P$; then $B_Q \subseteq A_P$.

Example 6 Let S be a multiplicative set of an integral domain A. Then A_S is a flat A−module. Indeed, let $x \in A_S$, $x = \frac{a}{s}$ with $a \in A$ and $s \in S$. Since $sx = a \in A$, then $s \in (A :_A x)$ with s an invertible element in A_S. Then $(A :_A x)A_S = A_S$. By Example 2, $A_S = A_\mathscr{S}$ with $\mathscr{S} = \{sA; s \in S\}$. For each $s \in S$, $sA.A_S = sA_S = A_S$.

Example 7 In this example, we give an integral domain A with an over ring B that is not a flat A−module. The domains $A = \mathbb{Z}[\sqrt{5}] \subset \mathbb{Z}[\frac{1+\sqrt{5}}{2}] = B$ have the same quotient field $\mathbb{Q}(\sqrt{5})$. The golden number $x = \frac{1+\sqrt{5}}{2}$ is integral over $\mathbb{Z}$ and satisfies $x^2 - x - 1 = 0$. Then B is a free $\mathbb{Z}$−module with basis $\{1, x\}$. The ideal $(A :_A x) = (2, 1 + \sqrt{5})A$. Indeed, $2x = 1 + \sqrt{5} \in A$ and $(1 + \sqrt{5})x = \frac{1}{2}(1+\sqrt{5})^2 = 3 + \sqrt{5} \in A$. Conversely, let $y = a + b\sqrt{5} \in (A :_A x)$ with $a, b \in \mathbb{Z}$. Then $yx = \frac{1}{2}(a + b\sqrt{5})(1 + \sqrt{5}) = \frac{1}{2}(a + 5b + (a + b)\sqrt{5}) \in A = \mathbb{Z}[\sqrt{5}]$, so $a + 5b \in 2\mathbb{Z}$ and $a + b \in 2\mathbb{Z}$. Since $a + 5b = a + b + 4b$, then $yx \in A \iff a + b \in 2\mathbb{Z} \iff a$ and b have the same parity. If $a = 2m$ and $b = 2n$ are even, then $y = 2(m + n\sqrt{5}) \in (2, 1 + \sqrt{5})A$. If $a = 2m + 1$ and $b = 2n + 1$ are odd, then $y = 2(m + n\sqrt{5}) + (1 + \sqrt{5}) \in (2, 1 + \sqrt{5})A$. We have the equality $(A :_A x) = (2, 1+\sqrt{5})A$. The ideal $(A :_A x)B = (2, 1+\sqrt{5})B = (2, 2x)B = 2B$. Suppose that $1 \in 2B$ and then $1 = 2(a + bx)$ with $a, b \in \mathbb{Z}$. Since B is a free $\mathbb{Z}$−module with basis $\{1, x\}$, then $1 = 2a$, which is absurd. We conclude that B is an over ring of A but B is not a flat A−module.

We generalize now Proposition 2.7.

Proposition 2.11 *Let A be an integral domain and $\mathscr{S}$ a nonempty family of nonzero integral ideals of A stable by multiplication. If the ring $A[X]$ is almost principal, then the ring $A_\mathscr{S}[X]$ is also almost principal.*

Proof Let K be the common quotient field of the domains A and $A_\mathscr{S}$. Let $0 \neq f(X) \in K[X]$. There exists $0 \neq s \in A$ such that $sg(X) \in A[X]$ for each $g(X) \in K[X]$ satisfying $f(X)g(X) \in A[X]$. Let $g(X) \in K[X]$ be such that $f(X)g(X) \in A_\mathscr{S}[X]$. There exists $I \in \mathscr{S}$ such that $If(X)g(X) \subseteq A[X]$ and then $f(X)(Ig(X)) \subseteq A[X]$. So $s(Ig(X)) \subseteq A[X]$ and then $sg(X) \subseteq A_\mathscr{S}[X]$. $\square$

Proposition 2.12 *Let $A = \bigcap\limits_{\alpha \in \Gamma} A_\alpha$ be an intersection of finite character of over rings of an integral domain A. If for each $\alpha \in \Gamma$, the ring $A_\alpha[X]$ is almost principal, then the ring $A[X]$ is also almost principal.*

Proof Let $0 \neq f(X) \in A[X]$. By hypothesis, there exist $\alpha_1, \ldots, \alpha_n \in \Gamma$ such that for each $\alpha \in \Gamma \setminus \{\alpha_1, \ldots, \alpha_n\}$, $c(f)A_\alpha = A_\alpha$. For $1 \leq i \leq n$, there exists $s_i \in A_{\alpha_i}$ such that $s_i g(X) \in A_{\alpha_i}[X]$ for each $g(X) \in K[X]$ satisfying $f(X)g(X) \in A_{\alpha_i}[X]$, where K is the quotient field of A. By multiplying by a convenient nonzero element of A, we may suppose that $s_i \in A$. Let $0 \neq s_1 \ldots s_n = s \in A$. Then $sg(X) \in A_{\alpha_i}[X]$ for all $i \in \{1, \ldots, n\}$ and $g(X) \in K[X]$ satisfying $f(X)g(X) \in A_{\alpha_i}[X]$. Let $g(X) \in K[X]$ be such that $f(X)g(X) \in A[X]$. We will show that $sg(X) \in A[X]$. For $\alpha \in \Gamma \setminus \{\alpha_1, \ldots, \alpha_n\}$, we have $c(f)A_\alpha = A_\alpha$ and then $c(g)A_\alpha = (c(f)A_\alpha)(c(g)A_\alpha) = c(fg)A_\alpha$, by the content formula. But $f(X)g(X) \in A[X] \subseteq A_\alpha[X]$ and then $c(fg) \subseteq A_\alpha$. So $c(g) \subseteq A_\alpha$ and then $g(X) \in A_\alpha[X]$. So $sg(X) \in \bigcap\limits_{\alpha \in \Gamma} A_\alpha[X] = (\bigcap\limits_{\alpha \in \Gamma} A_\alpha)[X] = A[X]$. □

3 Characterization of A[X] Almost Principal by the Ideals f(X)K[X]∩ A[X]

The results of this section are due to Hamann et al. (1988). In the sequel, A is an integral domain with quotient field K. We start by examining the ideal $f(X)K[X] \cap A[X]$ itself. The following proposition shows that it is far from being equal to the principal ideal $f(X)A[X]$.

Proposition 3.1 (Dobbs et al. 1992) *Let $0 \neq f(X) \in A[X]$. Then $f(X)K[X] \cap A[X] = f(X)A[X]$ if and only if $c(f)_v = A$.*

Proof " $\Longrightarrow$ " It suffices to show that $(c(f))^{-1} = A$. Let $u \in (c(f))^{-1}$. Then $uc(f) \subseteq A$, so $uf(X) \in A[X]$ and then $uf(X) \in f(X)K[X] \cap A[X] = f(X)A[X]$. By simplification, $u \in A[X] \cap K = A$.

" $\Longleftarrow$ " The inclusion $f(X)A[X] \subseteq f(X)K[X] \cap A[X]$ is clear. Conversely, let $g(X) \in f(X)K[X] \cap A[X]$. Then $g(X) = f(X)h(X)$ with $h(X) \in K[X]$. By the content formula, there exists $n \in \mathbb{N}^*$ such that $c(f)^{n+1}c(h) = c(f)^n c(fh) = c(f)^n c(g) \subseteq A$. Then $\big(c(f)^{n+1}c(h)\big)_v \subseteq A$. By Proposition 11.1 of Chap. 1, $\big((c(f)_v)^{n+1}c(h)\big)_v \subseteq A$. Since $\big(c(f)\big)_v = A$, then $\big(c(h)\big)_v \subseteq A$, so $h(X) \in A[X]$. Then $g(X) = f(X)h(X) \in f(X)A[X]$. □

Example Let a be a nonzero noninvertible element of A and $f(X) = a + aX \in A[X]$. Then $c(f)_v = (aA)_v = aA \neq A$, so $f(X)K[X] \cap A[X] \neq f(X)A[X]$.

Corollary 3.2 (Lucas 2012) *Let $f(X) \in A[X]$ be a nonconstant polynomial and $I = f(X)K[X] \cap A[X]$. If the content $c(f)$ is a principal ideal of A, then I is a principal ideal of $A[X]$.*

Proof Put $c(f) = aA$ with $0 \neq a \in A$. Then $f(X) = ag(X)$ with $0 \neq g(X) \in A[X]$ and $c(f) = ac(g)$, so $c(g) = A$. Since $f(X)$ and $g(X)$ are associated with $K[X]$, then $f(X)K[X] = g(X)K[X]$. So $I = f(X)K[X] \cap A[X] = g(X)K[X] \cap A[X] = g(X)A[X]$, by the preceding proposition.   □

Example 1 The converse of the corollary is false. Let $f(X) \in A[X]$ be a nonconstant polynomial such that $c(f) \subset c(f)_v = A$. By Proposition 3.1, $f(X)K[X] \cap A[X] = f(X)A[X]$ is principal. But $c(f)$ is not principal since $c(f) \neq c(f)_v$. To have a concrete example, recall Example 1 at the beginning of Sect. 11, Chap. 1. Let s and t be the two indeterminates over a field k, $A = k[s, t]$, and $I = (s, t)A$. Then $I_v = A$. Consider now the polynomial $f(X) = s + tX \in A[X]$. Then $c(f) = I \subset I_v = A$.

Example 2 The ideal $f(X)K[X] \cap A[X]$ may be principal but different from $f(X)A[X]$. Indeed, with the notations of the proof of the corollary, suppose that $a \notin U(A)$. Then $I = g(X)A[X] \neq f(X)A[X]$ since $f(X)$ and $g(X)$ are not associated with $A[X]$.

Example 3 Let A be an integral domain with quotient field K and $t \in K$ an element that it is not quasi-integral over A, but for which there exists $0 \neq a \in A$ such that $at^i \in A$ for an infinity of natural integers. Put $t = \frac{b}{c}$ with $b, c \in A \setminus (0)$. By Proposition 1.5, the ideal $P = (X - t)K[X] \cap A[X] = (cX - b)K[X] \cap A[X]$ is not almost principal. By Proposition 1.7, it is not finitely generated. It is not even generated by a family of polynomials of bounded degrees.

In the following proposition, we characterize when an ideal of the form $f(X)K[X] \cap A[X]$ is almost principal. The polynomial $f(X)$ plays a crucial role.

Proposition 3.3 *Let $f(X) \in A[X]$ be a nonconstant polynomial and $I = f(X)K[X] \cap A[X]$. The following assertions are equivalent:*

1. *The ideal I is almost principal.*
2. *There exists $0 \neq s \in A$ such that for each $g(X) \in K[X]$ satisfying $fg \in A[X]$, we have $sg \in A[X]$.*
3. *There exists $0 \neq s \in A$ such that $sI \subseteq f(X)A[X]$.*

Proof "(1) $\Longrightarrow$ (2)" There exist $0 \neq s \in A$ and a nonconstant polynomial $f'X) \in I$ such that $sI \subseteq f'(X)A[X]$. Since $f(X) \in I$, then $sf(X) \in f'(X)A[X]$, so $f'(X)$ divides $f(X)$ in $K[X]$. Since $f'(X) \in I \subseteq f(X)K[X]$, then $f(X)$ divides $f'(X)$ in $K[X]$. Then $f(X)$ and $f'(X)$ are associated with $K[X]$. There

exist $a, b \in A \setminus (0)$ such that $af(X) = bf'(X)$. Let $g(X) \in K[X]$ be
such that $f(X)g(X) \in A[X]$. Then $f(X)g(X) \in f(X)K[X] \cap A[X] = I$,
so $sf(X)g(X) \in sI \subseteq f'(X)A[X]$, and then $bsf(X)g(X) \in bf'(X)A[X]$
$= af(X)A[X]$. By simplification, $bsg(X) \in aA[X] \subseteq A[X]$ with $0 \neq bs \in A$
an independent element of $g(X)$.

$''(2) \implies (3)''$ Let $h(X) \in I$, $h(X) = f(X)g(X)$ with $g(X) \in K[X]$. Then
$f(X)g(X) = h(X) \in I \subseteq A[X]$. By the hypothesis (2), $sg(X) \in A[X]$. Then
$sh(X) = f(X)sg(X) \in f(X)A[X]$. So $sI \subseteq f(X)A[X]$.

$''(3) \implies (1)''$ Clear.   □

Notations Let $f(X) \in A[X]$ be a nonconstant polynomial. Consider:

(1) The set $G_f = \{h(X) \in K[X]; \ f(X)h(X) \in A[X]\}$, a fractional ideal of $A[X]$.
    Indeed, the stability of G_f by addition and multiplication by the elements of
    $A[X]$ is clear. Moreover, by definition, $f(X)G_f \subseteq A[X]$.

(2) The set $H_f = \bigcup \{c(h); \ h(X) \in G_f\}$, a A−sub-module of K.
    Indeed, we use (1). For each $a \in A$ and $h(X) \in G_f$, $ah(X) \in G_f$ and $ac(h) = c(ah)$. Let $h_1(X), h_2(X) \in G_f$, and $k = deg \ h_1(X)+1$. Then $h(X) = h_1(X)+ X^k h_2(X) \in G_f$ and $c(h) = c(h_1) + c(h_2)$.

The following corollary shows that there is a link between the almost principal
ideals of $A[X]$ of the form $f(X)K[X] \cap A[X]$ and the fractional ideals of A.

Corollary 3.4 *Let $f(X) \in A[X]$ a nonconstant polynomial and $I = f(X)K[X] \cap A[X]$. The following assertions are equivalent:*

1. *The ideal I is almost principal.*
2. *There exists $0 \neq s \in A$ such that $sG_f \subseteq A[X]$.*
3. *The A−sub-module H_f of K is a fractional ideal of A.*

Proof We rewrite condition (2) of the proposition in terms of G_f and H_f.   □

To prove the principal result of this section, we need the following two lemmas.

Lemma 3.5

1. *Let $f_1(X)$ and $f_2(X) \in A[X]$ be the two nonconstant polynomials.*
   Put $f(X) = f_1(X)f_2(X)$, $I = f(X)K[X]\cap A[X]$, and $I_i = f_i(X)K[X]\cap A[X]$, $i = 1, 2$.
   The ideal I is almost principal if and only if the ideals I_1 and I_2 are almost principal.
2. *Let $f(X)$ and $f_1(X) \in A[X]$ be the two nonconstant polynomials. Consider the ideals $I = f(X)K[X] \cap A[X]$ and $I_1 = f_1(X)K[X] \cap A[X]$.*

If $I \subseteq I_1$ and if the ideal I is almost principal, then I_1 is almost principal.

Proof

(1) We use Proposition 3.3 (2).

" $\Longrightarrow$ " Let $0 \neq s \in A$ be such that for each $h(X) \in K[X]$ satisfying
$f(X)h(X) \in A[X]$, $sh(X) \in A[X]$. Let $h(X) \in K[X]$ satisfying
$f_1(X)h(X) \in A[X]$.
Then $f(X)h(X) = f_2(X)(f_1(X)h(X)) \in A[X]$, so $sh(X) \in A[X]$. We
conclude that the ideal I_1 is almost principal. By symmetry, the ideal I_2 is
also almost principal.

" $\Longleftarrow$ " There exist $s_1, s_2 \in A \setminus (0)$ such that for each $h(X) \in K[X]$, if
$f_1(X)h(X) \in A[X]$, then $s_1h(X) \in A[X]$, and if $f_2(X)h(X) \in A[X]$, then
$s_2h(X) \in A[X]$.
Let $h(X) \in K[X]$ be such that $f(X)h(X) \in A[X]$. Then $f_1(X)(f_2(X)h(X))$
$\in A[X]$. So $s_1(f_2(X)h(X)) \in A[X]$ and then $s_2s_1h(X) \in A[X]$. So I is an
almost principal ideal.

(2) Since $f(X) \in I \subseteq I_1 \subseteq f_1(X)K[X]$, then $f(X) = f_1(X)f_2(X)$ with
$f_2(X) \in K[X]$. Let $0 \neq a \in A$ be such that $af_2(X) \in A[X]$. Then $af(X) =$
$f_1(X)(af_2(X))$, and the ideal $I = f(X)K[X] \cap A[X] = (af(X))K[X] \cap A[X]$
is almost principal by hypothesis. By (1), the ideal I_1 is almost principal. $\square$

Lemma 3.6 *Let J be an ideal of $A[X]$. If the ideal $I = JK[X] \cap A[X]$ is almost
principal, then the ideal J is also almost principal.*

Proof By definition, there exist $0 \neq s \in A$ and a nonconstant polynomial $f(X) \in I$
such that $sI \subseteq f(X)A[X]$. Then $f(X) = f_1(X)g_1(X) + \ldots + f_n(X)g_n(X)$ where
the polynomials $f_1(X), \ldots, f_n(X) \in J$ and $g_1(X), \ldots, g_n(X) \in K[X]$. Let $0 \neq$
$a \in A$ be such that $ag_1(X), \ldots, ag_n(X) \in A[X]$. Then $0 \neq as \in A$ and $af(X) \in$
$J \subseteq I$ is a nonconstant polynomial. Since $J \subseteq I$, then $asJ \subseteq asI \subseteq af(X)A[X]$.
So J is an almost principal ideal. $\square$

The following proposition is the principal result of this section. It shows that the
ideals of the form $f(X)K[X] \cap A[X]$, with $f(X) \in A[X]$ irreducible in $K[X]$, are
alone sufficient to decide if the ring $A[X]$ is almost principal.

Proposition 3.7 *The following assertions are equivalent:*

1. *The polynomial ring $A[X]$ is almost principal.*
2. *For each nonconstant polynomial $f(X) \in A[X]$, the ideal $f(X)K[X] \cap A[X]$ is
   almost principal.*
3. *For each polynomial $f(X) \in A[X]$ irreducible in $K[X]$, the ideal $f(X)K[X] \cap
   A[X]$ is almost principal in $A[X]$.*

Proof The two implications "(1) $\Longrightarrow$ (2) $\Longrightarrow$ (3)" are clear.
"(3) $\Longrightarrow$ (1)" Let I be a nonzero ideal of $A[X]$ such that $IK[X] \neq K[X]$. Since
$K[X]$ is a PID, $IK[X] = f(X)K[X]$ with $f(X) \in I$ a nonconstant polynomial.
Then $f(X) = f_1(X) \ldots f_n(X)$, where $f_1(X), \ldots, f_n(X) \in K[X]$ are irreducible
polynomials. For $1 \leq i \leq n$, there exists $0 \neq a_i \in A$ such that $a_i f_i(X) \in A[X]$.

Put $0 \neq a = a_1 \ldots a_n \in A$ and then $af(X) = a_1 f_1(X) \ldots a_n f_n(X)$ with $f_i(X)K[X] = a_i f_i(X)K[X]$ and $af(X)K[X] = f(X)K[X]$. By hypothesis, the ideals $\left(a_i f_i(X)K[X]\right) \cap A[X]$, $1 \leq i \leq n$, are almost principal. By Lemma 3.5 (1), the ideal $\left(af(X)K[X]\right) \cap A[X]$ is almost principal. Then the ideal $\left(IK[X]\right) \cap A[X] = \left(af(X)K[X]\right) \cap A[X]$ is almost principal. By the preceding lemma, the ideal I is almost principal. $\qquad\Box$

The UFD property of the domain $K[X]$ has played an important role in the proof of the preceding proposition. If the field K is algebraically closed, the irreducible polynomials of $K[X]$ are linear. Then we have the following corollary.

Corollary 3.8 *Suppose that the field K is algebraically closed. Then the ring $A[X]$ is almost principal if and only if for each $a, b \in A \setminus (0)$, the ideal $(aX - b)K[X] \cap A[X]$ is almost principal.*

Proof The irreducible polynomials of $K[X]$ are linear. $\qquad\Box$

In the general case, we do not know if the validity of the property almost principal for the linear polynomials implies that of all the polynomials.

4 Applications to Extensions of Integral Domains

The results of the two following sections are due to Lucas (2012). We fix in these two sections an integral domain A with quotient field K.

Proposition 4.1 *Let $A \subseteq B$ be an extension of integral domains with a conductor $(A : B) \neq (0)$ and $f(X) \in A[X]$ a nonconstant polynomial.*
The following assertions are equivalent:
1. *The ideal $I = f(X)K[X] \cap A[X]$ is almost principal in the ring $A[X]$.*
2. *The ideal $J = f(X)K[X] \cap B[X]$ is almost principal in the ring $B[X]$.*
3. *The ideal $I.B[X]$ is almost principal in the ring $B[X]$.*

Proof We fix an element $0 \neq a \in (A : B)$.

$"(1) \implies (2)"$ By Proposition 3.3, there exists $0 \neq s \in A$ such that $sI \subseteq f(X)A[X]$. Since $aJ \subseteq f(X)K[X] \cap A[X] = I$, then $(sa)J \subseteq sI \subseteq f(X)A[X] \subseteq f(X)B[X]$ with $0 \neq sa \in B$. Then the ideal J is almost principal in the ring $B[X]$.

$"(2) \implies (3)"$ Since $I \subseteq J$, then $I.B[X] \subseteq J = f(X)K[X] \cap B[X]$. By Proposition 3.3, there exists $0 \neq s \in B$ such that $sJ \subseteq f(X)B[X]$. Then $s(IB[X]) \subseteq sJ \subseteq f(X)B[X]$ with $f(X) \in I \subseteq IB[X]$. Then the ideal $I.B[X]$ is almost principal in the ring $B[X]$.

$"(3) \implies (1)"$ By hypothesis, there exist $0 \neq b \in B$ and $g(X) \in IB[X]$ a nonconstant polynomial such that $bIB[X] \subseteq g(X)B[X]$. In particular, $bI \subseteq g(X)B[X]$ and $ag(X) \in IaB[X] \subseteq IA[X] = I$. Then $a^2 bI \subseteq ag(X)aB[X] \subseteq$

$ag(X)A[X]$ with $0 \neq a^2b \in A$ and $ag(X) \in I$ a nonconstant polynomial. Then I is almost principal in $A[X]$. $\square$

Examples The implications $"(1) \Longrightarrow (3)"$ and $"(2) \Longrightarrow (3)"$ are true without the hypothesis $(A : B) \neq (0)$. Indeed, the second is clear from the preceding proof. For the other one, by Proposition 3.3, there exists $0 \neq s \in A$ such that $sI \subseteq f(X)A[X]$. Then $sIB[X] \subseteq f(X)B[X]$ with $f(X) \in I \subseteq IB[X]$. So $IB[X]$ is almost principal in the domain $B[X]$.

The following first example shows that the implications $"(2) \Longrightarrow (1)"$ and $"(3) \Longrightarrow (1)"$ are in general false. The second one shows that the implication $"(1) \Longrightarrow (2)"$ is in general false.

(1) Let s and t be the two indeterminates over a field F. The domain $A = F\left[s; \{st^{2^n}, n \in \mathbb{N}\}\right]$ has a quotient field $K = F(s, t)$ and an integral closure $B = F[st^n; n \in \mathbb{N}]$. See Example 2 of Proposition 2.5. Let $f(X) = sX - st \in A[X]$. The ideal $I = f(X)K[X] \cap A[X] = (X - t)K[X] \cap A[X]$ is not almost principal in $A[X]$. By Proposition 2.5, the ring $B[X]$ is almost principal. Its ideals $f(X)K[X] \cap B[X]$ and $IB[X]$ are then almost principal.

(2) Let s and t be the two indeterminates over a field F. We consider the domains $B = F\left[s; \{st^{2^n}, n \in \mathbb{N}\}\right]$ and $A = F[s, st] \subset B \subset F[s, t]$, with quotient field $K = F(s, t)$. By Example 2 of Corollary 1.8, the ring $A[X]$ is almost principal. Let $f(X) = sX - st \in A[X]$. The ideal $I = f(X)K[X] \cap A[X]$ is almost principal in $A[X]$. But the ideal $J = f(X)K[X] \cap B[X] = (X-t)K[X] \cap B[X]$ is not almost principal in $B[X]$.

Let A be an integral domain with quotient field K and $f(X) \in A[X]$ a nonconstant polynomial such that $f(X)K[X] \cap A[X]$ is almost principal in $A[X]$. The preceding examples show that if B is an over ring of A, the ideal $f(X)K[X] \cap B[X]$ is not necessarily almost principal in $B[X]$. However, the following proposition shows that there exists a class of over rings B of A for which the property is true. This is the class of the rings of fractions of A.

Proposition 4.2 *Let S be a multiplicative set of A and $f(X) \in A[X]$ a nonconstant polynomial. Let $I = f(X)K[X] \cap A[X]$ and $J = f(X)K[X] \cap A_S[X]$. Then $J = I.A_S[X]$.*

Moreover, if the ideal I is almost principal in the ring $A[X]$, then the ideal J is almost principal in the ring $A_S[X]$.

Proof

(1) Since $I \subseteq J$, then $I.A_S[X] \subseteq J$. Conversely, let $h(X) \in J$, $h(X) = f(X)g(X)$ with $g(X) \in K[X]$ and $h(X) \in A_S[X]$. There exists $t \in S$ such that $th(X) \in A[X]$ and then $f(X)tg(X) \in A[X]$. So $h(X) = f(X)tg(X).\frac{1}{t} \in \left(f(X)K[X] \cap A[X]\right)A_S[X] = I.A_S[X]$. Then $J \subseteq I.A_S[X]$, so $J = I.A_S[X]$.

(2) By Proposition 3.3, there exists $0 \neq a \in A$ such that $ag(X) \in A[X]$ for each $g(X) \in K[X]$ satisfying $f(X)g(X) \in A[X]$. Let $g(X) \in K[X]$ be such that $f(X)g(X) \in A_S[X]$. There exists $s \in S$ such that $sf(X)g(X) \in A[X]$. Then $asg(X) \in A[X]$, so $ag(X) \in \frac{1}{s}A[X] \subseteq A_S[X]$. By Proposition 3.3, the ideal $J = f(X)K[X] \cap A_S[X]$ is almost principal in the ring $A_S[X]$. □

Example Let s and t be the two indeterminates over a field F. The domains $A = F[s, st]$ and $B = F[s; \{st^{2^n}, n \in \mathbb{N}\}]$ satisfy $A \subset B \subset F[s, t]$ and have the same quotient field $K = F(s, t)$. Let $f(X) = sX - st \in A[X]$. The ideal $I = f(X)K[X] \cap A[X]$ is almost principal in $A[X]$, but $J = f(X)K[X] \cap B[X] = (X - t)K[X] \cap B[X]$ is not almost principal in $B[X]$. In fact, B is not a ring of fractions of A. Indeed, suppose that there exists a multiplicative set S of A such that $B = S^{-1}A$. Then $F[s; \{st^{2^n}, n \in \mathbb{N}\}] = S^{-1}F[s, st]$. Since the units of the domain $F[s; \{st^{2^n}, n \in \mathbb{N}\}]$ are the nonzero constants, then $S \subseteq F^*$. So $F[s; \{st^{2^n}, n \in \mathbb{N}\}] = F[s, st]$, which is absurd.

We apply the preceding proposition to the localized by prime ideals of an integral domain.

Corollary 4.3 *Let $f(X) \in A[X]$ be a nonconstant polynomial such that the ideal $I = f(X)K[X] \cap A[X]$ is almost principal in the ring $A[X]$. For each prime ideal P of A, the ideal $f(X)K[X] \cap A_P[X] - I.A_P[X]$ is almost principal in the ring $A_P[X]$.*

Recall from Sect. 10, Chap. 1, that a ring A has a finite character if each nonzero element of A belongs to at most a finite number of maximal ideals.

Corollary 4.4 *Suppose that the domain A has a finite character. Let $f(X) \in A[X]$ be a nonconstant polynomial and $I = f(X)K[X] \cap A[X]$.*

The ideal I is almost principal in the ring $A[X]$ if and only if for each maximal ideal M of A, the ideal $f(X)K[X] \cap A_M[X] = I.A_M[X]$ is almost principal in the ring $A_M[X]$.

Proof " $\Longrightarrow$ " By the preceding corollary.
" $\Longleftarrow$ " If $c(f) = A$, by Proposition 3.1, the ideal $I = f(X)A[X]$ is almost principal. Suppose that $c(f) \neq A$. By hypothesis, there exist a finite number of maximal ideals $M_1, \ldots, M_n$ of A containing $c(f)$. For each $M \in Max(A) \setminus \{M_1, \ldots, M_n\}$, $c(f) \not\subseteq M$, then $c(f)A_M = A_M$. By Proposition 3.1, the ideal $f(X)K[X] \cap A_M[X] = f(X)A_M[X]$ and then $I \subseteq f(X)A_M[X]$. By Proposition 3.3, for $1 \leq i \leq n$, there exists $0 \neq a_i \in A_{M_i}$ such that $a_i(f(X)K[X] \cap A_{M_i}[X]) \subseteq f(X)A_{M_i}[X]$ and then $a_i I \subseteq f(X)A_{M_i}[X]$. We may suppose that $0 \neq a_i \in A$. Let $0 \neq a = a_1 \ldots a_n \in A$. Then $aI \subseteq f(X)A_{M_i}[X]$ for $1 \leq i \leq n$. For each $M \in Max(A)$, $aI \subseteq f(X)A_M$. Then $aI \subseteq \bigcap\{fA_M[X]; M \in Max(A)\} = f(\bigcap\{A_M[X]; M \in Max(A)\}) = fA[X]$. So the ideal I is almost principal in the ring $A[X]$. □

We do not know if the equivalence in the preceding corollary is true for any integral domain.

Definition 4.5 An integral domain A is said to have a finite t-character if each nonzero element of A belongs to at most a finite number of t-maximal ideals of A.

Example 1 A Mori domain A has a finite t-character. Note first that the t-operation and the v-operation are identical over A. Indeed, let I be a nonzero fractional ideal of A. There exists a nonzero finitely generated fractional ideal $J \subseteq I$ of A such that $I^{-1} = J^{-1}$. Then $I_v = (I^{-1})^{-1} = (J^{-1})^{-1} = J_v = J_t$, since J is finitely generated. Since $J \subseteq I$, then $J_t \subseteq I_t \subseteq I_v$. So $I_v = J_t \subseteq I_t \subseteq I_v$, then $I_t = I_v$. Suppose that there exists a nonzero element $a \in A$ that belongs to an infinity $P_0, P_1, \ldots$ of distinct t-maximal ideals of A, i.e, integral maximal divisorial ideals of A. By Example 4 at the beginning of Sect. 11, Chap. 1, for each $n \in \mathbb{N}$, $I_n = P_0 \cap \ldots \cap P_n$ is an integral divisorial ideal of A containing a, and then $a(A : I_n) \subseteq A$ and $a(A : I_n)$ is an integral divisorial ideal of A. Since $(I_n)_n$ is a decreasing sequence, then $(a(A : I_n))_n$ is increasing. But A is a Mori domain. There exists $n \in \mathbb{N}$ such that $a(A : I_n) = a(A : I_{n+1})$ and then $(A : I_n) = (A : I_{n+1})$, so $A : (A : I_n) = A : (A : I_{n+1})$. Since the I_k are divisorial, then $I_n = I_{n+1}$, so $P_0 \cap \ldots \cap P_n = P_0 \cap \ldots \cap P_n \cap P_{n+1}$. In particular, $P_0 \ldots P_n \subseteq P_{n+1}$. By Proposition 11.3 of Chap. 1, P_{n+1} is a prime ideal of A. There exists $i \le n$ such that $P_i \subseteq P_{n+1}$. By t-maximality of the P_k, $P_i = P_{n+1}$, which is absurd.

Example 2 In a Prufer domain A, each nonzero integral ideal is a t-ideal. Then $t - Max(A) = Max(A)$. So A has a finite t-character if and only if it has a finite character. But in general, the two notions are different.

Example 3 (Kabbaj and Mimouni 2007) Let A be an integral domain. Then A has a finite t-character if and only if the polynomial ring $A[X]$ has a finite t-character. Indeed, let $0 \ne a \in A$ and $\{M_\alpha\}_{\alpha \in \Gamma}$ the family of the t-maximal ideals of A containing a. By Proposition 11.10 (2) of Chap. 1, $\{M_\alpha[X]\}_{\alpha \in \Gamma}$ is the family of the t-maximal ideals of $A[X]$ containing a. Then Γ is finite. Conversely, let $f(X)$ be a nonzero noninvertible element of $A[X]$ and $\{Q_\alpha\}_{\alpha \in \Gamma}$ the family of the t-maximal ideals of $A[X]$ containing $f(X)$. Let $\Gamma_1 = \{\alpha \in \Gamma; \ Q_\alpha \cap A = (0)\}$ and $\Gamma_2 = \{\alpha \in \Gamma; \ Q_\alpha \cap A = q_\alpha \ne (0)\}$. Let $\alpha \in \Gamma_1$, K the quotient field of A and $S = A \setminus (0)$. Then $S^{-1}Q_\alpha$ is a nonzero prime ideal of $S^{-1}A[X] = K[X]$. Then it is maximal and contains $f(X)$. Since $K[X]$ is a PID, by Example 2 of Proposition 10.1, Chap. 1, it has a finite character. So the family $\{Q_\alpha\}_{\alpha \in \Gamma_1}$ is finite, and then Γ_1 is finite. Let $\alpha \in \Gamma_2$. By Proposition 11.10 (1) of Chap. 1, q_α is a t-maximal ideal of A and $Q_\alpha = q_\alpha[X]$. Let a be the leading coefficient of $f(X)$. Then $0 \ne a \in q_\alpha$. Since A has a finite t-character, the family $\{q_\alpha\}_{\alpha \in \Gamma}$ is finite. Since $Q_\alpha = q_\alpha[X]$ for each $\alpha \in \Gamma_2$, the family $\{Q_\alpha\}_{\alpha \in \Gamma_2}$ is finite, and then Γ_2 is finite.

We show now the analogue of Corollary 4.4 for domains with finite t−character.

Corollary 4.6 *Suppose that the domain A has a finite t−character. Let $f(X) \in A[X]$ be a nonconstant polynomial and $I = f(X)K[X] \cap A[X]$.*

The ideal I is almost principal in $A[X]$ if and only if for each t−maximal ideal M of A, the ideal $f(X)K[X] \cap A_M[X] = I.A_M[X]$ is almost principal in $A_M[X]$.

Proof "$\Longrightarrow$" By Corollary 4.3.

" $\Longleftarrow$" If $c(f) = A$, by Proposition 3.1, $I = f(X)A[X]$ is almost principal. Suppose that $c(f) \neq A$. By hypothesis, there exist a finite number $M_1, \ldots, M_n$ of t−maximal ideals of A containing $c(f)$. For each $M \in t - Max(A) \setminus \{M_1, \ldots, M_n\}$, $c(f) \not\subseteq M$, then $c(f)A_M = A_M$. By Proposition 3.1, $f(X)K[X] \cap A_M[X] = f(X)A_M[X]$, and then we have $I \subseteq f(X)A_M[X]$. By Proposition 3.3, for $1 \leq i \leq n$, there exists $0 \neq a_i \in A_{M_i}$ such that $a_i\big(f(X)K[X] \cap A_{M_i}[X]\big) \subseteq f(X)A_{M_i}[X]$, and then $a_i I \subseteq f(X)A_{M_i}[X]$. We may suppose that $a_i \in A$. Let $0 \neq a = a_1 \ldots a_n \in A$. Then $aI \subseteq f(X)A_{M_i}[X]$ pour $1 \leq i \leq n$. For each $M \in Max(A)$, $aI \subseteq f(X)A_M$, and then $aI \subseteq \bigcap\{f(X)A_M[X]; \ M \in t - Max(A)\} = f(X)\big(\bigcap\{A_M[X]; \ M \in t - Max(A)\}\big) = f(X)A[X]$. So I is almost principal in $A[X]$. □

We define in the sequel a particular class of Mori domains.

Definition 4.7 An integral domain A is said to be strong Mori domain if it has a finite t−character, and for each $M \in t - Max(A)$, the ring A_M is Noetherian.

Other characterizations of strong Mori domains will be given in Exercise 18.

Corollary 4.8 *If A is a strong Mori domain, then the ring $A[X]$ is almost principal.*

Proof Let $f(X) \in A[X]$ be a nonconstant polynomial and $I = f(X)K[X] \cap A[X]$. For each $M \in t - Max(A)$, the ring A_M is Noetherian. By Corollary 1.8, the domain $A_M[X]$ is almost principal, and then its ideal $f(X)K[X] \cap A_M[X]$ is almost principal. Since A has a finite t−character, by the preceding corollary, I is almost principal in $A[X]$. □

5 The Almost Principal Ideals of the Form f(X)K[X]∩A[X] in the Ring A[X]

We have always shown in Sect. 3 that the ideals of the form $f(X)K[X] \cap A[X]$ allow us to see if the ring $A[X]$ is almost principal. This kind of ideals deserves the merit to be studied deeply. Recall from Sect. 3 that for $f(X) \in A[X]$ a nonconstant polynomial, the set $G_f = \big\{h(X) \in K[X]; \ f(X)h(X) \in A[X]\big\}$ is a fractional ideal of $A[X]$. The following proposition, whose proof is very short, will be used in many situations in the sequel.

Proposition 5.1 *Let* $f(X) \in A[X]$ *be a nonconstant polynomial and* $I = f(X)K[X] \cap A[X]$. *Suppose that there exists a nonzero integral ideal* J *of* A *such that* $c(f)c(h) \subseteq (J : J)$ *for each* $h(X) \in G_f$. *Then the ideal* I *is almost principal.*

Proof Fix two nonzero elements $a \in c(f)$ and $b \in J$. By hypothesis, for each $h(X) \in G_f$, $ah(X) \in (J : J)[X]$ and $b(J : J)[X] \subseteq J[X]$. Then $bah(X) \in b(J : J)[X] \subseteq J[X] \subseteq A[X]$. So $(ba)G_f \subseteq A[X]$ with $0 \neq ba \in A$. By Corollary 3.4, I is almost principal. $\square$

Corollary 5.2 *Let* $f(X) \in A[X]$ *be a nonconstant polynomial and* $I = f(X)K[X] \cap A[X]$. *Suppose that there exists a nonzero integral ideal* J *of* A *such that* $(c(f)^n : c(f)^n) \subseteq (J : J)$ *for each* $n \in \mathbb{N}^*$. *Then the ideal* I *is almost principal.*

Proof Let $h(X) \in G_f$. By the content formula, there exists $n \in \mathbb{N}^*$ such that $c(f)^n c(fh) = c(f)^n c(f)c(h)$ with $c(fh) \subseteq A$. Then $c(f)c(h) \subseteq (c(f)^n : c(f)^n) \subseteq (J : J)$. By the preceding proposition, the ideal I is almost principal. $\square$

To establish the converse of Proposition 5.1, we need the following lemma. We recall from Sect. 3 that for $f(X) \in A[X]$ a nonconstant polynomial, the set $H_f = \bigcup \{c(h);\ h(X) \in G_f\}$ is a A–sub-module of K.

Lemma 5.3 *Let* $f(X) \in A[X]$ *be a nonconstant polynomial. For each* $m \in \mathbb{N}^*$ *and* $s \in H_f^m$, *there exists a polynomial* $g(X) \in G_f$ *such that* $s \in c(g)^m$; *i.e,* $H_f^m = \bigcup c(g)^m, g \in G_f$.

Proof Let $m \in \mathbb{N}^*$ and $s \in H_f^m$, $s = a_1 + \ldots + a_n$ where each a_i is a product of m elements of H_f. Put $a_i = a_{i,1}a_{i,2} \ldots a_{i,m}$ with $a_{i,j} \in H_f$. Let $g_1(X), \ldots, g_r(X) \in G_f$ be such that $a_{i,j}, 1 \leq i \leq n, 1 \leq j \leq m$, belongs to some $c(g_k), 1 \leq k \leq r$. Put $n_k = deg\, g_k(X), 1 \leq k \leq r$, and $g(X) = g_1(X) + X^{n_1+1}g_2(X) + \ldots + X^{n_1+n_2+\ldots+n_{r-1}+r-1}g_r(X) \in G_f$. Then $c(g) = c(g_1) + \ldots + c(g_r)$, so $a_1, \ldots, a_n \in c(g)^m$, and then $s \in c(g)^m$. $\square$

Proposition 5.4 *Let* $f(X) \in A[X]$ *be a polynomial of degree* $k \geq 1$. *For each* $h(X) \in G_f$, *we have* $c(f)c(h) \subseteq (H_f^k : H_f^k)$.

Proof Let $h(X) \in G_f$ and $s \in H_f^k$. By the preceding lemma, there exists $g(X) \in G_f$ such that $s \in c(g)^k$. Let $t(X) = h(X) + X^n g(X)$ where $n = deg\, h(X) + 1$. Then $t(X) \in G_f$ and $c(t) = c(h) + c(g)$, so $s \in c(t)^k$. By the content formula, $c(t)^k c(ft) = c(t)^k c(t)c(f) = c(t)^k (c(h) + c(g))c(f) \supseteq c(t)^k c(h)c(f)$. Since $c(ft) \subseteq A$ because $t \in G_f$, then we have $c(t)^k c(h)c(f) \subseteq c(t)^k \subseteq H_f^k$. But $s \in c(t)^k$, then $sc(f)c(h) \subseteq H_f^k$. This is true for each $s \in H_f^k$. Then $c(f)c(h) \subseteq (H_f^k : H_f^k)$. $\square$

Now, we are ready to state the converse of Proposition 5.1.

Corollary 5.5 *Let $f(X) \in A[X]$ be a nonconstant polynomial and $I = f(X)K[X] \cap A[X]$. Then I is almost principal if and only if there exists a nonzero integral ideal J of A such that $c(f)c(h) \subseteq (J : J)$ for each $h(X) \in G_f$.*

Proof " $\Longleftarrow$" By Proposition 5.1.
" $\Longrightarrow$" By Corollary 3.4, since the ideal I is almost principal, H_f is a fractional ideal of A. There exists $0 \neq a \in A$ such that $a H_f \subseteq A$. Let $k = deg\ f(X) \in \mathbb{N}^*$. The ideal $J = a^k H_f^k$ of A in integral and nonzero. By the preceding proposition, for each $h(X) \in G_f, c(f)c(h) \subseteq \left(H_f^k : H_f^k\right) = \left(a^k H_f^k : a^k H_f^k\right) = (J : J)$. $\qquad\square$

Lemma 5.6 *Let J_1 and J_2 be the two fractional ideals of an integral domain A. Then $(J_1 : J_1) \cap (J_2 : J_2) \subseteq (J_1 + J_2 : J_1 + J_2)$.*

Proof Let $x \in (J_1 : J_1) \cap (J_2 : J_2)$ and $y \in J_1 + J_2, y = y_1 + y_2$ with $y_1 \in J_1$ et $y_2 \in J_2$. Then $xy = xy_1 + xy_2 \in J_1 + J_2$. So $x \in (J_1 + J_2 : J_1 + J_2)$. $\qquad\square$

Proposition 5.7 *Let $f(X) \in A[X]$ be a nonconstant polynomial. Suppose that the ideal $I = f(X)K[X] \cap A[X]$ is almost principal. There exists a biggest nonzero integral ideal J of A (may be A itself) such that $c(f)c(h) \subseteq (J : J)$ for each $h(X) \in G_f$.*

Proof Let $\mathscr{F}$ be the set of the nonzero integral ideals J of A such that $c(f)c(h) \subseteq (J : J)$ for each $h(X) \in G_f$. By Corollary 5.5, $\mathscr{F}$ is nonempty. Let $(J_\alpha)_{\alpha \in \Lambda}$ be a chain of elements of $\mathscr{F}$ and $J = \bigcup_{\alpha \in \Lambda} J_\alpha$. For each $x \in J$, there exists $\alpha \in \Lambda$ such that $x \in J_\alpha$. For each $h(X) \in G_f, xc(f)c(h) \subseteq J_\alpha \subseteq J$. Then $c(f)c(h) \subseteq (J : J)$. So $J \in \mathscr{F}$ and bounds the family $(J_\alpha)_{\alpha \in \Lambda}$. By Zorn's lemma, $(\mathscr{F}, \subseteq)$ has a maximal element J. It is the biggest element of $\mathscr{F}$. Indeed, if $J' \in \mathscr{F}$, then $J + J' \in \mathscr{F}$ because by the preceding lemma, for each $h(X) \in G_f, c(f)c(h) \subseteq (J : J) \cap (J' : J') \subseteq (J+J' : J+J')$. By the maximality of J in $\mathscr{F}, J+J' = J$, then $J' \subseteq J$. $\qquad\square$

In the following corollary, we give some properties of the ideal J defined in the preceding proposition.

Corollary 5.8 *Let $f(X) \in A[X]$ be a nonconstant polynomial such that the ideal $I = f(X)K[X] \cap A[X]$ is almost principal and J the biggest nonzero integral ideal of A satisfying $c(f)c(h) \subseteq (J : J)$ for each $h(X) \in G_f$. Then J is divisorial and satisfies the equality $J^{-1} = (J : J)$.*

Proof

(1) We have $(J : J) \subseteq (J_v : J_v)$. Indeed, for each $x \in (J : J), xJ \subseteq J$, then $xJ_v \subseteq J_v$, so $x \in (J_v : J_v)$. For each $h(X) \in G_f, c(f)c(h) \subseteq (J : J) \subseteq (J_v : J_v)$. By the preceding proposition, we necessarily have $J_v \subseteq J$, then $J_v = J$ and J is divisorial.

(2) The integral ideal JJ^{-1} of A satisfies $(J : J) \subseteq (JJ^{-1} : JJ^{-1})$. Indeed, for each element $x \in (J : J)$, $xJ \subseteq J$, then $xJJ^{-1} \subseteq JJ^{-1}$ so $x \in (JJ^{-1} : JJ^{-1})$. For each $h(X) \in G_f$, $c(f)c(h) \subseteq (J : J) \subseteq (JJ^{-1} : JJ^{-1})$. By the preceding proposition, we necessarily have $JJ^{-1} \subseteq J$ and then $J^{-1} \subseteq (J : J) \subseteq (A : J) = J^{-1}$. So $J^{-1} = (J : J)$. $\qquad\qquad\square$

Notations Let A be an integral domain with quotient field K and I a nonzero integral ideal of A. Then $A \subseteq (I : I) \subseteq I^{-1} \subseteq T_A(I) \subseteq K$, where $T_A(I) = \bigcup_{n:1}^{\infty}(A : I^n)$ is the Nagata transform of A with respect the ideal I. It is introduced in Example 1 of Lemma 2.10. Remark that $(I : I)$ is the biggest over ring of A in which I is an ideal. Indeed, let $x, y \in (I : I)$. Then $(xy)I = x(yI) \subseteq xI \subseteq I$, so $xy \in (I : I)$. Moreover, $I \subseteq A \subseteq (I : I)$ and $I(I : I) = I$ since $1 \in (I : I)$, and then I is an ideal of $(I : I)$. Let B be an over ring of A in which I is an ideal of B. Then $IB = I$, so $B \subseteq (I : I)$.

Beside the ideal J in the preceding corollary satisfying $J^{-1} = (J : J)$, we give another example satisfying the same equality.

Example 1 Let k be a field, $A = k[[X, Y]]$ and $I = (X, Y)$, the maximal ideal of A. By Example 1 at the beginning of Sect. 11, Chap. 1, we have $I^{-1} = A$. Then $A = (I : I) = I^{-1}$.

Example 2 The inclusions $(I : I) \subseteq I^{-1} \subseteq T_A(I)$ may be strict, and I^{-1} is not always a ring. Indeed, let k be a field and X, Y, Z three indeterminates over k. Consider the sub-ring $A = k[Z, XZ, YZ]$ of $k[X, Y, Z]$ and its ideal $I = ZA$. Then $X = \frac{XZ}{Z}$ and $Y = \frac{YZ}{Z}$ are elements of the quotient field of A. They belong to $I^{-1} = (A : I) = (A : ZA)$ since $XZ, YZ \in A$. But $XY \notin I^{-1}$ since $XYZ \notin A$. Then I^{-1} is not stable by multiplication, so it is not a ring. Since $(I : I)$ and $T_A(I)$ are rings, then $(I : I) \neq I^{-1} \neq T_A(I)$.

Example 3 We may have $A \neq (I : I) = I^{-1}$. Let k be a field, $A = k[[X^2, X^3]] \subset k[[X]]$, and $I = (X^2, X^3)A = X^2k[[X]]$. Indeed, the direct inclusion is clear. Conversely, let $f = \sum_{i:0}^{\infty} a_i X^i \in k[[X]]$, $f = \sum_{i:0}^{\infty} a_{2i} X^{2i} + X \sum_{i:0}^{\infty} a_{2i+1} X^{2i} = f_1 + X f_2$ with $f_1, f_2 \in A$. Then $X^2 f = X^2 f_1 + X^3 f_2 \in (X^2, X^3)A = I$. Note that the quotient field of A is equal to $k((X))$. Indeed, it suffices to show that $frac(A)$ contains $k[[X]]$. Let $f \in k[[X]]$, $f = f_1 + X f_2$ with $f_1, f_2 \in A$ and $X = \frac{X^3}{X^2} \in frac(A)$. Then $f \in frac(A)$. We will show that $(I : I) = I^{-1} = k[[X]]$. It suffices to prove that $k[[X]] \subseteq (I : I) \subseteq I^{-1} \subseteq k[[X]]$. For the first inclusion, it suffices to show that $X^2k[[X]] \subseteq I$ and $X^3k[[X]] \subseteq I$. But an element f of $k[[X]]$ is of the form $f = f_1 + X f_2$ with $f_1, f_2 \in A$, and then $X^3 f = X^3 f_1 + (X^2)^2 f_2 \in I$. For the last inclusion, let $f \in I^{-1} \subseteq frac(A) = k((X))$. Then $fI \subseteq A$, so

$X^2 f$ and $X^3 f \in A = k[[X^2, X^3]] \subseteq k[[X]]$. Then the order of f is ≥ -2. Put $f = aX^{-2} + bX^{-1} + \ldots$. Since $X^2 f = a + bX + \ldots \in k[[X^2, X^3]]$, then $b = 0$, and since $X^3 f = aX + bX^2 + \ldots \in k[[X^2, X^3]]$, then $a = 0$. So the order of f is positive, then $f \in k[[X]]$.

Example 4 Let A be an integral domain with quotient field K, $f(X) \in A[X]$ a nonconstant polynomial and $I = f(X)K[X] \cap A[X]$. Then $\big(A : c(I)\big) = \big(c(I) : c(I)\big)$, where $c(I)$ is the ideal of A formed by the coefficients of the polynomials in I. Indeed, let $q \in \big(A : c(I)\big)$. Then $qc(I) \subseteq A$. For each $h(X) \in I, h(X) = f(X)g(X)$ with $g(X) \in K[X]$ and $qh(X) \in A[X]$. Then $qh(X) \in f(X)K[X] \cap A[X] = I$, so $qc(h) = c(qh) \subseteq c(I)$. Then $qc(I) \subseteq c(I)$, so $q \in \big(c(I) : c(I)\big)$.

Example 5 (Bass 1963) Let A be an integral domain and I any nonzero fractional ideal of A. The integral ideal II^{-1} satisfies $(A : II^{-1}) = (II^{-1} : II^{-1})$ is a ring. Indeed, it suffices to show the first inclusion. Let $x \in (A : II^{-1})$. Then $x(II^{-1}) \subseteq A$, so $xI^{-1} \subseteq (A : I) = I^{-1}$, and then $xI^{-1}I \subseteq I^{-1}I$. So $x \in (II^{-1} : II^{-1})$.

Example 6 (Fontana et al. 1993) The inclusion $I^{-1} \subseteq T_A(I)$ may be strict. Let $A = \mathbb{Z} + X\mathbb{Q}[X]$ and $I = X\mathbb{Q}[X]$. Then $I^2 = X^2\mathbb{Q}[X]$. So $\frac{1}{X} \in (A : I^2) \subseteq T_A(I)$ since $\frac{1}{X}I^2 = X\mathbb{Q}[X] \subseteq A$. But $\frac{1}{X} \notin I^{-1}$ since $\frac{1}{X}I = \mathbb{Q}[X] \not\subseteq A$.

Proposition 5.9 (Huckaba and Papick 1982) *Let A be an integral domain with quotient field K and I a nonzero integral ideal of A.*
The following assertions are equivalent:

1. *The dual (or inverse) I^{-1} of I is a sub-ring of K.*
2. $I^{-1} = (I_v : I_v)$.
3. $I^{-1} = (II^{-1} : II^{-1})$.

Proof "(1) $\Longrightarrow$ (2)" Since I^{-1} is a ring, $I^{-1}(I^{-1}I_v) = (I^{-1}I^{-1})I_v = I^{-1}I_v = I^{-1}(A : I^{-1}) \subseteq A$. Then $I^{-1}I_v \subseteq (A : I^{-1}) = I_v$, so $I^{-1} \subseteq (I_v : I_v)$. Conversely, $I_v \subseteq A$, then $(I_v : I_v) \subseteq (A : I_v) = (I_v)^{-1} = (I^{-1})_v = I^{-1}$ since I^{-1} is divisorial, by Example 5 at the beginning of Sect. 11, Chap. 1. So $I^{-1} = (I_v : I_v)$.
"(2) $\Longrightarrow$ (3)" By Bass example, $(A : II^{-1}) = (II^{-1} : II^{-1})$. It suffices to show that $(II^{-1})^{-1} = I^{-1}$. By hypothesis, $I^{-1} = (I_v : I_v)$ is a ring, and then $I \subseteq II^{-1}$, so $(II^{-1})^{-1} \subseteq I^{-1}$. Conversely, let $u \in I^{-1}$ and $x \in II^{-1}$, $x = \sum a_i u_i$, with $a_i \in I$ and $u_i \in I^{-1}$. Then $ux = \sum a_i(uu_i)$, with $uu_i \in I^{-1}$ since I^{-1} is a ring. Then $ux \in II^{-1} \subseteq A$. So $u \in (A : II^{-1}) = (II^{-1})^{-1}$ and then $I^{-1} \subseteq (II^{-1})^{-1}$. So we have the equality $I^{-1} = (II^{-1} : II^{-1})$.
"(3) $\Longrightarrow$ (1)" We have $I^{-1} = (II^{-1} : II^{-1})$ is a ring.   □

Example 1 Let X, Y, Z be the three indeterminates over a field k, $A = k[Z, XZ, YZ]$ and $I = ZA$. By Example 2 after Corollary 5.8, $I^{-1} = \frac{1}{Z}A$ is not a ring. The equalities (2) and (3) in the preceding proposition are not satisfied. In fact, since I is principal, $I_v = I$ and $II^{-1} = A$. Then $(I_v : I_v) = (I : I) = \frac{1}{Z}Z(A : A) = (A : A) = A \neq I^{-1}$ and $(II^{-1} : II^{-1}) = (A : A) = A \neq I^{-1}$.

Example 2 Let A be a valuation domain and P a nonprincipal prime ideal of A. Then $A_P = (P : P) = P^{-1}$ is a ring.

Indeed, since $(P : P) \subseteq P^{-1}$, it suffices to show that $A_P \subseteq (P : P)$ and $P^{-1} \subseteq A_P$. For the first inclusion, it suffices to show that if $s \in A \setminus P$, then $\frac{1}{s} \in (P : P)$. Since A is a valuation domain, $P \subseteq sA$, then $\frac{1}{s}P \subseteq A$. But $s(\frac{1}{s}P) = P$, which is a prime ideal of A. Then $\frac{1}{s}P \subseteq P$, so $\frac{1}{s} \in (P : P)$. Then we have the inclusion $A_P \subseteq (P : P)$.

Let $x \in P^{-1}$. Suppose that $x \notin A_P$. Since A_P is a valuation domain, $\frac{1}{x} \in A_P$. In fact, $\frac{1}{x} \in PA_P = P$, since P is divided. For each $y \in P$, $y = (yx)\frac{1}{x}$ with $yx \in PP^{-1} \subseteq A$. Then $P = (\frac{1}{x})$ is a principal ideal, which is absurd. Then $x \in A_P$ so $P^{-1} \subseteq A_P$.

Corollary 5.10 *Let A be an integral domain and $(I_n)_{n \in \mathbb{N}^*}$ a decreasing sequence of integral divisorial ideals of A satisfying $I_n^{-1} = (I_n : I_n)$ for each $n \in \mathbb{N}^*$. Suppose that $I = \bigcap\limits_{n \in \mathbb{N}^*} I_n \neq (0)$. Then I is divisorial and satisfies the equality $I^{-1} = (I : I)$.*

Proof For each $n \in \mathbb{N}^*$, $I_v \subseteq (I_n)_v = I_n$, and then $I_v \subseteq \bigcap\limits_{n \in \mathbb{N}^*} I_n = I$, so $I_v = I$. For each $n \in \mathbb{N}^*$, $I_n^{-1} = (I_n : I_n)$ is a ring. Since the sequence $(I_n)_{n \in \mathbb{N}^*}$ is decreasing, the sequence $(I_n^{-1})_{n \in \mathbb{N}^*}$ is increasing, and then $J = \bigcup\limits_{n \in \mathbb{N}^*} I_n^{-1}$ is a ring. On the other hand, for each $n \in \mathbb{N}^*$, $I \subseteq I_n$, and then $I_n^{-1} \subseteq I^{-1}$. So $J = \bigcup\limits_{n \in \mathbb{N}^*} I_n^{-1} \subseteq I^{-1}$. Since I^{-1} is a fractional ideal, then J is a nonzero fractional ideal of A. It is divisorial and $J^{-1} = (A : \bigcup\limits_{n \in \mathbb{N}^*} I_n^{-1}) = \bigcap\limits_{n \in \mathbb{N}^*} (A : I_n^{-1}) = \bigcap\limits_{n \in \mathbb{N}^*} (I_n)_v = \bigcap\limits_{n \in \mathbb{N}^*} I_n = I$. Then $I^{-1} = (J^{-1})^{-1} = J_v = J$ is a ring. By the preceding proposition, $I^{-1} = (I_v : I_v) = (I : I)$. $\square$

Lemma 5.11 *Let A be an integral domain and I a nonzero integral ideal of A.*

1. *The sequence $(I^n(A : I^n))_{n \in \mathbb{N}^*}$ is decreasing.*
2. *For each $n \in \mathbb{N}^*$, $\sqrt{I^n(A : I^n)} = \sqrt{I(A : I)}$.*
3. *For each $n \in \mathbb{N}^*$, $\sqrt{(I^n(A : I^n))_v} = \sqrt{(I(A : I))_v}$.*

Proof

(1) Let I and J be the two nonzero ideals of A. Since $I.J(A : IJ) \subseteq A$, then $J(A : IJ) \subseteq (A : I)$, so $IJ(A : IJ) \subseteq I(A : I)$. By exchanging I with I^n and J with I in the preceding inclusion, we find $I^{n+1}(A : I^{n+1}) \subseteq I^n(A : I^n)$.

(2) Since $I(A : I) \subseteq A$, then $I^n(A : I)^n \subseteq A$, so $(A : I)^n \subseteq (A : I^n)$, and then $(I(A : I))^n \subseteq I^n(A : I^n)$. By (1), $(I(A : I))^{n+1} \subseteq I^{n+1}(A : I^{n+1}) \subseteq I^n(A : I^n) \subseteq \ldots \subseteq I(A : I)$. By taking the radicals, we find $\sqrt{I(A : I)} \subseteq \sqrt{I^n(A : I^n)} \subseteq \sqrt{I(A : I)}$, then $\sqrt{I^n(A : I^n)} = \sqrt{I(A : I)}$ for each $n \in \mathbb{N}^*$.

(3) By the computation done in (2), we have $(I(A : I))^{n+1} \subseteq I^n(A : I^n) \subseteq I(A : I)$.

Then $\left((I(A : I))^{n+1}\right)_v \subseteq \left(I^n(A : I^n)\right)_v \subseteq \left(I(A : I)\right)_v$. By Proposition 11.1, Chap. 1, we have $\left((I(A : I))^{n+1}\right)_v = \left(((I(A : I))_v)^{n+1}\right)_v$ that contains $((I(A : I))_v)^{n+1}$.

By taking the radicals, we obtain $\sqrt{((I(A : I))_v)^{n+1}} \subseteq \sqrt{\left(I^n(A : I^n)\right)_v} \subseteq \sqrt{\left(I(A : I)\right)_v}$. Then $\sqrt{(I(A : I))_v} \subseteq \sqrt{\left(I^n(A : I^n)\right)_v} \subseteq \sqrt{\left(I(A : I)\right)_v}$, so $\sqrt{\left(I^n(A : I^n)\right)_v} = \sqrt{\left(I(A : I)\right)_v}$. □

Proposition 5.12 *Let $f(X) \in A[X]$ be a nonconstant polynomial and $I = f(X)K[X] \cap A[X]$. For each $n \in \mathbb{N}^*$, let $J_n = c(f)^n(A : c(f)^n)$. Suppose that $\bigcap_{n \in \mathbb{N}^*} (J_n)_v \neq (0)$. Then the ideal I is almost principal in the ring $A[X]$.*

Proof Let $J = \bigcap_{n \in \mathbb{N}^*} (J_n)_v \neq (0)$. By the preceding lemma, the sequence $(J_n)_{n \in \mathbb{N}^*}$ is decreasing. It is the same for the sequence $\left((J_n)_v\right)_{n \in \mathbb{N}^*}$. By Bass example, $J_n^{-1} = (J_n : J_n)$ is a ring. By Proposition 5.9, $((J_n)_v : (J_n)_v) = J_n^{-1} = ((J_n)_v)^{-1}$. By Corollary 5.10, the ideal J is divisorial and $J^{-1} = (J : J)$. We have $\left(c(f)^n : c(f)^n\right) \subseteq (J_n : J_n)$ since $\left(c(f)^n : c(f)^n\right)J_n = \left(c(f)^n : c(f)^n\right)c(f)^n(A : c(f)^n) \subseteq c(f)^n(A : c(f)^n) = J_n$. On the other hand, for each $n \in \mathbb{N}^*$, $J \subseteq (J_n)_v$, and then $J^{-1} \supseteq ((J_n)_v)^{-1} = J_n^{-1}$. So, for each $n \in \mathbb{N}^*$, $(A : J_n) \subseteq (A : J) = (J : J)$, and then $\left(c(f)^n : c(f)^n\right) \subseteq (J_n : J_n) = J_n^{-1} \subseteq (J : J)$. By Corollary 5.2, I is almost principal. □

We do not know if the converse of the preceding proposition is true.

Corollary 5.13 *Let $f(X) \in A[X]$ be a nonconstant polynomial and $I = f(X)K[X] \cap A[X]$. If $c(f)^n(A : c(f)^n) = c(f)(A : c(f))$ for an infinity of natural integers n, then the ideal I is almost principal in the ring $A[X]$.*

Proof Let $J_n = c(f)^n(A : c(f)^n)$. By Lemma 5.11 (1), the sequence $(J_n)_{n \in \mathbb{N}^*}$ is decreasing. Since, $J_n = J_1$ for an infinity of integers, this sequence is constant. By the preceding proposition, the ideal I is almost principal. □

Corollary 5.14 *Let A be an integral domain such that $J(A : J)$ is a radical ideal for each nonzero integral ideal J of A. Then the ring $A[X]$ is almost principal.*

Proof Let $f(X) \in A[X]$ be a nonconstant polynomial and $I = f(X)K[X] \cap A[X]$. By Lemma 5.11, for each $n \in \mathbb{N}^*$, $\sqrt{c(f)^n(A : c(f)^n)} = \sqrt{c(f)(A : c(f))}$, and then $c(f)^n(A : c(f)^n) = c(f)(A : c(f))$. By the preceding corollary, the ideal I is almost principal. We conclude by Proposition 3.7.   □

Corollary 5.15 *Let a and b be the two nonzero elements of A and $f(X) \in A[X]$ a nonconstant polynomial such that $c(f) = (a, b)$. If the element $t = \frac{a}{b}$ of K is quasi-integral over A, the ideal $I = f(X)K[X] \cap A[X]$ is almost principal in the ring $A[X]$.*

Proof There exists $0 \neq d \in A$ such that $dt^n \in A$ for each $n \in \mathbb{N}$. Let $i, j, n \in \mathbb{N}$ be such that $i + j = n$. Then $\frac{d}{b^n} a^i b^j = d(\frac{a}{b})^i = dt^i \in A$. Since $c(f)^n$ is generated by the set $\{a^i b^j; \ i + j = n\}$, then $\frac{d}{b^n} \in (A : c(f)^n)$ and $d \in b^n(A : c(f)^n) \subseteq$

$$c(f)^n(A : c(f)^n) = J_n. \text{ Then } \bigcap_{n:1}^{\infty}(J_n)_v \neq (0). \text{ We conclude by Proposition 5.12.}$$

□

Note that in the preceding corollary, the hypotheses on the polynomial $f(X)$ do not bound neither its degree nor the number of its nonzero coefficients.

Example 1 Let a and b be the two nonzero elements of A and $f(X) = aX - b \in A[X]$. If one of the elements $\frac{a}{b}$ or $\frac{b}{a}$ of K is quasi-integral over A, the ideal $I = f(X)K[X] \cap A[X]$ is almost principal in $A[X]$.

Example 2 Let s, t, u be the three indeterminates over a field F, $A = F[s; \{st^{2^n}, n \in \mathbb{N}\}]$ and $K = F(s, t)$ its quotient field. Let $f(X) = sX - st \in A[X]$. We know that the ideal $I = f(X)K[X] \cap A[X] = (sX - st)K[X] \cap A[X] = (X - t)K[X] \cap A[X]$ is not almost principal in $A[X]$. Consider the ring $B = A[u; \{ut^i, i \in \mathbb{N}^*\}]$ and its quotient field $K(u)$. The ideal $f(X)K(u)[X] \cap B[X]$ is almost principal in $B[X]$. Indeed, the content $c(f) = (s, st)B$ with the element $\frac{st}{s} = t$ is quasi-integral over the ring B since $ut^i \in B$ for each $i \in \mathbb{N}^*$. We conclude by the preceding corollary.

The conductor $(A : B) = (0)$ since $u \in B$ and for each $0 \neq f(s, t) \in A$, $f(s, t)u \notin A$.

Proposition 5.16 *Let J be an ideal of $A[X]$ such that $JK[X] \neq K[X]$. Suppose that J contains a polynomial $f(X)$ such that $\bigcap_{n:1}^{\infty}(c(f))^n \neq (0)$. Then J is almost principal.*

Proof By Lemma 3.6, it suffices to show that the ideal $I_1 = JK[X] \cap A[X]$ is almost principal. Consider the ideal $I = f(X)K[X] \cap A[X] \subseteq I_1$. Since $I_1 = f_1(X)K[X] \cap A[X]$ where $f_1(X) \in J$ a nonconstant polynomial, by Lemma 3.5, it suffices to show that I is almost principal. Let $0 \neq a \in \bigcap_{n:1}^{\infty}(c(f))^n$. For each

$n \in \mathbb{N}^*$, let $J_n = c(f)^n \big(A : c(f)^n\big)$. Then $a = a.1 \in J_n$. So $\bigcap_{n:1}^{\infty} J_n \neq (0)$. By
Proposition 5.12, the ideal I is almost principal.   □

6 Applications to Semi-Normal Integral Domains

Definition 6.1 Let A be an integral domain with quotient field K. We say that A is semi-normal if for each $x \in K$ satisfying $x^2, x^3 \in A$, the element $x \in A$.

Examples

(1) An integrally closed integral domain is semi-normal.
(2) Let X be an indeterminate over a field K. Then $K[X^2, X^3]$ is not semi-normal. The goal of the following example is to construct a semi-normal domain that is not integrally closed.
(3) Let t be an indeterminate over a field K and $A = K\big[t(t-1), t^2(t-1)\big] \subset K[t]$. Since $t = \frac{t^2(t-1)}{t(t-1)}$, the quotient field of A is $K(t)$, and since t is a root of the polynomial $f(X) = X^2 - X - t(t-1) \in A[X]$, the extension $A \subset K[t]$ is integral. But $K[t]$ is integrally closed, and then the integral closure of A is equal to $K[t]$. So A is not integrally closed. We will show that $A = \{f(t) \in K[t]; f(0) = f(1)\}$. The direct inclusion is clear. Note first that for each $n \in \mathbb{N}^*$, $t^n(t-1) \in A$. Indeed, the property is clear for $n = 1$ and $n = 2$. Suppose that $n \geq 3$ and that the property is true until the order n. We multiply the equality $t^2 = t(t-1) + t$ by $t^{n-1}(t-1)$, and we obtain $t^{n+1}(t-1) = t(t-1).t^{n-1}(t-1) + t^n(t-1) \in A$. We deduce that for two integers $1 \leq k < n$, $t^n - t^k \in A$. Indeed, $t^n - t^k = t^k(t^{n-k} - 1) = t^k(t-1)(t^{n-k-1} + t^{n-k-2} + \ldots + t + 1) = \sum_{i:0}^{n-k-1} t^{k+i}(t-1) \in A$. Let now $f(t) = a_0 + a_1 t + \ldots + a_n t^n \in K[t]$ such that $f(0) = f(1)$ with $n \geq 2$. Then $a_0 = a_0 + a_1 + \ldots + a_n$, so $a_n = -a_1 - a_2 - \ldots - a_{n-1}$. Then $f(t) = a_0 - a_1(t^n - t) - a_2(t^n - t^2) - \ldots - a_{n-1}(t^n - t^{n-1}) \in A$. We will show that A is semi-normal. Let $f(t) \in K(t)$ be such that $f(t)^2, f(t)^3 \in A$. Then $f(t)$ is integral over A, so $f(t) \in K[t]$. Since $f(0)^2 = f(1)^2$ and $f(0)^3 = f(1)^3$, then $\big(f(0) - f(1)\big)^3 = f(0)^3 - 3f(0)^2 f(1) + 3f(0)f(1)^2 - f(1)^3 = -3f(0)^2 f(1) + 3f(0)f(1)^2 = -3f(1)^3 + 3f(0)^3 = 0$. So $f(0) = f(1)$, and then $f \in A$.

Proposition 6.2 (Rush 1980) *Let A be an integral domain with quotient field K. The following assertions are equivalent:*

1. The domain A is semi-normal.

2. *For each $x \in K$, if there exists $n \in \mathbb{N}^*$ such that $x^n, x^{n+1}, \ldots \in A$, then $x \in A$.*
3. *If $x \in K$ and $m, n \in \mathbb{N}^*$ are two relatively prime integers such that $x^m, x^n \in A$, then $x \in A$.*

Proof "(1) $\Longrightarrow$ (2)" Let $x \in K$ and $n \in \mathbb{N}^*$ such that $x^n, x^{n+1}, \ldots \in A$. If $n \geq 2$, then $(x^{n-1})^2, (x^{n-1})^3 \in A$ since $3(n-1) > 2(n-1) \geq n$. Since A is semi-normal, then $x^{n-1} \in A$, so $x^{n-1}, x^n, x^{n+1}, \ldots \in A$. By induction, we show that $x \in A$.

"(2) $\Longrightarrow$ (3)" Let $x \in K$ and $m, n \in \mathbb{N}^*$ be the two relatively prime integers such that $x^m, x^n \in A$. By Bézout identity, there exist $u, v \in \mathbb{Z}$ such that $um + vn = 1$, and then $\frac{u}{n} + \frac{v}{m} = \frac{1}{mn}$. For each integer $k \geq mn$, $\frac{uk}{n} - (-\frac{vk}{m}) = \frac{k}{mn} \geq 1$, there exists a relative integer t such that $-\frac{vk}{m} \leq t \leq \frac{uk}{n}$. The integers $u' = uk - nt$ and $v' = vk + mt$ are positive and satisfy $u'm + v'n = (uk - nt)m + (vk + mt)n = (um + vn)k = k$. Then $x^k = (x^m)^{u'}(x^n)^{v'} \in A$ so $x \in A$.

"(3) $\Longrightarrow$ (1)" Clear. $\square$

Example (Fontana et al. (1993)) Let X, Y, Z be indeterminates over a field K and $A = K\big[\{X^n Z, n \in \mathbb{N}\}; \{Y^n Z, n \in \mathbb{N}\}\big] \subset K[X, Y, Z]$. The quotient field of A is $K(X, Y, Z)$ since $X = \frac{XZ}{Z}$ and $Y = \frac{YZ}{Z}$. The domain A is not semi-normal since for each integer $n \geq 2$, $(XYZ)^n = (X^n Z)(Y^n Z^{n-1}) \in A$ but $XYZ \notin A$.

Proposition 6.3 (Fontana et al. (1993)) *Let A be a semi-normal integral domain and I a nonzero integral ideal of A such that $(A : I)$ is a ring. Then $(A : I) = (A : \sqrt{I}) = (\sqrt{I} : \sqrt{I})$.*

Proof The inclusions $(\sqrt{I} : \sqrt{I}) \subseteq (A : \sqrt{I}) \subseteq (A : I)$ are true without a supplementary condition on A. Indeed, the first one is in notations after Corollary 5.8. On the other hand, $I \subseteq \sqrt{I}$, and then $(A : \sqrt{I}) \subseteq (A : I)$. We only need to show that $(A : I) \subseteq (\sqrt{I} : \sqrt{I})$. Let $x \in (A : I)$, which is a ring. Then $x^n \in (A : I)$ for each $n \in \mathbb{N}^*$. Let $y \in \sqrt{I}$. Then $y^m \in I$ pour $m \gg 0$. So, for $m \gg 0$, $(xy)^m = x^m y^m \in (A : I)I \subseteq A$. By the preceding proposition, since A is semi-normal, then $xy \in A$. Also, for $m \gg 0$, $(xy)^{m+1} = y(x^{m+1}y^m) \in y(A : I)I \subseteq yA \subseteq \sqrt{I}$, and then $xy \in \sqrt{I}$. So $x \in (\sqrt{I} : \sqrt{I})$. Then $(A : I) \subseteq (\sqrt{I} : \sqrt{I})$. $\square$

Example (Fontana et al. 1993) The hypothesis "A is semi-normal" is necessary in the proposition. Let X be an indeterminate over a field K and $A = K[[X^2, X^5]] \subset K[[X]]$. The domain A is local with maximal ideal (X^2, X^5) and of quotient field $K((X))$. It is not semi-normal since $X^2, X^5 \in A$ but $X \notin A$. Consider the ideal $I = (X^4, X^5)$ of A. For each integer $n \geq 4$, $X^n \in I$. Indeed, it suffices to multiply X^4 and X^5 by X^{2n} with $n \in \mathbb{N}^*$. In particular, $X^2, X^5 \in \sqrt{I}$. Then $(X^2, X^5) \subseteq \sqrt{I}$, and since (X^2, X^5) is a maximal ideal of A, then $\sqrt{I} = (X^2, X^5)$.

We will show that $(A : I) = K[[X]]$ is a ring. Let $f(X) \in K[[X]]$. A monomial of $f(X)$ is of either the form aX^{2i} or aX^{2i+1} with $a \in K$ and $i \in \mathbb{N}$. But $aX^{2i} X^4 = aX^{2(i+2)} = a(X^2)^{i+2}, aX^{2i+1} X^4 = a(X^2)^i X^5, aX^{2i} X^5 = a(X^2)^i X^5$, and $aX^{2i+1}X^5 = a(X^2)^{i+3}$. Then $X^4f(X)$ and $X^5f(X) \in K[[X^2, X^5]] = A$. So $f(X) \in (A : I)$. Then $K[[X]] \subseteq (A : I)$. Conversely, if $f(X) \in (A : I) \subseteq K((X))$, then $X^4f(X)$ and $X^5f(X) \in A = K[[X^2, X^5]] \subseteq K[[X]]$. The order $\omega(f(X)) \geq -4$. If we put $f(X) = a_0X^{-4} + a_1X^{-3} + a_2X^{-2} + a_3X^{-1} + a_4 + a_5X + \ldots$, then $X^4f(X) = a_0 + a_1X + a_2X^2 + a_3X^3 + a_4X^4 + a_5X^5 + \ldots \in K[[X^2, X^5]]$ (1) and $X^5f(X) = a_0X + a_1X^2 + a_2X^3 + a_3X^4 + a_4X^5 + a_5X^6 + \ldots \in K[[X^2, X^5]]$ (2). By (1), $a_1 = a_3 = 0$, and by (2), $a_0 = a_2 = 0$, then $f(X) = a_4 + a_5X + \ldots \in K[[X]]$. We obtain the equality $(A : I) = K[[X]]$.

We will show that $(\sqrt{I} : \sqrt{I}) = K[[X^2, X^3]]$. Let $f(X) \in K[[X^2, X^3]]$. A monomial of $f(X)$ is of the form aX^{2i+3j} with $a \in K$ and $i, j \in \mathbb{N}$. If $j = 2l$, $aX^{2i+3j}X^2 = aX^{2(i+3l)}X^2 = a(X^2)^{i+3l}X^2$, and if $j = 2l + 1$, $aX^{2i+3j}X^2 = aX^{2(i+3l)}X^5 = a(X^2)^{i+3l}X^5$. Then we have $X^2f(X) \in \sqrt{I}$. In the same way, if $j = 2l$, $aX^{2i+3j}X^5 = aX^{2(i+3l)}X^5 = a(X^2)^{i+3l}X^5$, and if $j = 2l+1$, $aX^{2i+3j}X^5 = aX^{2(i+3l+3)}X^2 = a(X^2)^{i+3l+3}X^2$. Then $X^5f(X) \in \sqrt{I}$. So $K[[X^2, X^3]] \subseteq (\sqrt{I} : \sqrt{I})$. Conversely, let $f(X) \in (\sqrt{I} : \sqrt{I}) \subseteq K((X))$. Then $X^2f(X)$ and $X^5f(X) \in \sqrt{I} = (X^2, X^5)$, ideal of the domain $A = K[[X^2, X^5]]$. There exist $g(X), h(X) \in K[[X^2, X^5]]$ such that $X^2f(X) = X^2g(X) + X^5h(X)$. Then $f(X) = g(X) + X^3h(X) \in K[[X^2, X^5]] + X^3K[[X^2, X^5]] \subseteq K[[X^2, X^3]]$. It is clear that $(A : I) = K[[X]] \neq (\sqrt{I} : \sqrt{I}) = K[[X^2, X^3]]$.

Corollary 6.4 *Let A be a semi-normal integral domain and I a nonzero integral ideal of A. Put $J = \sqrt{I(A : I)}$. For each $n \in \mathbb{N}^*$, $(I^n : I^n) \subseteq (J : J)$.*

Proof For each $n \in \mathbb{N}^*$, let $J_n = I^n(A : I^n)$. By Bass example, $(A : J_n) = (J_n : J_n)$ is a ring. By Lemma 5.11 (2), $\sqrt{I^n(A : I^n)} = \sqrt{I(A : I)}$, and then $\sqrt{J_n} = J$. We also have $(I^n : I^n) \subseteq (J_n : J_n)$ since $(I^n : I^n)J_n = (I^n : I^n)I^n(A : I^n) \subseteq I^n(A : I^n) = J_n$. Since A is semi-normal, by the preceding proposition, $(A : J_n) = (A : \sqrt{J_n}) = (\sqrt{J_n} : \sqrt{J_n})$, then $(A : J_n) = (A : J) = (J : J)$. So $(I^n : I^n) \subseteq (J_n : J_n) = (A : J_n) = (A : J) = (J : J)$. Then $(I^n : I^n) \subseteq (J : J)$. □

The semi-normality of an integral domain A implies that the polynomial ring $A[X]$ is almost principal.

Proposition 6.5 *Let A be a semi-normal integral domain. Then the polynomial ring $A[X]$ is almost principal.*

Proof Let K be the quotient field of A, $f(X) \in A[X]$ a nonconstant polynomial, $I = f(X)K[X] \cap A[X]$ and $J = \sqrt{c(f)(A : c(f))}$. By the preceding corollary, for each $n \in \mathbb{N}^*$, $(c(f)^n : c(f)^n) \subseteq (J : J)$. By Corollary 5.2, the ideal I is almost principal. We conclude by Proposition 3.7. □

Example 1 Let t be an indeterminate over a field K. By Example 3 of Definition 6.1, the domain $A = K[t(t-1), t^2(t-1)]$ is semi-normal. Then the polynomial ring $A[X]$ is almost principal. Since A is not integrally closed, the preceding proposition improves Proposition 2.5.

Example 2 Let s and t be the two indeterminates over a field F and $A = F[s, \{st^{2^n}, n \in \mathbb{N}\}]$. The quotient field of A is $F(s, t)$. The ring $A[X]$ is not almost principal. The domain A is not semi-normal. Indeed, $(st^3)^2 = s^2 t^6 = (st^2)(st^{2^2}) \in A$ and $(st^3)^3 = s^3 t^9 = s(st^{2^3})(st) \in A$. But $st^3 \notin A$ since $\phi(3) = 2 > 1$, where $\phi(n)$ is the number of 1 in the development of the natural integer n in the binary basis.

Definition 6.6 Let A be an integral domain with quotient field K and $n \geq 2$ an integer. We say that A is $n-$root closed if for each $x \in K$ satisfying $x^n \in A$, we have $x \in A$. If A is $n-$root closed for each $n \in \mathbb{N}^*$, we say that A is root closed.

Example An integrally closed domain is radically closed.

The following lemma and its examples study the link between the semi-normal and radically closed properties.

Lemma 6.7 *Let A be an integral domain. If A is $n-$root closed for some integer $n \geq 2$, then it is semi-normal.*

Proof Let K be the quotient field of A and $x \in K$ such that $x^2, x^3 \in A$. Since each integer $m \geq 2$ can be written $m = 2i + 3j$ with $i, j \in \mathbb{N}$, then $x^n \in A$, so $x \in A$. □

Example 1 Let t be an indeterminate over the field $\mathbb{R}$. By Example 3 of Definition 6.1, the domain $A = \mathbb{R}[t(t-1), t^2(t-1)] = \{f(t) \in \mathbb{R}[t]; \ f(0) = f(1)\}$ is semi-normal. But for each $n \in \mathbb{N}^*$, the domain A is not $2n-$root closed. Indeed, let $f(t) = -2t + 1 \in \mathbb{R}[t] \subseteq \mathbb{R}(t)$, the quotient field of A. Then $f^{2n} = (4t^2 + 1 - 4t)^n = (4t(t-1) + 1)^n \in A$. Since $f(0) = 1 \neq f(1) = -1$, then $f \notin A$.

Example 2 In many steps:

(1) Let $d \geq 2$ be a square-free integer. The quotient field of the domain $A = \mathbb{Z}[\sqrt{d}]$ is $\mathbb{Q}(\sqrt{d}) = \{a + b\sqrt{d}; \ a, b \in \mathbb{Q}\}$. Its integral closure is $\overline{A} = \{a + b\sqrt{d}; \ a, b \in \mathbb{Q}, \ 2a = u \in \mathbb{Z}, \ 2b = v \in \mathbb{Z}, \ u^2 - dv^2 \equiv 0 \ (mod \ 4)\}$. Indeed, note that A is integral over $\mathbb{Z}$. Let $x \in \overline{A}, x = a + b\sqrt{d}$ with $a, b \in \mathbb{Q}$. If $b = 0$, then $x = a$ is an element of $\mathbb{Q}$ integral over $\mathbb{Z}$ and then $x \in \mathbb{Z}$, and we have the result. If $b \neq 0, x \notin \mathbb{Q}$. Let $P(X) = (X - (a + b\sqrt{d}))(X - (a - b\sqrt{d})) = X^2 - 2aX + a^2 - b^2 d \in \mathbb{Q}[X]$ a monic polynomial of degree 2. It is then the minimal polynomial of $x = a + b\sqrt{d}$ over $\mathbb{Q}$. Since x is integral over $\mathbb{Z}$, then it is a root of a monic polynomial $f(X) \in \mathbb{Z}[X]$. There exists a monic polynomial $Q(X) \in \mathbb{Q}[X]$ such that $P(X)Q(X) = f(X)$. By Lemma 2.4, $P(X) \in \mathbb{Z}[X]$, then $u = 2a \in \mathbb{Z}$ and $a^2 - b^2 d \in \mathbb{Z}$. So $(2a)^2 - (2b)^2 d = 4(a^2 - b^2 d) \in 4\mathbb{Z}$, and then $(2b)^2 d \in \mathbb{Z}$. Since $d \in \mathbb{Z}$ is square-free, the

denominator of the rational number $2b$ is equal to 1, and then $v = 2b \in \mathbb{Z}$. Finally, $u^2 - dv^2 = (2a)^2 - (2b)^2 d = 4(a^2 - b^2 d) \in 4\mathbb{Z}$, and then $u^2 - dv^2 \equiv 0 \ (mod \ 4)$. Conversely, let $x = a + b\sqrt{d}$ with $a, b \in \mathbb{Q}$, $2a = u \in \mathbb{Z}$, $2b = v \in \mathbb{Z}$, and $u^2 - dv^2 \equiv 0 \ (mod \ 4)$. Then $x = a + b\sqrt{d}$ is a root of the polynomial $P(X) = X^2 - 2aX + (a^2 - b^2 d) = X^2 - uX + \frac{u^2 - v^2 d}{4} \in \mathbb{Z}[X]$. Then x is integral over $\mathbb{Z}$ and so over A.

(2) Suppose that $d \equiv 1 \ (mod \ 4)$. Then $\overline{A} = \{\frac{u}{2} + \frac{v}{2}\sqrt{d}$, with $u, v \in \mathbb{Z}$ having the same parity $\} \neq A$. Indeed, let $y = a + b\sqrt{d} \in \overline{A}$ with $a, b \in \mathbb{Q}$. By (1), $2a = u \in \mathbb{Z}$, $2b = v \in \mathbb{Z}$, and $u^2 - dv^2 \equiv 0 \ (mod \ 4)$. We examine all the possibilities for u and v.

u	Even	Even	Odd	Odd
v	Even	Odd	Even	Odd
$u^2 - dv^2 \equiv$	0	3	1	0(mod 4)

Conversely, let u and $v \in \mathbb{Z}$ having the same parity and $y = \frac{u}{2} + \frac{v}{2}\sqrt{d}$. If u and v are even, then $y \in A \subseteq \overline{A}$. If $u = 2k + 1$ and $v = 2l + 1$ are odd integers with k and $l \in \mathbb{Z}$, then $y = k + l\sqrt{d} + \frac{1 + \sqrt{d}}{2}$. Since $x = \frac{1 + \sqrt{d}}{2}$ satisfies $x^2 - x - \frac{d-1}{4} = 0$ with $\frac{d-1}{4} \in \mathbb{Z}$, then x is integral over $\mathbb{Z}$ and $y \in \overline{A}$. The element $x = \frac{1 + \sqrt{d}}{2} \in \overline{A} \setminus A$.

(3) Let $d \geq 2$ be a square-free integer with $d \equiv 1 \ (mod \ 4)$. Then $\{1, \frac{1 + \sqrt{d}}{2}\}$ is a basis for the $\mathbb{Z}$−module $\overline{A}$. In particular, $\overline{A} = \mathbb{Z}[x]$ with $x = \frac{1 + \sqrt{d}}{2}$. Indeed, let $y \in \overline{A}$. Then $y = \frac{u}{2} + \frac{v}{2}\sqrt{d}$ with u and $v \in \mathbb{Z}$ are integers having the same parity. If $u = 2a$ and $v = 2b$ are even integers with $a, b \in \mathbb{Z}$, then $y = a + b\sqrt{d} = (a - b).1 + 2b\frac{1 + \sqrt{d}}{2} = (a - b).1 + vx$. If $u = 2k + 1$ and $v = 2l + 1$ are odd integers with k and $l \in \mathbb{Z}$, then $y = k + l\sqrt{d} + \frac{1 + \sqrt{d}}{2} = (k - l).1 + (2l + 1)\frac{1 + \sqrt{d}}{2} = (k - l).1 + (2l + 1)x$.

Let $a, b \in \mathbb{Z}$ such that $a.1 + b\frac{1 + \sqrt{d}}{2} = 0$. Then $2a + b + b\sqrt{d} = 0$, so $b = a = 0$. Then $\{1, x\}$ is a basis of the $\mathbb{Z}$−module $\overline{A}$. The rest is clear.

(4) Let $d \geq 2$ be a square-free integer with $d \equiv 1 \ (mod \ 4)$. The domain $A = \mathbb{Z}[\sqrt{d}\,]$ is semi-normal but not integrally closed. Indeed, by (3), $\overline{A} = \{a + bx;\ a, b \in \mathbb{Z}\}$ with $x = \frac{1 + \sqrt{d}}{2}$ satisfying $x^2 - x - \frac{d-1}{4} = 0$. Let $\alpha \in \overline{A}$, $\alpha = a + bx = a + b\frac{1 + \sqrt{d}}{2} = a + \frac{b}{2} + \frac{b}{2}\sqrt{d}$ with $a, b \in \mathbb{Z}$. We see that $\alpha \in A = \mathbb{Z}[\sqrt{d}\,]$ if and only if the integer b is even. We suppose in the sequel that $\alpha \in \overline{A} \setminus A$, then b is an odd integer. We have $\alpha^2 = (a + bx)^2 = a^2 + b^2 x^2 + 2abx = a^2 + b^2\left(x - \frac{1-d}{4}\right) + 2abx = (2a + b)bx + a^2 - b^2\frac{1-d}{4} = (2a + b)(a + bx) - 2a^2 - ab + a^2 - b^2\frac{1-d}{4} = (2a + b)\alpha - a(a + b) - b^2\frac{1-d}{4} = T\alpha - N$.

The integer $T = 2a + b$ is odd. Since the integer $a(a + b)$ is always even, then $N = a(a + b) + b^2\frac{1-d}{4}$ has the same parity as the integer $\frac{1-d}{4}$. Then

$\alpha^2 = T\alpha - N = T(a+bx) - N = Tbx + Ta - N$ with Tb an odd integer, so $\alpha^2 \notin A$. Then A is semi-normal.

(5) Angermuller (1983). An integral domain may be radically closed but not integrally closed. It may be semi-normal but not n−radically closed with $n \geq 2$ an integer. Let $d \geq 2$ be a square-free integer and $A = \mathbb{Z}[\sqrt{d}\,]$.

If $d \equiv 5 \ (mod \ 8)$, for each $\alpha \in \overline{A}$, $\alpha^3 \in A$, then A is not 3−radically closed.

If $d \equiv 1 \ (mod \ 8)$, then A is radically closed but not integrally closed.

In both cases, the domain A is semi-normal.

Indeed, in both cases, $d \equiv 1 \ (mod \ 4)$. By (4), A is semi-normal but not integrally closed. By (3), $\overline{A} = \mathbb{Z}[x] = \{a + bx; \ a, b \in \mathbb{Z}\}$ where $x = \frac{1+\sqrt{d}}{2}$ satisfying $x^2 - x - \frac{d-1}{4} = 0$. Let $\alpha \in \overline{A} \setminus A$. Then $\alpha = a + bx$ with $a, b \in \mathbb{Z}$ and b is odd. With the notations of (4), $\alpha^2 = T\alpha - N$ where $T = 2a + b$ is an odd integer and $N = a(a+b) + b^2 \frac{1-d}{4}$ is an integer with the same parity as the integer $\frac{1-d}{4}$.

(a) If $d \equiv 5 \ (mod \ 8)$, then $d = 5 + 8k$ with $k \in \mathbb{Z}$, so $\frac{d-1}{4} = 1 + 2k$ is odd and N is also odd. We have $\alpha^3 = T\alpha^2 - N\alpha = T(T\alpha - N) - N\alpha = (T^2 - N)\alpha - TN = (T^2 - N)(a+bx) - TN = (T^2 - N)a - TN + (T^2 - N)bx \in A$ since $T^2 - N$ is even. Then A is not 3−radically closed.

(b) If $d \equiv 1 \ (mod \ 8)$, then $d = 1 + 8k$ with $k \in \mathbb{Z}$, so $\frac{d-1}{4} = 2k$ and then N is even. We will prove by induction that for each integer $n \geq 2$, there exists an even integer N_n and an odd integer T_n such that $\alpha^n = T_n\alpha + N_n$. Since $\alpha^2 = T\alpha - N$, the result is true for $n = 2$. On the other hand, $\alpha^{n+1} = (T_n\alpha + N_n)\alpha = T_n\alpha^2 + N_n\alpha = T_n(T\alpha - N) + N_n\alpha = (T_nT + N_n)\alpha - T_nN$ where T_nN is even since N is even. But $T_nT + N_n$ is odd since T and T_n are odd and N_n is even. To conclude, note that for each integer $n \geq 2$, $\alpha^n = T_n\alpha + N_n = T_n(a+bx) + N_n = T_na + N_n + T_nbx \notin A$ since T_nb is odd.

Corollary 6.8 *Let A be an integral domain that is n−radically closed for each integer $n \geq 2$. Then the polynomial ring $A[X]$ is almost principal.*

In Proposition 1.5, the reason for which the polynomial ring $A[X]$ is not almost principal is the existence of an element $t \in K$ that is not quasi-integral, but there is $0 \neq a \in A$ such that $at^i \in A$ for infinitely many natural integers i.

Example (Anderson and Zafrullah 2007) An example of Arnold type does not exist in a radically closed domain. Let A be an integral domain that is radically closed, K its quotient field, and $t \in K^*$. Then t is quasi-integral over A if and only if there exists $0 \neq a \in A$ such that $at^n \in A$ for infinitely many natural integers n. Indeed, let $m \in \mathbb{N}^*$ be any integer. We will prove that $at^m \in A$. Let $n \geq m$ be an integer such that $at^n \in A$. Then $(at^m)^n = a^{n-m}(at^n)^m \in A$. Since A is radically closed, $at^m \in A$. Then t is quasi-integral over A.

Lemma-Definition 6.9 *Let A be an integral domain.*

1. *The intersection of all the over rings of A that are semi-normal is the smallest semi-normal over ring of A. It is called the semi-normalization of A.*
2. *Let $n \geq 2$ be an integer. The intersection of all the over rings of A that are n−radically closed is the smallest n−radically closed over ring of A. It is called the n−radical closure of A.*

Proof Clear. ☐

Example 1 Let t be an indeterminate over a field K. The domains $K[t^2, t^3] \subset K[t]$ have the same quotient field $K(t)$. The domain $K[t^2, t^3]$ is not semi-normal. But $K[t]$ is integrally closed and then semi-normal. Let S be the semi-normalization of $K[t^2, t^3]$. Then $S \subseteq K[t]$. Conversely, since $t^2, t^3 \in K[t^2, t^3] \subset S$, then $t \in S$, so $K[t] \subseteq S$, and then $S = K[t]$.

Example 2 Let $d \geq 2$ be a square-free integer such that $d \equiv 1 \ (mod \ 8)$. The domains $A = \mathbb{Z}[2\sqrt{d}] \subset B = \mathbb{Z}[\sqrt{d}]$ have the same quotient field $K = \mathbb{Q}(\sqrt{d}) = \{a + b\sqrt{d}; \ a, b \in \mathbb{Q}\}$. Let R be the 2−radical closure of A. Since B is radically closed, $R \subseteq B$. Conversely, let $y = a + b\sqrt{d} \in B$ with $a, b \in \mathbb{Z}$. Then $y^2 = a^2 + b^2 d + ab2\sqrt{d} \in A \subseteq R$, so $y \in R$. Then $R = B$.

Corollary 6.10

1. *Let A be an integral domain and S its normalization. If the conductor $(A : S) \neq (0)$, the polynomial ring $A[X]$ is almost principal.*
2. *Let A be an integral domain, $n \geq 2$ an integer and R the n−radical closure of A. If the conductor $(A : R) \neq (0)$, the polynomial ring $A[X]$ is almost principal.*

Proof By Proposition 6.5 and Corollary 6.8, the rings $S[X]$ and $R[X]$ are almost principal. By Proposition 2.2, since the conductors $(A : S) \neq (0)$ and $(A : R) \neq (0)$, the ring $A[X]$ is almost principal. ☐

Other Proof Let A be an integral domain with quotient field K and semi-normalization S such that the conductor $C = (A : S) \neq (0)$. Let $f(X) \in A[X]$ be a nonconstant polynomial and $I = f(X)K[X] \cap A[X]$. Let $c_A(f)$ be the content of $f(X)$ in A and $c_S(f)$ its content in S. Let $J = \sqrt{c_S(f)(S : c_S(f))}$ be the radical in S. By the proof of Proposition 6.5, for each $n \in \mathbb{N}^*$, $(c_S(f)^n : c_S(f)^n) \subseteq (J : J)$. Note that CJ is a nonzero common integral ideal of A and S and $(J : J) \subseteq (CJ : CJ)$. Also $(c_A(f)^n : c_A(f)^n) \subseteq (c_S(f)^n : c_S(f)^n)$. Then for each $n \in \mathbb{N}^*$, $(c_A(f)^n : c_A(f)^n) \subseteq (CJ : CJ)$. By Corollary 5.2, the ideal I is almost principal in $A[X]$. We conclude by Proposition 3.7. ☐

Example 1 Let t be an indeterminate over a field K. The semi-normalization of the domain $K[t^2, t^3]$ is $K[t]$. Since the extension $K[t^2, t^3] \subset K[t]$ is integral and $K[t]$ is integrally closed, and then $K[t]$ is the integral closure of $K[t^2, t^3]$. Each integer

$n \geq 2$ may be written $n = 2i + 3j$ with $i, j \in \mathbb{N}$, then $t^2 \in \left(K[t^2, t^3] : K[t]\right)$. By the preceding corollary, the polynomial ring $K[t^2, t^3][X]$ is almost principal.

Example 2 Let $d \geq 2$ be a square-free integer such that $d \equiv 1 \ (mod \ 8)$. The 2−radical closure of $\mathbb{Z}[2\sqrt{d}\,]$ is $\mathbb{Z}[\sqrt{d}\,]$. Note that $2 \in \left(\mathbb{Z}[2\sqrt{d}\,] : \mathbb{Z}[\sqrt{d}\,]\right)$ since for each $a, b \in \mathbb{Z}$, $2(a+b\sqrt{d}) = a+b2\sqrt{d} \in \mathbb{Z}[2\sqrt{d}\,]$. By the preceding corollary, the polynomial ring $\mathbb{Z}[2\sqrt{d}\,][X]$ is almost principal. By Example 2 of Lemma 6.7, the integral closure of $\mathbb{Z}[\sqrt{d}\,]$ is $\mathbb{Z}[x]$, with $x = \frac{1+\sqrt{d}}{2}$, and $\{1, x\}$ is a basis of the $\mathbb{Z}$−module $\mathbb{Z}[x]$. The extension $\mathbb{Z}[2\sqrt{d}\,] \subset \mathbb{Z}[\sqrt{d}\,]$ is integral since for each $y \in \mathbb{Z}[\sqrt{d}\,]$, $y = a+b\sqrt{d}$ with $a, b \in \mathbb{Z}$, and then $y^2 = a^2+b^2d+ab2\sqrt{d} \in \mathbb{Z}[2\sqrt{d}\,]$. Then the integral closure of $\mathbb{Z}[2\sqrt{d}\,]$ is $\mathbb{Z}[x]$. Since $4x = 2 + 2\sqrt{d} \in \mathbb{Z}[2\sqrt{d}\,]$, then $4 \in \left(\mathbb{Z}[2\sqrt{d}\,] : \mathbb{Z}[x]\right)$.

In the two preceding examples, the conductor $(A : \overline{A}) \neq (0)$. To show the importance of the preceding corollary, we must give an example of an integral domain A with $(A : \overline{A}) = (0)$ but $(A : S) \neq (0)$, where $\overline{A}$ is the integral closure of A and S is its semi-normalization A. To do this, we need the following two propositions.

Proposition 6.11 (Brewer et al. 1979) *Let A be a semi-normal integral domain. The polynomial ring $A[X]$ is semi-normal.*

Proof Let K be the quotient field of A. Then $K(X)$ is the quotient field of $A[X]$. Suppose that $A[X]$ is not semi-normal. There exists $f \in K(X) \setminus A[X]$ such that $f^2, f^3 \in A[X]$. Since f is integral over $A[X] \subseteq K[X]$ and $K[X]$ is integrally closed, then $f \in K[X]$. Choose a minimal counterexample f, i.e, $f = a_0 + a_1 + \ldots + a_n X^n$ of minimal degree n, and among all the counterexamples of degree n, f has the biggest initial part in $A[X]$, i.e, $a_0, a_1, \ldots, a_{k-1} \in A$ but $a_k \notin A$. Since $f^2, f^3 \in A[X]$, then $a_0^2, a_0^3 \in A$, so $a_0 \in A$. Note that for each $r \in A$ such that $ra_k \in A$, $rf \in A[X]$ since $(rf)^2 = r^2 f^2 \in A[X]$ and $(rf)^3 = r^3 f^3 \in A[X]$ with $ra_0, ra_1, \ldots, ra_{k-1}, ra_k \in A$. The coefficient of X^k in f^2 is $2a_0a_k$+terms in A, and then $2a_0a_k \in A$. The coefficient of X^k in f^3 est $3a_0^2a_k$+terms in A, and then $3a_0^2a_k \in A$. By the preceding step, $2a_0f, 3a_0^2f \in A[X]$. We have $f = a_0 + Xg$ with $g \in K[X]$, then $(Xg)^2 = (f - a_0)^2 = f^2 - 2a_0f + a_0^2 \in A[X]$ and $(Xg)^3 = (f - a_0)^3 = f^3 - a_0^3 - 3a_0f^2 + 3a_0^2f \in A[X]$. So $g^2, g^3 \in A[X]$ with $deg \ g < deg \ f$. By the minimality of f as a counterexample, $g \in A[X]$, and then $f = a_0 + Xg \in A[X]$, which is absurd. Then $A[X]$ is semi-normal. $\square$

Proposition 6.12 *Let A be an integral domain with integral closure $\overline{A}$. Then the integral closure of the polynomial domain $A[X]$ is $\overline{A[X]} = \overline{A}[X]$. In particular, if the domain A is integrally closed, then $A[X]$ is integrally closed.*

Proof Since $A \subset A[X]$, then $\overline{A} \subset \overline{A[X]}$, and since $X \in A[X] \subseteq \overline{A[X]}$, then $\overline{A}[X] \subseteq \overline{A[X]}$. Conversely, let K be the quotient field of A. Then $K(X)$ is the quotient field of $A[X]$. If an element f of $K(X)$ is integral $A[X]$, then it is also integral over $K[X]$. But $K[X]$ is integrally closed, and then $f \in K[X]$. So $\overline{A[X]} \subseteq K[X]$. Let $P \in \overline{A[X]}$. There exists a monic polynomial $Q(Y) = F_0 + F_1 Y + \ldots + F_{m-1} Y^{m-1} + Y^m \in A[X][Y]$ such that $Q(P) = 0$. Let $r > max\{deg\, P;\ deg\, F_i,\ 0 \le i \le m-1\}$ a natural integer and $P_1(X) = P(X) - X^r \in K[X]$. Then P_1 is a root of the polynomial $Q_1(Y) = Q(Y + X^r) = G_0 + G_1 Y + \ldots + G_{m-1} Y^{m-1} + Y^m \in A[X][Y]$. So $G_0 + G_1 P_1 + \ldots + G_{m-1} P_1^{m-1} + P_1^m = 0$; then $-P_1(G_1 + \ldots + G_{m-1} P_1^{m-2} + P_1^{m-1}) = G_0 \in A[X]$ and $G_0 = Q_1(0) = Q(X^r) = F_0 + F_1 X^r + \ldots + F_{m-1} X^{r(m-1)} + X^{rm}$. Since $-P_1(X)$ is monic in X and G_0 is monic of degree rm by the choice of r, then $G_1 + \ldots + G_{m-1} P_1^{m-2} + P_1^{m-1}$ is also monic in $K[X]$. By Lemma 2.4, $P_1(X) \in \overline{A[X]}$, and then $P(X) \in \overline{A[X]}$. □

Example Let A be a semi-normal integral domain with integral closure $\overline{A} \ne A$ such that the conductor $(A : \overline{A}) = (0)$. By the preceding two propositions, if t is an indeterminate over A, the domain $A[t]$ is semi-normal of integral closure $\overline{A[t]} = \overline{A}[t]$. The conductor $(A[t] : \overline{A}[t]) = (0)$. Indeed, let $f = a_0 + a_1 t + \ldots + a_n t^n \in A[t]$ be such that $f.\overline{A}[t] \subseteq A[t]$. Then $f.\overline{A} \subseteq A[t]$. For $0 \le i \le n$, $a_i.\overline{A} \subseteq A$, and then $a_i \in (A : \overline{A}) = (0)$ so $f = 0$. Since the extension $A[t^2, t^3] \subset A[t]$ is integral, the integral closure of $A[t^2, t^3]$ is equal to $\overline{A}[t]$, and the conductor $A([t^2, t^3] : \overline{A}[t]) \subseteq (A[t] : \overline{A}[t]) = (0)$ is zero. Let S be the semi-normalization of the domain $A[t^2, t^3]$. Since $A[t]$ is semi-normal, then $S \subseteq A[t]$. Conversely, since $t^2, t^3 \in A[t^2, t^3] \subseteq S$, then $t \in S$, so $A[t] \subseteq S$, and then $S = A[t]$. It is clear that $t^2 \in (A[t^2, t^3] : A[t])$ and then $(A[t^2, t^3] : S) \ne (0)$. By Corollary 6.10, the polynomial ring $A[t^2, t^3][X]$ is almost principal.

7 The Almost Principal Polynomial Rings and the v–Operation

The results of this section are due to Hamann et al. (1988). In the sequel, we fix an integral domain A with quotient field K.

Notation Let $f(X) \in A[X]$ be a nonconstant polynomial. Then $I = f(X)K[X] \cap A[X]$ and $C = \left(f(X)A[X] : I\right)_{A[X]}$ are two integral ideals of $A[X]$ containing $f(X)$.

Lemma 7.1

1. *The ideal I is almost principal in $A[X]$ if and only if $C \cap A \ne (0)$.*
2. *We have the equality $I^{-1} = \frac{1}{f(X)} C$, i.e, $f(X) I^{-1} = C$.*
3. *If for each integer $n \in \mathbb{N}^*$, $I^{-n} = (I^{-1})^n$, then $T_{A[X]}(I) = A\left[\frac{1}{f(X)} C\right]$, where $T_{A[X]}(I)$ is the Nagata transform of $A[X]$ with respect the ideal I and $I^{-n} = (I^n)^{-1} = (A[X] : I^n)$.*

4. *Suppose that $f(X)$ is irreducible in $K[X]$ and C contains a polynomial nondivisible by $f(X)$ in $K[X]$. Then I is almost principal in $A[X]$.*
5. *If $f(X)$ is irreducible in $K[X]$ and I is not almost principal, then $C \subseteq I$.*

Proof

(1) $C \cap A \neq (0) \iff$ There exists $0 \neq s \in A$ such that $sI \subseteq f(X)A[X] \iff$ The ideal I is almost principal in $A[X]$. In the last equivalence, we used Proposition 3.3.

(2) $"\subseteq"$ Since $I^{-1}I \subseteq A[X]$, $(f(X)I^{-1})I \subseteq f(X)A[X]$ with $f(X)I^{-1} \subseteq II^{-1} \subseteq A[X]$. Then $f(X)I^{-1} \subseteq (f(X)A[X] : I)_{A[X]} = C$.

   $"\supseteq"$ Let $g(X) \in C$. Then $g(X) \in A[X]$ and $g(X)I \subseteq f(X)A[X]$, so $\frac{g(X)}{f(X)}I \subseteq A[X]$. Then $\frac{g(X)}{f(X)} \in (A[X] : I) = I^{-1}$, so $g(X) \in f(X)I^{-1}$. Then $C \subseteq f(X)I^{-1}$.

(3) Recall from Lemma 2.10 that $T_{A[X]}(I) = \bigcup_{n:1}^{\infty} I^{-n} = \bigcup_{n:1}^{\infty} (I^{-1})^n$ is an over ring of $A[X]$. By (2), $I^{-1} = \frac{1}{f(X)}C \subseteq A\left[\frac{1}{f(X)}C\right]$, and then $(I^{-1})^n \subseteq A\left[\frac{1}{f(X)}C\right]$. So $T_{A[X]}(I) \subseteq A\left[\frac{1}{f(X)}C\right]$. Conversely, since $\frac{1}{f(X)}C = I^{-1} \subseteq T_{A[X]}(I)$, then $A\left[\frac{1}{f(X)}C\right] \subseteq T_{A[X]}(I)$.

(4) Let $g(X) \in C$ be a polynomial nondivisible by $f(X)$ in $K[X]$. Then $f(X)$ and $g(X)$ are relatively prime in the PID $K[X]$. By Bézout identity, there exist two polynomials $h_1(X)$ and $h_2(X) \in K[X]$ such that $1 = h_1(X)f(X) + h_2(X)g(X)$. Let $0 \neq s \in A$ be such that $sh_1(X)$ and $sh_2(X) \in A[X]$. Then $0 \neq s = sh_1(X)f(X) + sh_2(X)g(X) \in C \cap A$ since $f(X)$ and $g(X) \in C$. By (1), the ideal I is almost principal in $A[X]$.

   We can also observe that $f(X)K[X]$ is a maximal ideal of $K[X]$ and $g(X) \notin f(X)K[X]$. Then $f(X)K[X] + g(X)K[X] = K[X]$ and we finish like before.

(5) By (4), since I is not almost principal in $A[X]$, $C \subseteq f(X)K[X]$. So $C \subseteq f(X)K[X] \cap A[X] = I$. $\square$

Example Let A be an integral domain and I a nonzero integral ideal of A. The equalities $I^{-n} = (I^{-1})^n$, $n \in \mathbb{N}^*$ are not always true. Let $A = \mathbb{Z} + X\mathbb{Q}[X]$ and $I = X\mathbb{Q}[X]$. The quotient field of A is $\mathbb{Q}(X)$. For each $n \in \mathbb{N}^*$, $I^n = X^n\mathbb{Q}[X]$. We have $I^{-1} = (A : I) = \mathbb{Q}[X]$. Indeed, $I\mathbb{Q}[X] = X\mathbb{Q}[X] \subset A$, and then $\mathbb{Q}[X] \subseteq I^{-1}$. Conversely, let $f(X) \in I^{-1} \subseteq \mathbb{Q}(X)$. Then $f(X)I = f(X)X\mathbb{Q}[X] \subseteq A = \mathbb{Z} + X\mathbb{Q}[X] \subset \mathbb{Q}[X]$, so $f(X)X \in \mathbb{Q}[X]$, and then $f(X) \in \frac{1}{X}\mathbb{Q}[X]$. Suppose that $f(X) \notin \mathbb{Q}[X]$ and then $f(X) = \frac{g(X)}{X}$ with $g(X) \in \mathbb{Q}[X]$ and $g(0) \neq 0$. Let p be a prime number sufficiently large in such a way that $\frac{g(0)}{p} \notin \mathbb{Z}$. Since $Xf(X)\mathbb{Q}[X] \subseteq A = \mathbb{Z} + X\mathbb{Q}[X]$, then $Xf(X)\frac{1}{p} \in \mathbb{Z} + X\mathbb{Q}[X]$, so $g(X)\frac{1}{p} \in \mathbb{Z} + X\mathbb{Q}[X]$, and then $\frac{g(0)}{p} \in \mathbb{Z}$, which is absurd. Then $f(X) \in \mathbb{Q}[X]$ and $I^{-1} = \mathbb{Q}[X]$. For each $n \in \mathbb{N}^*$, $(I^{-1})^n = \mathbb{Q}[X]$. We will show that for each integer $n \geq 2$, $I^{-n} = (A : I^n) = \frac{1}{X^{n-1}}\mathbb{Q}[X]$. Indeed, $\frac{1}{X^{n-1}}\mathbb{Q}[X]I^n = \frac{1}{X^{n-1}}\mathbb{Q}[X]X^n\mathbb{Q}[X] = X\mathbb{Q}[X] \subset A$.

Then $\frac{1}{X^{n-1}}\mathbb{Q}[X] \subseteq I^{-n}$. Conversely, let $f(X) \in I^{-n} \subseteq \mathbb{Q}(X)$. Then $f(X)I^n \subseteq A$, so $X^n f(X)\mathbb{Q}[X] \subseteq \mathbb{Z} + X\mathbb{Q}[X] \subset \mathbb{Q}[X]$. Then $X^n f(X) \in \mathbb{Q}[X]$, so $f(X) \in \frac{1}{X^n}\mathbb{Q}[X]$. Suppose that $f(X) \notin \frac{1}{X^{n-1}}\mathbb{Q}[X]$ and then $f(X) = \frac{g(X)}{X^n}$ with $g(X) \in \mathbb{Q}[X]$ and $g(0) \neq 0$. Let p be a prime number sufficiently large in such a way that $\frac{g(0)}{p} \notin \mathbb{Z}$. Since $X^n f(X)\mathbb{Q}[X] \subseteq \mathbb{Z} + X\mathbb{Q}[X]$, and then $X^n f(X)\frac{1}{p} \in \mathbb{Z} + X\mathbb{Q}[X]$, so $g(X)\frac{1}{p} \in \mathbb{Z} + X\mathbb{Q}[X]$, then $\frac{g(0)}{p} \in \mathbb{Z}$, which is absurd. So $f(X) \in \frac{1}{X^{n-1}}\mathbb{Q}[X]$ and $I^{-n} = (A : I^n) = \frac{1}{X^{n-1}}\mathbb{Q}[X] \neq (I^{-1})^n = \mathbb{Q}[X]$.

Lemma 7.2 *We have the equality* $I^{-1} \cap K[X] = (I : I)$.

Proof "$\subseteq$" Let $h \in I^{-1} \cap K[X]$. Then $hI \subseteq A[X]$. Since $h \in K[X]$ and $I \subseteq f K[X]$, then $hI \subseteq hf K[X] \subseteq f K[X]$. So $hI \subseteq f K[X] \cap A[X] = I$, and then $h \in (I : I)$.
"$\supseteq$" By the notations following Corollary 5.8, since I is an integral ideal, then $(I : I) \subseteq I^{-1}$. Let $h \in (I : I)$. Then $hI \subseteq I \subseteq f K[X]$. Since $f \in I$, then $hf \in f K[X]$, so $h \in K[X]$. Then $(I : I) \subseteq K[X]$. So $(I : I) \subseteq I^{-1} \cap K[X]$. $\square$

In the following proposition, we suppose that the polynomial $f(X) \in A[X]$ is irreducible in $K[X]$. We give a characterization of the ideal $I = f(X)K[X] \cap A[X]$ to be almost principal in terms of I^{-1} and $(I : I)$.

Proposition 7.3 (Hamann et al. (1988) and Houston and Zafrullah (2005)) *Let $f(X) \in A[X]$ be an irreducible polynomial in $K[X]$ and $I = f(X)K[X] \cap A[X]$. The following assertions are equivalent:*

1. *The ideal I is almost principal in the ring $A[X]$.*
2. $I^{-1} K[X] = \frac{1}{f}K[X]$, *i.e,* $C K[X] = K[X]$.
3. $I^{-1} \nsubseteq K[X]$.
4. I^{-1} *is not a ring.*
5. $I^{-1} \neq (I : I)$.
6. *There exists $g(X) \in A[X] \setminus I$ such that $g(X)I \subseteq f(X)A[X]$.*
7. *There exists $0 \neq a \in A$ such that $I = (fA[X] : a)_{A[X]}$.*
8. *There exists $\psi(X) \in K(X)$ such that $I = (A[X] : \psi(X))_{A[X]}$.*

Proof "(1) $\Longrightarrow$ (2)" By Lemma 7.1 (1), since the ideal I is almost principal, $C \cap A \neq (0)$. The ideal C of $A[X]$ contains some nonzero constants, so $C K[X] = K[X]$.
"(2) $\Longrightarrow$ (3)" By (2), $\frac{1}{f} \in I^{-1}K[X]$. Put $\frac{1}{f} = u_1 h_1 + \ldots + u_n h_n$ with $u_1, \ldots, u_n \in I^{-1}$ and $h_1, \ldots, h_n \in K[X]$. Let $0 \neq a \in A$ be such that $ah_1, \ldots, ah_n \in A[X]$. Then $\frac{a}{f} = u_1(ah_1) + \ldots + u_n(ah_n) \in I^{-1}$. Since $f(X)$ is not a constant, $\frac{a}{f} \notin K[X]$. Then $I^{-1} \nsubseteq K[X]$.

$"(3) \implies (4)"$ Let $u \in I^{-1} \setminus K[X]$. By Lemma 7.1 (2), $I^{-1} = \frac{1}{f}C \subseteq \frac{1}{f}A[X]$ and then $u = \frac{g(X)}{f(X)}$ with $g(X) \in A[X]$. Suppose that $u^2 \in I^{-1}$, then $fu^2 \in II^{-1} \subseteq A[X]$. But $fu^2 = \frac{g^2}{f}$, and then f divides g^2 in $K[X]$. Since f is irreducible, then f divides g, so $u = \frac{g(X)}{f(X)} \in K[X]$, which is absurd. Then I^{-1} is not stable by multiplication.

$"(4) \implies (5)"$ By the notations following Corollary 5.8, $(I : I)$ is a ring. Since I^{-1} is not a ring, then $I^{-1} \neq (I : I)$.

$"(5) \implies (6)"$ By hypothesis, $(I : I) \subset I^{-1}$. Let $u \in I^{-1} \setminus (I : I)$. Then $uf \in A[X]$, so $u = \frac{g}{f}$ with $g \in A[X]$. Then $gI = fuI \subseteq fI^{-1}I \subseteq fA[X]$, so $gI \subseteq fA[X]$. By the preceding lemma, $I^{-1} \cap K[X] = (I : I)$, and then $u = \frac{g}{f} \notin K[X]$, so $g \notin fK[X]$. By definition of the ideal I, $g \notin I$, and then $g \in A[X] \setminus I$.

$"(6) \implies (1)"$ By hypothesis, $g(X) \in (f(X)A[X] : I)_{A[X]} = C$ and $g(X) \in A[X] \setminus I$. By definition of I, $g(X) \notin f(X)K[X]$. Then C contains a polynomial nondivisible by $f(X)$ in $K[X]$. By Lemma 7.1 (4), I is almost principal.

$"(1) \implies (7)"$ By Proposition 3.3, since I is almost principal, there exists $0 \neq a \in A$ such that $aI \subseteq fA[X]$, and then $I \subseteq (fA[X] : a)_{A[X]}$. Conversely, let $h \in (fA[X] : a)_{A[X]}$. Then $ah \in fA[X]$, so $h \in f\frac{1}{a}A[X] \subseteq fK[X]$. Then $h \in fK[X] \cap A[X] = I$. So we have $I = (fA[X] : a)_{A[X]}$.

$"(7) \implies (8)"$ By hypothesis, $I = (fA[X] : a)_{A[X]} = \{g(X) \in A[X]; \ ag(X) \in fA[X]\} = \{g(X) \in A[X]; \ g(X)\frac{a}{f(X)} \in A[X]\} = (A[X] : \frac{a}{f(X)})_{A[X]}$ with $\frac{a}{f(X)} \in K(X)$.

$"(8) \implies (6)"$ By hypothesis, $I = (A[X] : \psi(X))_{A[X]}$ with $\psi(X) \in K(X)$. Suppose that $\psi(X) \in K[X]$. There exists $0 \neq a \in A$ such that $a\psi(X) \in A[X]$. Then $a \in (A[X] : \psi(X))_{A[X]} = I = fK[X] \cap A[X] \subseteq fK[X]$, which is absurd since f is not constant. Then $\psi(X) \notin K[X]$. Since $f(X) \in I = (A[X] : \psi(X))_{A[X]}$, then $f(X)\psi(X) \in A[X]$, so $\psi(X) = \frac{g(X)}{f(X)}$ with $g(X) \in A[X] \setminus f(X)K[X]$, since $\psi(X) \notin K[X]$. In particular, $g(X) \notin I$. We have $gI = f(\psi I) \subseteq fA[X]$. Then $g(X) \in A[X] \setminus I$ and $gI \subseteq fA[X]$. $\qquad\square$

The following proposition is in the same way as the preceding one.

Proposition 7.4 *Let $f \in A[X]$ be a nonconstant polynomial and $I = fK[X] \cap A[X]$. Then $(I : I) \neq A[X]$ if and only if there exists $g \in I \setminus fA[X]$ such that $gI \subseteq fA[X]$.*

Proof $"\implies"$ We always have the inclusion $A[X] \subseteq (I : I)$. Let $h \in (I : I) \setminus A[X]$. Then $hI \subseteq I$. So $g = hf \in I \setminus fA[X]$. On the other hand, $gI = fhI \subseteq fI \subseteq fA[X]$.

$"\impliedby"$ By hypothesis, $g(X) \in I \subseteq fK[X]$, and then $g(X) = f(X)h(X)$ with $h(X) \in K[X]$. Since $g \notin fA[X]$, then $h \notin A[X]$. On the other hand, $gI \subseteq fA[X]$ with $g = fh$. By simplification, $hI \subseteq A[X]$. Since $hI \subseteq hfK[X] \subseteq fK[X]$, then $hI \subseteq fK[X] \cap A[X] = I$. So $h \in (I : I) \setminus A[X]$, then $(I : I) \neq A[X]$. $\qquad\square$

Notation Let A be a ring and I an ideal of $A[X]$. Let $I^{ec} = (IA[[X]]) \cap A[X]$, an ideal of the ring $A[X]$ containing I. The following two examples show that the ideals I and I^{ec} may be equal or different.

Example 1 Let J be an ideal of A and $I = J[X] = JA[X]$. Then $IA[[X]] = JA[[X]]$ and $I^{ec} = (IA[[X]]) \cap A[X] = (JA[[X]]) \cap A[X] = I$. Indeed, an element of the intersection is a polynomial of the form $f(X) = a_0 + a_1 X + \ldots + a_n X^n = b_1 g_1(X) + \ldots + b_m g_m(X)$ with $a_k \in A$, $b_k \in J$ and $g_k(X) \in A[[X]]$.

Put $g_k(X) = \displaystyle\sum_{j:0}^{\infty} c_{kj} X^j$ with $c_{kj} \in A$ for $1 \le k \le m$ and $j \in \mathbb{N}$. For $0 \le i \le n$,

$a_i = \displaystyle\sum_{k:1}^{m} b_k c_{ki} \in J$. Then $f(X) = a_0 + a_1 X + \ldots + a_n X^n \in J[X] = I$.

Example 2 Let A be an integral domain and $f(X) \in A[X]$ a nonconstant polynomial such that its constant term is invertible in A. Then $I = f A[X] \neq A[X]$. But $I^{ec} = (f A[[X]]) \cap A[X] = A[[X]] \cap A[X] = A[X] \neq I$.

The following two lemmas are due to Hamann et al. (1988).

Lemma 7.5 *Let A be an integral domain with quotient field K, $f(X) \in A[X]$, and $I = f(X)K[X] \cap A[X]$. Then $I^{ec} = I$ if and only if $(f A[X])^{ec} = f A[X]$.*

Proof Note first that $(f A[X])^{ec} = (f A[[X]]) \cap A[X] \subseteq f A[[X]]$.

" $\Longrightarrow$ " Since $f \in I$, then $(f A[X])^{ec} \subseteq I^{ec} = I \subseteq f K[X]$. Then $(f A[X])^{ec} \subseteq (f K[X]) \cap (f A[[X]]) = f(K[X] \cap A[[X]]) = f A[X]$. So $(f A[X])^{ec} = f A[X]$.

" $\Longleftarrow$ " Let $g \in I^{ec} = (IA[[X]]) \cap A[X] \subseteq IA[[X]]$. Then $g = g_1 h_1 + \ldots + g_n h_n$ with $g_1, \ldots, g_n \in I \subseteq f K[X]$ and $h_1, \ldots, h_n \in A[[X]]$. Let $0 \neq a \in A$ be such that $ag_1, \ldots, ag_n \in f A[X]$. Then $ag = (ag_1)h_1 + \ldots + (ag_n)h_n \in (f A[X])A[[X]] = f A[[X]]$. So $ag \in f A[[X]] \cap A[X] = (f A[X])^{ec} = f A[X]$, and then $g \in (f K[X]) \cap A[X] = I$. So $I^{ec} = I$. $\square$

Notation Let A be a ring and I an ideal of $A[X]$. Let $\overline{I} = \displaystyle\bigcap_{n:0}^{\infty}(I + X^n A[X])$, the closure of I for the X−adic topology in the ring $A[X]$. It is clear that for each $f \in A[X]$, $f.\overline{I} \subseteq \overline{f I}$ and if $I \subseteq J$ are two ideals of $A[X]$, then $\overline{I} \subseteq \overline{J}$.

Lemma 7.6 *Let A be an integral domain and $f(X) \in A[X]$. Then $(f A[X])^{ec} = \overline{f A[X]}$.*

Proof " $\subseteq$ " Let $g(X) \in (fA[X])^{ec} = (fA[[X]]) \cap A[X]$, $g(X) = f(X)h(X)$ with $h(X) = \sum_{i:0}^{\infty} a_i X^i \in A[[X]]$. For each $n \in \mathbb{N}^*$, $h_n(X) = \sum_{i:0}^{n-1} a_i X^i \in A[X]$ and $h'_n(X) = \sum_{i:n}^{\infty} a_i X^{i-n} \in A[[X]]$. Then $h(X) = h_n(X) + X^n h'_n(X)$, so $g(X) = f(X)h(X) = f(X)h_n(X) + X^n f(X) h'_n(X)$, and then $X^n f(X)h'_n(X) = g(X) - f(X)h_n(X) \in A[X]$ and even $f(X)h'_n(X) \in A[X]$. Then $g(X) = f(X)h_n(X) + X^n f(X)h'_n(X) \in f(X)A[X] + X^n A[X]$.

So $g(X) \in \bigcap_{n:0}^{\infty} \left(f(X)A[X] + X^n A[X] \right) = \overline{fA[X]}$.

" $\supseteq$ " Particular case: $f(0) \neq 0$. Let $h \in \overline{fA[X]}$. For each $n \in \mathbb{N}$, $h = fh_n + X^n l_n$ with $h_n, l_n \in A[X]$. For $m > n \geq 1$ be two integers, $fh_n + X^n l_n = fh_m + X^m l_m$, and then $f(h_n - h_m) = X^n (X^{m-n} l_m - l_n) \in X^n A[X]$. Since A is an integral domain and $f(0) \neq 0$, then $h_n - h_m \in X^n A[X]$, and then h_n and h_m have the same coefficient of X^k for $0 \leq k \leq n - 1$. Let a_0 be the constant coefficient shared by all the h_n, $n \geq 1$; a_1 the coefficient of X shared by all the h_n, $n \geq 2$; ...; and a_k the coefficient of X^k shred by all the h_n, $n \geq k + 1$.

Let $h' = a_0 + a_1 X + \ldots \in A[[X]]$. We will show that $h = fh' \in (fA[[X]]) \cap A[X] = (fA[X])^{ec}$. We must show that for each $k \in \mathbb{N}$, the coefficients of X^k in h and in fh' are the same. Since $h = fh_{k+1} + X^{k+1} l_{k+1}$, the coefficient of X^k in h is the same as the coefficient of X^k in fh_{k+1}. And the coefficient of X^k in fh' is the same as the coefficient of X^k in $f(a_0 + a_1 X + \ldots + a_k X^k)$. But by construction of a_i, $h_{k+1} = a_0 + a_1 X + \ldots + a_k X^k + X^{k+1} l(X)$ with $l(X) \in A[X]$. Then the coefficient of X^k in $f(a_0 + a_1 X + \ldots + a_k X^k)$ is the same as the coefficient of X^k in fh_{k+1}.

General case: $f(X) = X^k f'(X)$ with $k \geq 0$, $f'(X) \in A[X]$ and $f'(0) \neq 0$. Then $(fA[X])^{ec} = (fA[[X]]) \cap A[X] = \left(X^k f'A[[X]] \right) \cap A[X] = X^k (f'A[[X]] \cap A[X])$ and $\overline{fA[X]} = \bigcap_{n:0}^{\infty} (fA[X] + X^n A[X]) = \bigcap_{n:k}^{\infty} (fA[X] + X^n A[X]) = \bigcap_{n:k}^{\infty} (X^k f'A[X] + X^n A[X]) = X^k \bigcap_{n:k}^{\infty} (f'A[X] + X^{n-k} A[X]) = X^k \bigcap_{n:0}^{\infty} (f'A[X] + X^n A[X])$. By the particular case, we have $(f'A[X])^{ec} = \overline{f'A[X]}$. This means that $f'A[[X]] \cap A[X] = \bigcap_{n:0}^{\infty} (f'A[X] + X^n A[X])$. Multiplying by X^k, we obtain $(fA[X])^{ec} = \overline{fA[X]}$. □

In the following theorem, the nonconstant polynomial $f(X) \in A[X]$ is not necessarily irreducible in $K[X]$. We give some necessary conditions for the ideal $I = f(X)K[X] \cap A[X]$ to be almost principal. But they are not sufficient.

Theorem 7.7 *Let A be an integral domain with quotient field K, $f(X) \in A[X]$ a nonconstant polynomial, and $I = f(X)K[X] \cap A[X]$. We consider the following properties:*

1. *The ideal I is almost principal in the ring $A[X]$.*
2. *The ideal I is divisorial in the ring $A[X]$.*
3. *If $I^{ec} = I$, then $\overline{I} = I$.*

Then (1) $\Longrightarrow$ (2) $\Longrightarrow$ (3).

Proof "(1) $\Longrightarrow$ (2)" By Proposition 3.3, there is $0 \neq s \in A$ such that $sI \subseteq fA[X]$. Let v be the v−operation on $A[X]$. Then $sI_v = (sI)_v \subseteq (fA[X])_v = fA[X]$. So $I_v \subseteq f\frac{1}{s}A[X] \subseteq fK[X]$, and then $I_v \subseteq fK[X] \cap A[X] = I$, so $I_v = I$.

"(2) $\Longrightarrow$ (3)" By hypothesis, $I^{ec} = I$. By Lemma 7.5, $(fA[X])^{ec} = fA[X]$. By the preceding lemma, $\overline{fA[X]} = (fA[X])^{ec} = f(X)A[X]$.

Let $u(X) \in I^{-1} \subseteq K(X)$. Then $u(X)f(X) \in I^{-1}I \subseteq A[X]$. There exists $h(X) \in A[X]$ such that $u(X) = \frac{h(X)}{f(X)}$. Then $h(X)I = f(X)u(X)I \subseteq f(X)I^{-1}I \subseteq f(X)A[X]$. By the preceding step, $h(X)\overline{I} \subseteq \overline{h(X)I} \subseteq \overline{f(X)A[X]} = f(X)A[X]$. Then $\frac{h(X)}{f(X)}\overline{I} \subseteq A[X]$, so $u(X)\overline{I} \subseteq A[X]$. Then $I^{-1}\overline{I} \subseteq A[X]$, so $I \subseteq (A[X] : I^{-1}) = (I^{-1})^{-1} = I_v = I$, and then $\overline{I} = I$. $\square$

Notation Let A be an integral domain. A maximal element in the set of the divisorial proper integral ideals of A is said to be maximal divisorial or v−maximal. The set of these ideals is denoted by $v - Max(A)$. It is included in $Spec(A)$. Indeed, let P be a maximal divisorial ideal of A. Let $b, c \in A$ such that $bc \in P$ and $c \notin P$. The integral ideals $I = bA + P$ and $J = cA + P$ satisfy $IJ = (bA+P)(cA+P) = bcA + bP + cP + P^2 \subseteq P$, and then $P \subset J \subseteq (P : I) \cap A$. Since $(P : I) \cap A$ is an integral divisorial ideal, as intersection of two divisorial ideals, and strictly contains P, and then $(P : I) \cap A = A$, so $A \subseteq P : I$, and then $I \subseteq P$ so $b \in P$. $\square$

Other Proof We have the inclusions $P \subset J \subseteq J_v \subseteq A$. By the v-maximality of P, we have $J_v = A$. On the other hand, $bJ = b(cA + P) = bcA + bP \subseteq P$, and then $(bJ)_v \subseteq P_v = P$. But $(bJ)_v = bJ_v = bA$. Then $bA \subseteq P$ so $b \in P$. $\square$

Example 1 In contrast to $t - Max(A)$, the set $v - Max(A)$ may be empty. Indeed, let (A, M) be a nondiscrete valuation domain of rank 1. Since $v - Max(A) \subseteq Spec(A) = \{(0), M\}$, it suffices to show that M is not divisorial. Let K be the quotient field of A and ω a valuation on K with values in $\mathbb{R}$ that defines the ring A. We will show that $M^{-1} = A$. It is clear that $A \subseteq M^{-1} = (A : M)$. If $x \in K \setminus A$, then $\omega(x) < 0$. There exists $y \in M$ such that $\omega(y) < -\omega(x)$ then $\omega(xy) < 0$ and $xy \notin A$. So $x \notin (A : M) = M^{-1}$. Then $M^{-1} = A$ and $M_v = A \neq M$.

Other Proof By Example 3 at the beginning of Sect. 11, Chap. 1, the maximal ideal of a valuaion domain is divisorial if and only if it is principal. In our case, M is not principal.

Example 2 Let A be a Mori domain. Each proper divisorial ideal I of A is contained in at least a $v-$maximal ideal of A. Indeed, if I is not $v-$maximal, then it is properly contained in proper divisorial ideal $I_1, \ldots$ Since A is a Mori domain, the process must stop.

Lemma 7.8 (Gabelli and Roitman 2004) *Let A be an integral domain and I a divisorial integral proper ideal of A. Then I is $v-$maximal if and only if for each $x \in I^{-1} \setminus A$, $I = (A : xA)_A$.*

Proof Note first that if J a divisorial integral proper ideal of A, then $A \subset J^{-1}$, and for each $x \in J^{-1} \setminus A$, we have $J \subseteq (x^{-1}A) \cap A \subset A$, so $(x^{-1}A) \cap A$ is a divisorial integral proper ideal of A and $(x^{-1}A) \cap A = (A : xA)_A$.
Indeed, we have $A \subseteq J^{-1}$ but not the equality. In the contrary case, $J = J_v = (J^{-1})^{-1} = A^{-1} = A$, which is absurd. Let $x \in J^{-1} \setminus A$. Then $xJ \subseteq J^{-1}J \subseteq A$, so $J \subseteq x^{-1}A$. Then $J \subseteq (x^{-1}A) \cap A \subseteq A$. We cannot have $(x^{-1}A) \cap A = A$. In the contrary case, $A \subseteq x^{-1}A$, and then $xA \subseteq A$, so $x \in A$, which ist absurd. Then $J \subseteq (x^{-1}A) \cap A \subset A$. The integral ideal $(x^{-1}A) \cap A$ is divisorial, as intersection of two divisorial ideals. We have $(x^{-1}A) \cap A = (A : xA) \cap A = (A : xA)_A$.
" $\Longrightarrow$" We have $A \subset I^{-1}$ and for each $x \in I^{-1} \setminus A$, $I \subseteq (x^{-1}A) \cap A \subset A$ with $(x^{-1}A) \cap A$ a divisorial integral proper ideal of A. Since I is a $v-$maximal ideal of A, then $I = (x^{-1}A) \cap A = (A : xA)_A$.
" $\Longleftarrow$" Let J be a divisorial integral proper ideal of A containing I. Then $J^{-1} \subseteq I^{-1}$ and $A \subset J^{-1}$. Let $x \in J^{-1} \setminus A \subseteq I^{-1} \setminus A$. Then $I \subseteq J \subseteq (x^{-1}A) \cap A \subset A$. By hypothesis, $(x^{-1}A) \cap A = (A : xA)_A = I$. Then $I = J$ and I is a $v-$maximal ideal. $\qquad\square$

The following corollary is a partial converse of Theorem 7.7 and more precisely for the implication $"(1) \Longrightarrow (2)"$.

Corollary 7.9 (Houston and Zafrullah 2005) *Let A be an integral domain with quotient field K, $f(X) \in A[X]$ an irreducible polynomial in $K[X]$, and $I = f(X)K[X] \cap A[X]$. If I is $v-$maximal in $A[X]$, then it is almost principal.*

Proof By the preceding lemma, there exists $0 \neq \psi(X) \in K(X)$ such that $I = \big(A[X] : \psi(X)\big)_{A[X]}$. By Proposition 7.3, the ideal I is almost principal. $\qquad\square$

Lemma 7.10 (J.T. Arnold) *Let c and d be the two nonzero elements of A and $f(X) = cX - d \in A[X]$. Suppose that the element $y = \frac{c}{d}$ of K is not quasi-integral over A. Then $\big(f(X)A[X]\big)^{ec} = f(X)A[X]$.*

Proof Let $0 \neq g \in \left(f(X)A[X]\right)^{ec} = \left(f(X)A[[X]]\right) \cap A[X]$, $g(X) = f(X)h(X)$

with $h(X) = \sum_{i:0}^{\infty} a_i X^i \in A[[X]]$. Then $g(X) = (cX-d)h(X) = d(yX-1)h(X) =$

$d\left(-a_0 - a_1 X - a_2 X^2 - \ldots + ya_0 X + ya_1 X^2 + \ldots\right)$. Let n be the degree of $g(X)$. Then the terms of degrees $\geq n+1$, in the preceding expression, are all zero. So $a_{n+1} = ya_n$, $a_{n+2} = ya_{n+1}, \ldots$. We will show by induction that $a_n y^k = a_{n+k}$ for each $k \in \mathbb{N}^*$. The property is true for the order 1. Suppose that it is true until the order k. Then $a_n y^{k+1} = (a_n y^k)y = a_{n+k}y = a_{n+k+1}$. Since the $a_i \in A$, then $a_n y^k \in A$ for each $k \in \mathbb{N}^*$. But y is not quasi-integral over A, and then $a_n = 0$. From the preceding relations, we deduce that $0 = a_n = a_{n+1} = a_{n+2} = \ldots$. Then $h(X) = a_0 + a_1 X + \ldots + a_{n-1} X^{n-1} \in A[X]$ and $g(X) = f(X)h(X) \in f(X)A[X]$. $\qquad\square$

Corollary 7.11 *Let c and d be the two nonzero elements of A and $f(X) = cX-d \in A[X]$. Suppose that the element $y = \frac{c}{d}$ of K is not quasi-integral over A, but there exist $0 \neq a \in A$ and an infinity of natural integers n such that $ay^n \in A$. Then the ideal $I = f(X)K[X] \cap A[X]$ is not almost principal, is not divisorial, $I^{ec} = I$ and $\bar{I} \neq I$.*

Proof By the preceding lemma, $\left(f(X)A[X]\right)^{ec} = f(X)A[X]$. By Lemma 7.5, $I^{ec} = I$. On the other hand, $f(X) = cX - d = d(\frac{c}{d}X - 1) = d(yX - 1)$, and then $I = f(X)K[X] \cap A[X] = (yX - 1)K[X] \cap A[X]$. For each integer $n \geq 1$, $ay^n X^n - a = a(y^n X^n - 1) = a(yX-1)\left(y^{n-1} X^{n-1} + y^{n-2} X^{n-2} + \ldots + yX + 1\right) \in (yX - 1)K[X]$. If $ay^n \in A$, then $ay^n X^n - a \in (yX - 1)K[X] \cap A[X] = I$, so $a = -(ay^n X^n - a) + ay^n X^n \in I + X^n A[X]$. This is true for an infinity of natural integers n. Then $0 \neq a \in \bigcap_{n:1}^{\infty}(I + X^n A[X]) = \bar{I}$. Since $f(X)$ is not constant, then $a \notin I$, so $\bar{I} \neq I$. By Theorem 7.7, I is neither divisorial nor almost principal. $\qquad\square$

In the preceding corollary, we do not give necessary and sufficient conditions on the element $y = \frac{c}{d}$ of K for the ideal $(cX - d)K[X] \cap A[X]$ to be almost principal. But we have some conditions that allow the construction of a class of ideals that are not almost principal in $A[X]$.

Example 1 Let s and t be the two indeterminates over a field F and $A = F\left[s; \{st^{2^n}, n \in \mathbb{N}\}\right]$. The quotient field of A is $K = F(s, t)$. The polynomial $f(X) = stX - s \in A[X]$ satisfies the hypotheses of the preceding corollary. The ideal $I = f(X)K[X] \cap A[X] = (tX - 1)K[X] \cap A[X]$ is not almost principal, is not divisorial, $I^{ec} = I$ and $\bar{I} \neq I$.

Example 2 The conditions over the element $t = \frac{c}{d}$ in the preceding corollary are not necessary in order that the ideal $I = (cX - d)K[X] \cap A[X]$ is not almost principal.

Let $\{s_i;\ i \in \mathbb{N}^*\} \cup \{t\}$ be indeterminates over a field F and $A = F\big[\{s_i,\ i \in \mathbb{N}^*\}; \{s_i t^i, i \in \mathbb{N}^*\}\big] \subset B = F\big[\{s_i,\ i \in \mathbb{N}^*\}; t\big]$, domains with quotient field $K = F(\{s_i,\ i \in \mathbb{N}^*\}; t)$ since $t = \frac{s_1 t}{t}$. A monomial $s_1^{i_1} \ldots s_n^{i_n} t^j$ of B belongs to A if and only if $j = i_1' + 2i_2' + \ldots + ni_n'$ with $0 \le i_r' \le i_r$ for each $r \in \{1, 2, \ldots, n\}$. A polynomial of B belongs to A if and only if all its monomials belong to A. Let $c = s_1 t$ and $d = s_1 \in A$, $f(X) = cX - d \in A[X]$ and $I = (cX - d)K[X] \cap A[X]$. Then $t = \frac{c}{d}$. Suppose that there exists a monomial $a = s_1^{i_1} \ldots s_n^{i_n} t^j \in A$ such that $at^k \in A$ for an infinity of positive integers k. Since $at^k = s_1^{i_1} \ldots s_n^{i_n} t^{j+k} \in A$, then $j + k = i_1' + 2i_2' + \ldots + ni_n'$ with $0 \le i_r' \le i_r$ for each $r \in \{1, 2, \ldots, n\}$. But this is impossible for k sufficiently large. For each $n \in \mathbb{N}^*$, let $h_n(X) = s_n\big(t^{n-1}X^{n-1} + t^{n-2}X^{n-2} + \ldots + tX + 1\big) \in K[X]$. Then $f(X)h_n(X) = (cX - d)h_n(X) = d(\frac{c}{d}X - 1)h_n(X) = s_1(tX - 1)h_n(X) = s_1 s_n(t^n X^n - 1) = s_1(s_n t^n X^n - s_n) \in A[X]$, so $f(X)h_n(X) \in f(X)K[X] \cap A[X] = I$. Suppose that I is almost principal. By Proposition 3.3, there exists $0 \ne b \in A$ such that $bI \subseteq f(X)A[X]$. For each $n \in \mathbb{N}^*$, $bf(X)h_n(X) \in f(X)A[X]$, and then $bh_n(X) \in A[X]$. Then for each $i, n \in \mathbb{N}^*$ with $i < n$, $bs_n t^i \in A$. In particular, $bs_{2m}t^m \in A$ for each $m \in \mathbb{N}^*$. We may suppose that $b = s_1^{i_1} \ldots s_n^{i_n} t^j$ is a monomial of A where $j = i_1' + 2i_2' + \ldots + ni_n'$ with $0 \le i_r' \le i_r$ for each $r \in \{1, 2, \ldots, n\}$. Choose $m > i_1 + 2i_2 + \ldots + ni_n$. Since $bs_{2m}t^m = s_1^{i_1} \ldots s_n^{i_n} s_{2m} t^{j+m} \in A$, then $m = i_1'' + 2i_2'' + \ldots + ni_n'' + 2ml$ with $l \in \{0; 1\}$ and $0 \le i_r'' \le i_r$ for each $r \in \{1, 2, \ldots, n\}$. The value $l = 1$ is obviously impossible, and the value $l = 0$ is also impossible by the choice of $m > i_1 + 2i_2 + \ldots + ni_n \ge i_1'' + 2i_2'' + \ldots + ni_n''$. Then I is not almost principal.

We end this section by the following result.

Proposition 7.12 (Houston and Zafrullah 2005) *Let $f(X) \in A[X]$ be an irreducible polynomial in $K[X]$, $I = f(X)K[X] \cap A[X]$, and $\mathscr{I}$ the ideal of A formed by the constant terms of the elements of I. If $\mathscr{I}^{-1} = A$ and $I^{-1} \ne A[X]$, then the ideal I is almost principal.*

Proof Suppose that I is not almost principal. By Proposition 7.3, $I^{-1} \subseteq K[X]$. Let $g(X) \in I^{-1}$ and a its constant term. Since $g(X)I \subseteq A[X]$, then $a\mathscr{I} \subseteq A$, so $a \in (A : \mathscr{I}) = \mathscr{I}^{-1} = A \subset A[X] \subseteq I^{-1}$, and then $g(X) - a \in I^{-1}$. So $(g(X) - a)I \subseteq A[X]$, then $\frac{1}{X}(g(X) - a)I \subseteq A[X]$. So $\frac{g(X) - a}{X} \in (A[X] : I) = I^{-1}$. By the preceding step, the coefficient of X in $g(X)$ belongs to A. By induction, we show that $g(X) \in A[X]$. Then $I^{-1} = A[X]$, which is absurd. $\square$

Example (Houston and Zafrullah 2005) Let s and t be the two indeterminates over a field F, $A = F\big[s; \{st^{2^n}, n \in \mathbb{N}\}\big] \subset F[s, t]$, and $K = F(s, t)$ the quotient field of A. By Proposition 1.5, $P = (X - t)K[X] \cap A[X]$ is a prime ideal of $A[X]$ that is not almost principal. Note that $P = (sX - st)K[X] \cap A[X]$ with $f(X) = sX - st \in A[X]$ an irreducible polynomial in $K[X]$. We will show that $P^{-1} = A[X]$. By the preceding proposition, it suffices to show that $\mathscr{I}^{-1} = A$ where $\mathscr{I}$ is the ideal of A formed by the constant terms of the elements of P. For each $n \in \mathbb{N}^*$, the polynomial $sX^{2^n} - st^{2^n} = (X - t)\big(sX^{2^n-1} + stX^{2^n-2} + \ldots + st^{2^n-2}X + $

$st^{2^n-1}) \in (X - t)K[X] \cap A[X] = P$. Then the ideal $\mathscr{L} = (st^{2^n}; n \in \mathbb{N}) \subseteq \mathscr{I}$, so $A \subseteq \mathscr{I}^{-1} \subseteq \mathscr{L}^{-1}$. It suffices to show that $\mathscr{L}^{-1} \subseteq A$. Let $x \in \mathscr{L}^{-1}$. Then $xst \in A$, so $x = \frac{a}{st}$ with $a \in A$. It suffices to show that $a \in stA$. We may suppose that $a = s^i t^j$ is a monomial with $i \geq \phi(j)$, where $\phi(j)$ is the number of 1 in the development in binary basis of the integer j. By definition, for each $n \in \mathbb{N}^*$, $xst^{2^n} \in A$, and then $\frac{a}{st}st^{2^n} = s^i t^{2^n+j-1} \in A$. So $i \geq \phi(2^n + j - 1)$. For n sufficiently large, $\phi(2^n 1 + j - 1) = 1 + \phi(j - 1)$, then $i \geq 1 + \phi(j - 1)$, so $i - 1 \geq \phi(j-1)$, and then $s^{i-1}t^{j-1} \in A$. So $a = s^i t^j = st(s^{i-1}t^{j-1}) \in stA$. Note that $P_v = (P^{-1})^{-1} = A[X] \neq P$ since P is a prime ideal of $A[X]$. Then the P is not divisorial. This was also noted in Example 1 of Proposition 7.11.

8 Generators of the Ideal f(X)K[X]∩A[X]

The results in this section are due to, Hamann et al. (1988). We fix an integral domain A with quotient field K. Let $f(X) \in A[X]$ be a nonconstant polynomial. The ideal $f(X)K[X] \cap A[X]$ is not in general generated by $f(X)$ and is not principal. It is not even almost principal. We purpose to find a system of generators for this ideal.

Proposition 8.1 *Let* $f(X) = a_0 + a_1 X + \ldots + a_n X^n \in A[X]$ *be a polynomial of degree* $n \geq 1$ *with* $a_0 \neq 0$. *The ideal* $I = f(X)K[X] \cap A[X]$ *of* $A[X]$ *is generated by its elements of degree* n *if and only if for each element* $g(X) \in I$, $g(0) \in \bigcap_{i:1}^{n}(a_0 A : a_i)_A$.

Proof "$\Longrightarrow$" The polynomials of degree n in I are of the form $tf(X) \in I$ with $t \in K^*$. Let $g(X) \in I$, $g(X) = h_1(X)t_1 f(X) + \ldots + h_s(X)t_s f(X)$ with $h_1(X), \ldots, h_s(X) \in A[X]$, $t_1, \ldots, t_s \in K^*$, and $t_1 f(X), \ldots, t_s f(X) \in I \subseteq A[X]$. Then $t_j a_i \in A$ for $1 \leq j \leq s$ and $0 \leq i \leq n$ and $g(0) = h_1(0)t_1 a_0 + \ldots + h_s(0)t_s a_0 = (h_1(0)t_1 + \ldots + h_s(0)t_s)a_0$. For $1 \leq i \leq n$, $a_i g(0) = (h_1(0)(t_1 a_i) + \ldots + h_s(0)(t_s a_i))a_0 \in a_0 A$. Then $g(0) \in (a_0 A : a_i)$ for $1 \leq i \leq n$.

"$\Longleftarrow$" Suppose that I is not generated by its elements of degree n. Let $g(X) \in I$ be a polynomial of minimal degree m that is not a $A[X]$−linear combination of polynomials of degree n in I, $g(X) = f(X)h(X)$ with $h(X) \in K[X]$. Let $h(0) = b \in K$. By hypothesis, $a_0 b = g(0) \in \bigcap_{i:1}^{n}(a_0 A : a_i)_A$. For $1 \leq i \leq n$, $a_0 b a_i \in a_0 A$, and then $ba_i \in A$ so $bf(X) \in A[X] \cap f(X)K[X] = I$. Since $g(X) - bf(X) = (h(X) - b)f(X) = Xh'(X)f(X)$ with $h'(X) \in K[X]$ and $g(X) - bf(X) \in I \subseteq A[X]$, then $h'(X)f(X) \in f(X)K[X] \cap A[X] = I$ with $deg(h'(X)f(X)) \leq m - 1$. By minimality of m, $h'(X)f(X)$ is a $A[X]$−linear combination of elements in I of degree n. It is the same for $g(X) = bf(X) + Xh'(X)f(X)$, which is absurd. □

Remark 8.2 With the notations of the preceding proposition, we suppose else that the ideal I is generated by its elements of degree n. By Proposition 1.7, it is almost principal.

Example Let s and t be the two indeterminates over a field F, $A = F\big[s; \{st^{2^n}, n \in \mathbb{N}\}\big] \subset F[s, t]$, and $K = F(s, t)$ the quotient field of A. The ideal $I = (X - t)K[X] \cap A[X] = (sX - st)K[X] \cap A[X]$ is not almost principal. Then it is not generated by its linear elements.

Corollary 8.3 *Let* $f(X) = a_0 + a_1 X + \ldots + a_n X^n \in A[X]$ *be a polynomial of degree* $n \geq 1$ *with* $a_0 \neq 0$ *and* $\phi : A[X] \longrightarrow A$ *the homomorphism, defined by* $\phi(g(X)) = g(0)$ *for each* $g(X) \in A[X]$. *The ideal* $I = f(X)K[X] \cap A[X]$ *is generated by its elements of degree* n *if and only if* $\phi(I) = \bigcap_{i:1}^{n} (a_0 A : a_i)_A$.

Proof By the preceding proposition, the ideal I is generated by its elements of degree n if and only if $\phi(I) \subseteq \bigcap_{i:1}^{n} (a_0 A : a_i)_A$. The converse inclusion is always true without a supplementary hypothesis. Indeed, let $b \in \bigcap_{i:1}^{n} (a_0 A : a_i)_A$. Then $b \in A$ and $b a_i \in a_0 A$ for each $i = 1, \ldots, n$. So $bf(X) \in a_0 A[X]$. Then $\frac{b}{a_0} f(X) \in A[X] \cap f(X)K[X] = I$ with $\phi\big(\frac{b}{a_0} f(X)\big) = \frac{b}{a_0} a_0 = b$. □

Example Let $f(X) = a_0 + a_1 X \in A[X]$ with $a_0 a_1 \neq 0$. The ideal $I = f(X)K[X] \cap A[X]$ is generated by its linear elements if and only if $\phi(I) = (a_0 A : a_1)_A$.

Proposition 8.4 *Let* $f(X) = a_0 + a_1 X + \ldots + a_n X^n \in A[X]$ *be a polynomial of degree* $n \geq 1$, $I = f(X)K[X] \cap A[X]$ *and* M *the* A−*module formed by* 0 *and the polynomials of degree* n *in* I. *Then* $\mu\big(\bigcap_{i:0}^{n}(a_n A : a_i)_A\big) = \mu(M)$, *where* μ *is the minimal number of generators.*

Proof The A−module $M = \{t f(X) \in I; \ t \in K\}$. Let $\{t_\alpha \in A; \alpha \in \Lambda\}$ be a family of generators for the ideal $\bigcap_{i:0}^{n}(a_n A : a_i)_A$ of A. For $\alpha \in \Lambda$, we put $f_\alpha(X) = \frac{t_\alpha}{a_n} f(X) \in K[X]$. For each $i = 0, \ldots, n$ and $\alpha \in \Lambda$, $t_\alpha a_i \in a_n A$, then $\frac{t_\alpha}{a_n} a_i \in A$, so $f_\alpha(X) = \frac{t_\alpha}{a_n} f(X) \in A[X] \cap f(X)K[X] = I$, and then $f_\alpha(X) \in M$. We will show that the family $\{f_\alpha(X); \alpha \in \Lambda\}$ generates the A−module M. Let $0 \neq g(X) \in M$,

$g(X) = sf(X) \in I$ with $0 \neq s \in K$. The element $t = sa_n \in \bigcap\limits_{i:0}^{n}(a_n A : a_i)_A$ since

$sf(X) = sa_0 + sa_1 X + \ldots + sa_n X^n \in I \subseteq A[X]$, and then $sa_i \in A$, so $sa_i a_n \in a_n A$, and then $ta_i \in a_n A$. So $t = \sum\limits_{\alpha \in \Lambda} b_\alpha t_\alpha$, where the $b_\alpha \in A$ are almost zero.

Then $g(X) = sf(X) = \frac{sa_n}{a_n} f(X) = \frac{t}{a_n} f(X) = \sum\limits_{\alpha \in \Lambda} \frac{b_\alpha t_\alpha}{a_n} f(X) = \sum\limits_{\alpha \in \Lambda} b_\alpha f_\alpha(X)$.

Conversely, let $\{f_\alpha(X); \alpha \in \Lambda\}$ be a family of generators of the A−module M. For each $\alpha \in \Lambda$, $f_\alpha(X) = s_\alpha f(X)$ with $s_\alpha \in K^*$. Put $t_\alpha = s_\alpha a_n \in K^*$. Then

$t_\alpha \in \bigcap\limits_{i:0}^{n}(a_n A : a_i)_A$. Indeed, since $s_\alpha f(X) = f_\alpha(X) \in I \subseteq A[X]$, then $s_\alpha a_i \in A$

for $0 \leq i \leq n$, so $s_\alpha a_i a_n \in a_n A$, and then $t_\alpha a_i \in a_n A$. So $t_\alpha \in (a_n A : a_i)_A$ for each $i = 0, \ldots, n$. We will show that the family $\{t_\alpha; \alpha \in \Lambda\}$ generates the ideal

$\bigcap\limits_{i:0}^{n}(a_n A : a_i)_A$ of A. Let t be a nonzero element of this ideal. Then $ta_i \in a_n A$ for

each $i = 0, \ldots, n$, and then $tf(X) \in a_n A[X]$, so $\frac{t}{a_n} f(X) \in A[X] \cap f(X)K[X] = I$

is a polynomial of degree n. Then $\frac{t}{a_n} f(X) = \sum\limits_{\alpha \in \Lambda} b_\alpha f_\alpha(X)$, where the $b_\alpha \in A$ are

almost zero. Then $\frac{t}{a_n} f(X) = \sum\limits_{\alpha \in \Lambda} b_\alpha s_\alpha f(X)$, so $\frac{t}{a_n} = \sum\limits_{\alpha \in \Lambda} b_\alpha s_\alpha$, and then $t =$

$\sum\limits_{\alpha \in \Lambda} b_\alpha a_n s_\alpha = \sum\limits_{\alpha \in \Lambda} b_\alpha t_\alpha$.   □

Corollary 8.5 *Let* $f(X) = a_0 + a_1 X + \ldots + a_n X^n \in A[X]$ *be a polynomial of degree* $n \geq 1$ *with* $a_0 \neq 0$, $I = f(X)K[X] \cap A[X]$ *and* $\phi : A[X] \longrightarrow A$ *the homomorphism, defined by* $\phi(g(X)) = g(0)$ *for each* $g(X) \in A[X]$. *Suppose that*

$\phi(I) \subseteq \bigcap\limits_{i:1}^{n}(a_0 A : a_i)_A$. *Then the minimal number of generators of degree* n *of the*

ideal I *is* $\mu_{A[X]}(I) = \mu\left(\bigcap\limits_{i:0}^{n}(a_n A : a_i)_A\right)$.

Proof By the proof of Corollary 8.3, the inclusion $\phi(I) \subseteq \bigcap\limits_{i:1}^{n}(a_0 A : a_i)_A$ is in fact

an equality. By this corollary, the ideal I is generated by the set N of its elements of degree n. The A−module formed by 0 and the polynomials of degree n in I are $M = \{0\} \cup N$. The minimal number $\mu_{A[X]}(I)$ of generators of degree n of the ideal I is equal to the minimal number $\mu(M)$ of generators of the A−module M. Indeed, it is clear that a family of generators of the A−module M generates the ideal I. Conversely, let $\{t_\alpha f(X) \in I; \ \alpha \in \Lambda\} \subseteq N$ be a family of generators of the ideal I. Let $0 \neq g(X) = tf(X) \in M \subseteq I$ with $t \in K^*$. Then $g(X) = tf(X) =$

$\sum_{\alpha \in \Lambda} h_\alpha(X) t_\alpha f(X)$, where the $h_\alpha(X) \in A[X]$ are almost zero. By simplification,

$t = \sum_{\alpha \in \Lambda} h_\alpha(X) t_\alpha$, and then $t = \sum_{\alpha \in \Lambda} h_\alpha(0) t_\alpha$, where the $h_\alpha(0) \in A$ are almost zero.

Then we have $g(X) = tf(X) = \sum_{\alpha \in \Lambda} h_\alpha(0)(t_\alpha f(X))$. By the preceding corollary,

$$\mu_{A[X]}(I) = \mu(M) = \mu\left(\bigcap_{i:0}^{n} (a_n A : a_i)_A\right). \qquad \qquad \Box$$

Example Let $f(X) = a_0 + a_1 X \in A[X]$ with $a_0 a_1 \neq 0$. Since $(a_1 A : a_0)_A \cap (a_1 A : a_1)_A = (a_1 A : a_0)_A \cap A = (a_1 A : a_0)_A$, then the ideals $(a_1 A : a_0)_A$ and $(a_0 A : a_1)_A$ determine the minimal number of generators for the ideal $I = f(X)K[X] \cap A[X]$.

9 Prestable Ideals

Lemma 9.1 *Let A be an integral domain. Its integral closure $\overline{A} = \bigcup \{(I : I);\ I$ nonzero finitely generated fractional ideal of $A\}$. In particular, A is integrally closed if and only if for each nonzero finitely generated fractional ideal I of A, we have $(I : I) = A$.*

Proof Let I be a nonzero finitely generated fractional ideal of A. Then $I = y_1 A + \ldots + y_n A$ with $y_i \neq 0$. Let $x \in (I : I)$, and then $xI \subseteq I$. Put $xy_i = \sum_{j:1}^{n} a_{ij} y_j$ with $a_{ij} \in A$, and then $\sum_{j:1}^{n} (\delta_{ij} x - a_{ij}) y_j = 0$ for $1 \leq i \leq n$. Consider the n equations $\sum_{j:1}^{n} (\delta_{ij} x - a_{ij}) y_j = 0$ for $1 \leq i \leq n$, with coefficients in the quotient field K of A, where the indeterminates are $Y_1, \ldots, Y_n$. This system has at least a nontrivial solution $y_1, \ldots, y_n$. Then its determinant is zero and can be written $x^n + a_{n-1} x^{n-1} + \ldots + a_1 x + a_0 = 0$ with $a_0, \ldots, a_{n-1} \in A$. Then x is integral over A. Conversely, let $x \in \overline{A}$. The A−sub-module $I = A[x]$ of K is finitely generated. Then it is a nonzero finitely generated fractional ideal of A. Since $xI \subseteq I$, then $x \in (I : I)$. $\qquad \Box$

Example Let k be a field, $A = k[[X^2, X^3]]$, and $I = (X^2, X^3)A$. By Example 3 in notations after Corollary 5.8, $A \subset k[[X]] = (I : I) \subseteq \overline{A}$, the integral closure of A. But $k[[X]]$ is a PID, and then it is integrally closed. So $\overline{A} = (I : I) = k[[X]]$. We can also observe that $k[[X]] = A[X]$. Indeed, the converse inclusion is clear.

Let $f = \sum_{i:0}^{\infty} a_i X^i \in k[[X]]$, $f = \sum_{i:0}^{\infty} a_{2i} X^{2i} + X \sum_{i:0}^{\infty} a_{2i+1} X^{2i} = f_1 + X f_2$ with $f_1, f_2 \in A$. Then $f \in A[X]$. Since X is integral over A as a root of the polynomial $t^2 - X^2 \in A[t]$, the extension $A \subset k[[X]]$ is integral. We finish as before.

Lemma 9.2 *Let A be an integral domain, I a nonzero finitely generated integral ideal of A, and S a multiplicative set of A. The domain $(I : I) = T$ satisfies the equality $S^{-1}T = (S^{-1}I : S^{-1}I)$.*

Proof " $\subseteq$ " Let $x \in S^{-1}T$, $x = \frac{t}{s}$ with $t \in T$ and $s \in S$. For each $y \in S^{-1}I$, $y = \frac{a}{s'}$ with $a \in I$ and $s' \in S$, and then $xy = \frac{ta}{ss'} \in S^{-1}I$ since $ta \in I$ because I is an ideal of T. Then $x \in (S^{-1}I : S^{-1}I)$.
" $\supseteq$ " Put $I = (a_1, \ldots, a_n)$. Let $x \in (S^{-1}I : S^{-1}I)$. Then $x(S^{-1}I) \subseteq S^{-1}I$, so $xI \subseteq S^{-1}I$. There exists $s \in S$ such that $sxa_1, \ldots, sxa_n \in I$. Then $sxI = (sxa_1, \ldots, sxa_n) \subseteq I$ and $sx \in (I : I) = T$, so $x = \frac{sx}{s} \in S^{-1}T$. $\square$

Definition 9.3 Let A be an integral domain and I a fractional ideal of A. We say that I is invertible if there exists a fractional ideal J of A such that $IJ = A$. We say that J is an inverse of I in A.

Example 1 Each nonzero principal fractional ideal is invertible.

Example 2 (Schanuel)

(i) Let A be an integral domain with quotient field K and $a \in K^*$ such that $a^2, a^3 \in A$. Consider the fractional ideals $I = (a^2, 1 + aX)$ and $J = (a^2, 1 - aX)$ of $A[X]$. Then $IJ = (a^4, a^2 + a^3 X, a^2 - a^3 X, 1 - a^2 X^2) \subseteq A[X]$. Since $X^4 a^4 + (1 + a^2 X^2)(1 - a^2 X^2) = 1$, then $IJ = A[X]$. So I and J are invertible in $A[X]$.

(ii) Let A be an integral domain with quotient field K and $t \in K \setminus A$. Suppose that $t^n \in A$ for n sufficiently large. By exchanging t by a convenient power of t, we may suppose that $t^n \in A$ for each integer $n \geq 2$. Consider the fractional ideals: $I = (1 + tX, 1 + tX + t^2 X^2)$ and $J = (1 - tX + t^2 X^2, 1 - tX)$ of the domain $A[X]$. Then $IJ = (1 + t^3 X^3, 1 - t^2 X^2, 1 + t^2 X^2 + t^4 X^4, 1 - t^3 X^3) \subseteq A[X]$. Conversely, $1 = -t^2 X^2(1 + t^3 X^3) + t^2 X^2(1 - t^2 X^2) + (1 + t^2 X^2 + t^4 X^4) - t^2 X^2(1 - t^3 X^3) \in IJ$. Then $IJ = A[X]$ so I and J are invertible.

Lemma 9.4 *Let A be an integral domain and I an invertible fractional ideal of A. The unique inverse of I is $I^{-1} = (A : I)$. Moreover, I is divisorial.*

Proof Let J be a fractional ideal of A such that $IJ = A$. Then $J \subseteq (A : I) = I^{-1}$. Since $II^{-1} \subseteq A$, then $(JI)I^{-1} \subseteq J$. But $IJ = A$, and then $I^{-1} \subseteq J$. So $J = I^{-1}$. For the last assertion, the unique inverse of I^{-1} is $(I^{-1})^{-1} = I_v$. But $II^{-1} = A$, and then $I_v = I$. $\square$

Example 1 Let k be a field, $A = k[[X^2, X^3]] \subseteq k[[X]]$ and $I = (X^2, X^3)A$. By Example 3 in notations after Corollary 5.8, $I = X^2k[[X]]$ and $I^{-1} = k[[X]]$. The product $II^{-1} = X^2k[[X]] = I \neq A$. So I is not invertible.

Example 2 Let I be an invertible fractional ideal of an integral domain A. Then $(I : I) = A$. Indeed, the converse inclusion is clear. Let $x \in (I : I)$, and then $xI \subseteq I$. Since $xII^{-1} \subseteq II^{-1}$, then $xA \subseteq A$ so $x \in A$.

Example 3 A Prufer domain A is integrally closed. Indeed, a nonzero finitely generated fractional ideal I of A is invertible. By the preceding example, $(I : I) = A$. By Lemma 9.1, the integral closure $\overline{A} = A$.

The following lemma extends a result of Helms (1935).

Lemma 9.5 *Let A be an integral domain having a finite number of maximal ideals and I a nonzero fractional ideal of A.*

1. *The ideal I is invertible if and only if it is principal.*
2. *Suppose that A is local and I is principal. For each family G of generators for I, there exists $a \in G$ such that $I = aA$.*

Proof

(1) Let $M_1, \ldots, M_n$ be all the maximal ideals of A. Since $II^{-1} = A$, for each $k \in \{1, \ldots, n\}$, there exist $a_k \in I$ and $b_k \in I^{-1}$ such that $a_k b_k \in A \setminus M_k$. By maximality, for $j \neq k$, $M_j \not\subseteq M_k$. Choose $\lambda_{jk} \in M_j \setminus M_k$. Put $\lambda_k = \prod_{j:1, j \neq k}^{n} \lambda_{jk}$.
Then $\lambda_k \in M_j$ for $j \neq k$, but $\lambda_k \notin M_k$ since M_k is prime. Let $a = \lambda_1 a_1 + \ldots + \lambda_n a_n \in I$ and $b = \lambda_1 b_1 + \ldots + \lambda_n b_n \in I^{-1}$. Then $ab = \sum_{1 \leq i, j \leq n} \lambda_i \lambda_j a_i b_j$. (1)
Since $a_i b_j \in II^{-1} \subseteq A$, $\lambda_i \lambda_j a_i b_j \in M_k$ if $i \neq k$ or $j \neq k$. But $\lambda_k \lambda_k a_k b_k \notin M_k$. There is exactly one term on the right side on (1) that is not in M_k. Then $ab \notin M_k$. We conclude that ab does not belong to any maximal ideal of A. It is invertible in A. Then $aA \subseteq I = abI \subseteq aI^{-1}I = aA$. So $I = aA$.

(2) Let M be the unique maximal ideal of A. Put $I = Ac$ with $c \neq 0$. Then $c = \alpha_1 a_1 + \ldots + \alpha_n a_n$ with $\alpha_1, \ldots, \alpha_n \in A$ and $a_1, \ldots, a_n \in G$. Also, for $1 \leq i \leq n$, $a_i = \beta_i c$ with $\beta_1, \ldots, \beta_n \in A$. Then $c = \sum_{i:1}^{n} \alpha_i a_i = \sum_{i:1}^{n} \alpha_i \beta_i c = c(\sum_{i:1}^{n} \alpha_i \beta_i)$.

So $\sum_{i:1}^{n} \alpha_i \beta_i = 1$ with $\alpha_i \beta_i \in A$. There exists at least one index i such that $\alpha_i \beta_i \notin M$. Then $\alpha_i \beta_i$ is invertible in A. So β_i is invertible in A. Then $I = Ac = Aa_i$. □

Example 1 The (2) of the lemma is not true in general if A is not local. Indeed, $\mathbb{Z} = (1)$ and $\mathbb{Z} = (2; 3)$. But $\mathbb{Z} \neq (2)$ and $\mathbb{Z} \neq (3)$.

Example 2 We give an example of a nonprincipal invertible integral ideal.

The quotient field of the domain $A = \mathbb{Z} + i\mathbb{Z}\sqrt{5}$ is $K = \mathbb{Q} + i\mathbb{Q}\sqrt{5}$. Consider the integral ideal $I = (3; 1 + i\sqrt{5}) = (2 - i\sqrt{5}; 1 + i\sqrt{5})$ of A.

We start by determining $I^{-1} = A : I$. Let $x = a + ib\sqrt{5} \in K$ with $a, b \in \mathbb{Q}$. Then $x \in I^{-1}$ if and only if $3x = 3a + 3ib\sqrt{5} \in A$ and $(1 + i\sqrt{5})x = (1 + i\sqrt{5})(a + ib\sqrt{5}) \in A$. The first condition is equivalent to $a' = 3a \in \mathbb{Z}, b' = 3b \in \mathbb{Z}$. The second condition can be stated $(1 + i\sqrt{5})(a + ib\sqrt{5}) \in A \iff a - 5b \in \mathbb{Z}$ and $a + b \in \mathbb{Z} \iff \frac{a'}{3} - 5\frac{b'}{3} \in \mathbb{Z}$ and $\frac{a'}{3} + \frac{b'}{3} \in \mathbb{Z} \iff a' - 5b' \in 3\mathbb{Z}$ and $a' + b' \in 3\mathbb{Z}$ $\iff a' - 5b' \equiv 0 \ (mod \ 3\mathbb{Z})$ and $a' + b' \equiv 0 \ (mod \ 3\mathbb{Z})$.

For example, $-\frac{1}{3} + \frac{i}{3}\sqrt{5} \in I^{-1}$ since $-1 - 5 = -6 \equiv 0 \ (mod \ 3\mathbb{Z})$ and $-1 + 1 = 0$. Also, $\frac{2}{3} + \frac{i}{3}\sqrt{5} \in I^{-1}$ since $2 - 5 = -3 \equiv 0 \ (mod \ 3\mathbb{Z})$ and $2 + 1 = 3$. Note that $(2 - i\sqrt{5})(\frac{2}{3} + \frac{i}{3}\sqrt{5}) + (1 + i\sqrt{5})(-\frac{1}{3} + \frac{i}{3}\sqrt{5}) = \frac{1}{3}(4 + 5) + \frac{1}{3}(-5 - 1) = 3 - 2 = 1$. Then $II^{-1} = A$, so I is invertible. Let $N : K = \mathbb{Q} + i\mathbb{Q}\sqrt{5} \longrightarrow \mathbb{R}+$ be the map defined by $N(z) = |z|^2$. If $z \in A$, then $N(z) \in \mathbb{N}$ and N is multiplicative: $N(zz') = N(z)N(z')$. We will show that $z \in U(A)$ if and only if $N(z) = 1$. Indeed, if there exists $z' \in A$ such that $zz' = 1$, then $N(z)N(z') = 1$, so $N(z) = 1$. Conversely, let $z = a + ib\sqrt{5} \in A$ be such that $N(z) = a^2 + 5b^2 = 1$. Then $b = 0$ and $a^2 = 1$ so $a = +1$ and $z = \pm 1$. In particular, $U(A) = \{\pm 1\}$.

We will show that 3 and $1 + i\sqrt{5}$ are nonassociated irreducible elements in A. Let $z, z' \in A$ be such that $zz' = 3$. Then $N(z)N(z') = 9$ so $N(z), N(z') \in \{1; 3; 9\}$. If $N(z) = 3$ with $z = a + ib\sqrt{5}, a, b \in \mathbb{Z}$, then $a^2 + 5b^2 = 3$ so $b = 0$ and $a^2 = 3$, which is impossible. Then either $N(z) = 1$ or $N(z') = 1$ so either $z \in U(A)$ or $z' \in U(A)$.

Let $z, z' \in A$ be such that $zz' = 1 + i\sqrt{5}$. Then $N(z)N(z') = 6$ so $N(z), N(z') \in \{1; 2; 3; 6\}$. If either $N(z)$ or $N(z') = 2$ or 3, we will have an equation of the form $a^2 + 5b^2 = 2$ or 3, and then $b = 0$ and $a^2 = 2$ or 3, which is impossible. Then $N(z) = 1$ or $N(z') = 1$ so $z \in U(A)$ or $z' \in U(A)$.

Suppose that $I = (3; 1 + i\sqrt{5}) = Ax$ is an integral principal ideal. Then x divides 3 and $1 + i\sqrt{5}$ in A. But these two elements are nonassociated and irreducible in A. Then $x \in U(A) = \{\pm 1\}$ and $1 \in I$. Put $1 = 3(a + ib\sqrt{5}) + (1 + i\sqrt{5})(c + id\sqrt{5})$ with $a, b, c, d \in \mathbb{Z}$. Then we have the equalities $1 = 3a + c - 5d$ and $0 = 3b + c + d$. By subtraction, we get $1 = 3(a - b - 2d)$, which is absurd. We conclude that I is not principal.

Lemma 9.6 *Let A be an integral domain and I a nonzero fractional ideal of A. Then I is invertible if and only if it is finitely generated and locally principal; i.e, for each $M \in Max(A)$, the ideal IA_M is principal.*

Proof " $\Longrightarrow$ " Since $II^{-1} = A$, there exist $x_1, \ldots, x_n \in I$ and $y_1, \ldots, y_n \in I^{-1}$ such that $x_1y_1 + \ldots + x_ny_n = 1$. For each $x \in I$, $x = x_1(xy_1) + \ldots + x_n(xy_n)$ with $xy_i \in II^{-1} = A$. Then $I = (x_1, \ldots, x_n)$. On the other hand, by hypothesis, $II^{-1} = A$. For each $M \in Max(A)$, we have $(IA_M)(I^{-1}A_M) = A_M$. Then ideal IA_M is invertible in A_M. By the preceding lemma, IA_M is principal.

" $\Longleftarrow$ " Suppose that I is not invertible. Then $II^{-1} \neq A$. There exists $M \in Max(A)$ such that $II^{-1} \subseteq M$. Since IA_M is principal, there exists $a \in I$ such that $IA_M = aA_M$. In particular, $I \subseteq aA_M$, then $\frac{1}{a}I \subseteq A_M$. Since I is finitely generated, there exists $s \in A \setminus M$ such that $\frac{s}{a}I \subseteq A$. Then $\frac{s}{a} \in (A : I) = I^{-1}$, so $s \in aI^{-1} \subseteq II^{-1} \subseteq M$, which is absurd. $\qquad \square$

Example 1 In a Noetherian integral domain, the nonzero locally principal fractional ideals are invertible.

Example 2 We give an example of a nonzero finitely generated integral ideal that it is not invertible. Let A be an integral domain with quotient field $K \neq A$. There exists a noninvertible $0 \neq a \in A$. We will show that the integral ideal $I = (a, X)$ of $A[X]$ is not invertible. It is clear that $A[X] \subseteq (A[X] : I) = I^{-1}$. Conversely, let $u \in I^{-1} \subseteq K(X)$. Then $g = au \in A[X]$ so $u = \frac{g}{a}$. Also, $h = uX \in A[X]$, then $h = \frac{g}{a}X$ so $Xg = ah$. Then all the coefficients of g are divisible by a in A. Then $u = \frac{g}{a} \in A[X]$ and $I^{-1} = A[X]$. Finally, $II^{-1} = IA[X] = I = (a, X)$. Suppose that $(a, X) = A[X]$. There exist $f, s \in A[X]$ such that $1 = af + Xs$ and then $1 = af(0)$ with $f(0) \in A$, which is absurd.

Example 3 In a Prufer domain A, each nonzero finitely generated prime ideal P is maximal. Indeed, it suffices to show that if $a \in A \setminus P$, then $(a, P) = A$. Let $x \in (a, P)^{-1}$. Then $ax \in A$ and $xP \subseteq A$. So $axP \subseteq P$. Since P is a prime ideal of A, then $xP \subseteq P$, so $xPP^{-1} \subseteq PP^{-1}$. Since P is invertible, we have $xA \subseteq A$, and then $x \in A$. So $(a, P)^{-1} = A$, and then $(a, P)^{-1}(a, P) = (a, P)$. But (a, P) is invertible, and then $(a, P) = A$.

Lemma 9.7 *Let $A \subseteq B$ an extension of integral domains admitting a nonzero common ideal that is invertible in A. Then $A = B$.*

Proof Let $I \neq (0)$ be this common ideal of A and B. The domains A and B have the same quotient field. By hypothesis, $I(A : I) = A$. There exist $a_1, \ldots, a_n \in I$ and $x_1, \ldots, x_n \in (A : I)$ such that $1 = a_1x_1 + \ldots + a_nx_n$. For each $b \in B$, we have $b = (ba_1)x_1 + \ldots + (ba_n)x_n$ with $ba_1, \ldots, ba_n \in I$. Then $b \in I(A : I) = A$. So $B \subseteq A$. $\qquad \square$

Example The integral domains $\mathbb{R} + X\mathbb{C}[X] \subset \mathbb{C}[X]$ share the maximal ideal $X\mathbb{C}[X]$, but they are different.

Lemma 9.8 *Let $K \subseteq A$ be an extension of rings with K a field. If $dim_K(A) < \infty$, then $|Max(A)| \leq dim_K(A)$.*

Proof Since $dim_K(A) < \infty$, the extension $K \subseteq A$ is integral. Then $dim\ A = dim\ K = 0$. By Eakin-Nagata theorem, the ring A is Noetherian. See Example 1, Corollary 3.7, Chap. 1. By Lemma 7.1, Chap. 2, $Spec(A) = Max(A) = Min(A) = \{M_1, \ldots, M_k\}$ is finite. By CRT, $A/(M_1 \ldots M_k) \simeq (A/M_1) \times \ldots \times (A/M_k)$. See Theorem 7.7, Chap. 2. Denote the field $A/M_i = L_i$, for $1 \leq i \leq k$. The natural homomorphism $K \longrightarrow A/M_i = L_i$ is one to one, since $K \cap M_i = (0)$. We have $1 \leq [L_i : K] = dim_K(A/M_i) \leq dim_K(A) < \infty$. Then $L_1 \times \ldots \times L_k$ is a K−vector space of finite dimension. Since $A/(M_1 \ldots M_k)$ is a direct sum of k nonzero sub-vector spaces on K that are, respectively, isomorphic to $L_1, \ldots, L_k$, then $k \leq dim_K\big(A/(M_1 \ldots M_k)\big) \leq dim_K(A) < \infty$. This means that $|Max(A)| \leq dim_K(A) < \infty$. $\qquad\qquad\square$

Example $dim_{\mathbb{C}}\mathbb{C}[X] = \aleph_0$, $Max(\mathbb{C}[X]) = \{(X - \alpha); \ \alpha \in \mathbb{C}\}$, of cardinal $2^{\aleph_0} > \aleph_0$.

Lemma 9.9 *Let $A \subseteq B$ be an extension of rings such that B is generated as A−module by n elements. For each prime ideal P of A, there exist at most n different prime ideals in B over P.*

Proof Particular case: A is a field. Since the A−module B is generated by n elements, $dim_A(B) \leq n < \infty$. By the preceding lemma, $|Max(B)| \leq n$. Since A is a field and the extension $A \subseteq B$ is integral, $dim\ B = dim\ A = 0$. So $Spec(B) = Max(B)$ with cardinal $\leq n$. All the prime ideals of B are over the unique prime ideal (0) of A.

General case: Let P be a prime ideal of A. By Example 2 of Remark 9.4, Chap. 2, $(PB) \cap A = P$. Consider the multiplicative set $S = A \setminus P$ of both A and B. We have the extension $A_P = S^{-1}A \subseteq S^{-1}B$ with $S^{-1}B$ a A_P−module generated by n elements. The ideal $S^{-1}(PB)$ of $S^{-1}B$ satisfies $S^{-1}(PB) \cap A_P = S^{-1}(PB) \cap S^{-1}A = S^{-1}((PB) \cap A) = S^{-1}P = PA_P$. Then the extension $A_P/PA_P \subseteq S^{-1}B/S^{-1}(PB)$ with $S^{-1}B/S^{-1}(PB)$ a A_P/PA_P−module generated by n elements. Since A_P/PA_P is a field, by the particular case, there exist at most n prime ideals in $S^{-1}B/S^{-1}(PB)$ over (0) in the field A_P/PA_P. Let $(Q_\lambda)_{\lambda \in \Lambda}$ be the family of the prime ideals of B over P. For each $\lambda \in \Lambda$, $S \cap Q_\lambda = (S \cap A) \cap Q_\lambda = S \cap P = \emptyset$ and $PB \subseteq Q_\lambda$. Then $S^{-1}Q_\lambda/S^{-1}(PB)$ is a prime ideal of $S^{-1}B/S^{-1}(PB)$ and $\big(S^{-1}Q_\lambda/S^{-1}(PB)\big) \cap \big(A_P/PA_P\big) = \big(S^{-1}Q_\lambda/S^{-1}(PB)\big) \cap \big(S^{-1}A/S^{-1}P\big) = S^{-1}(Q_\lambda \cap A)/S^{-1}P = S^{-1}P/S^{-1}P = (0)$. Then $|\Lambda| \leq n$. $\qquad\qquad\square$

Example 1 We do not have the uniqueness in the lemma. Indeed, the ring $\mathbb{Z}[i]$ is a $\mathbb{Z}$−module generated by $\{1, i\}$. The elements $2 \pm i$ are irreducible in $\mathbb{Z}[i]$. Indeed, let $z, z' \in \mathbb{Z}[i]$ be such that $2 \pm i = zz'$. Then $5 = |2 \pm i|^2 = |z|^2|z'|^2$. Since the integral solutions of the equation $a^2 + b^2 = 1$ are $(\pm 1, 0)$ and $(0, \pm 1)$, then either z or $z' \in \{\pm 1, \pm i\} = U(\mathbb{Z}[i])$. The ideals $(2 \pm i)\mathbb{Z}[i]$ are prime and different since the elements $2 + i$ and $2 - i$ are not associated. Moreover, $(2 + i)\mathbb{Z}[i] = \{(2+i)(a+bi); a, b \in \mathbb{Z}\} = \{2a-b+i(a+2b); a, b \in \mathbb{Z}\}$. Then $(2+i)\mathbb{Z}[i] \cap \mathbb{Z} = \{2a - b + i(a + 2b); a, b \in \mathbb{Z}, a = -2b\} = \{-5b; b \in \mathbb{Z}\} = 5\mathbb{Z}$. Also, we have $(2 - i)\mathbb{Z}[i] \cap \mathbb{Z} = 5\mathbb{Z}$.

Example 2 The condition "B is a finitely generated $A-$module" cannot be weaken to "the extension $A \subseteq B$ is integral". Indeed, let X be any infinite set and A the ring of the maps $X \longrightarrow \mathbb{F}_2$, endowed with the natural operations. The map $\phi : \mathbb{F}_2 \longrightarrow A$, defined by $\phi(0) = 0$ and $\phi(1) = 1$, is a homomorphism one to one. Then $\mathbb{F}_2 \subset A$. For each $f \in A$, $f^2 - f = 0$ since for each $x \in \mathbb{F}_2$, $x^2 - x = 0$. The extension $\mathbb{F}_2 \subset A$ is then integral. In particular, $dim(A) = dim(\mathbb{F}_2) = 0$. All the prime ideals of A are maximal. There exist an infinity of prime ideals of A over (0) in $\mathbb{F}_2$. Indeed, for each $x \in X$, the set $P_x = \{f \in A; \ f(x) = 0\}$ is a prime ideal of A. Let $f, g \in A$ be such that $fg \in P_x$. Then $f(x)g(x) = 0$, so either $f(x) = 0$ or $g(x) = 0$, so $f \in P_x$ or $g \in P_x$. All the ideals P_x are maximal and over (0) in $\mathbb{F}_2$.

Lemma 9.10 *Let $A \subseteq B$ be an extension of rings such that B is a finitely generated $A-$module, M a maximal ideal of A, and $S = A \setminus M$.*

1. *There exists a finite number $M'_1, \ldots, M'_n$ of maximal ideals of B containing M.*
2. *For each $M' \in Max(B) \setminus \{M'_1, \ldots, M'_n\}$, $M' \cap S \neq \emptyset$.*
3. *The ring $S^{-1}B$ admits a finite number of maximal ideals.*

Proof

(1) Let $M' \in Max(B)$ such that $M \subseteq M'$. Then $M \subseteq M' \cap A$, so $M = M' \cap A$ since M is a maximal ideal of A. By the preceding lemma, the number of prime ideals of B over M in A is finite.
(2) Let $M' \in Max(B) \setminus \{M'_1, \ldots, M'_n\}$. Suppose that $M' \cap S = \emptyset$. Then $M' \cap A \subseteq M$. Since the extension $A \subseteq B$ is integral, by the GU property, there exists $P \in Spec(B)$ such that $M' \subseteq P$ and $P \cap A = M$. Then $M' = P$ and $M \subseteq P = M'$, which contradicts the choice of M'.
(3) The only maximal ideals of B disjoint with S are $M'_1, \ldots, M'_n$. Then the only maximal ideals of $S^{-1}B$ are $S^{-1}M'_1, \ldots, S^{-1}M'_n$. $\square$

Proposition-Definition 9.11 *Let A be an integral domain and I an integral ideal of A. The following assertions are equivalent:*

1. *For each prime ideal P of A, there exist $n \in \mathbb{N}^*$ and $a \in I^n$ such that $I^{2n}A_P = aI^nA_P$.*
2. *For each maximal ideal M of A, there exist $n \in \mathbb{N}^*$ and $a \in I^n$ such that $I^{2n}A_M = aI^nA_M$.*

We say that the ideal I is prestable if it is finitely generated and satisfies the two preceding equivalent conditions.

Proof Let P be a prime ideal of A. There exists a maximal ideal M of A containing P. Since $A \setminus M \subseteq A \setminus P$, then $A_M \subseteq A_P$. Let $n \in \mathbb{N}^*$ and $a \in I^n$ such that $I^{2n} A_M = a I^n A_M$. Then $I^{2n} A_P = a I^n A_P$. $\qquad\qquad\qquad\qquad\qquad\qquad\square$

Example The integral principal ideals are prestable.

The prestable ideals of a domain A have many characterizations. In the following theorem, we study their relations with the ideals of the form $f(X)K[X] \cap A[X]$.

Theorem 9.12 (Lucas 2012) *Let A be an integral domain with quotient field K and integral closure $\overline{A}$ and J a nonzero finitely generated integral ideal of A. The following assertions are equivalent:*

1. *For each $f(X) \in A[X]$ such that $c(f) = J$, the ideal $f(X)K[X] \cap A[X]$ of $A[X]$ contains a polynomial with content equal to A.*
2. *There exists $f(X) \in A[X]$ such that $c(f) = J$ and the ideal $f(X)K[X] \cap A[X]$ of $A[X]$ contains a polynomial with content equal to A.*
3. *The ideal $J . \overline{A}$ is invertible in the domain $\overline{A}$.*
4. *There exists $m \in \mathbb{N}^*$ such that J^m is invertible in the domain $T_m = (J^m : J^m)$.*
5. *There exists $m \in \mathbb{N}^*$ such that for each $M \in Max(A)$, there exists $t \in J^m$ satisfying $J^{2m} A_M = t J^m A_M$.*
6. *The ideal J is prestable.*

Proof $"(1) \Longrightarrow (2)"$ Clear.

$"(2) \Longrightarrow (3)"$ By hypothesis, there exists $f(X) \in A[X]$ such that $c(f) = J$, and there exists $g(X) \in K[X]$ such that $f(X)g(X) \in A[X]$ and $c(fg) = A$. By the content formula, there exists $m \in \mathbb{N}^*$ such that $c(f)^m c(f)c(g) = c(f)^m c(fg) = c(f)^m$. Then $c(f)c(g) \subseteq (c(f)^m : c(f)^m) \subseteq \overline{A}$, by Lemma 9.1. Then $c(g) \subseteq (\overline{A} : c(f))$ so $\overline{A} = c(fg)\overline{A} \subseteq c(f)c(g)\overline{A} \subseteq c(f)(\overline{A} : c(f))\overline{A} \subseteq \overline{A}$. Then $c(f)c(g)\overline{A} = \overline{A}$, so $J \overline{A}.c(g)\overline{A} = \overline{A}$. Then the ideal $J \overline{A}$ is invertible in the domain $\overline{A}$.

$"(4) \Longrightarrow (3)"$ By hypothesis, there exists $m \in \mathbb{N}^*$ such that the ideal J^m is invertible in the domain $(J^m : J^m) = T_m \subseteq \overline{A}$. There exists a fractional ideal T of T_m such that $J^m T = T_m$. Then $J \overline{A}(J^{m-1}T)\overline{A} = \overline{A}$. So the ideal $J \overline{A}$ is invertible in the domain $\overline{A}$.

$"(3) \Longrightarrow (1)"$ and $"(3) \Longrightarrow (4)"$ Let $f(X) = a_0 + \ldots + a_n X^n \in A[X]$ be such that $c(f) = J$. By hypothesis, the ideal $J \overline{A}$ is invertible in the domain $\overline{A}$, and then $(\overline{A} : c(f)\overline{A})c(f) \overline{A} = \overline{A}$. There exist $b_0, \ldots, b_n \in (\overline{A} : c(f)\overline{A}) \subseteq K$ such that $a_0 b_0 + \ldots + a_n b_n = 1$. Let $g(X) = \displaystyle\sum_{i:0}^{n} b_i X^{n-i} \in K[X]$. The coefficients of $g(X)f(X)$ belong to $(\overline{A} : c(f)\overline{A})c(f)\overline{A} = \overline{A}$, and the coefficient of X^n is equal to 1, and then $c(gf) = \overline{A}$. Consider the multiplicative sets $N = \{h(X) \in A[X]; \ c(h) = A\}$ of $A[X]$ and $N' = \{h(X) \in \overline{A}[X]; \ c(h) = \overline{A}\}$

of $\overline{A}[X]$. By Proposition 1.4, Chap. 2, N' is the saturation of N in $\overline{A}[X]$. Since $g(X)f(X) \in N'$, there exists $h(X) \in \overline{A}[X]$ such that $g(X)f(X)h(X) \in N$. Then $g(X)f(X)h(X) \in f(X)K[X] \cap A[X]$ and $c(gfh) = A$. So we have the implication $"(3) \Longrightarrow (1)"$.

By the content formula, $c(f)^m c(f)c(gh) = c(f)^m c(fgh) = c(f)^m$ with $m \in \mathbb{N}^*$. Then $c(f)c(gh) \subseteq \left(c(f)^m : c(f)^m \right) = T_m$, so $Jc(gh) \subseteq T_m$. Then $c(gh) \subseteq (T_m : J)$. But $A = c(fgh) \subseteq c(f)c(gh) = Jc(gh) \subseteq J(T_m : J) \subseteq T_m$, and then $T_m \subseteq JT_m(T_m : J) \subseteq T_m$. So $JT_m(T_m : J) = T_m$. The ideal JT_m is then invertible in the domain T_m so the ideal $J^m = J^m T_m$ is invertible in T_m. Then we have the implication $"(3) \Longrightarrow (4)"$.

$"(4) \Longrightarrow (5)"$ Let $m \in \mathbb{N}^*$ be such that the ideal J^m is invertible in the domain $(J^m : J^m) = T_m$. Since J is a finitely generated ideal of A, then J^m is also finitely generated. Put $J^m = b_1 A + \ldots + b_n A$. Since $(T_m : J^m)J^m = T_m$, there exist $t_1, \ldots, t_n \in (T_m : J^m) \subseteq K$ such that (*) $t_1 b_1 + \ldots + b_n t_n = 1$. Since $A \subseteq A[t_i b_j; 1 \le i, j \le n] \subseteq T_m$ because $t_i b_j \in (T_m : J^m)J^m = T_m$, and then J^m is an ideal of the domain $A[t_i b_j; 1 \le i, j \le n]$ and is invertible in it. Indeed, let $I = (t_1, \ldots, t_n)A[t_i b_j; 1 \le i, j \le n]$. Then $IJ^m = A[t_i b_j; 1 \le i, j \le n]$. The direct inclusion is clear and the inverse inclusion follows from (*). By Lemmas 9.7 and 9.1, $A[t_i b_i; 1 \le i \le n] = T_m = (J^m : J^m) \subseteq \overline{A}$. The elements $t_i b_j, 1 \le i, j \le n$ are then integral over A so $T_m = A[t_i b_j; 1 \le i, j \le n]$ is a finitely generated A−module. Let M be a maximal ideal of A and $S = A \setminus M$. By Lemma 9.10 (3), $S^{-1}T_m$ has a finite number of maximal ideals. Since the ideal J^m is invertible in the domain T_m, the ideal $S^{-1}J^m$ is invertible in the domain $S^{-1}T_m$. By Lemma 9.5, the ideal $S^{-1}J^m$ is principal in the domain $S^{-1}T_m$. There exists $t \in J^m$ such that $S^{-1}J^m = t(S^{-1}T_m)$. By Lemma 9.2, $S^{-1}T_m = S^{-1}(J^m : J^m) = \left(S^{-1}J^m : S^{-1}J^m \right) = (J^m A_M : J^m A_M)$. Then $J^{2m}A_M = S^{-1}J^{2m} = J^m(S^{-1}J^m) = J^m t(S^{-1}T_m) = tJ^m(J^m A_M : J^m A_M) \subseteq tJ^m A_M \subseteq J^{2m}A_M$. So $J^{2m}A_M = tJ^m A_M$.

$"(5) \Longrightarrow (6)"$ Clear.

$"(6) \Longrightarrow (3)"$ Since the ideal J is finitely generated, then the ideal $J.\overline{A}$ is also finitely generated. By Lemma 9.6, it remains to show that $J.\overline{A}$ is locally principal. Let $M' \in Max(\overline{A})$, $M = M' \cap A \in Max(A)$, and $S = A \setminus M$. By hypothesis, there exist $k \in \mathbb{N}^*$ and $0 \ne t \in J^k$ such that $J^{2k}A_M = tJ^k A_M$. Then $(t^{-1}J^k A_M)(J^k A_M) = J^k A_M$, so $t^{-1}J^k A_M \subseteq (J^k A_M : J^k A_M)$, and then $J^k A_M \subseteq t(J^k A_M : J^k A_M) \subseteq J^k(J^k A_M : J^k A_M) \subseteq J^k A_M$. So $J^k A_M = t(J^k A_M : J^k A_M)$. By Lemma 9.1, $(J^k A_M : J^k A_M)$ is a sub-ring of $\overline{A_M} = \overline{S^{-1}A} = S^{-1}\overline{A} \subseteq (A \setminus M')^{-1}\overline{A} = \overline{A}_{M'}$. Then $J^k \overline{A}_{M'} = t \overline{A}_{M'}$, so $J^k \overline{A}_{M'}$ is a principal ideal of $\overline{A}_{M'}$. Then it is invertible. So the ideal $J \overline{A}_{M'}$ is also invertible. By Lemma 9.5 (1), since the domain $\overline{A}_{M'}$ is local, the ideal $J \overline{A}_{M'}$ is principal. $\qquad\square$

Example 1 Let A be an integrally closed integral domain and J a nonzero finitely generated integral ideal of A. Then J is prestable if and only if it is invertible in A.

Example 2 Let B be an over ring of an integral domain of A with B a finitely generated A−module. Let I be a nonzero finitely generated integral ideal of A. Then I is prestable in A if and only if IB is prestable in B.
Indeed, since the extension $A \subseteq B$ is integral, then $\overline{A} = \overline{B}$. So $I\,\overline{A} = I\,\overline{B} = (IB)\overline{B}$.
Then $I\,\overline{A}$ is invertible in $\overline{A}$ if and only if $(IB)\overline{B}$ is invertible in $\overline{A} = \overline{B}$.

Example 3 An integral domain A is prestable ; i.e, all its nonzero finitely generated integral ideals are prestable, if and only if its integral closure $\overline{A}$ is a Prufer domain. Indeed, suppose that A is prestable. Let $\mathscr{I}$ be any nonzero finitely generated integral ideal of $\overline{A}$, $\mathscr{I} = (x_1, \ldots, x_n)\overline{A}$, $x_i \neq 0$, $x_i = \frac{a_i}{b}$ with $b, a_1, \ldots, a_n \in A \setminus (0)$. By hypothesis, the ideal $I = (a_1, \ldots, a_n)A$ is prestable in A. Then the ideal $I\,\overline{A} = \mathscr{I}$ is invertible in $\overline{A}$. So A is a Prufer domain. Conversely, suppose that A is a Prufer domain. Let I be any nonzero finitely generated integral ideal of A. By hypothesis, the ideal $I\,\overline{A}$ is invertible in $\overline{A}$. By the preceding theorem, I is prestable in A.

The following proposition connects the property almost principal of an ideal I of $A[X]$ and the contents of the polynomials in I.

Proposition 9.13 (T.G. Lucas; 212) *Let A be an integral domain, $f(X) \in A[X]$ a nonconstant polynomial, and $I = f(X)K[X] \cap A[X]$. Suppose that there exists $g(X) \in I$ such that $c(g)^{-1} = A$. Then I is almost principal. In particular, if I contains a polynomial with content equal to A, then I is almost principal.*

Proof Put $g(X) = f(X)l(X)$ with $l(X) \in K[X]$. There exists $0 \neq s \in A$ such that $sl(X) \in A[X]$. We will apply Proposition 3.3. Let $h(X) \in K[X]$ be such that $f(X)h(X) \in A[X]$. There exists $0 \neq a \in A$ such that $ah(X) \in A[X]$. The ideal $(a, g(X))sh(X)f(X) \subseteq f(X)A[X]$ since the following polynomials $ash(X)f(X) = s(ah(X))f(X) \in f(X)A[X]$ and $g(X)sh(X)f(X) = f(X)l(X)sh(X)f(X) = (sl(X))(f(X)h(X))f(X) \in f(X)A[X]$. By Proposition 11.1, Chap. 1, $((a, g(X))_v sh(X)f(X))_v = ((a, g(X))sh(X)f(X))_v \subseteq f(X)A[X]$. By Lemma 11.9, Chap. 1, since $c(g)_v = A$, then $(a, g(X))_v = A[X]$. So $sh(X)f(X)A[X] \subseteq f(X)A[X]$, and then $sh(X) \in A[X]$. $\square$

To prove the following corollary, we use the same techniques as in the proof of Theorem 9.12.

Corollary 9.14 (T.G. Lucas; 212) *Let A be an integral domain and $f(X)$ and $g(X) \in A[X]$ such that $c(f) = c(g)$. If the ideal $I = f(X)K[X] \cap A[X]$ contains a polynomial with content equal to A, then the ideal $J = g(X)K[X] \cap A[X]$ is almost principal.*

Proof Let $h(X) \in K[X]$ be such that $f(X)h(X) \in I$ and $c(fh) = A$. There exists $n \in \mathbb{N}^*$ such that $c(f)^n c(f)c(h) = c(f)^n c(fh) = c(f)^n$. Then $c(f)c(h) \subseteq \left(c(f)^n : c(f)^n\right) = T$. But $A = c(fh) \subseteq c(f)c(h) \subseteq T$, and then $T \subseteq c(f)T.c(h)T \subseteq T$ so $c(f)T.c(h)T = T$. The ideal $c(g)T = c(f)T$ is invertible in the ring T. Then $\left(T : c(g)T\right).c(g)T = T$. Put $g(X) = a_0 + a_1 X + \ldots + a_n X^n \in A[X]$. There exist $b_0, \ldots, b_n \in \left(T : c(g)T\right) \subseteq K$ such that $b_0 a_0 + \ldots + b_n a_n = 1$. Let $t(X) = \sum_{i:0}^{n} b_i X^{n-i} \in \left(T : c(g)T\right)[X] \subseteq K[X]$. The coefficients of the polynomial $g(X)t(X)$ belong to $c(g)T.(T : c(g)T) = T$, and the coefficient of X^n is 1, and then $c_T(gt) = T$. Let $\overline{A}$ be the integral closure of A. By Lemma 9.1, $T = \left(c(f)^n : c(f)^n\right) \subseteq \overline{A}$. Consider the multiplicative sets $N = \{r(X) \in A[X];\ c(r) = A\}$ of $A[X]$ and $N' = \{r(X) \in \overline{A}[X];\ c(r) = \overline{A}\}$ of $\overline{A}[X]$. By Proposition 1.4, Chap. 2, N' is the saturation of N in the domain $\overline{A}[X]$. Since $c_{\overline{A}}(gt) = \overline{A}$, then $gt \in N'$. There exists $v(X) \in N$ such that $vgt \in N$. Then $vgt \in g(X)K[X] \cap A[X] = J$ with $c(vgt) = A$. By the preceding the proposition, the ideal J is almost principal.   □

Corollary 9.15 (T.G. Lucas ; 212) *Let A be an integral domain with an integral closure $\overline{A}$ a Prufer domain. Then the polynomial ring $A[X]$ is almost principal.*

Proof We will apply Proposition 3.7. Let $f(X) \in A[X]$ be a nonconstant polynomial and $I = f(X)K[X] \cap A[X]$ where K is the quotient field of A. It suffices to show that the ideal I is almost principal. Consider the ideal $J = c(f)$ of A. Since A is a Prufer domain, $J\overline{A}$ is invertible in $\overline{A}$. By Theorem 9.12, there exists $g(X) \in A[X]$ such that $c(g) = J = c(f)$, and the ideal $g(X)K[X] \cap A[X]$ contains a polynomial with content equal to A. By the preceding corollary, the ideal I is almost principal.   □

10   When Is the Ideal f(X)K[X]∩ A[X] Maximal?

This section is devoted to characterize the maximal ideals of $A[X]$ that are over (0) in an integral domain A. The following proposition shows that the ideals of the form $f(X)K[X] \cap A[X]$ can characterize the integrally closed domains. This is a very useful result.

Proposition 10.1 (Querré 1980) *Let A be an integral domain with quotient field K. The following assertions are equivalent:*

1. *The domain A is integrally closed.*
2. *For each polynomials $f(X), g(X) \in K[X] \setminus (0)$ (or even just $f(X), g(X) \in A[X] \setminus (0)$), $\left(c(f)c(g)\right)_v = c(fg)_v$.*
3. *For each polynomial $0 \neq f(X) \in A[X]$, $(fK[X]) \cap A[X] = f.c(f)^{-1}[X]$.*

4. For each $f(X), g(X) \in A[X] \setminus (0)$ and $0 \neq a \in A$, if $c(fg) \subseteq aA$, and then $c(f)c(g) \subseteq aA$.

Proof "(1) $\Longrightarrow$ (2)" Let $(V_\alpha)_{\alpha \in \Lambda}$ be the family of over-valuation domains of A. Since A is integrally closed, $A = \bigcap \{V_\alpha; \ \alpha \in \Lambda\}$. Let $f, g \in K[X] \setminus (0)$. Since $c(fg) \subseteq c(f)c(g)$, then $c(fg)_v \subseteq (c(f)c(g))_v$. Conversely, it suffices to show that $c(f)c(g) \subseteq c(fg)_v$. Recall from Sect. 11, Chap. 1, that for each nonzero fractional ideal I of A, $I_v = \bigcap\{xA; \ x \in K, I \subseteq xA\}$. It suffices to show that for each $x \in K$ satisfying $c(fg) \subseteq xA$, then $c(f)c(g) \subseteq xA$. Since $xA = x(\bigcap\{V_\alpha; \ \alpha \in \Lambda\}) = \bigcap\{xV_\alpha; \ \alpha \in \Lambda\}$, it suffices to show that $c(f)c(g) \subseteq xV_\alpha$ for each $\alpha \in \Lambda$. But for the valuation domains V_α, $c(f)c(g) \subseteq c(f)V_\alpha.c(g)V_\alpha = c(fg)V_\alpha \subseteq (xA)V_\alpha = xV_\alpha$. See Example 4, Theorem 11.5, Chap. 1.

"(2) $\Longrightarrow$ (3)" Let $0 \neq f \in A[X]$. It suffices to show that if $g \in K[X]$, then $fg \in A[X]$ if and only if $g \in c(f)^{-1}[X]$. By (2), $fg \in A[X] \Longleftrightarrow c(fg) \subseteq A \Longleftrightarrow c(fg)_v \subseteq A \Longleftrightarrow (c(f)c(g))_v \subseteq A \Longleftrightarrow c(f)c(g) \subseteq A \Longleftrightarrow c(g) \subseteq c(f)^{-1} \Longleftrightarrow g \in c(f)^{-1}[X]$.

"(3) $\Longrightarrow$ (1)" Let $u \in K$ be an integral element over A. There exists a monic polynomial $f(X) \in A[X]$ such that $f(u) = 0$. Then $f(X) = (X - u)g(X)$ with $g(X) \in K[X]$ a monic polynomial. Let $0 \neq a \in A$ be such that $ag(X) \in A[X]$. By (3), we have $f \in (gK[X]) \cap A[X] = (agK[X]) \cap A[X] = ag.c(ag)^{-1}[X] = ag.(ac(g))^{-1}[X] = ag.\frac{1}{a}c(g)^{-1}[X] = g.c(g)^{-1}[X]$. Then $(X - u)g \in gc(g)^{-1}[X]$ so $X - u \in c(g)^{-1}[X]$, and then $u \in c(g)^{-1}$. Since g is monic, $1 \in c(g)$ and $u = u.1 \in c(g)^{-1}c(g) \subseteq A$. So the domain A is integrally closed.

"(4) $\Longrightarrow$ (1)" Let $u \in K$ be an integral element over A. There exists a monic polynomial $f(X) \in A[X]$ such that $f(u) = 0$. Then $f = (X - u)g$ with $g(X) \in K[X]$ a monic polynomial. Let $0 \neq a \in A$ be such that $a(X - u) \in A[X]$ and $ag(X) \in A[X]$. Then $a(X-u)ag(X) = a^2 f(X) \in a^2 A[X]$, so $c(a(X-u)ag) \subseteq a^2 A$. By (4), $c(a(X - u))c(ag) \subseteq a^2 A$. Since $au \in c(a(X - u))$ and $a \in c(ag)$ because g is monic, then $a^2 u \in a^2 A$. So $u \in A$. This shows that A is integrally closed.

"(2) $\Longrightarrow$ (4)" Let $f(X), g(X) \in A[X] \setminus (0)$ and $0 \neq a \in A$ such that $c(fg) \subseteq aA$. By condition (2), $c(f)c(g) \subseteq (c(f)c(g))_v = c(fg)_v \subseteq (aA)_v = aA$.   □

Example The (u, u^{-1}) Lemma: Let (A, M) be an integrally closed local integral domain and K its quotient field. Let $u \in K$ and $g(X) \in A[X]$ be such that $g(u) = 0$ and $c(g) = A$. Then either $u \in A$ or $u^{-1} \in A$.

Indeed, $u = \frac{a}{b}$ with $a, b \in A \setminus (0)$. By condition (3) of the preceding proposition, the ideal $I = (X - u)K[X] \cap A[X] = (bX - a)K[X] \cap A[X] = (bX - a).(b, a)^{-1}[X] = (X - u).(1, u)^{-1}[X]$. On the other hand, $g(X) = (X - u)h(X)$ with $h(X) \in K[X]$, and then $g(X) \in I$. So $h(X) \in (1, u)^{-1}[X]$, and then $c(h) \subseteq (1, u)^{-1}$, so $c(h) \subseteq A$ and $uc(h) \subseteq A$. If $c(h) = A$, then $uA = uc(h) = c(uh) \subseteq A$, so $u \in A$. If $c(h) \neq A$, then $c(h) \subseteq M$. So

$A = c(g) = c\big((X - u)h(X)\big) \subseteq (1, u)c(h) \subseteq (1, u)M = M + uM$. There exist $\alpha, \beta \in M$ such that $1 = \alpha + u\beta$, then $u^{-1} = \alpha u^{-1} + \beta$, so $u^{-1}(1 - \alpha) = \beta$. But $1 - \alpha$ is invertible in the local domain A, and then $u^{-1} = \frac{\beta}{1-\alpha} \in M \subset A$.

Proposition 10.2 *A minimal prime ideal over a t−ideal is a t−prime.*

Proof Let P be a minimal prime ideal over a t−ideal I of an integral domain A. By Lemma 3.11 of Chap. 2, for each $x \in P$, there exist $n \in \mathbb{N}^*$ and $s \in A \setminus P$ such that $sx^n \in I$. If J is a nonzero finitely generated subideal of P, there are $n \in \mathbb{N}^*$ and $s \in A \setminus P$ such that $sJ^n \subseteq I$. Then $s(J^n)_t = (sJ^n)_t \subseteq I_t = I \subseteq P$, so $(J^n)_t \subseteq P$ because $s \notin P$ and P is prime. By Proposition 11.1 of Chap. 1, $(J_t)^n \subseteq (J^n)_t \subseteq P$, then $J_t \subseteq P$. So $P_t = \bigcup\{J_v = J_t;\ J$ a nonzero finitely generated subideal of $P\} \subseteq P$. □

Example The height 1 prime ideals of an integral domain are t−prime.
Indeed, they are minimal over the nonzero principal integral ideals that they contain.

In the following proposition, we characterize the nonzero prime ideals of $A[X]$ over (0) in an integral domain A.

Proposition 10.3 *Let A be an integral domain with quotient field K.*

1. *The nonzero prime ideals of $A[X]$ over (0) in A are exactly the ideals of the form $P = f(X)K[X] \cap A[X]$ with $f(X) \in A[X]$ irreducible in $K[X]$. Moreover, $A[X]_P$ is a DVR of rank 1.*
2. *A nonzero prime ideal of $A[X]$ over (0) in A is of height 1, and then a t−prime.*

Proof

(1) Let P be a nonzero prime ideal of $A[X]$ such that $P \cap A = (0)$. Then $PK[X] = f(X)K[X]$ with $0 \neq f(X) \in P \subset A[X]$ and $P \subseteq PK[X] \cap A[X] = f(X)K[X] \cap A[X]$. Since $P \cap A = (0)$, $f(X)$ is not constant. Let $h(X) \in f(X)K[X] \cap A[X]$, $h(X) = f(X)l(X)$ with $l(X) \in K[X]$. There exists $0 \neq a \in A$ such that $al(X) \in A[X]$. Then $ah(X) = f(X)al(X) \in P$. But $a \notin P$, and then $h(X) \in P$. So $f(X)K[X] \cap A[X] = P$. It remains to show that $f(X)$ is irreducible in $K[X]$. Let $g(X)$ and $h(X) \in K[X]$ be such that $f(X) = g(X)h(X)$. There exist $a, b \in A \setminus (0)$ such that $ag(X), bh(X) \in A[X]$. Then $ag(X)bh(X) = abf(X) \in P$. For example, $ag(X) \in P \subseteq f(X)K[X]$, and then $f(X)$ divides $g(X)$ in $K[X]$ with $deg\ g(X) \leq deg\ f(X)$. Then $f(X)$ and $g(X)$ are associated with $K[X]$ and $h(X) \in K^*$. So $f(X)$ is irreducible in $K[X]$. □

Other Proof Let P be a nonzero prime ideal of $A[X]$ over (0) in A. Since $K[X] = S^{-1}A[X]$ with $S = A \setminus (0)$, then $P \cap S = \emptyset$ and $S^{-1}P$ is a nonzero prime ideal of $K[X]$. Then $S^{-1}P = f(X)K[X]$ with $f(X) \in P \subset A[X]$ and irreducible in $K[X]$. So $f(X)K[X] \cap A[X] = S^{-1}P \cap A[X] = P$. Indeed, an element of the intersection can be written $\frac{g(X)}{s} = h(X)$ with $g(X) \in P$, $0 \neq s \in A$ and $h(X) \in A[X]$. Then $sh(X) = g(X) \in P$ with $s \notin P$ because $P \cap A = (0)$, so $h(X) \in P$. Finally, by

Lemma 10.8, Chap. 1, $A[X]_P = (A[X]_S)_{P_S} = K[X]_{f(X)K[X]}$. Since $f(X)$ is an irreducible element of the UFD $K[X]$, then $K[X]_{f(X)K[X]}$ is a DVR of rank 1.

Conversely, let $f(X) \in A[X]$ be an irreducible polynomial in $K[X]$. Then $f(X)K[X]$ is maximal ideal in $K[X]$, so $P = f(X)K[X] \cap A[X]$ is prime in $A[X]$. Since $P \subseteq f(X)K[X]$ with $f(X)$ nonconstant, then $P \cap A = (0)$.

(2) Let P be a nonzero prime ideal of $A[X]$ such that $P \cap A = (0)$. Suppose that there exists a prime ideal Q of $A[X]$ such that $(0) \subset Q \subset P$. The trace over A of these three prime ideals of $A[X]$ is (0), which is absurd.   □

Other Proof By (1), $P = f(X)K[X] \cap A[X]$ with $f(X) \in A[X]$ an irreducible polynomial in $K[X]$. We will show that P is a minimal prime over the principal ideal $f(X)A[X]$. Then it is a $t-$prime, by the preceding proposition. Indeed, let Q be a prime ideal of $A[X]$ such that $f(X)A[X] \subseteq Q \subseteq P$. Then $Q \cap A = (0)$. As in (1), we show that $(QK[X]) \cap A[X] = Q$. Since $f(X) \in Q$, then $f(X)K[X] \subseteq QK[X]$. So $P = f(X)K[X] \cap A[X] \subseteq QK[X] \cap A[X] = Q$, and then $P = Q$.   □

We need some results on the $t-$operation before giving the main theorem of this section. Recall that for an integral domain A, we mean by $\mathscr{F}(A)$ the set of the nonzero fractional ideals of A.

Definition 10.4 Let A be an integral domain and $I \in \mathscr{F}(A)$. We say that the ideal I is $t-$invertible if there exists $J \in \mathscr{F}(A)$ such that $(IJ)_t = A$.

Example An invertible ideal is $t-$invertible.

Proposition 10.5 *Let A be an integral domain and $I \in \mathscr{F}(A)$. Then I is $t-$invertible if and only if I_t is $t-$invertible.*

Proof The ideal I is $t-$invertible $\iff \exists J \in \mathscr{F}(A)$ such that $(IJ)_t = A \iff \exists J \in \mathscr{F}(A)$ such that $(I_tJ)_t = A \iff I_t$ is $t-$invertible.   □

Proposition 10.6 *Let A be an integral domain and $I, J \in \mathscr{F}(A)$ be such that $(IJ)_t = A$. Then $J_t = I^{-1}$.*

Proof We have $I^{-1} \subseteq (I^{-1})_t = (I^{-1}A)_t = (I^{-1}(IJ)_t)_t = (I^{-1}IJ)_t \subseteq (AJ)_t = J_t$. Conversely, by Proposition 11.1 of Chap. 1, $IJ_t \subseteq (IJ_t)_t = (IJ)_t = A$, and then $J_t \subseteq (A : I) = I^{-1}$. So $J_t = I^{-1}$.   □

Corollary 10.7 *Let A be an integral domain and $I \in \mathscr{F}(A)$. Then I is $t-$invertible if and only if $(II^{-1})_t = A$.*

Proof If I is $t-$invertible, there exists $J \in \mathscr{F}(A)$ such that $(IJ)_t = A$. By the preceding proposition, $J_t = I^{-1}$. By Proposition 11.1 of Chap. 1, $A = (IJ)_t = (IJ_t)_t = (II^{-1})_t$.   □

Example If I is $t-$invertible, then I^{-1} is also $t-$invertible. The converse is true if I is finitely generated. Indeed, by the preceding corollary, $(I^{-1}(I^{-1})^{-1})_t = A$, and then $(I^{-1}I_v)_t = A$. So $(I^{-1}I)_t = (I^{-1}I_t)_t = (I^{-1}I_v)_t = A$. Then I is t-invertible.

Lemma 10.8 *Let A be an integral domain and $I, J \in \mathscr{F}(A)$.*
Then $(IJ)_t = \bigcup \{(I_0J_0)_v; \ (0) \neq I_0 \subseteq I, (0) \neq J_0 \subseteq J$ finitely generated fractional ideals $\}$.

Proof Since I_0J_0 is finitely generated, the inverse inclusion is clear. Conversely, let $L = (a_1, \ldots, a_n)$ be a nonzero finitely generated subideal of IJ, $a_i = b_{i,1}c_{i,1} + \ldots + b_{i,m_i}c_{i,m_i}$, with $b_{i,j} \in I$ and $c_{i,j} \in J$. Let $I_0 = (b_{i,j}; \ 1 \le i \le n, 1 \le j \le m_i) \subseteq I$ and $J_0 = (c_{i,j}; \ 1 \le i \le n, 1 \le j \le m_i) \subseteq J$, nonzero finitely generated ideals and $L \subseteq I_0J_0$. Then $L_v \subseteq (I_0J_0)_v$. So $(IJ)_t \subseteq \bigcup \{(I_0J_0)_v; \ (0) \neq I_0 \subseteq I, (0) \neq J_0 \subseteq J$ finitely generated fractional ideals $\}$. □

Lemma 10.9 *Let A be an integral domain and I a fractional t-invertible ideal of A. There exists a nonzero finitely generated fractional ideal $F \subseteq I$ such that $I_t = F_v$.*

Proof There exists $J \in \mathscr{F}(A)$ such that $(IJ)_t = A$. By the preceding lemma, there exists nonzero finitely generated fractional ideals $I_0 \subseteq I$ and $J_0 \subseteq J$ such that $1 \in (I_0J_0)_v$. Then $A \subseteq (I_0J_0)_v \subseteq (IJ)_t = A$. So $(I_0J_0)_t = (I_0J_0)_v = A$. By Proposition 10.6, $(I_0)_t = J_0^{-1}$ and $I_t = J^{-1}$. Since $J_0 \subseteq J$, then $J^{-1} \subseteq J_0^{-1}$. So $I_t \subseteq (I_0)_t$, and since $I_0 \subseteq I$, then $(I_0)_t \subseteq I_t$. So $I_t = (I_0)_t = (I_0)_v$. □

Lemma 10.10 (Houston and Zafrullah 1989) *Let A be an integral domain and P a t-prime t-invertible ideal of A. Then P is t-maximal in A.*

Proof Suppose that P is not t-maximal. By Proposition 11.2, Chap. 1, there exists a t-maximal ideal M of A such that $P \subset M$. Let $a \in M \setminus P$. By the preceding lemma, since P is t-invertible, there exists a nonzero finitely generated ideal $F \subseteq P$ of A such that $P = F_v = F_t$. We will show that $(a, F)^{-1} = A$. Let $x \in (a, F)^{-1}$, and then $xa \in A$ and $xF \subseteq A$. So $axF \subseteq F \subseteq P$. Since P is a prime ideal of A, then $xF \subseteq P$. So $xF_t \subseteq P_t$, and then $xP \subseteq P$. So $x(PP^{-1})_t \subseteq (PP^{-1})_t$. But P is t-invertible, and then $(PP^{-1})_t = A$ and $x \in A$. So $(a, F)^{-1} = A$ and $(a, F)_v = A$. Since $(a, F)_v \subseteq M_t = M$, then $M = A$, which is absurd. Then P is t-maximal in A. □

Example The converse of the preceding lemma is false. Let A be a nondiscrete valuation domain of rank 1. Its maximal ideal M is t-maximal. See Example of Proposition 11.3, Chap. 1. Suppose that M is t-invertible. By Lemma 10.9, there exists a nonzero finitely generated subideal F of M such that $M_t = F_v$. Since A is a valuation domain, F is principal. Then $M = F$ is a principal ideal, which is absurd. Then M is not t-invertible.

Lemma 10.11 *Let A be an integral domain.*

1. *Let P be a t−maximal ideal of A. Then either $PP^{-1} = P$ or P is t−invertible.*
2. *Let I be a nonzero fractional ideal of A. Then I is t−invertible if and only if for each $P \in t - Max(A)$, $II^{-1} \nsubseteq P$.*

Proof

(1) Since $A \subseteq (A : P) = P^{-1}$, then $P \subseteq PP^{-1} \subseteq (PP^{-1})_t \subseteq A$. Suppose that P is not t−invertible. Then $(PP^{-1})_t \subset A$. Since P is t−maximal, then $P = PP^{-1} = (PP^{-1})_t$. So $P = PP^{-1}$.

(2) " ⟹" Suppose that there exists $P \in t - Max(A)$ such that $II^{-1} \subseteq P$. By hypothesis, $A = (II^{-1})_t \subseteq P_t = P$. Then $P = A$, which is absurd since P is a prime ideal, by Proposition 11.3 of Chapter 1.

" ⟸" Suppose that $(II^{-1})_t \neq A$. By Proposition 11.2 of Chap. 1, there exists $P \in t - Max(A)$ such that $(II^{-1})_t \subseteq P$. Then $II^{-1} \subseteq (II^{-1})_t \subseteq P$, which is absurd. □

The notions of t−maximality and t−inversion are equivalent for some classes of prime ideals, as the following proposition shows.

Proposition 10.12 (Houston and Zafrullah 1989) *Let A be an integral domain with quotient field K, $f(X) \in A[X]$ irreducible in K[X] and $P = f(X)K[X] \cap A[X]$. Then P is a t−maximal ideal of A[X] if and only if it is t−invertible. In this case, there exists $g(X) \in P$ such that $c(g)_v = A$ and the ideal P is almost principal in A[X].*

Proof By Proposition 10.3, P is a nonzero t−prime ideal of A[X] over (0) in A.

" ⟸" By Lemma 10.10, since P is t−prime and t−invertible, then it is t−maximal.

" ⟹" Let c(P) be the ideal of A formed by the coefficients of all the elements of P. See notations after Proposition 10.3 of Chap. 1. Since $P \neq (0)$, then $c(P) \neq (0)$ and $P \subseteq c(P)[X]$. If $P = c(P)[X]$, then $P \cap A = c(P) \neq (0)$, which is absurd. Then $P \subset c(P)[X] \subseteq (c(P)[X])_t \subseteq A[X]$. But P is t−maximal, and then $(c(P)[X])_t = A[X]$. By Corollary 11.7 of Chap. 1, we have $(c(P)[X])_t = (c(P))_t[X]$, and then $(c(P))_t = A$. There exists a nonzero finitely generated ideal $F = (a_1, \ldots, a_k) \subseteq c(P)$ of A such that $1 \in F_v$. Let $g_1(X), \ldots, g_k(X) \in P$ such that a_i is a coefficient of $g_i(X)$, $1 \leq i \leq k$. Put $r_1 = deg\ g_1, \ldots, r_k = deg\ g_k$ and $g = g_1 + X^{r_1+1}g_2 + X^{r_1+r_2+2}g_3 + \ldots + X^{r_1+\ldots+r_{k-1}+k-1}g_k \in P \subseteq A[X]$. Then $a_1, \ldots, a_k$ are coefficients of $g(X)$, so $F \subseteq c(g)$ and $1 \in c(g)_v$. Then $c(g)_v = A$. By Proposition 9.13, the ideal P is almost principal. By Proposition 3.3, there exists $0 \neq a \in A$ such that $aP \subseteq fA[X]$, and then $\frac{a}{f}P \subseteq A[X]$, so $\frac{a}{f} \in P^{-1}$. Since $P \cap A = (0)$, then $\frac{a}{f}f = a \notin P$. So $P^{-1}P \neq P$. By the preceding lemma, P is t−invertible. □

Example (Houston and Zafrullah 1989) Let A be an integral domain with quotient field K and $f(X) \in A[X]$ an irreducible polynomial in $K[X]$. Suppose that the ideal $P = f(X)K[X] \cap A[X]$ is maximal in $A[X]$. Then P is invertible.

Indeed, by Proposition 10.3, P is a t−ideal, and then it is t−maximal. By the preceding proposition, P is t−invertible, and then $(PP^{-1})_t = A[X]$. So $PP^{-1} \neq P$. Since $A[X] \subseteq (A[X] : P) = P^{-1}$, then $P \subset PP^{-1} \subseteq A[X]$. Since P is a maximal ideal of $A[X]$, then $PP^{-1} = A[X]$. So P is invertible.

Proposition 10.13 *Let A be an integral domain and I an integral t−ideal of A. Then $I = \bigcap \{IA_M; \ M \in t - Max(A)\}$. In particular, $A = \bigcap \{A_M; \ M \in t - Max(A)\}$.*

Proof Let $0 \neq x \in \bigcap \{A_M; \ M \in t - Max(A)\}$. For each $M \in t - Max(A)$, there exists $s_M \in A \setminus M$ such that $s_M x \in I$. Since $(\frac{1}{x}I)_t = \frac{1}{x}I_t = \frac{1}{x}I$, then $J = A \cap \frac{1}{x}I$ is an integral t−ideal, as an intersection of two t−ideals, and $s_M \in J$ for each $M \in t - Max(A)$. Suppose that $J \neq A$. By Proposition 11.2 of Chap. 1, there exists $M_0 \in t - Max(A)$ such that $J \subseteq M_0$, and then $s_{M_0} \in J \subseteq M_0$, which is absurd. Then $J = A$ and $A \subseteq \frac{1}{x}I$, so $xA \subseteq I$, and then $x \in I$.   □

Example Let I and J be the two nonzero integral ideals of an integral domain A. If $IA_M = JA_M$ for each $M \in t - Max(A)$, then $I_t = J_t$.

Indeed, by symmetry, it suffices to show that $I \subseteq J_t$. But by hypothesis, for each $M \in t - Max(A)$, $I \subseteq JA_M \subseteq J_t A_M$. Then $I \subseteq \bigcap \{J_t A_M; \ M \in t - Max(A)\} = J_t$.

The following lemma establishes the relationship between the v−operation and the t−operation over an integral domain A with their analogues in a ring of fractions of A.

Lemma 10.14 (Kang 1989; Zafrullah 1978) *Let A be an integral domain, S a multiplicative set of A, and I a nonzero fractional ideal of A.*

1. *We have $(I^{-1})_S \subseteq (I_S)^{-1}$. If I is finitely generated, then $(I^{-1})_S = (I_S)^{-1} = ((I_v)_S)^{-1}$.*
2. *If I is finitely generated, then $(I_S)_v = ((I_v)_S)_v$.*
3. *We have the equality $(I_S)_t = ((I_t)_S)_t$. In particular, $(I_t)_S \subseteq (I_S)_t$.*

Proof

(1) Since $(I^{-1})_S I_S = (I^{-1}I)_S \subseteq A_S$, then $(I^{-1})_S \subseteq (I_S)^{-1}$.

Suppose now that I is finitely generated. Let $x \in (I_S)^{-1}$. Then $xI_S \subseteq A_S$, and in particular, $xI \subseteq A_S$. Since I is finitely generated, there exists $s \in S$ such that $sxI \subseteq A$, and then $sx \in I^{-1}$, so $x \in (I^{-1})_S$. Then we have the inverse inclusion, so the equality $(I_S)^{-1} = (I^{-1})_S$. On the other hand, $I \subseteq I_v$, and then $I_S \subseteq (I_v)_S$, so $((I_v)_S)^{-1} \subseteq (I_S)^{-1}$. In the same way, $(I^{-1})_S(I_v)_S = (I^{-1}I_v)_S \subseteq ((I^{-1}I_v)_v)_S = ((I^{-1}I)_v)_S \subseteq (A_v)_S = A_S$, and then $(I^{-1})_S \subseteq ((I_v)_S)^{-1}$. Then we have the inclusions $((I_v)_S)^{-1} \subseteq (I_S)^{-1} = (I^{-1})_S \subseteq ((I_v)_S)^{-1}$. So $(I^{-1})_S = (I_S)^{-1} = ((I_v)_S)^{-1}$.

(2) By (1), $(I_S)^{-1} = ((I_v)_S)^{-1}$. Then $((I_S)^{-1})^{-1} = (((I_v)_S)^{-1})^{-1}$, so $(I_S)_v = ((I_v)_S)_v$.

(3) Since $I \subseteq I_t$, then $I_S \subseteq (I_t)_S$, so $(I_S)_t \subseteq ((I_t)_S)_t$. Conversely, let $y \in (I_t)_S$, $y = \frac{x}{s}$ with $x \in I_t$ and $s \in S$. By definition of I_t, there exists a nonzero finitely generated fractional ideal $J \subseteq I$ of A such that $x \in J_v$. By (2), since J is finitely generated, $y = \frac{x}{s} \in (J_v)_S \subseteq ((J_v)_S)_v = (J_S)_v$. But J_S is a nonzero finitely generated fractional ideal of A_S with $J_S \subseteq I_S$, and then $(J_S)_v \subseteq (I_S)_t$ so $y \in (I_S)_t$. We conclude that $(I_t)_S \subseteq (I_S)_t$ and then $((I_t)_S)_t \subseteq ((I_S)_t)_t = (I_S)_t$. So $(I_S)_t = ((I_t)_S)_t$.   □

Example 1 Let S be a multiplicative set of an integral domain A. If $\mathscr{I}$ is a t−ideal of A_S, then $\mathscr{I} \cap A$ is a t−ideal of A.

Indeed, let F a nonzero finitely generated subideal of $\mathscr{I} \cap A$. By (2), $F_v \subseteq (F_v)_S \subseteq (F_S)_v \subseteq \mathscr{I}$, and then $F_v \subseteq \mathscr{I} \cap A$. So $(\mathscr{I} \cap A)_t = \mathscr{I} \cap A$.

Example 2 Let I be a fractional t−invertible ideal of an integral domain A and S a multiplicative set of A. Then I_S is t−invertible in A_S.

Indeed, by (3), $\left(I_S(I^{-1})_S\right)_t = \left((II^{-1})_S\right)_t = \left(((II^{-1})_t)_S\right)_t = (A_S)_t = A_S$.

We will use the preceding lemma in the following particular case.

Corollary 10.15 *Let A be an integral domain, P a prime ideal of A, and I a nonzero fractional ideal of A.*

1. *We have the equality $(I_t A_P)_t = (I A_P)_t$.*
2. *If I is finitely generated, then $I^{-1} A_P = (I A_P)^{-1}$.*

Lemma 10.16 *Let A be a ring and $f(X) = a_0 + a_1 X + \ldots + a_n X^n \in A[X]$. Then $f(X)$ is invertible in $A[X]$ if and only if a_0 is invertible in A and $a_1, \ldots, a_n$ are nilpotent.*

Proof "$\Longleftarrow$" Since $a_1, \ldots, a_n$ are nilpotent, then $f_2 = a_1 X + \ldots + a_n X^n$ is also nilpotent. But the sum of an invertible element and a nilpotent element is invertible. Then $f = a_0 + a_1 X + \ldots + a_n X^n = a_0 + f_2$ is invertible in $A[X]$.   □

Other Proof $f = a_0 + a_1 X + \ldots + a_n X^n$ is invertible in the ring formal series $A[[X]]$. Its inverse is $f^{-1} = a_0^{-1}(1 + \alpha + \alpha^2 + \ldots)$ with $\alpha = -a_0^{-1}(a_1 X + \ldots + a_n X^n)$. Since α is nilpotent, as sum of nilpotent elements, then there exists $k \in \mathbb{N}^*$ such that $\alpha^k = 0$. So $f^{-1} = a_0^{-1}(1 + \alpha + \alpha^2 + \ldots + \alpha^{k-1}) \in A[X]$.

"$\Longrightarrow$" Let $g(X) = b_0 + b_1 X + \ldots + b_m X^m$ be the inverse of $f(X)$ in $A[X]$. Then $1 = fg = a_0 b_0 + \ldots + a_n b_m X^{n+m}$, so a_0 is invertible in A. Let b_0 be its inverse. If f is constant, there is nothing to do. Suppose that $n \geq 1$. We will prove by induction on r that $a_n^{r+1} b_{m-r} = 0$ for $0 \leq r \leq m$. Since $a_n b_m = 0$, the property is satisfied for $r = 0$. Suppose that $a_n^{s+1} b_{m-s} = 0$ for $0 \leq s \leq r < m$. By multiplying the equality $fg = 1$ by a_n^{r+1}, we obtain $f(a_n^{r+1} g) = a_n^{r+1}$. By the induction hypothesis, $a_n^{r+1}g = a_n^{r+1}b_0 + a_n^{r+1}b_1X + \ldots + a_n^{r+1}b_{m-r-1}X^{m-r-1}$, and then $a_n^{r+2}b_{m-r-1} = 0$.   □

Other Proof for the Induction The coefficient of $X^{n+m-(r+1)}$ in fg is $0 = \sum a_i b_j$ (*), and the sum is taken for $i + j = n + m - (r + 1)$, $0 \le i \le n$, $0 \le j \le m$. We will prove that for $0 \le i < n$, $j = m - s$ with $0 \le s \le r$. To do this $j = m - (-n + r + 1 + i)$, then $s = (i - n) + r + 1 \le r$ and $0 \le m - j = s$. We multiply (*) by a_n^{r+1} and we call (**) the new equality. Since $0 \le s \le r$, then $r + 1 \ge s + 1$ and $a_n^{s+1}b_{m-s} = 0$, so $a_n^{r+1}b_{m-s} = 0$. The equality (**) is reduced to the only term corresponding to $i = n$ and can be written $a_n^{r+2}b_{m-(r+1)} = 0$. This finishes the proof of the induction.

In particular, for $r = m$, $a_n^{m+1}b_0 = 0$. Since b_0 is invertible, $a_n^{m+1} = 0$, then a_n is nilpotent and $a_n X^n$ is nilpotent. So $f_1 = a_0 + a_1 X + \ldots + a_{n-1}X^{n-1} = f - a_n X^n$ is invertible, as the sum of an invertible element and a nilpotent element. By the preceding step, a_{n-1} is nilpotent. By induction, we show that the coefficients $a_n, a_{n-1}, \ldots, a_1$ are nilpotent.   □

Lemma 10.17 (Chang 2015) *Let A be an integral domain, I an integral ideal of A, and $f(X) = a_0 + a_1 X + \ldots + a_n X^n \in A[X]$. Then $f A[X] + I[X] = A[X]$ if and only if $a_0 A + I = A$ and there exists $k \in \mathbb{N}^*$ such that we have $(a_1, \ldots, a_n)^k \subseteq I$.*

Proof We use the preceding lemma: $f A[X] + I[X] = A[X] \iff A[X]/I[X] \simeq (A/I)[X] = (\bar{f}) \iff \bar{f} = \bar{a}_0 + \bar{a}_1 X + \ldots + \bar{a}_n X^n$ is invertible in $(A/I)[X] \iff \bar{a}_0$ is invertible in A/I and $\bar{a}_1, \ldots, \bar{a}_n$ are nilpotent in $A/I \iff A/I = (\bar{a}_0)$ and there exists $s \in \mathbb{N}^*$ such that $a_1^s, \ldots, a_n^s \in I \iff a_0 A + I = A$ and there exists $k \in \mathbb{N}^*$ such that $(a_1, \ldots, a_n)^k \subseteq I$.   □

Lemma 10.18 (Chang 2015) *Let A be an integral domain with quotient field K and $f(X) = a_0 + a_1 X + \ldots + a_n X^n \in A[X]$ an irreducible polynomial in $K[X]$. Then $\sqrt{f A[X]}$ is a maximal ideal of $A[X]$ if and only if a_0 is invertible in the ring A and $(a_1, \ldots, a_n) \subseteq P$, for each nonzero prime ideal P of A. In this case, $\sqrt{f A[X]} = f A[X]$.*

Proof $"\Longrightarrow"$ Let P be any nonzero prime ideal of A. Suppose that $P \subseteq \sqrt{f A[X]}$ and let $0 \ne p \in P$. There exists $n \in \mathbb{N}^*$ such that $p^n \in f A[X]$, which is absurd, since the polynomial f is not constant. Then $P \not\subseteq \sqrt{f A[X]}$. Since $\sqrt{f A[X]}$ is a maximal ideal of $A[X]$, then $\sqrt{f A[X]} + P[X] = A[X]$. So $f A[X] + P[X] = A[X]$. Indeed, $1 = g(X) + h(X)$ with $g(X) \in \sqrt{f A[X]}$ and $h(X) \in P[X]$. There exists $n \in \mathbb{N}^*$ such that $g^n \in f A[X]$. So $1 = g^n + \sum_{i:0}^{n-1} C_n^i g^i h^{n-i} \in f A[X] + P[X]$. By the preceding lemma, $a_0 A + P = A$, and there exists $k \in \mathbb{N}^*$ such that $(a_1, \ldots, a_n)^k \subseteq P$. Since P is a prime ideal of A, $a_0 \notin P$ and $(a_1, \ldots, a_n) \subseteq P$. This is for each nonzero prime ideal P of A. Then $a_0 \in U(A)$.

$"\Longleftarrow"$ Since a_0 is invertible, $c(f) = A$. By Proposition 3.1, $f K[X] \cap A[X] = f A[X]$. Let Q be any prime ideal of $A[X]$ containing $f(X)$. Suppose that $Q \cap$

$A \neq (0)$. By hypothesis, $(a_1, \ldots, a_n) \subseteq Q \cap A$. Since a_0 is invertible, then $a_0 A + Q \cap A = A$. By the preceding lemma, $A[X] = fA[X] + (Q \cap A)[X] \subseteq Q$, which is absurd since Q is prime. Then $Q \cap A = (0)$ so $QK[X] \neq K[X]$ because $QK[X]$ is a principal ideal of $K[X]$ generated by an element of Q. Since $fK[X] \subseteq QK[X] \subset K[X]$ and $fK[X]$ is a maximal ideal of $K[X]$, then $fK[X] = QK[X]$. So $Q \subseteq fK[X] \cap A[X] = fA[X] \subseteq Q$, and then $Q = fA[X]$. We conclude that the only prime ideal of $A[X]$ containing f is $fA[X]$. Then $fA[X]$ is a maximal ideal and $\sqrt{fA[X]} = fA[X]$. □

Now, we can state the main result of this section.

Theorem 10.19 (Chang 2015) *Let A be an integrally closed integral domain and $f(X) = a_0 + a_1 X + \ldots + a_n X^n \in A[X]$ an irreducible polynomial in $K[X]$ with $a_0 \neq 0$. The ideal $M = f(X)K[X] \cap A[X]$ is maximal in $A[X]$ if and only if $\left(\frac{a_1}{a_0}, \ldots, \frac{a_n}{a_0}\right)A \subseteq P$ for each nonzero prime ideal P of A.*
In this case, $M = \frac{f(X)}{a_0}A[X]$ is a principal ideal of the ring $A[X]$.

Proof " $\implies$ " By hypothesis, M is a maximal ideal of $A[X]$ over (0) in A. By the example of Proposition 10.12, M is invertible in $A[X]$, so $MM^{-1} = A[X]$. By Proposition 10.1, since A is integrally closed, $M = f.c(f)^{-1}[X]$. Let the fractional ideal $c(f)^{-1} = I$. Then $M = f.I[X]$. By Corollary 11.7 of Chap. 1, $M^{-1} = (fI[X])^{-1} = f^{-1}(I[X])^{-1} = f^{-1}I^{-1}[X]$. Then $A[X] = MM^{-1} = I[X]I^{-1}[X] \subseteq II^{-1}[X] \subseteq A[X]$, so $II^{-1}[X] = A[X]$ and $II^{-1} = A$. The fractional ideal $c(f)^{-1} = I$ is $(t-)$invertible. By the example of Corollary 10.7, since $c(f)$ is finitely generated, it is also $t-$invertible. Let $P \in t - Max(A)$ and $S = A \setminus P$. Since M is a maximal ideal of $A[X]$ disjoint with S because $M \cap A = (0)$, then $S^{-1}M = S^{-1}(fK[X] \cap A[X]) = fK[X] \cap A_P[X]$ is a maximal ideal of $S^{-1}A[X] = A_P[X]$. By Lemma 10.11 (2), $c(f)c(f)^{-1} \nsubseteq P$, and then the integral ideal $c(f)c(f)^{-1}$ of A contains an invertible element of the ring A_P. Then $c(f)c(f)^{-1}A_P = A_P$ so $c(f)A_P c(f)^{-1}A_P = A_P$. Then the ideal $c(f)A_P$ is invertible in the local ring A_P. By Lemma 9.5 (2), the ideal $c(f)A_P$ is principal, and there exists $i \in \{0, \ldots, n\}$ such that $c(f)A_P = a_i A_P$. By Corollary 10.15 (1), $c(f)A_P \subseteq c(f)_t A_P \subseteq \left(c(f)_t A_P\right)_t = \left(c(f)A_P\right)_t = \left(a_i A_P\right)_t = a_i A_P = c(f)A_P$. Then $c(f)_t A_P = c(f)A_P = a_i A_P$. By Corollary 10.15 (2), $S^{-1}M = S^{-1}(f.c(f)^{-1}A[X]) = f.c(f)^{-1}A_P[X] = f(c(f)A_P)^{-1}[X] = \frac{f}{a_i}A_P[X]$ is a maximal ideal of $A_P[X]$. Then it is equal to its radical. By the preceding lemma, $\frac{a_0}{a_i}$ is invertible in A_P, and then $a_i A_P = a_0 A_P$, so $c(f)_t A_P = a_0 A_P$. This is for each $P \in t - Max(A)$. By Proposition 10.13, $c(f)_t = a_0 A \subseteq c(f) \subseteq c(f)_t$. Then $c(f) = a_0 A$, $c(f)^{-1} = \frac{1}{a_0}A$, and $M = f.c(f)^{-1}A[X] = \frac{f}{a_0}A[X]$ is a maximal ideal of $A[X]$, and then it is equal to its radical. By the preceding lemma, $\left(\frac{a_1}{a_0}, \ldots, \frac{a_n}{a_0}\right)A \subseteq P$ for each nonzero prime ideal P of A.
" $\impliedby$ " Let $h(X) = \frac{1}{a_0}f(X) = 1 + \frac{a_1}{a_0}X + \ldots + \frac{a_n}{a_0}X^n$. By hypothesis, $\frac{a_1}{a_0}, \ldots, \frac{a_n}{a_0} \in A$, and then $h(X) \in A[X]$ so $h(X)A[X] \subseteq f(X)K[X] \cap A[X] = M$. Conversely, let $g(X) \in M \subset A[X]$. Then $g(X) = f(X)l(X)$ with $l(X) \in K[X]$. So

$g(X) = \frac{1}{a_0} f(X) a_0 l(X) = h(X) a_0 l(X)$. Since $h(0) = 1$, then $h(X)$ is invertible in the formal power series ring $A[[X]]$. Then $\frac{g(X)}{h(X)} = a_0 l(X) \in A[[X]] \cap K[X] = A[X]$ and $g(X) = h(X) a_0 l(X) \in h(X) A[X]$. So $M = h(X) A[X]$. By Proposition 10.3, M is a prime ideal of $A[X]$. Then $M = h(X) A[X] = \sqrt{h(X) A[X]}$ with $h(X) = \frac{1}{a_0} f(X) \in A[X]$ an irreducible polynomial in $K[X]$. By the preceding lemma, the ideal M is maximal.   □

Example (Chang 2015) This example shows that the hypothesis A integrally closed is necessary in the preceding theorem. Let (A, P) be a local integral domain of dimension 1 which integral closure $\overline{A}$ is local with maximal ideal Q satisfying $P. \overline{A} \subset Q$. This is the case, for example, if $A = F[[t^2, t^3]]$ where t is an indeterminate over a field F. In fact, the maximal ideal of A is $P = (t^2, t^3) A$, and by the example of Lemma 9.1, the integral closure of A is $\overline{A} = F[[t]]$. Since the maximal ideal of $\overline{A}$ is $Q = t F[[t]]$, then $P. \overline{A} = (t^2, t^3) F[[t]] = t^2 F[[t]] \subset t F[[t]] = Q$.

Let's return to the general case. Since $A \subset \overline{A}$ is an integral extension of local rings, then $Q \cap A = P$. Let $u \in Q \backslash P. \overline{A}, u = \frac{b}{a}$ with $a, b \in A \backslash (0)$. Let $f(X) = a + bX \in A[X]$, an irreducible polynomial in $K[X]$. Note that $u = \frac{b}{a} \notin P$, the unique nonzero prime ideal of A. We will show that the ideal $M = f(X) K[X] \cap A[X]$ is maximal in $A[X]$, but it is not principal. Indeed, $\overline{A}$ is integrally closed of dimension 1 and $u = \frac{b}{a} \in Q$, the unique nonzero prime ideal of $\overline{A}$. By the preceding theorem, the ideal $f(X) K[X] \cap \overline{A}[X]$ is maximal in $\overline{A}[X]$. Since the extension $A[X] \subset \overline{A}[X]$ is integral, the ideal $M = f(X) K[X] \cap A[X] = (f(X) K[X] \cap \overline{A}[X]) \cap A[X]$ is maximal in $A[X]$. Suppose that $M = f(X) K[X] \cap A[X] = h(X) A[X]$ is principal. Then $h(X) \in f(X) K[X]$, so $h(X) K[X] \subseteq f(X) K[X]$. We also have $f(X) \in h(X) A[X]$, and then $f(X) K[X] \subseteq h(X) K[X]$. So $f(X) K[X] = h(X) K[X]$, and then $f(X) = \alpha h(X)$ with $\alpha \in K^*$. So $h(X) K[X] \cap A[X] = f(X) K[X] \cap A[X] = h(X) A[X]$. By Proposition 3.1, $c_A(h)_t = c_A(h)_v = A$. Since $ht(P) = dim(A) = 1$, then P is a t-ideal. Then $c_A(h) \nsubseteq P$, the unique maximal ideal of A. Then $c_A(h) = A$. We have $(a, b) A = c_A(f) = \alpha c_A(h) = \alpha A$, a fractional principal ideal of the local ring A. By Lemma 9.5 (2), either $(a, b) A = bA$ and in this case, $a \in bA$, and then $\frac{a}{b} \in A$, so $1 = \frac{a}{b} \frac{b}{a} = \frac{a}{b} u \in Q$, which is absurd. Or $(a, b) A = aA$, and then $b \in aA$, so $u = \frac{b}{a} \in A \cap Q = P$, which is absurd.

Corollary 10.20 (Chang 2015) *Let A be an integral domain with quotient field K and integral closure $\overline{A}$ and $f(X) = a_0 + a_1 X + \ldots + a_n X^n \in A[X]$ an irreducible polynomial in $K[X]$ with $a_0 \neq 0$. The following assertions are equivalent:*

1. *The ideal $f(X) K[X] \cap A[X]$ is maximal in $A[X]$.*
2. *The ideal $f(X) K[X] \cap \overline{A}[X]$ is maximal in $\overline{A}[X]$.*
3. *$(\frac{a_1}{a_0}, \ldots, \frac{a_n}{a_0}) \overline{A} \subseteq Q$ for each nonzero prime ideal Q of $\overline{A}$.*

Moreover, if a_0 is invertible in A, the ideal $f(X) K[X] \cap A[X]$ is maximal in $A[X]$ if and only if $(a_1, \ldots, a_n) A \subseteq P$ for each nonzero prime ideal P of A.

Proof "(1) $\Longleftrightarrow$ (2)" The extension of rings $A[X] \subseteq \overline{A}[X]$ is integral. By Proposition 10.3, the ideal $f(X)K[X] \cap \overline{A}[X]$ of $\overline{A}[X]$ is prime. Since $f(X)K[X] \cap A[X] = (f(X)K[X] \cap \overline{A}[X]) \cap A[X]$, we have the equivalence.

"(2) $\Longleftrightarrow$ (3)" Since $\overline{A}$ is integrally closed, the equivalence follows from the theorem.

Suppose that a_0 is invertible in A. Then $\left(\frac{a_1}{a_0}, \ldots, \frac{a_n}{a_0}\right)A = (a_1, \ldots, a_n)A$.

" $\Longrightarrow$ " Let P be nonzero prime ideal of A. Since the extension of rings $A \subseteq \overline{A}$ is integral, there exists $Q \in spec(\overline{A})$ such that $Q \cap A = P$. By (3), $(a_1, \ldots, a_n)\overline{A} \subseteq Q$. Then $(a_1, \ldots, a_n)A \subseteq ((a_1, \ldots, a_n)\overline{A}) \cap A \subseteq Q \cap A = P$.

" $\Longleftarrow$ " Let Q be a nonzero prime ideal of $\overline{A}$. Then $(0) \subset Q$. Since the extension of rings $A \subseteq \overline{A}$ is integral, by the INC property, $Q \cap A \neq (0)$. By hypothesis, $(a_1, \ldots, a_n)A \subseteq Q \cap A$. Then $(a_1, \ldots, a_n)\overline{A} \subseteq Q$. By (3), the ideal $f(X)K[X] \cap A[X]$ is maximal in $A[X]$. $\square$

Corollary 10.21 (Chang 2015) *Let A be an integral domain with quotient field K, B an over ring of A, and $f(X) = a_0 + a_1 X + \ldots + a_n X^n \in A[X]$ an irreducible polynomial in $K[X]$ with $a_0 \neq 0$. If the ideal $f(X)K[X] \cap A[X]$ is maximal in $A[X]$, the ideal $f(X)K[X] \cap B[X]$ is maximal in $B[X]$.*

Proof The integral closures of the domains $A \subseteq B$ satisfy $\overline{A} \subseteq \overline{B} \subseteq K$. If x is an element of $\overline{A}$ belonging to each nonzero prime ideal of $\overline{A}$, then x belongs to each nonzero prime ideal of $\overline{B}$. Indeed, let Q be a nonzero prime ideal of $\overline{B}$. Then $(0) \subset Q$. Since the extension $B \subseteq \overline{B}$ is integral, by the INC property, $Q \cap B \neq (0)$. Since B is an over ring of A, $Q \cap A \neq (0)$, and then $Q \cap \overline{A} \neq (0)$. So $x \in Q \cap \overline{A} \subseteq Q$. By hypothesis, the ideal $f(X)K[X] \cap A[X]$ is maximal in $A[X]$. By the preceding corollary, $\left(\frac{a_1}{a_0}, \ldots, \frac{a_n}{a_0}\right)\overline{A} \subseteq P$ for each nonzero prime ideal P of $\overline{A}$. By the preceding step, $\left(\frac{a_1}{a_0}, \ldots, \frac{a_n}{a_0}\right)\overline{B} \subseteq Q$ for each nonzero prime ideal Q of $\overline{B}$. Again, by the preceding corollary, the ideal $f(X)K[X] \cap B[X]$ is maximal in $B[X]$. $\square$

We will show by an example in the following section that the converse of the preceding corollary is false.

11 Application to Goldman Rings

Definition 11.1 Let A be an integral domain. The pseudo-radical of A is the intersection of all its nonzero prime ideals.

Example 1 If A is an integral domain of dimension 1, its pseudo-radical and its Jacobson radical are identical.

Example 2 Let A be an integral domain with pseudo-radical $J \neq (0)$. Then each nonzero prime ideal P of A contains a prime ideal of height 1. Indeed, if $0 \neq a \in J$, then P contains a prime ideal Q minimal over aA. We will show that $ht(Q) = 1$. Let R be a prime ideal of A such that $(0) \subset R \subseteq Q$. By hypothesis, $a \in R$, and then $R = Q$.

Proposition 11.2 (Chang 2015) *Let A be an integral domain with quotient field K and pseudo-radical $I \neq (0)$ and $f(X) = 1 + aX \in A[X]$ with $0 \neq a \in I$. Then the ideal $f(X)K[X] \cap A[X] = f(X)A[X]$ is maximal in the ring $A[X]$.*

Proof By Proposition 3.1, since $c(f) = A$, the ideal $f(X)K[X] \cap A[X] = f(X)A[X]$. Let P be a nonzero prime ideal of A. By definition of the pseudo-radical, $a \in I \subseteq P$. By Corollary 10.20, the ideal $f(X)K[X] \cap A[X]$ is maximal in the ring $A[X]$.   □

In the following proposition, we show that if the pseudo-radical of an integral domain A is nonzero, then the quotient field of A is a simple extension of A.

Proposition 11.3 (Gilmer 1966) *Let A be an integral domain with quotient field K. The following assertions are equivalent:*

1. *The pseudo-radical of A is nonzero.*
2. *There exists $0 \neq c \in A$ such that $K = A[\frac{1}{c}]$.*
3. *There exist $x_1, \ldots, x_n \in K^*$ such that $K = A[x_1, \ldots, x_n]$.*

Proof "(1) $\implies$ (2)" Let c be a nonzero element of the pseudo-radical of A and $S = \{c^n; n \in \mathbb{N}\}$. Let $y \in K^*$, $y = \frac{a}{b}$ with $a, b \in A \setminus (0)$. Suppose that $bA \cap S = \emptyset$. There exists a prime ideal P of A such that $bA \subseteq P$ and $P \cap S = \emptyset$. In particular, $P \neq (0)$ and $c \notin P$, which is absurd. Then $bA \cap S \neq \emptyset$. Let $n \in \mathbb{N}$ be such that $c^n \in bA$, $c^n = bt$ with $t \in A$, so $y = \frac{a}{b} = \frac{at}{c^n} \in A[\frac{1}{c}]$. Then $K = A[\frac{1}{c}]$.

"(2) $\implies$ (3)" Clear. "(3) $\implies$ (1)" Put $x_i = \frac{b_i}{c}$, $1 \leq i \leq n$, with $c, b_1, \ldots, b_n \in A \setminus (0)$. It is clear that $K = A[\frac{1}{c}]$. Suppose that there exists a nonzero prime ideal P of A that does not contain c. Then $P \cap S = \emptyset$ where $S = \{c^n; n \in \mathbb{N}\}$. Then $S^{-1}P$ is a prime ideal of the ring $S^{-1}A = A[\frac{1}{c}] = K$. But K is a field, and then $S^{-1}P = (0)$, so $P = (0)$, which is absurd. Then c is a nonzero element in the pseudo-radical of A.   □

The equivalence in the preceding proposition defines a new class of integral domains.

Definition 11.4 An integral domain satisfying the equivalent conditions of the preceding proposition is called Goldman domain.

Example 1 An integral domain A with finite prime spectrum is a Goldman domain. Indeed, let $P_1, \ldots, P_n$ be the nonzero prime ideals of A. For $1 \leq i \leq n$, choose $0 \neq x_i \in P_i$. Then $0 \neq x_1 \ldots x_n \in P_1 \cap \ldots \cap P_n$.

Example 2 An over ring B of a Goldman domain A is a Goldman domain.

Indeed, let K be the quotient field of A. There exists $0 \neq c \in A \subseteq B$ such that $K = A[\frac{1}{c}]$. Then $K = B[\frac{1}{c}]$.

Example 3 The ring $\mathbb{Z}$ is not a Goldman domain.
Indeed, $\bigcap p\mathbb{Z} = (0)$, where p runs the set of the prime numbers.

Example 4 The polynomial ring over an integral domain A is never a Goldman domain. Indeed, suppose that there exists $0 \neq f(X) \in A[X]$ such that $K(X) = A[X][\frac{1}{f(X)}]$ where K is the quotient field of A. Necessarily $f(X) \notin A$, and then $1 + f(X)$ is nonzero and noninvertible in $A[X]$. There exist $n \in \mathbb{N}^*$ and $g(X) \in A[X]$ such that $\frac{1}{1+f(X)} = \frac{g(X)}{f(X)^n}$, and then $f(X)^n = (1+f(X))g(X)$. In the nonzero quotient ring $A[X]/(1 + f(X))$, $\overline{f(X)} = \overline{-1}$. But $\overline{f(X)}^n = \overline{0}$, and then $\overline{(-1)}^n = \overline{0}$, so $\overline{1} = \overline{0}$, which is absurd in nonzero ring.

Lemma 11.5 *Let B be an over ring of an integral domain A with B integral over A.*

1. If J is the pseudo-radical of B and I is the pseudo-radical of A, then $I = J \cap A$.
2. The ideal J is zero if and only if the ideal I is zero.

Proof

(1) Let $x \in J \cap A$ and P a nonzero prime ideal of A. There exists a nonzero prime ideal Q of B over P. Then $x \in Q \cap A = P$. So $x \in I$. The inverse inclusion is true even if the extension $A \subseteq B$ is not integral. Let Q be a nonzero prime ideal of B and $0 \neq x \in Q$. Since B is an over ring of A, there exists $0 \neq a \in A$ such that $ax \in A$. Then $0 \neq ax \in Q \cap A$ so $Q \cap A$ is a nonzero prime ideal of A. By definition of the pseudo-radical, $I \subseteq Q \cap A$. Then $I \subseteq J \cap A$.

(2) By (1), if $J = (0)$, then $I = (0)$. Conversely, suppose that there exists $0 \neq x \in J$. Let $x^n + a_{n-1}x^{n-1} + \ldots + a_0 = 0$ be a relation of integral dependence of x over A of minimal degree. Then $a_0 \neq 0$ and $a_0 = -x(x^{n-1} + a_{n-1}x^{n-2} + \ldots + a_1) \in J \cap A = I = (0)$, which is absurd.   □

Let A be an integral domain with nonzero pseudo-radical. The following proposition shows how this reflects on the ideal theory of the domain $A[X]$.

Proposition 11.6 (Gilmer 1966) *The following assertions are equivalent for an integral domain A:*

1. A is a Goldman domain; i.e, the pseudo-radical of A is nonzero.
2. There exists a linear polynomial $f(X) \in A[X]$ such that the ideal $f(X)A[X]$ is maximal of height 1 in the domain $A[X]$.
3. The domain $A[X]$ has a height 1 maximal ideal.
4. The domain $A[X]$ has a height 1 maximal ideal over (0) in A.

Proof "(1) $\implies$ (2)" Let a be a nonzero element in the pseudo-radical of A and $f(X) = 1 + aX \in A[X]$. By Proposition 11.2, the ideal $f(X)K[X] \cap A[X] = f(X)A[X]$ is maximal in $A[X]$. Its height is one, by Proposition 10.3.

"(2) $\implies$ (3)" Clear. "(3) $\implies$ (4)" Let M be a maximal ideal of $A[X]$ of height 1 and $P = M \cap A$. Then $P[X] \subseteq M$. Since $A[X]/P[X] \simeq (A/P)[X]$ is not a field, the ideal $P[X]$ is not maximal. So $(0) \subseteq P[X] \subset M$. But $ht(M) = 1$, then $P[X] = (0)$, so $P = (0)$.

"(4) $\implies$ (1)" Let M be a maximal ideal of $A[X]$ such that $M \cap A = (0)$ and K the quotient field of A. By Proposition 10.3, $M = f(X)K[X] \cap A[X]$ with $f(X) = a_0 + a_1 X + \ldots + a_n X^n \in A[X]$ an irreducible polynomial in $K[X]$. By Corollary 10.20, for each nonzero prime ideal Q of the integral closure $\overline{A}$ of A, $\left(\frac{a_1}{a_0}, \ldots, \frac{a_n}{a_0}\right)\overline{A} \subseteq Q$. In particular, the pseudo-radical of $\overline{A}$ is nonzero. By the preceding lemma, the pseudo-radical of A is also nonzero. Then A is a Goldman domain. $\square$

Corollary 11.7 (Chang 2015) *Let A be a PID. Then A has an infinity of nonassociated irreducible elements if and only if the height of all the maximal ideals of $A[X]$ is 2.*

Proof We know that $dim\ A[X] = 2$. Let $(p_\lambda)_{\lambda \in \Lambda}$ be a system that represents the irreducible elements of A. The pseudo-radical of A is $\bigcap_{\lambda \in \Lambda} p_\lambda A \neq (0)$ if and only if Λ is finite. By the preceding proposition, Λ is infinite if and only if A is not a Goldman domain if and only if all the maximal ideals of $A[X]$ are of height 2. $\square$

Example 1 All the maximal ideals of the domain $\mathbb{Z}[X]$ are of height 2.

Example 2 (Chang 2015) We purpose to give a counterexample to Corollary 10.21. Let $f(X) = a_0 + a_1 X + \ldots + a_n X^n \in \mathbb{Z}[X]$ a primitive polynomial ; i.e, the elements $a_0, a_1, \ldots, a_n$ are relatively prime, such that $f(X)$ is irreducible in $\mathbb{Q}[X]$. Suppose that there exist some prime numbers that divide $a_1, \ldots, a_n$ but do not divide a_0. Denote them $p_1, \ldots, p_r$. Let $S = \mathbb{Z} \setminus p_1\mathbb{Z} \cup \ldots \cup p_r\mathbb{Z}$ and $A = S^{-1}\mathbb{Z}$. Since the content $c_{\mathbb{Z}}(f) = \mathbb{Z}$, then $c_A(f) = A$. By Proposition 3.1, the ideals $P = f\mathbb{Q}[X] \cap \mathbb{Z}[X] = f\mathbb{Z}[X]$ and $M = f\mathbb{Q}[X] \cap A[X] = fA[X]$. The only nonzero prime ideals of A are the $S^{-1}(p_i\mathbb{Z})$, $1 \leq i \leq r$. They all contain $a_1, \ldots, a_n$. Since $a_0 \in S$, it is invertible in A. By Lemma 10.18, M is a maximal ideal of $A[X]$. By Proposition 10.3, the ideal P is prime in $\mathbb{Z}[X]$ of height 1, by the principal ideal theorem. It is not maximal by the preceding example. Then we have a counterexample to Corollary 10.21.

For a concrete example, we may take $f(X) = 10 + 15X + 45X^2 + 3X^4$. It is irreducible in $\mathbb{Q}[X]$ by Eisenstein criteria with $p = 5$. Then $p_1 = 3$ and $A = \mathbb{Z}_{(3)}$.

Problems

Exercise 1 Let $z = a + ib \in \mathbb{C}$ with $a, b \in \mathbb{R}$. Show that z is integral over $\mathbb{Q}$ if and only if a and b are integral over $\mathbb{Q}$.

Exercise 2 Give the integral closure A' of the domain $A = \{a + 2ib;\ a, b \in \mathbb{Z}\}$.

Exercise 3

(1) Let $A \subset B$ be an extension of rings. Suppose that $B \setminus A$ is multiplicatively stable. Show that A is integrally closed in B.
(2) Let A be an integrally closed domain. Let b and c be the two elements of A satisfying $b^2 = c^3$. Show that there exists a unique element $x \in A$ such that $b = x^3$ and $c = x^2$. Hint: Take $x = \frac{b}{c}$.

Exercise 4 Let A be the integral closure of $\mathbb{Z}$ in $\mathbb{C}$. Using the equalities $2 = \sqrt{2}\sqrt{2} = \sqrt[4]{2}\sqrt[4]{2}\sqrt[4]{2}\sqrt[4]{2} = \ldots$, show that A is not UFD. Show that A is not Noetherian.

Exercise 5

(a) Show that for each $r \in \mathbb{Q}$, the number $e^{ir\pi}$ is integral over $\mathbb{Z}[i]$.
(b) Let $r \in \mathbb{Q}$ be such that $\cos r\pi$ and $\sin r\pi \in \mathbb{Q}$. Show that there exists $m \in \mathbb{Z}$ such that $r = \frac{m}{2}$.

Exercise 6 Let A be a ring, $(Aut\ A, \circ)$ the group of the automorphisms of A and G a finite subgroup of $Aut\ A$:

(1) Let $\sigma \in Aut\ A$. Show that the map $\tilde{\sigma} : A[X] \longrightarrow A[X]$, defined for $P = \sum_{i:0}^{n} a_i X^i$ by $\tilde{\sigma}(P) = \sum_{i:0}^{n} \sigma(a_i) X^i$, is an automorphism of the ring $A[X]$ that extends σ.
(2) Show that the set $A^G = \{x \in A;\ \sigma(x) = x$ for each $\sigma \in G\}$ is a sub-ring of A.
(3) Show that for each $x \in A$, the polynomial $P_x(X) = \prod_{\sigma \in G} (X - \sigma(x)) \in A^G[X]$.
(4) Show that the extension $A^G \subseteq A$ is integral.

Exercise 7 Let K be a field and $A = K + X^2 K[[X]]$. Show that $K[[X]]$ is an over ring of A, and the extension $A \subset K[[X]]$ is integral. Deduce the integral closure of A.

Exercise 8 (Houston 1994) Let A be an integral domain with quotient field K. Suppose that K contains an element t that is not quasi-integral over A, but there exists $0 \neq a \in A$ such that $at^i \in A$ for an infinity of natural integers i. Show that the complete integral closure A^* of A is not completely integrally closed.

Exercise 9 (Anderson and Nour El Abidine 1999) Let $A \subseteq B$ be an extension of integral domains and $R = A + XB[[X]]$:

(1) Show that the following two conditions are equivalent:

    (a) R satisfies the acc for the principal ideals.

    (b) Each increasing sequence of the form $b_1 A \subseteq b_2 A \subseteq \ldots$, with $b_n \in B$ is stationary.

(2) Show that if A is a field, R satisfies the acc for principal ideals.

(3) Suppose that $qf(A) \subseteq B$. Show that R satisfies the acc for the principal ideals if and only if A is a field. What about the domain $\mathbb{Z} + X\mathbb{Q}[[X]]$? Give an example.

(4) Suppose that $U(B) \cap A = U(A)$ and B satisfies the acc for the principal ideals. Show that R satisfies the acc for the principal ideals. What is the converse?

(5) Suppose that R satisfies the acc for the principal ideals. Show that $U(B) \cap A = U(A)$ and A satisfies the acc for the principal ideals.

(6) Suppose that B satisfies the acc for the principal ideals. Deduce that R satisfies the acc for the principal ideals if and only if $U(B) \cap A = U(A)$.

(7) Examine the converse of (5) using the domains $A = \mathbb{Z}$ and $B = \mathbb{Z} + Y\mathbb{Q}[Y]$ and the sequence $\left(\frac{XY}{2^n}\right)_{n\in\mathbb{N}}$.

Exercise 10 (Anderson et al. (1995) and Querré (1980)) Let A be an integral domain:

(1) Show that A is integrally closed if and only if for each integral ideal I of $A[X]$ over (0) in A, we have $I_v = \left(c(I)[X]\right)_v = c(I)_v[X]$, where $c(I)$ is the ideal of A formed by the coefficients of all the elements of I.

(2) Show that A is integrally closed if and only if for each integral ideal I of $A[X]$ over (0) in A, we have $I_t = c(I)_t[X]$.

(3) Suppose that the ring A is integrally closed and I is an integral ideal of $A[X]$ with nonzero trace in A:

    (a) Show that if I is divisorial (resp. a t−ideal) in $A[X]$, then $I \cap A$ is divisorial (resp. a t−ideal) in A and $I = (I \cap A)[X]$.

    (b) Show that if I is invertible (resp. a t−invertible t−ideal, resp. a v−invertible divisorial ideal) in $A[X]$, then $I \cap A$ is invertible (resp. a t−invertible t−ideal, resp. a v−invertible divisorial ideal) in A.

(4) Suppose that the ring A is integrally closed. Show that for each divisorial fractional ideal (resp. t−ideal) J of $A[X]$ is of the form $J = \frac{f(X)}{g(X)}L[X]$ with $f(X), g(X) \in A[X]$ and L a divisorial ideal (resp. a t−ideal) of A.

Exercise 11 Let (A, M) be a valuation domain and I a nonzero integral ideal of A:

(1) (a) Show that if I is not invertible, then $I = IM$.

    (b) Show that if I is not divisorial, there exists $a \in A$ such that $I = aM$.

(2) Suppose that I is not invertible. Show that $P = II^{-1}$ is a prime ideal of A, $(I : I) = A_P$ and $IA_P = I$.

Exercise 12 Let A be an integral domain and I an ideal of $A[X]$ satisfying $I \cap A = (0)$. Show that there exist $f(X) \in I$ and $0 \neq a \in A$ such that for each $g(X) \in I$, there exist $q(X) \in A[X]$ and $m \in \mathbb{N}$ satisfying $a^m g(X) = f(X)q(X)$.

Exercise 13 (Khurana and Kumar 2005) Let A be a PID:

(1) Let P be a nonzero prime ideal of $A[X]$ such that $P \cap A = (0)$. Show that $P = f(X)A[X]$ with $f(X) \in A[X]$ a nonconstant irreducible polynomial.
(2) Let M be a maximal ideal of $A[X]$ such that $M \cap A \neq (0)$. Show that $M = pA[X] + f(X)A[X]$ with $p \in A$ an irreducible element and $f(X) \in A[X]$ irreducible modulo the ideal $pA[X]$.
(3) Show that the following conditions are equivalent:

   (a) The domain A admits an infinity of nonassociated irreducible elements.
   (b) For each maximal ideal M of $A[X]$, $M \cap A \neq (0)$.
   (c) For each maximal ideal M of $A[X]$, there exist $p \in A$ irreducible and $f(X) \in A[X]$ irreducible modulo $pA[X]$ such that $M = pA[X] + f(X)A[X]$.
   (d) No maximal ideal of $A[X]$ is principal.

(4) Suppose that A admits an infinity of nonassociated irreducible elements. Show that an ideal M of $A[X]$ is maximal if and only if it is of the form $M = pA[X] + f(X)A[X]$ with $p \in A$ an irreducible element and $f(X) \in A[X]$ an irreducible polynomial modulo $pA[X]$.
   In this case, show that the height of all the maximal ideals of $A[X]$ is 2.
(5) Show that an ideal P of $A[X]$ is prime if and only if either P is maximal or $P = f(X)A[X]$ with $f(X)$ an irreducible polynomial of $A[X]$ that is not invertible modulo $pA[X]$ for some irreducible element $p \in A$.
(6) Show that the height of all the maximal ideals of $A[X]$ is 2 if and only if A admits an infinity of nonassociated irreducible elements.

Exercise 14 The goal of this exercise is to determine the prime ideals of the domains $\mathbb{Z}[X]$ and $\mathbb{Z}[i]$ and the irreducible elements of $\mathbb{Z}[i]$. Let p be a prime number and $s : \mathbb{Z}[X] \longrightarrow (\mathbb{Z}/p\mathbb{Z})[X]$ the natural homomorphism. We say that the polynomial $f(X)$ of $\mathbb{Z}[X]$ is irreducible modulo p if $s(f(X))$ is irreducible in the ring $(\mathbb{Z}/p\mathbb{Z})[X]$:

(1) Let $\phi : A \longrightarrow B$ be a homomorphism onto between rings. Show that ϕ induces a bijection between the set of the prime ideals of A containing $ker\ \phi$ and the set of the prime ideals of B.
(2) Show that the elements of the three following families are prime ideals of the domain $\mathbb{Z}[X]$: $F_1 = \{p\mathbb{Z}[X];\ p$ a prime number$\}$; $F_2 = \{f(X)\mathbb{Z}[X];\ f(X) \in \mathbb{Z}[X]$ an irreducible primitive polynomial in $\mathbb{Q}[X]\}$; $F_3 = \{p\mathbb{Z}[X] + f(X)\mathbb{Z}[X];\ p$ a prime number and $f(X) \in \mathbb{Z}[X]$ a polynomial irreducible modulo $p\}$. Hint: use (1).

(3) Let P be a nonzero prime ideal of $\mathbb{Z}[X]$:

   (a) Show that $P \cap \mathbb{Z}$ is either (0) or equal to $p\mathbb{Z}$ with p a prime number.
   (b) If $P \cap \mathbb{Z} = (0)$, show that $P \in F_2$.
   (c) Suppose that $P \cap \mathbb{Z} = p\mathbb{Z}$ with p a prime number. If $P \neq p\mathbb{Z}[X]$, show that there exists $f(X) \in \mathbb{Z}[X]$ irreducible modulo p such that $P = p\mathbb{Z}[X] + f(X)\mathbb{Z}[X]$.
   Hint: use (1).
   (d) Deduce that the set of the nonzero prime ideals of $\mathbb{Z}[X]$ is $F_1 \cup F_2 \cup F_3$.

(4) Let p be a prime number. Show that the following assertions are equivalent:

   (a) The polynomial $X^2 + 1$ of $\mathbb{Z}[X]$ is not irreducible modulo p.
   (b) Either $p = 2$ or the multiplicative group $(\mathbb{Z}/p\mathbb{Z})^*$ contains an element of order 4.
   (c) Either $p = 2$ or $p \equiv 1 \ (mod\ 4)$.
   Hint: for the implication (c)$\Longrightarrow$(a), if $p = 4n + 1$ and u is a generator of the multiplicative group $(\mathbb{Z}/p\mathbb{Z})^*$, show that u^n is a root of $X^2 + 1$.

(5) Show that the set of the prime ideals of $\mathbb{Z}[X]$ containing $X^2 + 1$ is $\{(X^2 + 1)\mathbb{Z}[X]\} \cup \{p\mathbb{Z}[X] + a(X^2 + 1)\mathbb{Z}[X]; \ p$ a prime number, $p > 2$ and $p \not\equiv 1 \ (mod\ 4), \ 1 \leq a \leq p - 1\} \cup \{p\mathbb{Z}[X] + a(X - b)\mathbb{Z}[X]; \ p$ a prime number, $p = 2$ or $p \equiv 1 \ (mod\ 4), 1 \leq a, b \leq p - 1, \ \bar{b}$ root of $X^2 + 1$ in $\mathbb{Z}/p\mathbb{Z}\}$.

(6) Show that $\mathbb{Z}[i] \simeq \mathbb{Z}[X]/(X^2 + 1)$.

(7) Show that for each nonzero prime ideal of the domain $\mathbb{Z}[i] = \{a + ib; \ a, b \in \mathbb{Z}\}$ belongs to one of the two following classes:
   $C_1 = \{p\mathbb{Z}[i]; \ p$ a prime number, $p \not\equiv 1 \ (mod\ 4)$ and $p > 2\}$.
   $C_2 = \{p\mathbb{Z}[i] + z\mathbb{Z}[i]; \ p$ a prime number, $p \equiv 1 \ (mod\ 4)$ or $p = 2, z \in \mathbb{Z}[i]$ and p divides $|z|^2$ but does not divide $z\}$. Hint: use (1).

(8) Give the group of unities of the domain $\mathbb{Z}[i]$.

(9) Show that an element of $\mathbb{Z}[i]$ is irreducible if and only if its complex conjugate is irreducible.

(10) Let $\mathscr{P}$ be a system of representative of the irreducible elements of $\mathbb{Z}[i]$. Show that $\mathscr{P}$ is equivalent to the set of the nonzero prime ideals of $\mathbb{Z}[i]$.

(11) Let p be a prime number that is not irreducible in $\mathbb{Z}[i]$:

   (a) Show by an example that such a prime number exists.
   (b) Show that if t is an irreducible element of $\mathbb{Z}[i]$ that divides p, then $p = |t|^2$.

(12) Let P be a prime ideal of $\mathbb{Z}[i]$ in the class C_2, so of the form $P = p\mathbb{Z}[i] + z\mathbb{Z}[i]$, with p a prime number, $p \equiv 1 \ (mod\ 4)$ or $p = 2, z \in \mathbb{Z}[i]$ and p divides $|z|^2$ but does not divide z:

   (a) Show that there exists an irreducible element t of $\mathbb{Z}[i]$ such that $p = |t|^2$.
   (b) Show that $P = \alpha\mathbb{Z}[i]$ with $\alpha \in \{t, \bar{t}\}$.

(13) Show that each irreducible element of $\mathbb{Z}[i]$ is in one of the following families:
$I_1 = \{pi^n;\ n = 0, 1, 2, 3,\ p \text{ a prime number}, p > 2,\ p \not\equiv 1(mod\ 4)\}$.
$I_2 = \{z \in \mathbb{Z}[i];\ |z|^2 \text{ is a prime number}, |z|^2 \equiv 1(mod\ 4) \text{ or } |z|^2 = 2\}$.

Exercise 15 (Barucci 1986) Let A be an integral domain, $\mathscr{F}(A)$ the set of the nonzero fractional ideals of A, and $I \in \mathscr{F}(A)$. We say that I is strong if $II^{-1} = I$. If I is strong and divisorial, we say that it is strongly divisorial. We denote by $D_f(A)$ the set of the strongly divisorial fractional ideals of A:

(1) Show that a strong ideal is necessarily integral and a maximal ideal is either strong or invertible.
(2) Let $I \in \mathscr{F}(A)$. Show that I is strong if and only if $I^{-1} = (I : I)$. Deduce that if I is strong, then I^{-1} is an over ring of A.
(3) Show that if $I \in \mathscr{F}(A)$ satisfies $I^{-1} = A$, then it is strong. Deduce that $D_f(A) \neq \emptyset$.
(4) Let $I \in \mathscr{F}(A)$. Show that I is strong if and only if there exists a nonzero integral ideal J of A such that $I = JJ^{-1}$. Deduce that for each $J \in \mathscr{F}(A)$, JJ^{-1} is strong.
(5) Show that a nonzero ideal of A is strongly divisorial if and only if it is the conductor of A in some over ring.
   Deduce that if A is a valuation domain, the set of the strongly divisorial ideals of A is $\{A\} \cup Spec(A) \setminus (0)$.
(6) Show that the complete integral closure of A is $A^c = \bigcup_{I \in \mathscr{F}(A)} (I : I) = \bigcup_{I \in D_f(A)} I^{-1}$.
   Deduce that A is completely integrally closed if and only if A is the only strongly divisorial ideal of A.
(7) Suppose that A is completely integrally closed. Let $J \in \mathscr{F}(A)$ not invertible. Show that $I = JJ^{-1}$ is strong but not divisorial.
(8) Suppose that the conductor $(A : A^c) \neq (0)$. Let $(I_\alpha)_{\alpha \in \Lambda}$ be a family of strongly divisorial ideals of A. Show that $\bigcap_{\alpha \in \Lambda} I_\alpha \neq (0)$.
(9) Let $I \in \mathscr{F}(A)$. Show that I is strongly divisorial if and only if there exists a nonzero integral ideal J of A such that $I = (JJ^{-1})_v$. Deduce that for each $J \in \mathscr{F}(A)$, the ideal $(JJ^{-1})_v$ is strongly divisorial.
(10) Show that the map α defined by $\alpha(B) = (A : B)$ is an increasing bijection between the set of the over rings of A of the form $B = I^{-1}, I \in \mathscr{F}(A)$, and the set $D_f(A)$.

Exercise 16 Let A be an integral semi-normal domain and B an over ring of A. Show that the conductor $(A : B)$ is a radical ideal of A and B.

Exercise 17 (Fanggui and McCasland (1997) and Park (2001)) Let A be an integral domain and w the $w-$operation on A, defined in Exercise 27 of Chap. 1:

(1) (a) Let I be a nonzero integral ideal of A. Show that $I_w = A$ if and only if I contains a GV-ideal of A.
   (b) Deduce that for each $I \in GV(A)$, $I_w = A$.
   (c) Show that each prime ideal P of A satisfies either $P_w = P$ or $P_w = A$.
(2) Let I be a nonzero integral ideal of A. Show that $I_t = A$ if and only if there exists a nonzero finitely generated subideal J of I such that $J^{-1} = A$.
(3) Show that $w - Max(A) = t - Max(A)$.
(4) Let $I \in \mathscr{F}(A)$. Show that $I_w A_P = I A_P$ for each $w-$prime ideal P of A.
(5) Show that for each $I \in \mathscr{F}(A)$, $I_w = \cap\{I A_P; \; P \in t - Max(A)\}$.
(6) Let I be an invertible fractional ideal of A and M a nonzero fractional ideal of A:

   (a) Show that if M is a fractional $w-$ideal, then IM is a $w-$ideal.
   (b) Show that if $M \in \mathscr{F}(A)$, then $(IM)_w = I M_w$.

(7) Let $M \in \mathscr{F}(A)$ and $J \in GV(A)$. Show that $(JM)_w = M_w$.

Exercise 18 (Fanggui and McCasland 1997) Let A be an integral domain. We say that A is a SM-ring (strong Mori) if it satisfies the acc for the proper integral $w-$ideals:

(1) Show that a SM-ring A is a Mori domain. Deduce $w - Max(A) = v - Max(A)$.
(2) Let P be a maximal element in the set of all the integral $w-$ideals of A that are not $w-$finite. Show that P is a prime ideal of A.
(3) Show that the following assertions are equivalent

   (a) A is a SM-ring.
   (b) Each $w-$ideal of A is $w-$finite.
   (c) Each prime $w-$ideal of A is $w-$finite.

(4) Show that if A is a SM-ring, then A_P is Noetherian for each $P \in w - Max(A)$.
(5) Let S be a multiplicative set of A:

   (a) Show that if $J \in GV(A)$, then $J_S \in GV(A_S)$.
   (b) Show that the trace on A of a fractional $w-$ideal of A_S is a $w-$ideal of A.
   (c) Show that if A is a MS-ring, so is A_S.

(6) Show that A is a SM-ring if and only if:

   (a) For each $P \in w - Max(A)$, the ring A_P is Noetherian.
   (b) Each nonzero element of A belongs to a finite number of $w-$maximal ideals.

Exercise 19 (Kabbaj and Mimouni 2021) Let A be a ring and I and J be the two ideals of A. We say that I and J are linked if there exists an ideal $L \subseteq I \cap J$ of A such that $I = (L :_A J)$ and $J = (L :_A I)$. In this case, we say that I and J are linked over L:

(1) Show that I and J are linked if and only if they are linked over the ideal IJ.
    Application: Let X, Y, and Z be the indeterminates over a field k. Using the
    element $X^2Y^2Z^2$, deduce that the ideals $I = (X^2, Y^2)$ and $J = (X^3, Y^3, Z^3)$
    are not linked in the polynomial ring $k[X, Y, Z]$.
(2) Show that two comaximal ideals I and J (i.e., $I + J = A$) are linked.
(3) Show that two primary ideals with incomparable radicals are linked.
(4) Deduce that two incomparable prime ideals are linked.
(5) Show that ideals of the form $P \subseteq M$ with P prime and M maximal are linked
    if and only if $PM \neq M$. Hint: Distinguish the cases: $P = M$ and $P \neq M$.

Application: Let X, Y, and Z be the indeterminates over a field k. Deduce that the
ideals (X, Y) and (X, Y, Z) are linked in the polynomial ring $k[X, Y, Z]$.

Solutions

Exercise 1 $" \Longleftarrow "$ Since i is integral over $\mathbb{Q}$, then $z = a + ib$ is integral over $\mathbb{Q}$.
$" \Longrightarrow "$ There exist $c_0, \ldots, c_{n-1} \in \mathbb{Q}$ such that $c_0 + c_1 z + \ldots + c_{n-1} z^{n-1} + z^n = 0$.
By taking the complex conjugate, $c_0 + c_1 \bar{z} + \ldots + c_{n-1} \bar{z}^{n-1} + \bar{z}^n = 0$, and then $\bar{z}$
is integral over $\mathbb{Q}$. So $a = \frac{1}{2}(z + \bar{z})$ and $b = -\frac{1}{2}i(z - \bar{z})$ are integral over $\mathbb{Q}$ since i
is integral over $\mathbb{Q}$.

Exercise 2 The domain $\mathbb{Z}[i]$ is an over ring of A since $a + ib = \frac{2a + 2ib}{2}$. The
extension $A \subset \mathbb{Z}[i]$ is integral since $(a + ib)^2 = a^2 - b^2 + 2iab \in A$. The domain
$\mathbb{Z}[i]$ is integrally closed since it is UFD. Then $A' = \mathbb{Z}[i]$.

Exercise 3

(1) Suppose that there exists $x \in B \setminus A$ an integral element over A. There exist
    $a_0, \ldots, a_{n-1} \in A$ such that $x^n + a_{n-1}x^{n-1} + \ldots + a_1 x + a_0 = 0$. We may suppose
    that n is the smallest. Then $x(x^{n-1} + a_{n-1}x^{n-2} + \ldots + a_1) = -a_0 \in A$. Since
    $x \in B \setminus A$ and $B \setminus A$ is multiplicatively stable, then $x^{n-1} + a_{n-1}x^{n-2} + \ldots + a_1 =$
    $c \in A$. The equality $x^{n-1} + a_{n-1}x^{n-2} + \ldots + a_1 - c = 0$ contradicts the
    minimality of n.
(2) Existence: Note that $b = 0$ if and only if $c = 0$. In this case, take $x = 0$. Suppose
    that $b \neq 0$ and $c \neq 0$. Since $b^2 = c^3$, then $(\frac{b}{c})^2 = c$. The element $x = \frac{b}{c} \in K$,
    the quotient field of A, satisfies $x^2 - c = 0$. Then x is integral over A. But A is
    integrally closed, so $x \in A$. On the other hand, $b = \frac{b^3}{b^2} = \frac{b^3}{c^3} = (\frac{b}{c})^3 = x^3$ and
    $c = \frac{c^3}{c^2} = \frac{b^2}{c^2} = (\frac{b}{c})^2 = x^2$.

Uniqueness: Let $x, y \in A$ be such that $b = x^3 = y^3$ and $c = x^2 = y^2$. Then
$(x - y)^3 = x^3 - y^3 - 3x^2 y + 3xy^2 = -3x^2 y + 3xy^2 = -3y^3 + 3x^3 = 3(x^3 - y^3) = 0$.
Since A is integral, $x - y = 0$, and then $x = y$.

Exercise 4

(1) For each integer $n \geq 1$, $\sqrt[n]{2} \in A \setminus U(A)$. Indeed, if $\frac{1}{\sqrt[n]{2}} \in A$, then $(\frac{1}{\sqrt[n]{2}})^n = \frac{1}{2} \in A$. So $\frac{1}{2} \in \mathbb{Q} \setminus \mathbb{Z}$ is an integral element over $\mathbb{Z}$, which is absurd since $\mathbb{Z}$ is integrally closed. These equalities contradict the uniqueness of the factorization in an UFD.

(2) Let $\alpha_n = 2^{\frac{1}{2^n}} \in A$ and $I_n = (\alpha_n)$. Since $\alpha_{n+1}^2 = \alpha_n$, then $I_n \subseteq I_{n+1}$. If $I_{n+1} = I_n$, then $\alpha_{n+1} = b\alpha_{n+1}^2$ with $b \in A$. So $\frac{1}{\alpha_{n+1}} \in A$. Since $\alpha_{n+1}^{2^{n+1}} = 2$, then $\frac{1}{2} \in A$ with $\frac{1}{2} \in \mathbb{Q}$. This is absurd because $\mathbb{Z}$ is integrally closed.

Exercise 5

(a) Put $x = e^{ir\pi}$ and $r = \frac{n}{m}$ with $n, m \in \mathbb{Z}$, $m \geq 1$. Then $x^{2m} = e^{2in\pi} = 1 \in \mathbb{Z}$. So x is integral over $\mathbb{Z}[i]$.

(b) Let $x = e^{ir\pi} = \cos r\pi + i \sin r\pi \in \mathbb{Q} + i\mathbb{Q} = qf(\mathbb{Z}[i])$. By (a), x is integral over $\mathbb{Z}[i]$, and since $\mathbb{Z}[i]$ is integrally closed, $x \in \mathbb{Z}[i]$. So $\cos r\pi$, $\sin r\pi \in \mathbb{Z}$. But $\cos^2 r\pi + \sin^2 r\pi = 1 \iff$ either ($\cos^2 r\pi = 1$ and $\sin^2 r\pi = 0$) or ($\cos^2 r\pi = 0$ and $\sin^2 r\pi = 1$) $\iff$ either ($\cos r\pi = \pm 1$ and $\sin r\pi = 0$) or ($\cos r\pi = 0$ and $\sin r\pi = \pm 1$) $\iff r\pi = m\frac{\pi}{2}$ with $m \in \mathbb{Z} \iff r = \frac{m}{2}$ with $m \in \mathbb{Z}$.

Exercise 6

(1) If $P = \sum a_i X^i$ and $Q = \sum b_i X^i$, then $PQ = \sum c_n X^n$ with $c_n = \sum_{i+j=n} a_i b_j$. So $\tilde{\sigma}(P) = \sum \sigma(a_i) X^i$, $\tilde{\sigma}(Q) = \sum \sigma(b_i) X^i$, $\tilde{\sigma}(PQ) = \sum \sigma(c_n) X^n = \sum \sigma(\sum_{i+j=n} a_i b_j) X^n = \sum \sum_{i+j=n} \sigma(a_i) \sigma(b_j) X^n = \tilde{\sigma}(P)\tilde{\sigma}(Q)$. The other properties are clear.

(2) For each $\sigma \in G$, $\sigma(0) = 0$, and $\sigma(1) = 1$, and then $0, 1 \in A^G$. Let $x, y \in A^G$, and then $\sigma(x - y) = \sigma(x) - \sigma(y) = x - y$ and $\sigma(xy) = \sigma(x)\sigma(y) = xy$, so $x - y, xy \in A^G$.

(3) Let $\tau \in G$, $\tilde{\tau}(P_x(X)) = \prod_{\sigma \in G} (X - \tau \circ \sigma(x))$. But the map $\phi : G \longrightarrow G$, defined by $\phi(\sigma) = \tau \circ \sigma$, is bijective, and then $\tau(P_x(X)) = \prod_{\sigma \in G} (X - \tau \circ \sigma(x)) = \prod_{\rho \in G} (X - \rho(x)) = P_x(X)$. Then the coefficients of $P_x(X)$ are fixed by τ, so $P_x(X) \in A^G[X]$.

(4) Let $x \in A$. The polynomial $P_x(X) = \prod_{\sigma \in G} (X - \sigma(x)) \in A^G[X]$ is monic. For $\sigma = id_A$, $X - \sigma(x) = X - x$ is a factor of $P_x(X)$, and then $P_x(x) = 0$. So x is integral over A^G.

Exercise 7 For each $f \in K[[X]]$, $f = \frac{X^2 f}{X^2}$ with $X^2, X^2 f \in A$. Then $K[[X]]$ is an over ring of A. Each element $f \in K[[X]]$ can be written $f = a + Xg$ with $g \in K[[X]]$, and then $f^2 = a^2 + X^2 g^2 + 2aXg$. So $f^2 - 2af = a^2 + X^2 g^2 + 2aXg - 2a^2 - 2aXg = -a^2 + X^2 g^2 \in A$, and then f is integral over A. Then the extension $A \subset K[[X]]$ is integral. Since $K[[X]]$ is integrally closed, the integral closure of A is $K[[X]]$.

Exercise 8 For each $i \in \mathbb{N}$, there is an integer $j \geq i$ such that $at^j \in A$. Then $(at^i)^j = a^{j-i}(at^j)^i \in A$. Let $\bar{A}$ be the integral closure of A. Then $at^i \in \bar{A} \subseteq A^*$ for each $i \in \mathbb{N}$. So $t \in (A^*)^* \setminus A^*$.

Exercise 9

(1) "$(a) \implies (b)$" Let $b_1, b_2, \ldots \in B$ be such that $b_1 A \subseteq b_2 A \subseteq \ldots$. Then the elements $b_1 X, b_2 X, \ldots \in R$, and we have successively $b_1 XA \subseteq b_2 XA \subseteq \ldots$ and $b_1 XR \subseteq b_2 XR \subseteq \ldots$. Since R satisfies the acc for principal ideals, there exists $k_0 \in \mathbb{N}$ such that for each $k \geq k_0$, $b_k XR = b_{k+1} XR$, and then $b_k R = b_{k+1} R$. So $b_k a = b_{k+1}$ with $a \in A$, then $b_k A = b_{k+1} A$.

"$(b) \implies (a)$" Let $f_1, f_2, \ldots \in R \setminus (0)$ be such that $f_1 R \subseteq f_2 R \subseteq \ldots$. Then $ord(f_{k+1}) \leq ord(f_k)$, and we can suppose that all the f_k have the same order, by omitting a finite number of terms of the sequence. Let b_k be the initial coefficient of f_k, and put $f_k = f_{k+1} g_k$ with $g_k \in R$. Then $g_k(0) \in A$ and $b_k = b_{k+1} g_k(0)$. The increasing sequence $b_1 A \subseteq b_2 A \subseteq \ldots$ is by hypothesis stationary. There exists $k_0 \in \mathbb{N}$ such that for each $k \geq k_0$, $b_k A = b_{k+1} A$. The coefficient of proportionality $g_k(0)$ is invertible in A, and g_k is invertible in R, and then $f_k R = f_{k+1} R$.

(2) Let $(0) \subset b_1 A \subseteq b_2 A \subseteq \ldots$ be an increasing sequence with $0 \neq b_k \in B$. Put $b_k = b_{k+1} a$ with $0 \neq a \in A$. Then $b_{k+1} = b_k \frac{1}{a}$ with $\frac{1}{a} \in A$, so $b_k A = b_{k+1} A$.

(3) "$\Longleftarrow$" By (2). "$\implies$" Let $0 \neq a \in A$, and then $\frac{1}{a} \in qf(A) \subseteq B$ and $\frac{1}{a} A \subseteq \frac{1}{a^2} A \subseteq \frac{1}{a^3} A \subseteq \ldots$ since $\frac{1}{a^n} = \frac{1}{a^{n+1}} a$. By (1), there exists $k \in \mathbb{N}^*$ such that $\frac{1}{a^k} A = \frac{1}{a^{k+1}} A$, and then $A = \frac{1}{a} A$, so $\frac{1}{a} \in A$ and A is a field.

The domain $R = \mathbb{Z} + X\mathbb{Q}[[X]]$ does not satisfy the acc for principal ideals since $qf(\mathbb{Z}) = \mathbb{Q}$ and $\mathbb{Z}$ is not a field. More precisely, the sequence $\left(\frac{X}{2^n} R\right)_{n \in \mathbb{N}}$ is strictly increasing since if $\frac{X}{2^{n+1}} = \frac{X}{2^n}(a + Xf)$ with $a \in \mathbb{Z}$ and $f \in \mathbb{Q}[[X]]$, and then $a = \frac{1}{2} \notin \mathbb{Z}$, which is absurd.

(4) Let $b_1 A \subseteq b_2 A \subseteq \ldots$ be an increasing sequence with $0 \neq b_k \in A$. Then $b_1 B \subseteq b_2 B \subseteq \ldots$ is an increasing sequence of principal ideals of B. By hypothesis, there exists $k_0 \in \mathbb{N}^*$ such that $b_k B = b_{k+1} B$ for each $k \geq k_0$, and then $\frac{b_k}{b_{k+1}} \in U(B)$. But $b_k A \subseteq b_{k+1} A$, and then $\frac{b_k}{b_{k+1}} \in A$. So $\frac{b_k}{b_{k+1}} \in U(B) \cap A = U(A)$, and $b_k A = b_{k+1} A$. By (1), R satisfies the acc for the principal ideals.

The converse is false: Let B be a ring that does not satisfy the acc for principal ideals, containing a field A. Then $U(B) \cap A = U(A)$, and by (2), $R = A + XB[[X]]$ satisfies the acc for the principal ideals.

(5) First method: By (1), A satisfies the acc for the principal ideals.
Second method: $R = A + XB[[X]]$, and then $A \simeq R/XB[[X]]$ satisfies the acc for the principal ideals.
Let $a \in U(B) \cap A$. Then $\frac{1}{a} \in B$, and we have the increasing sequence $\frac{1}{a}A \subseteq \frac{1}{a^2}A \subseteq \frac{1}{a^3}A \subseteq \ldots$ since $\frac{1}{a^n} = \frac{1}{a^{n+1}}a$. By (1), it is stationary. There exists k such that $\frac{1}{a^k}A = \frac{1}{a^{k+1}}A$, and then $A = \frac{1}{a}A$. So $\frac{1}{a} \in A$, and then $a \in U(A)$. So $U(B) \cap A = U(A)$.

(6) $'' \Longrightarrow ''$ By (5). $'' \Longleftarrow ''$ By (4).

(7) $R = A + XB[[X]] = \mathbb{Z} + XB[[X]]$; $\frac{XY}{2^n} \in R$ since $\frac{Y}{2^n} \in B = \mathbb{Z} + Y\mathbb{Q}[Y]$. The sequence $(\frac{XY}{2^n})_{n \in \mathbb{N}}$ is increasing since $\frac{XY}{2^n} = \frac{XY}{2^{n+1}}2$ with $2 \in \mathbb{Z} \subset R$. It is strictly increasing because if not, $\frac{XY}{2^{n+1}} = \frac{XY}{2^n}f$ with $f \in R$. But then $f = \frac{1}{2} \notin R = \mathbb{Z} + XB[[X]]$, which is absurd. Then R does not satisfy the acc for the principal ideals. We have $U(B) \cap A = U(\mathbb{Z} + Y\mathbb{Q}[Y]) \cap \mathbb{Z} = U(\mathbb{Z}) \cap \mathbb{Z} = U(\mathbb{Z}) = U(A)$ and $A = \mathbb{Z}$ is Noetherian.

Exercise 10

(1) The equality $(c(I)[X])_v = c(I)_v[X]$ follows from Corollary 11.7 of Chap. 1.
$'' \Longrightarrow ''$ Since $I \subseteq c(I)[X]$, then $I_v \subseteq (c(I)[X])_v$. Conversely, I_v is the intersection of the principal fractional ideals containing I. It suffices to show that if $I \subseteq \frac{a(X)}{b(X)}A[X]$ with $a(X)$ and $b(X) \in A[X] \setminus (0)$, then $c(I)[X] \subseteq \frac{a(X)}{b(X)}A[X]$. We may suppose that $a(X)$ and $b(X)$ are relatively prime in $K[X]$. Necessarily, $a(X) \in A$. Indeed, let $0 \neq c \in I \cap A$, $c = \frac{a(X)}{b(X)}d(X)$ with $d(X) \in A[X]$. Then $cb(X) = a(X)d(X)$, and since $b(X)$ must divide $d(X)$ in $K[X]$, for degrees reasons, $a(X)$ is a constant. The inclusion $I \subseteq \frac{a(X)}{b(X)}A[X]$ implies that a divides $b(X)g(X)$ in $A[X]$ for each $g(X) \in I$, and then $c(bg) \subseteq aA$. Since A is integrally closed, by Proposition 10.1, $c(b)c(g) \subseteq aA$, then $c(b)c(I) \subseteq aA$. So $b(X)c(I)[X] \subseteq aA[X]$, and then $c(I)[X] \subseteq \frac{a}{b(X)}A[X]$.
$'' \Longleftarrow ''$ By Proposition 10.1, it suffices to show that for each $f(X), g(X) \in A[X] \setminus (0)$ and for each $0 \neq a \in A$, the inclusion $c(fg) \subseteq aA$ implies $c(f)c(g) \subseteq aA$. But $c(fg) \subseteq aA$ implies that $af(X)A[X] + f(X)g(X)A[X] \subseteq aA[X]$ so $aA[X] + g(X)A[X] \subseteq \frac{a}{f(X)}A[X]$. Since the ideal $aA[X] + g(X)A[X]$ has a nonzero trace in A, by hypothesis, $(aA[X] + g(X)A[X])_v = c(a, g)_v[X]$. Then $c(g)[X] \subseteq c(a, g)[X] \subseteq c(a, g)_v[X]$
$= (aA[X] + g(X)A[X])_v \subseteq \frac{a}{f(X)}A[X]$. So $f(X)c(g)[X] \subseteq aA[X]$, and then $c(f)c(g) \subseteq aA$.

(2) $'' \Longrightarrow ''$ Since $I \subseteq c(I)[X]$, then $I_t \subseteq (c(I)[X])_t$. Conversely, consider the ideal $J = \bigcup F_v \subseteq I_t$ of $A[X]$, where F runs the set of the nonzero finitely generated subideals of I such that $F \cap A \neq (0)$. Then $c(I) = \bigcup c(F)$. Indeed, if $0 \neq a \in A \cap I$ and $f \in I$, we take $F = (a, f)$. By (1), $F_v = c(F)_v[X]$. Then $c(I)[X] = \bigcup c(F)[X] \subseteq \bigcup c(F)_v[X] = \bigcup F_v = J \subseteq I_t$. So $(c(I)[X])_t \subseteq I_t$. By Corollary 11.7, Chap. 1, $I_t = (c(I)[X])_t = c(I)_t[X]$.

" $\Longleftarrow$ " By Proposition 10.1, it suffices to show that for all $f(X), g(X) \in A[X]\backslash$ (0) and for each $0 \neq a \in A$, the inclusion $c(fg) \subseteq aA$ implies $c(f)c(g) \subseteq aA$. But $c(fg) \subseteq aA$ implies that $af(X)A[X] + f(X)g(X)A[X] \subseteq aA[X]$ so $aA[X] + g(X)A[X] \subseteq \frac{a}{f(X)}A[X]$. Since the ideal $aA[X] + g(X)A[X]$ has a nonzero trace in A, by hypothesis, $\left(aA[X] + g(X)A[X]\right)_t = c(a, g)_t[X]$. Then $c(g)[X] \subseteq c(a, g)[X] \subseteq c(a, g)_t[X]$
$= \left(aA[X] + g(X)A[X]\right)_t \subseteq \frac{a}{f(X)}A[X]$. So $f(X)c(g)[X] \subseteq aA[X]$, then $c(f)c(g) \subseteq aA$.

(3)

(a) Let $*$ be either the $v-$operation or the $t-$operation. By (1) and (2), $I = I^* = c(I)^*[X]$, and then $I \cap A = c(I)^*$ is a $*-$ideal. Moreover, $I = (I \cap A)[X]$.

(b) Let $*$ be the identity or the $v-$operation or the $t-$operation. By Corollary 11.7, Chap. 1, $I^{-1} = \left(((I \cap A)[X])\right)^{-1} = (I \cap A)^{-1}[X]$, and then $II^{-1} = (I \cap A)[X](I \cap A)^{-1}[X] \subseteq (I \cap A)(I \cap A)^{-1}[X] \subseteq A[X]$. By Corollary 11.7, Chap. 1, $(II^{-1})^* \subseteq \left((I \cap A)(I \cap A)^{-1}\right)^*[X] \subseteq A[X]$. When I is $*-$invertible, $(II^{-1})^* = A[X]$, and then $\left((I \cap A)(I \cap A)^{-1}\right)^* = A$.

(4) By Proposition 2.5, since A is integrally closed, then $A[X]$ is almost principal. By Theorem 1.10, $J = \frac{f(X)}{g(X)}I$ with $f(X), g(X) \in A[X]$ and I is an integral ideal of $A[X]$ with nonzero trace in A. By (3-a), $I = (I \cap A)[X]$, and then $J = \frac{f(X)}{g(X)}(I \cap A)[X]$. Moreover, if I is divisorial (resp. a $t-$ideal) in $A[X]$, then $I \cap A$ is divisorial (resp. a $t-$ideal) in A.

Exercise 11

(1) (a) It suffices to show the inclusion $I \subseteq IM$. Let $a \in I$. Since I is not invertible, it is not principal, and then $I \neq aA$. Let $b \in I \setminus aA$. Then $aA \subset bA$. So $a = b\alpha$ with α not invertible, and then $\alpha \in M$ and $a \in IM$.

(b) Let $a \in I_v \setminus I$. Then $I \subset aA \subseteq I_v$, so $I_v \subseteq aA \subseteq (I_v)_v = I_v$. Then $I_v = aA$. On the other hand, for each $x \in I$, $x = a\alpha$ with α not invertible in A, and then $\alpha \in M$. So $I \subseteq aM$. Suppose that $I \neq aM$ and choose $b \in M$ such that $ab \notin I$. Then $I \subset abA$. So $aA = I_v \subseteq abA$. By simplification, $A \subseteq bA$, and then $bA = A$, so $b \in U(A)$, which is absurd since $b \in M$. Then $I = aM$.

(2) (a) The integral ideal II^{-1} is proper. Its radical $P = \sqrt{II^{-1}}$ is a prime ideal of A. Suppose that $II^{-1} \neq P$ and choose $a \in P \setminus II^{-1}$. There exists $n \in \mathbb{N}^*$ such that $a^n \notin II^{-1}$ but $a^{n+1} \in II^{-1}$. Then $II^{-1} \subset a^n A$, so $\frac{1}{a^n}II^{-1} \subset A$. Then $\frac{1}{a^n}I^{-1} \subseteq (A:I) = I^{-1}$. So $\frac{1}{a^n}II^{-1} \subseteq II^{-1}$, and then $II^{-1} \subseteq a^nII^{-1} \subset a^{2n}A \subseteq a^{n+1}A$. So $a^{n+1} \notin II^{-1}$, which is absurd. Then $II^{-1} = P$ is a prime ideal of A.

(b) First case: I is not divisorial. By (1-b), there exists $0 \neq a \in A$ such that $I = aM$. Then $I_v = aM_v$, so M is not divisorial. Since $M \subset M_v \subseteq A$, then $M_v = A$, so $M^{-1} = A$. Since $I^{-1} = (aM)^{-1} = \frac{1}{a}M^{-1}$, then $P =$

$II^{-1} = aM\frac{1}{a}M^{-1} = MM^{-1} = M$. So $(I : I) = (aM : aM) = a\frac{1}{a}(M : M) = (M : M) = A = A_M = A_P$. For the inclusion $(M : M) \subseteq A$, if $x \in (M : M)$, then $xM \subseteq M$, so $xM_v \subseteq M_v$, then $xA \subseteq A$ so $x \in A$. Second case: I is divisorial. Let $s \in A \setminus P$. Then $P \subseteq sA$, so $\frac{1}{s}II^{-1} \subseteq A$, then $\frac{1}{s}I \subseteq (A : I^{-1}) = (I^{-1})^{-1} = I_v = I$. So $\frac{1}{s} \in (I : I)$, then $A_P \subseteq (I : I)$. Conversely, let $x \in (I : I)$. Then $xI \subseteq I$, so $xII^{-1} \subseteq II^{-1}$, and then $xP \subseteq P$. Suppose that $x \notin A_P$. Since A_P is a valuation domain, $\frac{1}{x} \in A_P$ and more precisely, $\frac{1}{x} \in PA_P = P$ because P is divided. Then $1 = x\frac{1}{x} \in xP \subseteq P$, which is absurd. Then $x \in A_P$ and $A_P = (I : I)$.

(c) Since I is an ideal of the ring $(I : I) = A_P$, then $IA_P = I$.

Exercise 12 The result is clear if $I = (0)$. Suppose that $I \neq (0)$. Choose a nonzero polynomial $f(X) \in I$ of minimal degree $n \geq 1$. Let a be its leading coefficient. Let $g(X) \in I$. The Euclidean division of $g(X)$ by $f(X)$ in $A[\frac{1}{a}][X]$ is of the form $g(X) = f(X)q_1(X) + r_1(X)$ where $q_1(X), r_1(X) \in A[\frac{1}{a}][X]$ with either $r_1(X) = 0$ or $deg\ r_1(X) < n$. Multiplying by a convenient power of a, we find $a^m g(X) = f(X)q(X) + r(X)$ with $q(X)$ and $r(X) \in A[X]$. Then $r(X) = a^m g(X) - f(X)q(X) \in I$. By the choice of $f(X)$ in I, we necessarily have $r(X) = 0$ and $a^m g(X) = f(X)q(X)$.

Exercise 13

(1) Let $0 \neq f(X) \in P$ of minimum degree. Since P is prime and $A[X]$ is UFD, we can suppose that $f(X)$ is irreducible in $A[X]$. Let K be the quotient field of A. Let $g(X) \in P$. The Euclidean division of $g(X)$ by $f(X)$ in $K[X]$ is $g(X) = f(X)q_1(X) + r_1(X)$ with $q_1(X), r_1(X) \in K[X]$ and either $r_1(X) = 0$ or $deg\ r_1(X) < deg\ f(X)$. Let $0 \neq a$ be the lcm of the denominators of the coefficients of $q_1(X)$ and $r_1(X)$. Then $a \in A$, $q_1(X) = \frac{q(X)}{a}$ and $r_1(X) = \frac{r(X)}{a}$ with $q(X), r(X) \in A[X]$ and $deg\ r(X) = deg\ r_1(X) < deg\ f(X)$. So $ag(X) = f(X)q(X) + r(X)$, and then $r(X) = ag(X) - f(X)q(X) \in P$. If $r(X) \neq 0$, then $deg\ r(X) \geq deg\ f(X)$, which is absurd. Then $r(X) = 0$ and $ag(X) = f(X)q(X) \in f(X)A[X]$, which is a prime ideal since $f(X)$ is irreducible in the UFD $A[X]$. Then a or $g(X) \in f(X)A[X]$. Since $f(X)A[X] \cap A \subseteq P \cap A = (0)$, then $a \notin f(X)A[X]$, so $g(X) \in f(X)A[X]$, then $P = f(X)A[X]$.

(2) Since $M \cap A \neq (0)$, then M contains a nonzero noninvertible constant. Then it contains an irreducible element p of A. Then $pA[X] \subset pA[X] + XA[X] \subset A[X]$. Indeed, in the contrary case, there exist $u(X), v(X) \in A[X]$ such that $pu(X) + Xv(X) = 1$, then $pu(0) = 1$, which is absurd. So $pA[X]$ is not a maximal ideal of $A[X]$, and then $pA[X] \neq M$. So $pA[X] \subset M$ and $M/pA[X]$ is a nonzero prime ideal of the quotient ring $A[X]/pA[X] \simeq (A/pA)[X]$. Since A/pA is a field, then $(A/pA)[X]$ is a PID. There exists $f(X) \in M$ such that $M/pA[X] = \bar{f}(X)A[X]/pA[X]$. Then $M = pA[X] + f(X)A[X]$ with $\bar{f}(X)$ irreducible in $A[X]/pA[X]$.

(3) $"(a) \Longrightarrow (b)"$ Suppose that there exists $M \in Max(A[X])$ such that $M \cap A = (0)$. By (1), there exists $f(X) \in A[X]$ a nonconstant irreducible polynomial such that $M = f(X)A[X]$. Put $f(X) = a_0 + a_1 X + \ldots + a_n X^n$, $a_n \neq 0$, $n \geq 1$. Let $p \in A$ be an irreducible element. Then $p \notin M$ since $M \cap A = (0)$, so $M + pA[X] = A[X]$. There exist $u(X)$ and $v(X) \in A[X]$ such that $u(X)f(X) + pv(X) = 1$. We reduce modulo $pA[X]$, we find $\bar{u}(X)\bar{f}(X) = 1$ in $(A/pA)[X]$, and then $\bar{f}(X) = \bar{a}_0 + \bar{a}_1 X + \ldots + \bar{a}_n X^n$ is a constant of A/pA. Then $a_1, \ldots, a_n$ are divisible by p in A. This is for an infinity of nonassociated irreducible elements p. Since A is UFD, then $a_1 = \ldots = a_n = 0$ and $f = a_0 \in A$, which is absurd.

$"(b) \Longrightarrow (c)"$ By (2). $"(c) \Longrightarrow (d)"$ Suppose that there exist $M \in Max(A[X])$ and $g(X) \in A[X]$ such that $M = g(X)A[X]$. By hypothesis, $M = pA[X] + f(X)A[X]$ with $p \in A$ an irreducible element and $f(X) \in A[X]$ an irreducible polynomial modulo $pA[X]$. Since $M \neq A[X]$, then $g(X) \notin U(A[X]) = U(A)$. Put $p = g(X)u(X)$ with $u(X) \in A[X]$ and then $g(X)$ and $u(X) \in A$, and since p is irreducible, then $u(X) \in U(A)$ and $g(X)$ is associated to p in A. Then $pA[X] = g(X)A[X] = M = pA[X] + f(X)A[X]$. So $f(X) \in pA[X]$ and $\bar{f}(X) = \bar{0}$ in $A[X]/pA[X]$, which is absurd.

$"(d) \Longrightarrow (a)"$ Suppose that A admits a finite number of nonassociated irreducible elements $p_1, \ldots, p_n$. Let $f(X) = 1 - p_1 \ldots p_n X$ and $I = f(X)A[X] \neq A[X]$. There exists $M \in Max(A[X])$ such that $I \subseteq M$. But M is not principal, and then by (1) and (2), $M = pA[X] + g(X)A[X]$ with $p \in A$ irreducible and $g(X) \in A[X]$ irreducible modulo $pA[X]$. Put $f(X) = pu(X) + g(X)v(X)$ with $u(X)$ and $v(X) \in A[X]$. In $(A/pA)[X]$, we have $\bar{1} = \bar{f}(X) = \bar{g}(X)\bar{v}(X)$, which is impossible since $g(X)$ is irreducible modulo $pA[X]$.

(4) $" \Longrightarrow "$ By the preceding question. $" \Longleftarrow "$ Put $B = A[X]/pA[X] \simeq (A/pA)[X]$: a PID since A/pA is a field. On the other hand, $A[X]/M \simeq \frac{A[X]/pA[X]}{M/pA[X]} = B/\bar{f}(X)B$ is a field since by hypothesis, $\bar{f}(X)$ is irreducible in the PID B. Then M is maximal.

Finally, we have the chain of prime ideals: $(0) \subset pA[X] \subset M = pA[X] + f(X)A[X]$. Indeed, since $f(X)$ is irreducible modulo $pA[X]$, then $f(X) \notin pA[X]$. So $ht\, M \geq 2$, and since $dim\, A[X] = 2$, then $ht\, M = 2$.

(5) $" \Longleftarrow "$ An irreducible element of the UFD $A[X]$ generates a prime ideal.

$" \Longrightarrow "$ Let P be a prime ideal of $A[X]$ that is not maximal.

First case: $P \cap A \neq (0)$. Put $P \cap A = pA$ and then $0 \neq pA[X] \subseteq P$ with $pA[X]$ prime. Since P is not maximal, then $pA[X] = P$.

Second case: $P \cap A = (0)$. By (1), $P = f(X)A[X]$ with $f(X) \in A[X]$ a nonconstant irreducible polynomial. There exists $M \in Max(A[X])$ such that $P \subset M$. By (1) and (2), M has two types. If $M = g(X)A[X]$ with $g(X) \in A[X]$ a nonconstant irreducible polynomial, then $g(X)$ divides $f(X)$, so $g(X)$ and $f(X)$ are associated and $P = M$ is a maximal ideal, which is absurd. Then $M = pA[X] + f(X)A[X]$ with p an irreducible element of A and $g(X) \in A[X]$ an irreducible polynomial modulo $pA[X]$. In the ring $A[X]/pA[X]$, $\bar{g}(X)$ divides $\bar{f}(X)$, and then $\bar{f}(X)$ is not invertible.

(6) $" \Longleftarrow "$ By (4). $" \Longrightarrow "$ Suppose that A admits a finite number of nonassociated irreducible elements. By (3-d), $A[X]$ admits a maximal principal ideal $M = f(X)A[X]$ with $f(X)$ an irreducible polynomial. Suppose that there exists a prime ideal P of $A[X]$ such that $(0) \subset P \subset M$. By (5), $P = g(X)A[X]$ with $g(X)$ an irreducible polynomial. Then $f(X)$ divides $g(X)$, so these two polynomials are associated. Then $P = M$, which is absurd. We conclude that $ht\ M = 1$, which contradicts the hypothesis.

Exercise 14

(1) Since $B \simeq A/ker\ \phi$, the set of the prime ideals of B is equivalent to the set of the prime ideals of $A/ker\ \phi$, which is equivalent to the set of the prime ideals of A containing $ker\ \phi$.

(2) – The prime numbers p are irreducible in $\mathbb{Z}$, so in $\mathbb{Z}[X]$. Then the $p\mathbb{Z}[X]$ are prime ideals of the UFD $\mathbb{Z}[X]$.

    – A primitive polynomial $f(X) \in \mathbb{Z}[X]$ irreducible in $\mathbb{Q}[X]$ is also irreducible in $\mathbb{Z}[X]$. Then $f(X)\mathbb{Z}[X]$ is a prime ideal of the domain $\mathbb{Z}[X]$.

    – Let $I = p\mathbb{Z}[X] + f(X)\mathbb{Z}[X] = (p, f(X))$ with p a prime number and $f(X) \in \mathbb{Z}[X]$ a polynomial that is irreducible modulo p. Let $s : \mathbb{Z}[X] \longrightarrow (\mathbb{Z}/p\mathbb{Z})[X]$ be the natural homomorphism. It is onto with kernel $ker\ s = p\mathbb{Z}[X] \subseteq I$ and $s(I) = s(f)(\mathbb{Z}/p\mathbb{Z})[X]$. By hypothesis, $s(f)$ is irreducible in $(\mathbb{Z}/p\mathbb{Z})[X]$. So $s(I)$ is a prime ideal of $(\mathbb{Z}/p\mathbb{Z})[X]$. By (1), I is a prime ideal of $\mathbb{Z}[X]$.

(3)

  (a) $P \cap \mathbb{Z}$ is a prime ideal of $\mathbb{Z}$.

  (b) By (1) of the preceding exercise, $P = f(X)A[X]$ with $f(X) \in \mathbb{Z}[X]$ a nonconstant irreducible polynomial. It is necessarily primitive in $\mathbb{Z}[X]$ and then irreducible in $\mathbb{Q}[X]$.

  (c) We have $P \cap \mathbb{Z} = p\mathbb{Z}$ and $P \neq p\mathbb{Z}[X]$. Let $s : \mathbb{Z}[X] \longrightarrow (\mathbb{Z}/p\mathbb{Z})[X]$ be the natural homomorphism. It is onto with kernel $ker\ s = p\mathbb{Z}[X] \subset P$. By (1), $s(P)$ is a nonzero prime ideal of the PID $(\mathbb{Z}/p\mathbb{Z})[X]$. There exists $f \in \mathbb{Z}[X]$ that is irreducible modulo p such that $s(P) = s(f(X))(\mathbb{Z}/p\mathbb{Z})[X]$. By (2), $p\mathbb{Z}[X] + f(X)\mathbb{Z}[X]$ is a prime ideal of $\mathbb{Z}[X]$ containing $ker\ s$. Since P is also a prime ideal of $\mathbb{Z}[X]$ containing $ker\ s$ and the two ideals have the same image by s, then by (1), we have $P = p\mathbb{Z}[X] + f(X)\mathbb{Z}[X]$.

  (d) Clear.

(4) $"(a) \Longrightarrow (b)"$ Let $a \in \mathbb{Z}/p\mathbb{Z}$ be such that $a^2 = -1$. If $p \neq 2$, then $-1 \neq 1$ in $\mathbb{Z}/p\mathbb{Z}$ and $a^4 = 1$, so the order of a is 4.

$"(b) \Longrightarrow (c)"$ Let $a \in (\mathbb{Z}/p\mathbb{Z})^*$ be an element of order 4. Then 4 divides $p - 1$, so $p \equiv 1 (mod\ 4)$.

$"(c) \Longrightarrow (a)"$ Let u be a generator of the cyclic group $(\mathbb{Z}/p\mathbb{Z})^*$. Then $u^{4n} = 1$, so $(u^{2n} - 1)(u^{2n} + 1) = 0$. But $u^{2n} - 1 \neq 0$ since the order of u is $4n$. Then $u^{2n} + 1 = 0$, so u^n is a root of the polynomial $X^2 + 1$.

(5) – It is clear that no member of F_1 contains $1 + X^2$.

  – Let $P = f(X)\mathbb{Z}[X]$ be a member of F_2 containing $1 + X^2$. The polynomial $f(X) \in \mathbb{Z}[X]$ is primitive and irreducible in $\mathbb{Q}[X]$. Then $f(X)$ is irreducible in $\mathbb{Z}[X]$ and $f(X)$ divides $1 + X^2$, with $1 + X^2$ irreducible in $\mathbb{Z}[X]$. Then $f(X)$ and $1 + X^2$ are associated with $\mathbb{Z}[X]$, so $P = f(X)\mathbb{Z}[X] = (1 + X^2)\mathbb{Z}[X]$.

  – Let $P = p\mathbb{Z}[X] + f(X)\mathbb{Z}[X]$ be a member of F_3 containing $1 + X^2$. Then p is a prime number and $f(X) \in \mathbb{Z}[X]$ is irreducible modulo p. Put $1 + X^2 = pg(X) + f(X)h(X)$ with $g(X)$ and $h(X) \in \mathbb{Z}[X]$. By reducing modulo p, we obtain $1 + X^2 = \bar{f}(X)\bar{h}(X)$ with $\bar{f}(X)$ irreducible in $(\mathbb{Z}/p\mathbb{Z})[X]$. We distinguish two cases according to that $1 + X^2$ is or not irreducible in $(\mathbb{Z}/p\mathbb{Z})[X]$. We use the preceding question.
  First case: $p \neq 2$ and $p \not\equiv 1(mod\ 4)$. Then $1 + X^2$ is irreducible in $(\mathbb{Z}/p\mathbb{Z})[X]$. There exists a, $0 < a \leq p - 1$ such that $\bar{f}(X) = \bar{a}(1 + X^2)$. Then $f(X) = a(1 + X^2) + pl(X)$ with $l(X) \in \mathbb{Z}[X]$ and $P = p\mathbb{Z}[X] + a(1 + X^2)\mathbb{Z}[X]$.
  Second case: $p = 2$ or $p \equiv 1(mod\ 4)$. Then $1 + X^2$ is not irreducible in $(\mathbb{Z}/p\mathbb{Z})[X]$. There exist $1 \leq a, b \leq p - 1$ such that $1 + \bar{b}^2 = 0$ in $\mathbb{Z}/p\mathbb{Z}$ and $\bar{f}(X) = \bar{a}(X - \bar{b})$. Then $f(X) = a(X - b) + pl(X)$ with $l(X) \in \mathbb{Z}[X]$. So $P = p\mathbb{Z}[X] + a(X - b)\mathbb{Z}[X]$.

(6) The map $\phi : \mathbb{Z}[X] \longrightarrow \mathbb{Z}[i]$, defined by $\phi(f(X)) = f(i)$, is a homomorphism onto with $X^2 + 1 \in ker\ \phi$. Let $f(X) \in \mathbb{Z}[X]$. We have $f(X) = q(X)(X^2 + 1) + r(X)$ with $q(X), r(X) \in \mathbb{Z}[X]$ since $X^2 + 1$ is monic, $r(X) = aX + b$. Then $f(X) \in ker\ \phi \iff f(i) = 0 \iff r(i) = ai + b = 0 \iff a = b = 0 \iff r(X) = 0 \iff f(X) = q(X)(X^2 + 1)$. So $ker\ \phi = (X^2 + 1)\mathbb{Z}[X]$ and $\mathbb{Z}[i] \simeq \mathbb{Z}[X]/(X^2 + 1)$.

(7) The map $\phi : \mathbb{Z}[X] \longrightarrow \mathbb{Z}[i]$, defined by $\phi(f(X)) = f(i)$, is a homomorphism onto with kernel $ker\ \phi = (1 + X^2)\mathbb{Z}[X]$. The prime ideals of $\mathbb{Z}[i]$ correspond to the prime ideals of $\mathbb{Z}[X]$ containing $1 + X^2$. Then they are (0); $\{p\mathbb{Z}[i]\}$, p a prime > 2 and $p \not\equiv 1(mod\ 4)$ and $\{p\mathbb{Z}[i] + a(i - b)\mathbb{Z}[i]\}$, $p = 2$ or p a prime, $p \equiv 1(mod\ 4)$ and $1 \leq a, b \leq p - 1$, $\bar{b}$ a root of $1 + X^2$ in $\mathbb{Z}/p\mathbb{Z}$. In the last case, let $z = a(i - b) = ai - ab$ is not divisible by p in $\mathbb{Z}[i]$ since p does not divide a in $\mathbb{Z}$. But there exists $c \in \mathbb{Z}$ such that $1 + b^2 = pc$ and then $|z|^2 = a^2 + (ab)^2 = a^2(1 + b^2) = a^2 cp$, so p divides $|z|^2$.

(8) $U(\mathbb{Z}[i]) = \{\pm 1; \pm i\}$.

(9) Suppose that z is irreducible. If $\bar{z} = z_1 z_2$, then $z = \bar{z}_1 \bar{z}_2$ so $\bar{z}_1$ or $\bar{z}_2 \in U(\mathbb{Z}[i])$. So z_1 or $z_2 \in U(\mathbb{Z}[i])$.

(10) Let $\theta : \mathcal{P} \longrightarrow Spec(\mathbb{Z}[i]) \setminus (0)$ be the map defined by $\theta(t) = t\mathbb{Z}[i]$. If $t\mathbb{Z}[i] = t'\mathbb{Z}[i]$, and then t and t' are associated, so equal. Since $\mathbb{Z}[i]$ is a PID, a nonzero prime ideal of $\mathbb{Z}[i]$ is of the form $t\mathbb{Z}[i]$ with t an irreducible element and we may suppose that $t \in \mathcal{P}$. We conclude that θ is a bijection.

(11) (a) Examples: $2 = 1^2 + 1^2 = (1 - i)(1 + i); 5 = 1^2 + 2^2 = (1 - 2i)(1 + 2i);$ $17 = 1^2 + 4^2 = (1 - 4i)(1 + 4i)$.

(b) Let t be an irreducible element of $\mathbb{Z}[i]$ that divides p, such that a t exists since $\mathbb{Z}[i]$ is UFD. Put $p = ty$, y is not invertible since p is not irreducible. Since $p^2 = |t|^2|y|^2$, then $|t|^2, |y|^2 \in \{1, p, p^2\}$. But t and y are not invertible, and then $|t|^2 \neq 1$ and $|y|^2 \neq 1$. So $|t|^2 = p$.

(12) (a) Since p does not divide z, then $z \notin p\mathbb{Z}[i]$. So $(0) \subset p\mathbb{Z}[i] \subset p\mathbb{Z}[i] + z\mathbb{Z}[i] = P$. But in a PID, each nonzero prime ideal is maximal, and then $p\mathbb{Z}[i]$ is not prime so p is not irreducible in $\mathbb{Z}[i]$. By (11-b), there exists an irreducible element t of $\mathbb{Z}[i]$ such that $p = |t|^2$.

(b) $t\bar{t} = |t|^2 = p$ divides $|z|^2 = z\bar{z}$. Since $\mathbb{Z}[i]$ is UFD, one of the irreducible elements t or $\bar{t}$ divides z. Then $z \in \alpha\mathbb{Z}[i]$ with $\alpha \in \{t, \bar{t}\}$, and $\alpha\mathbb{Z}[i]$ is a prime ideal. Then $P = p\mathbb{Z}[i] + z\mathbb{Z}[i] \subseteq \alpha\mathbb{Z}[i]$ so $P = \alpha\mathbb{Z}[i]$ since in a PID, each nonzero prime ideal is maximal.

(13) By (10), an irreducible element of $\mathbb{Z}[i]$ belongs to one the following two families obtained using the prime ideals of $\mathbb{Z}[i]$ and the unities of $\mathbb{Z}[i]$:

$p, -p, -ip, ip$, where p is a prime number, $p \not\equiv 1 (mod\ 4)$ and $p > 2$.

$\alpha \in \mathbb{Z}[i]$, with $|\alpha|^2$ prime, $|\alpha|^2 \equiv 1 (mod\ 4)$ or $|\alpha|^2 = 2$.

Exercise 15

(1) Let $I \in \mathscr{F}(A)$ be a strong ideal. Then $I = II^{-1} \subseteq A$. Then I is integral. Let $M \in Max(A)$. Then $M = M.1 \subseteq M(A : M) = MM^{-1} \subseteq A$. Either $M = MM^{-1}$, and in this case, M is strong. Or $MM^{-1} = A$, and in this case, M is invertible.

(2) $" \Longrightarrow "$ Since I is strong, then by (1), it is integral. So $(I : I) \subseteq I^{-1}$. But $II^{-1} = I$, and then $I^{-1} \subseteq (I : I)$. So $I^{-1} = (I : I)$.

$" \Longleftarrow "$ We have $II^{-1} = I(I : I) \subseteq I$. Conversely, $I = I.1 \subseteq I(I : I) = II^{-1}$. Then $II^{-1} = I$.

If I is strong, then $I^{-1} = (I : I)$ is an over ring of A, by notations after Corollary 5.8.

(3) We have $II^{-1} = IA = I$. Then I is strong. Since $A^{-1} = A$, then $A \in D_f(A)$.

(4) $" \Longrightarrow "$ By (1), if I is strong, then it is integral and $I = II^{-1}$.

$" \Longleftarrow "$ Let J be a nonzero integral ideal of A and $I = JJ^{-1}$. By Bass example 5 of Corollary 5.8, we have $(A : JJ^{-1}) = (JJ^{-1} : JJ^{-1})$. Then $I^{-1} = (I : I)$. By (2), I is strong.

Let $J \in \mathscr{F}(A)$. There is $0 \neq d \in A$ such that $dJ \subseteq A$. Then $JJ^{-1} = (dJ)(dJ)^{-1}$ is strong.

(5) $" \Longrightarrow "$ Let I be a strongly divisorial ideal of A. By (2), $B = I^{-1} = (I : I)$ is an over ring of A. The conductor $(A : B) = A : (A : I) = I_v = I$ because I is divisorial.

$" \Longleftarrow "$ Let B be an over ring of A and $I = (A : B)$, the conductor. By Lemma 2.1, I is an ideal of A and B. Then $(I : I) = (A : B) : I = (A : IB) = (A : I) = I^{-1}$. By (2), I is strong in A. Let $0 \neq a \in I$. Then $aB \subseteq A$, so $B \in \mathscr{F}(A)$. We have $I = (A : B) = B^{-1}$, a divisorial ideal of A, by Example 5 at the beginning of Sect. 11, Chap. 1.

Suppose now that A is a valuation domain. The over rings of A are the localizations A_P, $P \in Spec(A)$. The strongly divisorial ideals of A are the nonzero conductors $(A : A_P)$, $P \in Spec(A)$. If P is the maximal ideal of A, then $(A : A_P) = (A : A) = A$. If P is nonmaximal, then $(A : A_P) = P$. Indeed, since $PA_P = P \subseteq A$, then $P \subseteq (A : A_P)$. Conversely, let $x \in (A : A_P)$, and then $xA_P \subseteq A$, so $x \in A$. Suppose that $x \notin P$. Then $\frac{1}{x^2} \in A_P$ and $\frac{1}{x} = x\frac{1}{x^2} \in xA_P \subseteq A$. So $x \in U(A)$ and $A_P \subseteq \frac{1}{x}A = A$. Then $A_P = A$ and P is the maximal ideal of A, which is absurd. So $x \in P$ and $(A : A_P) = P$.

(6) Let $x \in A^c$. There exists $0 \neq d \in A$ such that $dx^i \in A$ for each $i \in \mathbb{N}$. Let I be the A−sub-module of $K = qf(A)$ generated by $\{x^i; i \in \mathbb{N}\}$. Then $dI \subseteq A$, so $I \in \mathscr{F}(A)$. Since $xI \subseteq I$, then $x \in (I : I)$. Conversely, let $I \in \mathscr{F}(A)$ and $x \in (I : I)$. Then $xI \subseteq I$. So $x^2I \subseteq xI \subseteq I, \ldots$. By induction, $x^iI \subseteq I$ for each $i \in \mathbb{N}$. Then $\{x^i; i \in \mathbb{N}\} \subseteq (I : I)$. Since $(I : I) \in \mathscr{F}(A)$, there exists $0 \neq d \in A$ such that $d(I : I) \subseteq A$. Then $dx^i \in A$ for each $i \in \mathbb{N}$. So $x \in A^c$. We have shown the equality $A^c = \bigcup_{I \in \mathscr{F}(A)} (I : I)$.

Let $I \in D_f(A)$. By (2), $I^{-1} = (I : I)$. Conversely, let $F \in \mathscr{F}(A)$. By notations after Corollary 5.8, $(F : F)$ is an over ring of A. By (5), the conductor $I = A : (F : F) \in D_f(A)$. We have $(F : F) \subseteq (F : F)_v = A : (A : (F : F)) = (A : I) = I^{-1}$. Then we have shown the equality
$$\bigcup_{I \in D_f(A)} I^{-1} = \bigcup_{I \in \mathscr{F}(A)} (I : I).$$

Now, if A is the only strongly divisorial ideal of A, then $A^c = A^{-1} = A$. Conversely, suppose that $A^c = \bigcup_{I \in D_f(A)} I^{-1} = A$. For each $I \in D_f(A)$, the over ring $I^{-1} = A$. Since I is divisorial, $I = I_v = (I^{-1})^{-1} = A^{-1} = A$.

(7) By (4), I is strong. By Bass example 5 of Corollary 5.8, we have $(A : JJ^{-1}) = (JJ^{-1} : JJ^{-1})$. Then $I^{-1} = (I : I) = A$, by (6) since A is completely integrally closed. So $I_v = (I^{-1})^{-1} = A^{-1} = A$. Then I is not divisorial.

(8) By (6), $\bigcup_{\alpha \in \Lambda} I_\alpha^{-1} \subseteq A^c$. Then $(0) \neq (A : A^c) \subseteq (A : \bigcup_{\alpha \in \Lambda} I_\alpha^{-1}) = \bigcap_{\alpha \in \Lambda} (A : I_\alpha^{-1}) = \bigcap_{\alpha \in \Lambda} (I_\alpha)_v = \bigcap_{\alpha \in \Lambda} I_\alpha$. So $\bigcap_{\alpha \in \Lambda} I_\alpha \neq (0)$.

(9) " $\Longrightarrow$ " By definition, since I is strong, $I = II^{-1}$. Since I is divisorial, $I = I_v = (II^{-1})_v$. " $\Longleftarrow$ " By hypothesis, $I = (JJ^{-1})_v$ is divisorial. By Bass example 5 of Corollary 5.8, we have $(A : JJ^{-1}) = (JJ^{-1} : JJ^{-1})$. Then $I^{-1} = (I : I)$. By (2), I is strong. Then I is strongly divisorial. Let $J \in \mathscr{F}(A)$, there exists $0 \neq d \in A$ such that $dJ \subseteq A$. Then $(JJ^{-1})_v = (dJ(dJ)^{-1})_v$ is strongly divisorial.

(10) The map α takes its values in $D_f(A)$ because the conductor of A in an over ring is a strongly divisorial ideal by (5). It is clear that α is increasing.

Let $I \in D_f(A)$; i.e, I is a strongly divisorial ideal of A. Then $I = I_v = A :$
$(A : I) = (A : I^{-1})$. Since I is strong, then $I^{-1} = (I : I)$ is an over ring of
A by (2). So $I = \alpha(I^{-1})$ and α is surjective.

Let $B_1 = I_1^{-1}$ and $B_2 = I_2^{-1}$ be two over rings of A, with $I_1, I_2 \in \mathscr{F}(A)$,
such that $\alpha_1(B_1) = \alpha_2(B_2)$. Then $(A : B_1) = (A : B_2)$, so $A : (A : I_1^{-1}) =$
$A : (A : I_2^{-1})$, i.e., $(I_1^{-1})_v = (I_2^{-1})_v$. By Example 5 at the beginning of
Sect. 11, Chap. 1, the ideals of the form I^{-1}, with $I \in \mathscr{F}(A)$, are divisorial.
So $B_1 = B_2$ and α is injective.

Exercise 16 By Lemma 2.1, the conductor $(A : B)$ is an ideal of A and B. It
suffices to show that $(A : B)$ is a radical ideal of B. Let $t \in B$ such that $t^2 \in (A : B)$.
We must show that $t \in (A : B)$. We have $t^2 B \subseteq A$ and $t^3 B = t^2(tB) \subseteq t^2 B \subseteq A$.
For each $x \in B$, $(tx)^2 = t^2 x^2 \in A$ and $(tx)^3 = t^3 x^3 \in A$. Since A is semi-normal,
for each $x \in B$, $tx \in A$. Then $t \in (A : B)$.

Exercise 17

(1) (a) Since $I \subseteq A$, then $I_w \subseteq A$. So $I_w = A \Longleftrightarrow A \subseteq I_w \Longleftrightarrow 1 \in I_w \Longleftrightarrow$
$\exists J \in GV(A)$ such that $1 \in (I : J) \Longleftrightarrow \exists J \in GV(A)$ such that $J \subseteq I$.

    (b) Clear (c) Suppose that $P_w \neq P$. Let $x \in P_w \setminus P$. Then $x \in A$, and there
exists $J \in GV(A)$ such $x \in (P : J)$. Since $xJ \subseteq P$, then $J \subseteq P$. By (a),
$P_w = A$.

(2) We have $I_t = A \Longleftrightarrow 1 \in I_t \Longleftrightarrow$ There exists a nonzero finitely generated
subideal J of I such that $1 \in J_v = (J^{-1})^{-1} = (A : J^{-1}) \Longleftrightarrow J^{-1} \subseteq A \Longleftrightarrow$
$J^{-1} = A$.

(3) "$\subseteq$" Let $Q \in w - Max(A)$. Then $Q \subseteq Q_t \subseteq A$. By Exercise 27 (8-c), Chap. 1,
each t-ideal is a w-ideal. Then Q_t is a w-ideal. By the w-maximality of Q,
either $Q_t = Q$ or $Q_t = A$. Suppose that $Q_t = A$. By (2), there exists $J \subseteq Q$ a
nonzero finitely generated ideal such that $J^{-1} = A$. Then $J \in GV(A)$. By (1-
a), $Q_w = A$. So $Q = Q_w = A$, which is absurd. Then $Q_t = Q$. This means that
Q is a proper integral t-ideal of A. By Proposition 11.2, Chap. 1, there exists
$M \in t - Max(A)$ such that $Q \subseteq M$. Since each t-ideal is a w-ideal, then M
is a w-ideal. By the w-maximality of Q, we have $Q = M \in t - Max(A)$.
"$\supseteq$" Let $Q \in t - Max(A)$. By Exercise 27 (8-c), Chap. 1, Q is a w-ideal. By
Exercise 27 (4), Chap. 1, there exists $M \in w - Max(A)$ such that $Q \subseteq M$. Then
$Q \subseteq M = M_w \subseteq M_t \subseteq A$. By the t-maximality of Q, we have either $M_t = A$
or $M_t = Q$. Suppose that $M_t = A$. By (2), there exists $(0) \neq J \subseteq M$ a finitely
generated ideal such that $J^{-1} = A$. Then $J \in GV(A)$. By (1), $M_w = A$. So
$M = M_w = A$, which is absurd. Then $M_t = Q$. So $Q = M \in w - Max(A)$.

(4) The inclusion $IA_P \subseteq I_w A_P$ is clear. Conversely, let $x \in I_w A_P$, $x = \frac{a}{s}$ with
$a \in I_w$ and $s \in A \setminus P$. Then $sx = a \in I_w$. By definition, there exists $J \in$
$GV(A)$ such that $sx \in (I : J)$. So $sxJ \subseteq I$. By (1-a), since $P_w = P \subset A$,
$J \nsubseteq P$. Let $t \in J \setminus P$. Then $sxt \in I$ and $x \in \frac{1}{st} I \subseteq IA_P$. So $I_w A_P \subseteq IA_P$.

(5) Let $I \in \mathscr{F}(A)$. By Exercise 27 (6), Chap. 1, we have $I_w = \cap\{I_w A_P;\ P \in w - Max(A)\}$

$= \cap\{I A_P;\ P \in w - Max(A)\}$, by (4).

$= \cap\{I A_P;\ P \in t - Max(A)\}$, by (3).

(6) (a) Let $x \in (IM)_w$. There exists $J \in GV(A)$ such that $x \in (IM : J)$. Since $xJ \subseteq IM$ and I is invertible, then $xI^{-1}J \subseteq M$. So $xI^{-1} \subseteq (M : J) \subseteq M_w = M$. Then $x \in IM$ and $(IM)_w = IM$.

(b) By (a), since M_w is a w-ideal, then $I M_w$ is a w-ideal. By Exercise 27 (7), Chap. 1, we have $I M_w = (I M_w)_w = (IM)_w$.

(7) By Exercise 27 (7), Chap. 1 and (1-b), we have $(JM)_w = (J_w M)_w = (AM)_w = M_w$.

Exercise 18

(1) By Exercise 27 (8-b), Chap. 1, each v-ideal of A is a w-ideal. Then a SM-ring satisfies the acc on divisorial ideals. So it is a Mori domain. By Example 1 of Definition 4.5, in a Mori domain, the star operations v and t are identical. By the preceding exercise, $w - Max(A) = t - Max(A) = v - Max(A)$.

(2) Since A is w-finite, $P \neq A$. Suppose that P is not prime. There exist $r, m \in A \setminus P$ such that $rm \in P$. Then $P \subset P + rA \subseteq (P + rA)_w \subseteq A$. Then $(P + rA)_w$ is w-finite. Exercise 27 (9), Chap. 1, $(P + rA)_w = (x_1 + ra_1, \ldots, x_n + ra_n)_w$ with $x_1, \ldots, x_n \in P$ and $a_1, \ldots, a_n \in A$. Let $B = (P :_A r)$, and then $P \subset B$ because of m. So $P \subset B \subseteq B_w \subseteq A$. Then B_w is w-finite. Then $B_w = (c_1, \ldots, c_k)_w$ with $c_1, \ldots, c_k \in B$. We have $(x_1, \ldots, x_n, rc_1, \ldots, rc_k) \subseteq P$. Then $(x_1, \ldots, x_n, rc_1, \ldots, rc_k)_w \subseteq P_w = P$. Conversely, let $x \in P \subset (P + rA)_w = (x_1 + ra_1, \ldots, x_n + ra_n)_w$. There exist $J = (d_1, \ldots, d_l) \in GV(A)$ such that $xJ \subseteq (x_1 + ra_1, \ldots, x_n + ra_n)$.

For $1 \leq t \leq l$, $xd_t = \sum_{i:1}^{n} s_{it}(x_i + ra_i) = \sum_{i:1}^{n} s_{it}x_i + r(\sum_{i:1}^{n} s_{it}a_i)$ with $s_{it} \in A$. The element $y_t = \sum_{i:1}^{n} s_{it}a_i \in A$ and $ry_t = xd_t - \sum_{i:1}^{n} s_{it}x_i \in P$.

So $y_t \in (P : r) = B \subseteq B_w = (c_1, \ldots, c_k)_w$. There exists $J_t \in GV(A)$ such that $y_t J_t \subseteq (c_1, \ldots, c_k)$, $1 \leq t \leq l$. Then $ry_t J_1 \ldots J_l \subseteq (rc_1, \ldots, rc_k)$, $1 \leq t \leq l$. On the other hand, $xd_t = ry_t + \sum_{i:1}^{n} s_{it}x_i$, $1 \leq t \leq l$. Then

$J_1 \ldots J_l Jx \subseteq J_1 \ldots J_l(xd_1, \ldots, xd_l) \subseteq J_1 \ldots J_l(ry_1, \ldots, ry_l, x_1, \ldots, x_n) \subseteq (rc_1, \ldots, rc_k, x_1, \ldots, x_n)$. So $x \in \big((rc_1, \ldots, rc_k, x_1, \ldots, x_n) : J_1 \ldots J_l J\big)$ with $J_1 \ldots J_l J \in GV(A)$, by Exercise 27 (1), Chap. 1. We conclude that $P = (rc_1, \ldots, rc_k, x_1, \ldots, x_n)_w$, which is absurd since P is not w-finite. Then P is a prime ideal.

(3) "(a) $\Longrightarrow$ (b)" Let I be an integral w-ideal of A. Let $0 \neq x_1 \in I$. If $(x_1)_w = (x_1) \subset I$, there exists $x_2 \in I \setminus (x_1)$. We have $(x_1)_w \subset (x_1, x_2)_w \subseteq I$. If $(x_1, x_2)_w \subset I$, there exists $x_3 \in I \setminus (x_1, x_2)_w$. We have $(x_1)_w \subset (x_1, x_2)_w \subset$

$(x_1, x_2, x_3)_w \subseteq I, \ldots$. Since A is a SM-ring, the process must stop. So there exists $n \in \mathbb{N}^*$ such that $I = (x_1, \ldots, x_n)_w$.

$''(b) \implies (c)''$ Clear. $''(c) \implies (a)''$ Let $\mathscr{F}$ be the set of the integral w−ideals of A that are not w−finite. Suppose that $\mathscr{F} \neq \emptyset$. Then $(\mathscr{F}, \subseteq)$ is inductive. Indeed, let $(I_\alpha)_{\alpha \in \Lambda}$ be a totally ordered family of elements of $\mathscr{F}$ and $I = \bigcup_{\alpha \in \Lambda} I_\alpha$. Then I is a w−ideal. To see this, let $x \in I_w$. Since w has a finite character, there exists an ideal $J = (a_1, \ldots, a_n) \subseteq I$ such that $x \in J_w$. There exists $\alpha_0 \in \Lambda$ such that $a_1, \ldots, a_n \in I_{\alpha_0}$. So $x \in J_w \subseteq (I_{\alpha_0})_w = I_{\alpha_0} \subseteq I$. Then $I_w = I$. Moreover, I is not w−finite. Indeed, if $I = (a_1, \ldots, a_n)_w$ with $a_1, \ldots, a_n \in I$, there exists $\alpha_0 \in \Lambda$ such that $a_1, \ldots, a_n \in I_{\alpha_0}$. So $I_{\alpha_0} = (a_1, \ldots, a_n)_w$, which is absurd. We conclude that I bounds the family in $\mathscr{F}$. By Zorn's lemma $(\mathscr{F}, \subseteq)$ admits a maximal element P. By (2), P is a prime ideal of A, which contradicts (c).

(4) A prime ideal of A_P is of the form QA_P with Q a prime ideal of A contained in P. By (3), $Q_w = (q_1, \ldots, q_n)_w$ with $q_1, \ldots, q_n \in Q$. By the preceding exercise, $QA_P = Q_wA_P = (q_1, \ldots, q_n)_wA_P = (q_1, \ldots, q_n)A_P$, a finitely generated ideal of A_P. We conclude by Cohen's theorem.

(5) (a) Since J is finitely generated, so is J_S. By Lemma 10.14 (1), $(J_S)^{-1} = (J^{-1})_S = A_S$. Then $J_S \in GV(A_S)$.

   (b) Let I be a fractional w−ideal of A_S and $x \in (I \cap A)_w \subseteq A$. There exists $J \in GV(A)$ such that $x \in (I \cap A : J)$. Since $xJ \subseteq I \cap A$, then $xJ_S \subseteq (I \cap A)_S \subseteq I$. So $x \in (I : J_S)$ with $J_S \in GV(A_S)$, by (a). Then $x \in I_w = I$. So $x \in A \cap I$. Then $(A \cap I)_w = A \cap I$.

   (c) Note first that if I is an integral ideal of A_S, then $(I \cap A)_S = I$. Indeed, the direct inclusion is clear since the elements of S are invertible in A_S. Conversely, let $x \in I$, $x = \frac{a}{s}$ with $a \in A$ and $s \in S$. Then $a = sx \in I \cap A$, so $x = \frac{a}{s} \in (I \cap A)_S$.

Now, let $(I_n)_n$ be an increasing sequence of integral w−ideals of A_S. By (b), $(I_n \cap A)_n$ in an increasing sequence of integral w−ideals of A. Since A is a SM-ring, the sequence $(I_n \cap A)_n$ is stationary. Then the sequence $(I_n)_n = ((I_n \cap A)_S)_n$ is stationary.

6) $'' \implies ''$ (a) Follows from (4). (b) By Exercise 17 (3), $w - Max(A) = t - Max(A)$. By Example 1 of Definition 4.5, a Mori domain has a finite character.

   $'' \impliedby ''$ By (3), it suffices to show that each nonzero w−ideal I of A is w−finite. Let $0 \neq x \in I$. By (b), there are a finite number $P_1, \ldots, P_n \in w - Max(A)$ containing x. By (a), the domains $A_{P_1}, \ldots, A_{P_n}$ are Noetherian. There are finite subsets $F_1, \ldots, F_n$ of I such that $IA_{P_i} = (F_i)A_{P_i}$, $1 \leq i \leq n$. Consider the finite set $F = \{x\} \cup F_1 \cup \ldots \cup F_n \subseteq I$. Then $IA_{P_i} = (F)A_{P_i}$, $1 \leq i \leq n$. For each $P \in w - Max(A) \setminus \{P_1, \ldots, P_n\}$, $x \notin P$, then x is invertible in A_P. So $IA_P = A_P = (F)A_P$. By Exercise 17 (4), for each $P \in w - Max(A)$, $IA_P = (F)A_P = (F)_wA_P$. By Exercise 17 (5), $I = (F)_w$.

Exercise 19

(1) There exists an ideal $L \subseteq I \cap J$ of A such that $I = (L :_A J)$ et $J = (L :_A I)$.
Then $IJ = J(L :_A J) \subseteq L$. So $I \subseteq (IJ :_A J) \subseteq (L :_A J) = I$, and then
$I = (IJ :_A J)$. By symmetry, $J = (IJ :_A I)$. Then I and J are linked over
IJ.
Application:
$IJ = (X^5, X^2Y^3, X^2Z^3, Y^2X^3, Y^5, Y^2Z^3)$ and $X^2Y^2Z^2I = (X^4Y^2Z^2, X^2Y^4Z^2) \subseteq IJ$.
Then $X^2Y^2Z^2 \in (IJ :_A I) \setminus J$. So $(IJ :_A I) \neq J$,and then I and J are not
linked.

(2) $(IJ :_A J) = (IJ :_A J)A = (IJ :_A J)(I + J) = (IJ :_A J)I + (IJ :_A J)J \subseteq I + IJ = I$. Then $(IJ :_A J) \subseteq I$. The inverse inclusion is clear. So
$I = (IJ :_A J)$. By symmetry, $J = (IJ :_A I)$. Then I and J are linked over
IJ.

(3) Let I and J be the two primary ideals such that the prime ideals $P = \sqrt{I}$ and
$Q = \sqrt{J}$ are incomparable. Then $J(IJ :_A J) \subseteq IJ \subseteq I$ with $J \not\subseteq P$. Since I
is primary, $(IJ :_A J) \subseteq I$. The inverse inclusion is clear. Then $I = (IJ :_A J)$.
By symmetry, $J = (IJ :_A I)$. Then I and J are linked over IJ.

(4) The radical of a prime ideal is the equal to the ideal itself. We conclude by (3).

(5) Prime case: $P = M$. We must show the equivalence $(M^2 :_A M) = M \Longleftrightarrow M^2 \neq M$.
Since $(M^2 :_A M)$ is an ideal of A containing the maximal ideal M, then it is
equal to M or A. But $(M^2 :_A M) = A \Longleftrightarrow 1 \in (M^2 :_A M) \Longleftrightarrow M \subseteq M^2 \Longleftrightarrow M^2 = M$. Then $(M^2 :_A M) = M \Longleftrightarrow M^2 \neq M$.
Second case: $P \subset M$. Let $x \in M \setminus P$. Then $x(PM :_A M) \subseteq M(PM :_A M) \subseteq PM \subseteq P$. Since P is a prime ideal and $x \notin P$, then $(PM :_A M) \subseteq P$. The
other inclusion is clear. Then $P = (PM :_A M)$.
Since $(PM :_A P)$ is an ideal of A containing the maximal ideal M, then it is
equal to M or A. But $(PM :_A P) = A \Longleftrightarrow 1 \in (PM :_A P) \Longleftrightarrow P \subseteq PM \Longleftrightarrow PM = P$. Then $(PM :_A P) = M \Longleftrightarrow PM \neq P$. So P and M are
linked if and only if $PM \neq P$.
Application: We have the ideals $P = (X, Y) \subset M = (X, Y, Z)$ in the ring
$k[X, Y, Z]$. Since $k[X, Y, Z]/(X, Y) \simeq k[Z]$ is an integral domain, the ideal P
is prime, and since $k[X, Y, Z]/(X, Y, Z) \simeq k$ is a field, the ideal M is maximal.
On the other hand, $PM = (X^2, XY, XZ, Y^2, YZ)$. Since $X, Y \in P \setminus PM$, then
$PM \neq P$. So P and M are linked.

References

D.F. Anderson, D.N. El Abidine, Factorization in integral domain III. J. Pure Appl. Algebra **135**,
    107–127 (1999)
D.D. Anderson, M. Zafrullah, Pseudo-almost integral elements. Commun. Algebra **35**, 1127–1131
    (2007)

D.D. Anderson, D.J. Kwak, M. Zafrullah, Agreeable domains. Commun. Algebra **23**(13), 4861–4883 (1995)

G. Angermuller, On the root and integral closure of Noetherian domains of dimension 1. J. Algebra **83**, 437–441 (1983)

J.T. Arnold, J.W. Brewer, On flat overrings, ideal transforms and generalized transforms of a commutative ring. J. Algebra **18**, 254–263 (1971)

V. Barucci, Strongly divisorial ideals and complete integral closure of an integral domain. J. Algebra **99**, 132–142 (1986)

H. Bass, On the ubiquity of Gorenstein rings. Math. Zeitschr. **82**, 8–28 (1963)

J.W. Brewer, D.S. Costa, K. McCrimmon, Seminormality and root closure in polynomial rings and algebraic curves. J. Algebra **58**, 217–226 (1979)

G.W. Chang, Uppers to zero in polynomial rings which are maximal ideals. Bull. Korean Math. Soc. **52**(2), 525–530 (2015)

D.E. Dobbs, E.G. Houston, T.G. Lucas, M. Roitman, M. Zafrullah, On t-linked overrings. Commun. Algebra **20**(5), 1463–1488 (1992)

W. Fanggui, R.L. McCasland, On w−modules over strong Mori domains. Commun. Algebra **25**(4), 1285–1306 (1997)

M. Fontana, J.A. Huckaba, I.J. Papick, M. Roitman, Prufer domains and endomorphism rings of their ideals. J. Algebra **157**, 489–516 (1993)

S. Gabelli, M. Roitman, Maximal divisorial ideals and t-maximal ideals. JP Journal of Algebra, Number Theory and Applications **4**(2), 323–336 (2004)

R. Gilmer, The pseudo-radical of a commutative ring. Pacific J. Math. **19**(2), 275–284 (1966)

E. Hamann, E. Houston, J.L. Johnson, Properties of uppers to zero in $R[X]$. Pacific J. Math. **135**(1), 65–79 (1988)

A. Helms, Algebraische Geometrie. Math. Ann. **111**, 438–458 (1935)

E.G. Houston, Prime t-ideals in $R[X]$, in *Commutative Ring Theory*, ed. by P.J. Cahen, D.L. Costa, M. Fontana. Lecture Notes in Pure and Applied Mathematics, vol. 153 (Dekker, New York, 1994), pp. 163–170

E.G. Houston, M. Zafrullah, On t-invertibility II. Commun. Algebra **17**(8), 1955–1969 (1989)

E.G. Houston, M. Zafrullah, UMV-domains, in *Arithmetical Properties of Commutative Rings and Modules*, ed. by S. Chapman. Lecture Notes in Pure and Applied Mathematics, vol. 241 (Chapman, Hall/CRC, London, 2005), pp. 304–315

J.A. Huckaba, I.J. Papick, When the dual of an ideal is a ring. Manuscripta Math. **37**, 67–85 (1982)

J.L. Johnson, Three topological properties from Noetherian rings. Canad. J. Math. **34**(3), 525–534 (1982)

S. Kabbaj, A. Mimouni, t−class semigroup of integral domains. J. Reine Angew. Math. **612**, 213–229 (2007)

S. Kabbaj, A. Mimouni, Linkage of ideals in integral domains. Algebras and Representation Theory **24**, 799–809 (2021)

B.G. Kang, Prufer v-multiplication domains and the ring $R[X]_{N_v}$. J. Algebra **123**(1), 151–170 (1989)

D. Khurana, C. Kumar, The maximal and prime ideals of $R[x]$, R a PID. Resonance, 67–71 February (2005)

T.G. Lucas, Almost principal ideals in R[X]. Arab. J. Math. **1**, 97–111 (2012)

M.H. Park, Group rings and semigroup rings over strong Mori domains. J. Pure Applied Algebra **163**, 301–318 (2001)

J. Querré, Idéaux divisoriels d'un anneau de polynômes. J. Algebra **64**, 270–284 (1980)

D.E. Rush, Seminormality. J. Algebra **67**, 377–384 (1980)

M. Zafrullah, On finite conductor domains. Manuscripta Math. **24**(2), 191–204 (1978)

Chapter 4
The SFT and t-SFT Rings

1 Krull Dimension of the Formal Power Series Ring

In this section, the results are essentially due to Arnold (1973), but we follow the plan in the article of Roitman (2015). All the rings considered are commutative with identity. The multiplicative sets may contain 0 and do contain necessarily the identity.

Proposition 1.1 *Let A be a ring and $I \neq A$ un ideal of A. The following conditions are equivalent:*

1. *There exists a strictly increasing infinite sequence of prime ideals of A containing I.*
2. *There exists a decreasing sequence $S_1 \supseteq S_2 \supseteq \ldots$ of multiplicative sets of A such that $I \cap S_1 = \emptyset$ and elements $s_i \in S_i$, $i \in \mathbb{N}^*$, such that $S_{i+1} + As_i \subseteq S_{i+1}$ for each $i \in \mathbb{N}^*$.*
3. *There exists a decreasing sequence $S_1 \supseteq S_2 \supseteq \ldots$ of multiplicative sets of A such that $I \cap S_1 = \emptyset$ and elements $s_i \in S_i$, $i \in \mathbb{N}^*$, such that $S_{i+1} + As_i \subseteq S_i$ for each $i \in \mathbb{N}^*$.*

The inclusions $S_1 \supseteq S_2 \supseteq \ldots$ in (2) and (3) are all necessarily strict.

Proof "(1) $\Longrightarrow$ (2)" Let $P_1 \subset P_2 \subset \ldots$ be a strictly increasing infinite sequence of prime ideals of A containing the ideal I. Then the $S_i = A \setminus P_i$, $i \in \mathbb{N}^*$, are multiplicative sets of A with $S_1 \supset S_2 \supset \ldots$ and $S_1 \cap I = \emptyset$. For each $i \in \mathbb{N}^*$, $S_i + P_i \subseteq S_i$. Because, in the contrary case, there exist $i \in \mathbb{N}^*$, $s_i \in S_i$ and $p_i \in P_i$ such that $s_i + p_i \in P_i$, then $s_i \in P_i$, which is absurd. For each $i \in \mathbb{N}^*$, choose an element $s_i \in P_{i+1} \setminus P_i = P_{i+1} \cap S_i$. Then $s_i \in S_i$ and $S_{i+1} + As_i \subseteq S_{i+1} + P_{i+1} \subseteq S_{i+1}$. So $S_{i+1} + As_i \subseteq S_{i+1}$ for each $i \in \mathbb{N}^*$.

"(2) $\Longrightarrow$ (3)" With the same notations of (2), we have $S_{i+1} + As_i \subseteq S_{i+1} \subseteq S_i$, then $S_{i+1} + As_i \subseteq S_i$, for each $i \in \mathbb{N}^*$.

© The Author(s), under exclusive license to Springer Nature Switzerland AG 2022   281
A. Benhissi, *Chain Conditions in Commutative Rings*,
https://doi.org/10.1007/978-3-031-09898-7_4

"(3) $\Longrightarrow$ (1)" Since $I \cap S_1 = \emptyset$, there exists a prime ideal P_1 of A such that $I \subseteq P_1$ and $P_1 \cap S_1 = \emptyset$. We have $(P_1 + As_1) \cap S_2 = \emptyset$, because in the contrary case, there exist $p_1 \in P_1$ and $a \in A$ such that $p_1 + as_1 = s_2' \in S_2$, then $s_2' - as_1 = p_1 \in P_1$. But $s_2' - as_1 \in S_2 + As_1 \subseteq S_1$ with $P_1 \cap S_1 = \emptyset$, which is absurd. Then $(P_1 + As_1) \cap S_2 = \emptyset$. There exists a prime ideal P_2 of A such that $P_1 + As_1 \subseteq P_2$ and $S_2 \cap P_2 = \emptyset$. Since $s_1 \in P_2$ and $s_1 \in S_1$, then $s_1 \notin P_1$, so $P_1 \subset P_2$. We continue the construction of the P_i, $i \in \mathbb{N}^*$, by induction.

We will now show that the inclusions $S_1 \supseteq S_2 \supseteq \ldots$ are strict. For example, in (2), $s_i \in S_i \setminus S_{i+1}$. In the contrary case, $s_i \in S_{i+1}$, then $0 = s_i + (-s_i) \in S_{i+1} + As_i \subseteq S_{i+1}$, so $0 \in S_1 \cap I = \emptyset$, which is absurd. $\square$

Corollary 1.2 *The following conditions are equivalent to a ring A:*

1. *There exists a strictly increasing infinite sequence of prime ideals of A.*
2. *There exists a decreasing sequence $S_1 \supseteq S_2 \supseteq \ldots$ of multiplicative sets of A that do not contain 0 and elements $s_i \in S_i$, $i \in \mathbb{N}^*$, such that $S_{i+1} + As_i \subseteq S_{i+1}$ for each $i \in \mathbb{N}^*$.*
3. *There exists a decreasing sequence $S_1 \supseteq S_2 \supseteq \ldots$ of multiplicative sets of A that do not contain 0 and elements $s_i \in S_i$, $i \in \mathbb{N}^*$, such that $S_{i+1} + As_i \subseteq S_i$ for each $i \in \mathbb{N}^*$.*

The inclusions $S_1 \supseteq S_2 \supseteq \ldots$ in (2) and (3) are all necessarily strict.

Proof Take $I = (0)$. $\square$

In the sequel, A is a ring.

Notation Let $f = \displaystyle\sum_{i:0}^{\infty} a_i X^i \in A[[X]]$. For each $m \in \mathbb{N}$, let $cont_m f = (a_0, \ldots, a_m)A$ be the ideal of A generated by the $m+1$ first coefficients $a_0, \ldots, a_m$ of f.

Definition 1.3 Let $a = (a_0, a_1, \ldots)$ be an infinite sequence of elements of A and $u = (u(n))_{n \in \mathbb{N}}$ a sequence of natural integers strictly increasing from certain rank. Let $I(a, u)$ be the set of the formal power series $g(X) \in A[[X]]$ such that there exists an integer $n_0 \in \mathbb{N}$ satisfying $cont_{u(n)}g \subseteq (a_0, \ldots, a_{n-1})A$, for each integer $n \geq n_0$.

Lemma 1.4 *With the preceding notations, $I(a, u)$ is an ideal of the ring $A[[X]]$.*

Proof It is clear that $0 \in I(a, u)$ and $I(a, u)$ is stable by addition. Let $h \in A[[X]]$ and $g \in I(a, u)$ and n_0 the natural integer corresponding to g in the preceding definition. For each $n \geq n_0$, $cont_{u(n)}(gh) \subseteq cont_{u(n)}(g) \subseteq (a_0, \ldots, a_{n-1})A$, then $gh \in I(a, u)$. So $I(a, u)$ is an ideal of $A[[X]]$. $\square$

Definition 1.5 Let $a = (a_0, a_1, \ldots)$ be an infinite sequence of elements of the ring A and $u = (u(n))_{n \in \mathbb{N}}$ a sequence of natural integers strictly increasing from certain rank. We define $S(a, u)$ as the set of formal power series $g \in A[[X]]$ such that there exist natural integers $e \geq 1$ and n_0 and a sequence $v = (v(n))_{n \in \mathbb{N}}$ of natural integers strictly increasing from certain rank satisfying the following conditions:

(1) $\forall n \geq n_0, \, v(n) \leq eu(n)$ (2) $g \in f_e + I$,

with $f_e = f_e(a, v) = \sum_{i:0}^{\infty} a_i^e X^{v(i)}$ and the ideal $I = I(a, v)$ as in the definition 1.3

(the parameters e, n_0, and v depend on g).

Lemma 1.6 *Condition (2) of the preceding definition is equivalent to each one of the following two conditions:*

(1') *For n sufficiently big,* $g \equiv a_n^e X^{v(n)} \, mod \, (X^{v(n)+1}, a_0, \ldots, a_{n-1}) A[[X]].$

(2') *Put* $g = \sum_{i:0}^{\infty} b_i X^i$. *For n sufficiently big:*

(a) $b_i \in (a_0, \ldots, a_{n-1}) A$ *for* $i < v(n)$, *i.e.,* $cont_{v(n)-1} g \subseteq (a_0, \ldots, a_{n-1}) A$.
(b) $b_{v(n)} \equiv a_n^e \, mod \, (a_0, \ldots, a_{n-1}) A$.

Proof "(2) $\implies$ (1')" We have $g = f_e + h$ with $f_e = \sum_{i:0}^{\infty} a_i^e X^{v(i)}$ and

$h = \sum_{i:0}^{\infty} b_i X^i \in I(a, v)$. For n sufficiently big, $cont_{v(n)} h \subseteq (a_0, \ldots, a_{n-1}) A$,

then $g = a_n^e X^{v(n)} + \sum_{i:0}^{n-1} a_i^e X^{v(i)} + \sum_{i:n+1}^{\infty} a_i^e X^{v(i)} + \sum_{i:0}^{v(n)} b_i X^i + \sum_{i:v(n)+1}^{\infty} b_i X^i$. Note

that $v(n+1) > v(n)$, then $v(n+1) \geq v(n) + 1$. All the sums belong to $(X^{v(n)+1}, a_0, \ldots, a_{n-1}) A[[X]]$.

Then $g \equiv a_n^e X^{v(n)} \, mod \, (X^{v(n)+1}, a_0, \ldots, a_{n-1}) A[[X]]$.

"(1') $\implies$ (2')" For n sufficiently big, $g(X) = a_n^e X^{v(n)} + X^{v(n)+1} h(X) + \sum_{i:0}^{n-1} a_i h_i(X)$ with $h(X)$ and $h_i(X) \in A[[X]]$.

(a) $cont_{v(n)-1} g = cont_{v(n)-1} (\sum_{i:0}^{n-1} a_i h_i) \subseteq (a_0, \ldots, a_{n-1}) A.$

(b) Put $h_i(X) = \sum_{j:0}^{\infty} h_{ij} X^j$, for $0 \leq i \leq n-1$, then $b_{v(n)} = a_n^e + \sum_{i:0}^{n-1} a_i h_{iv(n)}$, so
$b_{v(n)} \equiv a_n^e \, mod \, (a_0, \ldots, a_{n-1}) A.$

$"(2') \implies (2)"$ We will show that $g - f_e \in I(a, v)$ with $g = \sum_{i:0}^{\infty} b_i X^i$ and $f_e = \sum_{i:0}^{\infty} a_i^e X^{v(i)}$. For n sufficiently big, $cont_{v(n)}(g - f_e) = cont_{v(n)-1}(g - f_e) + (b_{v(n)} - a_n^e)A$. By (a) and the expression of f_e, the first ideal is included in $(a_0, \ldots, a_{n-1})A$. By (b), the second ideal is also included in $(a_0, \ldots, a_{n-1})A$.   □

Lemma 1.7 *Let* $a = (a_0, a_1, \ldots)$ *be an infinite sequence of elements of* A *and* $u = (u(n))_{n \in \mathbb{N}}$ *a sequence of natural integers strictly increasing from certain rank. Then* $S(a, u)$ *is stable by multiplication and does not contain the series* $f(a, u) = f_1(a, u) = \sum_{i:0}^{\infty} a_i X^{u(i)}$.

Proof Let g_1 and $g_2 \in S(a, u)$. Conditions (1) and (2) of Definition 1.5 are satisfied by g_i for e_i and v_i, $i = 1, 2$. Put $e = e_1 + e_2$ and $v = v_1 + v_2$. Then v is strictly increasing from certain rank and for n sufficiently big, $v(n) = v_1(n) + v_2(n) \leq e_1 u(n) + e_2 u(n) = eu(n)$ and $g_1 \in a_n^{e_1} X^{v_1(n)} + (X^{v_1(n)+1}, a_0, \ldots, a_{n-1})A[[X]]$ and $g_2 \in a_n^{e_2} X^{v_2(n)} + (X^{v_2(n)+1}, a_0, \ldots, a_{n-1})A[[X]]$, then $g_1 g_2 \in a_n^{e_1+e_2} X^{v_1(n)+v_2(n)} + (X^{v_1(n)+v_2(n)+1}, a_0, \ldots, a_{n-1})A[[X]] = a_n^e X^{v(n)} + (X^{v(n)+1}, a_0, \ldots, a_{n-1})A[[X]]$. So $g_1 g_2 \in S(a, u)$. To see that $f(a, u) = \sum_{i:0}^{\infty} a_i X^{u(i)} \in S(a, u)$, we take $e = 1$, $n_0 = 0$ and $v = u$.   □

Lemma 1.8 *Let* $a = (a_0, a_1, \ldots)$ *be an infinite sequence of elements of the ring* A *and* $u = (u(n))_{n \in \mathbb{N}}$ *and* $u' = (u'(n))_{n \in \mathbb{N}}$ *be two sequences of natural integers strictly increasing from certain ranks.*

1. *If* $u'(n) \leq u(n)$ *for* n *sufficiently big, then* $S(a, u') \subseteq S(a, u)$.

2. *If* $u'(n) < u(n)$ *for* n *sufficiently big, then* $f = f_1(a, u) = \sum_{n:0}^{\infty} a_n X^{u(n)} \in I(a, u')$.

3. *If* $\lim\limits_{n \longrightarrow +\infty} \dfrac{u(n)}{u'(n)} = +\infty$, *then* $S(a, u') + A[[X]]f \subseteq S(a, u')$.

Proof

(1) Let $g \in S(a, u')$. There exist natural integers $e \geq 1$ and n_0 and a sequence $v = (v(n))_{n \in \mathbb{N}}$ natural integers strictly increasing from certain rank such that for each $n \geq n_0$, $v(n) \leq eu'(n)$ and $g \in f_e + I$ with $f_e = f_e(a, v) = \sum_{n:0}^{\infty} a_n^e X^{v(n)}$ and $I = I(a, v)$. Since $v(n) \leq eu'(n) \leq eu(n)$ for n sufficiently big, then $g \in S(a, u)$.

(2) We have $f = f_1(a, u) = \sum_{i:0}^{\infty} a_i X^{u(i)}$. For n a sufficiently big integer, $u'(n) < u(n)$, then $cont_{u'(n)} f \subseteq cont_{u(n)-1} f = (a_0, \ldots, a_{n-1})A$. So $f \in I(a, u')$.

(3) Let $g \in S(a, u')$. There exists an integer $e \geq 1$ and a sequence of natural integers $v = (v(n))_{n \in \mathbb{N}}$ strictly increasing from certain rank such that for n sufficiently big, $v(n) \leq eu'(n)$ and $g \in f_e(a, v) + I(a, v)$. Since $\lim_{n \to +\infty} \frac{u(n)}{u'(n)} = \infty$, then $v(n) < u(n)$ for n sufficiently big. By (2), $f_1(a, u) \in I(a, v)$. Then $g + A[[X]]f_1(a, u) \subseteq f_e(a, v) + I(a, v) + A[[X]]I(a, v) = f_e(a, v) + I(a, v)$ since $I(a, v)$ is an ideal of $A[[X]]$. By Definition 1.5, $g + A[[X]]f_1(a, u) \subseteq S(a, u')$. Then $S(a, u') + A[[X]]f \subseteq S(a, u')$. $\qquad\square$

By the end of the year 1960, R. Gilmer and some of his students were interested in the study of the ring of formal power series. To study the Krull dimension of this ring, J.T. Arnold, one of Gilmer's students, introduced a notion of almost Noetherian that he called the SFT property.

Definition 1.9 The ideal I of the ring A is called SFT-ideal if there exist a finitely generated ideal $F \subseteq I$ of A and an integer $k \geq 1$ such that for each $x \in I$, $x^k \in F$. The ring A satisfies the SFT property if all its ideals are SFT.

Example 1 Every finitely generated ideal is an SFT-ideal and every Noetherian ring is an SFT ring.

Example 2 (Coykendall 2006 and Dutta 2006) Let $A = \mathbb{Z} + 2X\mathbb{Z}[X] \subset \mathbb{Z}[X]$. Note that $X \notin A$ but $2X \in A$. An element of the ideal $I = (2X^n; \ n \in \mathbb{N})$ of A is of the form $f = \sum_{i:1}^{r} 2X^{n_i} f_i$ with $n_i \in \mathbb{N}$ and $f_i \in A$. Then $f^2 = \sum_{i:1}^{r}(2X^{n_i} f_i)^2 + 2\sum_{i \neq j}(2X^{n_i} f_i)(2X^{n_j} f_j) = 2\sum_{i:1}^{r} 2X^{2n_i} f_i^2 + 2\sum_{i \neq j} 2X^{n_i+n_j} 2 f_i f_j \in 2A \subseteq I$. So I is an SFT-ideal of A. Suppose that I is finitely generated, $I = (2, \ldots, 2X^n)$. Put $2X^{n+1} = \sum_{i:0}^{n} 2X^i f_i$ with $f_i \in A$, $f_i = a_i + 2Xg_i$, where $a_i \in \mathbb{Z}$ and $g_i \in \mathbb{Z}[X]$. Then $X^{n+1} = \sum_{i:0}^{n} X^i(a_i + 2Xg_i)$, so $X^{n+1} - \sum_{i:0}^{n} a_i X^i = 2\sum_{i:0}^{n} X^{i+1} g_i$. By identification, the coefficient of X^{n+1} is of the form $1 = 2b$ with $b \in \mathbb{Z}$, which is absurd.

Example 3 (Coykendall 2006 and Dutta 2006) Let $\{Y, X_i; \ i \in \mathbb{N}^*\}$ be indeterminates over the field $\mathbb{F}_2$ and $A = \mathbb{F}_2[Y, X_1, X_2, \ldots, \frac{Y}{X_1}, \frac{Y}{X_1^2}, \ldots, \frac{Y}{X_2}, \frac{Y}{X_2^2}, \ldots]$. Consider the ideals $I = (Y, \frac{Y}{X_1}, \frac{Y}{X_1^2}, \ldots, \frac{Y}{X_2}, \frac{Y}{X_2^2}, \ldots)$ and $F = (Y) \subseteq I$ of A. The generators of I satisfy $\left(\frac{Y}{X_k^n}\right)^2 = \frac{Y}{X_k^{2n}} Y \in F$. Since the characteristic of A is 2, for each $f \in I$, $f^2 \in F$. Then I is an SFT-ideal of A.

Example 4 Let $(K_i)_{i \in \mathbb{N}}$ be a sequence of fields and $A = \prod_{i:0}^{\infty} K_i$ the product ring. For $f = (a_i)_{i \in \mathbb{N}} \in A$, the support of f is the set $S(f) = \{i \in \mathbb{N}; a_i \neq 0\}$. The subset I of A of the elements with finite supports is an ideal of A. Suppose that I is an SFT-ideal. There exist $k \in \mathbb{N}^*$ and $f_1, \ldots, f_n \in I$ such that for each $f \in I$, $f^k \in (f_1, \ldots, f_n)$. Then for each $f \in I$, $S(f) = S(f^k) \subseteq S(f_1) \cup \ldots \cup S(f_n)$, which is a fixed finite set. Since the elements of I may have finite supports, as large as we want, then I is not an SFT-ideal.

Example 5 Let (A, M) be a valuation domain and I a nonzero proper ideal of A. Then I is an SFT if and only if $I^2 \neq I$. In particular, M is SFT if and only if it is principal. For example, if A is a nondiscrete valuation domain of rank 1, M is not SFT.

Indeed, suppose that $I^2 = I$. By induction, for each $k \in \mathbb{N}^*$, $I^k = I$. Since I is an SFT-ideal, there exist $n \in \mathbb{N}^*$ and a finitely generated ideal $F \subseteq I$ such that $x^n \in F$ for each $x \in I$. Since A is a valuation domain, then $F = aA$ with $a \neq 0$. The ideal I^n is generated by the set $\{x^n; \ x \in I\}$. Then $I = I^n \subseteq F = aA \subseteq I$, so $I = aA$. Since $I^2 = I$, there exists $b \in A$ such that $a = a^2 b$, then $ab = 1$. So $I = A$, which is absurd.

Conversely, if $I^2 \neq I$, there exists $a \in I \setminus I^2$. Since the ideals are comparable in a valuation domain, then $I^2 \subset aA \subseteq I$. So I is an SFT-ideal. Suppose that M is SFT not principal. Let $a \in M$. Since M is not principal, then $aA \subset M$. Let $b \in M \setminus aA$. Then $aA \subset bA$, so $a = bc$ with $c \in M$. Then $M = M^2$, which is absurd.

Example 6 (Coykendall 2006 and Dutta 2006) Let I be an SFT-ideal of a ring A and $n \in \mathbb{N}^*$. Then I^n is an SFT-ideal. Indeed, there exist $k \in \mathbb{N}^*$ and a finitely generated ideal $F \subseteq I$ such that $x^k \in F$ for each $x \in I$. Let $x \in I^n \subseteq I$. Then $x^k \in F$, so $x^{kn} \in F^n$. Since F^n is finitely generated included in I^n, we have the result.

Lemma 1.10 *Let I be an ideal that is not SFT in a ring A. There exists an infinite sequence $a_0, a_1, \ldots$ of elements of I such that $a_{n+1}^{n+1} \notin (a_0, \ldots, a_n)$ for each $n \in \mathbb{N}$.*

Proof Construction by induction. We start by any element $a_0 \in I$. There exists $a_1 \in I$ such that $a_1 \notin (a_0)$. There exists $a_2 \in I$ such that $a_2^2 \notin (a_0, a_1), \ldots$. $\square$

Lemma 1.11 *Let $a = (a_0, a_1, \ldots)$ be an infinite sequence of elements of a ring A such that $a_{n+1}^{n+1} \notin (a_0, \ldots, a_n)$ for each $n \in \mathbb{N}$. Then for each sequence $u = (u(n))_{n \in \mathbb{N}}$ of natural integers strictly increasing from certain rank, $0 \notin S(a, u)$.*

Proof By Lemma 1.6 (2'), if $g = \sum_{i:0}^{\infty} b_i X^i \in S(a, u)$, there exist an integer $e \geq 1$ and a sequence $v = (v(n))_{n \in \mathbb{N}}$ of natural integers strictly increasing from certain rank, such that $b_{v(n)} \equiv a_n^e \, mod(a_0, \ldots, a_{n-1})A$ for n sufficiently big. For $n \geq e$ sufficiently big, $a_n^n \notin (a_0, \ldots, a_{n-1})A$, then $a_n^e \notin (a_0, \ldots, a_{n-1})A$, so $b_{v(n)} \notin (a_0, \ldots, a_{n-1})A$. Then $b_{v(n)} \neq 0$ and $g \neq 0$.   □

Remark 1.12 With the preceding notations, g has an infinity of nonzero coefficients $b_{v(n)}$. In particular, $1 \notin S(a, u)$.

Let A be a ring. If there is a chain $P_0 \subset P_1 \subset \ldots \subset P_n$ of $n+1$ prime ideals of A, but there is not a chain of $n+2$ prime ideals, we say that the Krull dimension of A is equal to n and we denote $dim\ A = n$. If not, we say that A has an infinite dimension and we denote $dim\ A = \infty$. It is well known that if A has a Krull dimension n, then the dimension of the polynomials ring $A[X]$ is bounded by $n + 1$ and $2n + 1$. The formal power series ring $A[[X]]$ behaves in a very different way. The importance of the SFT property is clear in the following theorem. Indeed, if A is not an SFT-ring, not only the lengths of the chains of prime ideals in $A[[X]]$ are not bounded, but $A[[X]]$ admits a strictly increasing infinite chain of prime ideals.

Theorem 1.13 *If the ring A does not satisfy the SFT property, there exists an infinite strictly increasing sequence of prime ideals in $A[[X]]$. In particular, $dim\ A[[X]] = \infty$.*

Proof Since A does not have the SFT property, by Lemma 1.10, there exists an infinite sequence $a = (a_0, a_1, \ldots)$ of elements of A such that $a_{n+1}^{n+1} \notin (a_0, \ldots, a_n)$ for each $n \in \mathbb{N}$. For integers $i \geq 1$ and $n \geq 0$, we define $u_i(n) = n^{n-i}$ if $n \geq i$ and $u_i(n) = 0$ if $0 \leq n < i$. For each $i \geq 1$, $u_i = (u_i(n))_{n \in \mathbb{N}}$ is a sequence of natural integers strictly increasing from the rank i. By Lemma 1.11, for each $i \geq 1$, the multiplicative set $S_i = S(a, u_i)$ of $A[[X]]$ does not contain 0. Note also that for $n > i + 1$, $\frac{u_i(n)}{u_{i+1}(n)} = \frac{n^{n-i}}{n^{n-i-1}} = n \longrightarrow \infty$. By Lemma 1.8, $S_{i+1} = S(a, u_{i+1}) \subseteq S(a, u_i) = S_i$ and $S(a, u_{i+1}) + A[[X]]f_i \subseteq S(a, u_{i+1})$ with $f_i = f_1(a, u_i) = \sum_{k:0}^{\infty} a_k X^{u_i(k)}$. By Lemma 1.7, $f_i \in S(a, u_i)$. Then the sequence $(S_i)_{i \in \mathbb{N}^*}$ satisfies condition (2) of Corollary 1.2. So $A[[X]]$ contains an infinite strictly increasing sequence of prime ideals.   □

The technical difficulty in the proof of the preceding theorem becomes legendary. We start by constructing a strictly decreasing sequence of multiplicative sets having certain properties. Then, we define by induction a strictly increasing sequence of prime ideals. This strategy is used in other situations. This is the case to show that the fragmented rings have infinite dimension. We prove later that the converse of the preceding theorem is false.

Lemma 1.14 *Let I be an ideal of a ring A and p a prime number. For each $k \in \mathbb{N}^*$ and $f = \sum_{i:0}^{\infty} a_i X^i \in I[[X]]$, $f^{p^k} = \sum_{i:0}^{\infty} a_i^{p^k} X^{ip^k} + pg(X)$ with $g(X) \in I[[X]]$.*

Proof Note first that for $f, g \in I[[X]]$, $(f + g)^p = f^p + g^p + ph$ with $h \in I[[X]]$. Indeed, $(f + g)^p = f^p + g^p + \sum_{i:1}^{p-1} C_p^i f^i g^{p-i}$ with p dividing C_p^i. Suppose by induction that $(f_1 + \ldots + f_{n-1})^p = f_1^p + \ldots + f_{n-1}^p + ph$ with $h \in I[[X]]$. Then $(f_1 + \ldots + f_{n-1} + f_n)^p = (f_1 + \ldots + f_{n-1})^p + f_n^p + ph' = f_1^p + \ldots + f_{n-1}^p + ph + f_n^p + ph' = f_1^p + \ldots + f_n^p + ph''$ with $h, h', h'' \in I[[X]]$. We will show now that if $f = \sum_{i:0}^{\infty} a_i X^i \in I[[X]]$, then $f^p = \sum_{i:0}^{\infty} a_i^p X^{ip} + pg$ with $g \in I[[X]]$. Indeed, let $l \in \mathbb{N}$, $f = \sum_{i:0}^{l} a_i X^i + X^{l+1} g$ with $g \in I[[X]]$. Then $f^p = \sum_{i:0}^{l} a_i^p X^{ip} + X^{p(l+1)} g^p + ph$ with $h \in I[[X]]$. The coefficient of X^l in the sum of the two first expressions is a_i^p if $l = ip$ and 0 if not. The coefficient of X^l in f^p is of the form $a_i^p + pb_i$ if $l = ip$ and pb_i if not, with $b_i \in I$. Then $f^p = \sum_{i:0}^{\infty} a_i^p X^{ip} + pl(X)$ with $l(X) \in I[[X]]$. Suppose by induction that $f^{p^k} = \sum_{i:0}^{\infty} a_i^{p^k} X^{ip^k} + pg_k(X)$ with $g_k(X) \in I[[X]]$. Then $f^{p^{k+1}} = (\sum_{i:0}^{\infty} a_i^{p^k} X^{ip^k} + pg_k)^p = (\sum_{i:0}^{\infty} a_i^{p^k} X^{ip^k})^p + (pg_k)^p + pg = \sum_{i:0}^{\infty} a_i^{p^{k+1}} X^{ip^{k+1}} + pg' + (pg_k)^p + pg = \sum_{i:0}^{\infty} a_i^{p^{k+1}} X^{ip^{k+1}} + pg_{k+1}$ with $g, g', g_{k+1} \in I[[X]]$. $\square$

Example Let A be a ring with characteristic equal to a prime number p. For each $f = \sum_{i:0}^{\infty} a_i X^i \in A[[X]]$, we have $f^p = \sum_{i:0}^{\infty} a_i^p X^{pi}$.

Lemma 1.15 *Let I be a nil ideal with bounded index of a ring A, i.e., there exists $k \in \mathbb{N}^*$ such that for each $x \in I$, $x^k = 0$:*

1. *There exists an integer $m \geq 2$ such that $mI^m = (0)$.*
2. *$I[[X]]$ is a nil ideal of bounded index of $A[[X]]$.*

Proof

(1) Suppose that there exist some integers $r, s_1, \ldots, s_t \geq 1$ such that for each $a_1, \ldots, a_t \in I$, we have $r a_1^{s_1} \ldots a_t^{s_t} = 0$ (this is always true for $r = t = 1$ and $s_1 = k$).

Suppose that $s_1 \geq 2$. For each $b_0, b_1, \ldots, b_t \in I$, we have $r(b_0 + b_1)^{s_1} b_2^{s_2} \ldots b_t^{s_t} = 0$, then $r b_0^{s_1-2} (b_0 + b_1)^{s_1} b_2^{s_2} \ldots b_t^{s_t} = 0$, so

$$r b_0^{s_1-2} \left(\sum_{j:0}^{s_1} C_{s_1}^j b_0^{s_1-j} b_1^j \right) b_2^{s_2} \ldots b_t^{s_t} = 0.$$

Then $\sum_{j:0}^{s_1} r C_{s_1}^j b_0^{2s_1-2-j} b_1^j b_2^{s_2} \ldots b_t^{s_t} = 0$.

For $0 \leq j \leq s_1$, put $\alpha_j = r C_{s_1}^j b_0^{2s_1-2-j} b_1^j b_2^{s_2} \ldots b_t^{s_t}$. Then $\alpha_0 + \ldots + \alpha_{s_1} = 0$. For $0 \leq j \leq s_1 - 2$, we have $2s_1 - 2 - j = s_1 + (s_1 - 2 - j) \geq s_1$, then $\alpha_j = 0$. We also have $\alpha_{s_1} = b_0^{s_1-2} (r b_1^{s_1} b_2^{s_2} \ldots b_t^{s_t}) = 0$. Then $\alpha_{s_1-1} = 0$, so $r s_1 b_0^{s_1-1} b_1^{s_1-1} b_2^{s_2} \ldots b_t^{s_t} = 0$. Then by adding an element, we can reduce two exponents. After a finite number of steps, we can reduce all the exponents to 1. We find two integers m_1 and $m_2 \geq 1$ such that for each $a_1, \ldots, a_{m_2} \in I$, we have $m_1 a_1 \ldots a_{m_2} = 0$. Let $m = m_1 m_2$, then $m I^m = (0)$.

(2) Let $k \in \mathbb{N}^*$ be such that $x^k = 0$ for each $x \in I$ and $m \geq 2$ be an integer satisfying $m I^m = 0$. Let $f = \sum_{i:0}^{\infty} a_i X^i \in (I[[X]])^m \subseteq I^m[[X]]$. Then $a_i \in I^m \subseteq I$. By the preceding lemma, for each prime number p,

$$f^{p^k} = \sum_{i:0}^{\infty} a_i^{p^k} X^{ip^k} + pg = pg \text{ with } g \in I^m[[X]]. \text{ Indeed, since } p^k > k,$$

$a_i^{p^k} = 0$. Put $m = p_1^{e_1} \ldots p_t^{e_t}$ where the p_i are different prime numbers and the integers $e_i \geq 1$. Then $f^{p_i^k} = p_i g_i$ with $g_i \in I^m[[X]], 1 \leq i \leq t$. Let $n = p_1^k e_1 + \ldots + p_t^k e_t$. Then $f^n = (f^{p_1^k})^{e_1} \ldots (f^{p_t^k})^{e_t} = (p_1 g_1)^{e_1} \ldots (p_t g_t)^{e_t} = p_1^{e_1} \ldots p_t^{e_t} g_1^{e_1} \ldots g_t^{e_t} = mg$ with $g = g_1^{e_1} \ldots g_t^{e_t} \in I^m[[X]]$. Then $f^n \in m I^m[[X]] = 0$, so $f^n = 0$. We conclude that for each $g \in I[[X]], g^{mn} = 0$. $\qquad \square$

Example Let k be a field, $B = k[X_1, X_2, \ldots]$ the polynomial ring in infinitely countably variables with coefficients in k, $I = (X_1, X_2^2, X_3^3, \ldots)$, $A = B/I$, and $\mathscr{I} = (\overline{X_1}, \overline{X_2}, \ldots)$. Then $\mathscr{I} \subseteq Nil(A)$. Since $\overline{X_i}^i = 0$ but $\overline{X_i}^{i-1} \neq 0$ for each integer $i \geq 2$, then the index of $\mathscr{I}$ is not bounded. Note also $\mathscr{I}^n \neq (0)$ for each $n \in \mathbb{N}^*$.

Let I be an ideal of a ring A. In the polynomial case, $I[X] = I A[X]$. But for the formal power series, the ideals $I[[X]]$ and $I A[[X]]$ are always different.

Lemma 1.16 *Let I be an ideal of a ring A. Then $I A[[X]] \subseteq I[[X]]$. Moreover, if I is finitely generated, we have the equality.*

Proof Since $I \subseteq I[[X]]$, then $I A[[X]] \subseteq I[[X]]$. Suppose that $I = (a_1, \ldots, a_n)$ is finitely generated. Let $f = \sum_{i:0}^{\infty} b_i X^i \in I[[X]]$. For each $i \in \mathbb{N}$, $b_i = \sum_{j:1}^{n} \beta_{ij} a_j$ with $\beta_{ij} \in A$. Then $f = \sum_{i:0}^{\infty} \left(\sum_{j:1}^{n} \beta_{ij} a_j \right) X^i = \sum_{j:1}^{n} \left(\sum_{i:0}^{\infty} \beta_{ij} X^i \right) a_j \in I A[[X]]$. $\qquad \square$

Example Let (A, M) be a nondiscrete valuation domain of rank one. It may be defined by a valuation v with group G, dense in $\mathbb{R}$. There exists a sequence $(\alpha_i)_{i \in \mathbb{N}}$ of elements in G that decreases strictly to 0. For each $i \in \mathbb{N}$, there exists $a_i \in M$ such that $v(a_i) = \alpha_i$. Let $f = \sum_{i:0}^{\infty} a_i X^i \in M[[X]]$. Suppose that $f \in MA[[X]]$, there exist $b \in M$ and $g = \sum_{i:0}^{\infty} c_i X^i \in A[[X]]$ such that $f = bg$. For each $i \in \mathbb{N}$, $a_i = bc_i$, then $\alpha_i = v(bc_i) = v(b) + v(c_i) \geq v(b)$, a strictly positive constant, which is absurd. Then $M[[X]] \neq MA[[X]]$.

Proposition 1.17 *If I is an SFT-ideal of a ring A, then $\sqrt{I[[X]]} = \sqrt{I A[[X]]}$. Moreover, there exists a finitely generated ideal $F \subseteq I$ of A such that $\sqrt{I[[X]]} = \sqrt{F A[[X]]}$.*

Proof By definition, there exists a finitely generated ideal $F \subseteq I$ of A such that I/F is a nil ideal of bounded index. By Lemma 1.15, $(I/F)[[X]] = I[[X]]/F[[X]]$ is a nil ideal of bonded index k in the ring $A[[X]]/F[[X]]$. For each series $f \in I[[X]]$, $f^k \in F[[X]]$. By the preceding lemma, $\sqrt{I[[X]]} = \sqrt{F[[X]]} = \sqrt{F A[[X]]} \subseteq \sqrt{I A[[X]]} \subseteq \sqrt{I[[X]]}$. So $\sqrt{I[[X]]} = \sqrt{I A[[X]]} = \sqrt{F A[[X]]}$. $\qquad \square$

Theorem 1.18 *The following conditions are equivalent to a ring A:*

1. *The ring A satisfies the SFT property.*
2. *For each ideal I of A, $\sqrt{I[[X]]} = \sqrt{I A[[X]]}$.*
3. *For each prime ideal P of A, $P[[X]] = \sqrt{P A[[X]]}$.*

Proof "$(1) \Longrightarrow (2)$" The preceding proposition. "$(2) \Longrightarrow (3)$" Clear.

"$(3) \Longrightarrow (2)$" Suppose that there exists an ideal I of A such that $\sqrt{I[[X]]} \neq \sqrt{I A[[X]]}$. Then $\sqrt{I[[X]]} \nsubseteq \sqrt{I A[[X]]}$, so $I[[X]] \nsubseteq \sqrt{I A[[X]]}$. There exists a prime ideal Q of $A[[X]]$ containing $I A[[X]]$ but not containing $I[[X]]$. Let $P = Q \cap A$: a prime ideal of A with $I \subseteq P$. Then $I[[X]] \subseteq P[[X]]$. Since Q does not contain $I[[X]]$, it does not contain $P[[X]]$. But Q contains $P A[[X]]$, then $P[[X]] \nsubseteq \sqrt{P A[[X]]}$, so $P[[X]] \neq \sqrt{P A[[X]]}$.

"$(2) \Longrightarrow (1)$" Suppose that the ring A does not have the SFT property. By Lemma 1.10, there exists an infinite sequence $a_0, a_1, \ldots$ of elements of A such that $a_{n+1}^{n+1} \notin (a_0, \ldots, a_n) A$ for each $n \in \mathbb{N}$. Let I be the ideal of A generated by the $a_n, n \in \mathbb{N}$, and $f = \sum_{i:0}^{\infty} a_i X^i \in I[[X]]$. If $f \in \sqrt{IA[[X]]}$, there exists an integer $k \geq 1$ such that $f^k \in IA[[X]]$. There exists a finitely generated ideal $I_0 \subseteq I$ of A such that $f^k \in I_0 A[[X]]$. There exists an integer n_0 such that $f^k \in (a_0, \ldots, a_{n_0}) A[[X]]$. For each integer $n \geq max(k, n_0)$, the series

$$f^{n+1} = (\sum_{i:0}^{\infty} a_i X^i)^{n+1} \in (a_0, \ldots, a_n) A[[X]].$$ In the ring $(A/(a_0, \ldots, a_n))[[X]]$,

$0 = \bar{f}^{n+1} = (\bar{a}_{n+1} X^{n+1} + \ldots)^{n+1} = \bar{a}_{n+1}^{n+1} X^{(n+1)^2} + \ldots$, then $\bar{a}_{n+1}^{n+1} = 0$, so $a_{n+1}^{n+1} \in (a_0, \ldots, a_n) A$, which is absurd. □

The SFT property over a ring is completely determined by its prime spectrum. Then we have an analogue to Cohen theorem.

Proposition 1.19 *A ring A satisfies the SFT property if and only if all its prime ideals are SFT-ideals.*

Proof By Proposition 1.17, for each prime ideal P of A, we have $P[[X]] = \sqrt{PA[[X]]}$. By Theorem 1.18, the ring A has the SFT property. □

The preceding proposition may be shown without using formal power series. A maximal element in the set of the ideals which are not SFT is prime. By Zorn lemma, such ideal exists if the ring A does not have the SFT property.

Example (Condo et al. 1996) Let $A = \mathbb{F}_2[Y_i; i \in \mathbb{N}]/(Y_i^2; i \in \mathbb{N})$. A prime ideal $\mathcal{P}$ of $\mathbb{F}_2[Y_i; i \in \mathbb{N}]$ containing $(Y_i^2; i \in \mathbb{N})$ contains all the $Y_i, i \in \mathbb{N}$. But $\mathbb{F}_2[Y_i; i \in \mathbb{N}]/(Y_i; i \in \mathbb{N}) \simeq \mathbb{F}_2$ is a field, then $(Y_i; i \in \mathbb{N})$ is a maximal ideal of $\mathbb{F}_2[Y_i; i \in \mathbb{N}]$, so $\mathcal{P} = (Y_i; i \in \mathbb{N})$. Then the prime spectrum of A is reduced to the singleton $\{\mathcal{M} = (\bar{Y}_i; i \in \mathbb{N})\}$ and $dim(A) = 0$. An element of $\mathcal{M}$ is of the form $f = f_0 \bar{Y}_0 + \ldots + f_n \bar{Y}_n$ with $f_i \in A$, then $f^2 = f_0^2 \bar{Y}_0^2 + \ldots + f_n^2 \bar{Y}_n^2 = 0$. So $\mathcal{M}$ is an SFT-ideal and A is an SFT-ring.

Proposition 1.20 (Condo et al. 1996) *Let A be an SFT-ring. Then $Min(A[[X]]) = \{\mathcal{M}[[X]]; \mathcal{M} \in Min(A)\}$.*

Proof "$\supseteq$" Let $\mathcal{M} \in Min(A)$ and $P \in spec(A[[X]])$ such that $P \subseteq \mathcal{M}[[X]]$. Then $P \cap A \subseteq \mathcal{M}$, so $P \cap A = \mathcal{M}$. Since A is an SFT-ring, by Theorem 1.18, $\mathcal{M}[[X]] = \sqrt{\mathcal{M}.A[[X]]} \subseteq P$, then $\mathcal{M}[[X]] = P$. So $\mathcal{M}[[X]] \in Min(A[[X]])$.

"$\subseteq$" Let $P \in Min(A[[X]])$ and $\mathcal{M} = P \cap A$. By Theorem 1.18, $\mathcal{M}[[X]] = \sqrt{\mathcal{M}.A[[X]]} \subseteq P$, then $\mathcal{M}[[X]] = P$. We will show that $\mathcal{M} \in Min(A)$. Let $\mathcal{P} \in spec(A)$ with $\mathcal{P} \subseteq \mathcal{M}$. Then $\mathcal{P}[[X]] \subseteq \mathcal{M}[[X]] = P$, so $\mathcal{P}[[X]] = P = \mathcal{M}[[X]]$. Then $\mathcal{P} = \mathcal{M}$. □

The SFT property is sufficient for the Krull dimension of the ring of formal power series with coefficients in a zero-dimensional ring to be finite.

Proposition 1.21 *Let A be zero-dimensional SFT-ring. Then $dim\ A[[X]] = 1$.*

Proof (Condo et al. 1996) $dim A[[X]] = sup\{coht P; P \in Min(A[[X]])\} = sup\{dim(A[[X]]/P); P \in Min(A[[X]])\} = sup\{dim(A[[X]]/\mathcal{M}[[X]]); \mathcal{M} \in Min(A)\} = sup\{dim(A/\mathcal{M})[[X]]; \mathcal{M} \in Min(A)\} = 1$, since all the prime ideals of A are maximal. □

Example For $A = \mathbb{F}_2[Y_i; i \in \mathbb{N}]/(Y_i^2; i \in \mathbb{N})$, we have $dim\ A = 0$ and $dim\ A[[X]] = 1$.

Corollary 1.22 (Condo et al. 1996) *Let A be a zero-dimensional SFT-ring. The prime ideals of $A[[X]]$ are of two types $\mathcal{M}[[X]] \subset \mathcal{M} + XA[[X]]$, corresponding to $\mathcal{M} \in Spec(A)$, and there is no other proper inclusion between the members of $spec(A[[X]])$.*

Example If $A = \mathbb{F}_2[Y_i; i \in \mathbb{N}]/(Y_i^2; i \in \mathbb{N})$, the only prime ideal of A is $\mathcal{M} = (\bar{Y}_i; i \in \mathbb{N})$. Then the only prime ideals of $A[[X]]$ are $\mathcal{M}[[X]] \subset \mathcal{M} + XA[[X]]$.

Lemma 1.23 *For any ring A, $dim\ A[[X]] \geq dim\ A + 1$.*

Proof Let $\mathcal{P}_0 \subset \mathcal{P}_1 \subset \ldots \subset \mathcal{P}_n$ be a chain of length n of prime ideals of A. Then $\mathcal{P}_0[[X]] \subset \mathcal{P}_1[[X]] \subset \ldots \subset \mathcal{P}_n[[X]] \subset \mathcal{P}_n + XA[[X]]$ is a chain of length $n+1$ of prime ideals of $A[[X]]$. Then $dim\ A[[X]] \geq dim\ A + 1$. □

It follows from the preceding lemma that if $dim\ A = \infty$, then $dim\ A[[X]] = \infty$. In the sequel, we suppose that $dim\ A < \infty$. On the other hand, Noetherian rings constitute the simplest class among the SFT-rings. With the help of the generalized principal ideal theorem (cf. Sect. 8 of Chap. 2), we can compute the Krull dimension of the formal power series ring with coefficients in a Noetherian ring. In this case, the formal power series ring behaves like the polynomial ring. To do this, we first need the following lemma.

Lemma 1.24 *Let A be a ring, I a proper ideal of A, and P a prime ideal of A minimal over I. Then $P + (X_1, \ldots, X_n)A[[X_1, \ldots, X_n]]$ is a prime ideal minimal over the ideal $I + (X_1, \ldots, X_n)A[[X_1, \ldots, X_n]]$ in the ring $A[[X_1, \ldots, X_n]]$.*

Proof Let Q be a prime ideal of $A[[X_1, \ldots, X_n]]$ such that $I + (X_1, \ldots, X_n)A[[X_1, \ldots, X_n]] \subseteq Q \subseteq P + (X_1, \ldots, X_n)A[[X_1, \ldots, X_n]]$. Intersecting with A, we obtain $I \subseteq Q \cap A \subseteq P$, then $Q \cap A = P$, so $P \subseteq Q$. Since $X_1, \ldots, X_n \in Q$, then $Q = P + (X_1, \ldots, X_n)A[[X_1, \ldots, X_n]]$. So we have the result. □

Theorem 1.25 *Let A be a Noetherian ring with finite Krull dimension n. Then $dim\ A[[X]] = n + 1$.*

Proof Let M be a maximal ideal of $A[[X]]$. There exists a maximal ideal $\mathcal{M}$ of A such that $M = \mathcal{M} + XA[[X]]$. Let $k = ht\ \mathcal{M} \leq dim\ A = n$. Since A is Noetherian, by the generalized principal ideal theorem, there exist $a_1, \ldots, a_k \in \mathcal{M}$ such that $\mathcal{M}$ is minimal over the ideal $(a_1, \ldots, a_k)A$. By the preceding lemma, $M = \mathcal{M} + XA[[X]]$ is minimal over the ideal $(a_1, \ldots, a_k, X)A[[X]]$. Since A

is Noetherian, $A[[X]]$ is also Noetherian, by Corollary 5.2 of Chap. 1. Again by the generalized principal ideal theorem, $ht\ M \leq k+1 \leq n+1$. This is for each maximal ideal M of $A[[X]]$. Then $dim\ A[[X]] \leq n + 1$. $\square$

By induction, we have the following corollary.

Corollary 1.26 *Let A be a Noetherian ring of finite Krull dimension n. Then for each positive integer m, we have $dim\ A[[X_1, \ldots, X_m]] = n + m$.*

2 Transfer of the SFT Property to the Amalgamation

Let $f : A \longrightarrow B$ be a homomorphism of ring and J an ideal of B. Recall from Sect. 7, Chap. 1, that the set $A \bowtie^f J = \{(a, f(a) + j); a \in A, j \in J\}$ is a subring of the product ring $A \times B$, called the amalgamation of A with B along J with respect to f. The goal of this section is to find necessary and sufficient conditions in order that $A \bowtie^f J$ be an SFT-ring. The results are due to Louartiti (2013).

Lemma 2.1 (Arnold 1973) *A homomorphic image of an SFT-ring is an SFT-ring. In particular, the quotient of an SFT-ring is an SFT-ring.*

Proof Let $f : A \longrightarrow B$ be a homomorphism onto between rings such that A has the SFT property. Let J be an ideal of B. The ideal $I = f^{-1}(J)$ of A satisfies $J = f(I)$. By hypothesis, there exist $n \in \mathbb{N}^*$ and a finitely generated ideal $F \subseteq I$ of A such that $x^n \in F$ for each $x \in I$. Since f is onto, $F' = f(F)$ is a finitely generated ideal of B included in $f(I) = J$. For each $y \in J$, there exists $x \in I$ such that $y = f(x)$, then $y^n = f(x^n) \in f(F) = F'$. $\square$

Lemma 2.2 *The sum of two SFT-ideals is an SFT-ideal.*

Proof Let I and J be two SFT-ideals of a ring A. There exist $n, n' \in \mathbb{N}^*$ and two finitely generated ideals $F \subseteq I$ and $F' \subseteq J$ of A such that for each $x \in I$ and $y \in J$, $x^n \in F$ and $y^{n'} \in F'$. Then $(x + y)^{n+n'} = \displaystyle\sum_{i:0}^{n+n'} C_{n+n'}^i x^i y^{n+n'-i} =$

$\left(\displaystyle\sum_{i:0}^{n} C_{n+n'}^i x^i y^{n-i}\right)y^{n'} + \left(\displaystyle\sum_{i:n+1}^{n+n'} C_{n+n'}^i x^{i-n} y^{n+n'-i}\right)x^n \in F' + F$. Since $F + F'$ is a finitely generated ideal of A included in $I + J$, then $I + J$ is an SFT-ideal. $\square$

Lemma 2.3 *Let $f : A \longrightarrow B$ be a homomorphism of rings, J an ideal of B, and I an ideal of A.*

1. *If $I \bowtie^f J$ is an SFT-ideal of the ring $A \bowtie^f J$, then I is an SFT-ideal of A.*
2. *Suppose that J is an SFT-ideal of the ring $f(A) + J$. Then $I \bowtie^f J$ is an SFT-ideal of the ring $A \bowtie^f J$ if and only if I is an SFT-ideal of the ring A.*

Proof

(1) There exist $n \in \mathbb{N}^*$ and a finitely generated ideal $F = \big((a_i, f(a_i) + j_i); \ 1 \le i \le m\big) \subseteq I \bowtie^f J$ of the ring $A \bowtie^f J$, $a_1, \ldots, a_m \in I$ and $j_1, \ldots, j_m \in J$, such that for each $x \in I \bowtie^f J$, $x^n \in F$. The ideal $F' = (a_1, \ldots, a_m)A$ is finitely generated and included in I. For each $a \in I$, $(a, f(a)) \in I \bowtie^f J$, then $(a^n, f(a^n)) = (a, f(a))^n \in F$. So $a^n \in F'$. Then I is an SFT-ideal of the ring A.

(2) The direct way follows from (1). For the inverse way, since J is an SFT-ideal of the ring $f(A) + J$, there exist $n \in \mathbb{N}^*$ and a finitely generated ideal $F = (j_1, \ldots, j_m) \subseteq J$ of $f(A) + J$ such that $b^n \in F$ for each $b \in J$. The finitely generated ideal $\mathscr{F} = \big((0, j_l); \ 1 \le l \le m\big)$ of $A \bowtie^f J$ is included in $I \bowtie^f J$. Since I is an SFT-ideal of A, there exist $n' \in \mathbb{N}^*$ and a finitely generated ideal $F' = (i_1, \ldots, i_s) \subseteq I$ of A such that $a^{n'} \in F'$ for each $a \in I$. Consider the finitely generated ideal $\mathscr{I} = \big((i_1, f(i_1)); \ldots; (i_s, f(i_s))\big)$ of $A \bowtie^f J$ included in $I \bowtie^f J$. Any element of $I \bowtie^f J$ is of the form $x = (i, f(i) + j)$ with $i \in I$ and $j \in J$. Then $x = (i, f(i)) + (0, j)$, a sum of two elements of $I \bowtie^f J$, with $(i, f(i))^{n'} = (i^{n'}, f(i^{n'})) \in \mathscr{I}$ and $(0, j)^n = (0, j^n) \in \mathscr{F}$. Indeed,

$$i^{n'} = \sum_{k:1}^{s} a_k i_k \text{ with } a_k \in A, \text{ then } f(i^{n'}) = \sum_{k:1}^{s} f(a_k) f(i_k) \text{ and } (i^{n'}, f(i^{n'})) =$$

$$\sum_{k:1}^{s} (a_k i_k, f(a_k) f(i_k)) = \sum_{k:1}^{s} (a_k, f(a_k))(i_k, f(i_k)) \text{ with } (a_k, f(a_k)) \in A \bowtie^f$$

J and $(i_k, f(i_k)) \in \mathscr{I}$. Also, we have $j^n = \sum_{k:1}^{m} (f(a'_k) + j'_k) j_k$ with $a'_k \in A$ and

$j'_k \in J$. Then $(0, j^n) = \sum_{k:1}^{m} (0, (f(a'_k) + j'_k) j_k) = \sum_{k:1}^{m} (a'_k, f(a'_k) + j'_k)(0, j_k)$

with $(a'_k, f(a'_k) + j'_k) \in A \bowtie^f J$ and $(0, j_k) \in \mathscr{F}$. As in the proof of the preceding lemma, we show that $x^{n+n'} \in \mathscr{I} + \mathscr{F}$. Since $\mathscr{I} + \mathscr{F}$ is a finitely generated ideal of $A \bowtie^f J$ included in $I \bowtie^f J$, then $I \bowtie^f J$ is an SFT-ideal of $A \bowtie^f J$. □

Remark 2.4 With the notations of the preceding lemma, J is an SFT-ideal of $f(A) + J$ if and only if J is an SFT-ideal of B. More generally, let $A \subseteq B$ be an extension of rings with a common ideal J. Then J is an SFT-ideal of A if and only if J is an SFT-ideal of B. Indeed, for the first way, there exist a finitely generated ideal $F \subseteq J$ of A and $n \in \mathbb{N}^*$ such that for each $x \in J$, $x^n \in F$. Since FB is a finitely generated ideal of B with $F \subseteq FB \subseteq J$, we have the result. Conversely, there exist $a_1, \ldots, a_s \in J$ and $n \in \mathbb{N}^*$ such that for each $x \in J$, $x^n \in (a_1, \ldots, a_s)B$.

Put $x^n = b_1 a_1 + \ldots + b_s a_s$ with $b_1, \ldots, b_s \in B$. Then $x^{2n} = (\sum_{i:1}^{s} b_i a_i)^2 =$

$\sum_{i:1}^{s} (a_i b_i^2) a_i + \sum_{i \neq j} (2 b_i b_j a_i) a_j \in (a_1, \ldots, a_s) A$ since $a_i b_i^2$ and $2 b_i b_j a_i \in J \subseteq A$.

Now, we are ready to prove the main result of this section.

Theorem 2.5 *Let $f : A \longrightarrow B$ be a homomorphism of rings and J an ideal of B. Then $A \bowtie^f J$ is an SFT-ring if and only if A and $f(A) + J$ are SFT-rings.*

Proof "$\Longrightarrow$" By Proposition 7.3 of Chap. 1, $(A \bowtie^f J)/(f^{-1}(J) \times (0)) \simeq f(A) + J$ and $(A \bowtie^f J)/((0) \times J) \simeq A$. By Lemma 2.1, the rings $f(A) + J$ and A are SFT. We can also remark that if we denote $p_A : A \bowtie^f J \subseteq A \times B \longrightarrow A$ and $p_B : A \bowtie^f J \subseteq A \times B \longrightarrow B$ the canonical projections, then $p_A(A \bowtie^f J) = A$ and $p_B(A \bowtie^f J) = f(A) + J$. We conclude by Lemma 2.1.

"$\Longleftarrow$" By Proposition 1.19, it suffices to show that the prime ideals of $A \bowtie^f J$ are SFT. By Remark 7.4 of Chap. 1, the prime ideals $\mathscr{P}$ of $A \bowtie^f J$ have two types:

First case: $\mathscr{P} = P \bowtie^f J$ with P a prime ideal of A. By hypothesis, J is an SFT-ideal of $f(A) + J$ and P is an SFT-ideal of A. By Lemma 2.3 (2), $\mathscr{P}$ is an SFT-ideal of $A \bowtie^f J$.

Second case: $\mathscr{P} = \overline{Q}^f = \{(a, f(a) + j); \ a \in A, \ j \in J, \ f(a) + j \in Q\}$ with Q a prime ideal of the ring B. Since $f(A) + J$ is an SFT-subring of B, then $Q_0 = Q \cap (f(A) + J)$ is an SFT-ideal of $f(A) + J$. There exist $k_0 \in \mathbb{N}^*$, $a_1, \ldots, a_n \in A$, $j_1, \ldots, j_n \in J$ with $f(a_i) + j_i \in Q$, $1 \leq i \leq n$, such that for each $x \in Q_0$, $x^{k_0} \in (f(a_i) + j_i, 1 \leq i \leq n)(f(A) + J)$ (1).

The ideal $F_0 = ((a_i, f(a_i) + j_i); \ 1 \leq i \leq n)$ of $A \bowtie^f J$ is finitely generated and included in $\overline{Q}^f = \mathscr{P}$. The first projection $p_A : A \bowtie^f J \subseteq A \times B \longrightarrow A$ is a homomorphism onto, then $p_A(\overline{Q}^f)$ is an ideal of A. Since $f^{-1}(J)$ is also an ideal of A, then $I = f^{-1}(J) \cap p_A(\overline{Q}^f)$ is an SFT-ideal of A. There exist $k_1 \in \mathbb{N}^*$ and elements $a_{n+1}, \ldots, a_m \in I$ such that for each $x \in I$, $x^{k_1} \in (a_{n+1}, \ldots, a_m) A$.   (2).

By definition of I, the elements $f(a_{n+1}), \ldots, f(a_m) \in J$ and there exist $j_{n+1}, \ldots, j_m \in J$ such that the elements $(a_{n+1}, f(a_{n+1}) + j_{n+1}); \ \ldots; \ (a_m, f(a_m) + j_m) \in \overline{Q}^f$. The ideal $F_1 = ((a_{n+1}, f(a_{n+1}) + j_{n+1}); \ \ldots; \ (a_m, f(a_m) + j_m))$ of $A \bowtie^f J$ is finitely generated and included in $\overline{Q}^f = \mathscr{P}$. Since J is an ideal of the subring $f(A) + J$ of B and Q is an ideal of B, then $Q_1 = J \cap Q$ is an SFT-ideal of $f(A) + J$. There exist $k_2 \in \mathbb{N}^*$ and elements $j_{m+1}, \ldots, j_l \in Q_1$ such that for each $x \in Q_1$, $x^{k_2} \in (j_{m+1}, \ldots, j_l)(f(A) + J)$ (3).

The ideal $F_2 = ((0, j_{m+1}); \ \ldots; \ (0, j_l))$ of $A \bowtie^f J$ is finitely generated and is included in $\overline{Q}^f = \mathscr{P}$. The ideal $F = F_0 + F_1 + F_2$ of the ring $A \bowtie^f J$ is finitely generated and is included in $\overline{Q}^f = \mathscr{P}$.

We will show that for each $y \in \mathscr{P}$, $y^{k_0 k_1 k_2} \in F$. Indeed, $y = (a, f(a) + j)$ with $a \in A$, $j \in J$, and $f(a) + j \in Q$, then $f(a) + j \in Q \cap (f(A) + J) = Q_0$.

By (1), $\left(f(a)+j\right)^{k_0} \in \left(f(a_i)+j_i;\ 1 \le i \le n\right)(f(A)+J)$.

Put $\left(f(a)+j\right)^{k_0} = \sum_{i:1}^{n} \left(f(a_i)+j_i\right)\left(f(b_i)+j_i'\right)$ with $b_i \in A$ and $j_i' \in J$ for $1 \le i \le n$.

Then $f\left(a^{k_0} - \sum_{i:1}^{n} a_i b_i\right) \in J$, so the element $\alpha = a^{k_0} - \sum_{i:1}^{n} a_i b_i \in f^{-1}(J)$.

Then $y^{k_0} = (a, f(a)+j)^{k_0} = \left(a^{k_0}, (f(a)+j)^{k_0}\right) = \left(\alpha + \sum_{i:1}^{n} a_i b_i, \sum_{i:1}^{n} (f(a_i)+$

$j_i)\left(f(b_i)+j_i'\right)\right) = \sum_{i:1}^{n} \left(a_i b_i, \left(f(a_i)+j_i\right)\left(f(b_i)+j_i'\right)\right) + (\alpha, 0) = \sum_{i:1}^{n} \left(a_i, f(a_i)+\right.$

$j_i\right)\left(b_i, f(b_i)+j_i'\right) + (\alpha, 0)$.

Since $\sum_{i:1}^{n} \left(a_i, f(a_i)+j_i\right)\left(b_i, f(b_i)+j_i'\right) \in F_0$, then $y^{k_0} = f_0 + (\alpha, 0)$ with $f_0 \in F_0$ (4).

Since $(\alpha, 0) = y^{k_0} - f_0 \in \mathscr{P} = \overline{Q}^f$, then $\alpha \in p_A(\overline{Q}^f)$, so $\alpha \in f^{-1}(J) \cap p_A(\overline{Q}^f) = I$. By (2), $\alpha^{k_1} \in (a_{n+1}, \ldots, a_m)A$. Put $\alpha^{k_1} = \sum_{i:n+1}^{m} a_i a_i'$ with $a_{n+1}', \ldots, a_m' \in A$.

Then $(\alpha, 0)^{k_1} = (\alpha^{k_1}, 0) = \sum_{i:n+1}^{m} (a_i a_i', 0) = \sum_{i:n+1}^{m} \left(a_i, f(a_i)+j_i\right)\left(a_i', f(a_i')\right) +$

$(0, e)$ with $e = -\sum_{i:n+1}^{m} \left(f(a_i)+j_i\right)f(a_i') = -f\left(\sum_{i:n+1}^{m} a_i a_i'\right) - \sum_{i:n+1}^{m} j_i f(a_i') =$

$-f(\alpha^{k_1}) - \sum_{i:n+1}^{m} j_i f(a_i') = -f(\alpha)^{k_1} - \sum_{i:n+1}^{m} j_i f(a_i') \in J$. So $e \in J$.

The element $f_1 = \sum_{i:n+1}^{m} \left(a_i, f(a_i)+j_i\right)\left(a_i', f(a_i')\right) \in F_1$ and $(\alpha, 0)^{k_1} = f_1 +$

$(0, e)$ (5).

By definition of $\overline{Q}^f$, since $(0, e) = (\alpha, 0)^{k_1} - f_1 \in \overline{Q}^f$, then $e \in Q$, so $e \in J \cap Q = Q_1$. By (3), $e^{k_2} \in (j_{m+1}, \ldots, j_l)(f(A)+J)$. Put $e^{k_2} = \sum_{i:m+1}^{l} (f(c_i)+$

$j_i')j_i$ with $c_i \in A$ and $j_i' \in J$. Then $(0, e)^{k_2} = \left(0, \sum_{i:m+1}^{l} (f(c_i)+j_i')j_i\right) =$

$\sum_{i:m+1}^{l} \left(0, (f(c_i)+j_i')j_i\right) = \sum_{i:m+1}^{l} \left(c_i, f(c_i)+j_i'\right)(0, j_i) \in F_2$. So $(0, e)^{k_2} \in F_2$ (6).

Conclusion: By (4), $y^{k_0} = f_0 + (\alpha, 0)$ with $f_0 \in F_0$. Then $(y^{k_0})^{k_1} = f_0' + (\alpha, 0)^{k_1}$ with $f_0' \in F_0$. By (5), $y^{k_0 k_1} = f_0' + f_1 + (0, e)$ with $f_1 \in F_1$. Then $(y^{k_0 k_1})^{k_2} = f_1' + (0, e)^{k_2}$ with $f_1' \in F_0 + F_1$. But by (6), $(0, e)^{k_2} \in F_2$. Then $y^{k_0 k_1 k_2} \in F_0 + F_1 + F_2 = F$. $\square$

Corollary 2.6 *Let A be a ring, I an ideal of A, $s : A \longrightarrow A/I$ the canonical homomorphism, and J an ideal of A/I. Then $A \bowtie^s J$ is an SFT-ring if and only if A is an SFT-ring.*

Proof We have $s(A) + J = A/I + J = A/I$. By Lemma 2.1, if A is an SFT-ring, so is A/I. $\square$

Corollary 2.7 *Let A be a ring, I an ideal of A, and $i : A \longrightarrow A$ the identity homomorphism. Then $A \bowtie^i I$ is an SFT-ring if and only if A is an SFT-ring.*

Proof We have $i(A) + I = A + I = A$. $\square$

3 Transfer of the SFT Property to the Polynomial Ring

The results of this section are due to Park (2019).

Lemma 3.1 *Let I be an ideal of a ring A. Then $I[X]$ is an SFT-ideal of $A[X]$ if and only if I is an SFT-ideal of A.*

Proof " $\Longrightarrow$ " There exist $n \in \mathbb{N}^*$ and $f_1(X), \ldots, f_s(X) \in I[X]$ such that for each $f(X) \in I[X]$, $f^n \in (f_1, \ldots, f_s)A[X]$. Put $a_i = f_i(0) \in I$, for $1 \leq i \leq n$. For each $x \in I$, $x^n \in (a_1, \ldots, a_n)A$.

" $\Longleftarrow$ " There exist $m \in \mathbb{N}^*$ and a finitely generated ideal $F \subseteq I$ of A such that for each $x \in I$, $x^m \in F$. For each $\bar{x}$ in the ideal I/F of the ring A/F, $\bar{x}^m = 0$. By Lemma 1.15, $(I/F)[X]$ is a nil ideal of bounded index of the ring $(A/F)[X]$. There exists $n \in \mathbb{N}^*$ such that for each $f \in I[X]$, $\bar{f}^n = 0$ in $(A/F)[X]$, then $f^n \in F[X] = FA[X]$. Since $FA[X]$ is a finitely generated ideal of $A[X]$ included in $I[X]$, then $I[X]$ is an SFT-ideal. $\square$

Example Let (A, M) be a nondiscrete valuation domain of rank one. By the preceding lemma, the ideal $M[X]$ is not SFT. The ideal $M + XA[X]$ is also non-SFT. Indeed, suppose that there exist $n \in \mathbb{N}^*$ and a finitely generated ideal $\mathscr{F} \subseteq M + XA[X]$ of $A[X]$ such that $f^n \in \mathscr{F}$ for each $f \in M + XA[X]$. The set $F = \{f(0); \ f \in \mathscr{F}\}$ is a finitely generated ideal of A included in M and for each $a \in M$, $a^n \in F$, which is absurd.

Lemma 3.2 *Let A be an SFT-integral domain and $0 \neq f(X) \in A[X]$.*
Then $f(X).c(f)^{-1}[X]$ is an SFT-integral ideal of the domain $A[X]$.

Proof The integral ideal $c(f)c(f)^{-1}$ of A is SFT. There exist $k \in \mathbb{N}^*$ and a finitely generated ideal $F \subseteq c(f)c(f)^{-1}$ of A such that $a^k \in F$ for each $a \in c(f)c(f)^{-1}$. There exists a finitely generated fractional ideal $J \subseteq c(f)^{-1}$ of A such that $F \subseteq c(f)J$. Indeed, if $f(X) = a_0 + a_1X + \ldots + a_nX^n \in A[X]$ and $F = (b_1, \ldots, b_s)$, then

$$b_i = \sum_{j:0}^{n} a_jc_{ij} \text{ with } c_{ij} \in c(f)^{-1} \text{ for } 1 \leq i \leq s. \text{ We take } J = (c_{ij}; 1 \leq i \leq s, 0 \leq$$

$j \leq n)$. Since $F \subseteq c(f)J$, then $F[X] \subseteq c(f).J[X]$. For each element $\overline{a}$ in the ideal $c(f)c(f)^{-1}/F$ of the ring A/F, $\overline{a}^k = 0$. By Lemma 1.15, $(c(f)c(f)^{-1}/F)[X]$ is a nil ideal of bounded index of the ring $(A/F)[X]$. There exists $m \in \mathbb{N}^*$ such that for each $h \in (c(f)c(f)^{-1})[X]$, $\overline{h}^m = 0$ in $(A/F)[X]$, then $h^m \in F[X]$. In particular, for each $g \in c(f)^{-1}[X]$, $(fg)^m \in F[X] \subseteq c(f).J[X]$. Then $(fg)^{m+1} = fg(fg)^m \in (gc(f)).fJ[X] \subseteq A[X].fJ[X] = fJ[X]$ with $fJ[X] = fJ.A[X]$ a finitely generated ideal of $A[X]$ included in $f.c(f)^{-1}[X]$. Then $f.c(f)^{-1}[X]$ is an SFT-integral ideal of the ring $A[X]$. $\square$

Proposition 3.3 *Let A be an SFT-integral domain that is integrally closed of dimension one. Then $A[X]$ is an SFT-domain.*

Proof By Proposition 1.19, it suffices to show that the prime ideals of $A[X]$ are SFT. Let $(0) \neq Q \in spec(A[X])$ and $P = Q \cap A$. Then $P[X] \subseteq Q$.

First case: $P \neq (0)$. Since $dim\, A = 1$, P is a maximal ideal of A. Either $Q = P[X]$, then it is SFT by Lemma 3.1, or $P[X] \subset Q$. In the last case, $Q/P[X]$ is an ideal of the PID $A[X]/P[X] \simeq (A/P)[X]$. There exists $f(X) \in Q$ such that $Q/P[X] = (\overline{f})$, then $Q = P[X] + fA[X]$ is an SFT-ideal by Lemma 2.2.

Second case: $P = (0)$. By Proposition 26.3, Chap. 3, $Q = f(X)K[X] \cap A[X]$ with $f(X) \in Q$ and K the quotient field of A. By Proposition 26.1, Chap. 3, since A is integrally closed, $Q = f(X).c(f)^{-1}[X]$ is an SFT-ideal of $A[X]$, by the preceding lemma. $\square$

Corollary 3.4 *Let A be an SFT-integral domain. Suppose that for each prime ideal P of A, the domain A/P is integrally closed. Then $A[X]$ is an SFT-domain.*

Proof Let $(0) \neq Q \in spec(A[X])$ and $P = Q \cap A$. Then $P[X] \subseteq Q$.

First case: $Q = P[X]$, it is an SFT-ideal, by Lemma 3.1.

Second case: $P[X] \subset Q$. Then $Q/P[X]$ is a nonzero prime ideal of $A[X]/P[X] \simeq (A/P)[X]$ with $(Q/P[X]) \cap (A/P) = (Q \cap A)/P = P/P = (0)$. By Proposition 26.3 (1) of Chap. 3, $Q/P[X] = (\overline{f(X)}K[X]) \cap (A/P)[X]$ where $f(X) \in Q$ and K is the quotient field of A/P. By Proposition 26.1 of Chap. 3, since A/P is integrally closed, $Q/P[X] = \overline{f}.c(\overline{f})^{-1}[X]$. By Lemma 2.1, since A is an SFT-ring, so is the ring A/P. By Lemma 3.2, $Q/P[X]$ is an SFT-ideal of $A[X]/P[X]$. We will show that Q is an SFT-ideal of $A[X]$. By Lemma 3.1, $P[X]$ is an SFT-ideal of $A[X]$. There exist $k_1 \in \mathbb{N}^*$ and a finitely generated ideal J_1 of $A[X]$ included in $P[X] \subset Q$ such that $g^{k_1} \in J_1$ for each $g \in P[X]$. Also, there exist $k_2 \in \mathbb{N}^*$ and a finitely generated ideal $\mathscr{I}$ of $(A/P)[X]$ included in

$Q/P[X]$ such that $\overline{g}^{k_2} \in \mathscr{I}$ for each $\overline{g} \in Q/P[X]$. The ideal $\mathscr{I}$ is the image by the canonical homomorphism onto $\pi : A[X] \longrightarrow A[X]/P[X]$ of a finitely generated ideal J_2 of $A[X]$ included in Q. Let $k = k_1 k_2 \in \mathbb{N}^*$ and $J = J_1 + J_2$, a finitely generated ideal of $A[X]$ included in Q. We will show that for each $g \in Q$, $g^k \in J$. Indeed, $\overline{g}^{k_2} \in \mathscr{I}$, then $\overline{g}^{k_2}$ is the class of an element of J_2 modulo $P[X]$. Then $g^{k_2} \in P[X] + J_2$. Put $g^{k_2} = g_1 + g_2$ with $g_1 \in P[X]$ and $g_2 \in J_2$, then $g_1^{k_1} \in J_1$. So

$$g^k = g^{k_1 k_2} = (g^{k_2})^{k_1} = (g_1 + g_2)^{k_1} = g_1^{k_1} + \sum_{i:0}^{k_1-1} C_{k_1}^i g_1^i g_2^{k_1-i} \in J_1 + J_2 = J. \quad \square$$

Corollary 3.5 *Let A be an SFT-Prufer domain. Then $A[X]$ is an SFT-domain.*

Proof For each $P \in spec(A)$, the domain A/P is a Prufer, then integrally closed. We conclude by the preceding corollary. $\qquad\square$

4   J. Coykendall Example (2002)

In the first section, we showed that if a ring A does not have the SFT property, then $dim\ A[[X]] = \infty$. There are some natural questions. If A is an SFT-ring with finite dimension, is the dimension of $A[[X]]$ finite? Does the SFT property pass to the formal power series ring? The following example, introduced and studied by J. Coykendall, answers negatively the last question. It also shows that the SFT property is not sufficient for the formal power series ring to be of finite dimension.

Notation Consider the monoid algebra $\mathbb{F}_2[\mathbb{Q}+]$, its maximal ideal $M = \left\{ \sum_{i:0}^{n} \epsilon_i t^{\alpha_i};\ n \in \mathbb{N},\ \epsilon_i \in \mathbb{F}_2,\ \alpha_i \in \mathbb{Q}^* + \right\} = (t^{\alpha};\ \alpha > 0)$, and the domains $V = \mathbb{F}_2[\mathbb{Q}+]_M$ and $V_1 = \mathbb{F}_2 + tV \subseteq V$. The ring V_1 is an example of the classical construction $"K + M"$.

Example Let $n \geq 2$ be an integer. Then $t^{\frac{1}{n}} \in V \setminus V_1$. Indeed, suppose that $t^{\frac{1}{n}} \in V_1$. Then $t^{\frac{1}{n}} = \epsilon + t t^{\alpha} \frac{f}{g}$ with $\epsilon \in \mathbb{F}_2, \alpha \geq 0, f, g \in \mathbb{F}_2[\mathbb{Q}+] \setminus M$. So $g t^{\frac{1}{n}} = g \epsilon + t^{\alpha+1} f$. If $\epsilon = 0$, by identification, we find $\frac{1}{n} = \alpha + 1$, which is absurd. If $\epsilon = 1$, then $g(1 + t^{\frac{1}{n}}) = t^{\alpha+1} f$, and by identification, we find $0 = \alpha + 1$, which also absurd.

Proposition 4.1 *V is a nondiscrete valuation domain of rank one.*

Proof To show that V is a valuation domain, it suffices to show that its principal ideals are comparable. This is also equivalent to show that if f and g are two nonzero elements of $\mathbb{F}_2[\mathbb{Q}+]$, then fV and gV are comparable, since the elements of $\mathbb{F}_2[\mathbb{Q}+] \setminus M$ are invertible in V. We have $f = t^r f_1$ and $g = t^s g_1$ with $r, s \in \mathbb{Q}+$ and $f_1, g_1 \in \mathbb{F}_2[\mathbb{Q}+] \setminus M$. For example, $r \leq s$. Then $g = t^s g_1 = \frac{t^r f_1 t^{s-r} g_1}{f_1} = f \frac{t^{s-r} g_1}{f_1} \in$

fV, so $gV \subseteq fV$. The extension $\mathbb{F}_2[t] \subset \mathbb{F}_2[\mathbb{Q}+]$ is integral since a monomial of the form $f = t^{\frac{p}{q}}$, with $p, q \in \mathbb{N}^*$, satisfies $f^q = t^p \in \mathbb{F}_2[t]$. Since $\mathbb{F}_2[t]$ is a PID, $dim\ \mathbb{F}_2[\mathbb{Q}+] = dim\ \mathbb{F}_2[t] = 1$. Since $V = \mathbb{F}_2[\mathbb{Q}+]_M$, $dim\ V = ht\ M \leq 1$. But $(0) \subset M$, then $ht\ M = 1$ and $dim\ V \leq 1$. The domain V is not a field. Indeed, in the contrary case, $\frac{1}{t} \in V$, then $\frac{1}{t} = \frac{f}{g_1}$ with $f \in \mathbb{F}_2[\mathbb{Q}+]$ and $g_1 \in \mathbb{F}_2[\mathbb{Q}+] \setminus M$. Then $g_1 = tf \in M$, which is absurd. We conclude that $dim\ V = 1$. To show that V is a nondiscrete valuation domain, it suffices to show that its maximal ideal MV is not principal. In the contrary case, $MV = fV$ with $f \in M$, $f = t^r f_1, r \in \mathbb{Q}^*+$, and $f_1 \in \mathbb{F}_2[\mathbb{Q}+] \setminus M$. Let $n \in \mathbb{N}^*$ be such that $\frac{1}{n} < r$. Since $t^{\frac{1}{n}} \in M$, there exist $g \in V, g = \frac{t^s h_1}{k_1}$ with $s \in \mathbb{Q}+$ and $h_1, k_1 \in \mathbb{F}_2[\mathbb{Q}+] \setminus M$ such that $t^{\frac{1}{n}} = fg$. Then $t^{\frac{1}{n}} k_1 = t^{r+s} f_1 h_1$. By identification of the least degree, we find $\frac{1}{n} = r + s$, which is absurd.   □

Proposition 4.2 *The integral closure of V_1 is V. In particular, V_1 is not a valuation domain and $dim\ V_1 = dim\ V = 1$.*

Proof Since the domains $V_1 \subset V$ share the nonzero ideal tV, they have the same quotient field. Since V is integrally closed, it contains the integral closure of V_1. Conversely, let $f \in V$, $f = t^r \frac{f_1}{g_1}$ with $r \in \mathbb{Q}+$ and $f_1, g_1 \in \mathbb{F}_2[\mathbb{Q}+] \setminus M$. We distinguish two cases: If $r > 0$, put $r = \frac{p}{q}$ with $p, q \in \mathbb{N}^*$, then $f^q = t^p (\frac{f_1}{g_1})^q \in tV \subseteq V_1$. So f is integral over V_1. If $r = 0$, $f = \frac{f_1}{g_1}$. Since the constant terms of f_1 and g_1 are equal to 1, then $f - 1 = \frac{f_1 - g_1}{g_1}$ is integral over V_1, by the first case. The same is true for f. The ring V_1 is not a valuation domain since it is not integrally closed.   □

Contrary to the case of V, the domain V_1 is SFT.

Proposition 4.3 *V_1 is a non-Noetherian SFT-domain.*

Proof

(1) Since $V_1/tV \simeq \mathbb{F}_2$ is a field, the ideal tV is maximal in V_1. Since $V_1 \subset V$ is an integral extension, the contraction $Spec(V) \longrightarrow Spec(V_1)$ is onto. Then the prime ideals of V_1 are traces of prime ideals of V. Since $Spec(V)$ is reduced to two elements, then $Spec(V_1) = \{(0), tV\}$. It suffices to show that tV is a SFT-ideal of V_1. Let the principal ideal $F = tV_1$ of V_1. Then $F = tV_1 \subseteq tV$ and $(tV)^2 = t(tV) \subseteq tV_1 = F$.
(2) Suppose that the ring V_1 is Noetherian. The ideal tV is finitely generated in V_1. Put $tV = (tf_1, \ldots, tf_s)V_1$ with $f_1, \ldots, f_s \in V$, then $V = (f_1, \ldots, f_s)V_1$. So V is a finitely generated V_1-module. By Eakin-Nagata theorem (see Sect. 3 of Chap. 1), V is also Noetherian, which is absurd. Then V_1 is not Noetherian.
                                                                                                                              □

In the following proposition, we show that V_1 is an example of a non-Noetherian SFT-integral domain of dimension one that it is not integrally closed such that $V_1[X]$ is an SFT-ring.

Proposition 4.4 *[Park 2019] $V_1[X]$ is an SFT-domain.*

Proof The only prime ideals of V_1 are (0) and tV.

Let Q be a nonzero prime ideal of $V_1[X]$. Then $Q \cap V_1 \in \{0, tV\}$.

First case: $Q \cap V_1 = tV$. Then $(tV)[X] \subseteq Q$. Since $V_1[X]/(tV)[X] \simeq (V_1/tV)[X] \simeq \mathbb{F}_2[X]$ is a PID, the ideal $Q/(tV)[X] = (\overline{f})$ with $f(X) \in Q$. Then $Q = (tV)[X] + fV_1[X]$. By Lemma 3.1, $(tV)[X]$ is an SFT-ideal of $V_1[X]$. So Q is an SFT-ideal of $V_1[X]$, as a sum of two SFT-ideals.

Second case: $Q \cap V_1 = (0)$. By Proposition 26.3 of Chap. 3, $Q = (lK[X]) \cap V_1[X]$ with $l(X) \in V_1[X] \subset V[X]$, where K is the common quotient field of V_1, V and $\mathbb{F}_2[\mathbb{Q}+]$. Since V is a valuation domain, $l(X) = al'(X)$ with $0 \neq a \in V$ and $l'(X) \in V[X]$ is a polynomial having at least one coefficient equal to 1. Put $l'(X) = l_0 + l_1 X + \ldots + l_n X^n \in V[X]$, $l_i = \frac{1}{c}(\epsilon_i + t^{\alpha_i} b_i)$ with $\epsilon_i \in \mathbb{F}_2$, $\alpha_i > 0$, $c \in \mathbb{F}_2[\mathbb{Q}+] \setminus M$ and $b_i \in \{0\} \cup (\mathbb{F}_2[\mathbb{Q}+] \setminus M)$. Put $c_i = \epsilon_i + t^{\alpha_i} b_i \in \mathbb{F}_2[\mathbb{Q}+]$ for $0 \leq i \leq n$ and $f(X) = c_0 + c_1 X + \ldots + c_n X^n$. One of the c_i is equal to c, so invertible in V. Then the content $c_V(f) = V$. Note that $fK[X] = l'K[X] = lK[X]$. Let $k \in \mathbb{N}^*$ be an integer such that $2^k \alpha_i \geq 1$ for $0 \leq i \leq n$. Since the characteristic of V is 2, then $c_i^{2^k} = \epsilon_i^{2^k} + t^{2^k \alpha_i} b_i^{2^k} = \epsilon_i + t^{2^k \alpha_i} b_i^{2^k} \in \mathbb{F}_2 + tV = V_1$ and the polynomial $g = f^{2^k} = c_0^{2^k} + c_1^{2^k} X^{2^k} + \ldots + c_n^{2^k} X^{2^k n} \in V_1[X]$ with $c_V(g) = c_V(f^{2^k}) = c_V(f)^{2^k} = V$ since V is a valuation domain. Then $g = f^{2^k} \in fK[X] \cap V_1[X] = lK[X] \cap V_1[X] = Q$. It suffices to show that for each $r(X) \in Q$, $r^{2^k} \in gV_1[X] \subseteq Q$. Indeed, $r = fh$ with $h \in K[X]$. Then $gh^{2^k} = f^{2^k} h^{2^k} = (fh)^{2^k} = r^{2^k} \in Q \subset V_1[X]$. Since V is a valuation domain, $c_V(g)c_V(h^{2^k}) = c_V(gh^{2^k})$. But $c_V(g) = V$, then $c_V(h^{2^k}) = c_V(gh^{2^k}) \subseteq V_1$, so $h^{2^k} \in V_1[X]$. Then $r^{2^k} = (fh)^{2^k} = gh^{2^k} \in gV_1[X] \subseteq Q$, so Q is an SFT-ideal of $V_1[X]$. $\square$

5 Transfer of the SFT Property to the Formal Power Series Ring

The results of this section are due to Coykendall (2002). Along it, we consider the monoid algebra $\mathbb{F}_2[\mathbb{Q}+]$, its maximal ideal $M = (t^\alpha; \ \alpha > 0)$, and the domains $V = \mathbb{F}_2[\mathbb{Q}+]_M$ and $V_1 = \mathbb{F}_2 + tV \subset V$. The goal is to prove that the SFT property does not pass to the formal power series ring.

Lemma 5.1 *Let A be an integral domain, $a, b \in A$ with $a \neq 0$, $n \in \mathbb{N}^*$ and $f(X) \in A[[X]]$. If $(a + X^n)f(X) = b$, then $f(X) \in A[[X^n]]$.*

Proof We have $f(X) = \frac{b}{a+X^n} = \frac{b}{a} \cdot \frac{1}{1+\frac{X^n}{a}}$. Since $\frac{1}{1+X} = \sum_{i:0}^{\infty} (-1)^i X^i$, then we

have $f(X) = \sum_{i:0}^{\infty} \frac{(-1)^i b}{a^{i+1}} X^{ni} \in K[[X^n]] \cap A[[X]] = A[[X^n]]$, where K is the

quotient field of A.   □

Lemma 5.2 *For all $n \in \mathbb{N}^*$ and $\alpha \in \mathbb{Q}^*+$, the principal ideal $I = (t^\alpha + X^n)V[[X]]$ has a trace (0) over V.*

Proof Suppose that $I \cap V \neq (0)$. There exist $f(X) \in V[[X]]$ and $0 \neq \omega \in V$ such that $(t^\alpha + X^n)f(X) = \omega$. By the preceding lemma, $f(X) \in V[[X^n]]$. Without a loss of generality, we may suppose that $n = 1$, then $(t^\alpha + X)f(X) = \omega$. The constant term of $f(X)$ is not zero since $\omega \neq 0$. Then $f(X) = t^{\beta_0} u_0 + \epsilon_1 t^{\beta_1} u_1 X + \epsilon_2 t^{\beta_2} u_2 X^2 + \ldots \in V[[X]]$ with $\epsilon_1, \epsilon_2, \ldots \in \mathbb{F}_2 = \{0, 1\}$, $\beta_0, \beta_1, \ldots \in \mathbb{Q}+$, and $u_0, u_1, \ldots \in U(V)$. The equality $(t^\alpha + X)f(X) = \omega$ becomes $t^{\alpha+\beta_0} u_0 +$

$\left(\epsilon_1 t^{\alpha+\beta_1} u_1 + t^{\beta_0} u_0\right) X + \sum_{i:2}^{\infty} \left(\epsilon_i t^{\alpha+\beta_i} u_i + \epsilon_{i-1} t^{\beta_{i-1}} u_{i-1}\right) X^i = \omega$. Since $\epsilon_1 t^{\alpha+\beta_1} u_1 =$

$t^{\beta_0} u_0$, then $\epsilon_1 = 1$. Since $\epsilon_i t^{\alpha+\beta_i} u_i = \epsilon_{i-1} t^{\beta_{i-1}} u_{i-1}$ for $i \geq 2$, we show by induction that $\epsilon_i = 1$ for each $i \in \mathbb{N}^*$. By taking the valuations in the domain V in the preceding equalities, we see that $\alpha + \beta_i = \beta_{i-1}$ for $i \in \mathbb{N}^*$. Then $i\alpha + \beta_i = \beta_0$ for each $i \in \mathbb{N}^*$, which is absurd since $\alpha > 0$ and the $\beta_i \geq 0$.   □

The infinite product defined in the following lemma is specific to this situation.

Lemma 5.3 *For each $m \in \mathbb{N}$, let $f_m(X) = t^{1/2^m} + X^{2^m} \in V[X]$. For each $n \in \mathbb{N}$,*

the element $g_n(X) = \prod_{m:n}^{\infty} f_m(X)$ is well defined in the domain $V[[X]]$, and for each

integer $k \geq n$, $f_k(X)$ divides $g_n(X)$ in $V[[X]]$.

Proof Let's start by precising the definition of the infinite product: $(b_{10} + b_{11} X + \ldots + b_{1n} X^n + \ldots)(b_{20} + b_{21} X + \ldots + b_{2n} X^n + \ldots) \ldots (b_{m0} + b_{m1} X + \ldots + b_{mn} X^n + \ldots) \ldots = c_0 + c_1 X + \ldots + c_n X^n + \ldots$.

By definition, $c_n = \sum_{\substack{j_1+j_2+\ldots+j_k=n \\ 0<j_1 \leq j_2 \leq \ldots \leq j_k}} \left(\sum_{i_1,i_2,\ldots,i_k \text{ distinct}} (b_{i_1 j_1} b_{i_2 j_2} \ldots b_{i_k j_k} \prod_{i \neq i_1,i_2,\ldots,i_k} \right.$

$b_{i0}))$. It is well possible that in this expression, certain infinite products or sums are not well defined. In our case, it is the infinite product of the monomials $\{t^{\alpha_i}\}_{i \in \mathbb{N}}$.

Put $\prod_{i:0}^{\infty} t^{\alpha_i} = t^{\alpha_0 + \alpha_1 + \cdots}$ when the numerical series $\sum_{i:0}^{\infty} \alpha_i$ converges in $\mathbb{R}$. To prove

the first part of the lemma, it suffices to prove that each coefficient of $g_n(X)$ is an element of V. The constant term of $f_n(X)$ is $c_0 = t^{\frac{1}{2^n} + \frac{1}{2^{n+1}} + \cdots} = t^{\frac{1}{2^{n-1}}} \in V$. For each integer $n \geq 1$, the terms in the infinite sums involved in c_n are almost zero and the infinite products are powers of t which represent subseries of $\frac{1}{2^n} + \frac{1}{2^{n+1}} + \cdots$, then convergent. So $g_n(X) \in V[[X]]$.

For the second part of the lemma, we must show that for each integer $k \geq n$, $f_k(X)$ divides $g_n(X)$ in $V[[X]]$. Let $g(X) = f_n(X) \ldots f_{k-1}(X) f_{k+1}(X) f_{k+2}(X) \ldots$. By the same method of the preceding step, we show that $g(X)$ is a well-defined element of $V[[X]]$, and by identification of the coefficients, we see that $g(X) f_k(X) = g_n(X)$. $\qquad \square$

Example (Coykendall 2005) The series $g_0(X) = \displaystyle\prod_{m:0}^{\infty} f_m(X) = (t + X)(t^{1/2} +$ $X^2)(t^{1/4} + X^4) \ldots (t^{1/2^m} + X^{2^m}) \ldots$ is well defined in the domain $V[[X]]$. Since the characteristic of V is 2, for each integer $m \geq 1$, $t^{1/2^m} + X^{2^m} = (t^{1/2^m} + X)^{2^m}$. The product $(t+X)(t^{1/4}+X)^2(t^{1/16}+X)^4 \ldots (t^{1/2^{2m}}+X)^{2^m} \ldots$ is not even defined. In fact, since each factor of the product contains a linear term, the coefficient of X in the product is not defined. Note that the infinite product is not associative.

Our strategy in the following proposition is to collect information on the formal power series ring over the non-SFT-domain V and then to transport them the formal power series ring over the SFT-domain V_1.

Proposition 5.4 *There exists a prime ideal $\mathscr{P}$ of the domain $V[[X]]$ included in the ideal $(MV)[[X]]$ with trace (0) over V such that $\{f(0); \; f(X) \in \mathscr{P}\} = MV$.*

Proof Let $\mathscr{I}$ be the ideal of $V[[X]]$ generated by the family $\{g_n(X); \; n \in \mathbb{N}^*\}$, where $g_n(X) = \displaystyle\prod_{m:n}^{\infty} f_m(X)$ with $f_m(X) = t^{1/2^m} + X^{2^m}$ for each $m \in \mathbb{N}$. Let v be the natural valuation of V. For each real number $\alpha > 0$, there exists $f(X) \in \mathscr{I}$ such that $v(f(0)) < \alpha$. Indeed, let $N \in \mathbb{N}^*$ be such that $\frac{1}{2^N} < \alpha$. The constant term of the element $g_{N+1}(X) = \displaystyle\prod_{m:N+1}^{\infty} f_m(X)$ of $\mathscr{I}$ is $t^{\frac{1}{2^N}}$, and its valuation is $\frac{1}{2^N} < \alpha$. Let $I = \{f(0); \; f(X) \in \mathscr{I}\}$; it is an ideal of V. We will show that $I = MV$. Recall that $M = (t^{\alpha}; \; \alpha > 0)\mathbb{F}_2[\mathbb{Q}+]$. Since $g_n(0) = t^{\frac{1}{2^{n-1}}} \in M$ for each $n \in \mathbb{N}^*$, we have the first inclusion. Conversely, let $\alpha \in \mathbb{Q}^*+$. There exists $f(X) \in \mathscr{I}$ such that $\beta = v(f(0)) < \alpha$. The element $g(X) = t^{\alpha-\beta} f(X) \in \mathscr{I}$, then $g(0) \in I$ and $g(0) = t^{\alpha-\beta} f(0) = t^{\alpha} t^{-\beta} f(0)$. Since $t^{-\beta} f(0)$ is invertible in V, then $t^{\alpha} \in I$. So $MV \subseteq I$. We will show that $\mathscr{I} \cap V = (0)$. Let $\omega \in \mathscr{I} \cap V$, $\omega = r_1(X) g_{k_1}(X) + \ldots + r_m(X) g_{k_m}(X)$ with $r_1(X), \ldots, r_m(X) \in V[[X]]$ and $k_1, \ldots, k_m \in \mathbb{N}^*$. Let $k > max\{k_1, \ldots, k_m\}$ be an integer. By the preceding lemma, $f_k(X)$ divides $g_{k_i}(X)$ in $V[[X]]$ for $1 \leq i \leq m$. Then $\omega = r(X) f_k(X)$ with $r(X) \in V[[X]]$. By Lemma 5.2, since $f_k(X) = t^{1/2^k} + X^{2^k}$, then $\omega = 0$. Since the ideal $\mathscr{I}$ of $V[[X]]$ is disjoint with the multiplicative set $V \setminus (0)$, it is included in a prime ideal $\mathscr{P}$ of $V[[X]]$ disjoint with $V \setminus (0)$, so it has a trace (0) on V. Moreover, $MV = \{f(0); \; f(X) \in \mathscr{I}\} \subseteq \{f(0); \; f(X) \in \mathscr{P}\}$. Since $\mathscr{P}$ is prime, it does not contain any invertible element of $V[[X]]$. Then the ideal $\{f(0); \; f(X) \in \mathscr{P}\}$ of V is proper. But MV is a maximal ideal of the valuation domain V, then $MV = \{f(0); \ f(X) \in \mathscr{P}\}$. Note that $X \notin \mathscr{P}$. Indeed, in the contrary case, $\mathscr{P}$ contains the constant terms of the generators $\{g_n(X), \ n \in \mathbb{N}^*\}$ of $\mathscr{I}$. Then $\{t^{1/2^n}, n \in \mathbb{N}\} \subseteq \mathscr{P} \cap V = (0)$, which is absurd.

For the inclusion $\mathscr{P} \subseteq (MV)[[X]]$, the equality $\{f(0); \ f(X) \in \mathscr{P}\} = MV$ shows that the constant terms of the elements of $\mathscr{P}$ belong to MV. Suppose that for each natural integer $k \leq N$ and for each series in $\mathscr{P}$, the coefficient of X^k belongs to MV. Let $f(X) = a_0 + a_1 X + \ldots + a_N X^N + a_{N+1} X^{N+1} + \ldots \in \mathscr{P}$. We want to show that the coefficient $a_{N+1} \in MV$. Since $\mathscr{P}$ is a prime ideal of $V[[X]]$, a_0 is not invertible in V, then $v(a_0) > 0$. If $a_0 = 0$, since $X \notin \mathscr{P}$, then $a_1 + a_2 X + \ldots + a_N X^{N-1} + a_{N+1} X^N + \ldots \in \mathscr{P}$, so $a_{N+1} \in MV$. If $a_0 \neq 0$, there exists a series $g(X) = b_0 + b_1 X + \ldots + b_N X^N + b_{N+1} X^{N+1} + \ldots \in \mathscr{P}$ such that $v(b_0) = \frac{1}{2} v(a_0)$. Indeed, let $\alpha = \frac{1}{2} v(a_0) \in \mathbb{Q}^*+$. By the preceding step, there exists $g_1(X) \in \mathscr{I} \subseteq \mathscr{P}$ such that $v(g_1(0)) = \beta < \alpha$. Take $g(X) = t^{\alpha-\beta} g_1(X) \in \mathscr{P}$. Then $g(0) = t^{\alpha-\beta} g_1(0)$ and $v(g(0)) = \alpha - \beta + v(g_1(0)) = \alpha - \beta + \beta = \alpha = \frac{1}{2} v(a_0)$. So $b_0 \neq 0$ so $v\left(\frac{a_0}{b_0}\right) = v(a_0) - v(b_0) = \frac{1}{2} v(a_0) = v\left(t^{\frac{v(a_0)}{2}}\right)$. Then $a_0 = b_0 t^{\frac{v(a_0)}{2}}$ modulo a multiplicative unity of V. Without loss of generality, we can suppose that this unity is equal to 1. The series $t^{\frac{v(a_0)}{2}} g(X) - f(X) = \left(t^{\frac{v(a_0)}{2}} b_1 - a_1\right) X + \ldots + \left(t^{\frac{v(a_0)}{2}} b_N - a_N\right) X^N + \left(t^{\frac{v(a_0)}{2}} b_{N+1} - a_{N+1}\right) X^{N+1} + \ldots \in \mathscr{P}$. Since $X \notin \mathscr{P}$, then $t^{\frac{v(a_0)}{2}} b_1 - a_1 + \left(t^{\frac{v(a_0)}{2}} b_2 - a_2\right) X + \ldots + \left(t^{\frac{v(a_0)}{2}} b_N - a_N\right) X^{N-1} + \left(t^{\frac{v(a_0)}{2}} b_{N+1} - a_{N+1}\right) X^N + \ldots \in \mathscr{P}$. By the induction hypothesis, $t^{\frac{v(a_0)}{2}} b_{N+1} - a_{N+1} \in MV$. Since $t^{\frac{v(a_0)}{2}} b_{N+1} \in MV$ because $t^{\frac{v(a_0)}{2}} \in M$ and $b_{N+1} \in V$, then $a_{N+1} \in MV$. $\quad\square$

The proof of the preceding proposition relies on the following points:

(1) The ideal $\mathscr{I}$ contains series whose constant terms have valuations that are as small as we want.
(2) The trace of $\mathscr{I}$ over V is (0) since traces over V of the principal ideals $f_m V[[X]]$, $m \in \mathbb{N}$, are (0).
(3) Any ideal disjoint with a multiplicative set may be enlarged to a prime ideal again disjoint with this set.

Lemma 5.5 (Condo and Coykendall 1999) *Let A be a ring, $\mathscr{I}$ an SFT radical ideal of $A[[X]]$, and $\{f_i(X); \ i \in \mathbb{N}\}$ a sequence of elements of $\mathscr{I}$. Then*

$$\sum_{i:0}^{\infty} f_i(X) X^i \in \mathscr{I}.$$

Proof The expression $f(X) = \sum_{i:0}^{\infty} f_i(X) X^i$ defines an element of $A[[X]]$. Indeed, for each $n \in \mathbb{N}$, the coefficient of X^n in $f(X)$ is the same as that of X^n in $\sum_{i:0}^{n} f_i(X) X^i$, so it is well defined in A. By Proposition 1.17, since $\mathscr{I}$ is an SFT-ideal

of $A[[X]]$, then $\mathscr{I}[[Y]] \subseteq \sqrt{\mathscr{I}.A[[X, Y]]}$. Let $g(X, Y) = \sum_{i:0}^{\infty} f_i(X)Y^i \in \mathscr{I}[[Y]]$.
There exists $n \in \mathbb{N}^*$ such that $(g(X, Y))^n \in \mathscr{I}.A[[X, Y]]$. Put $(g(X, Y))^n = \sum_{i:0}^{k} h_i(X)r_i(X, Y)$ with $h_i(X) \in \mathscr{I}$ and $r_i(X, Y) \in A[[X, Y]]$ for $1 \leq i \leq k$.
Let $\psi : A[[X, Y]] \longrightarrow A[[X]]$ be the homomorphism defined by $\psi(l(X, Y)) = l(X, X)$ for each $l(X, Y) \in A[[X, Y]]$. By applying ψ to the preceding equality, we find $(g(X, X))^n = \sum_{i:0}^{k} h_i(X)r_i(X, X) \in \mathscr{I}$. But $g(X, X) = \sum_{i:0}^{\infty} f_i(X)X^i = f(X)$, then $(f(X))^n \in \mathscr{I}$, so $f(X) \in \mathscr{I}$ since $\mathscr{I}$ is a radical ideal. $\qquad\square$

Intuitively, the conclusion of the preceding lemma says that under some hypotheses, an infinite sum of elements of $\mathscr{I}$ is again in $\mathscr{I}$, when it is well defined in $A[[X]]$.

Example 1 Let A be a ring. Each finitely generated ideal $\mathscr{I} = (g_1, \ldots, g_n)$ of $A[[X]]$ satisfies the conclusion of the lemma. Indeed, let $(f_i(X))_{i \in \mathbb{N}}$ be a sequence of elements of I. Put $f_i = \sum_{j:1}^{n} r_{ij}g_j$ with $r_{ij}(X) \in A[[X]]$, then $\sum_{i:0}^{\infty} f_i X^i = \sum_{i:0}^{\infty}(\sum_{j:1}^{n} r_{ij}g_j)X^i = \sum_{j:1}^{n}(\sum_{i:0}^{\infty} r_{ij}X^i)g_j \in \mathscr{I}$.

Example 2 The finiteness condition in the preceding example is not necessary. Let A be a ring. The maximal ideals of $A[[X]]$ satisfy the conclusion of the lemma. Indeed, let M be a maximal ideal of A, $(f_i(X))_{i \in \mathbb{N}}$ a sequence of elements of $M + XA[[X]]$, and $f(X) = \sum_{i:0}^{\infty} f_i(X)X^i$. Put $f_i(X) = a_i + Xg_i(X)$ with $a_i \in M$ and $g_i(X) \in A[[X]]$ for each $i \in \mathbb{N}$. Then $f(X) = \sum_{i:0}^{\infty}(a_i + Xg_i)X^i = a_0 + X\big(g_0 + \sum_{i:1}^{\infty} X^{i-1}(a_i + Xg_i)\big) \in M + XA[[X]]$.

Example 3 Let (A, M) be a nondiscrete valuation domain of rank one. Then $MA[[X]] \subset M[[X]]$. Let $f = \sum_{i:0}^{\infty} a_i X^i \in M[[X]] \setminus MA[[X]]$. The ideal $MA[[X]]$ of $A[[X]]$ does not satisfy the conclusion of the lemma. Indeed, $a_i \in MA[[X]]$ for each $i \in \mathbb{N}$, but $f \notin MA[[X]]$.

Example 4 (Fields 1971) Let K be a field with zero characteristic, $n \geq 2$ an integer, and $A = K[Y_i, i \in \mathbb{N}]/(Y_i^n, i \in \mathbb{N})$. The ideal $\mathscr{I} = (\overline{Y_i}, i \in \mathbb{N})A[[X]]$ does not have the conclusion of the lemma. Indeed, put $a_i = \overline{Y_i} \in \mathscr{I}$ for each $i \in \mathbb{N}$. It is clear that for each $i \in \mathbb{N}$, $a_i^n = 0$. Suppose that $f(X) = \sum_{i:0}^{\infty} a_i X^i \in \mathscr{I}$, then

$$f(X) = \sum_{i:0}^{s} f_i(X)a_i \text{ with } f_i(X) \in A[[X]] \text{ and } s \in \mathbb{N}. \text{ Since the } a_i \text{ are nilpotent,}$$

then $f(X)$ is also nilpotent. There exists $m \in \mathbb{N}^*$ such that $f^m = 0$. Then $a_0^m = 0$, so $m \geq n$. Put $f^m = \sum_{k:0}^{\infty} b_k X^k$. Let $t_1 = \inf\{k \in \mathbb{N}; \text{ the terms of the coefficient}$ b_k of X^k in f^m do not admit a_0^n as a factor $\}$. Since $f^m = 0$, the coefficient of X^{t_1} in f^m is zero so $0 = b_{t_1} = c a_0^{n-1} a_1^{m-(n-1)} + (\text{terms admitting } a_0^n \text{ as a factor})$ where $c \in \mathbb{N}^*$. Since the characteristic of A is zero, c is invertible in $K \subset A$, then $a_0^{n-1} a_1^{m-(n-1)} = 0$. By the definition of A, we must have $m - (n-1) \geq n$. We repeat the process in the following way. Let $t_2 = \inf\{k \in \mathbb{N}; \text{ the terms of the}$ coefficient b_k of X^k in f^m do not admit neither a_0^n nor a_1^n as a factor $\}$. Then $0 = b_{t_2} = b a_0^{n-1} a_1^{n-1} a_2^{m-2(n-1)} + (\text{terms admitting either } a_0^n \text{ or } a_1^n \text{ as a factor})$ where $b \in \mathbb{N}^*$, then $b a_0^{n-1} a_1^{n-1} a_2^{m-2(n-1)} = 0$, so $m - 2(n-1) \geq n$. By induction, for each $k \in \mathbb{N}$, we must have $m - k(n-1) \geq n$, then $m \geq n + k(n-1)$, which is absurd. Note that in this example, all the coefficients of the series f are nilpotent with bounded index. But f itself is not nilpotent.

The importance of the prime ideal $\mathscr{P}$ of the domain $V[[X]]$, constructed in Proposition 5.4, comes from the preceding lemma, as it will be shown in the following proposition.

Proposition 5.6 *The prime ideal $\mathscr{P}$ of $V[[X]]$ constructed in Proposition 5.4 is not SFT.*

Proof The prime ideal $\mathscr{P}$ of $V[[X]]$ satisfies the following properties:

(1) $\mathscr{P} \cap V = (0)$ (2) $\mathscr{P} \subseteq (MV)[[X]]$ (3) $\{f(0); f(X) \in \mathscr{P}\} = MV.$

Let ω be the order of the formal power series over $V[[X]]$. Fix an element $f(X) = a_0 + a_1 X + a_2 X^2 + \ldots \in \mathscr{P}$ with $a_0 \neq 0$. Such an element exists by (3) since $MV \neq (0)$. We can also take a generator $g_n(X), n \in \mathbb{N}^*$, of the ideal $\mathscr{I} \subseteq \mathscr{P}$, whose constant term is $t^{\frac{1}{2^{n-1}}}$. By (2), the series $f_1(X) = a_1 + a_2 X + a_3 X^2 + \ldots \in (MV)[[X]]$, and by (3), there exists a series $h_1(X) = a_1 + b_2 X + b_3 X^2 + \ldots \in \mathscr{P}$. Then $f_1(X) - h_1(X) = (a_2 - b_2)X + (a_3 - b_3)X^2 + \ldots \in (MV)[[X]]$. In particular, $\omega(f_1(X) - h_1(X)) \geq 1$. Let $f_2(X) = (a_2 - b_2) + (a_3 - b_3)X + \ldots \in (MV)[[X]]$. By (3), there exists a series $h_2(X) = (a_2 - b_2) + c_3 X + c_4 X^2 + \ldots \in \mathscr{P}$. Then $f_2(X) - h_2(X) = d_3 X + d_4 X^2 + \ldots$ and $f_1(X) - h_1(X) = X f_2(X) = X(f_2(X) - h_2(X)) + X h_2(X)$, so $f_1(X) - h_1(X) - X h_2(X) = X(f_2(X) - h_2(X)) = d_3 X^2 + d_4 X^3 + \ldots$ then $\omega(f_1(X) - h_1(X) - X h_2(X)) \geq 2$. By induction, we construct a sequence $\{h_i(X), i \in \mathbb{N}^*\}$ of elements of $\mathscr{P}$ such that $f_1(X) =$

$\sum\limits_{i:1}^{\infty} X^{i-1}h_i(X)$. Then $f(X) - \sum\limits_{i:1}^{\infty} X^i h_i(X) = f(X) - X f_1(X) = a_0 \notin \mathscr{P}$ since
$\mathscr{P} \cap V = (0)$. By Lemma 5.5, $\mathscr{P}$ is not an SFT-ideal of $V[[X]]$.   □

Remark 5.7 It follows from the preceding proposition that the domain $V[[X]]$ is not SFT. But the result is more general and its proof more simple. In fact, if (A, M) is a nondiscrete valuation domain of rank one, the prime ideals $M[[X]]$ and $M + XA[[X]]$ of $A[[X]]$ are not SFT. Indeed, suppose that there exist $n \in \mathbb{N}^*$ and a finitely generated ideal $\mathscr{F}$ of $A[[X]]$ included in $M[[X]]$ (resp. $M + XA[[X]]$) such that $f^n \in \mathscr{F}$ for each $f \in M[[X]]$ (resp. $M + XA[[X]]$). The ideal $F = \{f(0); f \in \mathscr{F}\}$ of A is finitely generated and is included in M and for each $x \in M$, $x^n \in F$. Then M is an SFT-ideal of A, which is absurd.

In our case, $\mathscr{P} \subset (MV)[[X]] \subset MV + XA[[X]]$. The first inclusion is strict since $\mathscr{P} \cap V = (0)$ and $(MV)[[X]] \cap V = MV \neq (0)$.

The following corollary shows that the SFT property does not pass to the formal power series ring. The conclusion of the preceding proposition transmits from V to V_1 since the conductor $(V_1 : V) \neq (0)$.

Corollary 5.8 *The formal power series ring $V_1[[X]]$ is not SFT.*

Proof Let $\mathscr{P}$ be the prime ideal of $V[[X]]$ constructed in Proposition 5.4. In the proof of Proposition 5.6, we constructed a sequence $\{h_i(X); i \in \mathbb{N}^*\}$ of elements of $\mathscr{P}$ such that $g(X) = \sum\limits_{i:1}^{\infty} h_i(X)X^{i-1} \in V[[X]] \setminus \mathscr{P}$. Let $P = \mathscr{P} \cap V_1[[X]]$, a prime ideal of $V_1[[X]]$. Since $V_1 = \mathbb{F}_2 + tV$, if $l(X) \in V[[X]]$, then $tl(X) \in V_1[[X]]$. So if $l(X) \in \mathscr{P}$, then $tl(X) \in P$. So $tg(X) = \sum\limits_{i:1}^{\infty} th_i(X)X^{i-1} \in V_1[[X]]$ with $th_i(X) \in P$ for each $i \in \mathbb{N}^*$. Suppose that $tg(X) \in P \subseteq \mathscr{P}$. Since $g(X) \notin \mathscr{P}$, then $t \in \mathscr{P}$, which is absurd since $\mathscr{P} \cap V = (0)$. By Lemma 5.5, P is not an SFT-ideal of $V_1[[X]]$.   □

Lemma 5.9 *Let I be an ideal of a ring A. The following assertions are equivalent:*

1. *I is an SFT-ideal of A.*
2. *$I[[X]]$ is an SFT-ideal of $A[[X]]$.*
3. *$I + XA[[X]]$ is an SFT-ideal of $A[[X]]$.*

Proof "(1) $\implies$ (2)" There exist $m \in \mathbb{N}^*$ and a finitely generated ideal $F \subseteq I$ of A such that for each $x \in I$, $x^m \in F$. For each $\bar{x}$ in the ideal I/F of the ring A/F, $\bar{x}^m = 0$. By Lemma 1.15, $(I/F)[[X]]$ is a nil ideal of bounded index of the ring $(A/F)[[X]]$. There exists $n \in \mathbb{N}^*$ such that for each $f \in I[[X]]$, $\bar{f}^n = 0$ in $(A/F)[[X]]$, then $f^n \in F[[X]] = FA[[X]]$ since F is finitely generated. Since

$FA[[X]]$ is a finitely generated ideal of $A[[X]]$ included in $I[[X]]$, then $I[[X]]$ is an SFT-ideal of $A[[X]]$.

$''(1) \implies (3)''$ There exist $n \in \mathbb{N}^*$ and a finitely generated ideal $F \subseteq I$ of A such that for each $x \in I$, $x^n \in F$. The ideal $(F, X)A[[X]]$ is finitely generated and included in $I + XA[[X]]$. For each $f \in I + XA[[X]]$, $f = a + Xg$ with $a \in I$ and $g \in A[[X]]$, then $f^n = a^n + \sum_{i:0}^{n-1} C_n^i a^i (Xg)^{n-i} \in (F, X)A[[X]]$. Then $I + XA[[X]]$ is an SFT-ideal of $A[[X]]$.

$''(2) \implies (1)''$ and $''(3) \implies (1)''$ There exist $n \in \mathbb{N}^*$ and a finitely generated ideal $\mathscr{F}$ of $A[[X]]$ included in $I[[X]]$ (resp. $I + XA[[X]]$) such that $f^n \in \mathscr{F}$ for each $f(X) \in I[[X]]$ (resp. $I + XA[[X]]$). The ideal $F = \{f(0); f(X) \in \mathscr{F}\}$ of A is finitely generated and is included in I and for each $x \in I$, $x^n \in F$. Then I is an SFT-ideal of A. $\square$

Corollary 5.10 *For each* $n \in \mathbb{N}^*$, $V_1[[X_1, \dots, X_n]]$ *is not an SFT-domain.*

Proof The case $n = 1$ follows from Corollary 5.8. By the preceding lemma, if A is a non-SFT-ring, then so is $A[[X]]$. We continue by induction. $\square$

Proposition 5.11 *If* A *is an SFT-ring of zero dimensional, then* $A[[X]]$ *is an SFT-ring of dimension one.*

Proof By Corollary 1.22, the only prime ideals of $A[[X]]$ are of the form $M[[X]]$ and $M + XA[[X]]$ with $M \in spec(A)$. They are SFT, by the preceding lemma. $\square$

Example The formal power series ring $\left(\mathbb{F}_2[Y_i; i \in \mathbb{N}]/(Y_i^2; i \in \mathbb{N})\right)[[X]]$ is SFT.

6   The I-Adic Topology on Rings

Let A be a ring and I an ideal of A. The sequence $(I^n)_{n \in \mathbb{N}}$ is decreasing with first term $I^0 = A$. The map $v : A \longrightarrow \mathbb{N} \cup \{+\infty\}$, defined by $v(x) = sup\{n \in \mathbb{N}; x \in I^n\}$ for each $x \in A$, is called the order function. It is clear that $v(x) = \infty$ if and only if $x \in \bigcap_{n:0}^{\infty} I^n$ and $v(-x) = v(x)$. We will show that $v(x + y) \geq min\{v(x), v(y)\}$ and $v(xy) \geq v(x) + v(y)$. Indeed, since $x \in I^{v(x)}$ and $y \in I^{v(y)}$, then $x + y \in I^{min\{v(x),v(y)\}}$, so $v(x + y) \geq min\{v(x), v(y)\}$. Also, $xy \in I^{v(x)}I^{v(y)} = I^{v(x)+v(y)}$, then $v(xy) \geq v(x) + v(y)$. Note that v is not necessarily a valuation over the ring A. For $x, y \in A$, let $d(x, y) = e^{-v(x-y)}$ with the convention that $e^{-\infty} = 0$. Then $d(x, y) \in \mathbb{R}+$, $d(x, y) = e^{-v(x-y)} = e^{-v(y-x)} = d(y, x)$, $d(x, y) = 0 \iff x - y \in \bigcap_{n:0}^{\infty} I^n$. We will prove the ultrametric inequality $d(x, y) \leq sup\{d(x, z), d(z, y)\}$. We have $x - y = (x - z) + (z - y)$, then $v(x - y) \geq min\{v(x - z), v(z - y)\}$, so $-v(x - y) \leq -min\{v(x - z), v(z - y)\} = sup\{-v(x - z), -v(z - y)\}$. Then $d(x, y) = e^{-v(x-y)} \leq e^{sup\{-v(x-z), -v(z-y)\}} = sup\{e^{-v(x-z)}, e^{-v(z-y)}\} = sup\{d(x, z), d(z, y)\}$. In general, d is not a distance since the separation axiom is not usually satisfied. More precisely, d is a distance over A if and only if $\bigcap_{n:0}^{\infty} I^n = (0)$. In this case, the topology defined by d over A is called the I-adic topology.

Example 1 Let B be a ring, $A = B[X]$ (resp. $B[[X]]$) and $I = XB[X]$ (resp. $XB[[X]]$). Then $\bigcap_{n:0}^{\infty} I^n = (0)$. In this case, d is a distance, called (X)-adic over $B[X]$ (resp. $B[[X]]$).

Example 2 Let A be an integral domain and $I = pA$ a nonzero principal prime ideal such that $\bigcap_{n:0}^{\infty} p^n A = (0)$. The order function v associated to I is a valuation on A, called the (p)-adic valuation. Indeed, since $\bigcap_{n:0}^{\infty} I^n = (0)$, then $v(x) = \infty$ if and only if $x = 0$. Let $x, y \in A \setminus (0)$. We have $v(xy) \geq v(x) + v(y)$. Suppose that the inequality is strict. Put $v(x) = m$, $v(y) = n$ and $v(xy) = k \in \mathbb{N}$ with $m + n < k$. Then $x = p^m a$, $y = p^n b$ and $xy = p^k c$ with $a, b, c \in A$ not divisible by p. Since $p^{m+n} ab = p^k c$, then $ab = p^{k-(m+n)} c \in pA$. But pA is a prime ideal of A, then either $a \in pA$ or $b \in pA$. So p divides either a or b, which is absurd. So $v(xy) = v(x) + v(y)$.

Example 3 Let A be a UFD and p an irreducible element of A. Then pA is a prime ideal of A and $\bigcap_{n:0}^{\infty} p^n A = (0)$. The (p)-adic valuation v is extended to the quotient field of A. Its valuation domain is $A_{(p)}$, the localized of A by (p).

Example 4 Take $A = \mathbb{Z}^2$, the product ring.

(a) Let I be the ideal of A generated by $(1, 0)$. For each $n \in \mathbb{N}^*$, $(1, 0)^n = (1, 0)$, then $I^n = I$. So $\bigcap_{n:0}^{\infty} I^n = I \neq (0)$.

(b) Let I be the ideal of A generated by $(1, 0)$, J the ideal generated by $(0, 2)$ and $T = I + J$. Then for each $x \in I$ and $y \in J$, $xy = 0$, for each $n \in \mathbb{N}^*$, $I^n = I$ and the sequence $(J^n)_n$ decreases strictly. Then the sequence $(T^n)_n$ decreases strictly but $\bigcap_{n:0}^{\infty} T^n \neq (0)$.

Remarks 6.1 Let A be a ring, I an ideal of A satisfying $\bigcap\limits_{n:0}^{\infty} I^n = (0)$, d the I-adic

distance over A, and $(x_n)_{n \in \mathbb{N}}$ a sequence of elements of A.

(1) By the ultrametric inequality, the sequence $(x_n)_{n \in \mathbb{N}}$ is Cauchy if and only if
    $\lim\limits_{n \longrightarrow +\infty} d(x_n, x_{n+1}) = 0$.
(2) The sequence $(x_n)_{n \in \mathbb{N}}$ is Cauchy if and only if for each $p \in \mathbb{N}$, there exists
    $q \in \mathbb{N}$ such that for each $n, m \geq q$, $x_n - x_m \in I^p$ (resp. for each $n \geq q$,
    $x_{n+1} - x_n \in I^p$).
(3) The sequence $(x_n)_{n \in \mathbb{N}}$ converges to an element $x \in A$ if and only if for each
    $p \in \mathbb{N}$, there exists $q \in \mathbb{N}$ such that for each integer $n \geq q$, $x_n - x \in I^p$.

Example The numerical sequence with general term $x_n = 1 + \frac{1}{2} + \ldots + \frac{1}{n}$ diverges,
so it is not Cauchy. However, $x_{n+1} - x_n = \frac{1}{n+1} \longrightarrow 0$ when $n \longrightarrow +\infty$.

Lemma 6.2 *Let A be a ring, I an ideal of A satisfying $\bigcap\limits_{n:0}^{\infty} I^n = (0)$, and d the*
I-adic distance over A. For each real number $r > 0$, there exists $n \in \mathbb{N}$ such that
$I^n \subseteq \mathcal{B}(0, r)$, the open ball for d with center 0 and radius r.
    Conversely, for each $n \in \mathbb{N}$, there exists a real number $r > 0$ such that $\mathcal{B}(0, r)$
$\subseteq I^n$.

Proof

(1) Let $r > 0$ be a real number. Then $x \in \mathcal{B}(0, r) \iff d(0, x) < r \iff e^{-v(x)} <$
    $r \iff -v(x) < Log\, r \iff v(x) > -Log\, r = Log \frac{1}{r}$. Let $n = sup\{0, 1 +$
    $E(Log \frac{1}{r})\} \in \mathbb{N}$. By definition, for each $x \in I^n$, $v(x) \geq n > Log \frac{1}{r}$, then
    $x \in \mathcal{B}(0, r)$. So $I^n \subseteq \mathcal{B}(0, r)$.
(2) Let $n \in \mathbb{N}$ and $x \in A$. Then $x \in I^n \iff v(x) \geq n \iff -v(x) \leq -n \iff$
    $e^{-v(x)} \leq e^{-n} \iff d(0, x) \leq e^{-n} \iff x \in \mathcal{B}_f(0, e^{-n})$, the closed ball of
    center 0 and radius e^{-n}. In particular, the open ball $\mathcal{B}(0, e^{-n}) \subseteq I^n$.   □

Let A be a ring and I an ideal of A satisfying $\bigcap\limits_{n:0}^{\infty} I^n = (0)$. The I-adic topology

over A may be defined by another ideal $I' \neq I$. More precisely, we have the
following proposition.

Proposition 6.3 *[Bounded difference] Let A be a ring and I and J be two ideals*
of A satisfying $\bigcap\limits_{n:0}^{\infty} I^n = \bigcap\limits_{n:0}^{\infty} J^n = (0)$. The I-adic and J-adic topologies over A
are identical if and only if there exist $n_1, n_2 \in \mathbb{N}^*$ such that $I^{n_1} \subseteq J$ and $J^{n_2} \subseteq I$.
Moreover, if the ring A is Noetherian, then these two topologies are identical if and
only if $\sqrt{I} = \sqrt{J}$.

Proof

(1) " $\Longrightarrow$ " By the preceding lemma, there exists a real number $r > 0$ such that $\mathscr{B}_J(0, r) \subseteq J$, where $\mathscr{B}_J(0, r)$ is the open ball with center 0 and radius r for the J-adic topology. Since the two topologies are identical, there exists a real number $r' > 0$ such that $\mathscr{B}_I(0, r') \subseteq \mathscr{B}_J(0, r)$. By the preceding lemma, there exists $n_1 \in \mathbb{N}^*$ such that $I^{n_1} \subseteq \mathscr{B}_I(0, r')$. Then $I^{n_1} \subseteq J$. By symmetry, there exists $n_2 \in \mathbb{N}^*$ such that $J^{n_2} \subseteq I$.

" $\Longleftarrow$ " Let $r' > 0$ be a real number. By the preceding lemma, there exists $k \in \mathbb{N}^*$ such that $J^k \subseteq \mathscr{B}_J(0, r')$ and there exists a real number $r > 0$ such that $\mathscr{B}_I(0, r) \subseteq I^{n_1 k} \subseteq J^k \subseteq \mathscr{B}_J(0, r')$. By symmetry, every open ball for the I-adic topology contains an open ball with the same center for the J-adic topology. The two topologies are then identical.

(2) " $\Longrightarrow$ " There exist $n_1, n_2 \in \mathbb{N}^*$ such that $I^{n_1} \subseteq J$ and $J^{n_2} \subseteq I$. Then $\sqrt{I} = \sqrt{I^{n_1}} \subseteq \sqrt{J} = \sqrt{J^{n_2}} \subseteq \sqrt{I}$. So $\sqrt{I} = \sqrt{J}$.

" $\Longleftarrow$ " Suppose that A is Noetherian. There exists $k \in \mathbb{N}^*$ such that $(\sqrt{I})^k \subseteq I$. Indeed, if $\sqrt{I} = (x_1, \ldots, x_n)$ and $s \in \mathbb{N}^*$ is such that $x_1^s, \ldots, x_n^s \in I$, then $(\sqrt{I})^{sn} \subseteq I$. We take $k = sn$. Also, there exists $l \in \mathbb{N}^*$ such that $(\sqrt{J})^l \subseteq J$. Then $I^l \subseteq (\sqrt{I})^l = (\sqrt{J})^l \subseteq J$. Also, $J^k \subseteq I$. We conclude by (1). $\qquad \square$

In the following section, we give an example of two distinct ideals defining the same topology in a non-Noetherian ring.

Theorem 6.4 *Let A be a ring. The formal power series ring $A[[X_1, \ldots, X_n]]$, endowed with its $(X_1, \ldots, X_n)A[[X_1, \ldots, X_n]]$-adic topology, is the completion of the polynomial ring $A[X_1, \ldots, X_n]$ for its proper $(X_1, \ldots, X_n)A[X_1, \ldots, X_n]$-adic topology.*

Proof The ideal $I = (X_1, \ldots, X_n)A[[X_1, \ldots, X_n]]$ satisfies $\bigcap_{k:1}^{\infty} I^k = (0)$.

The $(X_1, \ldots, X_n)$-adic topology over the ring $A[[X_1, \ldots, X_n]]$ induces the $(X_1, \ldots, X_n)$-adic topology over $A[X_1, \ldots, X_n]$. Indeed, for each $k \in \mathbb{N}$, $((X_1, \ldots, X_n)^k A[[X_1, \ldots, X_n]]) \cap A[X_1, \ldots, X_n] = (X_1, \ldots, X_n)^k A[X_1, \ldots, X_n]$. Let $f \in A[[X_1, \ldots, X_n]]$. Then $f = f_0 + f_1 + \ldots$ where f_i is a homogenous polynomial of degree i in $X_1, \ldots, X_n$. For each $k \in \mathbb{N}$, let $g_k = f_0 + f_1 + \ldots + f_k \in A[X_1, \ldots, X_n]$. Then $f - g_k = f_{k+1} + f_{k+2} + \ldots \in (X_1, \ldots, X_n)^{k+1}A[[X_1, \ldots, X_n]]$. So for two natural integers $k \geq l$, we have $f - g_k \in (X_1, \ldots, X_n)^{k+1}A[[X_1, \ldots, X_n]] \subseteq (X_1, \ldots, X_n)^l A[[X_1, \ldots, X_n]]$. Then $\lim_{k \to \infty} g_k = f$ for the $(X_1, \ldots, X_n)$-adic topology.

Conversely, let $(g_k)_{k \in \mathbb{N}}$ be a Cauchy sequence of elements of the ring $A[X_1, \ldots, X_n]$ for its $(X_1, \ldots, X_n)$-adic topology. By the Cauchy property, $g = g_0 + (g_1 - g_0) + (g_2 - g_1) + \ldots \in A[[X_1, \ldots, X_n]]$. By hypothesis, for each $l \in \mathbb{N}$, there exists $k_0 \in \mathbb{N}$, such that for each $k \geq k_0$, $g_{k+1} - g_k \in (X_1, \ldots, X_n)^{l+1}A[X_1, \ldots, X_n]$. Since $g = g_k + (g_{k+1} - g_k) + \ldots$, then

$$g - g_k = (g_{k+1} - g_k) + \ldots \in (X_1, \ldots, X_n)^{l+1} A[[X_1, \ldots, X_n]], \text{ so } \lim_{k \longrightarrow \infty} g_k = g.$$

$$\square$$

Even though $A[[X]]$ is the completion of $A[X]$ for the (X)-adic topology, the two rings do not share the same properties.

7 Infinite Product of Formal Power Series over a Nondiscrete Valuation Domain of Rank One

The results of this section are due to Kang and Park (1999).

Notation Let (A, M) be a nondiscrete valuation domain of rank one, defined by a valuation v taking its values in $\mathbb{R}$. For each $0 \neq \gamma \in M, \bigcap_{n:0}^{\infty} \gamma^n A = (0)$. Indeed, let x be an element of this intersection. For each $n \in \mathbb{N}$, $x = \gamma^n a_n$ with $a_n \in A$, then $v(x) = n v(\gamma) + v(a_n) \geq n v(\gamma) \longrightarrow +\infty$ when $n \longrightarrow +\infty$. Then $v(x) = +\infty$ and $x = 0$. The completion $\hat{A}$ of A for the γA-adic topology does not depend on $0 \neq \gamma \in M$. Indeed, let $0 \neq \gamma' \in M$ be another element with, for example, $v(\gamma) \leq v(\gamma')$. There exists $k \in \mathbb{N}^*$ such that $v(\gamma') \leq k v(\gamma) = v(\gamma^k)$. For each $n \in \mathbb{N}^*$, $\gamma'^n A \subseteq \gamma^n A$ and $\gamma^{kn} A \subseteq \gamma'^n A$. We conclude by Proposition 6.3. We admit that $\hat{A}$ is a valuation domain of rank one, with maximal ideal $M\hat{A}$, defined by a valuation, that we denote also by v, with the same group as that of A. The topology of $\hat{A}$ is the $\gamma \hat{A}$-adic topology. In the sequel, we keep the preceding notations.

Proposition 7.1 *Let $(a_i)_{i \in \mathbb{N}}$ be a sequence of elements of the domain $\hat{A}$. The sequence $\left(\sum_{i:0}^{n} a_i \right)_{n \in \mathbb{N}}$ is Cauchy in $\hat{A}$ if and only if $\lim_{i \longrightarrow +\infty} v(a_i) = +\infty$. In this case, we denote $\sum_{i:0}^{\infty} a_i$ its limit.*

Proof $" \Longrightarrow "$ Let $p \in \mathbb{N}$. By Remark 6.1 (2), there exists $n_0 \in \mathbb{N}$ such that for each integer $n > n_0$, $a_n = \sum_{i:0}^{n} a_i - \sum_{i:0}^{n-1} a_i \in \gamma^p \hat{A}$, then $v(a_n) \geq p v(\gamma)$. So $\lim_{i \longrightarrow \infty} v(a_i) = +\infty$.

"$\Longleftarrow$" Let $p \in \mathbb{N}$. There exists $n_0 \in \mathbb{N}$ such that for each integer $n \geq n_0$, $v(a_n) \geq v(\gamma^p)$, then $v\left(\sum_{i:0}^{n} a_i - \sum_{i:0}^{n-1} a_i\right) = v(a_n) \geq v(\gamma^p)$, so $\sum_{i:0}^{n} a_i - \sum_{i:0}^{n-1} a_i \in \gamma^p \hat{A}$. By Remark 6.1 (2), the sequence $\left(\sum_{i:0}^{n} a_i\right)_{n \in \mathbb{N}}$ is Cauchy. $\qquad\square$

Example Let $f(X) = \sum_{i:0}^{\infty} a_i X^i \in \hat{A}[[X]]$ and $\beta \in M$. Then the sequence $\left(\sum_{i:0}^{n} a_i \beta^i\right)_{n \in \mathbb{N}}$ converges in $\hat{A}$. Indeed, since $v(a_i \beta^i) = v(a_i) + i v(\beta) \geq i v(\beta) \longrightarrow +\infty$ when $i \longrightarrow \infty$, the sequence is Cauchy, then it converges in $\hat{A}$.

We denote $f(\beta) = \sum_{i:0}^{\infty} a_i \beta^i \in \hat{A}$ its limit.

Lemma 7.2 *Let $f(X), g(X) \in \hat{A}[[X]]$, and $\beta \in M$. Then $(fg)(\beta) = f(\beta)g(\beta)$ in $\hat{A}$.*

Proof The result is clear if $\beta = 0$. Suppose that $\beta \neq 0$. The domain $\hat{A}$ does not depend on γ, then we replace γ by $\beta \in M$. Put $f = \sum_{i:0}^{\infty} a_i X^i$, $g = \sum_{i:0}^{\infty} b_i X^i$, $a = f(\beta) \in \hat{A}$, and $b = g(\beta) \in \hat{A}$. For each $l \in \mathbb{N}$, there exists $N \geq l$ such that for each $n \geq N$, $(a_0 + \ldots + a_n \beta^n) - a \in \beta^l \hat{A}$ and $(b_0 + \ldots + b_n \beta^n) - b \in \beta^l \hat{A}$, then $a_0 + \ldots + a_n \beta^n \in a + \beta^l \hat{A}$ and $b_0 + \ldots + b_n \beta^n \in b + \beta^l \hat{A}$. Then $(a_0 + \ldots + a_n \beta^n)(b_0 + \ldots + b_n \beta^n) \in ab + \beta^l \hat{A}$. Since $n \geq l$, then $a_0 b_0 + (a_0 b_1 + a_1 b_0)\beta + \ldots + \left(\sum_{i+j=n} a_i b_j\right)\beta^n - ab \in \beta^l \hat{A}$. So $(fg)(\beta) = ab$. $\qquad\square$

Notation For each $i \in \mathbb{N}^*$, we consider a formal power series $f_i = \sum_{j:0}^{\infty} a_{ij} X^j \in A[[X]]$. Suppose that for each $i \in \mathbb{N}^*$, the constant term $0 \neq a_{i0} \in M$ and that there exists an element $0 \neq a \in M$ such that $\sum_{i:1}^{\infty} v(a_{i0}) \leq v(a)$ and for each $j \in \mathbb{N}^*$, $\lim_{i \rightarrow \infty} v(a_{ij}) = +\infty$. The numerical series, with positive terms, $\sum_{i:1}^{\infty} v(a_{i0})$ converges.

We will define a new notion of infinite product of formal power series. This allows us to study the prime spectrum of $A[[X]]$ when A is a nondiscrete valuation domain of rank one.

Definition 7.3 We define a formal power series denoted $\left(\prod_{i:1}^{\infty} f_i, a\right)$ of $\hat{A}[[X]]$ as follows:

1. The constant term of $\left(\prod_{i:1}^{\infty} f_i; a\right)$ is equal to a.

2. The coefficient of X is $a_{11}\dfrac{a}{a_{10}} + a_{21}\dfrac{a}{a_{20}} + \ldots + a_{n1}\dfrac{a}{a_{n0}} + \ldots \in \hat{A}$.

3. For each series $g \in \hat{A}[[X]]$, we denote $(g)_i$ the coefficient of X^i in g. Suppose that the coefficient of X^k in $\left(\prod_{i:1}^{\infty} f_i; a\right)$ is defined for each f_i and a and for each $k \le n-1$.

We define the coefficient of X^n in $\left(\prod_{i:1}^{\infty} f_i; a\right)$ by $\displaystyle\sum_{j:1}^{n} a_{1j}\left(\prod_{i:2}^{\infty} f_i; \frac{a}{a_{10}}\right)_{n-j} +$

$$\sum_{j:1}^{n}\sum_{k:2}^{\infty} a_{10}a_{20}\ldots a_{k-10}a_{kj}\left(\prod_{i:k+1}^{\infty} f_i; \frac{a}{a_{10}\ldots a_{k0}}\right)_{n-j}$$

$$= \sum_{j:1}^{n} a_{1j}\left(\prod_{i:2}^{\infty} f_i; \frac{a}{a_{10}}\right)_{n-j} + \sum_{k:2}^{\infty}\left[a_{10}a_{20}\ldots a_{k-10}\sum_{j:1}^{n} a_{kj}\left(\prod_{i:k+1}^{\infty} f_i; \frac{a}{a_{10}\ldots a_{k0}}\right)_{n-j}\right]$$

$$= a_{1n}\left(\prod_{i:2}^{\infty} f_i; \frac{a}{a_{10}}\right)_0 + a_{1n-1}\left(\prod_{i:2}^{\infty} f_i; \frac{a}{a_{10}}\right)_1 + \ldots + a_{11}\left(\prod_{i:2}^{\infty} f_i; \frac{a}{a_{10}}\right)_{n-1}$$

$$+ a_{10}\sum_{j:1}^{n} a_{2j}\left(\prod_{i:3}^{\infty} f_i; \frac{a}{a_{10}a_{20}}\right)_{n-j} + a_{10}a_{20}\sum_{j:1}^{n} a_{3j}\left(\prod_{i:4}^{\infty} f_i; \frac{a}{a_{10}a_{20}a_{30}}\right)_{n-j} + \ldots$$

$$+ a_{10}a_{20}\ldots a_{k-10}\sum_{j:1}^{n} a_{kj}\left(\prod_{i:k+1}^{\infty} f_i; \frac{a}{a_{10}a_{20}\ldots a_{k0}}\right)_{n-j} + \ldots.$$

Since $v\left(a_{n1}\dfrac{a}{a_{n0}}\right) = v(a_{n1}) + v(a) - v(a_{n0}) \ge v(a_{n1}) \ge 0$, then $a_{n1}\dfrac{a}{a_{n0}} \in A \subseteq \hat{A}$. By Proposition 7.1, since $\lim_{n\to+\infty} v(a_{n1}) = +\infty$, the coefficient of X is well defined in $\hat{A}$. For each $n \ge 1$, $v(a_{10}\ldots a_{1n}) = \displaystyle\sum_{i:1}^{n} v(a_{i0}) < \sum_{i:1}^{\infty} v(a_{i0}) \le v(a)$, then $\dfrac{a}{a_{10}\ldots a_{1n}} \in M$. The n infinite sums $\displaystyle\sum_{k:2}^{\infty} a_{10}a_{20}\ldots a_{k-10}a_{kj}\left(\prod_{i:k+1}^{\infty} f_i; \frac{a}{a_{10}\ldots a_{k0}}\right)_{n-j}$, $1 \le j \le n$, in the expression of the coefficient of X^n are well defined in $\hat{A}$ since the factor a_{kj} satisfies $\lim_{k\to+\infty} v(a_{kj}) = +\infty$.

Example We look for sufficient conditions in order that infinite product $\left(\prod_{i:1}^{\infty} f_i, a\right)$ of elements of $A[[X]]$ belongs to $A[[X]]$. The result will be used later.

For $f = \sum_{i:0}^{\infty} a_i X^i \in A[[X]]$, we denote $\delta(f) = \inf\{i \in \mathbb{N}^*;\ a_i \neq 0\}$. A sequence $(f_i)_{i \in \mathbb{N}^*}$ of elements of $A[[X]]$ is said to be upper triangular if $\delta(f_n) \geq n$ for each $n \in \mathbb{N}^*$.

We always remarked before that for each $n \geq 1$, $\frac{a}{a_{10}\cdots a_{1n}} \in M$. We will show by induction on the coefficient of X^k that for such sequence, the series $\left(\prod_{i:1}^{\infty} f_i,\ a\right) \in A[[X]]$. The constant term of $\left(\prod_{i:1}^{\infty} f_i;\ a\right)$ is $a \in A$ and the coefficient of X is $a_{11}\frac{a}{a_{10}} \in A$. Suppose that $n \geq 2$ and the coefficient of X^k belongs to A, for each $k \leq n - 1$ and for each upper triangular sequence. The coefficient of X^n in the formal power series $\left(\prod_{i:1}^{\infty} f_i;\ a\right)$ is $\sum_{j:1}^{n} a_{1j}\left(\prod_{i:2}^{\infty} f_i;\ \frac{a}{a_{10}}\right)_{n-j} +$

$$\sum_{j:1}^{n}\sum_{k:2}^{n} a_{10}a_{20}\cdots a_{k-10}a_{kj}\left(\prod_{i:k+1}^{\infty} f_i;\ \frac{a}{a_{10}\cdots a_{k0}}\right)_{n-j} \in A.$$

The following proposition shows that each f_i is a factor in the product $\left(\prod_{i:1}^{\infty} f_i;\ a\right)$ and the modification of the constant term changes the product. When there is no possible confusion, the product $\left(\prod_{i:1}^{\infty} f_i;\ a\right)$ will be simply denoted $\prod_{i:1}^{\infty} f_i$.

Proposition 7.4 *With the same notations and hypotheses:*

1. $\left(\prod_{i:1}^{\infty} f_i;\ a\right) = f_1\left(\prod_{i:2}^{\infty} f_i;\ \frac{a}{a_{10}}\right).$

2. $\left(\prod_{i:1}^{\infty} f_i;\ aa'\right) = \left(\prod_{i:1}^{\infty} f_i;\ a\right)a',\ \text{for each } a' \in A.$

Proof By identification of the coefficients in the two formal power series:

(1) The constant term of $f_1\left(\prod_{i:2}^{\infty} f_i;\ \frac{a}{a_{10}}\right)$ is $a_{10}\frac{a}{a_{10}} = a$, equal to that of the series

$$\left(\prod_{i:1}^{\infty} f_i;\ a\right).$$

The coefficient of X in the series $\left(\prod_{i:1}^{\infty} f_i;\ a\right)$ is $a_{11}\frac{a}{a_{10}} + a_{21}\frac{a}{a_{20}} + \ldots +$ $a_{n1}\frac{a}{a_{n0}} + \ldots$.

The coefficient of X in $f_1(\prod_{i:2}^{\infty} f_i; \ \frac{a}{a_{10}})$ is $(f_1)_1(\prod_{i:2}^{\infty} f_i; \ \frac{a}{a_{10}})_0 +$

$(f_1)_0(\prod_{i:2}^{\infty} f_i; \ \frac{a}{a_{10}})_1 = a_{11}\frac{a}{a_{10}} + a_{10}(a_{21}\frac{a}{a_{10}a_{20}} + \ldots + a_{n1}\frac{a}{a_{10}a_{n0}} + \ldots).$

So we have the equality of the coefficients of X.

For $n \geq 2$, the coefficient $(\prod_{i:1}^{\infty} f_i; \ a)_n = \sum_{j:1}^{n} a_{1j}(\prod_{i:2}^{\infty} f_i; \ \frac{a}{a_{10}})_{n-j} +$

$\sum_{k:2}^{\infty} a_{10}a_{20}\ldots a_{k-10} \sum_{j:1}^{n} a_{kj}(\prod_{i:k+1}^{\infty} f_i; \ \frac{a}{a_{10}\ldots a_{k0}})_{n-j} = \sum_{j:1}^{n} a_{1j}(\prod_{i:2}^{\infty} f_i; \ \frac{a}{a_{10}})_{n-j} +$

$a_{10}\Big[\sum_{j:1}^{n} a_{2j}(\prod_{i:3}^{\infty} f_i; \ \frac{a}{a_{10}a_{20}})_{n-j} + \sum_{k:3}^{\infty} a_{20}\ldots a_{k-10} \sum_{j:1}^{n} a_{kj}(\prod_{i:k+1}^{\infty} f_i; \ \frac{a}{a_{10}\ldots a_{k0}})_{n-j}\Big] =$

$\sum_{j:1}^{n} a_{1j}(\prod_{i:2}^{\infty} f_i; \ \frac{a}{a_{10}})_{n-j} + a_{10}(\prod_{i:2}^{\infty} f_i; \ \frac{a}{a_{10}})_n = \sum_{j:0}^{n} a_{1j}(\prod_{i:2}^{\infty} f_i; \ \frac{a}{a_{10}})_{n-j} =$

$\Big(f_1(\prod_{i:1}^{\infty} f_i; \ \frac{a}{a_{10}})\Big)_n.$

(2) The constant terms of the two series $(\prod_{i:1}^{\infty} f_i; \ aa')$ and $(\prod_{i:1}^{\infty} f_i; \ a)a'$ are equal to aa'.

Also, $(\prod_{i:1}^{\infty} f_i; \ aa')_1 = a_{11}\frac{aa'}{a_{10}} + a_{21}\frac{aa'}{a_{20}} + \ldots + a_{n1}\frac{aa'}{a_{n0}} + \ldots = a'(a_{11}\frac{a}{a_{10}} +$

$a_{21}\frac{a}{a_{20}} + \ldots + a_{n1}\frac{a}{a_{n0}} + \ldots) = a'(\prod_{i:1}^{\infty} f_i; \ a)_1.$ We have the equality of the coefficients of X in the two series.

Let $n \geq 2$ and suppose that for each $k \leq n - 1$, the coefficient of X^k is the same in the two members of the equality for each sequence of formal power series. Then

$(\prod_{i:1}^{\infty} f_i; \ aa')_n = \sum_{j:1}^{n} a_{1j}(\prod_{i:2}^{\infty} f_i; \ \frac{aa'}{a_{10}})_{n-j} + \sum_{j:1}^{n}\sum_{k:2}^{\infty} a_{10}a_{20}\ldots a_{k-10}a_{kj}(\prod_{i:k+1}^{\infty} f_i;$

$\frac{aa'}{a_{10}\ldots a_{k0}})_{n-j} = a'(\prod_{i:1}^{\infty} f_i; \ a)_n.$ $\square$

Let $f_1(X), \ldots, f_n(X) \in A[[X]]$ and $\beta \in M$. By Lemma 7.2, we have $(f_1 \ldots f_n)(\beta) = f_1(\beta) \ldots f_n(\beta)$. Then $v((f_1 \ldots f_n)(\beta)) = v(f_1(\beta)) + \ldots + v(f_n(\beta))$. In the following proposition, we show that the result is not true for an infinite product.

Proposition 7.5 *With the same hypotheses and notations, for each $\beta \in M$, we have*

$$v\left(\left(\prod_{i:1}^{\infty} f_i;\, a\right)(\beta)\right) = \sum_{i:1}^{\infty} v(f_i(\beta)) + v(a) - \sum_{i:1}^{\infty} v(a_{i0}).$$

Proof If $\beta = 0$, $v\left(\left(\prod_{i:1}^{\infty} f_i;\, a\right)(\beta)\right) = v(a)$ and $\sum_{i:1}^{\infty} v(f_i(\beta)) + v(a) - \sum_{i:1}^{\infty} v(a_{i0}) =$

$\sum_{i:1}^{\infty} v(a_{i0}) + v(a) - \sum_{i:1}^{\infty} v(a_{i0}) = v(a)$. Suppose that $\beta \neq 0$. By the density in $\mathbb{R}$ of

the values group of v, there exists $d \in A$ such that $\left(v(a) - \sum_{i:1}^{\infty} v(a_{i0})\right) - \frac{1}{2}v(\beta) <$

$v(d) \leq v(a) - \sum_{i:1}^{\infty} v(a_{i0})$. Then $v(a) - \sum_{i:1}^{\infty} v(a_{i0}) < v(d) + \frac{1}{2}v(\beta)$. Since

$\lim_{n \longrightarrow +\infty} \sum_{i:1}^{n} v(a_{i0}) = \sum_{i:1}^{\infty} v(a_{i0})$, there exists $N \in \mathbb{N}$ such that for each $n \geq N$,

$\sum_{i:1}^{\infty} v(a_{i0}) - \sum_{i:1}^{n} v(a_{i0}) < \frac{1}{2}v(\beta)$. Then $v(a) - \sum_{i:1}^{n} v(a_{i0}) = \left(v(a) - \sum_{i:1}^{\infty} v(a_{i0})\right) +$

$\left(\sum_{i:1}^{\infty} v(a_{i0}) - \sum_{i:1}^{n} v(a_{i0})\right) < v(d) + \frac{1}{2}v(\beta) + \frac{1}{2}v(\beta) = v(d) + v(\beta)$, so

$v(a) - v(a_{10} \ldots a_{n0}) - v(d) < v(\beta)$, then $v(\frac{a}{da_{10}\ldots a_{n0}}) < v(\beta)$. So $\beta = \frac{a}{da_{10}\ldots a_{n0}} \beta'$

with $\beta' \in M$. We have $v(d) + \sum_{i:1}^{\infty} v(a_{i0}) \leq v(a)$. Since $0 \neq a_{i0} \in M$, $1 \leq i \leq n$,

then $v(d) + \sum_{i:1}^{n} v(a_{i0}) < v(d) + \sum_{i:1}^{\infty} v(a_{i0}) \leq v(a)$, so $v(a) - v(d) - \sum_{i:1}^{n} v(a_{i0}) > 0$,

then $v(\frac{a}{da_{10}\ldots a_{n0}}) > 0$. So $\frac{a}{da_{10}\ldots a_{n0}} \in M$. Also, since $v(d) + \sum_{i:1}^{\infty} v(a_{i0}) \leq v(a)$, then

$\sum_{i:n+1}^{\infty} v(a_{i0}) \leq v(a) - v(d) - \sum_{i:1}^{n} v(a_{i0}) = v\left(\frac{a}{da_{10}\ldots a_{n0}}\right)$. These two conditions

allow us to define the infinite product $\left(\prod_{i:n+1}^{\infty} f_i;\, \frac{a}{da_{10}\ldots a_{n0}}\right) \in \hat{A}[[X]]$. For

each integer $n \geq N$, there exists $c_n \in \hat{A}$ such that $\left(\prod_{i:n+1}^{\infty} f_i;\, \frac{a}{da_{10}\ldots a_{n0}}\right)(\beta) =$

$\frac{a}{da_{10}\ldots a_{n0}} + \beta c_n = \frac{a}{da_{10}\ldots a_{n0}} + \frac{a}{da_{10}\ldots a_{n0}}\beta' c_n = \frac{a}{da_{10}\ldots a_{n0}}(1 + \beta' c_n)$.

Since $\hat{A}$ is a valuation domain with maximal ideal $M\hat{A}$, then $1 + \beta' c_n$ is invertible in the domain $\hat{A}$. So $v\left(\left(\prod_{i:n+1}^{\infty} f_i; \dfrac{a}{da_{10}\ldots a_{n0}}\right)(\beta)\right) = v\left(\dfrac{a}{da_{10}\ldots a_{n0}}\right).$

By Proposition 7.4, $\left(\prod_{i:n+1}^{\infty} f_i; \dfrac{a}{a_{10}\ldots a_{n0}}\right) = d\left(\prod_{i:n+1}^{\infty} f_i; \dfrac{a}{da_{10}\ldots a_{n0}}\right).$ Then

$$v\left(\left(\prod_{i:n+1}^{\infty} f_i; \dfrac{a}{a_{10}\ldots a_{n0}}\right)(\beta)\right) = v(d) + v\left(\left(\prod_{i:n+1}^{\infty} f_i; \dfrac{a}{da_{10}\ldots a_{n0}}\right)(\beta)\right) =$$

$$v(d) + v\left(\dfrac{a}{da_{10}\ldots a_{n0}}\right) = v(a) - \sum_{i:1}^{n} v(a_{i0}).$$ By Proposition 7.4 (1), we

have $\left(\prod_{i:1}^{\infty} f_i; a\right) = \left(\prod_{i:1}^{n} f_i\right)\left(\prod_{i:n+1}^{\infty} f_i; \dfrac{a}{a_{10}\ldots a_{n0}}\right).$ By Lemma 7.2, we have

$$\left(\prod_{i:1}^{\infty} f_i; a\right)(\beta) = \left(\prod_{i:1}^{n} f_i(\beta)\right)\left(\prod_{i:n+1}^{\infty} f_i; \dfrac{a}{a_{10}\ldots a_{n0}}\right)(\beta).$$ By the preceding

step, $v\left(\left(\prod_{i:1}^{\infty} f_i; a\right)(\beta)\right) = \sum_{i:1}^{n} v\left(f_i(\beta)\right) + v\left(\left(\prod_{i:n+1}^{\infty} f_i; \dfrac{a}{a_{10}\ldots a_{n0}}\right)(\beta)\right) =$

$$\sum_{i:1}^{n} v\left(f_i(\beta)\right) + v(a) - \sum_{i:1}^{n} v(a_{i0}),$$ for each integer $n \geq N.$

If $v\left(\left(\prod_{i:1}^{\infty} f_i; a\right)(\beta)\right) \neq \infty,$ by passage to the limit when $n \longrightarrow \infty,$ we find the result.

If $v\left(\left(\prod_{i:1}^{\infty} f_i; a\right)(\beta)\right) = \infty,$ then $\sum_{i:1}^{n} v\left(f_i(\beta)\right) = \infty.$ Then there exists at least one index i such that $v\left(f_i(\beta)\right) = \infty,$ which gives again the result. $\square$

8 The Zeros of Formal Power Series

The results of this section are due to Kang and Park (1999).

Definition 8.1 Let (A, M) be a nondiscrete valuation domain of rank one and $f(X) \in A[[X]].$ We define the set of the zeros of $f(X)$ by $Z(f) = \{\alpha \in M;\ f(\alpha) = 0 \text{ in } \hat{A}\}.$

Proposition 8.2 *With the notations and the hypotheses in the preceding section, we have the equality* $Z\left(\prod_{i:1}^{\infty} f_i\right) = \bigcup_{i:1}^{\infty} Z(f_i).$

Proof By Proposition 7.4, for each $n \in \mathbb{N}^*$, $\prod_{i:1}^{\infty} f_i = f_1 \ldots f_n \left(\prod_{i:n+1}^{\infty} f_i \right)$. By

Lemma 7.2, for each $\beta \in M$, we have $\left(\prod_{i:1}^{\infty} f_i \right)(\beta) = f_1(\beta) \ldots f_n(\beta) \left(\prod_{i:n+1}^{\infty} f_i \right)(\beta)$,

then $\bigcup_{i:1}^{n} Z(f_i) \subseteq Z \left(\prod_{i:1}^{\infty} f_i \right)$. Then $\bigcup_{i:1}^{\infty} Z(f_i) \subseteq Z \left(\prod_{i:1}^{\infty} f_i \right)$. Conversely, if $\beta \in$

$Z \left(\prod_{i:1}^{\infty} f_i \right)$, then $v \left(\left(\prod_{i:1}^{\infty} f_i \right)(\beta) \right) = \infty$. By the proof of Proposition 7.5, there exists

$i \in \mathbb{N}^*$ such that $f_i(\beta) = 0$, then $\beta \in Z(f_i)$. So $Z \left(\prod_{i:1}^{\infty} f_i \right) \subseteq \bigcup_{i:1}^{\infty} Z(f_i)$. $\qquad \square$

Let $f(X) = \sum_{i:0}^{\infty} a_i X^i \in A[[X]]$ and $\alpha \in Z(f)$. Then $f(\alpha) = a_0 + a_1 \alpha + a_2 \alpha^2 +$

$\ldots = 0$ in $\hat{A}$. By subtraction, we find $f(X) = a_1(X - \alpha) + a_2(X^2 - \alpha^2) + \ldots =$
$(X - \alpha)(a_1 + a_2(X + \alpha) + \ldots)$, then $X - \alpha$ divides $f(X)$ in the domain $\hat{A}[[X]]$.
Then we give the following definition.

Definition 8.3 Let $f(X) \in A[[X]]$ and $\alpha \in Z(f)$. We say that α is a zero for $f(X)$
with multiplicity $k \in \mathbb{N}^*$ and we denote $m(f, \alpha) = k$ if $(X - \alpha)^k$ divides $f(X)$ and
$(X - \alpha)^{k+1}$ does not divide $f(X)$ in the domain $\hat{A}[[X]]$.

Example If $f(X)$ is invertible in $A[[X]]$, then $Z(f) = \emptyset$. Indeed, if $X - \alpha$ divides
$f(X)$, then α divides $f(0)$ in $\hat{A}$, so α is invertible in $\hat{A}$. But $\alpha \in Z(f) \subseteq M \subseteq M\hat{A}$,
the maximal ideal of $\hat{A}$, which is absurd.

Lemma 8.4 *Let (A, M) be a nondiscrete valuation domain of rank one. There exist
a sequence $(b_n)_{n \in \mathbb{N}^*}$ of elements in M and a sequence $(g_n)_{n \in \mathbb{N}^*}$ of formal power
series in $A[[X]]$ such that $\{b_n; \ n \in \mathbb{N}^*\} \subseteq Z(g_i)$ and $m(g_i, b_n) = n^i$ for each
$i, n \in \mathbb{N}^*$.*

Proof Let v be a nondiscrete valuation with values in $\mathbb{R}$ and domain A, $(n_i)_{i \in \mathbb{N}^*}$
be a strictly increasing sequence of nonzero natural integers that are not divisible
by the characteristic of A, and $b_1, b_2, \ldots$ an infinite sequence of nonzero elements
in M such that $v(b_1) > v(b_2) > \ldots$ and $v(b_j) < \frac{e^{-j}}{n_j}$ for each $j \in \mathbb{N}^*$. Fix

$i \in \mathbb{N}^*$, then $\sum_{j:1}^{\infty} n_j j^i v(b_j) \leq \sum_{j:1}^{\infty} n_j j^i \frac{e^{-j}}{n_j} = \sum_{j:1}^{\infty} j^i e^{-j} < \infty$ (apply one

convergence criterion). Let $0 \neq b \in M$ be such that $\sum_{j:1}^{\infty} n_j j^i v(b_j) \leq v(b)$.

Then $\sum_{j:1}^{\infty} v((-b_j^{n_j})^{j^i}) \le v(b)$. The polynomial $-b_j^{n_j} + X^{n_j} \in A[X]$ and admits b_j a simple root. Indeed, its derivative $n_j X^{n_j-1}$ is not annihilated by b_j since the characteristic of A does not divide n_j. Let α be a root of $-b_j^{n_j} + X^{n_j}$ in A. Then $n_j v(\alpha) = n_j v(b_j)$, so $v(\alpha) = v(b_j)$, then $\{b_1, b_2, \ldots\} \cap Z(-b_j^{n_j} + X^{n_j}) = \{b_j\}$ since $v(b_1) > v(b_2) > \ldots$. Let $f_j(X) = (-b_j^{n_j} + X^{n_j})^{j^i} \in A[X]$. Then $\{b_1, b_2, \ldots\} \cap Z(f_j) = \{b_j\}$ and $m(f_j, b_j) = j^i$. For each $i \in \mathbb{N}^*$, the infinite product $g_i = (\prod_{j:1}^{\infty} f_j; b)$ is well defined and the sequence $(f_j)_{j \in \mathbb{N}^*}$ is upper triangular. Indeed, except the constant term, the first nonzero coefficient of f_j is that of X^{n_j} with $n_j \ge j$. By the example of Definition 7.3, the series $g_i \in A[[X]]$. By Proposition 8.2, $\{b_1, b_2, \ldots\} \subseteq \bigcup_{j:1}^{\infty} Z(f_j) = Z(g_i)$. Since $b_n \notin Z(f_j)$ for $j \ne n$, then $m(g_i, b_n) = m(f_n, b_n) = n^i$. $\square$

Let A be a ring. If $P_1 \subset P_2 \subset P_3$ are three prime ideals of the polynomial ring $A[X]$, then we cannot have $P_1 \cap A = P_2 \cap A = P_3 \cap A$. This phenomena may happen in formal power series ring $A[[X]]$.

Theorem 8.5 *Let A be a nondiscrete valuation domain of rank one.*

Then $dim\ A[[X]]_{A \setminus (0)} = \infty$. In fact, there exists an infinite strictly decreasing sequence of prime ideals in $A[[X]]$ with trace (0) over A.

Proof Let M be the maximal ideal of A. By Lemma 8.4, there exist a sequence $(b_n)_{n \in \mathbb{N}^*}$ of elements in M and a sequence $(g_n)_{n \in \mathbb{N}^*}$ of elements in $A[[X]]$ such that for each $i, n \in \mathbb{N}^*$, b_n is a zero for g_i of multiplicity n^i. For each $f \in A[[X]]$, we write $f(b_j, j^k) = 0$ when b_j is a zero for f of multiplicity at least equal to j^k. We say that f annihilates almost everywhere (a.e) over a subset S of A if $f(s) = 0$ in $\hat{A}$ for each $s \in S$, but perhaps not on a finite number of elements of S. For each $k \in \mathbb{N}^*$, put $A_k = \{(b_j, j^k); j \in \mathbb{N}^*\}$ and $I_k = \{f \in A[[X]]$, such that f annihilates a.e over $A_k\}$. Then I_k is an ideal of $A[[X]]$ containing g_k since $g_k(b_j, j^k) = 0$ for each $j \in \mathbb{N}^*$. It is proper since 1 does not have a zero and $I_{k+1} \subseteq I_k$ since if $f(b_j, j^{k+1}) = 0$, then $f(b_j, j^k) = 0$. Let P_1 be a prime ideal of $A[[X]]$ minimal over I_1. Suppose that $P_1 \cap A \ne (0)$ and let $0 \ne a \in P_1 \cap A$. By Lemma 3.11 of Chap. 2, $\sqrt{I_1 A[[X]]}_{P_1} = P_1 A[[X]]_{P_1}$. There exist $m \in \mathbb{N}^*$ and $h \in A[[X]] \setminus P_1$ such that $a^m h \in I_1$, i.e., $a^m h$ annihilates a.e on A_1, then h annihilates a.e on A_1, so $h \in I_1 \subseteq P_1$, which is absurd. Then $P_1 \cap A = (0)$. We have the inclusions $I_2 \subseteq I_1 \subseteq P_1$. Suppose that P_1 is minimal over I_2, then $\sqrt{I_2 A[[X]]}_{P_1} = P_1 A[[X]]_{P_1}$. Since $g_1 \in I_1 \subseteq P_1$, then there exist $l \in \mathbb{N}^*$ and $h \in A[[X]] \setminus P_1$ such that $hg_1^l \in I_2$, so hg_1^l annihilates a.e over A_2. There exists $N \in \mathbb{N}^*$ such that $hg_1^l(b_i, i^2) = 0$ for each $i \ge N$, then b_i is a zero for hg_1^l of multiplicity $\ge i^2$. On the other hand, since $h \notin P_1$, then $h \notin I_1$ so there exist an infinity of natural integers j such that

$h(b_j, j) \neq 0$ and since $m(g_1, b_j) = j$, then the multiplicity of b_j as a zero for hg_1^l is $< j + lj$. Then for an infinity of integers $j \geq N$, we have $j^2 \leq j + lj$, so $j < 1 + l$, which is absurd. We conclude that P_1 is not minimal over I_2. We consider a prime ideal $P_2 \subset P_1$ of $A[[X]]$ minimal over I_2. Suppose, by induction, the construction of a finite sequence $P_1, \ldots, P_n$ of prime ideals of $A[[X]]$ such that $P_1 \supsetneq P_2 \supsetneq \ldots \supsetneq P_n \supseteq I_n \supseteq I_{n+1}$. Suppose that P_n is minimal over I_{n+1}, then $\sqrt{I_{n+1}A[[X]]_{P_n}} = P_n A[[X]]_{P_n}$. Since $g_n \in I_n \subseteq P_n$, there exist $l \in \mathbb{N}^*$ and $h \in A[[X]] \setminus P_n$ such that $hg_n^l \in I_{n+1}$, i.e., hg_n^l annihilates a.e over A_{n+1}. There exists $N \in \mathbb{N}^*$ such that for each $i \geq N$, $hg_n^l(b_i, i^{n+1}) = 0$, then b_i is a zero for hg_n^l of multiplicity at least equal to i^{n+1}. On the other hand, since $h \notin P_n$, then $h \notin I_n$. There exists an infinity of natural integers j such that $h(b_j, j^n) \neq 0$, i.e., $m(h, b_j) < j^n$, but $m(g_n, b_j) = j^n$, then $m(hg_n^l, b_j) < j^n + lj^n$. So for an infinity of natural integers $j \geq N$, $j^{n+1} < j^n + lj^n$, then $j < 1 + l$, which is absurd. We conclude that P_n is not minimal over I_{n+1}. We consider a prime ideal $P_{n+1} \subset P_n$ of $A[[X]]$ minimal over I_{n+1}. By induction, we construct an infinite strictly decreasing chain of prime ideals $P_1 \supsetneq P_2 \ldots$ of $A[[X]]$. Since $P_1 \cap A = (0)$, then $P_n \cap A = (0)$ for each $n \in \mathbb{N}^*$. Then we have the result.   □

It follows from the preceding theorem that if A is a nondiscrete valuation domain of rank one, then $A[[X]]$ has an infinite strictly decreasing sequence of prime ideals and infinite strictly increasing sequence of prime ideals.

Consider the monoid algebra $\mathbb{F}_2[\mathbb{Q}+]$, its maximal ideal $M = (t^\alpha; \ \alpha > 0)$, and the domains $V = \mathbb{F}_2[\mathbb{Q}+]_M$ and $V_1 = \mathbb{F}_2 + tV \subset V$. The following corollary shows that the SFT property is not sufficient for the formal power series ring to be of finite Krull dimension.

Corollary 8.6 (Coykendall 2002) *The formal power series domain $V_1[[X]]$ has an infinite dimension.*

Proof By the preceding theorem, since V is a nondiscrete valuation domain of rank one, there exists an infinite strictly decreasing chain $P_1 \supsetneq P_2 \supsetneq \ldots$ of prime ideals in $V[[X]]$ such that $P_i \cap V = (0)$ for each $i \in \mathbb{N}^*$. It suffices to show that the chain $P_1 \cap V_1[[X]] \supseteq P_2 \cap V_1[[X]] \supseteq \ldots$ of prime ideals in $V_1[[X]]$ is strictly decreasing. Suppose that there exists $n \in \mathbb{N}^*$ such that $P_n \cap V_1[[X]] = P_{n+1} \cap V_1[[X]]$. Since $P_{n+1} \subset P_n$, there exists $p_n \in P_n \setminus P_{n+1}$. Then $tp_n \in P_n \cap V_1[[X]] = P_{n+1} \cap V_1[X]]$, so $tp_n \in P_{n+1}$. Since $t \in V$ and $P_{n+1} \cap V = (0)$, then $t \notin P_{n+1}$, which is absurd.   □

We do not know if the domain $V_1[[X]]$ has an infinite strictly increasing sequence of prime ideals.

Corollary 8.7 *For each integer $n \geq 1$, we have $\dim V_1[[X_1, \ldots, X_n]] = \infty$.*

Proof By Lemma 1.23, we have $\dim V_1[[X_1, \ldots, X_n]] \geq \dim V_1[[X_1, \ldots, X_{n-1}]] \geq \ldots \geq \dim V_1[[X_1]] = \infty$.   □

9 Infinite Product of Formal Power Series over an Integral Domain

The goal of this section is to construct a class of integral domains A, mostly larger than the class of nondiscrete valuation domains of rank one, such that we have $dim\ A[[X]]_{A\setminus(0)} = +\infty$. We extend also Coykendall's example to other domains for which the SFT property does not pass to the formal power series ring. The results are due to Coykendall and Dumitrescu (2006).

Notation Let A be an integral domain and $(f_n(X))_{n\in\mathbb{N}^*}$ a sequence of elements of $A[[X]]$ satisfying the two following conditions:

(1) $(f_n(X))_{n\in\mathbb{N}^*}$ is a echelon, i.e., $\lim\limits_{n\to\infty} \omega(f_n(X) - f_n(0)) = +\infty$, where ω is the order of the formal power series.

(2) There exists $0 \neq a \in \bigcap\limits_{n:1}^{\infty} f_1(0)\dots f_n(0)A$.

Note that for each $n \in \mathbb{N}^*$, $f_n(0) \neq 0$. We consider the sequence $(g_n(X))_{n\in\mathbb{N}^*}$ of formal power series defined by $g_n(X) = \frac{a}{f_1(0)\dots f_n(0)} f_1(X)\dots f_n(X) \in \frac{a}{f_1(0)\dots f_n(0)} A[[X]]$ with $\frac{a}{f_1(0)\dots f_n(0)} \in A$, for each $n \in \mathbb{N}^*$, by (2).

The following lemma shows that the infinite product of formal power series has a topological interpretation.

Lemma 9.1 *The sequence* $(g_n(X))_{n\in\mathbb{N}^*}$ *is Cauchy in the complete space* $A[[X]]$ *for its* (X)-*adic topology. We denote* $\lim\limits_{n\to+\infty} g_n(X) = \left(\prod\limits_{n:1}^{\infty} f_n,\ a\right) \in A[[X]]$ *and we call it the infinite product of the formal power series* $f_n(X)$, $n \in \mathbb{N}^*$, *with constant term* a.

Proof For each integer $n \geq 2$, we have $g_n(X) - g_{n-1}(X) = \frac{a}{f_1(0)\dots f_n(0)} f_1(X)$ $\dots f_n(X) - \frac{af_n(0)}{f_1(0)\dots f_{n-1}(0)f_n(0)} f_1(X)\dots f_{n-1}(X) = \frac{a}{f_1(0)\dots f_n(0)} f_1(X)\dots f_{n-1}(X)$ $\big(f_n(X) - f_n(0)\big)$. Since the sequence $(f_n(X))_{n\in\mathbb{N}^*}$ is an echelon, by Lemma 6.1 (2), $(g_n(X))_{n\in\mathbb{N}^*}$ is Cauchy. But the domain $A[[X]]$ is complete for its (X)-adic topology, by Theorem 6.4. Then $(g_n(X))_{n\in\mathbb{N}^*}$ converges. Let $g(X) = \lim\limits_{n\to+\infty} g_n(X)$. By definition, there exists $k \in \mathbb{N}^*$ such that for each integer $n \geq k$, $g_n(X) - g(X) \in XA[[X]]$, then $g_n(0) - g(0) = 0$. So $g(0) = g_n(0) = \frac{a}{f_1(0)\dots f_n(0)} f_1(0)\dots f_n(0) = a$. $\qquad\square$

Example Let A be an integral domain. The infinite product $\left(\prod\limits_{n:1}^{\infty}(1 + X^{2^{n-1}}),\ 1\right) =$
$\sum\limits_{n:0}^{\infty} X^n = \frac{1}{1-X}$. Indeed, for each $n \in \mathbb{N}^*$, $f_n(X) = 1 + X^{2^{n-1}}$ and $g_n(X) =$

$f_1(X) \ldots f_n(X) = (1+X)(1+X^2) \ldots + (1+X^{2^{n-1}}) = 1+X+X^2+\ldots+X^{2^n-1}$
since $deg\ g(X) = 1+2+2^2+\ldots+2^{n-1} = 2^n - 1$, and each natural integer is in a
unique way a sum of powers of 2. We can also use the identity $1 + \omega = \frac{1-\omega^2}{1-\omega}$ pour
$\omega \neq 1$. Then $g_n(X) = (1+X)(1+X^2) \ldots + (1+X^{2^{n-1}}) = \frac{1-X^2}{1-X}\frac{1-X^4}{1-X^2} \cdots \frac{1-X^{2^n}}{1-X^{2^{n-1}}} =$
$\frac{1-X^{2^n}}{1-X} = 1+X+X^2+\ldots+X^{2^n-1}$. Since $\sum_{i:0}^{\infty} X^i - g_n(X) = \sum_{i:2^n}^{\infty} X^i \in X^{2^n} A[[X]]$,

then $\lim_{n \longrightarrow +\infty} g_n(X) = \sum_{i:0}^{\infty} X^i$ for the (X)-adic topology.

Lemma 9.2 *The set* $I = \bigcup_{n:1}^{\infty} \frac{a}{f_1(0) \ldots f_n(0)} A$ *is an ideal of A and* $\left(\prod_{n:1}^{\infty} f_n, a\right) \in$
$I[[X]]$.

Proof

(1) By (2), for each $n \in \mathbb{N}^*$, $\frac{a}{f_1(0)\ldots f_n(0)} \in A$, then $\frac{a}{f_1(0)\ldots f_n(0)} A$ is an integral ideal
of A. Moreover, $\frac{a}{f_1(0)\ldots f_n(0)} = \frac{af_{n+1}(0)}{f_1(0)\ldots f_n(0)f_{n+1}(0)} \in \frac{a}{f_1(0)\ldots f_{n+1}(0)} A$. Then I is an
ideal of A, as the union of an increasing sequence of ideals of A.

(2) Let $g(X) - \left(\prod_{n:1}^{\infty} f_n, a\right) = \lim_{n \longrightarrow +\infty} g_n(X) = \sum_{i:0}^{\infty} a_i X^i \in A[[X]]$. Let $l \in \mathbb{N}$.
There exists $k \in \mathbb{N}^*$ such for each $n \geq k$, $g_n(X) - g(X) \in X^{l+1} A[[X]]$. The
coefficient a_l of X^l in $g(X)$ is equal to that of X^l in $g_n(X) \in \frac{a}{f_1(0)\ldots f_n(0)} A[[X]]$,
then $a_l \in \frac{a}{f_1(0)\ldots f_n(0)} A \subseteq I$. □

Lemma 9.3 *For each* $m \in \mathbb{N}^*$, $\left(\prod_{n:1}^{\infty} f_n, a\right) = f_1 \ldots f_m \left(\prod_{n:m+1}^{\infty} f_n, \frac{a}{f_1(0) \ldots f_m(0)}\right)$.

Proof By induction, it suffices to show the property for $m = 1$. Let
the series $g(X) = \left(\prod_{n:1}^{\infty} f_n, a\right)$, $g'(X) = \left(\prod_{n:2}^{\infty} f_n, \frac{a}{f_1(0)}\right)$, $g_n(X) =$
$\frac{a}{f_1(0)\ldots f_n(0)} f_1(X) \ldots f_n(X)$, and $g_n'(X) = \frac{a}{f_1(0)f_2(0)\ldots f_n(0)} f_2(X) \ldots f_n(X) \in$
$A[[X]]$. We will show that $g(X) = f_1(X)g'(X)$. We have $g(X) - f_1(X)g'(X) =$
$g(X) - g_n(X) + g_n(X) - f_1(X)g'(X) = g(X) - g_n(X) + f_1(X)g_n'(X) -$
$f_1(X)g'(X) = g(X) - g_n(X) + f_1(X)(g_n'(X) - g'(X))$. By definition,
$g(X) = \lim_{n \longrightarrow +\infty} g_n(X)$ and $g'(X) = \lim_{n \longrightarrow +\infty} g_n'(X)$ for the (X)-adic topology
of $A[[X]]$. Let $p \in \mathbb{N}^*$. By Remark 6.1 (3), there exists $q \in \mathbb{N}^*$ such that for each
integer $n \geq q$, $g(X) - g_n(X) \in X^p A[[X]]$ and $g'(X) - g_n'(X) \in X^p A[[X]]$, then

$g(X) - f_1(X)g'(X) \in X^p A[[X]]$. So we have $g(X) - f_1(X)g'(X) \in \bigcap_{p:1}^{\infty} X^p A[[X]]$

$= (0)$. Then $g(X) = f_1(X)g'(X)$.   □

Example Let A be an integral domain with quotient field K, $(a_n)_{n \in \mathbb{N}^*}$ a sequence of nonzero elements of A, and $(r_n)_{n \in \mathbb{N}^*}$ a sequence of nonzero natural integers. Suppose that $a_{n+1} = a_n^2$ and $r_{n+1} = 2r_n$ for n sufficiently big. Then the infinite product $f(X) = \left(\prod_{n:1}^{\infty} (1 + a_n X^{r_n}), \ 1 \right) \in K(X)$. Indeed, suppose that the equalities $a_{n+1} = a_n^2$ and $r_{n+1} = 2r_n$ are true from the rank N. Since $f(X) = (1 + a_1 X^{r_1}) \ldots (1 + a_{N-1} X^{r_{N-1}}) \left(\prod_{n:N}^{\infty} (1 + a_n X^{r_n}), \ 1 \right)$, we may suppose that $N = 1$. Then $a_n = a_{n-1}^2 = a_{n-2}^{2^2} = \ldots = a_1^{2^{n-1}}$ and $r_n = 2r_{n-1} = 2^2 r_{n-2} = \ldots = 2^{n-1} r_1$. So $f(X) = \left(\prod_{n:1}^{\infty} (1 + a_1^{2^{n-1}} X^{r_1 2^{n-1}}), \ 1 \right) = \left(\prod_{n:1}^{\infty} \left(1 + (a_1 X^{r_1})^{2^{n-1}} \right), \ 1 \right)$

$= \frac{1}{1 - a_1 X^{r_1}} \in K(X)$.

Lemma 9.4 *Let B be a ring and $a \in B$. Then $(X - a)B[[X]] \cap B \subseteq \bigcap_{n:1}^{\infty} a^n B$.*

Moreover, if B is an integral domain, we have the equality.

Proof Let $r \in (X - a)B[[X]] \cap B$. Then $r = (X - a)\left(\sum_{n:0}^{\infty} b_n X^n \right) = \sum_{n:0}^{\infty} b_n X^{n+1} -$

$\sum_{n:0}^{\infty} ab_n X^n = \sum_{n:1}^{\infty} b_{n-1} X^n - \sum_{n:0}^{\infty} ab_n X^n = -ab_0 + \sum_{n:1}^{\infty} (b_{n-1} - ab_n) X^n$. So $r =$

$-ab_0$ and for each $n \geq 1$, $b_{n-1} = ab_n$. Then $r = -ab_0 = -a^2 b_1 = -a^3 b_2 =$

$\ldots \in \bigcap_{n:1}^{\infty} a^n B$. Conversely, suppose now that B is an integral domain. If $a = 0$,

the equality is clear. Suppose that $a \neq 0$. Let $r \in \bigcap_{n:1}^{\infty} a^n B$, $r = a^n b_n$ with $b_n \in B$

for each $n \in \mathbb{N}$. Then $\frac{1}{X - a} = \frac{-1}{a(1 - \frac{X}{a})} = \frac{-1}{a} \sum_{n:0}^{\infty} (\frac{X}{a})^n = -\sum_{n:0}^{\infty} \frac{b_{n+1}}{r} X^n$. So $r =$

$-(X - a) \sum_{n:0}^{\infty} b_{n+1} X^n \in (X - a)B[[X]] \cap B$.   □

Example The integrity of B is not necessary to have the equality in the lemma. Let a be an idempotent element of any ring B. For example, $B = \mathbb{Z}^2$ and $a = (1, 0)$. Then $\bigcap\limits_{n:1}^{\infty} a^n B = aB$. Indeed, since $a = (X - a)\left(\sum\limits_{n:0}^{\infty} -aX^n\right)$, we have the result.

Definition 9.5 An integral domain A is said to be Archimedean if for each non-invertible element a of A, the intersection $\bigcap\limits_{n:1}^{\infty} a^n A = (0)$.

Example 1 An integral domain A satisfying the acc for principal ideals is Archimedean. Indeed, let $0 \neq a \in A$ be a non-invertible element. Suppose that there exists $0 \neq b \in \bigcap\limits_{n:1}^{\infty} a^n A$. For each $n \in \mathbb{N}$, there exists $0 \neq b_n \in A$ such that $b = a^n b_n$. Then $a^n b_n = b = a^{n+1} b_{n+1}$, so $b_n = a b_{n+1}$. The sequence $(b_n A)_{n \in \mathbb{N}}$ is increasing, so stationary. Let $n \in \mathbb{N}$ be such that $b_n A = b_{n+1} A$. Then b_n and b_{n+1} are associated, which is absurd since the coefficient of proportionality a is not invertible.

Example 2 A completely integrally closed integral domain A is Archimedean For example, a valuation domain of rank one is Archimedean. Indeed, let $0 \neq a \in A$ be a non-invertible element. Suppose that there exists $0 \neq b \in \bigcap\limits_{n:1}^{\infty} a^n A$. For each $n \in \mathbb{N}$, there exists $b_n \in A$ such that $b = a^n b_n$, then $b\frac{1}{a^n} = b_n \in A$. So $\frac{1}{a}$ is quasi-integral over A, which is absurd.

Example 3 Let A be a nondiscrete valuation domain of rank one, defined by a valuation with values group G dense in $\mathbb{R}$. Then A is Archimedean. Let $(\alpha_n)_{n \in \mathbb{N}}$ be a sequence of elements of G that decreases strictly to zero and $(a_n)_{n \in \mathbb{N}}$ be a sequence of elements of A such that $v(a_n) = \alpha_n$ for each $n \in \mathbb{N}$. The sequence $(a_n A)_{n \in \mathbb{N}}$ is strictly increasing.

Theorem 9.6 *Let $A \subseteq B$ be an extension of integral domains. Suppose that B is Archimedean and does not satisfy the acc for principal ideals and the conductor $(A : B) \neq (0)$. Then $A[[X]]$ is not an SFT-ring.*

Proof Let $(b_n B)_{n \in \mathbb{N}^*}$ be a strictly increasing sequence of nonzero principal ideals of B. Let $a_0 = b_0$ and $a_n = \frac{b_{n-1}}{b_n} \in B \setminus U(B)$ for each integer $n \geq 1$. Let $n, p \in \mathbb{N}^*$. Then $b_{n-1} = a_n b_n = a_n a_{n+1} b_{n+1} = a_n a_{n+1} a_{n+2} b_{n+2} = \ldots\ldots = a_n a_{n+1} a_{n+2} \ldots a_{n+p} b_{n+p} \in a_n a_{n+1} a_{n+2} \ldots a_{n+p} B$. So for each $n \geq 1, 0 \neq b_{n-1} \in \bigcap\limits_{p:1}^{\infty} a_n a_{n+1} a_{n+2} \ldots a_{n+p} B$. For each $n \geq 1$, we define the formal power series

$g_n(X) = \left(\prod\limits_{i:n}^{\infty} (X^i + a_i), \, b_{n-1} \right) \in B[[X]]$. Conditions (1) and (2) of the definition of

an infinite product are satisfied. By Lemma 9.2, for $n \in \mathbb{N}^*$, the coefficients of $g_n(X)$

belong to the following ideal: $I = \bigcup\limits_{p:1}^{\infty} \dfrac{b_{n-1}}{a_n a_{n+1} a_{n+2} \cdots a_{n+p}} B = \bigcup\limits_{p:1}^{\infty} b_{n+p} B =$

$\bigcup\limits_{p:1}^{\infty} b_p B$ of B, since the sequence $(b_p B)_{p \in \mathbb{N}^*}$ is increasing. By Lemma 9.3, for each

$n \geq 1$, $g_n(X) = (X^n + a_n) g_{n+1}(X)$. Note that the constant term of $g_{n+1}(X)$

is $b_n = \dfrac{b_{n-1}}{a_n}$. Then $\mathscr{I} = \bigcup\limits_{n:1}^{\infty} g_n(X) B[[X]]$ is an ideal of $B[[X]]$, as the union

of an increasing sequence of ideals of $B[[X]]$. Since $g_n(X) \in I[[X]]$ for each
$n \in \mathbb{N}^*$, then $\mathscr{I} \subseteq I[[X]]$. Let $\pi : B[[X]] \longrightarrow B$ be the homomorphism
defined by $\pi(f(X)) = f(0)$ for each $f(X) \in B[[X]]$. Then $\pi(\mathscr{I}) \subseteq I$.

Conversely, $I = \bigcup\limits_{p:1}^{\infty} b_p B$ with b_p the constant term of $g_{p+1}(X) \in \mathscr{I}$, then

$b_p \in \pi(\mathscr{I})$. So $I \subseteq \pi(\mathscr{I})$, then $I = \pi(\mathscr{I})$. Fix an element $0 \neq r \in I$. Then
$r = h_0(X) + X q_1(X)$ with $h_0(X) \in \mathscr{I}$ and $q_1(X) \in B[[X]]$. Since $\mathscr{I} \subseteq I[[X]]$,
then $q_1(X) \in I[[X]]$. So $q_1(X) = h_1(X) + X q_2(X)$ with $h_1(X) \in \mathscr{I}$ and
$q_2(X) \in B[[X]]$. Then $r = h_0(X) + X q_1(X) = h_0(X) + X h_1(X) + X^2 q_2(X)$.
By induction, we construct a sequence $(h_n(X))_{n \in \mathbb{N}}$ of elements of $\mathscr{I}$ such that

$r = \sum\limits_{n:0}^{\infty} X^n h_n(X)$. Let $0 \neq a \in (A : B)$. Then $ar = \sum\limits_{n:0}^{\infty} X^n (a h_n(X))$ with

$a h_n(X) \in A[[X]]$ for each $n \in \mathbb{N}$. Suppose that $A[[X]]$ is an SFT-ring. By
Lemma 5.5, since the ideal $\sqrt{(a h_n(X); \, n \in \mathbb{N}) A[[X]]}$ is SFT and radical, then

the element $ar = \sum\limits_{n:0}^{\infty} X^n (a h_n(X)) \in \sqrt{(a h_n(X); \, n \in \mathbb{N}) A[[X]]}$. Let $k \in \mathbb{N}^*$ be

such that $(ar)^k \in (a h_n(X); \, n \in \mathbb{N}) A[[X]] \subseteq \mathscr{I}$. Since $\mathscr{I} = \bigcup\limits_{n:1}^{\infty} g_n(X) B[[X]]$,

there exists $m \in \mathbb{N}^*$ such that $(ar)^k \in g_m(X) B[[X]]$. By Lemma 9.3, $g_m(X) = (X^m + a_m) g_{m+1}(X)$, then $0 \neq (ar)^k \in (X^m + a_m) B[[X]] \cap B$. By Lemma 5.1,
$(X^m + a_m) B[[X]] \cap B = (X^m + a_m) B[[X^m]] \cap B$. By Lemma 9.4, since B is an

Archimedean integral domain, $(X^m + a_m) B[[X^m]] \cap B = \bigcap\limits_{n:1}^{\infty} a_m^n B = (0)$, then

$(ar)^k = 0$, which is absurd.   □

We retrieve the Coykendall's example in the following corollary.

Corollary 9.7 *Let (V, M) be a nondiscrete valuation domain of rank one containing a field K and let $0 \neq x \in M$. Then the domain $W = K + xV$ is SFT but the formal power series domain $W[[X]]$ is not SFT.*

Proof Since $W/xV \simeq K$ is a field, xV is a maximal ideal of W. It is SFT since $(xV)^2 = x^2 V = x(xV) \subseteq xW \subseteq xV$. It is also the only nonzero prime ideal of W. Indeed, let P be a nonzero prime ideal of W and $0 \neq t \in P$. For each $v \in V$, since V is a nondiscrete valuation domain of rank one and $x \in M$, there exists $n \in \mathbb{N}^*$ such that $(xv)^n \in tV$. Then $(xv)^{n+1} = xv(xv)^n \in txV \subseteq tW \subseteq P$. But P is a prime ideal of W and $xv \in W$, then $xv \in P$. So $xV \subseteq P$, then $P = xV$ since xV is maximal. The domain $W = K + xV$ is SFT. By a preceding example, a nondiscrete valuation domain of rank one is Archimedean and does not satisfy the acc for the principal ideals. Since $0 \neq x \in (V : W)$, we can apply the preceding theorem. Then the domain $W[[X]]$ is not SFT. $\qquad\square$

The following proposition will be used later, but it is interesting in itself.

Proposition 9.8 *Let $A \subseteq B$ be an extension of integral domains, $(v_n)_{n \in \mathbb{N}^*}$ a sequence of discrete valuations positive over B and zero over $A \setminus (0)$, and $(g_i)_{i \in \mathbb{N}^*}$ a sequence of elements of B. Suppose that for each $i \in \mathbb{N}^*$, $v_n(g_i) = n^i$ for almost each $n \in \mathbb{N}^*$ (a.e), i.e., for each $n \in \mathbb{N}^*$ but not true for at most a finite number of them.*

Then $\dim B_{A \setminus (0)} = +\infty$. In fact, B admits an infinite strictly decreasing sequence of prime ideals with trace (0) over A.

Proof For each $k \in \mathbb{N}^*$, let $I_k = \{ f \in B; \ v_n(f) \geq n^k \ \text{a.e} \}$. Then I_k is an ideal of B containing g_k. Indeed, since $v_n(g_k) = n^k$ a.e, then $g_k \in I_k$. Let $f, g \in I_k$ and $h \in B$. If $v_n(f) \geq n^k$ and $v_n(g) \geq n^k$, then $v_n(f + g) \geq \min\{ v_n(f); v_n(g) \} \geq n^k$ and $v_n(fh) = v_n(f) + v_n(h) \geq v_n(f) \geq n^k$. Then $f + g, fh \in I_k$. The sequence $(I_k)_{k \in \mathbb{N}^*}$ is decreasing. Indeed, let $f \in I_{k+1}$. Since $v_n(f) \geq n^{k+1}$ a.e, then $v_n(f) \geq n^k$ a.e. Then $f \in I_k$. So $I_{k+1} \subseteq I_k$. Let us now mention the following two remarks:

(1) For each $k \in \mathbb{N}^*$, if P is a prime ideal of B minimal over I_k, then $P \cap A = (0)$. Indeed, by Lemma 3.11 of Chap. 2, $\sqrt{I_k B_P} = P B_P$. Suppose that there exists $0 \neq a \in P \cap A$. There exist $i \in \mathbb{N}^*$ and $s \in B \setminus P$ such that $a^i s \in I_k$. Then $n^k \leq v_n(a^i s) = v_n(a^i) + v_n(s) = v_n(s)$ a.e since v_n is zero over $A \setminus (0)$. Then $s \in I_k \subseteq P$, which is absurd. Then $P \cap A = (0)$.

(2) For each $k \in \mathbb{N}^*$, the ideals I_k and I_{k+1} cannot have a common minimal prime ideal P in B. Indeed, in the contrary case, by Lemma 3.11 of Chap. 2, $\sqrt{I_{k+1} B_P} = P B_P$. Since $g_k \in I_k \subseteq P$, there exist $i \in \mathbb{N}^*$ and $s \in B \setminus P$ such that $g_k^i s \in I_{k+1}$. Then $n^{k+1} \leq v_n(g_k^i s) = i v_n(g_k) + v_n(s) = i n^k + v_n(s)$ a.e, so $v_n(s) \geq n^{k+1} - i n^k = n^k(n - i) \geq n^k$ a.e. Then $s \in I_k \subseteq P$, which is absurd.

To finish the proof of the proposition, we start by prime ideal P_1 of B minimal over I_1. Since $I_2 \subseteq I_1 \subseteq P_1$, there exists a prime ideal $P_2 \subset P_1$ of B minimal over I_2, $\ldots$, etc. $\qquad\square$

Example (M. Laplaza) The domain E of analytic functions admits an infinite strictly decreasing sequence of prime ideals. Indeed, for $a \in \mathbb{C}$ and $f \in E$, we denote $ord_a(f)$ the multiplicity of a as a zero of f, then ord_a a discrete valuation positive over E and zero over $\mathbb{C}^*$. Fix an integer $i \in \mathbb{N}^*$. By Weierstrass theorem, there exists $g_i \in E$ with $ord_n(g_i) = n^i$ for each $n \in \mathbb{N}^*$. We conclude by the preceding proposition.

Lemma 9.9 *Let A be an integral domain and $0 \neq a \in A$ such that $\bigcap_{n:1}^{\infty} a^n A = (0)$.*

1. *The intersection $\bigcap_{n:1}^{\infty} (X - a)^n A[[X]] = (0)$.*
2. *Suppose that the ideal $(X-a)A[[X]]$ is prime. Let v be the $(X-a)$-adic valuation over $A[[X]]$. For each $0 \neq f(X) \in A[[X]]$, $v(f) = 0$ if and only if $f(a) \neq 0$ in the completion $\hat{A}$ of A for the (a)-adic topology.*

Proof

(1) Suppose that there exists $0 \neq f(X) = a_k X^k + \ldots \in \bigcap_{n:1}^{\infty} (X - a)^n A[[X]]$ with $a_k \neq 0$. For each $n \in \mathbb{N}^*$, $f(X) = (X - a)^n g_n(X)$ with $0 \neq g_n(X) = b_n X^{e_n} + \ldots \in A[[X]]$ where $b_n \neq 0$. The coefficient of the nonzero term of the least degree in $f(X)$ is $a_k = (-a)^n b_n$ for each $n \in \mathbb{N}^*$. Then $a_k \in \bigcap_{n:1}^{\infty} a^n A = (0)$, which is absurd.

(2) Let $0 \neq f(X) = \sum_{i:0}^{\infty} c_i X^i \in A[[X]]$. Recall that $v(f)$ is the biggest power of $X - a$ that divides $f(X)$ in $A[[X]]$. Since $\sum_{i:0}^{n+1} c_i a^i - \sum_{i:0}^{n} c_i a^i = c_{n+1} a^{n+1} \in a^{n+1} A$, the sequence $\left(\sum_{i:0}^{n} c_i a^i \right)_{n \in \mathbb{N}}$ is Cauchy in A, then it converges in $\hat{A}$. We denote $f(a) = \sum_{i:0}^{\infty} c_i a^i \in \hat{A}$ its limit. Suppose that $v(f) = 0$ and $f(a) = 0$. In $\hat{A}[[X]]$, $f(X) = f(X) - f(a) = \sum_{i:1}^{\infty} c_i (X^i - a^i) = (X - a)(c_1 + c_2(X + a) + \ldots) \in (X - a)A[[X]]$. Then $v(f) \geq 1$, which is absurd. Conversely, suppose that $v(f) \geq 1$. Then $f(X) = (X - a)g(X)$ with $g(X) \in A[[X]]$, so $f(a) = 0$ in $\hat{A}$, which is absurd. $\square$

Lemma 9.10 *Let A be a ring and $h(X) \in A[[X]]$ and $(h_n(X))_{n \in \mathbb{N}}$ a sequence of elements of $A[[X]]$ that converges to $h(X)$ for the (X)-adic topology. For each $a \in A$, if we denote $\hat{A}$ the completion of A for the aA-adic topology, the sequence $(h_n(a))_{n \in \mathbb{N}}$ converges in $\hat{A}$ to $h(a)$ for the $a\hat{A}$-adic topology.*

Proof Let $k \in \mathbb{N}$. There exists $N \in \mathbb{N}$ such that for each integer $n \geq N$, $h_n(X) - h(X) \in X^k A[[X]]$, then $h_n(a) - h(a) \in a^k \hat{A}$. The sequence $(h_n(a))_{n \in \mathbb{N}}$ converges to $h(a)$ in $\hat{A}$ for the $a\hat{A}$-adic topology. □

Theorem 9.11 *Let A be a local integral domain with zero characteristic. Suppose that there exists a strictly increasing sequence $(b_n A)_{n \in \mathbb{N}^*}$ of nonzero principal ideals of A satisfying:*

(1) *For each $n \in \mathbb{N}^*$, the ideal $(X - b_n)A[[X]]$ is prime and $\bigcap\limits_{k:1}^{\infty} b_n^k A = (0)$.*

(2) *For each $i \in \mathbb{N}^*$, the intersection $\bigcap\limits_{n:1}^{\infty} b_1^{1^i} b_2^{2^i} \ldots b_n^{n^i} A \neq (0)$.*

Then $\dim A[[X]]_{A \setminus (0)} = +\infty$.

Proof We look to apply Proposition 9.8 with the domain $B = A[[X]]$.

Fix an integer $n \in \mathbb{N}^*$. By Lemma 9.9, hypothesis (1) allows us to define the $(X - b_n)$-adic discrete valuation v_n over the domain $A[[X]]$. By Lemma 9.9 (2), for each $0 \neq f(X) \in A[[X]]$, $v_n(f) = 0$ if and only if $f(b_n) \neq 0$ in $\hat{A}$, the completion of A for the (b_n)-adic topology. In particular, the valuation v_n is zero on the set $A \setminus (0)$ and positive over $A[[X]]$. Fix an integer $i \in \mathbb{N}^*$ and consider an element $0 \neq a_i \in \bigcap\limits_{n:1}^{\infty} b_1^{1^{i+1}} b_2^{2^{i+1}} \ldots b_n^{n^{i+1}} A$. By Lemma 9.1, the infinite product

$$g_i(X) = \left(\prod\limits_{m:1}^{\infty} (X^m - b_m^m)^{m^i}; \, a_i \right) \in A[[X]] \text{ is well defined. By Lemma 9.3, } g_i(X) =$$

$$\left(X - b_1 \right) \left(X^2 - b_2^2 \right)^{2^i} \ldots \left(X^n - b_n^n \right)^{n^i} h(X) \text{ with } h(X) = \left(\prod\limits_{m:n+1}^{\infty} (X^m - b_m^m)^{m^i}; \, c_n \right) \in$$

$A[[X]]$, where $c_n = \pm \dfrac{a_i}{b_1^{1^{i+1}} b_2^{2^{i+1}} \ldots b_n^{n^{i+1}}} \in A \setminus (0)$. We will compute the valuation of each factor of $g_i(X)$ for the valuation v_n.

For $1 \leq k < n$, $v_n\left((X^k - b_k^k)^{k^i}\right) = 0$. Indeed, if not, by Lemma 9.9 (2), b_n will be a root of the polynomial $(X^k - b_k^k)^{k^i}$, then $b_n^k = b_k^k$. But by hypothesis, $b_k A \subset b_n A$, so $b_k = b_n c$ with $c \in M$, the maximal ideal of the local domain A. Then $b_n^k = b_n^k c^k$, so $c^k = 1$, then $c \in U(A)$, which is absurd.

On the other hand, b_n is a simple root of the polynomial $X^n - b_n^n$. Indeed, the derivative of this polynomial is $n X^{n-1}$. It does not annihilate on b_n since the characteristic of A is zero. Then $v_n(X^n - b_n^n) = 1$ and $v_n\left((X^n - b_n^n)^{n^i}\right) = n^i$.

By hypothesis, for each integer $m \geq n + 1$, $b_n A \subset b_m A$, then $b_n = b_m d_m$ with $d_m \in M$. Then $(b_n^m - b_m^m)^{m^i} = (b_m^m d_m^m - b_m^m)^{m^i} = b_m^{m^{i+1}} (d_m^m - 1)^{m^i} = b_m^{m^{i+1}} u_m$

with $u_m \in U(A)$. By definition, the infinite product $h(X) = \lim\limits_{p \longrightarrow +\infty} h_p(X)$ for the (X)-adic topology over $A[[X]]$, where $h_p(X) = \pm \dfrac{c_n}{b_{n+1}^{(n+1)^{i+1}} \cdots b_{n+p}^{(n+p)^{i+1}}} \big(X^{n+1} -$

$b_{n+1}^{n+1}\big)^{(n+1)^i} \cdots \big(X^{n+p} - b_{n+p}^{n+p}\big)^{(n+p)^i}$. Suppose that $v_n(h) \neq 0$. By Lemma 9.9 (2), $h(b_n) = 0$ in $\hat{A}$, the completion of A for the (b_n)-adic topology over A. By the preceding lemma, $\lim\limits_{p \longrightarrow +\infty} h_p(b_n) = h(b_n) = 0$. But we have

$$h_p(b_n) = \frac{\pm c_n}{b_{n+1}^{(n+1)^{i+1}} \cdots b_{n+p}^{(n+p)^{i+1}}} \big(b_n^{n+1} - b_{n+1}^{n+1}\big)^{(n+1)^i} \cdots \big(b_n^{n+p} - b_{n+p}^{n+p}\big)^{(n+p)^i} =$$

$\dfrac{\pm c_n}{b_{n+1}^{(n+1)^{i+1}} \cdots b_{n+p}^{(n+p)^{i+1}}} b_{n+1}^{(n+1)^{i+1}} u_{n+1} \cdots b_{n+p}^{(n+p)^{i+1}} u_{n+p} = \pm c_n u_{n+1} \cdots u_{n+p}$. Since

we have $\lim\limits_{p \longrightarrow +\infty} h_p(b_n) = 0$, for each $k \in \mathbb{N}^*$, there exists $N \in \mathbb{N}^*$ such that for each integer $p \geq N$, $h_p(b_n) \in b_n^k A$. But $u_{n+1}, \ldots, u_{n+p} \in U(A)$, then $c_n \in b_n^k A$.

So $c_n \in \bigcap\limits_{k:1}^{\infty} b_n^k A = (0)$, by hypothesis. Then $c_n = 0$, which is absurd. We conclude that $v_n(h) = 0$, then $v_n(g_i) = n^i$. The conditions of application of Proposition 9.8 are satisfied. $\square$

10 Completion of a Metric Space

Definition 10.1 Let (E, d) be a metric space. We call a completion of (E, d) a couple $\big((\hat{E}, \hat{d}); \phi\big)$ formed by a metric space $(\hat{E}, \hat{d})$ and an isometry $\phi : E \longrightarrow \hat{E}$ such that $\phi(E)$ is dense in $\hat{E}$.

We recall that an isometry is a map between metric spaces that preserve the distance, i.e., for each $x, y \in E$, $\hat{d}\big(\phi(x), \phi(y)\big) = d(x, y)$. In particular, ϕ is one to one since if $\phi(x) = \phi(y)$, then $d(x, y) = 0$, so $x = y$.

Theorem 10.2 *Each metric space has a completion.*

Proof Let (E, d) be a metric space and $C(E)$ the set of the Cauchy sequences of elements of E. Then $C(E) \neq \emptyset$ since it contains all the constant sequences. Note that if $(u_n)_{n \in \mathbb{N}}$ and $(v_n)_{n \in \mathbb{N}}$ are elements of $C(E)$, the numerical sequence $\big(d(u_n, v_n)\big)_{n \in \mathbb{N}}$ is Cauchy in the complete metric space $(\mathbb{R}, | \ |)$, so convergent. Indeed, $d(u_n, v_n) - d(u_m, v_m) \leq d(u_n, u_m) + d(u_m, v_m) + d(v_m, v_n) - d(u_m, v_m)$ $= d(u_n, u_m) + d(v_m, v_n)$. By symmetry, $d(u_m, v_m) - d(u_n, v_n) \leq d(u_n, u_m) + d(v_m, v_n)$. Then we have $|d(u_m, v_m) - d(u_n, v_n)| \leq d(u_n, u_m) + d(v_m, v_n)$. We endow $C(E)$ by the equivalence relation $\sim$ defined for $u = (u_n)_{n \in \mathbb{N}}$ and $v = (v_n)_{n \in \mathbb{N}}$ by $u \sim v$ if and only if $\lim\limits_{n \longrightarrow +\infty} d(u_n, v_n) = 0$. The transitivity follows from the triangular inequality. Let $d_1 : C(E) \times C(E) \longrightarrow \mathbb{R}$ be the map defined for each $u = (u_n)_{n \in \mathbb{N}}$ and $v = (v_n)_{n \in \mathbb{N}} \in C(E)$, by $d_1(u, v) = \lim\limits_{n \longrightarrow +\infty} d(u_n, v_n)$.

We will show that d_1 passes to the quotient by $\sim$, i.e., if $u' = (u'_n)_{n \in \mathbb{N}}$ and $v' = (v'_n)_{n \in \mathbb{N}} \in C(E)$ are such that $u \sim u'$ and $v \sim v'$, then $d_1(u, v) = d_1(u', v')$. Indeed, by the triangular inequality, $d(u_n, v_n) \le d(u_n, u'_n) + d(u'_n, v'_n) + d(v'_n, v_n)$. But $\lim\limits_{n \longrightarrow +\infty} d(u_n, u'_n) = \lim\limits_{n \longrightarrow +\infty} d(v_n, v'_n) = 0$. By passage to the limit, we find $d_1(u, v) \le d_1(u', v')$. By symmetry, $d_1(u', v') \le d_1(u, v)$. Then $d_1(u, v) = d_1(u', v')$. We consider the quotient set $\hat{E} = C(E)/\sim$ and the map $\hat{d} : \hat{E} \times \hat{E} \longrightarrow \mathbb{R}$ defined by $\hat{d}(\tilde{u}, \tilde{v}) = d_1(u, v) = \lim\limits_{n \longrightarrow +\infty} d(u_n, v_n)$ for each $u = (u_n)_{n \in \mathbb{N}}$ and $v = (v_n)_{n \in \mathbb{N}} \in C(E)$. Then $\hat{d}$ is a distance over $\hat{E}$. Let $\phi : E \longrightarrow \hat{E}$ be the map that associates to x the class of the constant sequence with general term x. Then ϕ is an isometry. Indeed, let $x, y \in E$. Put $\phi(x) = \tilde{u}$ and $\phi(y) = \tilde{v}$ with $u = (u_n)_{n \in \mathbb{N}}$ and $v = (v_n)_{n \in \mathbb{N}}$ the constant sequences with general terms x and y, respectively. Then $\hat{d}(\phi(x), \phi(y)) = \hat{d}(\tilde{u}, \tilde{v}) = d_1(u, v) = \lim\limits_{n \longrightarrow +\infty} d(u_n, v_n) = \lim\limits_{n \longrightarrow +\infty} d(x, y) = d(x, y)$.

We will show that $\phi(E)$ is dense in $\hat{E}$. It suffices to show that each element of $\hat{E}$ is the limit of a sequence of elements of $\phi(E)$. Let $\tilde{u} \in \hat{E}$ with $u = (u_n)_{n \in \mathbb{N}}$. We consider the sequence $(\phi(u_m))_{m \in \mathbb{N}} \subseteq \phi(E)$. Then $\hat{d}(\tilde{u}, \phi(u_m)) = \lim\limits_{n \longrightarrow +\infty} d(u_n, u_m)$. Since $(u_n)_{n \in \mathbb{N}}$ is Cauchy in (E, d), then $\lim\limits_{m \longrightarrow +\infty} \hat{d}(\tilde{u}, \phi(u_m)) = \lim\limits_{m \longrightarrow +\infty} \lim\limits_{n \longrightarrow +\infty} d(u_n, u_m) = 0$. So $(\phi(u_m))_{m \in \mathbb{N}}$ converges to $\tilde{u}$.

We finish by showing that the metric space $(\hat{E}, \hat{d})$ is complete. Let $(\widetilde{u_n})_{n \in \mathbb{N}}$ be a Cauchy sequence of $\hat{E}$. Since $\phi(E)$ is dense in $\hat{E}$, for each $n \in \mathbb{N}$, there exists $v_n \in E$ such that $\hat{d}(\widetilde{u_n}, \phi(v_n)) \le \frac{1}{2^n}$. We will show that the sequence $v = (v_n)_{n \in \mathbb{N}}$ is Cauchy in E, then $\tilde{v} \in \hat{E}$. Let $\epsilon > 0$ be a real number. There exists $N \in \mathbb{N}$ such that for each $n, m \ge N$, $\hat{d}(\widetilde{u_n}, \widetilde{u_m}) < \frac{\epsilon}{2}$. Then $d(v_n, v_m) = \hat{d}(\phi(v_n), \phi(v_m)) \le \hat{d}(\phi(v_n), \widetilde{u_n}) + \hat{d}(\widetilde{u_n}, \widetilde{u_m}) + \hat{d}(\widetilde{u_m}, \phi(v_m)) \le \frac{1}{2^n} + \frac{\epsilon}{2} + \frac{1}{2^m}$. There exists $M \in \mathbb{N}$ such that for each $n, m \ge M$, $\frac{1}{2^n} + \frac{1}{2^m} \le \frac{\epsilon}{2}$. Then for each $n, m \ge sup\{N, M\}$, $d(v_n, v_m) \le \epsilon$. The sequence $(\widetilde{u_n})_{n \in \mathbb{N}}$ converges to $\tilde{v}$ in $\hat{E}$. Indeed, $\hat{d}(\widetilde{u_n}, \tilde{v}) \le \hat{d}(\widetilde{u_n}, \phi(v_n)) + \hat{d}(\phi(v_n), \tilde{v}) \le \frac{1}{2^n} + \hat{d}(\phi(v_n), \tilde{v})$. Since $v = v_0, v_1, \ldots$ and $\phi(v_n)$ is the class of the constant sequence $v_n, v_n, \ldots$, then $\hat{d}(\phi(v_n), \tilde{v}) = \lim\limits_{m \longrightarrow +\infty} d(v_n, v_m)$. So $\hat{d}(\widetilde{u_n}, \tilde{v}) \le \frac{1}{2^n} + \lim\limits_{m \longrightarrow +\infty} d(v_n, v_m)$. Since the sequence $v = (v_n)_{n \in \mathbb{N}}$ is Cauchy, by passage to the limit, we find $\lim\limits_{n \longrightarrow +\infty} \hat{d}(\widetilde{u_n}, \tilde{v}) = 0$.   □

In the following example, we obtain the real numbers field $\mathbb{R}$ as a completion of the rational numbers field $\mathbb{Q}$, using the natural metric.

Example 1 Construction of the real numbers field.

The same process of completion can be used to construct the real numbers field. Let $C(\mathbb{Q})$ be the set of the Cauchy sequences $(a_n)_{n \in \mathbb{N}}$ of rational numbers for the absolute value, i.e., for each $m \in \mathbb{N}^*$, there exists $N \in \mathbb{N}$ such that if $n, n' \ge N$, then $|a_n - a_{n'}| \le \frac{1}{m}$. Then $C(\mathbb{Q})$ is a subring of the product ring $\mathbb{Q}^{\mathbb{N}}$, of all the rational sequences, for the natural addition and multiplication, i.e., $(a_n)_n + (b_n)_n =$

$(a_n + b_n)_n$ and $(a_n)_n(b_n)_n = (a_nb_n)_n$. Let M be the subset of $C(\mathbb{Q})$ formed by the sequences $(a_n)_{n\in\mathbb{N}}$ of rational numbers that converge to zero, i.e., for each $m \in \mathbb{N}^*$, there exists $N \in \mathbb{N}$ such that if $n \geq N$, then $|a_n| \leq \frac{1}{m}$. We will show that M is a maximal ideal of $C(\mathbb{Q})$. It is clear that M is a proper ideal of $C(\mathbb{Q})$. Let I be an ideal of $C(\mathbb{Q})$ containing strictly M and $a = (a_n)_{n\in\mathbb{N}} \in I \setminus M$. Since $(a_n)_{n\in\mathbb{N}}$ does not converge to zero, there exists a rational number $c > 0$ such that for each $k \in \mathbb{N}$, there exists $n_k \geq k$ for which $|a_{n_k}| \geq c$. Since $(a_n)_{n\in\mathbb{N}}$ is Cauchy, there exists $N \in \mathbb{N}$ such that $|a_n - a_{n_k}| \leq \frac{c}{2}$ for all integers $n, k \geq N$, then $|a_n| = |a_{n_k} + a_n - a_{n_k}| \geq |a_{n_k}| - |a_n - a_{n_k}| \geq \frac{c}{2}$. The sequence $(a_n^{-1})_{n\geq N}$ is Cauchy since $|a_n^{-1} - a_m^{-1}| = \frac{|a_n - a_m|}{|a_n a_m|} \leq \frac{4}{c^2}|a_n - a_m|$. Let $x = (x_n)_{n\in\mathbb{N}} \in C(\mathbb{Q})$ be any sequence. The sequence $y = (0, \ldots, 0, \frac{x_N}{a_N}, \ldots, \frac{x_n}{a_n}, \ldots) \in C(\mathbb{Q})$, as the product of two elements of the ring $C(\mathbb{Q})$. Since I is an ideal of $C(\mathbb{Q})$, then $ay \in I$. So $x = (x_0, \ldots, x_{N-1}, 0, \ldots) + (0, \ldots, 0, x_N, \ldots, x_n, \ldots) = (x_0, \ldots, x_{N-1}, 0, \ldots) + ay \in M + I \subseteq I$. Then $I = C(\mathbb{Q})$. The field denoted $\mathbb{R} = C(\mathbb{Q})/M = C(\mathbb{Q})/\sim = \hat{\mathbb{Q}}$, where $\sim$ is the equivalence relation defined in the proof of the preceding theorem and $\hat{\mathbb{Q}}$ is the completion of $\mathbb{Q}$ for the absolute value. Let $\phi : \mathbb{Q} \longrightarrow \mathbb{R}$ be the map that associates to each rational number q the class of the constant sequence with general term q. Then ϕ is a homomorphism one to one (isometry) between fields and $\phi(\mathbb{Q})$ is dense in $\mathbb{R}$ for the absolute value. We identify $\mathbb{Q}$ with $\phi(\mathbb{Q}) \subseteq \mathbb{R}$ and q with the equivalence class just defined in $\mathbb{R}$. Let $\hat{d}$ be the distance of $\mathbb{R} = \hat{\mathbb{Q}}$ defined from the absolute value of $\mathbb{Q}$. Let $\tilde{r} \in \mathbb{R}$ with $r = (q_n)_{n\in\mathbb{N}} \in C(\mathbb{Q})$. The sequence $(|q_n|)_{n\in\mathbb{N}} \in C(\mathbb{Q})$ is Cauchy in $(\mathbb{R}, \hat{d})$ and converges to an element of $\mathbb{R}$, denoted $|\tilde{r}|$ and called the absolute value of $\tilde{r}$. By definition, $\hat{d}(\tilde{r}, \tilde{0}) = \lim_{n \longrightarrow +\infty} |q_n - 0| = \lim_{n \longrightarrow +\infty} |q_n| = |\tilde{r}|$.

The following example comes from number theory. We use the idea of the preceding example for the construction of the p-adic numbers fields, by a modification of the metric.

Example 2 Construction of the p-adic numbers field.

In the sequel, p is a fixed prime number. For each integer $0 \neq a \in \mathbb{Z}$, we denote $\omega_p(a)$ the biggest integer m such that p^m divides a. For example, $\omega_5(35) = \omega_5(5.7) = 1$, $\omega_5(250) = \omega_5(2.5^3) = 3$, $\omega_2(96) = \omega_2(2^5.3) = 5$, and $\omega_2(97) = 0$. Note that for all $a, b \in \mathbb{Z} \setminus (0)$, $\omega_p(a_1a_2) = \omega_p(a_1) + \omega_p(a_2)$. Let $x \in \mathbb{Q}^*$, $x = \frac{a}{b}$ with $a, b \in \mathbb{Z}\setminus(0)$. We define $\omega_p(x) = \omega_p(a) - \omega_p(b)$. The value of $\omega_p(x)$ depends only on x and not on a and b. Indeed, if $x = \frac{ac}{bc}$, then $\omega_p(x) = \omega_p(ac) - \omega_p(bc) = \omega_p(a) - \omega_p(b)$. Note that $\omega_p(xy) = \omega_p(x) + \omega_p(y)$ for all $x, y \in \mathbb{Q}^*$. We define $|\ |_p : \mathbb{Q} \longrightarrow \mathbb{R}+$ by $|x|_p = \frac{1}{p^{\omega_p(x)}}$ if $x \neq 0$ and $|0|_p = 0$. The map $|\ |_p$ is a norm over $\mathbb{Q}$. Indeed, for the triangular inequality, if $x = 0$ or $y = 0$ or $x + y = 0$, this is trivial. Suppose that these three elements are nonzero. Put $x = \frac{a}{b}$ and $y = \frac{c}{d}$ and $\omega_p = \omega$. Then $x + y = \frac{ad+bc}{bd}$, so $\omega(x + y) = \omega(ad + bc) - \omega(b) - \omega(d)$. But $\omega(ad + bc) \geq min\{\omega(ad); \omega(bc)\}$. Then $\omega(x + y) \geq min\{\omega(a) + \omega(d); \omega(b) + \omega(c)\} - \omega(b) - \omega(d) = min\{\omega(a) - \omega(b); \omega(c) - \omega(d)\} = min\{\omega(x); \omega(y)\}$.

Then $| x + y|_p = p^{-\omega(x+y)} \leq p^{-min\{\omega(x);\,\omega(y)\}} = max\{p^{-\omega(x)};\, p^{-\omega(y)}\} = max\{|x|_p;\, |y|_p\}$. We say that the norm $|\ |_p$ is not Archimedean. It satisfies also the isosceles triangle principle, i.e., if $|x|_p \neq |y|_p$, then $|x - y|_p = max\{|x|_p, |y|_p\}$. Indeed, suppose, for example, that $|x|_p < |y|_p$. Then $|x - y|_p \leq max\{|x|_p, |y|_p\} = |y|_p$. On the other hand, $|y|_p = |x - (x - y)|_p \leq max\{|x|_p, |x - y|_p\}$. Since we cannot have $|y|_p \leq |x|_p$, then $max\{|x|_p, |x - y|_p\} = |x - y|_p |y|_p \leq |x - y|_p$. Then we have $|x - y|_p = |y|_p = max\{|x|_p, |y|_p\}$. Also, we have $|xy|_p = |x|_p |y|_p$ et $|\frac{1}{x}|_p = \frac{1}{|x|_p}$. Indeed, we have $|xy|_p = p^{-\omega(xy)} = p^{-\omega(x)-\omega(y)} = p^{-\omega(x)} p^{-\omega(y)} = |x|_p |y|_p$. Since $|1|_p = p^{-0} = 1$, if we take $y = \frac{1}{x}$, we find $|\frac{1}{x}|_p = \frac{1}{|x|_p}$.

A sequence (a_i) of rational numbers is Cauchy for $|\ |_p$ if for each real number $\epsilon > 0$, there exists $N \in \mathbb{N}$ such that for each integer $i, j \geq N$, $|a_i - a_j|_p < \epsilon$. Let $C_p(\mathbb{Q})$ be the set of the Cauchy sequences for the norm $|\ |_p$ and $\sim$ the equivalence relation defined by $|\ |_p$ on $C_p(\mathbb{Q})$ in the following way. Two sequences (a_i) and (b_i) of $C_p(\mathbb{Q})$ are equivalent if $\lim\limits_{i \to +\infty} |a_i - b_i|_p = 0$. Let $\mathbb{Q}_p = C_p(\mathbb{Q})/\sim$ be the quotient set. For each $a \in C_p(\mathbb{Q})$, let $\tilde{a}$ be the equivalence class of a. In particular, $\tilde{0}$ is the class of the zero sequence. A sequence (a_i) of rational numbers is a representative of $\tilde{0}$ if and only if $\lim\limits_{i \to +\infty} |a_i|_p = 0$. We will show that $\mathbb{Q}_p$ is a normed field.

Let $a = (a_i)$ be a Cauchy sequence of rational numbers, with class $\tilde{a} \neq \tilde{0}$. There exists a real number $\epsilon_0 > 0$ such that for each $N \in \mathbb{N}$, there exists an integer $i_N > N$ satisfying $|a_{i_N}|_p > \epsilon_0$. The sequence $(|a_i|_p)$, of modules, is constant nonzero from some rank, then it converges in $\mathbb{R}$. Indeed, we fix N sufficiently big in such a way that for each $i, i' > N$, $|a_i - a_{i'}|_p < \epsilon_0$. In particular, for each $i > N$, $|a_i - a_{i_N}|_p < \epsilon_0$. Then for each integer $i > N$, $|a_i - a_{i_N}|_p \neq |a_{i_N}|_p$. By the isosceles triangle principle, $|a_i|_p = |(a_i - a_{i_N}) + a_{i_N}|_p = max\{|a_i - a_{i_N}|_p, |a_{i_N}|_p\} = |a_{i_N}|_p$. Then for each $i > N$, $|a_i|_p = |a_{i_N}|_p$. We define $|\tilde{a}|_p = \lim\limits_{i \to +\infty} |a_i|_p$. Also, let $|\tilde{0}|_p = 0$. Let (a_i) and (b_i) be two representatives of the same equivalence class. Then $\lim\limits_{i \to +\infty} |a_i - b_i|_p = 0$ and $|a_i|_p = |(a_i - b_i) + b_i|_p \leq |a_i - b_i|_p + |b_i|_p$. Then $\lim\limits_{i \to \infty} |a_i|_p \leq \lim\limits_{i \to \infty} |b_i|_p$, and by symmetry, $\lim\limits_{i \to \infty} |a_i|_p = \lim\limits_{i \to \infty} |b_i|_p$.

Let $\tilde{a}$ and $\tilde{b} \in \mathbb{Q}_p$ be two elements; choose two representatives $(a_i) \in \tilde{a}$ and $(b_i) \in \tilde{b}$. We define $\tilde{a}.\tilde{b}$ as the class of the Cauchy sequence $(a_i b_i)$. This multiplication is well defined. Indeed, let (a'_i) and (b'_i) be other representatives of $\tilde{a}$ and $\tilde{b}$. Then $|a'_i b'_i - a_i b_i|_p = |a'_i(b'_i - b_i) + b_i(a'_i - a_i)|_p \leq max\{|a'_i|_p |b'_i - b_i|_p;\, |b_i|_p |a'_i - a_i|_p\}$. When $i \to +\infty$, the first expression tends to $|\tilde{a}|_p \lim |b'_i - b_i|_p = 0$ and the second tends to $|\tilde{b}|_p \lim |a'_i - a_i|_p = 0$. Then $(a'_i b'_i)$ and $(a_i b_i)$ are equivalent. In the same way, we define the sum of two equivalence classes by choosing two representatives and we define the sum termwise, then we show that the sum depends only on the equivalence classes. The opposite is defined in a clear way. The multiplicative converse is more complicated since there are zero terms in a Cauchy sequence. We start by the following two remarks:

(1) If (a_i) is a Cauchy sequence and (b_i) any sequence of rational numbers such that $\lim |a_i - b_i|_p = 0$, then (b_i) is also Cauchy.

Indeed, $|b_i - b_j|_p \leq |b_i - a_i|_p + |a_i - a_j|_p + |a_j - b_j|_p \longrightarrow 0$ when $i, j \longrightarrow +\infty$.

(2) If (a_i) and (b_i) are two representatives of the same nonzero class $\tilde{a}$, without zero term, then the Cauchy sequences $(\frac{1}{a_i})$ and $(\frac{1}{b_i})$ are equivalent.

Indeed, $|\frac{1}{a_i} - \frac{1}{b_i}|_p = |\frac{a_i - b_i}{a_i b_i}|_p = \frac{|a_i - b_i|_p}{|a_i|_p |b_i|_p}$ with $\lim |a_i - b_i|_p = 0$ and $\lim |a_i|_p = \lim |b_i|_p = |\tilde{a}|_p \neq 0$. Then $lim |\frac{1}{a_i} - \frac{1}{b_i}|_p = 0$.

Let now $\tilde{a}$ be a nonzero class and (a_i) a representative of $\tilde{a}$. We define a new sequence (a_i') by $a_i' = a_i$ if $a_i \neq 0$ and $a_i' = p^i$ if $a_i = 0$. Then $a_i' - a_i = 0$ if $a_i \neq 0$ and $a_i' - a_i = p^i$ if $a_i = 0$. Since $|p^i|_p = p^{-i} \longrightarrow 0$ when $i \longrightarrow \infty$, then $\lim_{i \longrightarrow +\infty} |a_i' - a_i|_p = 0$. By (1), (a_i') is Cauchy and the sequences (a_i) and (a_i') are equivalent. We conclude that a nonzero class $\tilde{a}$ has at least a representative (a_i') whose terms are all nonzero. We define $1/\tilde{a}$ as the class of $(\frac{1}{a_i'})$. By (2), the definition does not depend on the representative with nonzero terms.

Then the set $\mathbb{Q}_p$, of the equivalence classes of Cauchy sequences, is a field for the addition, the multiplication, and the converse previously defined. For example, for the distributivity: let (a_i), (b_i) and (c_i) be representatives of $\tilde{a}$, $\tilde{b}$ and $\tilde{c} \in \mathbb{Q}_p$. Then $\tilde{a}(\tilde{b} + \tilde{c})$ is the class of $(a_i(b_i + c_i)) = (a_i b_i + a_i c_i)$ and $\tilde{a}\tilde{b} + \tilde{a}\tilde{c}$ is the class of the same sequence. The map $|\ |_p$ is a norm on $\mathbb{Q}_p$. The axioms of the norm on $\mathbb{Q}_p$ follow from their analogues on $\mathbb{Q}$ by passage to the limit. We have $|\frac{1}{x}|_p = \frac{1}{|x|_p}$. The field $\mathbb{Q}$ may be identified to the subfield of $\mathbb{Q}_p$ formed by the equivalence classes of constant Cauchy sequences.

One can show that modulo a natural equivalence notion, the only absolute values on $\mathbb{Q}$ are the habitual absolute value and the p-adic absolute values, p a prime number.

11   Projective Limit of a Projective System of Rings

The notion of projective limit is an essential tool in mathematics. It appears in many particular cases in a simple form that does not justify a systematic study. However, it seems to be important to have general ideas on this notion which is involved in various questions. In our case, it allows us to give an algebraic aspect to the I-adic completion of a ring.

A preorder over a nonempty set is a reflexive and transitive relation. A set endowed with a preorder is said to be preordered.

Definition 11.1 Let I be a preordered set. A projective system of rings indexed by I consists of giving for each $i \in I$ a ring A_i and, for each couple $(i, j) \in I^2$ such that $i \leq j$, a homomorphism $f_{ij} : A_j \longrightarrow A_i$ such that if $i \leq j \leq k$ are three elements of I, then the following diagram is commutative:

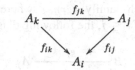

i.e., $f_{ik} = f_{ij} \circ f_{jk}$, and for each $i \in I$, $f_{ii} = id_{A_i}$. We denote $\left(A_i, f_{ij}\right)_{i,j \in I}$ such a system. The maps $\left(f_{ij}\right)_{i,j \in I}$ are called the homomorphisms of transition.

Example 1 Let I be a nonempty set endowed with the trivial order, in which each element is comparable only to itself. A projective system of rings indexed by I is a family $(A_i)_{i \in I}$ of rings. The only homomorphisms of transition are the identities $id_{A_i} : A_i \longrightarrow A_i, i \in I$.

Example 2 Let I be a preordered set and A a ring. For each $i \in I$, we consider the ring $A_i = A$, and for each couple $(i, j) \in I^2$ such that $i \leq j$, we consider the homomorphism $f_{ij} = id_A : A_j = A \longrightarrow A_i = A$. Then $\left(A_i, f_{ij}\right)_{i,j \in I}$ is a projective system of rings indexed by I. Indeed, if $i \leq j \leq k$ are three elements of I, the following diagram is commutative:

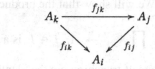

Definition 11.2 Let $\left(A_i, f_{ij}\right)_{i,j \in I}$ be a projective system of rings indexed by a preordered set I. A projective (or inverse) limit of this system consists of giving a ring A, and, for each $i \in I$, a homomorphism $f_i : A \longrightarrow A_i$ satisfying the two following conditions:

(1) If $i \leq j$, the following diagram is commutative:

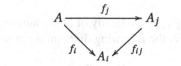

i.e., $f_i = f_{ij} \circ f_j$.

(2) For each ring B and each family $(g_i)_{i \in I}$ where $g_i : B \longrightarrow A_i$ is a homomorphism such that if $i \leq j$, the following diagram is commutative:

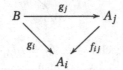

i.e., $g_i = f_{ij} \circ g_j$, then there exists a unique homomorphism $g : B \longrightarrow A$ such that for each $i \in I$, the following diagram is commutative:

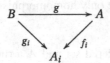

i.e., $g_i = f_i \circ g$.

Example 1 Let $\left(A_i, f_{ij} \right)_{i,j \in I}$ be a projective system of rings indexed by a set I endowed with the trivial order. We will show that the product ring $\prod_{j \in I} A_j$ endowed with the natural projections $f_i : \prod_{j \in I} A_j \longrightarrow A_i, i \in I$, is a projective limit of this system. The only homomorphisms of transition are the identities $id_{A_i} : A_i \longrightarrow A_i$, $i \in I$.

(1) It is clear that the following diagram is commutative:

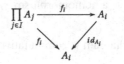

(2) Let B be a ring and $(g_i)_{i \in I}$ be a family of homomorphisms with $g_i : B \longrightarrow A_i$ such that for each $i \in I$, the following diagram is (trivially) commutative:

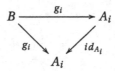

Suppose that there exists a homomorphism of rings $g : B \longrightarrow \prod_{j \in I} A_j$ such that for each $i \in I$, the following diagram is commutative:

i.e., $g_i = f_i \circ g$. Let $b \in B$ and $g(b) = (a_j)_{j \in I} \in \prod_{j \in I} A_j$. For each $i \in I$, $a_i = f_i \circ g(b) = g_i(b)$. Then $g(b) = (g_i(b))_{i \in I}$. We have the uniqueness of the homomorphism g (if it exists).

Conversely, the map $g : B \longrightarrow \prod_{j \in I} A_j$, defined by $g(b) = (g_i(b))_{i \in I}$ pour $b \in B$, is a homomorphism of rings that makes the preceding diagram commutative.

Example 2 Let I be a totally ordered set and A a ring. For each $i \in I$, we take the ring $A_i = A$ and for each couple $(i, j) \in I^2$ such that $i \leq j$, we take the homomorphism $f_{ij} = id_A : A_j = A \longrightarrow A_i = A$. We have always showed that $(A_i, f_{ij})_{i,j \in I}$ is a projective system of rings indexed by I. For each $i \in I$, we take the homomorphism $f_i = id_A : A \longrightarrow A_i = A$. We will show that $(A, f_i)_{i \in I}$ is a projective limit of this system.

(1) Let $i \leq j$ be two elements of I. It is clear that the following diagram is commutative:

$$
\begin{array}{ccc}
A & \xrightarrow{\;f_j\;} & A_j \\
 & \searrow f_i \quad \swarrow f_{ij} & \\
 & A_i &
\end{array}
$$

(2) Let B be ring and $(g_i)_{i \in I}$ a family of homomorphisms with $g_i : B \longrightarrow A_i$ such that if $i \leq j$, the following diagram is commutative:

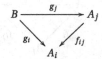

i.e., $g_i = f_{ij} \circ g_j$. Since $f_{ij} = id_A : A_j = A \longrightarrow A_i = A$, then $g_i = g_j$ for $i \leq j$. Since I is totally ordered, the family $(g_i)_{i \in I}$ is constant.

Suppose that there exists a homomorphism of rings $g : B \longrightarrow A$ such that for each $i \in I$, the following diagram is commutative:

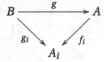

i.e., $g_i = f_i \circ g$. Since $f_i = id_A : A \longrightarrow A_i = A$, then $g = g_i$ for each $i \in I$. Then g takes the common value of the family $(g_i)_{i \in I}$. Then we have the uniqueness. For the existence, we take $g = g_i$ for each $i \in I$.

Example 3 Let C be a ring and $(A_i)_{i \in I}$ a family of subrings of C indexed by any set I (without a preorder). Suppose that for each couple $(i, j) \in I^2$, there exists $k \in I$ such that $A_k \subseteq A_i$ and $A_k \subseteq A_j$. We preorder I by supposing that $i \leq j$ if and only if $A_j \subseteq A_i$. Indeed, the reflexivity is clear. For the transitivity, if $i \leq j \leq k$, then $A_k \subseteq A_j \subseteq A_i$, so $A_k \subseteq A_i$, then $i \leq k$. With this preorder, $(I, \leq)$ is an increasing filter, i.e., for each couple $(i, j) \in I^2$, there exists $k \in I$ such that $i \leq k$ and $j \leq k$. If $i \leq j$ are two elements of I with $A_j \subseteq A_i$, let $f_{ij} : A_j \longrightarrow A_i$ be the canonical injection. We obtain a projective system of rings. Indeed, for each $i \in I$, $f_{ii} = id_{A_i} : A_i \longrightarrow A_i$ and if $i \leq j \leq k$ are three elements of I, the following diagram is trivially commutative:

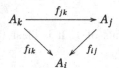

Let the subring $A = \bigcap_{i \in I} A_i$ of C and for each $i \in I$, $f_i : A \longrightarrow A_i$ be the canonical injection. We will show that $(A, f_i)_{i \in I}$ is a projective limit of the system $(A_i, f_{ij})_{i,j \in I}$.

(1) If $i \leq j$, the following diagram is trivially commutative:

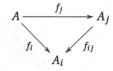

(2) Let B be a ring and $(g_i)_{i \in I}$ a family of homomorphisms with $g_i : B \longrightarrow A_i$ such that if $i \leq j$, the following diagram is commutative:

i.e., $g_i = f_{ij} \circ g_j$.

Let $i, j \in I$ be any two elements (not necessarily comparable for $\leq$). Since $(I, \leq)$ is an increasing filter, there exists $k \in I$ such that $i \leq k$ and $j \leq k$. Then $g_i = f_{ik} \circ g_k$ and $g_j = f_{jk} \circ g_k$. For each $x \in A$, $g_i(x) = f_{ik} \circ g_k(x) = g_k(x)$ and $g_j(x) = f_{jk} \circ g_k(x) = g_k(x)$. Then for each $i, j \in I$ and for each $x \in A$, $g_i(x) = g_j(x)$.

Suppose that there exists a homomorphism of rings $g : B \longrightarrow A$ such that for each $i \in I$, the following diagram is commutative:

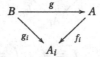

i.e., $g_i = f_i \circ g$.

Let $b \in B$. For each $i \in I$, $g_i(b) = f_i \circ g(b) = g(b)$. Then $g(b)$ takes the common value of the $g_i(b)$, $i \in I$. Then we have the uniqueness of g. For the existence, we take $g(b) = g_i(b)$ for each $b \in B$ and for each $i \in I$. The preceding diagram is commutative.

The following proposition shows that the projective limit of a projective system of rings (if it exists) is unique modulo an isomorphism.

Proposition 11.3 *Let* $\left(A_i, f_{ij}\right)_{i,j \in I}$ *be a projective system of rings indexed by a preordered set* I. *If* $\left(A, f_i\right)_{i \in I}$ *and* $\left(A', f'_i\right)_{i \in I}$ *are two projective limits of this system, the rings* A *and* A' *are isomorphic.*

Proof Since $\left(A, f_i\right)_{i \in I}$ is a projective limit of the system and $f'_i = f_{ij} \circ f'_j$, there exists a homomorphism $f' : A' \longrightarrow A$ such that for each $i \in I$, the diagram

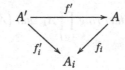

is commutative. Also, there exists a homomorphism $f : A \longrightarrow A'$ such that for each $i \in I$, the diagram

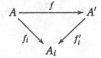

is commutative. We see that for each $i \in I$, the two following diagrams

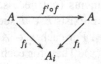

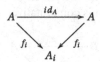

are commutative. Indeed, $f_i \circ f' \circ f = f_i' \circ f = f_i$. By the uniqueness in the definition of a projective limit, we have $f' \circ f = id_A$. Also, $f \circ f' = id_{A'}$. Then $A \simeq A'$.   □

Notation Let $\left(A_i, f_{ij}\right)_{i,j \in I}$ be a projective system of rings and $\left(A, f_i\right)_{i \in I}$ a projective limit of this system. We denote $\varprojlim \left(A_i, f_{ij}\right)_{i,j \in I} = A$, if there is no possible confusion for the homomorphisms $f_i, i \in I$.

We prove now the existence of the projective limit for a projective system of rings.

Proposition 11.4 *Let $\left(A_i, f_{ij}\right)_{i,j \in I}$ be a projective system of rings indexed by a preordered set I. Let A be the subset of $\prod_{k \in I} A_k$ of the elements $(x_k)_{k \in I}$ such that for each couple (i, j) of elements of I satisfying $i \leq j$, $f_{ij}(x_j) = x_i$. For each $i \in I$, let $f_i : A \longrightarrow A_i$ be the restriction to A of the projection $\prod_{k \in I} A_k \longrightarrow A_i$. Then A is a subring of the product ring $\prod_{k \in I} A_k$ and $\left(A, f_i\right)_{i \in I}$ is a projective limit of $\left(A_i, f_{ij}\right)_{i,j \in I}$.*

Proof It is clear that A is a subring of the product ring $\prod_{k \in I} A_k$.

(1) Let $i \leq j$ be two indexes in I. The following diagram is commutative:

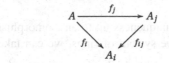

Indeed, let $x = (x_k)_{k \in I} \in A$. Then $f_{ij} \circ f_j(x) = f_{ij}(x_j) = x_i = f_i(x)$.

(2) Let B be a ring, and, for each $i \in I$, $g_i : B \longrightarrow A_i$ a homomorphism such that if $i \leq j$, the following diagram is commutative:

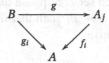

i.e., $g_i = f_{ij} \circ g_j$.

Suppose that there exists a homomorphism of rings $g : B \longrightarrow A$ such that for each $i \in I$, the following diagram is commutative:

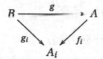

i.e., $g_i = f_i \circ g$. Let $y \in B$. Put $g(y) = x = (x_k)_{k \in I} \in A$. For each $i \in I$, $g_i(y) = f_i \circ g(y) = f_i(x) = x_i$. Then $g(y) = (g_k(y))_{k \in I}$. Then we have the uniqueness of g.

Conversely, the formula $g(y) = (g_k(y))_{k \in I}$ for each $y \in B$ defines a map $g : B \longrightarrow A$. Indeed, by hypothesis, if $i \leq j$, then $g_i = f_{ij} \circ g_j$, so $g_i(y) = f_{ij}(g_j(y))$, then $g(y) \in A$, by definition of A. It is also clear that g is a homomorphism of rings. Moreover, the following diagram is commutative:

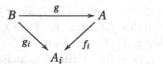

Indeed, for each $y \in B$, $f_i \circ g(y) = f_i(g(y)) = f_i((g_k(y))_{k \in I}) = g_i(y)$. Then $f_i \circ g = g_i$. □

Example Let I be a nonempty set endowed with the trivial order and $(A_i, f_{ij})_{i,j \in I}$ a projective system of rings indexed by I. The product ring $A = \prod_{k \in I} A_k$, endowed with the canonical projections $f_i : A \longrightarrow A_i$, $i \in I$, is a projective limit of this system. Indeed, if $i \leq j$, then $i = j$ with $f_{ii} = id_{A_i}$. We retrieve a preceding example.

Remark 11.5 Thanks to the uniqueness up to an isomorphism the projective limit, if $(A_i, f_{ij})_{i,j \in I}$ is a projective system of rings, we can take the subring A of the product ring $\prod_{k \in I} A_k$, endowed with the homomorphisms $f_i, i \in I$, defined in the preceding proposition, as a definition for the projective limit of this system.

Example 1 Let $(A_i, f_{ij})_{i,j \in I}$ and $(B_i, g_{ij})_{i,j \in I}$ be two projective systems of rings indexed by the same preordered set I. Suppose that for each $i \in I$, there exists an isomorphism of rings $f_i : A_i \longrightarrow B_i$ such that if $i \leq j$ are two indexes in I, the following diagram is commutative:

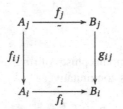

Then $\varprojlim (A_i, f_{ij})_{i,j \in I} \simeq \varprojlim (B_i, g_{ij})_{i,j \in I}$.

Indeed, we may suppose that $\varprojlim (A_i, f_{ij})_{i,j \in I} = A = \{(x_k)_{k \in I} \in \prod_{k \in I} A_k;\ f_{ij}(x_j) = x_i$ if $i \leq j\}$ and $\varprojlim (B_i, g_{ij})_{i,j \in I} = B = \{(y_k)_{k \in I} \in \prod_{k \in I} B_k;\ g_{ij}(y_j) = y_i$ if $i \leq j\}$. Let $\phi : A \longrightarrow B$, the map defined by $\phi(x) = (f_k(x_k))_{k \in I}$ for each $x = (x_k)_{k \in I} \in A$. Note that ϕ takes its values in B since if $i \leq j$ are two indexes in I, then $f_{ij}(x_j) = x_i$, so $f_i(f_{ij}(x_j)) = f_i(x_i)$. Since the diagram is commutative, then $g_{ij}(f_j(x_j)) = f_i(x_i)$. So $(f_k(x_k))_{k \in I} \in B$. It is clear that ϕ is a homomorphism onto since the f_k are isomorphisms. Let $y = (y_k)_{k \in I} \in B$. Then $g_{ij}(y_j) = y_i$ if $i \leq j$. Since the f_k are onto, for each $k \in I$, there exists $x_k \in A_k$ such that $f_k(x_k) = y_k$. Then $x = (x_k)_{k \in I} \in A$ since if $i \leq j$ are two indexes in I, then $g_{ij}(y_j) = y_i$, so $g_{ij}(f_j(x_j)) = f_i(x_i)$. Since the diagram is commutative, $f_i(f_{ij}(x_j)) = f_i(x_i)$. But f_i is an isomorphism, then $f_{ij}(x_j) = x_i$, so $x \in A$. Finally, $y = (y_k)_{k \in I} = (f_k(x_k))_{k \in I} = \phi(x)$. Then ϕ is onto.

Example 2 We will construct a projective system of rings indexed by the set $(\mathbb{N}, \leq)$, endowed by its habitual order. Let $A_0, A_1, \ldots$ be a sequence of rings and for each $i \in \mathbb{N}$, a homomorphism $f_{i,i+1} : A_{i+1} \longrightarrow A_i$. The other homomorphisms are obtained by composition from these ones, with $f_{ii} = id_{A_i} : A_i \longrightarrow A_i$. Note that if $i \leq j \leq k$ are three natural integers, the diagram of Definition 11.1, of the projective systems

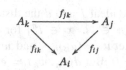

is commutative by construction, i.e., $f_{ik} = f_{ij} \circ f_{jk}$.

By Proposition 11.4, an element of the projective limit A of this system is a sequence $x_0, x_1, \ldots$ of elements of $A_0, A_1, \ldots$ respectively such that x_{i+1} is sent over x_i by the homomorphism $f_{i,i+1} : A_{i+1} \longrightarrow A_i$, for each $i \in \mathbb{N}$. In practice, to construct an element of A, we can start by an element $x_0 \in A_0$, then choose an element $x_1 \in A_1$ with image x_0, then choose an element $x_2 \in A_2$ with image $x_1, \ldots$, etc. If all the homomorphisms $f_{i,i+1}, i \in \mathbb{N}$, are onto, given an element $x_i \in A_i$, we can construct an element of A having x_i as i^{th} coordinate. For $j < i$, we take the successive images of x_i, and for $i + 1, i + 2, \ldots$, we choose the values of $x_{i+1}, x_{i+2}, \ldots$ by induction using the fact that the homomorphisms are onto.

12 Construction of the I-Adic Completion of a Ring

The goal of this section is an algebraic formulation of the I-adic completion of a ring by using the projective limit of a projective system of rings. Let A be a ring and I an ideal of A. For each $i \in \mathbb{N}$, the map $f_{i,i+1} : A/I^{i+1} \longrightarrow A/I^i$, which sends $\bar{x}$ on $\tilde{x}$, is well defined since if $\bar{x} = \bar{y}$ in A/I^{i+1}, then $x - y \in I^{i+1} \subseteq I^i$, so $\bar{x} = \bar{y}$ in A/I^i. It is also a homomorphism onto of rings. By the last example of the preceding section, these homomorphisms allow us to define a projective system $\left(A/I^i, \ f_{ij} \right)_{ij \in \mathbb{N}}$, indexed by $(\mathbb{N}, \leq)$. The model of the projective limit of this system defined in Proposition 11.4 is the subring $\mathscr{A}$ of the product ring $\prod_{k:0}^{\infty}(A/I^k)$ formed by the sequences $(x_k + I^k)_{k \in \mathbb{N}}$, $x_k \in A$, such that $f_{i,i+1}(x_{i+1} + I^{i+1}) = x_i + I^i$ or $x_{i+1} + I^i = x_i + I^i$ for each $i \in \mathbb{N}$.

Suppose now that $\bigcap_{n:0}^{\infty} I^n = (0)$. Consider the product ring $\prod_{n:0}^{\infty} A = A^{\mathbb{N}}$, of all the sequences of elements of A and its subring $C_I(A)$ of the Cauchy sequences for the I-adic topology, i.e., the sequences $(a_n)_{n \in \mathbb{N}}$ of elements of A such that for each $i \in \mathbb{N}$, there exists $N \in \mathbb{N}$ satisfying for each integers $n, m \geq N$, $a_n - a_m \in I^i$. Let $C_I^0(A)$ be the subset of $C_I(A)$ formed by the sequences that converge to 0, i.e., the sequences $(a_n)_{n \in \mathbb{N}}$ of elements of A such that for each $i \in \mathbb{N}$, there exists $N \in \mathbb{N}$ satisfying for each integer $n \geq N$, $a_n \in I^i$. Then $C_I^0(A)$ is an ideal of $C_I(A)$. Indeed, let $(a_n)_{n \in \mathbb{N}} \in C_I^0(A)$ and $(b_n)_{n \in \mathbb{N}} \in C_I(A)$. Let $i \in \mathbb{N}$. There exists $N \in \mathbb{N}$ satisfying for each integer $n \geq N$, $a_n \in I^i$, then $a_n b_n \in I^i$. The sequence $(a_n b_n)_{n \in \mathbb{N}}$ converges to 0. By Theorem 10.2, the completion of the ring A for the I-adic topology is the quotient ring $C_I(A)/\sim \, = C_I(A)/C_I^0(A)$. Indeed, let v be the

order function associated to the ideal I and d the distance associated to I, defined
in Sect. 6. Then $v(x) = sup\{n \in \mathbb{N}; \ x \in I^n\}$ for each $x \in A$ and $d(x, y) =$
$e^{-v(x-y)}$ for each $x, y \in A$. Let $u = (u_n)_{n \in \mathbb{N}}$ and $w = (w_n)_{n \in \mathbb{N}} \in C_I(A)$. Then
$u \sim w \iff \lim\limits_{n \to \infty} d(u_n, w_n) = 0$. But $d(u_n, w_n) = e^{-v(u_n - w_n)} = d(u_n - w_n, 0)$.
Then $u \sim w \iff \lim\limits_{n \to \infty} d(u_n - w_n, 0) = 0 \iff u - w \in C_I^0(A)$. The map $\phi :$
$A \longrightarrow C_I(A)/C_I^0(A)$, which sends an element $a \in A$ on the class modulo $C_I^0(A)$
of the constant sequence with general term a, is a homomorphism of rings (easy)
and one to one (Theorem 10.2). Let $C_I(A)/C_I^0(A) = \hat{A}^I$, or simply $\hat{A}$, if there is
no confusion.

Proposition 12.1 *With the preceding notations, we have an isomorphism of rings*

$$\hat{A}^I = C_I(A)/C_I^0(A) \simeq \varprojlim_i A/I^i.$$

Proof We may suppose that $\varprojlim_i A/I^i = \mathscr{A}$, the subring of $\prod\limits_{i:0}^{\infty}(A/I^i)$, defined in
Proposition 11.4 and formed by the sequences $(x_i + I^i)_{i \in \mathbb{N}}$ with $x_i \in A$ such that
$x_{i+1} + I^i = x_i + I^i$ in A/I^i for each $i \in \mathbb{N}$. We define the map $\phi : C_I(A)/C_I^0(A)$
$\longrightarrow \mathscr{A}$ in the following way. Let $\tilde{a} \in C_I(A)/C_I^0(A)$ with $a = (a_k)_{k \in \mathbb{N}} \in C_I(A)$.
For each $i \in \mathbb{N}$, there exists $N_i \in \mathbb{N}$ such that for each integer $n, m \geq N_i, a_n - a_m \in$
I^i, then $a_n + I^i = a_m + I^i$ in A/I^i. Let $b_i + I^i$ be the common value of the
$a_n + I^i$ for $n >> 0$, $b_i \in A$. Without loss of generality, we can suppose that
$N_0 \leq N_1 \leq N_2 \leq \ldots$. For each integer $n \geq N_{i+1} \geq N_i, a_n + I^{i+1} = b_{i+1} + I^{i+1}$,
then $a_n - b_{i+1} \in I^{i+1} \subseteq I^i$, so $a_n + I^i = b_{i+1} + I^i$, then $b_i + I^i = b_{i+1} + I^i$.
So $b_i + I^i = b_{i+1} + I^i$ in A/I^i for each $i \in \mathbb{N}$. We conclude that the sequence
$b = (b_k + I^k)_{k \in \mathbb{N}} \in \mathscr{A}$. The sequence b does not depend of the representative
a of $\tilde{a}$. Indeed, let $a' = (a'_k)_{k \in \mathbb{N}}$ be another representative of $\tilde{a}$. Let $i \in \mathbb{N}$. Since
$a - a' \in C_I^0(A)$, then for $n >> 0$, $a_n - a'_n \in I^i$, so $a_n + I^i = a'_n + I^i$. Put
$\phi(\tilde{a}) = b$. It is clear that the map ϕ defined is a homomorphism of rings. It is one to
one. Indeed, suppose that $\phi(\tilde{a}) = 0$. With the same notations, for each $i \in \mathbb{N}$ and for
each integer $n > N_i, a_n + I^i = 0$ in A/I^i, then $a_n \in I^i$. The sequence $a = (a_k)_{k \in \mathbb{N}}$
converges to zero, then $a \in C_I^0(A)$ so $\tilde{a} = 0$. The map ϕ is also onto. Indeed, let
$(b_k + I^k)_{k \in \mathbb{N}} \in \mathscr{A}$. Then $b_k \in A$ and $b_{k+1} + I^k = b_k + I^k$ in A/I^k for each $k \in \mathbb{N}$.
Let $i \in \mathbb{N}$. For each integer $k \geq i, b_{k+1} - b_k \in I^k \subseteq I^i$. By Remark 6.1 (2), the
sequence $b = (b_k)_{k \in \mathbb{N}}$ is Cauchy in A for the I-adic topology, then $b \in C_I(A)$ and
$\tilde{b} \in C_I(A)/C_I^0(A)$. We have $\phi(\tilde{b}) = (b_k + I^k)_{k \in \mathbb{N}}$. Indeed, let $i \in \mathbb{N}$. For each
integer $n \geq i, b_n - b_i = (b_n - b_{n-1}) + (b_{n-1} - b_{n-2}) + \ldots + (b_{i+1} - b_i) \in$
$I^{n-1} + I^{n-2} + \ldots + I^i = I^i$. Then $b_n + I^i = b_i + I^i$ for each integer $n \geq i$. Then
the common value of the $b_n + I^i, n \geq i$, is equal to $b_i + I^i$. So $\phi(\tilde{b}) = (b_k + I^k)_{k \in \mathbb{N}}$.
$\qquad \square$

Example 1 Let B be any ring, $A = B[X_1, \ldots, X_n]$ and $I = (X_1, \ldots, X_n)$. By Theorem 6.4, $B[[X_1, \ldots, X_n]] \simeq \varprojlim_i B[X_1, \ldots, X_n] / (X_1, \ldots, X_n)^i$.

Example 2 Let p be a prime number. The (p)-adic completion of $\mathbb{Z}$, denoted $\mathbb{Z}_p \simeq \varprojlim_i \mathbb{Z}/p^i \mathbb{Z}$, is an integral domain. Indeed, we can suppose that $\varprojlim_i \mathbb{Z}/p^i \mathbb{Z}$ is the model A defined in Proposition 11.4. For $i \leq j$ two natural integers, the corresponding homomorphism of transition of the projective system is f_{ij} : $\mathbb{Z}/p^j \mathbb{Z} \longrightarrow \mathbb{Z}/p^i \mathbb{Z}$, defined by $f_{ij}(\bar{x}) = \tilde{x}$ for each $\bar{x} \in \mathbb{Z}/p^j \mathbb{Z}$. Suppose that there exist two nonzero elements $x = (x_k + p^k \mathbb{Z})_{k \in \mathbb{N}}$ and $y = (y_k + p^k \mathbb{Z})_{k \in \mathbb{N}} \in A$ such that $xy = 0$. There exist $m, l \in \mathbb{N}$ such that $x_m \notin p^m \mathbb{Z}$ and $y_l \notin p^l \mathbb{Z}$. Since $0 = xy = (x_k y_k + p^k \mathbb{Z})_{k \in \mathbb{N}}$, then $x_{m+l} y_{m+l} \in p^{m+l} \mathbb{Z}$. So either p^m divides x_{m+l} or p^l divides y_{m+l}. Suppose, for example, that p^m divides x_{m+l}. By definition of the ring A, $f_{m,m+l}(x_{m+l} + p^{m+l} \mathbb{Z}) = x_m + p^m \mathbb{Z}$. By definition of the homomorphisms of transition, $f_{m,m+l}(x_{m+l} + p^{m+l} \mathbb{Z}) = x_{m+l} + p^m \mathbb{Z} = 0$ since p^m divides x_{m+l}. Then $x_m + p^m \mathbb{Z} = 0$, which is absurd since p^m does not divide x_m.

The following corollary is important in the computation.

Corollary 12.2 *Let I be an ideal of a ring A, $C_I(A)$ the ring of the Cauchy sequences of elements of A for the I-adic topology, and $C_I^0(A)$ the ideal of $C_I(A)$ formed by the sequences that converge to 0.*

1. *For each sequence $(a_k)_{k \in \mathbb{N}} \in C_I(A)$, there exists a sequence $(b_k)_{k \in \mathbb{N}} \in C_I(A)$ that defines the same class in the quotient ring $C_I(A) / C_I^0(A)$ and satisfies $b_k + I^k = b_{k+1} + I^k$ for each $k \in \mathbb{N}$.*
2. *Each element of $C_I(A)/C_I^0(A)$ may be represented by a sequence $(x_n)_{n \in \mathbb{N}} \in C_I(A)$ for which, there exists a sequence $(a_i)_{i \in \mathbb{N}}$ of elements of A satisfying $a_{i+1} \in I^i$ for each $i \in \mathbb{N}$ such that $x_n = a_n + a_{n-1} + \ldots + a_0$ for each $n \in \mathbb{N}^*$.*

Proof

(1) Let $(a_k)_{k \in \mathbb{N}} \in C_I(A)$. In the proof of the preceding proposition, we construct a sequence $(b_i)_{i \in \mathbb{N}}$ of elements of A that satisfies $b_i + I^i = b_{i+1} + I^i$ for each $i \in \mathbb{N}$. Moreover, if $i \in \mathbb{N}$, there exists $N_i \in \mathbb{N}$ (and we can suppose that $N_i \geq i$) such that for each integer $n \geq N_i$, $a_n + I^i = b_i + I^i$. We will prove first that the sequence $(a_k)_{k \in \mathbb{N}} - (b_k)_{k \in \mathbb{N}} \in C_I^0(A)$. Let $i \in \mathbb{N}$. For each integer $n \geq N_i$, $a_n + I^i = b_i + I^i = b_{i+1} + I^i \supseteq b_{i+1} + I^{i+1} = b_{i+2} + I^{i+1} \supseteq b_{i+2} + I^{i+2} = b_{i+3} + I^{i+2} \supseteq \ldots \supseteq b_n + I^{n-1}$, then $a_n - b_n \in I^i$. The sequence $(a_n - b_n)_{n \in \mathbb{N}}$ converges to zero for the I-adic topology. It remains to show that $(b_n)_{n \in \mathbb{N}} \in C_I(A)$. Let $i \in \mathbb{N}$. There exists $N \in \mathbb{N}$ such that for each integer $n \geq N$, $a_n - b_n, a_n - a_{n+1} \in I^i$. Then $b_n - b_{n+1} = (b_n - a_n) + (a_n - a_{n+1}) + (a_{n+1} - b_{n+1}) \in I^i$.

(2) By (1), an element of $C_I(A)/C_I^0(A)$ may be represented by a sequence $(x_n)_{n \in \mathbb{N}} \in C_I(A)$ satisfying $a_{n+1} = x_{n+1} - x_n \in I^n$ for each $n \in \mathbb{N}$. Then $x_n = (x_n - x_{n-1}) + (x_{n-1} - x_{n-2}) + (x_{n-2} - x_{n-3}) + \ldots + (x_1 - x_0) + x_0 = a_n + a_{n-1} + a_{n-2} + \ldots + a_1 + a_0$ with $a_{i+1} \in I^i$ for each $i \in \mathbb{N}$ and $a_0 = x_0$.

$\square$

Proposition 12.3 *Let A and B be two rings, I an ideal of A, J an ideal of B, and $f : A \longrightarrow B$ a homomorphism of rings such that $\bigcap\limits_{n:1}^{\infty} I^n = (0)$, $\bigcap\limits_{n:1}^{\infty} J^n = (0)$ and $f(I) \subseteq J$. There exists a canonical homomorphism of rings $\hat{f} : \hat{A}^I \longrightarrow \hat{B}^J$ that extends f.*

Moreover, if f is onto and $f(I) = J$, then $\hat{f}$ is onto.

Proof If $(x_n)_{n \in \mathbb{N}}$ is a Cauchy sequence in A for the I-adic topology, then $\left(f(x_n)\right)_{n \in \mathbb{N}}$ is a Cauchy in B for the J-adic topology. Indeed, let $i \in \mathbb{N}$. There exists $N \in \mathbb{N}$ such that for all integers $m, n \geq N$, $x_n - x_m \in I^i$, then $f(x_n) - f(x_m) = f(x_n - x_m) \in f(I^i) = f(I)^i \subseteq J^i$. Also, if the sequence$(x_n)_{n \in \mathbb{N}}$ converges to an element x in A for the I-adic topology, then $\left(f(x_n)\right)_{n \in \mathbb{N}}$ converges to $f(x)$ in B for the J-adic topology.

Let $x \in \hat{A}$. Since A is dense in $\hat{A}$, there exists a sequence $(x_n)_{n \in \mathbb{N}}$ of elements of A that converges to x in $\hat{A}$. Since $(x_n)_{n \in \mathbb{N}}$ is a Cauchy sequence in A, then $\left(f(x_n)\right)_{n \in \mathbb{N}}$ is a Cauchy sequence in B. But $\hat{B}$ is complete, let $y \in \hat{B}$ be its limit. The element y does not depend on the sequence $(x_n)_{n \in \mathbb{N}}$. Indeed, let $(x'_n)_{n \in \mathbb{N}}$ be another sequence of elements of A that converges to x in $\hat{A}$. Then $(x_n - x'_n)_{n \in \mathbb{N}}$ converges to zero in A. By the preceding step, $\left(f(x_n) - f(x'_n)\right)_{n \in \mathbb{N}} = \left(f(x_n - x'_n)\right)_{n \in \mathbb{N}}$ converges to $f(0) = 0$ in B. Then $\left(f(x_n)\right)_{n \in \mathbb{N}}$ and $\left(f(x'_n)\right)_{n \in \mathbb{N}}$ have the same limit in $\hat{B}$. We define a map $\hat{f} : \hat{A} \longrightarrow \hat{B}$, by $\hat{f}(x) = y$. By definition of $\hat{f}$, its restriction to A is equal to f. In particular, $\hat{f}(1) = 1$. We will show that $\hat{f}$ is a homomorphism of rings. Let $x, x' \in \hat{A}$ and $(x_n)_{n \in \mathbb{N}}$, $(x'_n)_{n \in \mathbb{N}}$ be two sequences of elements of A that converge to x and x', respectively. By definition, the sequences $\left(f(x_n)\right)_{n \in \mathbb{N}}$ and $\left(f(x'_n)\right)_{n \in \mathbb{N}}$ converge to $\hat{f}(x)$ and $\hat{f}(x')$, respectively. Since f is a homomorphism, $f(x_n) + f(x'_n) = f(x_n + x'_n)$ and $f(x_n)f(x'_n) = f(x_n x'_n)$ for each $n \in \mathbb{N}$. Then the sequences $\left(f(x_n + x'_n)\right)_{n \in \mathbb{N}}$ and $\left(f(x_n x'_n)\right)_{n \in \mathbb{N}}$ converge to $\hat{f}(x) + \hat{f}(x')$ and $\hat{f}(x)\hat{f}(x')$, respectively. On the other hand, the sequences $(x_n + x'_n)_{n \in \mathbb{N}}$ and $(x_n x'_n)_{n \in \mathbb{N}}$ converge to $x + x'$ and xx', respectively. By definition of $\hat{f}$, the sequences $\left(f(x_n + x'_n)\right)_{n \in \mathbb{N}}$ and $\left(f(x_n x'_n)\right)_{n \in \mathbb{N}}$ converge to $\hat{f}(x + x')$ and $\hat{f}(xx')$, respectively. By the uniqueness of the limit in $\hat{B}$, $\hat{f}(x) + \hat{f}(x') = \hat{f}(x + x')$ and $\hat{f}(x)\hat{f}(x') = \hat{f}(xx')$. Then $\hat{f}$ is a homomorphism of rings.

Suppose now that f is onto and $f(I) = J$. Let $y \in \hat{B}$. Since B is dense in $\hat{B}$, there exists a sequence $(y_n)_{n \in \mathbb{N}}$ of elements of B that converges to y. By the preceding corollary, there exists a sequence $(b_i)_{i \in \mathbb{N}}$ of elements of B such that $y_n = b_n + b_{n-1} + \ldots + b_0$ with $b_{i+1} \in J^i$ for each $i \in \mathbb{N}$. Since f is onto and $f(I^i) = f(I)^i = J^i$ for each $i \in \mathbb{N}$, we can find a sequence $(a_i)_{i \in \mathbb{N}}$ of elements of A such that $f(a_i) = b_i$ and $a_{i+1} \in I^i$ for each $i \in \mathbb{N}$. Put $x_n = a_n + a_{n-1} + \ldots + a_0 \in A$ for each $n \in \mathbb{N}$. The sequence $(x_n)_{n \in \mathbb{N}}$ is Cauchy in A. Indeed, let $i \in \mathbb{N}$. For each integer $n \geq i$, $x_{n+1} - x_n = a_{n+1} \in I^n \subseteq I^i$. But $\hat{A}$ is complete. Let $x \in \hat{A}$ be the limit of the sequence $(x_n)_{n \in \mathbb{N}}$. For each $n \in \mathbb{N}$, $f(x_n) = f(a_n) + f(a_{n-1}) + \ldots + f(a_0) = b_n + b_{n-1} + \ldots + b_0 = y_n$. By definition of $\hat{f}$, we have $\hat{f}(x) = y$. $\quad\square$

13 The *I*-Adic Completion of a Noetherian Ring

Proposition 13.1 *Let A be a ring and $J = (a_1, \ldots, a_n)A$ a finitely generated ideal of A satisfying $\bigcap_{n:1}^{\infty} J^n = (0)$. The completion $\hat{A}^J \simeq A[[X_1, \ldots, X_n]]/(X_1 - a_1, \ldots, X_n - a_n)$.*

In particular, if A is Noetherian, the completion $\hat{A}^J$ is also Noetherian for each ideal J of A satisfying $\bigcap_{n:1}^{\infty} J^n = (0)$.

Proof The map $\phi : A[X_1, \ldots, X_n] \longrightarrow A$, defined by $\phi\big(f(X_1, \ldots, X_n)\big) = f(a_1, \ldots, a_n)$ for each $f(X_1, \ldots, X_n) \in A[X_1, \ldots, X_n]$, is a A-homomorphism onto of rings such that $\phi(X_i) = a_i$ for each $1 \leq i \leq n$. The ideal $I = (X_1, \ldots, X_n)$ of $A[X_1, \ldots, X_n]$ satisfies $\phi(I) = J$. By Proposition 12.3, there exists a homomorphism onto of rings $\hat{\phi} : \widehat{A[X_1, \ldots, X_n]}^I \longrightarrow \hat{A}^J$ that extends ϕ. By Theorem 6.4, $\widehat{A[X_1, \ldots, X_n]}^I = A[[X_1, \ldots, X_n]]$. Then $\hat{\phi} : A[[X_1, \ldots, X_n]] \longrightarrow \hat{A}^J$ is a A- homomorphism onto of rings and $\hat{\phi}(X_i) = \phi(X_i) = a_i$ for $1 \leq i \leq n$. Then $\hat{\phi}(X_i - a_i) = 0$ for $1 \leq i \leq n$ so $(X_1 - a_1, \ldots, X_n - a_n) \subseteq \ker \hat{\phi}$. We will show by induction on n that $\ker \hat{\phi} = (X_1 - a_1, \ldots, X_n - a_n)$. If $n = 1$, $I = XA[X]$, $J = aA$ and $\hat{\phi} : A[[X]] \longrightarrow \hat{A}^J$ with $\hat{\phi}(X) = a$. Let $f(X) = \sum_{i:0}^{\infty} b_i X^i \in \ker \hat{\phi}$. Then $\hat{\phi}\big(f(X)\big) = 0$. The sequence with general term $x_m = b_0 + b_1 a + \ldots + b_m a^m$ is Cauchy in A for the J-adic topology. Indeed, let $k \in \mathbb{N}$. For each $m \geq k$, $x_{m+1} - x_m = b_{m+1} a^{m+1} \in J^{m+1} \subseteq J^k$. Let $f(a) = \sum_{i:0}^{\infty} b_i a^i \in \hat{A}^J$ be its limit. Note that $f(X)$ is the limit in $A[[X]]$ for the $XA[[X]]$-adic topology of the sequence with general term $f_m(X) = b_0 + b_1 X + \ldots + b_m X^m \in A[X]$ and $\phi\big(f_m(X)\big) = x_m$ for each $m \in \mathbb{N}$. By the proof of Proposition 12.3, $\hat{\phi}\big(f(X)\big) = f(a)$, then $f(a) = 0$. So $f(X) = f(X) - f(a) = \sum_{i:1}^{\infty} b_i(X^i - a^i)$. Since $X^i - a^i = (X - a)(X^{i-1} + X^{i-2}a + \ldots + a^{i-1})$ for each $i \in \mathbb{N}^*$, then $f(X) \in (X - a)A[[X]]$. Suppose that $n \geq 2$ and the property is true at the order $n - 1$. Note that $\hat{\phi}$ is the composite of the A-homomorphism $\psi : A[[X_1, \ldots, X_n]] \longrightarrow A[[X_1, \ldots, X_{n-1}]]$, defined by $\psi(X_i) = X_i$ for $1 \leq i \leq n - 1$, and $\psi(X_n) = a_n$ and the A-homomorphism $\theta : A[[X_1, \ldots, X_{n-1}]] \longrightarrow \hat{A}^J$, defined by $\theta(X_i) = a_i$ for $1 \leq i \leq n - 1$. By the induction hypothesis, $\ker \theta = (X_1 - a_1, \ldots, X_{n-1} - a_{n-1})$. Since $\hat{\phi} = \theta \circ \psi$, then $\ker \hat{\phi} = \hat{\phi}^{-1}(0) = \psi^{-1}\big(\theta^{-1}(0)\big) = \psi^{-1}(\ker \theta)$. Let $f(X_1, \ldots, X_n) \in \ker \hat{\phi}$. Then $\psi\big(f(X_1, \ldots, X_n)\big) \in \ker \theta$. For each $1 \leq i \leq n - 1$, there exists $h_i(X_1, \ldots, X_{n-1}) \in A[[X_1, \ldots, X_{n-1}]]$ such that $\psi\big(f(X_1, \ldots, X_n)\big) = $

$\sum\limits_{i:1}^{n-1} h_i(X_1, \ldots, X_{n-1})(X_i - a_i)$. By definition, for each $j \in \mathbb{N}$, there exists $f_j(X_1, \ldots, X_{n-1}) \in A[[X_1, \ldots, X_{n-1}]]$ such that we have $f(X_1, \ldots, X_n) = \sum\limits_{j:0}^{\infty} f_j(X_1, \ldots, X_{n-1})X_n^j = \sum\limits_{j:0}^{\infty} f_j(X_1, \ldots, X_{n-1})a_n^j + \sum\limits_{j:1}^{\infty} f_j(X_1, \ldots, X_{n-1})(X_n^j - a_n^j)$. Since $\sum\limits_{j:0}^{\infty} f_j(X_1, \ldots, X_{n-1})a_n^j = \psi(f(X_1, \ldots, X_n)) \in (X_1 - a_1, \ldots, X_{n-1} - a_{n-1})$ and for each $j \geq 1$, $X_n^j - a_n^j = (X_n - a_n)(X_n^{j-1} + X_n^{j-2}a_n + \ldots + a_n^{j-1}) \in (X_n - a_n)$, then $f(X_1, \ldots, X_n) \in (X_1 - a_1, \ldots, X_{n-1} - a_{n-1}, X_n - a_n)$. So $ker \, \hat{\phi} \subseteq (X_1 - a_1, \ldots, X_n - a_n)$. Then we have the isomorphism $\hat{A}^J \simeq A[[X_1, \ldots, X_n]]/(X_1 - a_1, \ldots, X_n - a_n)$. If A is Noetherian, by Corollary 5.2 of Chap. 1, the formal power series ring $A[[X_1, \ldots, X_n]]$ is also Noetherian. $\square$

Example 1 The completion of an integral domain is not always an integral domain. By the preceding proposition, the (6)-adic completion of $\mathbb{Z}$ is $\hat{\mathbb{Z}} \simeq \mathbb{Z}[[X]]/(X - 6)$. Let $f = \sum\limits_{i:0}^{\infty} a_i X^i \in \mathbb{Z}[[X]]$ be the formal power series defined by $a_0 = 0, a_1 = 1$ and for each integer $n \geq 2$, $a_n = -\sum\limits_{i:1}^{n-1} a_i a_{n-i}$. We will show that $(3+f)(-2+f) = -6 + X$, which is equivalent to $f^2 + f = X$. The constant term of $f^2 + f$ is zero, and the coefficient of X is 1. For each integer $n \geq 2$, the coefficient of X^n is $\sum\limits_{i:1}^{n-1} a_i a_{n-i} + a_n = 0$. The constant terms of the series $3 + f$ and $-2 + f$ are respectively 3 and -2. Then $3 + f$ and $-2 + f$ are not multiple of $X - 6$ in $\mathbb{Z}[[X]]$. Then the ideal $(X - 6)$ is not prime and $\mathbb{Z}[[X]]/(X - 6)$ is not an integral domain.

Example 2 Let A be an integral domain. The polynomials ring $A[X_1, \ldots, X_n]$ and its $(X_1, \ldots, X_n)$-adic completion $A[[X_1, \ldots, X_n]]$ are both integral domains.

Notation Let A be a ring and $f = \sum\limits_{i:0}^{\infty} a_i X^i \in A[[X]]$. The content $c(f) = (a_i, \, i \in \mathbb{N})$ of f is the ideal of A generated by the coefficients of f. We define $n_f = min\{n \in \mathbb{N}; \, c(f) = (a_0, \ldots, a_n)\}$ if $c(f)$ is finitely generated. In the contrary case, $n_f = \infty$.

Lemma 13.2 (Haouaoui and Benhissi 2016) *Let A be a ring, $f \in A[[X]]$, and $0 \neq g \in A[[X]]$ be such that $gf = 0$ and n_g is finite. If n_g is minimal, i.e.,*

$n_g = min\{n_h; \ 0 \neq h \in A[[X]], hf = 0\}$, then for each element $a \in A$ satisfying $ag(0) = 0$, we have $ag = 0$.

Proof Put $g = \sum_{i:0}^{\infty} b_i X^i$. Let $a \in A$ be such that $ab_0 = 0$. We have $c(ag) = ac(g) = a(b_0, \ldots, b_{n_g}) = (ab_0, ab_1, \ldots, ab_{n_g}) = (ab_1, \ldots, ab_{n_g})$ and the equality $ag = ab_0 + ab_1 X + \cdots + ab_{n_g} X^{n_g} + \cdots = ab_1 X + \cdots + ab_{n_g} X^{n_g} + \cdots = Xh$ with $h = \sum_{i:0}^{\infty} ab_{i+1} X^i$. Then $c(h) = c(Xh) = c(ag) = (ab_1, \ldots, ab_{n_g})$, so $n_h \leq n_g - 1$. Since $Xhf = agf = 0$ with X a regular element, then $hf = 0$. So $h = 0$ since n_g is minimal. Then $ag = 0$. $\square$

McCoy theorem asserts that for any ring A, if $f(X) \in A[X]$ is a zero divisor, there exists $0 \neq a \in A$ such that $af(X) = 0$. The following lemma shows that McCoy property is valid for for formal power series over Noetherian rings.

Lemma 13.3 *[Fields 1971] Let A be a Noetherian ring and $f \in A[[X]]$. Then f is a zero divisor if and only if there exists $0 \neq a \in A$ such that $af = 0$.*

Proof *(Haouaoui and Benhissi 2016)* Since A is Noetherian, for each $g \in A[[X]]$, n_g is finite. Let $0 \neq f = \sum_{i:0}^{\infty} a_i X^i$ be a zero divisor of $A[[X]]$. There exists $0 \neq g = \sum_{i:0}^{\infty} b_i X^i \in A[[X]]$ such that $gf = 0$. We may suppose that $n_g = min\{n_h; \ 0 \neq h \in A[[X]], hf = 0\}$. Since $a_0 b_0 = 0$ and n_g is minimal, by the preceding lemma, $a_0 g = 0$. Then $(f - a_0)g = 0$ so $a_1 b_0 = 0$. Again, by the preceding lemma, $a_1 g = 0$. By induction, we show that for each $n \in \mathbb{N}$, $a_n g = 0$. In particular, $b_{n_g} a_n = 0$ pour tout $n \in \mathbb{N}$. Then $b_{n_g} f = 0$ with $b_{n_g} \neq 0$. $\square$

The following two examples show that the result is not true in general.

Example 1 (Fields 1971) Let K be a field, Y, $\{X_i; \ i \in \mathbb{N}\}$ indeterminates over K, $R = K[Y, \{X_i; \ i \in \mathbb{N}\}]$, and I the ideal of R generated by the elements $X_0 Y$ and $\{X_i + Y X_{i+1}\}_{i \in \mathbb{N}}$. Let $A = R/I$; $a_0 = \bar{Y}$ and $b_i = \bar{X}_i$ for each $i \in \mathbb{N}$. Let $f = a_0 + X$ and $0 \neq g = \sum_{i:0}^{\infty} b_i X^i \in A[[X]]$. Then $fg = (a_0 + X)(\sum_{i:0}^{\infty} b_i X^i) = \sum_{i:0}^{\infty} a_0 b_i X^i + \sum_{i:0}^{\infty} b_i X^{i+1} = a_0 b_0 + \sum_{i:0}^{\infty} a_0 b_{i+1} X^{i+1} + \sum_{i:0}^{\infty} b_i X^{i+1} = a_0 b_0 + \sum_{i:0}^{\infty} (b_i + a_0 b_{i+1}) X^{i+1} = 0$. It is clear that a nonzero constant of A cannot annihilate $f = a_0 + X$.

Example 2 (Gilmer et al. 1975) Let K be a field of characteristic 2 and $(Y_i)_{i \in \mathbb{N}}$ a sequence of indeterminates over K. The ring $A = K[Y_i; \ i \in \mathbb{N}]/(Y_i^2; \ i \in \mathbb{N})$ is

also of characteristic 2. The formal power series $f = \sum\limits_{i:0}^{\infty} \bar{Y}_i X^i$ of $A[[X]]$ satisfies
$f^2 = 0$. Let $\bar{P} \in A$ be such that $\bar{P}f = 0$. Then $\bar{P}\bar{Y}_i = 0$, that means that $PY_i \in (Y_j^2; \ j \in \mathbb{N})$ for each $i \in \mathbb{N}$. Then each monomial of P is divisible by at least one Y_i^2, so $\bar{P} = 0$.

Lemma 13.4 (Hwang 2004) *Let A be a Noetherian ring and $a_1, \ldots, a_n \in A$. If Q is a prime ideal of the ring $A[[X_1, \ldots, X_n]]$ containing the elements $X_1 - a_1, \ldots, X_n - a_n$, then $ht(Q) \geq n$.*

Proof By induction. If $n = 1$, Q is a prime ideal of $A[[X]]$ containing $X - a$ with $a \in A$. By the preceding lemma, since A is Noetherian, the ring $A[[X]]$ has the McCoy property. Then $X - a$ is not a zero divisor in $A[[X]]$. Let $P \subseteq Q$ be a prime ideal of $A[[X]]$ minimal over the ideal $(X - a)A[[X]]$. By Example 16.1 of Theorem 16.2, Chap. 2, $ht(P) = 1$. Then $ht(Q) \geq 1$. Suppose that the property is true at the order $n - 1$. Let Q be a prime ideal of the ring $A[[X_1, \ldots, X_n]]$ containing $X_1 - a_1, \ldots, X_n - a_n$. Then $Q_0 = Q \cap A[[X_1, \ldots, X_{n-1}]]$ is a prime ideal of the ring $B = A[[X_1, \ldots, X_{n-1}]]$ containing $X_1 - a_1, \ldots, X_{n-1} - a_{n-1}$. Then $ht_B(Q_0) \geq n - 1$. By Lemma 1.16, since the ring $B[[X_n]]$ is Noetherian, $Q_0[[X_n]] = Q_0 B[[X_n]] \subset Q$. The inclusion is strict since $X_n - a_n \in Q \setminus Q_0[[X_n]]$ because the coefficient of X_n is $1 \in Q_0$. Since $ht(Q_0[[X_n]]) \geq n - 1$, then $ht(Q) \geq n$. $\qquad\square$

Lemma 13.5 *Let $n \in \mathbb{N}^*$, Λ a subset of $\{1, \ldots, n\}$, and P a prime ideal of a ring A. Then $P[[X_1, \ldots, X_n]] + (X_i, i \in \Lambda)$ is a prime ideal of $A[[X_1, \ldots, X_n]]$.*

Proof Let Λ^c be the complementary of Λ in the set $\{1, \ldots, n\}$. Then
$A[[X_1, \ldots, X_n]]/P[[X_1, \ldots, X_n]] + (X_i, i \in \Lambda) \simeq A[[X_1, \ldots, X_n]]/(X_i, i \in \Lambda)/P[[X_1, \ldots, X_n]] + (X_i, i \in \Lambda)/(X_i, i \in \Lambda) \simeq A[[X_i, i \in \Lambda^c]]/P[[X_i, i \in \Lambda^c]] \simeq (A/P)[[X_i, i \in \Lambda^c]]$, which is an integral domain. $\qquad\square$

Lemma 13.6 *Let A be a ring, $a_1, \ldots, a_n \in A$, and P an ideal of A. Then*
$$A[[X_1, \ldots, X_n]]/P[[X_1, \ldots, X_n]] + (X_1 - a_1, \ldots, X_n - a_n) \simeq$$
$$(A/P)[[X_1, \ldots, X_n]]/(X_1 - \bar{a}_1, \ldots, X_n - \bar{a}_n).$$

Proof Consider the composite of the following canonical homomorphisms: $\psi :$ $A[[X_1, \ldots, X_n]] \rightarrow (A/P)[[X_1, \ldots, X_n]] \rightarrow (A/P)[[X_1, \ldots, X_n]]/(X_1 - \bar{a}_1, \ldots, X_n - \bar{a}_n)$. Then ψ is a homomorphism onto and $P[[X_1, \ldots, X_n]] + (X_1 - a_1, \ldots, X_n - a_n) \subseteq \ker \psi$. Conversely, let $f = \sum a_{i_1 \ldots i_n} X_1^{i_1} \ldots X_n^{i_n} \in \ker \psi$. Its image in the ring $(A/P)[[X_1, \ldots, X_n]]$ is $\bar{f} = \sum \overline{a_{i_1 \ldots i_n}} X_1^{i_1} \ldots X_n^{i_n} \in (X_1 - \bar{a}_1, \ldots, X_n - \bar{a}_n)$. Put $\bar{f} = \bar{g}_1(X_1 - \bar{a}_1) + \ldots + \bar{g}_n(X_n - \bar{a}_n)$ with $g_1, \ldots, g_n \in A[[X_1, \ldots, X_n]]$ and $\bar{g}_1, \ldots, \bar{g}_n$ their respective images in the ring $(A/P)[[X_1, \ldots, X_n]]$. Then $f - g_1(X_1 - a_1) - \ldots - g_n(X_n - a_n) \in P[[X_1, \ldots, X_n]]$. So $f \in P[[X_1, \ldots, X_n]] + (X_1 - a_1, \ldots, X_n - a_n)$. $\qquad\square$

In the following theorem, we compute the Krull dimension of the completion of a ring for the *I*-adic topology, by using the formal power series.

Theorem 13.7 (Hwang 2004) *Let A be a Noetherian ring with finite dimension, I an ideal of A satisfying $\bigcap\limits_{n:1}^{\infty} I^n = (0)$, and $\hat{A}$ the completion of A for the I-adic topology. Then $\dim \hat{A} = \sup\{ht_A(M); \ M \in Max(A), I \subseteq M\} \le \dim A.$*

Proof Put $I = (a_1, \ldots, a_n)$. By Proposition 13.1, $\hat{A} \simeq A[[X_1, \ldots, X_n]]/(X_1 - a_1, \ldots, X_n - a_n)$. Let $\overline{Q_0} \subset \overline{Q_1} \subset \ldots \subset \overline{Q_l}$ be a chain of prime ideals of $\hat{A}$, $\overline{Q_i} = Q_i/(X_1 - a_1, \ldots, X_n - a_n)$ with Q_i a prime ideal of $A[[X_1, \ldots, X_n]]$ and $(X_1 - a_1, \ldots, X_n - a_n) \subseteq Q_0 \subset Q_1 \subset \ldots \subset Q_l$. By adding supplementary prime ideals, we may suppose that Q_0 is minimal over $(X_1 - a_1, \ldots, X_n - a_n)$ and Q_l is a maximal ideal of $A[[X_1, \ldots, X_n]]$. There exists a maximal ideal M of A such that $Q_l = M + (X_1, \ldots, X_n)$. Note that M contains the ideal $(a_1, \ldots, a_n) = I$. By Lemma 13.4, $ht(Q_0) \ge n$, and by the generalized principal ideal theorem (GPIT), $ht(Q_0) \le n$. Then $ht(Q_0) = n$ so $ht(Q_l) \ge n + l$. Put $ht(M) = k \le \dim A < \infty$. By GPIT, there exist $b_1, \ldots, b_k \in A$ such that M is minimal over the ideal $(b_1, \ldots, b_k)$. By Lemma 1.24, $Q_l = M + (X_1, \ldots, X_n)$ is minimal over $(b_1, \ldots, b_k) + (X_1, \ldots, X_n)$. Again, by GPIT, we have $ht(Q_l) \le k + n$. Let $P_0 \subset P_1 \subset \ldots \subset P_k = M$ be a chain of prime ideals of the ring A that realizes the height of the maximal ideal M. By Lemma 13.5, we have the following chain of prime ideals: $P_0[[X_1, \ldots, X_n]] \subset P_1[[X_1, \ldots, X_n]] \subset \ldots \subset P_k[[X_1, \ldots, X_n]] \subset P_k[[X_1, \ldots, X_n]] + (X_1) \subset P_k[[X_1, \ldots, X_n]] + (X_1, X_2) \subset \ldots \subset P_k[[X_1, \ldots, X_n]] + (X_1, \ldots, X_n) = M[[X_1, \ldots, X_n]] + (X_1, \ldots, X_n) = M.A[[X_1, \ldots, X_n]] + (X_1, \ldots, X_n) = Q_l$ of length $k + n$. Then $ht(Q_l) \ge k + n$ so $ht(Q_l) = k + n$. On the other hand, $ht(Q_l) \ge l + n$, then $l \le k = ht_A(M)$. Then we have shown that for each chain of prime ideals of $\hat{A}$ of length l, there exists a maximal ideal M of A containing $I = (a_1, \ldots, a_n)$ such that $l \le ht_A(M)$. We have the inequality $\dim \hat{A} \le \sup\{ht_A(M); \ M \in Max(A), I \subseteq M\}$.

We will show the inverse inequality by induction on $\dim A$. If $\dim A = 0$, for each $M \in Max(A)$, $ht(M) = 0 \le \dim \hat{A}$ and the property is satisfied. Suppose that $\dim A \ge 1$. Let M_0 be a maximal ideal of A containing I, $l = ht(M_0) \le \dim A < \infty$ and $P_0 \subset P_1 \subset \ldots \subset P_l = M_0$ a chain of prime ideals of A that realizes the height of M_0. It suffices to construct a chain of length l of prime ideals in $\hat{A}$. By the preceding lemma, the ring $B = A[[X_1, \ldots, X_n]]/P_1[[X_1, \ldots, X_n]] + (X_1 - a_1, \ldots, X_n - a_n) \simeq (A/P_1)[[X_1, \ldots, X_n]]/(X_1 - \bar{a}_1, \ldots, X_n - \bar{a}_n) = \widehat{A/P_1}$, the completion of the ring A/P_1 for the $(\bar{a}_1, \ldots, \bar{a}_n)$-adic topology. Since $\dim(A/P_1) < \dim A$, we can apply the induction hypothesis to the ring A/P_1. Then $\dim B = \sup\{ht(M/P_1); M \in Max A, P_1 \subseteq M, (\bar{a}_1, \ldots, \bar{a}_n) \subseteq \bar{M}\} = \sup\{ht(M/P_1); M \in Max A, P_1 + (a_1, \ldots, a_n) \subseteq M\}$. Since $M_0 \in Max(A)$ and $P_1 + (a_1, \ldots, a_n) \subseteq M_0$, then $\dim B \ge ht(M_0/P_1) = l - 1$. There exists a chain of prime ideals in $A[[X_1, \ldots, X_n]]$ of length $l - 1$ such that $P_1[[X_1, \ldots, X_n]] + (X_1 - a_1, \ldots, X_n - a_n) \subseteq Q_0 \subset Q_1 \subset \ldots \subset Q_{l-1}$. Suppose that Q_0 is minimal over the ideal $P_0[[X_1, \ldots, X_n]] + (X_1 - a_1, \ldots, X_n - a_n)$. Then the ideal $\overline{Q_0} = Q_0/P_0[[X_1, \ldots, X_n]]$ is prime minimal over $(X_1 - \tilde{a}_1, \ldots, X_n - \tilde{a}_n)$ in the ring $A[[X_1, \ldots, X_n]]/P_0[[X_1, \ldots, X_n]] \simeq (A/P_0)[[X_1, \ldots, X_n]]$. Let $b \in P_1 \setminus P_0$. Then $\tilde{b}$ is not a zero divisor in the ring A/P_0. Indeed, in the contrary case, there exists $0 \neq \tilde{c} \in A/P_0$ such that $\tilde{c}\tilde{b} = 0$, then $cb \in P_0$ so $c \in P_0$, which is absurd. The element $\tilde{b}$ of A/P_0 belongs to $\overline{Q_0}$. By Lemma 12.22 of Chap. 2, there exist $k \in \mathbb{N}^*$ and $h \in (A/P_0)[[X_1, \ldots, X_n]] \setminus \overline{Q_0}$ such that $h\tilde{b}^k \in (X_1 - \tilde{a}_1, \ldots, X_n - \tilde{a}_n)$. By taking $X_1 = \tilde{a}_1, \ldots, X_n = \tilde{a}_n$, we find $h(\tilde{a}_1, \ldots, \tilde{a}_n)\tilde{b}^k = 0$ in $\widehat{A/P_0}$, the completion of the ring A/P_0 for the $(\tilde{a}_1, \ldots, \tilde{a}_n)$-adic topology. But $\tilde{b}$ is not a zero divisor in $\widehat{A/P_0}$. Then $h(\tilde{a}_1, \ldots, \tilde{a}_n) = 0$ so $h \in (X_1 - \tilde{a}_1, \ldots, X_n - \tilde{a}_n) \subseteq \overline{Q_0}$, which is absurd. We conclude that Q_0 is not minimal over $P_0[[X_1, \ldots, X_n]] + (X_1 - a_1, \ldots, X_n - a_n)$. Let $Q_{-1} \subset Q_0$ be a prime ideal of $A[[X_1, \ldots, X_n]]$ minimal over $P_0[[X_1, \ldots, X_n]] + (X_1 - a_1, \ldots, X_n - a_n)$. Then we have the chain $Q_{-1} \subset Q_0 \subset Q_1 \subset \ldots \subset Q_{l-1}$ of length l of prime ideals in $A[[X_1, \ldots, X_n]]$ containing $(X_1 - a_1, \ldots, X_n - a_n)$. This gives a chain of length l in the ring $\hat{A} \simeq A[[X_1, \ldots, X_n]]/(X_1 - a_1, \ldots X_n - a_n)$. $\square$

Example 1 The inequality in the preceding theorem may be strict. Let's see the example of Corollary 8.8, Chap. 2. Let k be a field and $A = k[X]_{(X)}[t]$. Then A is a Noetherian ring with dimension 2, its ideal $I = (Xt - 1)$ is maximal with height 1, and its ideal (X, t) is maximal of height 2. It is clear that $\bigcap\limits_{n:1}^{\infty} I^n = (0)$. Let $\hat{A}$ be the completion of A for the I-adic topology. By the preceding theorem, $dim\ \hat{A} = 1 < dim\ A$.

Example 2 We generalize the preceding example in the following way. Let B be a DVR of rank one and maximal ideal $M = (p)$. The domain $B[Y]$ is Noetherian of dimension 2 and its ideal $(pY - 1)$ is maximal of height 1. Indeed, the element $pY - 1$ is irreducible since if $pY - 1 = (aY + b)c$ with $a, b, c \in B$, then $-bc = 1$, so c is invertible. Since $B[Y]$ is UFD, the ideal $(pY - 1)$ is prime of height one. The quotient field of B is $K = S^{-1}B$ with $S = \{p^n;\ n \in \mathbb{N}\}$. The map $\phi : B[Y] \longrightarrow K$, defined by $\phi(P(Y)) = P(\frac{1}{p})$ for each $P(Y) \in B[Y]$, is a homomorphism onto of rings since if $x \in K$, $x = \frac{b}{p^n}$ with $b \in B$ and $n \in \mathbb{N}$, then $x = \phi(bY^n)$. It is clear that $(pY - 1) \subseteq ker\ \phi$. Conversely, let $P(Y) \in ker\ \phi$. Then $P(\frac{1}{p}) = 0$, so $P(Y) = (Y - \frac{1}{p})Q(Y)$ with $Q(Y) \in K[Y]$. Then $P(Y) = (pY - 1)\frac{1}{p^n}f(Y)$ with $n \in \mathbb{N}^*$ and $f(Y) \in B[Y]$. Then $p^n P(Y) = (pY - 1)f(Y) \in (pY - 1)$. Since $(pY - 1)$ is a prime ideal and $p^n \notin (pY - 1)$ for degrees reasons, then $P(Y) \in (pY - 1)$. So $ker\ \phi = (pY - 1)$ and $B[Y]/(pY - 1) \simeq K$ is a field. Then $(pY - 1)$ is a maximal ideal of $B[Y]$ and $\bigcap\limits_{n:1}^{\infty}(pY - 1)^n = (0)$. Let $\widehat{B[Y]}$ be the completion of $B[Y]$ for the $(pY - 1)$-adic topology. By the preceding theorem, $dim\ \widehat{B[Y]} = 1 < dim\ B[Y] = 2$.

Corollary 13.8 *Let A be a Noetherian ring of finite dimension, I an ideal of A contained in the Jacobson radical of A and satisfying* $\bigcap\limits_{n:1}^{\infty} I^n = (0)$, *and* $\hat{A}$ *the completion of A for the I-adic topology. Then* $\dim \hat{A} = \dim A$.

14   The *I*-Adic Completion of a Valuation Domain

The results of this section are due to Kang and Park (1999).

Lemma 14.1 *Let (A, M) be a valuation domain and Q a prime ideal of $A[[X]]$. If $X \notin Q$ and if $Q \nsubseteq M[[X]]$, then Q is a principal ideal.*

Proof If $g = \sum\limits_{i:0}^{\infty} a_i X^i \in Q \setminus M[[X]]$, let $n(g) = \inf\{n \in \mathbb{N};\ a_n \notin M\}$. Choose f such that $n(f)$ be the smallest possible element in the set $\{n(g);\ g \in Q \setminus M[[X]]\}$. Then $a_0 = f(0) \neq 0$. Indeed, in the contrary case $f = Xh$ with $h \in A[[X]]$ and since $X \notin Q$, then $h \in Q \setminus M[[X]]$. But $n(h) = n(f) - 1$, which contradicts the minimality of $n(f)$. Then $a_0 \in M \setminus (0)$ since f cannot be invertible. Let v be a valuation with ring A. We will show that $v(a_0) = \min\{v(g(0));\ g \in Q\}$. In the contrary case, there exists $g = \sum\limits_{i:0}^{\infty} b_i X^i \in Q$ with $v(b_0) < v(a_0)$, then $a_0 = b_0 c$ with $c \in M$. If we denote $n(f) = n$, then $f - cg = X\big(\ldots + (u - cb_n)X^{n-1} + \ldots\big) \in Q$ where u is the first invertible coefficient of f. Since $X \notin Q$, the series $h = \ldots + (u - cb_n)X^{n-1} + \ldots \in Q$. But the coefficient of X^{n-1} in h is $u - cb_n \notin M$, which contradicts the minimality of $n(f) = n$. We will show that $Q = fA[[X]]$. Let $g \in Q$. There exists $c_1 \in A$ such that X divides $g - c_1 f$ in $A[[X]]$, then $g - c_1 f = Xg_1 \in Q$, so $g_1 \in Q$. Again, $g_1 - c_2 f = Xg_2 \in Q$ with $c_2 \in A$ and $g_2 \in Q$. Then $g = c_1 f + Xg_1 = c_1 f + X(c_2 f + Xg_2) = c_1 f + c_2 Xf + X^2 g_2$. By induction, we construct an infinite sequence $(c_i)_{i \in \mathbb{N}^*}$ of elements of A and a sequence $(g_i)_{i \in \mathbb{N}^*}$ of elements of $A[[X]]$ such that for each $n \in \mathbb{N}^*$, $g = \big(\sum\limits_{i:1}^{n-1} c_i X^{i-1}\big)f + X^n g_n$. If we put $h = \sum\limits_{i:1}^{\infty} c_i X^{i-1} \in A[[X]]$, then $g = hf$. $\square$

Corollary 14.2 *Let (A, M) be a valuation domain and Q a prime ideal of $A[[X]]$. If $Q \cap A = (0)$ and if $Q \nsubseteq M[[X]]$, then Q is a principal ideal.*

Proof If $X \in Q$, then $Q = Q \cap A + XA[[X]] = XA[[X]]$ is principal. If $X \notin Q$, since $Q \nsubseteq M[[X]]$, we apply the preceding lemma. $\square$

The hypothesis $Q \nsubseteq M[[X]]$ is necessary in the preceding lemma and corollary.

Example 1 Let (A, M) be a nondiscrete valuation domain of rank one. We admit that $M.A[[X]]$ is a prime ideal of $A[[X]]$. See Arnold and Brewer (1973). Since $M.A[[X]] \subseteq M[[X]]$, then $X \notin M.A[[X]]$. Suppose that $M.A[[X]]$ is principal. There exists $m \in M$ such that $M.A[[X]] = m.A[[X]]$. Let v be a valuation with values in $\mathbb{R}$ that defines A. If we take $m' \in M$ such that $v(m') < v(m)$, then $m' \notin m.A[[X]]$.

Example 2 (Seidenberg 1966) We need the following preliminary results.

(1) Let A be an integral domain and A^c its complete integral closure. Then $(A[[X]])^c \subseteq A^c[[X]]$. In particular, $A[[X]]$ is completely integrally closed if and only if A is completely integrally closed.

   Indeed, let $0 \neq f \in (A[[X]])^c$. There exists $0 \neq g \in A[[X]]$ such that $gf^n \in A[[X]]$ for each $n \in \mathbb{N}$. Let $K = qf(A)$. Then $qf(A[[X]]) \subseteq K((X))$. But $K[[X]]$ is a DVR of rank one, then it is completely integrally closed with quotient field $K((X))$. Then $f \in K[[X]]$. Put $f = b_t X^t + ...$ and $g = a_s X^s + ...$ with $b_i \in K, a_i \in A, a_s \neq 0$ and $t, s \in \mathbb{N}$. Since $gf^n = a_s b_t^n X^{s+nt} + ... \in A[[X]]$, then $a_s b_t^n \in A$ for each $n \in \mathbb{N}$, i.e., $b_t \in A^c$. Again, $f - b_t X^t \in (A[[X]])^c$, then $b_{t+1} \in A^c$. By induction, we see that all the coefficients of f belong to A^c, then $f \in A^c[[X]]$. So $(A[[X]])^c \subseteq A^c[[X]]$.

   In a particular case, if $A[[X]]$ is completely integrally closed, $A^c \subseteq (A[[X]])^c = A[[X]]$, then $A^c = A$. Conversely, if A is completely integrally closed, $(A[[X]])^c \subseteq A^c[[X]] = A[[X]]$, then $(A[[X]])^c = A[[X]]$.

(2) Let (A, M) be a nondiscrete valuation domain of rank one. We always admit in the first example that the ideal $M.A[[X]]$ is prime. But thanks to (1), we can easily prove that it is radical. Indeed, let v be the associated valuation to A and G its value group. We may suppose that G is a subgroup that is dense in $(\mathbb{R}, +)$. Let $0 \neq f \in A[[X]]$ and $n \in \mathbb{N}^*$ be such that $f^n \in M.A[[X]]$, $f^n = ag$ with $0 \neq a \in M$ and $g \in A[[X]]$. Since G is dense in $\mathbb{R}$, there exists $\alpha \in G$ such that $0 < \alpha \leq \frac{v(a)}{n}$. Let $b \in M$ be such that $v(b) = \alpha$. Then $v(b) \leq \frac{v(a)}{n}$, so $v(b^n) \leq v(a)$. The element $c = \frac{a}{b^n} \in A$ and $\left(\frac{f}{b}\right)^n = \frac{ag}{b^n} = cg$, then $\left(\frac{f}{b}\right)^n - cg = 0$. Since $cg \in A[[X]]$, the element $\frac{f}{b}$ of $qf(A[[X]])$ is integral over $A[[X]]$. But A is a valuation domain of rank one. By (1), $A[[X]]$ is completely integrally closed. Then $\frac{f}{b} \in A[[X]]$ and $f \in bA[[X]] \subseteq M.A[[X]]$.

(3) Let B be a completely integrally closed integral domain, $b \in B$ a non-invertible element, and P a prime ideal of B such that $P \subset bB$. Then $P = (0)$. Indeed, $\frac{1}{b}P \subseteq B$ and $b(\frac{1}{b}P) = P$. Since P is prime and $b \notin P$, then $\frac{1}{b}P \subseteq P$, so $P \subseteq bP \subseteq P$, then $bP = P$. By induction, $P = b^n P$ for each $n \in \mathbb{N}^*$. Then $P \subseteq \bigcap_{n:1}^{\infty} b^n B = (0)$ since B is an Archimedean domain. Then $P = (0)$.

(4) Let A be a nondiscrete valuation domain of rank one. By Theorem 8.5, there exists an infinite strictly decreasing sequence $(P_n)_{n \in \mathbb{N}^*}$ of prime ideals of $A[[X]]$ such that $P_n \cap A = (0)$ for each $n \in \mathbb{N}^*$. By (1), $A[[X]]$ is completely

integrally closed. By (3), no P_n is principal. Then we have a counterexample for the preceding corollary.

(5) To finish with this example, if A is an integral domain with quotient field $K \neq A$, then $A[[X]] \subseteq K[[X]]$, so $qf(A[[X]]) \subseteq K((X))$. But in general, we do not have equality. Suppose that A is completely integrally closed. Let a be a non-invertible nonzero element in A and (n_i) the sequence of natural integers defined by $n_0 = 1$ and $n_{r+1} = \sum_{i+j=r} n_i n_j$. The series $f = \sum_{i=1}^{\infty} \frac{n_{i-1}}{a^{2i-1}} X^i$ of $K[[X]]$ does not belong to $A[[X]]$ and satisfies $f^2 - af + X = 0$. Indeed, the coefficient of X in this expression is $-a\frac{n_0}{a} + 1 = 0$, and for each $r \geq 2$, the coefficient of X^r is $\sum_{i+j=r} \frac{n_{i-1}}{a^{2i-1}} \frac{n_{j-1}}{a^{2j-1}} - a\frac{n_{r-1}}{a^{2r-1}} = \frac{1}{a^{2r-2}} \sum_{i+j=r-2} n_i n_j - \frac{n_{r-1}}{a^{2r-2}} = 0$. Since $A[[X]]$ is completely integrally closed by (1), then $f \notin qf(A[[X]])$. So $f \in K((X)) \setminus qf(A[[X]])$.

Proposition 14.3 *Let A be valuation domain and $a \in A$. Then $(X - a)A[[X]]$ is a prime ideal if and only if a is not invertible in A*

Proof If a is invertible in A, $a - X$ is invertible in $A[[X]]$, then $(X - a)A[[X]] = A[[X]]$ is not prime. Suppose that a is not invertible. If $a = 0$, then $(X-a)A[[X]] = XA[[X]] \in spec(A[[X]])$. Suppose that $a \neq 0$.

Particular case: $\bigcap_{i:1}^{\infty} a^n A = (0)$. By Lemma 9.4, $(X-a)A[[X]] \cap A = \bigcap_{i:1}^{\infty} a^n A = (0)$. We even have $\sqrt{(X - a)A[[X]]} \cap A = (0)$. Indeed, let $b \in \sqrt{(X - a)A[[X]]} \cap A$. There exists $n \in \mathbb{N}^*$ such that $b^n \in (X - a)A[[X]] \cap A = (0)$, then $b = 0$. Let $Min(X - a)$ be the set of prime ideals of $A[[X]]$ that are minimal over the ideal $(X - a)A[[X]]$. We will show that there exists $Q \in Min(X - a)$ such that $Q \cap A = (0)$. Suppose that for each $Q \in Min(X - a)$, $Q \cap A \neq (0) \Longrightarrow \forall Q \in Min(X-a)$, $Q \cap A \not\subseteq \bigcap_{n:1}^{\infty} a^n A \Longrightarrow \forall Q \in Min(X-a)$, $\exists n \in \mathbb{N}^*$, $Q \cap A \not\subseteq a^n A$.

Since A is a valuation domain, for each $Q \in Min(X - a)$, there exists $n \in \mathbb{N}^*$ such that $a^n A \subseteq Q \cap A \Longrightarrow \forall Q \in Min(X - a)$, $a \in \sqrt{Q \cap A} = Q \cap A$. Then $0 \neq a \in \sqrt{(X - a)A[[X]]} \cap A = (0)$, which is absurd. We conclude that there exists $Q \in Min(X-a)$ such that $Q \cap A = (0)$. Since $X-a \in Q$, then $Q \not\subseteq M[[X]]$, where M is the maximal ideal of A. By Corollary 14.2, Q is principal. Put $Q = fA[[X]]$. There exists $g \in A[[X]]$ such that $X - a = fg$. Put $f = a_0 + a_1 X + \ldots$ and $g = b_0 + b_1 X + \ldots$, then $a_0 \in M$ since f cannot be invertible. By identification, $a_0 b_1 + a_1 b_0 = 1$, then $b_0 \notin M$ so g is invertible. Then $(X-a)A[[X]] = fA[[X]] = Q \in spec(A[[X]])$.

General case: $P = \bigcap_{i:1}^{\infty} a^n A$. Since A is a valuation domain, $P \in spec(A)$. By Lemma 9.4, $P \subseteq (X - a)A[[X]]$. For each $f = \sum_{n:0}^{\infty} p_n X^n \in P[[X]]$, $p_n = (X - a)g_n$, with $g_n \in A[[X]]$, then $f = (X - a)\sum_{n:0}^{\infty} g_n X^n$. Since the order of $g_n X^n$ is $\geq n$, then $\sum_{n:0}^{\infty} g_n X^n \in A[[X]]$, so $f \in (X - a)A[[X]]$. We conclude that $P[[X]] \subseteq (X-a)A[[X]]$. In the valuation domain A/P, $\bar{a} \neq 0$ since $a \notin \bigcap_{n:1}^{\infty} a^n A = P$, but $\bigcap_{n:1}^{\infty}(\bar{a}^n) = (0)$. In a particular case, $(X - \bar{a})(A/P)[[X]]$ is a prime ideal of $(A/P)[[X]] \simeq A[[X]]/P[[X]]$, then $(X - a)A[[X]]$ is a prime ideal of $A[[X]]$. □

Example 1 Let (A, M) be a valuation domain and $0 \neq a \in M$. The ideal $\bigcap_{i:1}^{\infty} a^n A$ introduced in the proof of the preceding proposition may be either zero or not. For example, if A is of dimension 1, it is Archimedean, then $\bigcap_{i:1}^{\infty} a^n A = (0)$ for each $a \in M$. Suppose that $dim\, A \geq 2$. Let $(0) \subset P \subset M$ be a prime ideal of A and $a \in M \setminus P$. For each $i \in \mathbb{N}^*$, $a^i \notin P$, then $P \subset a^i A$. So $(0) \neq P \subseteq \bigcap_{i:1}^{\infty} a^n A$.

Example 2 The result proved in the preceding proposition is not true for any ring. In Example 1 of Proposition 13.1, we showed that the ideal $(X - 6)\mathbb{Z}[[X]]$ is not prime.

Example 3 Let (A, M) be a valuation domain of rank one and $a \neq b$ two elements of M. The prime ideals $(X - a)A[[X]] \neq (X - b)A[[X]]$. Indeed, by Lemma 9.4, if $X - b \in (X - a)A[[X]]$, then $a - b = (X - b) - (X - a) \in (X - a)A[[X]] \cap A = \bigcap_{i:1}^{\infty} a^i A = (0)$ since A is Archimedean. Then $a = b$, which is absurd.

Example 4 Let A be a valuation domain of rank one and $a \in A$ a non-invertible element. Since A is Archimedean, $\bigcap_{i:1}^{\infty} a^i A = (0)$. By the preceding proposition, the ideal $(X - a)A[[X]]$ is prime. By Proposition 13.1, the completion of A for the (a)-adic topology is $\hat{A} \simeq A[[X]]/(X - a)$ is an integral domain.

Lemma 14.4 *Let A be an integral domain. If $P \subseteq Q$ are two nonzero principal prime ideals of A, then $P = Q$.*

Proof Put $P = pA$ and $Q = qA$ with p and q two irreducible elements of the domain A. Since q divides p, these elements are associated, then $P = Q$. $\square$

Theorem 14.5 *Let A be a valuation domain with finite dimension, I a nonzero ideal of A such that $\bigcap\limits_{i:1}^{\infty} I^i = (0)$, and $\hat{A}$ the completion of A for the I-adic topology. Then $dim\ \hat{A} = dim(A/I) + 1$.*

Proof The condition $\bigcap\limits_{i:1}^{\infty} I^i = (0)$ with $I \neq (0)$ implies $I^2 \neq I$. Let $a \in I \setminus I^2$. Then $I^2 \subset (a) \subseteq I$ with $\bigcap\limits_{i:1}^{\infty} a^i A = (0)$ since $a \in I$. By Proposition 6.3, the I-adic and (a)-adic topologies over A are identical. We may suppose that $I = (a)$. By Proposition 13.1, $\hat{A} \simeq A[[X]]/(X - a)$. Let $(0) \subset P_1 \subset \ldots \subset P_n = M$ be the prime spectrum of A. Suppose that $a \notin P_1$, then $a^i \notin P_1$ for each $i \in \mathbb{N}^*$. Since A is a valuation domain, $P_1 \subseteq \bigcap\limits_{i:1}^{\infty} a^i A = (0)$, which is absurd. Then $a \in P_1 \subseteq M$. By Proposition 14.3, the ideal $(X - a)A[[X]]$ is prime. Let Q be a prime ideal of $A[[X]]$ containing strictly the ideal $(X - a)A[[X]]$. By the preceding lemma, Q is not principal. Since $X - a \notin M[[X]]$, then $Q \nsubseteq M[[X]]$. By Corollary 14.2, $Q \cap A \neq (0)$, then $P_1 \subseteq Q \cap A$. So $X = a + (X - a) \in Q$, then $P_1 + XA[[X]] \subseteq Q$. Then we have $coht((X - a)A[[X]]) \leq coht(P_1 + XA[[X]]) + 1$ so $dim(A[[X]]/(X - a)) \leq dim(A[[X]]/P_1 + XA[[X]]) + 1$. But $A[[X]]/P_1 + XA[[X]] \simeq A/P_1$, then $dim\ \hat{A} \leq dim(A/P_1) + 1 = dim(A/I) + 1 = n$. On the other hand, $(X - a)A[[X]] \subseteq P_1 + XA[[X]] \subset P_2 + XA[[X]] \subset \ldots \subset P_n + XA[[X]]$. By Lemma 9.4, $((X-a)A[[X]]) \cap A = \bigcap\limits_{i:1}^{\infty} a^i A = (0)$. But, $(P_1 + XA[[X]]) \cap A = P_1 \neq (0)$. Then we have the strict inclusion $(X - a)A[[X]] \subset P_1 + XA[[X]]$. By passage in the quotient ring $A[[X]]/(X - a)$, we obtain a chain of length n of prime ideals $(\bar{0}) \subset P_1 + XA[[X]]/(X - a) \subset P_2 + XA[[X]]/(X - a) \subset \ldots \subset P_n + XA[[X]]/(X - a)$. Then $dim\ \hat{A} = n$. $\square$

Corollary 14.6 *Let A be a valuation domain with finite dimension, I a nonzero ideal of A such that $\bigcap\limits_{i:1}^{\infty} I^i = (0)$, and $\hat{A}$ the completion of A for the I-adic topology. Then $Spec(\hat{A}) = \{P\hat{A};\ P \in Spec(A)\}$. Moreover, $P_1\hat{A} \subseteq P_2\hat{A} \iff P_1 \subseteq P_2$ and $P_1\hat{A} = P_2\hat{A} \iff P_1 = P_2$.*

Proof With the notations of the proof of the preceding theorem, we may suppose that $\hat{A} = A[[X]]/(X - a)$. By Lemma 9.4, $A \cap (X - a)A[[X]] = \bigcap_{i:1}^{\infty} a^i A = (0)$, then $A \subseteq A[[X]]/(X - a)$. Let $(0) \subset P_1 \subset \ldots \subset P_n = M$ be the prime spectrum of A, then $a \in P_1$ and $(\bar{0}) \subset P_1 + XA[[X]]/(X - a) \subset P_2 + XA[[X]]/(X - a) \subset \ldots \subset P_n + XA[[X]]/(X - a)$ are prime ideals of $\hat{A}$. Conversely, if Q is a prime ideal of $A[[X]]$ containing strictly $(X - a)A[[X]]$, then $P_1 + XA[[X]] \subseteq Q$. In particular, $X \in Q$, then $Q = Q \cap A + XA[[X]]$, and $0 \neq a \in Q$. Put $P_i = Q \cap A$ with $1 \leq i \leq n$, then $Q = P_i + XA[[X]]$. So $Spec(\hat{A}) = \{\bar{0}, P_i + XA[[X]]/(X - a); 1 \leq i \leq n\}$. It suffices to show that $P_i + XA[[X]]/(X - a) = P_i\hat{A}$ for $1 \leq i \leq n$. Since $X = (X - a) + a$, then $\bar{X} = a \in P_1 \subseteq P_i$. So we have the first inclusion. The inverse inclusion is clear. The last assertion is also clear.   □

15 Generalities on the t-SFT Domains

The results in the remaining of this chapter are basically due to Kang and Park (2006).

Definitions 15.1 Let A be an integral domain and I a nonzero integral ideal of A. We say that I is a t-SFT ideal if there exist a nonzero finitely generated ideal $F \subseteq I$ and an integer $k \geq 1$ such that for each $a \in I_t$, $a^k \in F_v$.

We say that A is a t-SFT domain if all its nonzero ideals are t-SFT ideals.

Example 1 Each Mori domain is a t-SFT domain. In particular, each Krull domain is a t-SFT domain. See Sect. 12 of Chap. 1 for the definition of a Krull domain. Note first that by Example 1 of Definition 4.5, Chap. 1, the t-operation and the v-operation are identical over a Mori domain A. We will now show that A is a t-SFT domain. Let I be a nonzero integral ideal of A. There exists a nonzero finitely generated ideal $J \subseteq I$ of A such that $I^{-1} = J^{-1}$. Then $I_v = J_v$, so $I_t = J_v$. We take $k = 1$.

Example 2 Each SFT-integral domain is a t-SFT domain. Indeed, let I be a nonzero integral ideal. By hypothesis, I_t is an integral SFT-ideal. There exist a finitely generated ideal $F = (a_1, \ldots, a_n) \subseteq I_t$ and an integer $k \geq 1$ such that for each $a \in I_t$, $a^k \in F$. There exist nonzero finitely generated ideals $J_1, \ldots, J_n \subseteq I$ such that $a_i \in (J_i)_v$ for $1 \leq i \leq n$. The finitely generated ideal $J = J_1 + \ldots + J_n \subseteq I$ and we have $F = (a_1, \ldots, a_n) \subseteq (J_1)_v + \ldots + (J_n)_v \subseteq J_v$ and for each $a \in I_t$, $a^k \in F \subseteq J_v$.

Example 3 In a Prufer domain, a nonzero ideal is SFT if and only if it is t-SFT. This is the case for valuation domains. The proof in many steps:

(i) In a Prufer domain, each nonzero finitely generated fractional ideal is invertible, so divisorial, by Lemma 9.4 of Chap. 3.

(ii) In a Prufer domain A, each nonzero fractional ideal I is a t-ideal.
    Indeed, $I_t = \bigcup J_v$, where J runs the set of the nonzero finitely generated fractional ideals contained in I. By (i), $J_v = J$. Then $I_t = \bigcup J = I$.

(iii) Let I be a t-SFT ideal of a Prufer domain A. There exist a nonzero finitely generated ideal $F \subseteq I$ and an integer $k \geq 1$ such that for each $x \in I_t, x^k \in F_v$. By (ii), $I_t = I$ and $F_v = F_t = F$. Then I is an SFT-ideal.

Proposition 15.1 *1. A nonzero integral ideal I is t-SFT if and only if the ideal I_t is t-SFT.*

2. An integral domain A is t-SFT if and only if its integral t-ideals are t-SFT.

Proof

(1) " $\Longrightarrow$ " There exists a nonzero finitely generated ideal $F \subseteq I$ and an integer $k \geq 1$ such that for each $a \in I_t, a^k \in F_v$. Then $F \subseteq I \subseteq I_t$ and for each $a \in (I_t)_t = I_t, a^k \in F_v$. So I_t is t-SFT.
    " $\Longleftarrow$ " There exist a nonzero finitely generated ideal $F = (a_1, \ldots, a_n) \subseteq I_t$ and an integer $k \geq 1$ such that for each $a \in (I_t)_t = I_t, a^k \in F_v$. There exist nonzero finitely generated subideals $J_1, \ldots, J_n$ of I such that $a_i \in (J_i)_v$ for $1 \leq i \leq n$. The finitely generated ideal $J = J_1 + \ldots + J_n \subseteq I$ and $F = (a_1, \ldots, a_n) \subseteq (J_1)_v + \ldots + (J_n)_v \subseteq J_v$. For each $a \in I_t, a^k \in F_v \subseteq (J_v)_v = J_v$. Then I is a t-SFT ideal.

(2) " $\Longleftarrow$ " Let I be a nonzero integral ideal of A. By hypothesis, I_t is a t-SFT ideal of A. By (1), I is a t-SFT ideal. $\square$

The following proposition is the analogue of Cohen theorem for the t-SFT property and improves the preceding proposition.

Proposition 15.2 *An integral domain A is t-SFT if and only if its t-prime ideals are t-SFT.*

Proof Suppose that A is not a t-SFT domain. Let $\mathscr{F}$ be the set of the t-ideals of A that are not t-SFT. By the preceding proposition, $\mathscr{F} \neq \emptyset$. We will show that $\mathscr{F}$ is inductive for the inclusion. Let $(I_\lambda)_{\lambda \in \Lambda}$ be a totally ordered family of elements of $\mathscr{F}$. Then $I = \bigcup_{\lambda \in \Lambda} I_\lambda$ is a t-ideal. Indeed, let J be a nonzero finitely generated subideal of I. There exists $\lambda_0 \in \Lambda$ such that $J \subseteq I_{\lambda_0}$, then $J_v = J_t \subseteq (I_{\lambda_0})_t = I_{\lambda_0} \subseteq I$. The ideal I is not t-SFT because in the contrary case, there exists a finitely generated ideal $F \subseteq I$ and an integer $k \geq 1$ such that for each $a \in I = I_t, a^k \in F_v$. There exists $\lambda_0 \in \Lambda$ such that $F \subseteq I_{\lambda_0}$. For each $a \in I_{\lambda_0} \subseteq I, a^k \in F_v$. Then I_{λ_0} is a t-SFT ideal, which is absurd. Then I bounds $(I_\lambda)_{\lambda \in \Lambda}$ in $\mathscr{F}$. By Zorn lemma, $\mathscr{F}$ has a maximal element P. It is a t-ideal that is not t-SFT. It suffices to show that it is prime to have a contradiction. Suppose that there exist $a_1, a_2 \in A \setminus P$ such that $a_1 a_2 \in P$. Then $(P + a_i A)_t$ is a t-ideal of A containing strictly P, then $(P + a_i A)_t$ is a t-SFT ideal with $i = 1, 2$. By the preceding proposition, $P + a_i A$ is also a

t-SFT ideal. There exist a finitely generated ideal $J_i \subseteq P + a_i A$ and an integer $k_i \geq 1$ such that for each $x \in (P + a_i A)_t$, $x^{k_i} \in (J_i)_v$. The finitely generated ideal $J = J_1 J_2 \subseteq (P + a_1 A)(P + a_2 A) = P^2 + a_1 P + a_2 P + a_1 a_2 A \subseteq P$. Let $k = k_1 + k_2$. By Proposition 11.1 of Chap. 1, for each $x \in P_t = P$, $x^k = x^{k_1} x^{k_2} \in (J_1)_v (J_2)_v \subseteq \left((J_1)_v (J_2)_v \right)_v = (J_1 J_2)_v = J_v$. Then P is a t-SFT ideal, which is absurd. $\square$

16   Transfer of the t-SFT Property to the Polynomial Domain

In the following proposition, we extend Proposition 11.10 of Chap. 1 to the t-ideals non-necessarily t-maximal in the polynomial ring with coefficients in an integrally closed domain.

Proposition 16.1 (Houston et al. 1984 and Houston et al. 1984) *Let A be an integrally closed integral domain and $\mathscr{I}$ an integral t-ideal of the ring $A[X]$ such that $I = \mathscr{I} \cap A \neq (0)$. Then I is a t-ideal of the domain A and $\mathscr{I} = I[X]$.*

Proof

(1) The inclusion $I[X] \subseteq \mathscr{I}$ is clear. Fix a nonzero element $a \in I$. Let $g(X) \in \mathscr{I}$. For each $h \in (a, g)^{-1} \subseteq K(X)$, where K is the quotient field of A, $gh \in A[X]$ and $ah \in A[X]$. Then $h \in \frac{1}{a} A[X] \subseteq K[X]$ and $c(gh) \subseteq A$. By Proposition 10.1 of Chap. 3, since A is integrally closed, $\left(c(g)c(h) \right)_v = \left(c(gh) \right)_v \subseteq A$, then $c(g)c(h) \subseteq A$. So $c(g)h(X) \subseteq A[X]$. We conclude that $c(g)(a, g)^{-1} \subseteq A[X]$, then $c(g) \subseteq \left((a, g)^{-1} \right)^{-1} = (a, g)_v \subseteq \mathscr{I}_t = \mathscr{I}$. So $c(g) \subseteq \mathscr{I} \cap A = I$, then $g(X) \in I[X]$.

(2) By Corollary 11.7 (1) of Chap. 1, $I[X] = \mathscr{I} = \mathscr{I}_t = (I[X])_t = I_t[X]$, then $I_t = I$. So I is a t-ideal of A. $\square$

Example (Bass 1962) Hypothesis A is integrally closed and is necessary in the proposition. Indeed, let's see again Schanuel example of Definition 9.3, Chap. 3. Let A be an integral domain with quotient field K and $t \in K \setminus A$ such that $t^n \in A$ for each integer $n \geq 2$. The fractional ideals $I_1 = (1 + tX, 1 + tX + t^2 X^2)$ and $J_1 = (1 - tX, 1 - tX + t^2 X^2)$ of the domain $A[X]$ satisfy $I_1 J_1 = A[X]$. Let $0 \neq d \in A$ such that $dt \in A$. The fractional ideals $\mathscr{I} = dI_1$ and $\mathscr{J} = \frac{1}{d} J_1$ of the domain $A[X]$ are inverse to each other with $\mathscr{I}$ integral. By Lemma 9.4 of Chap. 3, $\mathscr{I}$ is a divisorial ideal in $A[X]$. Then it is a t-ideal since it is finitely generated. Since $0 \neq d^2 = d^2(1 + tX + t^2 X^2) - d(1 + tX)dtX \in \mathscr{I}$, then $I = \mathscr{I} \cap A \neq (0)$. Suppose else that A is local. Suppose that $\mathscr{I} = I[X]$, then $I[X]\mathscr{J} = A[X]$, so $Ic(\mathscr{J}) = A$ where $c(\mathscr{J})$ is the fractional ideal of A formed by the coefficients of all the elements in $\mathscr{J}$. Indeed, since $I\mathscr{J} \subseteq A[X]$, then $Ic(\mathscr{J}) \subseteq A$. Conversely, there exist $f_1, \ldots, f_n \in I[X]$ and $g_1, \ldots, g_n \in \mathscr{J}$ such that $1 = f_1 g_1 + \ldots + f_n g_n$, then $1 = f_1(0)g_1(0) + \ldots + f_n(0)g_n(0) \in Ic(\mathscr{J})$. So

$Ic(\mathcal{J}) = A$. Then the ideal I is invertible in the local domain A. By Lemma 9.5 of Chap. 3, I is principal. Put $I = aA$ with $0 \neq a \in A$. Then $\mathcal{I} = I[X] = aA[X]$ so $d(1 + tX, 1 + tX + t^2X^2) = aA[X]$, then $\frac{d}{a}(1 + tX, 1 + tX + t^2X^2) = A[X]$, so $\frac{d}{a} \in A$. On the other hand, $\mathcal{J} = \mathcal{I}^{-1} = \frac{1}{a}A[X]$, then $\frac{1}{a}(1 - tX, 1 - tX + t^2X^2) = \frac{1}{a}A[X]$, so $\frac{a}{d}(1 - tX, 1 - tX + t^2X^2) = A[X]$. In particular, $\frac{a}{d} \in A$ and $\frac{a}{d}t \in A$. We conclude that the element $u = \frac{a}{d}$ is invertible in A. Since $ut \in A$, then $t \in A$, which is absurd.

Lemma 16.2 *Let A be an integral domain and $(A_\alpha)_{\alpha \in \Lambda}$ a family of over-rings of A such that $A = \bigcap_{\alpha \in \Lambda} A_\alpha$. For each nonzero fractional ideal J of A, $J_v = \bigcap_{\alpha \in \Lambda} J_v A_\alpha$.*

Proof

(i) If J is a fractional ideal of A, then $A : J = \bigcap_{\alpha \in \Lambda} (A_\alpha : J A_\alpha)$.

Indeed, let K be the quotient field of A and $x \in K$. Then $x \in A : J \Longleftrightarrow xJ \subseteq A = \bigcap_{\alpha \in \Lambda} A_\alpha \Longleftrightarrow \forall \alpha \in \Lambda, xJ \subseteq A_\alpha \Longleftrightarrow \forall \alpha \in \Lambda, xJA_\alpha \subseteq A_\alpha \Longleftrightarrow \forall \alpha \in \Lambda, x \in (A_\alpha : JA_\alpha) \Longleftrightarrow x \in \bigcap_{\alpha \in \Lambda} (A_\alpha : JA_\alpha)$. Then we have the equality.

(ii) We apply (i) to the fractional ideal J^{-1} of A; we find $J_v = \bigcap_{\alpha \in \Lambda} (A_\alpha : (A : J)A_\alpha)$. It suffices to prove that $\bigcap_{\alpha \in \Lambda} (A_\alpha : (A : J)A_\alpha) = \bigcap_{\alpha \in \Lambda} J_v A_\alpha$. For each $\alpha \in \Lambda$, since $A \subseteq A_\alpha$, then $A : (A : J) \subseteq A_\alpha : (A : J) = A_\alpha : (A : J)A_\alpha$. So $J_v \subseteq A_\alpha : (A : J)A_\alpha$, then $J_v A_\alpha \subseteq A_\alpha : (A : J)A_\alpha$. So $\bigcap_{\alpha \in \Lambda} J_v A_\alpha \subseteq \bigcap_{\alpha \in \Lambda} (A_\alpha : (A : J)A_\alpha)$. Conversely, let $x \in \bigcap_{\alpha \in \Lambda} (A_\alpha : (A : J)A_\alpha)$. For each $\alpha \in \Lambda$, $x(A : J)A_\alpha \subseteq A_\alpha$, then $x(A : J) \subseteq A_\alpha$. So $x(A : J) \subseteq \bigcap_{\alpha \in \Lambda} A_\alpha = A$. Then $x \in (A : (A : J)) = J_v \subseteq \bigcap_{\alpha \in \Lambda} J_v A_\alpha$. Then we have the equality. $\square$

In the following theorem, we study the transfer of the t-SFT property to the polynomial ring. This allows us to construct new examples of t-SFT domains from old examples.

Theorem 16.3 *Let A be an integrally closed integral domain. Then $A[X]$ is a t-SFT domain if and only if A is a t-SFT domain.*

Proof "$\Longrightarrow$" Let P be a t-prime ideal of A. By Corollary 11.7 of Chap. 1, $(P[X])_t = P_t[X] = P[X]$, then $P[X]$ is a t-prime ideal of $A[X]$. By hypothesis, there exist an integer $k \geq 1$ and a nonzero finitely generated ideal $\mathcal{F}$ of $A[X]$ contained in $P[X]$ such that for each $f \in P[X]$, $f^k \in \mathcal{F}_v$. We can suppose that

$\mathscr{F} = F[X]$ with F a nonzero finitely generated ideal of A contained in P. For each $a \in P, a^k \in \mathscr{F}_v = (F[X])_v = F_v[X]$, then $a^k \in F_v$. So P is a t-SFT ideal of A.

"$\Longleftarrow$" By Proposition 15.2, it suffices to show that the t-prime ideals of $A[X]$ are t-SFT. Let $Q \in t - spec(A[X])$.

First case: $Q \cap A = (0)$. Let K be the quotient field of A. By Proposition 10.3 of Chap. 3, $Q = fK[X] \cap A[X]$ with $f(X) \in A[X]$ an irreducible polynomial in $K[X]$. By Proposition 10.1 of Chap. 3, since A is integrally closed, $Q = f.c(f)^{-1}[X]$. Since $c(f)c(f)^{-1}$ is an integral ideal of the t-SFT domain A, there exist an integer $k \geq 1$ and a nonzero finitely generated ideal $F \subseteq c(f)c(f)^{-1}$ of A such that for each $a \in (c(f)c(f)^{-1})_t$, $a^k \in F_v$. There exist a nonzero finitely generated fractional ideal $J \subseteq c(f)^{-1}$ of A such that $F \subseteq c(f)J$, then $F[X] \subseteq c(f).J[X]$. By Corollary 11.7 of Chap. 1, $F_v[X] = (F[X])_v \subseteq (c(f).J[X])_v$. For each element $\overline{a}$ in the ideal $(c(f)c(f)^{-1})_t/F_v$ of the ring A/F_v, $\overline{a}^k = 0$. By Lemma 1.15, $((c(f)c(f)^{-1})_t/F_v)[X]$ is a nil ideal with bounded index in the ring $(A/F_v)[X]$. There exists an integer $m \in \mathbb{N}^*$ such that for each $h(X) \in (c(f)c(f)^{-1})_t[X]$, $h^m \in F_v[X] \subseteq (c(f).J[X])_v$. For each $g(X) \in c(f)^{-1}[X]$, $fg \in (c(f)c(f)^{-1})[X] \subseteq (c(f)c(f)^{-1})_t[X]$, then $(fg)^m \in (c(f).J[X])_v$. So $(fg)^{m+1} = fg(fg)^m \in fg(c(f).J[X])_v = (gc(f).fJ[X])_v \subseteq (fJ[X])_v = (fJA[X])_v$. But $fJA[X]$ is a finitely generated ideal contained in $fc(f)^{-1}[X] = Q$. Then Q is a t-SFT ideal.

Second case: $Q \cap A = P \neq (0)$. By Proposition 16.1, P is a t-ideal of A and $Q = P[X]$. By hypothesis, there exist $k \in \mathbb{N}^*$ and a nonzero finitely generated ideal $F \subseteq P$ of A such that for each $a \in P, a^k \in F_v$. Let $(V_\alpha)_{\alpha \in \Lambda}$ be the family of over-valuation domains of A. Since A is integrally closed, $A = \bigcap_{\alpha \in \Lambda} V_\alpha$. We will show that

for each $f \in Q, f^k \in (F[X])_v$ with $F[X] = FA[X]$ is a nonzero finitely generated ideal of $A[X]$ contained in $P[X] = Q$. Put $f(X) = a_0 + a_1 X + \ldots + a_n X^n \in P[X] = Q$. By the preceding lemma and the content formula applied aux valuation domains, $c(f^k) \subseteq \bigcap_{\alpha \in \Lambda} c(f^k)V_\alpha = \bigcap_{\alpha \in \Lambda} c(f)^k V_\alpha = \bigcap_{\alpha \in \Lambda} (a_i^k; 0 \leq i \leq n)V_\alpha \subseteq \bigcap_{\alpha \in \Lambda} F_v V_\alpha = F_v$. Then $f^k \in F_v[X] = (F[X])_v$. $\qquad \square$

Corollary 16.4 *Let A be a t-SFT integrally closed integral domain. The polynomial ring $A[X_1, \ldots, X_n]$ is a t-SFT domain.*

Proof By Proposition 6.12 of Chap. 3, if A is integrally closed, then the polynomials domain $A[X]$ is also integrally closed. We finish by induction. $\qquad \square$

17 Transfer of the *t*-SFT Property to the *t*-Flat Over-Rings

The goal of this section is the transfer of the t-SFT property to a class of over-rings of an integral domain.

Proposition 17.1 *[Barucci et al. 1994] The following assertions are equivalent to an extension $A \subseteq B$ of integral domains:*

1. $I_v \subseteq (IB)_v$ *for each nonzero finitely generated fractional ideal I of A.*
2. $I_t \subseteq (IB)_t$ *(or $(I_t B)_t = (IB)_t$) for each nonzero fractional ideal I of A.*
3. $J \cap A$ *is a t-ideal for each nonzero t-fractional ideal J of B such that $J \cap A \neq (0)$.*
4. $(IB)_t \cap A$ *is a t-ideal of A for each nonzero fractional ideal I of A.*
5. $(IB)_v \cap A$ *is a t-ideal for each nonzero finitely generated fractional ideal I of A.*

Proof $"(1) \implies (2)"$ Let J be a nonzero finitely generated fractional ideal of A contained in I. By hypothesis, $J_v \subseteq (JB)_v \subseteq (IB)_t$. By definition of t, we have $I_t \subseteq (IB)_t$.

$"(2) \implies (1)"$ By hypothesis, $I_t \subseteq (IB)_t$. But I and IB are finitely generated, then $I_v \subseteq (IB)_v$.

$"(1) \implies (3)"$ Let I be a nonzero finitely generated integral ideal of A contained in J. By hypothesis, $I_v \subseteq (IB)_v$ with IB a nonzero finitely generated fractional ideal of B contained in J. Since J is a t-ideal, $(IB)_v \subseteq J$ and $I_v \subseteq J \cap A$. Then $J \cap A$ is a t-ideal of A.

$"(3) \implies (4)"$ Clear since $(0) \neq I \cap A \subseteq (IB)_t \cap A$.

$"(4) \implies (5)"$ Since I is finitely generated, then IB is also finitely generated. By hypothesis, $(IB)_v \cap A = (IB)_t \cap A$ is a t-ideal of A.

$"(5) \implies (1)"$ By hypothesis, $I_v \subseteq \big((IB)_v \cap A\big)_t = (IB)_v \cap A \subseteq (IB)_v$. $\square$

If the equivalent conditions of the preceding proposition are satisfied for an extension $A \subseteq B$ of integral domains, then $(I_t B)_t = (IB)_t$ for each nonzero fractional ideal I of A. In this case, we say that the t-operations over the domains A and B are compatible. The following two examples show it is not always the case even if B is an over-ring of A.

Example 1 Let's see again Example 3 of Proposition 11.10, Chap. 1. Let k be a field, $A = k[X, Y]$, and $I = (X, Y)$. We know that $I_t = A$. Let (B, M) be an over-valuation domain of A such that $M \cap A = I$. Then M is a t-deal of B, then $M_t = M$. Since $IB \subseteq M$, then $(IB)_t \subseteq M_t = M$. So $I_t = A \nsubseteq (IB)_t \subseteq M$.

Example 2 (Anderson 1988) Let k be a field and $A = k[X, XY] \subset B = k[X, Y]$. The two domains A and B have the same quotient field $k(X, Y)$ since $Y = \frac{XY}{X}$ is an element in the quotient field of A. The map $\phi : B \longrightarrow A$, defined by $\phi\big(f(X, Y)\big) = f(X, XY)$, is a homomorphism onto of rings. Then $A \simeq B/\ker(\phi)$ and the two domains A and B are Noetherian. Their ideals are finitely generated, then $t = v$ over A and B. Let $I = (X, XY)A$. Then $IB = (X, XY)B = XB$ is a principal ideal, then $(IB)_v = (IB)_t = XB$. We will show that $I^{-1} = A$. Let $f \in I^{-1} \subseteq k(X, Y)$,

$f = \frac{g}{h}$ with $g, h \in k[X, Y]$, $h \neq 0$, and since $k[X, Y]$ is UFD, we may suppose that g and h are relatively prime. Then $Xf, XYf \in A$. Since $(Xf)h = Xg$, then h divides X in $k[X, Y]$. But X is irreducible, then either $h \in k^*$ or h is associated to X. Suppose that $h = aX$ with $a \in k^*$. Then g is not divisible by X. Then it has a term non-divisible by X. So $XYf = \frac{1}{a}Yg \notin k[X, XY] = A$, which is absurd. Then $h = b \in k^*$ and $f = \frac{1}{b}g \in k[X, Y]$. Since $XYf \in k[X, XY]$, then $f \in k[X, XY] = A$. So $I^{-1} = A$, then $I_t = I_v = A$. In this example, $A = I_v \nsubseteq (IB)_v = XB$ and $A = I_t \nsubseteq (IB)_t = XB$.

Now, we will give some examples where the t-operations are compatible over the two domains of the extension.

Example 3 Let A be an integral domain. By Corollary 11.7, Chap. 1, for each nonzero fractional ideal I of A, $I_v[X] = I[X]_v = (IA[X])_v$ and $I_t[X] = I[X]_t = (IA[X])_t$. Then the t-operations are compatible over the domains of the extension $A \subset A[X]$.

Example 4 (Anderson and Ryckaert 1988) Let T be an integral domain of the form $T = K + M$ where K is a field and M a maximal ideal of T. Since $K \cap M = (0)$, we have the uniqueness in writing an element of T as a sum of an element of K and an element of M. If A is a subring of K, $B = A + M$ is a subring of T. This construction is very useful to produce counterexamples. An interesting particular case is $T = K[X]$ (resp. $K[[X]]$) and $M = XK[X]$ (resp. $XK[[X]]$), where K is a field. We will show in the general case that the t-operations are compatible on the domains of the extension $A \subseteq B$. To avoid the trivial case, we suppose that $M \neq (0)$ and $A \subset K$. If we denote F the quotient field of A, then $F \subseteq K$. Let I be a nonzero fractional ideal of A. Then $(I + M)^{-1} = I^{-1} + M$. Indeed, let $x \in I^{-1} \subseteq F$. Then $xI \subseteq A$, so $x(I + M) = xI + xM \subseteq A + M = B$. Then $x \in (I + M)^{-1}$ so $I^{-1} \subseteq (I+M)^{-1}$. Since $M(I+M) \subseteq M(F+M) \subseteq M(K+M) = MT = M \subseteq B$, then $M \subseteq (I + M)^{-1}$. So $I^{-1} + M \subseteq (I + M)^{-1}$. Conversely, let $x \in (I + M)^{-1}$. Then $x(I + M) \subseteq B$. In particular, $xI \subseteq B = A + M$. Let $0 \neq b \in I \subseteq F$. Then $xb = a + m'$ with $a \in A \subseteq F$ and $m' \in M$. So $x = \frac{a}{b} + \frac{m'}{b} = \alpha + m$ with $\alpha \in F$ and $m \in M$. It suffices to show that $\alpha \in I^{-1}$. Since $(\alpha + m)I = xI \subseteq B = A + M$, then $\alpha I \subseteq A + M$ since $mI \subseteq M$. But $\alpha I \subseteq F \subseteq K$ since $\alpha \in F$ and $I \subseteq F$. Then $\alpha I \subseteq (A + M) \cap F = A$ and $\alpha \in I^{-1}$.

The preceding equality applied to the fractional ideal I^{-1} of A, then to I gives $I_v + M = (I^{-1})^{-1} + M = (I^{-1} + M)^{-1} = ((I + M)^{-1})^{-1} = (I + M)_v$. On the other hand, $IB = I + M$ for each nonzero fractional ideal I of A. Indeed, $IB = I(A + M) = IA + IM = I + M$ since $I \subseteq F \subseteq K$, then the nonzero elements of I are invertible in T so $IM = M$. Then $(IB)_v = (I + M)_v = I_v + M$.

In the sequel, we will be interested in the extensions of domains of the form $A \subseteq A_{\mathscr{S}}$ where $\mathscr{S}$ is a multiplicative system of integral ideals of an integral domain A. We start by generalizing Lemma 10.14 of Chap. 3.

Proposition 17.2 (Querré 1975 and Kang 1989) *Let A be an integral domain, $\mathscr{S}$ a multiplicative system of integral ideals of A, and I and J two nonzero fractional ideals of A.*

1. *We have $(I : J)_{\mathscr{S}} \subseteq (I_{\mathscr{S}} : J_{\mathscr{S}})$. If J is finitely generated, then $(I : J)_{\mathscr{S}} = (I_{\mathscr{S}} : J_{\mathscr{S}}) = (I_{\mathscr{S}} : JA_{\mathscr{S}})$.*
2. *If J is finitely generated, then $J_v A_{\mathscr{S}} \subseteq (J_v)_{\mathscr{S}} \subseteq (J_{\mathscr{S}})_v = (JA_{\mathscr{S}})_v$.*
3. *If J is finitely generated, then $(J_v A_{\mathscr{S}})_v = (J_{\mathscr{S}})_v = (JA_{\mathscr{S}})_v$.*

Proof

(1) Let $x \in (I : J)_{\mathscr{S}}$ and $H \in \mathscr{S}$ be such that $xH \subseteq (I : J)$. If $y \in J_{\mathscr{S}}$ and $H' \in \mathscr{S}$ are such that $yH' \subseteq J$, then $xyHH' \subseteq (I : J)J \subseteq I$. Since $HH' \in \mathscr{S}$, then $xy \in I_{\mathscr{S}}$. This is for each $y \in J_{\mathscr{S}}$. Then $x \in (I_{\mathscr{S}} : J_{\mathscr{S}})$. We obtain the inclusion $(I : J)_{\mathscr{S}} \subseteq (I_{\mathscr{S}} : J_{\mathscr{S}})$.

Suppose now that $J = x_1 A + \ldots + x_n A$ is finitely generated. We have the inclusions $(I : J)_{\mathscr{S}} \subseteq (I_{\mathscr{S}} : J_{\mathscr{S}}) \subseteq (I_{\mathscr{S}} : JA_{\mathscr{S}})$ because $J \subseteq J_{\mathscr{S}}$ since $HJ \subseteq J$ for each $H \in \mathscr{S}$. We must show that $(I_{\mathscr{S}} : JA_{\mathscr{S}}) \subseteq (I : J)_{\mathscr{S}}$. Let $y \in (I_{\mathscr{S}} : JA_{\mathscr{S}})$. Then $yJA_{\mathscr{S}} \subseteq I_{\mathscr{S}}$. In particular, $yJ \subseteq I_{\mathscr{S}}$. There exist $H_1, \ldots, H_n \in \mathscr{S}$ such that $yx_i H_i \subseteq I, 1 \le i \le n$. Then $H = H_1 \ldots H_n \in \mathscr{S}$ and $yJH \subseteq I$, so $yH \subseteq (I : J)$ and $y \in (I : J)_{\mathscr{S}}$. We have the inclusion $(I_{\mathscr{S}} : JA_{\mathscr{S}}) \subseteq (I : J)_{\mathscr{S}}$.

(2) By (1), $J_v A_{\mathscr{S}} \subseteq (J_v)_{\mathscr{S}} = (A : (A : J))_{\mathscr{S}} \subseteq (A_{\mathscr{S}} : (A : J)_{\mathscr{S}}) = (A_{\mathscr{S}} : (A_{\mathscr{S}} : J_{\mathscr{S}})) = (J_{\mathscr{S}})_v = (A_{\mathscr{S}} : (A_{\mathscr{S}} : JA_{\mathscr{S}})) = (JA_{\mathscr{S}})_v$. Then $J_v A_{\mathscr{S}} \subseteq (J_v)_{\mathscr{S}} \subseteq (J_{\mathscr{S}})_v = (JA_{\mathscr{S}})_v$.

(3) We apply the v-operation to the inclusions $J_v A_{\mathscr{S}} \subseteq (J_{\mathscr{S}})_v = (JA_{\mathscr{S}})_v$; we find $(J_v A_{\mathscr{S}})_v \subseteq (J_{\mathscr{S}})_v = (JA_{\mathscr{S}})_v \subseteq (J_v A_{\mathscr{S}})_v$. Then we have the required equalities. □

Let $\mathscr{S}$ be a multiplicative system of integral ideals of an integral domain A. The following corollary shows that the t-operations are compatible on the domains $A \subset A_{\mathscr{S}}$.

Corollary 17.3 *Let A be an integral domain and $\mathscr{S}$ a multiplicative system of integral ideals of A. The following assertions are equivalent and satisfied:*

1. *We have $I_v \subseteq (IA_{\mathscr{S}})_v$ for each nonzero finitely generated fractional ideal I of A.*
2. *We have $(I_t A_{\mathscr{S}})_t = (IA_{\mathscr{S}})_t$ for each nonzero fractional ideal I of A.*
3. *If J is a t-ideal of $A_{\mathscr{S}}$ with $J \cap A \ne (0)$, then $J \cap A$ is a t-ideal of A.*

Proof The assertion (1) follows from (2) or (3) of the preceding proposition. The equivalence between the three assertions of the corollary follows from Proposition 17.1. □

Definition 17.4 Let B be an over-ring of an integral domain A. We say that B is a t-flat over-ring of A if for each $M \in t - Max(B)$, $B_M = A_{A \cap M}$.

Lemma 17.5 *(Kwak and Park 1995) Let* $A \subseteq B$ *be an extension of integral domains and* I *a nonzero integral t-ideal of* A *such that* $(IB)_t \neq B$. *There exists a t-prime ideal* P *of* A *containing* I *such that* $(PB)_t \neq B$.

Proof By Proposition 11.2 of Chap. 1, since $(IB)_t \neq B$, there exists $M \in t - Max(B)$ such that $(IB)_t \subseteq M$. Since $M \cap A$ is a prime ideal of A containing I, there exists a prime ideal P of A minimal over I and included in $M \cap A$. By Proposition 10.2 of Chap. 3, P is a t-prime ideal of A and $(PB)_t \subseteq M_t = M \subset B$. Then $(PB)_t \neq B$.   □

In the following proposition, we give various characterizations for t-flat over-rings of an integral domain.

Proposition 17.6 *[Kwak and Park 1995] Let* B *be an over-ring of an integral domain* A. *The following assertions are equivalent:*

1. *For each t-prime ideal* P *of* A, *either* $(PB)_t = B$ *or* $B \subseteq A_P$.
2. *For each element* $\frac{x}{y} \in B$ *with* $x, y \in A$, $y \neq 0$, $\left((yA :_A x)B\right)_t = B$.
3. B *is a t-flat over-ring of* A.
4. $B = \bigcap \left\{ A_{M \cap A}; \ M \in t - Max(B) \right\}$.
5. *There exists a multiplicative system* $\mathscr{S}$ *of nonzero integral ideals of* A *such that* $B = A_{\mathscr{S}}$ *and* $(IB)_t = B$ *for each* $I \in \mathscr{S}$.

Proof "(1) $\Longrightarrow$ (2)" Let $\frac{x}{y} \in B$ with $x, y \in A$, $y \neq 0$. Then $(yA :_A x)$ is a t-ideal of A. Indeed, let F be a nonzero finitely generated ideal of A contained in $(yA :_A x)$. Then $xF \subseteq yA$, so $xF_v \subseteq yA$, then $F_v \subseteq (yA :_A x)$. So $(yA :_A x)_t \subseteq (yA :_A x)$. Suppose that $\left((yA :_A x)B\right)_t \neq B$. By the preceding lemma, there exists a t-prime ideal P of A such that $(yA :_A x) \subseteq P$ and $(PB)_t \neq B$. By (1), $B \subseteq A_P$, then $\frac{x}{y} \in A_P$. Put $\frac{x}{y} = \frac{a}{s}$ with $a \in A$ and $s \in A \setminus P$. Since $sx = ya \in yA$, then $s \in (yA :_A x) \subseteq P$, which is absurd. Then $\left((yA :_A x)B\right)_t = B$.

"(2) $\Longrightarrow$ (3)" Let $M \in t - Max(B)$. We will show that $B_M = A_{M \cap A}$. Let $\frac{x}{y} \in B_M$ with $x \in B$ and $y \in B \setminus M$. Since B is an over-ring of A, $x = \frac{u}{s}$ and $y = \frac{v}{s}$ with $u, v, s \in A$, $s \neq 0$. By Proposition 11.1 of Chap. 1 and hypothesis (2),

$$\left((sA :_A u)(sA :_A v)B\right)_t = \left((sA :_A u)B.(sA :_A v)B\right)_t = \left(\left((sA :_A u)B\right)_t.\left((sA :_A v)B\right)_t\right)_t = B_t = B.$$

Then $(sA :_A u)(sA :_A v) \not\subseteq M \cap A$. Let $z \in (sA :_A u)(sA :_A v) \setminus M \cap A$. Since $(sA :_A u)(sA :_A v) \subseteq (sA :_A u) \cap (sA :_A v)$, then $zu \in sA$ and $zv \in sA$, so $zx \in A$ and $zy \in A$ and $zy \notin M \cap A$. Then $\frac{x}{y} = \frac{zx}{zy} \in A_{M \cap A}$ so $B_M \subseteq A_{M \cap A}$. The inverse inclusion is clear.

"(3) $\Longrightarrow$ (4)" By Proposition 10.13 of Chap. 3, $B = \bigcap \{B_M; \ M \in t - Max(B)\} = \bigcap \{A_{M \cap A}; \ M \in t - Max(B)\}$ because by (3), $B_M = A_{M \cap A}$ for each $M \in t - Max(B)$.

"(4) $\Longrightarrow$ (1)" Let P be a t-prime ideal of A such that $(PB)_t \neq B$. We will show that $B \subseteq A_P$. By Proposition 11.2 of Chap. 1, there exists $M \in t - Max(B)$ such that $(PB)_t \subseteq M$, then $P \subseteq M \cap A$. By (4), $B \subseteq A_{M \cap A} \subseteq A_P$.

"(4) $\Longrightarrow$ (5)" Let $\mathscr{S}$ be the set of the integral ideals I of A such that $I \not\subseteq M \cap A$ for each $M \in t - Max(B)$. Since $A \in \mathscr{S}$, then $\mathscr{S} \neq \emptyset$. We will show that $\mathscr{S}$ is stable by multiplication. Let $I, J \in \mathscr{S}$. Suppose that there exists $M \in t - Max(B)$ such that $IJ \subseteq M \cap A$. Since the ideal $M \cap A$ of A is prime, $I \subseteq M \cap A$ so $J \subseteq M \cap A$, which is absurd. Then $IJ \in \mathscr{S}$. Let $I \in \mathscr{S}$. For each $M \in t - Max(B)$, $I \not\subseteq M \cap A$, then $IB \not\subseteq M$, so $(IB)_t \not\subseteq M$. By Proposition 11.2 of Chap. 1, $(IB)_t = B$. We will show that $B = A_{\mathscr{S}}$. Let $x \in A_{\mathscr{S}}$. There exists $I \in \mathscr{S}$ such that $xI \subseteq A$, then $xIB \subseteq B$, so $x(IB)_t \subseteq B$. But $(IB)_t = B$, then $xB \subseteq B$ and $x \in B$. Conversely, let $x \in B = \bigcap \{A_{M \cap A}; \ M \in t - Max(B)\}$. Suppose that there exists $M \in t - Max(B)$ such that $(A :_A x) \subseteq M \cap A$. Since $x \in A_{M \cap A}$, $x = \frac{a}{s}$ with $a \in A$ and $s \in A \setminus M \cap A$. Then $sx = a \in A$, so $s \in (A :_A x) \subseteq M \cap A$, which is absurd. Then $(A :_A x) \in \mathscr{S}$ and since $(A :_A x)x \subseteq A$, then $x \in A_{\mathscr{S}}$.

"(5) $\Longrightarrow$ (4)" We will show the equality $B = \bigcap \{A_{M \cap A}; \ M \in t - Max(B)\}$. For each $M \in t - Max(B)$, $A_{M \cap A} \subseteq B_M$. By Proposition 10.13 of Chap. 3, $\bigcap \{A_{M \cap A}; \ M \in t - Max(B)\} \subseteq \bigcap \{B_M; \ M \in t - Max(B)\} = B$. Conversely, let $x \in B = A_{\mathscr{S}}$. There exists $I \in \mathscr{S}$ such that $xI \subseteq A$. Let $M \in t - Max(B)$. By definition of $\mathscr{S}$, $I \not\subseteq M \cap A$. Let $s \in I \setminus M \cap A$. Then $sx \in xI \subseteq A$ and $x = \frac{sx}{s} \in A_{A \cap M}$. Then $x \in \bigcap \{A_{M \cap A}; \ M \in t - Max(B)\}$. $\qquad \square$

Example 1 Let A be an integral domain. Each flat over-ring B of A is t-flat.

Indeed, let $x = \frac{a}{b} \in B$ with $a, b \in A$, $b \neq 0$. Then $(A :_A x) = (bA :_A a)$. Since B is flat, $B = (A :_A x)B = (bA :_A a)B$. Then $\big((bA :_A a)B\big)_t = B_t = B$. By the preceding proposition, B is t-flat. The converse of this example is false. Cf. Exercise 15.

Example 2 Let's see again Example 2.7 of Lemma 2.10, Chap. 3. Consider the domains $A = \mathbb{Z}[\sqrt{5}\,] \subset \mathbb{Z}\big[\frac{1+\sqrt{5}}{2}\big] = B$. Then B is an over-ring of A. Let $x = \frac{1+\sqrt{5}}{2} \in B$, the golden number. We have always shown that $(A :_A x) = (2, 1 + \sqrt{5})A$ and $(A :_A x)B = 2B \neq B$. Then B is not a flat over-ring of A. In fact, B is not even t-flat since $\big((A :_A x)B\big)_t = 2B \neq B$.

Lemma 17.7 *[Baghdadi 2002] The radical of an integral t-ideal is a t-ideal.*

Proof Let $(0) \neq I$ be an integral t-ideal. We will show that $(\sqrt{I}\,)_t = \sqrt{I}$. Let J be a nonzero finitely generated subideal of $\sqrt{I}$. There exists an integer $n \geq 1$ such that $J^n \subseteq I$ with J^n a finitely generated ideal. By Proposition 11.1 of Chap. 1, $(J_v)^n \subseteq (J^n)_v \subseteq I_t = I$, then $J_v \subseteq \sqrt{I}$, so $(\sqrt{I}\,)_t \subseteq \sqrt{I}$. $\qquad \square$

Example Let (A, M) be a nondiscrete valuation domain of rank one, defined by a valuation v with values in $\mathbb{R}$. Then $M.A[[X]]$ is a t-ideal of the domain $A[[X]]$.

Indeed, let $0 \neq m_0 \in M$ be a fixed element. For each $f \in M.A[[X]]$, there exist $m \in M$ and $f' \in A[[X]]$ such that $f = mf'$. Let $l \in \mathbb{N}^*$ be such that $lv(m) \geq v(m_0)$. Then $f^l = m_0(\frac{m^l}{m_0})f'^l \in m_0 A[[X]]$, so $M.A[[X]] \subseteq \sqrt{m_0 A[[X]]} \subseteq$

$\sqrt{MA[[X]]}$. By Example 2 (2) of Corollary 14.2, the ideal $M.A[[X]]$ is radical. Then $M.A[[X]] = \sqrt{m_0 A[[X]]}$ is the radical of a principal ideal. By the preceding lemma, $M.A[[X]]$ is a t-ideal.

Now, we are ready to prove the main result of this section.

Theorem 17.8 (Kang and Park 2006) *Let A be a t-SFT integral domain and B a t-flat over-ring of A. Then B is a t-SFT domain.*

Proof By Proposition 17.6, there exists a multiplicative system of nonzero integral ideals $\mathscr{S}$ of A such that $B = A_{\mathscr{S}}$ and $(IB)_t = B$ for each $I \in \mathscr{S}$. Let Q be a t-prime ideal of B and $P = Q \cap A$. By Corollary 17.3 (3), P is a t-prime ideal of A. We will show that $Q = P_{\mathscr{S}}$. Let $x \in Q \subseteq B$. There exists $I \in \mathscr{S}$ such that $xI \subseteq A$, then $xI \subseteq A \cap Q = P$. So $x \in P_{\mathscr{S}}$. Conversely, let $x \in P_{\mathscr{S}}$. There exists $I \in \mathscr{S}$ such that $xI \subseteq P$, then $xIB \subseteq PB$, so $x(IB)_t \subseteq (PB)_t \subseteq Q_t = Q$. But $(IB)_t = B$, then $xB \subseteq Q$ and $x \in Q$. Then we have the equality $Q = P_{\mathscr{S}}$. Since P is a t-ideal of A and A is a t-SFT domain, there exist a nonzero finitely generated ideal $F \subseteq P$ of A and an integer $k \geq 1$ such that for each $a \in P$, $a^k \in F_v$. Let $q \in Q$ be any fixed element. There exists $I \in \mathscr{S}$ such that $qI \subseteq P$. For each $a \in I$, $qa \in P$, then $(qa)^k \in F_v$. The ideal $J = \big((FB)_v :_B q^k\big)$ of B is a t-ideal. Indeed, let H be a nonzero finitely generated ideal of B included in J. Then $Hq^k \subseteq (FB)_v$, so $H_v q^k \subseteq (FB)_v$, then $H_v \subseteq \big((FB)_v :_B q^k\big) = J$. So $J_t = J$. By the preceding lemma, the radical $\sqrt{J}$ of J is a t-ideal of B. By Corollary 17.3, $F_v \subseteq (FB)_v$. Then for each $a \in I$, $(qa)^k \in (FB)_v$, so $a^k \in J$. Then $I \subseteq \sqrt{J}$, so $IB \subseteq \sqrt{J}$. Since $I \in \mathscr{S}$, $B = (IB)_t \subseteq (\sqrt{J})_t = \sqrt{J}$. Then $J = B$ so $q^k \in (FB)_v$. This is for each $q \in Q$. Since FB is finitely generated, then Q is a t-SFT ideal of B.   □

18 Chains Condition in the t-SFT Domains

Lemma 18.1 (Baghdadi 2002) *Let H and L be two nonzero integral ideals of an integral domain. Then $\sqrt{(HL)_t} = \sqrt{(\sqrt{H_t}\sqrt{L_t})_t}$.*

Proof Since $HL \subseteq \sqrt{H_t}\sqrt{L_t}$, then $\sqrt{(HL)_t} \subseteq \sqrt{(\sqrt{H_t}\sqrt{L_t})_t}$. Conversely, by Lemma 10.8 of Chap. 3, $(\sqrt{H_t}\sqrt{L_t})_t = \bigcup \{(IJ)_v; \ 0 \neq I \subseteq \sqrt{H_t}, 0 \neq J \subseteq \sqrt{L_t}$, I and J finitely generated ideals $\}$. For each couple (I, J) of such ideals, there exists an integer $n \geq 1$ such that $I^n \subseteq H_t$ and $J^n \subseteq L_t$. Since I^n and J^n are finitely generated, there exist nonzero finitely generated ideals $H_0 \subseteq H$ and $J_0 \subseteq L$ such that $I^n \subseteq (H_0)_v$ and $J^n \subseteq (J_0)_v$. By Proposition 11.1 of Chap. 1, $(IJ)^n = I^n J^n \subseteq (H_0)_v (J_0)_v \subseteq (H_0 J_0)_v \subseteq (HL)_t$. Then $IJ \subseteq \sqrt{(HL)_t}$. By Lemma 17.7, $\sqrt{(HL)_t}$ is a t-ideal. Since IJ is a finitely generated subideal of $\sqrt{(HL)_t}$, then $(IJ)_v \subseteq \sqrt{(HL)_t}$. So $(\sqrt{H_t}\sqrt{L_t})_t \subseteq \sqrt{(HL)_t}$, then $\sqrt{(\sqrt{H_t}\sqrt{L_t})_t} \subseteq \sqrt{(HL)_t}$.

□

Proposition 18.2 (Baghdadi 2002) *The following assertions are equivalent to an integral domain A:*

1. *For each t-prime ideal P of A, there exists a nonzero finitely generated ideal F of A such that $P = \sqrt{F_v}$.*
2. *For each radical t-ideal I of A, there exists a finitely generated ideal F of A such that $I = \sqrt{F_v}$.*
3. *The domain A satisfies the acc for the radical t-ideals.*

Proof "(1) $\Longrightarrow$ (2)" Let $\mathscr{F}$ be the set of the radical t-ideals I of A such that there exists no finitely generated ideal F of A satisfying $I = \sqrt{F_v}$. Suppose that $\mathscr{F} \neq \emptyset$. Then $\mathscr{F}$ is inductive for the inclusion. Indeed, let $(I_\lambda)_{\lambda \in \Lambda}$ be a totally ordered family of elements of $\mathscr{F}$. As in the proof of Proposition 15.2, $I = \bigcup_{\lambda \in \Lambda} I_\lambda$ is radical t-ideal.

Suppose that there exists a finitely generated ideal F of A such that $I = \sqrt{F_v}$. Then $F \subseteq I$, so there exists $\lambda_0 \in \Lambda$ such that $F \subseteq I_{\lambda_0}$. Since I_{λ_0} is a t-ideal, then $F_v \subseteq I_{\lambda_0}$, so $\sqrt{F_v} \subseteq I_{\lambda_0}$, then $\sqrt{F_v} = I_{\lambda_0}$, which is absurd. Then I bounds the family in $\mathscr{F}$. By Zorn's lemma, $\mathscr{F}$ has a maximal element P. It is a radical t-ideal such that there is no finitely generated ideal F of A satisfying $P = \sqrt{F_v}$. It suffices to show that P is prime to obtain a contradiction with (1). Suppose that there exist $a, b \in A \setminus P$ such that $ab \in P$. By Lemma 17.7, $\sqrt{(P + aA)_t}$ and $\sqrt{(P + bA)_t}$ are radical t-ideals strictly containing P. By maximality of P, there exist finitely generated ideals H and J of A such that $\sqrt{(P + aA)_t} = \sqrt{H_v}$ and $\sqrt{(P + bA)_t} = \sqrt{J_v}$. By the preceding lemma, we have the following equalities $\sqrt{((P + aA)(P + bA))_t} = \sqrt{(\sqrt{(P+aA)_t}\sqrt{(P+bA)_t})_t} = \sqrt{(\sqrt{H_v}\sqrt{J_v})_t} = \sqrt{(\sqrt{H_t}\sqrt{J_t})_t} = \sqrt{(HJ)_t} = \sqrt{(HJ)_v}$. On the other hand, $P \subseteq \sqrt{P + aA} \cap \sqrt{P + bA} = \sqrt{(P+aA)(P+bA)} \subseteq \sqrt{((P+aA)(P+bA))_t} = \sqrt{(P^2 + aP + bP + abA)_t} \subseteq \sqrt{P_t} = \sqrt{P} = P$, then $\sqrt{((P+aA)(P+bA))_t} = P$. So $P = \sqrt{(HJ)_v}$, which is absurd.

"(2) $\Longrightarrow$ (3)" Let (I_n) be an increasing sequence of radical t-ideals of A. As before, $I = \bigcup I_n$ is a radical t-ideal. By (2), there exists a nonzero finitely generated ideal J of A such that $I = \sqrt{J_v}$. Let n_0 be such that $J \subseteq I_{n_0}$. Then $J_v \subseteq (I_{n_0})_t = I_{n_0}$. So $I = \sqrt{J_v} \subseteq \sqrt{I_{n_0}} = I_{n_0} \subseteq I$, then $I = I_{n_0}$. So (I_n) is stationary from the rank n_0.

"(3) $\Longrightarrow$ (1)" Suppose that there exists a t-prime ideal P of A which is not of the form $P = \sqrt{F_v}$ with F a finitely generated ideal of A. Let $0 \neq x_1 \in P$. Then $\sqrt{(x_1)_v} \subset P$. Let $x_2 \in P \setminus \sqrt{(x_1)_v}$. Then $\sqrt{(x_1)_v} \subset \sqrt{(x_1, x_2)_v} \subset P$, ... By induction, we construct a strictly increasing sequence of radical t-ideals (Lemma 17.7) of A: contradicts (3). $\qquad\square$

Corollary 18.3 (Kang and Park 2006) *A t-SFT domain satisfies the acc for the radical t-ideals.*

Proof Let I be a radical t-ideal of a t-SFT integral domain A. There exist a nonzero finitely generated ideal $F \subseteq I$ of A and an integer $k \geq 1$ such that for each $a \in I$, $a^k \in F_v$, then $I \subseteq \sqrt{F_v}$. Conversely, $F \subseteq I$, then $F_v \subseteq I_t = I$, so $\sqrt{F_v} \subseteq \sqrt{I} = I$. Then $I = \sqrt{F_v}$. By the preceding proposition, we have the result.   □

19   Application to Formal Power Series

The results of this section are due to Dobbs and Houston (1995). We start by a very useful result.

Proposition 19.1 *Let I be a nonzero fractional ideal of an integral domain A.*

1. $(I.A[[X]])^{-1} = I^{-1}[[X]] = (I[[X]])^{-1}$.
2. $(I.A[[X]])_v = I_v[[X]] = (I[[X]])_v$.

Proof

(1) Let $u \in (I.A[[X]])^{-1}$. Then $uI \subseteq A[[X]]$. Let $0 \neq x \in I$, $x = \frac{a}{b}$ with $a, b \in A \setminus (0)$. Then $ux \in A[[X]]$, so $u \in \frac{b}{a} A[[X]] \subseteq K[[X]]$, where K

is the quotient field of A. Put $u = \sum_{i:0}^{\infty} u_i X^i \in K[[X]]$. For each $a \in I$,

$au = \sum_{i:0}^{\infty} a u_i X^i \in A[[X]]$, then $au_i \in A$ so $u_i \in (A : I) = I^{-1}$. Then

$u = \sum_{i:0}^{\infty} u_i X^i \in I^{-1}[[X]]$. Then we have the inclusion $(I.A[[X]])^{-1} \subseteq$

$I^{-1}[[X]]$. Let $u = \sum_{i:0}^{\infty} u_i X^i \in I^{-1}[[X]]$. For each $f = \sum_{i:0}^{\infty} a_i X^i \in I[[X]]$,

$uf = \sum_{n:0}^{\infty} b_n X^n$ with $b_n = \sum_{i+j=n} u_i a_j \in I^{-1}I \subseteq A$, so $uf \in A[[X]]$ and

$u \in (I[[X]])^{-1}$. Then we have the inclusion $I^{-1}[[X]] \subseteq (I[[X]])^{-1}$. On the other hand, $I.A[[X]] \subseteq I[[X]]$, so $(I[[X]])^{-1} \subseteq (I.A[[X]])^{-1}$. We obtain the inclusions $(I.A[[X]])^{-1} \subseteq I^{-1}[[X]] \subseteq (I[[X]])^{-1} \subseteq (I.A[[X]])^{-1}$. So we have equalities.

(2) By (1), $(I.A[[X]])^{-1} = I^{-1}[[X]] = (I[[X]])^{-1}$. Then $((I.A[[X]])^{-1})^{-1} = (I^{-1}[[X]])^{-1} = ((I[[X]])^{-1})^{-1}$, so $(I.A[[X]])_v = (I^{-1}[[X]])^{-1} = (I[[X]])_v$. But by (1), $(I^{-1}[[X]])^{-1} = (I^{-1})^{-1}[[X]] = I_v[[X]]$. Then we have the result.   □

Corollary 19.2 *Let I be a nonzero fractional ideal of an integral domain A. Then I is divisorial in A if and only if $I[[X]]$ is divisorial in $A[[X]]$.*

Proof By the preceding proposition, $I_v = I \iff I_v[[X]] = I[[X]] \iff (I[[X]])_v = I[[X]]$.   □

Corollary 19.3 *Let A be an integral domain and $\mathscr{I}$ an integral t-ideal of $A[[X]]$ such that $\mathscr{I} \cap A \neq (0)$. Then $\mathscr{I} \cap A$ is a t-ideal of A.*

Proof Let I be a nonzero (finitely generated) fractional ideal of A. By Proposition 19.1, $I_v \subset I_v[[X]] = (I.A[[X]])_v$. Then the t-operations are compatibles over the domains $A \subset A[[X]]$. By Proposition 17.1, if $\mathscr{I}$ is an integral t-ideal of $A[[X]]$ such that $\mathscr{I} \cap A \neq (0)$, then $\mathscr{I} \cap A$ is a t-ideal of A.   □

Other Proof Let J be a nonzero finitely generated subideal of $\mathscr{I} \cap A$. By Proposition 19.1, $J_v \subset J_v[[X]] = (J.A[[X]])_v$. Since $J.A[[X]]$ is a finitely generated subideal of $\mathscr{I}$, then $(J.A[[X]])_v \subseteq \mathscr{I}_t = \mathscr{I}$, so $J_v \subseteq \mathscr{I}$, then $J_v \subseteq \mathscr{I} \cap A$. This is for each nonzero finitely generated subideal J of $\mathscr{I} \cap A$, then $(\mathscr{I} \cap A)_t \subseteq \mathscr{I} \cap A$, so $(\mathscr{I} \cap A)_t = \mathscr{I} \cap A$.   □

Example 1 For each integral domain A, $XA[[X]]$ is a t-ideal and $XA[[X]] \cap A = (0)$.

Example 2 In contrast to the polynomial case, with the notions and the hypotheses of the preceding corollary, we do not always have $\mathscr{I} = (\mathscr{I} \cap A)[[X]]$. Indeed, let (A, M) be a nondiscrete valuation domain of rank one. We showed, in the example of Lemma 17.7, that $M.A[[X]]$ is a t-ideal of the domain $A[[X]]$. It is also clear that $(M.A[[X]]) \cap A = M$. But $M.A[[X]] \neq M[[X]]$.

In the following example, we show that we may have the equality $\mathscr{I} = (\mathscr{I} \cap A)[[X]]$, for a t-ideal $\mathscr{I}$ of $A[[X]]$, under some supplementary hypotheses.

Example 3 Let A be a Noetherian integral domain and Q a t-prime ideal of $A[[X]]$ such that $Q \cap A = \mathscr{P} \neq (0)$. Then $\mathscr{P}$ is a t-prime ideal of A and $Q = \mathscr{P}[[X]]$. Indeed, by the preceding corollary, $\mathscr{P}$ is t-prime in A. By Lemma 1.16, since A is Noetherian, $\mathscr{P}[[X]] = \mathscr{P}.A[[X]] \subseteq Q$. We will show the inverse inclusion. By hypothesis, there exists $0 \neq a \in \mathscr{P}$. Then $aQ^{-1} \subseteq QQ^{-1} \subseteq A[[X]]$, so $Q^{-1} \subseteq K[[X]]$ where K is the quotient field of A. Let $f = \sum_{i:0}^{\infty} b_i X^i \in Q$ and $k = \sum_{i:0}^{\infty} u_i X^i \in Q^{-1}$. Then $fk = b_0 u_0 + (b_0 u_1 + b_1 u_0)X + \ldots \in A[[X]]$. So $b_0 u_0 \in A$, then $b_0^2 u_1 \in A$ and by induction, $b_0^n u_{n-1} \in A$ for each $n \geq 1$. Let $c(k)$ be the sub-A-module of K generated by the coefficients of k. Since $ak \in aQ^{-1} \subseteq A[[X]]$, then $ac(k) \subseteq A$, so $c(k)$ is a fractional ideal of A. Since A is Noetherian, $c(k)$ is finitely generated. There exists $m \in \mathbb{N}^*$ (depending on k) such that $b_0^m c(k) \subseteq A$, then $b_0^m k \in A[[X]]$. By Corollary 5.2 of Chap. 1, the domain $A[[X]]$ is Noetherian. The fractional ideal Q^{-1} of $A[[X]]$ is then finitely generated by $k_1, \ldots, k_s$. There exists $r \in \mathbb{N}^*$ such that $b_0^r k_1, \ldots, b_0^r k_s \in A[[X]]$, then $b_0^r Q^{-1} \subseteq A[[X]]$. So $b_0^r \in (Q^{-1})^{-1} = Q_v = Q_t = Q$, then $b_0^r \in Q \cap A = \mathscr{P}$, so $b_0 \in \mathscr{P}$. We conclude that $b_1 X + b_2 X^2 + \ldots \in Q$. To continue, note that $(a, X)^{-1} = A[[X]]$. Indeed, let $t \in (a, X)^{-1} \subseteq K((X))$. Then $at \in A[[X]]$, so $t \in K[[X]]$. Also, $Xt \in A[[X]]$ with

$t \in K[[X]]$, then $t \in A[[X]]$. So $(a, X)^{-1} = A[[X]]$. Suppose now that $X \in Q$. Then $A[[X]] = (a, X)_v \subseteq Q_t = Q$, which is absurd since Q is a prime ideal of $A[[X]]$. Then $X \notin Q$ so $b_1 + b_2 X + \ldots \in Q$. We show as before that $b_1 \in \mathscr{P}$. By induction, we show that $f \in \mathscr{P}[[X]]$. Then $Q \subseteq \mathscr{P}[[X]]$ so we have the equality $Q = \mathscr{P}[[X]]$.

Lemma 19.4 *Let I be a nonzero fractional ideal of an integral domain A. Then $I_t.A[[X]] \subseteq (I.A[[X]])_t \subseteq I_t[[X]]$.*

Proof

(1) The inclusion $I_t.A[[X]] \subseteq (I.A[[X]])_t$ follows from the preceding corollary and Proposition 17.1 (compatibility of the t-operations over the domains $A \subset A[[X]]$).

   Other Proof Let $a \in I_t$. There exists a nonzero finitely generated subideal J of I such that $a \in J_v$. By Proposition 19.1, $J_v \subset J_v[[X]] = (JA[[X]])_v$. Since $JA[[X]]$ is a nonzero finitely generated subideal of $IA[[X]]$, then $(JA[[X]])_v \subseteq (IA[[X]])_t$, so $a \in (IA[[X]])_t$. Then $I_t \subseteq (IA[[X]])_t$, so $I_t.A[[X]] \subseteq (I.A[[X]])_t$.

(2) Let J be a nonzero finitely generated subideal of $I.A[[X]]$. There exists a nonzero finitely generated subideal F of I such that $J \subseteq F.A[[X]]$. By Proposition 19.1, $J_v \subseteq (F.A[[X]])_v = F_v[[X]] \subseteq I_t[[X]]$. Then $(I.A[[X]])_t \subseteq I_t[[X]]$. □

Example 1 The three inclusions may be equalities. If I is a principal ideal, then $I.A[[X]]$ is also principal. Then $I_t A[[X]] = IA[[X]] = (IA[[X]])_t = I[[X]] = I_t[[X]]$.

Example 2 The second inclusion of the lemma may be strict. Let (A, M) be a nondiscrete valuation domain of rank one. Then M et $M.A[[X]]$ are t-ideals. So $M_t A[[X]] = MA[[X]] = (MA[[X]])_t \subset M[[X]] = M_t[[X]]$.

We prove now the analogue of Theorem 1.18 for the t-SFT domains.

Theorem 19.5 (Kang and Park 2006) *The following assertions are equivalent to an integral domain A:*

1. *There exists a nonzero integral ideal I of A such that $I[[X]] \not\subseteq \sqrt{(IA[[X]])_t}$.*
2. *There exists $P \in t - Spec(A)$ such that $P[[X]] \neq \sqrt{(PA[[X]])_t}$.*
3. *A is not a t-SFT domain.*

Proof "(1) $\Longrightarrow$ (2)" Since $I[[X]] \not\subseteq \sqrt{(IA[[X]])_t}$, there exists a prime ideal Q of $A[[X]]$ minimal over $(IA[[X]])_t$ such that $I[[X]] \not\subseteq Q$. By Proposition 10.2 of Chap. 3, Q is a t-ideal. Let $P = Q \cap A$. Then $I \subseteq Q \cap A = P$, so $P \neq (0)$. By Corollary 19.3, $P \in t - Spec(A)$. Since $I[[X]] \subseteq P[[X]]$, then $P[[X]] \not\subseteq Q$. On the other hand, $P.A[[X]] \subseteq Q$, then $(P.A[[X]])_t \subseteq Q_t = Q$, so $\sqrt{(P.A[[X]])_t} \subseteq Q$, then $P[[X]] \neq \sqrt{(PA[[X]])_t}$.

$"(2) \implies (3)"$ Suppose that P is a t-SFT ideal of A. There exist a finitely generated ideal $C \subseteq P$ of A and an integer $k \geq 1$ such that for each $b \in P$, $b^k \in C_v$. Consider the quotient ring $\bar{A} = A/C_v$ and its ideal $\bar{P} = P/C_v$. In fact $C \subseteq P$, then $C_v \subseteq P_t = P$. For each $b \in P$, $\bar{b}^k = 0$. By Lemma 1.15, all the elements of $\bar{P}[[X]]$ are nilpotent in $\bar{A}[[X]]$. By using Proposition 19.1 and Lemma 19.4, $P[[X]] \subseteq \sqrt{C_v[[X]]} = \sqrt{(CA[[X]])_v} = \sqrt{(CA[[X]])_t} \subseteq \sqrt{(PA[[X]])_t} \subseteq \sqrt{P_t[[X]]} = \sqrt{P[[X]]} = P[[X]]$. Then $\sqrt{(PA[[X]])_t} = P[[X]]$, which is absurd. Then P is not a t-SFT ideal of A.

$"(3) \implies (1)"$ (A. Benhissi (2011)). By Proposition 15.1, there exists a t-ideal M of A that is not t-SFT. We can choose a sequence $(a_i)_{i\in\mathbb{N}}$ of elements of $M \setminus (0)$ such that for each integer $n \geq 1$, $a_n^n \notin (a_0, \dots, a_{n-1})_v$. Put $I_n = (a_0, a_1, \dots, a_n)$ and $I = \bigcup_{n:0}^{\infty} I_n$. It is clear that $\bigcup_{n:0}^{\infty} (I_n)_v \subseteq I_t$. Conversely, let J be a nonzero finitely generated subideal of I. There exists $n \in \mathbb{N}$ such that $J \subseteq I_n$, then $J_v \subseteq (I_n)_v$. This shows that $I_t \subseteq \bigcup_{n:0}^{\infty} (I_n)_v$, so we have the equality. Let $f = \sum_{i:0}^{\infty} a_i X^{i!} \in I[[X]]$. Suppose that $f \in \sqrt{(IA[[X]])_t}$. There exists an integer $k \geq 1$ such that $f^k \in (IA[[X]])_t$. There exists a nonzero finitely generated subideal J of $IA[[X]]$ such that $f^k \in J_v$. We can suppose that J is of the form $J = I_r A[[X]]$ with $r \in \mathbb{N}$. By Lemma 19.1, $f^k \in (I_r A[[X]])_v = (I_r)_v[[X]]$. For each integer $n > max(r, k)$, the coefficient of $X^{k(n!)}$ in f^k is $a_n^k \in (I_r)_v$. Indeed, $f = \sum_{i:0}^{n} a_i X^{i!} + X^{(n+1)!} g$ with

$g \in A[[X]]$. The coefficient of $X^{k(n!)}$ in f^k is equal to that of $X^{k(n!)}$ in $(\sum_{i:0}^{n} a_i X^{i!})^k$, which is a_n^k. Then for each integer $n > max(r, k)$, $a_n^n \in (I_r)_v \subseteq (I_{n-1})_v$, which is absurd. Then we have $f \notin \sqrt{(IA[[X]])_t}$, so $I[[X]] \nsubseteq \sqrt{(IA[[X]])_t}$. $\qquad\square$

Problems

Exercise 1

(1) Show that a ring A is Noetherian if and only if the following conditions are satisfied:

  (a) A is an SFT-ring.
  (b) For each $P \in Spec(A)$, $P.A[[X]] \in spec(A[[X]])$.

(2) Show that a ring A is SFT if and only if for each $P \in Spec(A)$, $P[[X]]$ is the only prime ideal of $A[[X]]$ minimal over $P.A[[X]]$.

Exercise 2 (Benhissi 2012) Let A be a ring. Consider the two following conditions:

(a) For each ideal I of A, $I[[X]] \subseteq \sqrt{IA[[X]]}$.   (b) A is an SFT-ring.

We want to give a new proof for the implication $(a) \implies (b)$.

(1) Let $a_0, a_1, \ldots$ be an infinite sequence of elements of A and $f = \sum_{i:0}^{\infty} a_i X^{i!} \in A[[X]]$. Show that for each integer $n \geq k \geq 1$, the coefficient of $X^{k(n!)}$ in f^k is a_n^k.

(2) Suppose that A is not an SFT-ring and consider an infinite sequence $a_0, a_1, \ldots$ of elements of A such that $a_{n+1}^{n+1} \notin (a_0, \ldots, a_n)$ for each $n \in \mathbb{N}$. Let $I = (a_0, a_1, \ldots)A$ and $f = \sum_{i:0}^{\infty} a_i X^{i!} \in I[[X]]$. Show that $f \notin \sqrt{IA[[X]]}$.

The last part of this exercise is due to Bakkari (2009)

Exercise 3 Let R be a ring and M a R-module. Recall from Sect. 7, Chap. 1, that the ring $R(+)M = \{(r, m); \ r \in R, m \in M\}$ is endowed with the operations $(r, m) + (r', m') = (r + r', m + m')$ and $(r, m)(r', m') = (rr', rm' + r'm)$.

(1) Give the expression of $(r, m)^k$ for $r \in R, m \in M$ and $k \in \mathbb{N}^*$.
(2) Express $dim(R(+)M)$ using $dim(R)$.
(3) Let S be a nonempty set of R and I the ideal generated by S. Show that the ideal of $R(+)M$ generated by the set $S \times \{0\}$ is $I(+)IM$.
(4) Show that $R(+)M$ is an SFT-ring if and only if R is an SFT-ring.
(5) Suppose that R is a Noetherian ring and the R-module M is not finitely generated. Show that the ring $R(+)M$ is SFT but not Noetherian.
(6) Show that $\mathbb{Z}(+)\mathbb{Q}$ is an SFT-ring but not Noetherian.

Exercise 4 (Toan and Kang 2018, 2020) Let A be a ring.

(1) Show that A is not an SFT-ring if and only if there exists an infinite sequence $a_0, a_1, \ldots$ of elements of A such that $a_m^m \notin (a_0, \ldots, a_{m-1})$ for each integer $m \geq 1$.
(2) Deduce that if A does not have a Noetherian spectrum, then it is not SFT.
(3) Show that if A is not an SFT-ring, then so is $A[[X]]$.
(4) Let I be an ideal of A that is not SFT. Suppose that I is the radical of a countably generated ideal. Show that there exists a sequence $(a_i)_{i \in \mathbb{N}}$ of elements of I such that $I = \sqrt{(a_0, a_1, \ldots)}$ and $a_m^m \notin (a_0, a_1, \ldots, a_{m-1})$ for each $m \in \mathbb{N}^*$.

Exercise 5 Let A be an integral domain and I a nonzero proper ideal that is idempotent ($I^2 = I$) of A. Show that I is not SFT.

Exercise 6 (Coykendall and Dutta 2013) Let A be a ring and I an ideal of A. We say that I is a VSFT-ideal (very strong finite type) if there exist $n \in \mathbb{N}^*$ and a finitely generated ideal $F \subseteq I$ of A such that $I^n \subseteq F$. We say that A is a VSFT-ring if all its ideals are VSFT.

(1) (a) Show that a VSFT-ideal is an SFT-ideal.
  (b) Show that a finitely generated ideal is a VSFT-ideal.
  (c) Let A be an integral domain with quotient field K, $x \in K$ a quasi-integral element over A, and M the A-sub-module of K generated by $\{1, x, x^2, \ldots\}$.
    Show that there exists $0 \neq a \in A$ such that aM is a VSFT-ideal of A.
  (d) Let $A = \mathbb{Z} + 2X\mathbb{Z}[X] \subset \mathbb{Z}[X]$. Show that the ideal $I = (2X^n; \ n \in \mathbb{N})$ of A is VSFT.
  (e) Show that Coykendall's example is a VSFT-ideal. See Sect. 4.
  (f) Show by an example that the notions SFT and VSFT are different.
(2) With the preceding notations, show that if I is a VSFT(resp. SFT)-ideal, then $\sqrt{I} = \sqrt{F}$. Deduce that if I is a radical VSFT (resp. SFT)-ideal, then $I = \sqrt{F}$ and if I is a VSFT-ideal that is not finitely generated, then F is not radical.
(3) Let I be a VSFT-ideal of A and $n \in \mathbb{N}^*$. Show that I^n is a VSFT-ideal.
(4) Show that A is a VSFT-ring if and only if its prime ideals are VSFT.
(5) Let I be a VSFT (resp. SFT)-ideal of A and F a finitely generated ideal of A such that $F \subseteq I \subseteq \sqrt{F}$. Show that there exists $n \in \mathbb{N}^*$ such that $I^n \subseteq F \subseteq I$ (resp. $\forall x \in I, x^n \in F$).
(6) Let I be an ideal of A such that $\sqrt{I}$ is a VSFT-ideal. Show that there exists $k \in \mathbb{N}^*$ such that $(\sqrt{I})^k \subseteq I$. Deduce that if the radical of an ideal I is finitely generated, then I is VSFT.
(7) Let $X, Y_1, Y_2, \ldots$ be indeterminates over the field $\mathbb{F}_2$.
    Consider the ring $A = \mathbb{F}_2(Y_1^2, Y_2^2, \ldots)[[X^2]][XY_1, XY_2, \ldots]$. Show that the ideal $M = (X^2, XY_1, XY_2, \ldots)A$ is SFT but not VSFT. Deduce that the ring A is SFT but not VSFT.
(8) Let A be a VSFT-integral domain and B an over-ring of A that is a flat A-module. Show that B is a VSFT-ring. Deduce that if A is a VSFT-integral domain and S a multiplicative set of A, then $S^{-1}A$ is a VSFT-ring.
(9) Show that a homomorphic image of a VSFT-ring is a VSFT-ring. Deduce that if I is an ideal of a VSFT-ring A, the quotient ring A/I is VSFT.
(10) Let $A \subseteq B$ be an extension of rings. Show that if I is a VSFT-ideal of A, then IB is a VSFT-ideal of B.
(11) Suppose that the ring A contains the field $\mathbb{Q}$ of rational numbers.
    Show that the notions SFT and VSFT are identical over A.
(12) Show that the notions SFT and VSFT coincide with over-valuation domains.
(13) Suppose that A is a radically closed integral domain. Let I be an SFT-ideal of A for which there exist $n \in \mathbb{N}^*$ and $0 \neq a \in I$ such that for each $x \in I$, $x^n \in aA$. Show that I is a VSFT-ideal.
(14) Let $A = \mathbb{Z} + 2X\mathbb{Z}[X]$. Show that the ideal $\mathscr{I} = (Y; Y + 2X^k, \ k \in \mathbb{N})$ of the polynomials ring $A[Y]$ is VSFT with $n = 2$ and $F = (2, Y)A[Y]$.

(15) Let $J \subseteq I$ be two ideals of A. Suppose that J is a reduction of I, i.e., there exists $n \in \mathbb{N}^*$ such that $I^{n+1} = JI^n$. Show that if J is finitely generated, then I is VSFT.

(16) Reconsider the example of (1-c). Show that $J = aA$ is a reduction of $I = aM$.

Exercise 7 (Condo and Coykendall 1999) Let A be a ring. Suppose that for each ideal $\mathscr{I}$ of $A[[X, Y]]$ and each sequence $(f_i)_{i \in \mathbb{N}}$ of elements of $\mathscr{I}$, the formal power series $\displaystyle\sum_{i:0}^{\infty} f_i Y^i \in \mathscr{I}$. Show that $A[[X]]$ is an SFT-ring.

Exercise 8 Let K be a field with zero characteristic, $n \geq 2$ an integer, and $A = K[Y_i, i \in \mathbb{N}]/(Y_i^n, i \in \mathbb{N})$. Show that A is a ring of dimension zero that is not SFT.

Exercise 9 Let A be an integral domain and $(f_n(X))_{n \in \mathbb{N}}$ and $(g_n(X))_{n \in \mathbb{N}}$ two sequences of elements of $A[[X]]$ that converge respectively to $f(X)$ and $g(X)$ for the (X)-adic topology of $A[[X]]$.

(1) Show that the sequences $(f_n(X) + g_n(X))_{n \in \mathbb{N}}$ and $(f_n(X)g_n(X))_{n \in \mathbb{N}}$ converge respectively to $f(X) + g(X)$ and $f(X)g(X)$.

(2) Suppose that $f_n(X)$ is invertible in $A[[X]]$ for each $n \in \mathbb{N}$. Show that $f(X)$ is invertible in $A[[X]]$ and $(f_n(X)^{-1})_{n \in \mathbb{N}}$ converges to $f(X)^{-1}$.

Exercise 10 Let A be a ring and $a_1, \ldots, a_n \in A$. Show that there exists one and only one A-homomorphism of rings $\phi : A[X_1, \ldots, X_n] \longrightarrow A$ such that $\phi(X_i) = a_i$ for $1 \leq i \leq n$. Moreover, $\ker \phi = (X_1 - a_1, \ldots, X_n - a_n)$.

Exercise 11 Let A be a ring, $f \in A[[X]]$ a formal power series, and $g \in A[X]$ a nonzero polynomial such that $fg = 0$. Show that there exists $0 \neq a \in A$ such that $af = 0$.

Exercise 12 (Kang and Park 2006) Let A be a t-SFT integrally closed domain.

(1) Show that each radical integral t-ideal of A is divisorial. Deduce that $t - Spec(A) = v - Spec(A)$.

(2) Let I be an integral ideal of A such that $I^{-1} = A$. Show that there exists a finitely generated ideal $J \subseteq I$ such that $J^{-1} = A$.

Exercise 13 (Kang and Park 2006) Let A be an integral domain and $(A_\lambda)_{\lambda \in \Lambda}$ a family of over-rings of A with finite character such that $A = \displaystyle\bigcap_{\lambda \in \Lambda} A_\lambda$.

(1) Show that for each nonzero fractional ideal I of A, $I^{-1} = \bigcap_{\lambda \in \Lambda}(IA_\lambda)^{-1}$. Deduce

that $\bigcap_{\lambda \in \Lambda}(IA_\lambda)_{v_\lambda} \subseteq I_v$, where v is the v-operation on A and v_λ the v-operation on A_λ.

(2) Show that if each A_λ is a t-SFT domain, so is A.

Exercise 14 (Anderson et al. 2017) Let A be an integral domain. Suppose that each nonzero finitely generated integral ideal of $A[[X]]$ is v- invertible. Show that the same is true for the domain A.

Exercise 15 Let T be an integral domain, $(0) \neq M$ a maximal ideal of T, and D a proper subring of the field $K = T/M$. Let $s : T \longrightarrow K = T/M$ be the canonical homomorphism and $A = s^{-1}(D)$.

$$
\begin{array}{ccc}
A = s^{-1}(D) & \xrightarrow{\ s/A\ } & D \\
i \downarrow & & \downarrow i \\
T & \xrightarrow{\ \ s\ \ } & K = T/M
\end{array}
$$

Show that

(1) M is a prime ideal of A and $A/M \simeq D$.
(2) T is an over-ring of A.
(3) For each prime ideal P of T other than M, we have $T_P = A_{P \cap A}$.
(4) $T_M = A_M$ if and only if K is the quotient field of D.
(5) T is a flat over-ring of A if and only if $qf(D) = K$.
(6) T is a t-flat over-ring of A if and only if either $qf(D) = K$ or $qf(D) \neq K$ and M are not a t-ideal of T. Give necessary and sufficient conditions in order that T be a t-flat over-ring of A that is not flat.
(7) Examine the case where T is of the form $T = K + M$ with K as field and $M \neq (0)$ a maximal ideal of T and $A = D + M$ with D a proper subring of K.

Give an example of a t-flat over-ring that is not flat using the domain $T = K[X, Y]$ and $M = (X, Y)$ with K a field and X and Y two indeterminates over K.

Exercise 16 (Hizem and Benhissi 2005) Let $A \subseteq B$ be an extension of rings and $R = A + XB[[X]]$.

(1) Show that the ring R endowed with its $XB[[X]]$-adic topology is the completion of $A + XB[X]$ for its $XB[X]$-adic topology .
(2) Show that for each ideal J of R containing $XB[[X]]$, there exists an ideal I of A such that $J = I + XB[[X]]$.
(3) Let I be an ideal of A. Show that the ideal $I + XB[[X]]$ is prime in R if and only if I is prime in A.

(4) Suppose that B is a finitely generated A-module. Let p be an ideal of A. Show that $P = p + XB[[X]]$ is finitely generated in R if and only if p is finitely generated.

(5) Show that the ring R is Noetherian if and only if A is a Noetherian ring and B is a finitely generated A-module.

(6) Examine the case of an extension $A = K \subset B = L$ of fields.

(7) Suppose that $A \subset B$ are distinct integral domains.

   (a) Show that R is never principal.
   (b) Show that R is never completely integrally closed. Deduce that R is never UFD.

Exercise 17 Let A be an integral domain and $0 \neq f(X) \in A[[X]]$.

(1) Suppose that $f(0) = a \neq 0$.

   (a) Show that for each $g(X) \in A[[X]]$, there exists $h(X) \in (A[\frac{1}{a}])[[X]]$ such that $g = fh$.
   (b) Deduce that the ideal $fA[[X]]$ is closed in $A[[X]]$ for the $XA[[X]]$-adic topology.

(2) Show that the result in (b) is true even if $f(0) = 0$.

Exercise 18 Let A be a ring. For each $f \in A[[X]]$, let $c(f)$ be the ideal of A generated by the coefficients of f. Show that for each $f, g \in A[[X]]$, $\sqrt{c(fg)} = \sqrt{c(f)c(g)}$.
   Hint: Use the prime ideals.

Exercise 19 Let K be a field.

(1) Is the element $X + X^2$ irreducible in $K[X]$? in $K[[X]]$?
(2) Give two polynomials of $K[X]$ relatively prime in $K[[X]]$ but not in $K[X]$.

Solutions

Exercise 1

(1) $"\Longrightarrow"$ (a) A Noetherian ring is clearly an SFT-ring.
(b) If A is Noetherian, for each $P \in Spec(A)$, $P.A[[X]] = P[[X]] \in Spec(A[[X]])$.
   $"\Longleftarrow"$ By Theorem 1.17, since A is an SFT-ring, for each $P \in Spec(A)$, we have $P[[X]] = \sqrt{P.A[[X]]}$. But by (b), $P.A[[X]] \in Spec(A[[X]])$, then $\sqrt{P.A[[X]]} = P.A[[X]]$. So $P[[X]] = P.A[[X]]$, then A is Noetherian.
(2) $"\Longrightarrow"$ Let Q be a prime ideal of $A[[X]]$ minimal over $P.A[[X]]$. Then $P \subseteq Q \cap A$. Since A is an SFT-ring, by Theorem 1.18, we have $P.A[[X]] \subseteq$

$P[[X]] \subseteq (Q \cap A)[[X]] = \sqrt{(Q \cap A).A[[X]]} \subseteq Q$. By minimality of Q over $P.A[[X]]$, we have $P[[X]] = Q$.

" $\Longleftarrow$ " Let $P \in Spec(A)$. Since $\sqrt{P.A[[X]]}$ is the intersection of all the prime ideals of $A[[X]]$ containing $P.A[[X]]$, then $\sqrt{P.A[[X]]} = P[[X]]$. By Theorem 1.18, A is SFT.

Exercise 2

(1) $f = \sum_{i:0}^{\infty} a_i X^{i!} = \sum_{i:0}^{n} a_i X^{i!} + X^{(n+1)!} g$ with $g \in A[[X]]$ and $(n+1)! > k(n!)$.

The coefficient of $X^{k(n!)}$ in f^k is equal to that of $X^{k(n!)}$ in $(\sum_{i:0}^{n} a_i X^{i!})^k$, which is a_n^k.

(2) The sequence $a_0, a_1, \ldots$ exists by Lemma 1.10. Let $I = (a_0, a_1, \ldots)A$ and $f = \sum_{i:0}^{\infty} a_i X^{i!} \in I[[X]]$. Suppose that $f \in \sqrt{IA[[X]]}$ and let $k \in \mathbb{N}^*$ be such that $f^k \in IA[[X]]$. There exists $n_0 \in \mathbb{N}^*$ such that $f^k \in (a_0, \ldots, a_{n_0})A[[X]]$. All the coefficients of f^k belong to $(a_0, \ldots, a_{n_0})A$. By (1), for each integer $n > max(n_0, k)$, the coefficient of $X^{k(n!)}$ in f^k is $a_n^k \in (a_0, \ldots, a_{n_0}) \subseteq (a_0, \ldots, a_{n-1})$ since $n_0 < n$. Then $a_n^n \in (a_0, \ldots, a_{n-1})$ since $k < n$, which is absurd.

Exercise 3

(1) We have $(r, m)^2 = (r^2, 2rm)$, $(r, m)^3 = (r^3, 3r^2 m)$. Suppose that $(r, m)^k = (r^k, kr^{k-1} m)$. Then $(r, m)^{k+1} = (r^k, kr^{k-1} m)(r, m) = (r^{k+1}, r^k m + kr^k m) = (r^{k+1}, (k+1)r^k m)$.

(2) By Lemma 8.9 (2), Chap. 1, we have $Spec(R(+)M) = \{\mathscr{P}(+)M; \mathscr{P} \in Spec(R)\}$. Then $dim(R(+)M) = dim(R)$.

(3) " $\subseteq$ " By Lemma 8.9 (1), Chap. 1, $I(+)IM$ is an ideal of $R(+)M$. It contains the set $S \times \{0\}$, then $< S \times \{0\} > \subseteq I(+)IM$.

" $\supseteq$ " An element of $I(+)IM$ is a finite sum of elements of the form (a, bm) with $a, b \in I$ and $m \in M$. But $(a, bm) = (a, 0) + (0, bm) = (a, 0) + (0, m)(b, 0)$. It suffices to show that an element of the form $(a, 0)$ with $a \in I$ belongs to $< S \times \{0\} >$. Put $a = \alpha_1 s_1 + \ldots + \alpha_n s_n$ with $\alpha_i \in R$ and $s_i \in S$.

Then $(a, 0) = (\sum_{i:1}^{n} \alpha_i s_i, 0) = \sum_{i:1}^{n} (\alpha_i s_i, 0) = \sum_{i:1}^{n} (\alpha_i, 0)(s_i, 0) \in < S \times \{0\} >$.

(4) " $\Longrightarrow$ " Let I be an ideal of R. Then $I(+)M$ is an ideal of $R(+)M$ since $IM \subseteq M$. Since $R(+)M$ is an SFT-ring, there exist an integer $k \geq 1$ and $x_1, \ldots, x_n \in I(+)M$ such that for each $x \in I(+)M$, $x^k \in (x_1, \ldots, x_n)$. Put $x_i = (r_i, m_i)$ with $r_i \in I$ and $m_i \in M$, $1 \leq i \leq n$. Then for each $a \in I$, $(a, 0) \in I(+)M$ and $(a, 0)^k = (a^k, 0)$, $(a, 0)^k \in (x_1, \ldots, x_n)R(+)M$, so $a^k \in (r_1, \ldots, r_n)R \subseteq I$. Then I i an SFT-ideal of R.

$" \Longleftarrow "$ Let $P \in Spec(R(+)M)$. By Lemma 8.9 (2), Chap. 1, $P = \mathscr{P}(+)M$ with $\mathscr{P} \in Spec(R)$. There exist a finite subset S of $\mathscr{P}$ and an integer $k \geq 1$ such that for each $x \in \mathscr{P}$, $x^k \in SR = I$. By (3), the ideal of $R(+)M$ generated by the finite set $S \times \{0\}$ of $\mathscr{P}(+)M = P$ is $I(+)IM \subseteq \mathscr{P}(+)\mathscr{P}M \subseteq \mathscr{P}(+)M = P$ and satisfies for each $y = (r, m) \in \mathscr{P}(+)M$, $y^{k+1} = (r^{k+1}, (k+1)r^k m) \in I(+)IM$. Then P is an SFT-ideal of $R(+)M$.

(5) Since R is Noetherian, then it is an SFT-ring. By the preceding question, $R(+)M$ is an SFT-ring. By Corollary 8.11 of Chap. 1, the ring $R(+)M$ is not Noetherian.

(6) By the example of Corollary 8.11, Chap. 1, the ring $\mathbb{Z}(+)\mathbb{Q}$ is not Noetherian.

Exercise 4

(1) $" \Longrightarrow "$ Lemma 1.10. $" \Longleftarrow "$ Suppose that the ideal $I = (a_0, a_1, \ldots)$ is SFT. There exist $k \in \mathbb{N}^*$ and a finitely generated ideal $F \subseteq I$ such that for each $x \in I$, $x^k \in F$. Let $n \in \mathbb{N}$ be such that $F \subseteq (a_0, \ldots, a_n)$. For each $m \in \mathbb{N}$, $a_m^k \in (a_0, \ldots, a_n)$. Let $m \geq max\{k, n+1\}$ be an integer. Then $a_m^m = a_m^{m-k} a_m^k \in (a_0, \ldots, a_n) \subseteq (a_0, \ldots, a_{m-1})$, which is absurd.

(2) Let $J_0 \subset J_1 \subset \ldots$ be strictly increasing sequence of radical ideals in A. Let $a_0 \in J_0$ and $a_i \in J_i \setminus J_{i-1}$ for each $i \geq 1$. Then $a_m^m \notin (a_0, \ldots, a_{m-1})$ since in the contrary case, $a_m \in \sqrt{(a_0, \ldots, a_{m-1})} \subseteq \sqrt{J_{m-1}} = J_{m-1}$, which is absurd.

(3) By Theorem 1.13, there exists an infinite strictly increasing sequence $P_0 \subset P_1 \subset \ldots$ of prime ideals of $A[[X]]$. Choose $f_0(X) \in P_0$ and $f_m(X) \in P_m \setminus P_{m-1}$ for each $m \in \mathbb{N}^*$. Since $f_m^m \notin P_{m-1}$ and $(f_0, \ldots, f_{m-1}) \subseteq P_{m-1}$, then $f_m^m \notin (f_0, \ldots, f_{m-1})$. So $A[[X]]$ is not an SFT-ring. We can also conclude by Lemma 5.9.

(4) We distinguish the following two cases:

   (i) $I = \sqrt{(a_0, a_1, \ldots, a_k)}$ is the radical of a finitely generated ideal. By omitting some generators, we can suppose that $a_i \notin \sqrt{(a_0, a_1, \ldots, a_{i-1})}$ for $1 \leq i \leq k$. Then $a_i^i \notin (a_0, a_1, \ldots, a_{i-1})$ for $1 \leq i \leq k$. Since I is not an SFT-ideal, there exists $a_{k+1} \in I$ such that $a_{k+1}^{k+1} \notin (a_0, a_1, \ldots, a_k)$. Also, there exists $a_{k+2} \in I$ such that $a_{k+2}^{k+2} \notin (a_0, \ldots, a_{k+1})$. By induction, we construct an infinite sequence $a_0, a_1, \ldots, a_k, a_{k+1}, \ldots$ of elements in I such that $a_m^m \notin (a_0, a_1, \ldots, a_{m-1})$ for each $m \in \mathbb{N}^*$.

   (ii) I is not the radical of a finitely generated ideal. By hypothesis, $I = \sqrt{(b_0, b_1, \ldots)}$ is the radical of a countably generated ideal. Take $i_0 = 0$ and $a_0 = b_{i_0}$. Since $I \neq \sqrt{(a_0)}$, there exists a smallest index $i_1 > i_0$ such that $b_{i_1} \notin \sqrt{(a_0)}$. Let $a_1 = b_{i_1}$. By induction, suppose constructed elements $a_0, \ldots, a_{k-1} \in I$ $(k \geq 2)$ such that $a_i \notin \sqrt{(a_0, a_1, \ldots, a_{i-1})}$ for $1 \leq i \leq k-1$. Since $I \neq \sqrt{(a_0, \ldots, a_{k-1})}$, let $i_k > \ldots > i_1 > i_0$ be the smallest index such that $b_{i_k} \notin \sqrt{(a_0, \ldots, a_{k-1})}$. Take $a_k = b_{i_k}$. By the choice of the i_k, $I = \sqrt{(a_0, a_1, \ldots)}$ and $a_m^m \notin (a_0, a_1, \ldots, a_{m-1})$ for each $m \in \mathbb{N}^*$.

Exercise 5 Since $I \neq A$, there exists $P \in spec(A)$ such that $I \subseteq P$. There exists an over-valuation domain V of A with maximal ideal M such that $M \cap A = P$. Suppose that I is an SFT-ideal of A. There exist $n \in \mathbb{N}^*$ and a finitely generated ideal $F \subseteq I$ of A such that $x^n \in F$ for each $x \in I$. Put $I' = IV \subseteq PV \subseteq M \subset V$ and $F' = FV$ a finitely generated ideal of V. The ideal I'^n is generated by the set $\{x^n; \ x \in I\}$. Then $I'^n \subseteq F' \subseteq I'$ so I' is a nonzero proper SFT-ideal of the valuation domain V. Then $I'^2 \neq I'$ so $I^2 \neq I$, which is absurd.

Exercise 6

(1) (a) Clear. (b) Take $n = 1$ and $F = I$. (c) There exists $0 \neq a \in A$ such that $ax^n \in A$ for each $n \in \mathbb{N}$. Then $I = aM$ is an integral ideal of A. The generators of I^2 have the form $(ax^i)(ax^j) = a(ax^{i+j}) \in aA \subseteq I$ with $i, j \in \mathbb{N}$. Then $I^2 \subseteq aA \subseteq I$, so I is a VSFT-ideal.

(d) The product of two generators of I is $(2X^n)(2X^m) = 2(2X^{n+m}) \in 2A$ with $n, m \in \mathbb{N}$. Then $I^2 \subseteq 2A \subseteq I$ so I is a VSFT-ideal.

(e) With the notations of Coykendall's example, $(tV)^2 = t^2V = t(tV) \subseteq tV_1 \subseteq tV$.

(f) Consider the ring $A = \mathbb{F}_2[Y_i; i \in \mathbb{N}]/(Y_i^2; i \in \mathbb{N})$. Its unique prime ideal $\mathcal{M} = (\bar{Y}_i; i \in \mathbb{N})$ is SFT. Then A is an SFT-ring. Suppose that there exist $n \in \mathbb{N}^*$ and a finitely generated ideal F of A such that $\mathcal{M}^n \subseteq F \subseteq \mathcal{M}$. A generator of F is a finite sum of monomials of the form $\bar{Y}_{t_1}\bar{Y}_{t_2}\ldots\bar{Y}_{t_s}$. Let k be the biggest integer such that $\bar{Y}_k$ is a factor in one of these monomials. Then $\bar{Y}_{k+1}\bar{Y}_{k+2}\ldots\bar{Y}_{k+n}$ is an element of $\mathcal{M}^n$ that is not F. Then $\mathcal{M}$ is not VSFT so A is not a VSFT-ring.

(2) Since $F \subseteq I$, then $\sqrt{F} \subseteq \sqrt{I}$. For each $x \in I$, $x^n \in F$, then $x \in \sqrt{F}$. So $\sqrt{I} = \sqrt{F}$. Suppose that F is radical. Then $I \subseteq \sqrt{I} = \sqrt{F} = F \subseteq I$, so $I = F$ is a finitely generated ideal, which is absurd.

(3) There exist $k \in \mathbb{N}^*$ and a finitely generated ideal F of A such that $I^k \subseteq F \subseteq I$. Consider a product of k elements of I^n of the form $x = (x_{11}\ldots x_{n1})(x_{12}\ldots x_{n2})\ldots(x_{1k}\ldots x_{nk})$ where $x_{ij} \in I$. Then $x = (x_{11}x_{12}\ldots x_{1k})(x_{21}x_{22}\ldots x_{2k})\ldots(x_{n1}x_{n2}\ldots x_{nk}) \in I^k\ldots I^k \subseteq F\ldots F = F^n$. Since an element of $(I^n)^k$ is a finite sum of elements of the preceding type, it belongs to F^n. Then $(I^n)^k \subseteq F^n$ with F^n a finitely generated ideal of A included in I^n. Then I^n is a VSFT-ideal.

(4) Suppose that A is not VSFT and consider an ideal I of A that is not VSFT. Let $\mathcal{F}$ be the set of the ideals of A containing I that are not VSFT. Let $\mathcal{C}$ be a chain of $(\mathcal{F}, \subseteq)$ and $M = \bigcup_{J \in \mathcal{C}} J$, an ideal of A containing I. Suppose that M is a VSFT-ideal. There exist $n \in \mathbb{N}^*$ and a finitely generated ideal $F \subseteq M$ of A such that $M^n \subseteq F$. There exists $J_0 \in \mathcal{C}$ such that $F \subseteq J_0$. Since $J_0^n \subseteq M^n \subseteq F$, then J_0 is VSFT, which is absurd. We conclude that M bounds $\mathcal{C}$ and $(\mathcal{F}, \subseteq)$ is inductive. By Zorn's lemma, $(\mathcal{F}, \subseteq)$ has a maximal element P. It suffices to show that P is a prime ideal of A to have a contradiction. Suppose that there exist $a, b \in A \setminus P$ such that $ab \in P$. By maximality of P, the ideals

(P, a) and (P, b) are VSFT. There exist $m, n \in \mathbb{N}^*$ and finitely generated ideals $F_a \subseteq (P, a)$ and $F_b \subseteq (P, b)$ such that $(P, a)^m \subseteq F_a$ et $(P, b)^n \subseteq F_b$. We may suppose that $F_a = (x_1, \ldots, x_s, a)$ and $F_b = (y_1, \ldots, y_t, b)$ with $x_1, \ldots, x_s, y_1, \ldots, y_t \in P$. The ideal $F = F_a F_b = (x_i y_j, a y_j, b x_i, ab)$ is finitely generated contained in P. Moreover, $P^{m+n} = P^m P^n \subseteq F_a F_b = F$. Then P is VSFT, which is absurd. Then P is prime.

(5) There exist $k \in \mathbb{N}^*$ and a finitely generated ideal F' of A such that $I^k \subseteq F' \subseteq I \subseteq \sqrt{F}$. The image $\overline{F'}$ of F' in the quotient ring A/F is a finitely generated nil ideal. Then $\overline{F'}$ is a nilpotent ideal. There exists $m \in \mathbb{N}^*$ such that $F'^m \subseteq F$. Then $I^{km} \subseteq F'^m \subseteq F \subseteq I$.

An analogue proof for the SFT property.

(6) There exist $n \in \mathbb{N}^*$ and a finitely generated ideal F of A such that $(\sqrt{I})^n \subseteq F \subseteq \sqrt{I}$. By (2), $\sqrt{\sqrt{I}} = \sqrt{F}$. Then $F \subseteq \sqrt{I} = \sqrt{F}$ with $\sqrt{I}$ a VSFT-ideal. By the preceding question, there exists $t \in \mathbb{N}^*$ such that $(\sqrt{I})^t \subseteq F$. Since F is finitely generated and contained in $\sqrt{I}$, there exists $s \in \mathbb{N}^*$ such that $F^s \subseteq I$. Then $(\sqrt{I})^{ts} \subseteq F^s \subseteq I$.

If $\sqrt{I}$ is finitely generated, then it is VSFT. By the preceding step, there exists $k \in \mathbb{N}^*$ such that $I^k \subseteq (\sqrt{I})^k \subseteq I$ with $(\sqrt{I})^k$ finitely generated. Then I is a VSFT-ideal.

(7) The ideal $F = X^2 A \subseteq M$. For each $i \in \mathbb{N}^*$, $(XY_i)^2 = X^2 Y_i^2 \in F$. Since *charact* $A = 2$, for each $f \in M$, $f^2 \in F$. Then M is an SFT-ideal of A.

We have $F \subseteq M \subseteq \sqrt{F}$. Suppose that M is a VSFT-ideal of A. By (5), there exists $n \in \mathbb{N}^*$ such that $M^n \subseteq F$. Let $k_1, \ldots, k_n$ be nonzero distinct natural integers, then $XY_{k_1}, \ldots, XY_{k_n} \in M$ but $(XY_{k_1}) \ldots (XY_{k_n}) \notin F$, which is absurd. Then M is not a VSFT-ideal and A is not a VSFT-ring.

The extension of rings $\mathbb{F}_2(Y_1^2, Y_2^2, \ldots)[[X^2]] \subset A$ is integral since for each $i \in \mathbb{N}^*$, $(XY_i)^2 = X^2 Y_i^2 \in \mathbb{F}_2(Y_1^2, Y_2^2, \ldots)[[X^2]]$. The ring $\mathbb{F}_2(Y_1^2, Y_2^2, \ldots)[[X^2]]$, of formal power series over a field, is a local integral domain with maximal ideal $X^2 \mathbb{F}_2(Y_1^2, Y_2^2, \ldots)[[X^2]]$ and its dimension is one. Then $dim\, A = 1$, and since $A/M \simeq \mathbb{F}_2(Y_1^2, Y_2^2, \ldots)$ is a field, the ideal M of A is maximal. It is the only maximal ideal of A. Indeed, let $f \in A \setminus M$. The term of f in the field $\mathbb{F}_2(Y_1^2, Y_2^2, \ldots)$ is nonzero. Since *charact*$(A) = 2$, f^2 is a formal power series in X^2 with nonzero constant term, then it is invertible. So f is invertible. Then M is the only nonzero prime ideal of A. The ring A is SFT.

(8) There exists a nonempty family $\mathscr{S}$ of nonzero integral ideals of A that is stable by multiplication such that $B = A_{\mathscr{S}}$ and $IB = B$ for each $I \in \mathscr{S}$. Let Q be a prime ideal of B and $P = Q \cap A$. By Example 2.5 of Lemma 2.10, Chap. 3, $Q = P_{\mathscr{S}}$. There exist $n \in \mathbb{N}^*$ and a finitely generated ideal F of A such that $P^n \subseteq F \subseteq P$. We will show that $Q^n \subseteq FB \subseteq Q$. Let $q_1, \ldots, q_n$ be any elements of Q. There exist $I_1, \ldots, I_n \in \mathscr{S}$ such that $q_i I_i \subseteq P$, $1 \leq i \leq n$. Then $q_1 \ldots q_n (I_1 \ldots I_n) \subseteq P^n \subseteq F$. Since $\mathscr{S}$ is stable by multiplication, $I = I_1 \ldots I_n \in \mathscr{S}$ and $q_1 \ldots q_n I \subseteq F$. Let $\mathscr{I} = (FB :_B q_1 \ldots q_n) = \{z \in B;\ q_1 \ldots q_n z \in FB\}$, an ideal of B containing I. Since $IB = B$, then $\mathscr{I} = B$.

In particular, $1 \in \mathscr{I}$, then $q_1 \ldots q_n \in FB$. This is for each $q_1, \ldots, q_n \in Q$. Since an element of Q^n is a finite sum of elements of the preceding type, then $Q^n \subseteq FB \subseteq Q$ with FB a finitely generated ideal of B. We conclude that Q is a VSFT-ideal of B.

(9) Let $f : A \longrightarrow B$ be a homomorphism onto of rings with A VSFT. Let P be a prime ideal of B. Then $f^{-1}(P)$ is a prime ideal of A. There exist $n \in \mathbb{N}^*$ and a finitely generated ideal $F = (a_1, \ldots, a_k)$ of A such that $\left(f^{-1}(P)\right)^n \subseteq F \subseteq f^{-1}(P)$. Let $b_1, \ldots, b_n$ be any elements of P. Since f is onto, there exist $x_1, \ldots, x_n \in A$ such that $f(x_i) = b_i$, $1 \leq i \leq n$. Then $x_1, \ldots, x_n \in f^{-1}(P)$ so $x_1 x_2 \ldots x_n \in \left(f^{-1}(P)\right)^n \subseteq F$. Put $x_1 \ldots x_n = \alpha_1 a_1 + \ldots + \alpha_k a_k$ with $\alpha_1, \ldots, \alpha_k \in A$. Then $b_1 \ldots b_n = f(x_1) \ldots f(x_n)$ $= f(x_1 \ldots x_n) = f(\alpha_1 a_1 + \ldots + \alpha_k a_k) = f(\alpha_1) f(a_1) + \ldots + f(\alpha_k) f(a_k) \in \left(f(a_1), \ldots, f(a_k)\right)$. Then $P^n \subseteq \left(f(a_1), \ldots, f(a_k)\right) B \subseteq P$. So P is a VSFT-ideal of the ring B.

(10) There exist $n \in \mathbb{N}^*$ and a finitely generated ideal F of A such that $I^n \subseteq F \subseteq I$. Then FB is a finitely generated ideal of B contained in IB. We will show that $(IB)^n \subseteq FB$. Let $y_1, \ldots, y_n$ be any elements of IB, $y_i = a_{i1}b_{i1} + \ldots + a_{in_i}b_{in_i}$ with $a_{ij} \in I$, $b_{ij} \in B$ and $n_i \in \mathbb{N}^*$ for $1 \leq i \leq n$ and $1 \leq j \leq n_i$. A typical term of the product $y_1 \ldots y_n$ is of the form $a_{1k_1}b_{1k_1} \ldots a_{nk_n}b_{nk_n} = (a_{1k_1} \ldots a_{nk_n})(b_{1k_1} \ldots b_{nk_n}) \in I^n B \subseteq FB$, with $1 \leq k_i \leq n_i$ for $1 \leq i \leq n$.

(11) Let I be an SFT-ideal of A. There exist $n \in \mathbb{N}^*$ and a finitely generated ideal $F \subseteq I$ of A such that $x^n \subseteq F$ for each $x \in I$. In the quotient ring A/F, the elements $\bar{x}$ of the ideal I/F satisfy $\bar{x}^n = 0$. By Lemma 1.15 (1), there exists $m \in \mathbb{N}^*$ such that $m(I/F)^m = (0)$, then $mI^m \subseteq F$. Since $\mathbb{Q} \subseteq A$, then $I^m \subseteq F \subseteq I$. So I is a VSFT-ideal.

(12) Let I be a nonzero SFT-ideal of a valuation domain A. There exist $n \in \mathbb{N}^*$ and $0 \neq a \in I$ such that $x^n \in aA$ for each $x \in I$. Since the ideal I^n is generated by the set $\{x^n; x \in I\}$, then $I^n \subseteq aA \subseteq I$.

(13) Let $x_1, \ldots, x_n \in I$. Then $x_i^n = ar_i$ with $r_i \in A$ for $1 \leq i \leq n$. So $\left(\frac{x_1 \ldots x_n}{a}\right)^n = r_1 \ldots r_n \in A$. Since A is radically closed, $\frac{x_1 \ldots x_n}{a} \in A$, then $x_1 \ldots x_n \in aA$. So $I^n \subseteq aA \subseteq I$.

(14) Since the product $(Y + 2X^k)(Y + 2X^l) = Y(Y + 2X^k + 2X^l) + 2(2X^{k+l}) \in (2, Y)A[Y]$, we have the result.

(15) Since $I^{n+1} = JI^n \subseteq J \subseteq I$, we have the result.

(16) For each $m, n \in \mathbb{N}$, $(ax^m)(ax^n) = a(ax^{m+n}) \in JI$, then $I^2 \subseteq JI$, so $I^2 = JI$.

Exercise 7 Suppose that $A[[X]]$ is not an SFT-ring. By Theorem 1.18, there exists $\mathscr{P} \in Spec(A[[X]])$ such that $\mathscr{P}[[Y]] \neq \sqrt{\mathscr{P}.A[[X, Y]]}$, then $\sqrt{\mathscr{P}.A[[X, Y]]} \subset \mathscr{P}[[Y]]$. There exists $Q \in Spec(A[[X, Y]])$ containing $\mathscr{P}.A[[X, Y]]$ but not $\mathscr{P}[[Y]]$. Let $f = \sum_{i:0}^{\infty} f_i(X)Y^i \in \mathscr{P}[[Y]] \setminus Q$. For each $i \in \mathbb{N}$, $f_i(X) \in \mathscr{P} \subset \mathscr{P}.A[[X, Y]]$. But $f \notin \mathscr{P}.A[[X, Y]]$, which contradicts the hypothesis.

Exercise 8 The only prime ideal of A is $P = (\overline{Y}_i, i \in \mathbb{N})$, then $dim\, A = 0$. Suppose that P is an SFT-ideal. By Theorem 1.18, $P[[X]] = \sqrt{PA[[X]]}$. Let

$$f(X) = \sum_{i:0}^{\infty} \overline{Y}_i X^i \in P[[X]]. \text{ There exists } n \in \mathbb{N}^* \text{ such that } f^n \in PA[[X]].$$

Put $f^n = \sum_{i:0}^{k} g_i \overline{Y}_i$ with $g_i(X) \in A[[X]]$. Since $\overline{Y}_0^n = \ldots = \overline{Y}_k^n = 0$, then f is nilpotent. This is absurd, by Example 5.4 of Lemma 5.5, Sect. 5.

Exercise 9

(1) Let $l \in \mathbb{N}$. There exists $N \in \mathbb{N}$ such that for each integer $n \geq N$, we have $f_n(X) - f(X) \in X^l A[[X]]$ and $g_n(X) - g(X) \in X^l A[[X]]$. Then $f_n(X) + g_n(X) - g(X) - f(X) = f_n(X) - f(X) + g_n(X) - g(X) \in X^l A[[X]]$ and $f_n(X)g_n(X) - g(X)f(X) = (f_n(X) - f(X))g_n(X) + f(X)(g_n(X) - g(X)) \in X^l A[[X]]$.

(2) Since $f(0) = f_n(0)$ for n sufficiently big, then $f(0)$ is invertible in A, so $f(X)$ is invertible in $A[[X]]$. Let $l \in \mathbb{N}$. There exists $N \in \mathbb{N}$ such that for each integer $n \geq N$, $f_n(X) - f(X) \in X^l A[[X]]$. Then $f(X)^{-1} - f_n(X)^{-1} = (f_n(X) - f(X))f(X)^{-1}f_n(X)^{-1} \in X^l A[[X]]$ since $f(X)^{-1}$ and $f_n(X)^{-1} \in A[[X]]$.

Exercise 10 We necessarily have $\phi(f(X_1, \ldots, X_n)) = f(a_1, \ldots, a_n)$ for each polynomial $f(X_1, \ldots, X_n) \in A[X_1, \ldots, X_n]$. For the last assertion, we proceed by induction. If $n = 1$, $\phi : A[X] \longrightarrow A$ with $\phi(X) = a$, then $(X - a) \subseteq ker\, \phi$. Conversely, let $f(X) = b_0 + b_1 X + \ldots + b_m X^m \in ker\, \phi$. Since $f(a) = 0$, then $f(X) = f(X) - f(a) = b_1(X - a) + (X^2 - a^2) + \ldots + b_m(X^m - a^m)$. Since $X^i - a^i = (X - a)(X^{i-1} + X^{i-2}a + \ldots + a^{i-1})$ for $1 \leq i \leq n$, then $f(X) \in (X - a)$. Suppose that $n \geq 2$ and the property is true at the order $n - 1$. Note that ϕ is the composite of the homomorphism $\psi : A[X_1, \ldots, X_n] \longrightarrow A[X_1, \ldots, X_{n-1}]$, defined by $\psi(X_i) = X_i$ for $1 \leq i \leq n - 1$ and $\psi(X_n) = a_n$ and the homomorphism $\theta : A[X_1, \ldots, X_{n-1}] \longrightarrow A$, defined by $\theta(X_i) = a_i$ for $1 \leq i \leq n-1$. By the induction hypothesis, $ker\, \theta = (X_1 - a_1, \ldots, X_{n-1} - a_{n-1})$. Since $\phi = \theta \circ \psi$, then $ker\, \phi = \phi^{-1}(0) = \psi^{-1}(\theta^{-1}(0)) = \psi^{-1}(ker\, \theta)$. Let $f(X_1, \ldots, X_n) \in ker\, \phi$. Then $\psi(f(X_1, \ldots, X_n)) \in ker\, \theta$. There exist $h_i(X_1, \ldots, X_{n-1}) \in A[X_1, \ldots, X_{n-1}]$ for $1 \leq i \leq n - 1$ such that $\psi(f(X_1, \ldots, X_n)) = \sum_{i:1}^{n-1} h_i(X_1, \ldots, X_{n-1})(X - a_i)$. There exist $f_j(X_1, \ldots, X_{n-1}) \in A[X_1, \ldots, X_{n-1}]$, $0 \leq j \leq d$, such that we have $f(X_1, \ldots, X_n) = \sum_{j:0}^{d} f_j(X_1, \ldots, X_{n-1})X_n^j = \sum_{j:0}^{d} f_j(X_1, \ldots, X_{n-1})a_n^j +$

$$\sum_{j:1}^{d} f_j(X_1, \ldots, X_{n-1})(X_n^j - a_n^j) = \psi(f(X_1, \ldots, X_n)) + \sum_{j:1}^{d} f_j(X_1, \ldots, X_{n-1})$$

$(X_n - a_n)(X_n^{j-1} + X_n^{j-2}a_1 + \ldots + a_n^{j-1}) \in (X_1 - a_1, \ldots, X_{n-1} - a_{n-1}, X_n - a_n)$.
Then $ker\ \phi \subseteq (X_1 - a_1, \ldots, , X_n - a_n)$. The inverse inclusion is clear.

Exercise 11 Put $f = \sum_{i:0}^{\infty} a_i X^i$ and $g = \sum_{i:0}^{m} b_i X^i$ with $b_m \neq 0$. If $deg\ g(X) = 0$,
then $b_0 f(X) = 0$. Suppose that $deg\ g(X) \geq 1$. If all the coefficients of $f(X)$
annihilate $g(X)$, then each nonzero coefficient of $g(X)$ annihilates $f(X)$ and we
have the result. If not, there exists $p \in \mathbb{N}$ such that $a_p g(X) \neq 0$ and $a_0 g(X) =$
$a_1 g(X) = \ldots = a_{p-1} g(X) = 0$. Then $0 = f(X)g(X) = (\sum_{i:p}^{\infty} a_i X^i)g(X) =$
$a_p b_0 X^p + \ldots$, so $a_p b_0 = 0$. Then $a_p g(X) = a_p(b_1 X + b_2 X^2 + \ldots + b_m X^m) =$
$X a_p(b_1 + b_2 X + \ldots + b_m X^{m-1})$. Put $g'(X) = a_p(b_1 + b_2 X + \ldots + b_m X^{m-1}) \in A[X]$,
then $0 \neq a_p g(X) = X g'(X)$, so $g'(X) \neq 0$, $deg\ g'(X) < deg\ g(X)$, and since
$0 = a_p(g(X)f(X)) = X g'(X)f(X)$, then $g'(X)f(X) = 0$. We replace $g(X)$ by
$g'(X)$ and we repeat the operation. After a finite number of steps, we find a nonzero
constant $a \in A$ such that $af(X) = 0$.

Exercise 12

(1) a) Let I be a radical integral t ideal of A. Since A is a t-SFT domain, there
    exist a nonzero finitely generated ideal $F \subseteq I$ and an integer $k \geq 1$ such
    that for each $a \in I$, $a^k \in F_v$. Let (V_α) be the family of the over-valuation
    domains of A. Since A is integrally closed, $A = \bigcap_\alpha V_\alpha$. By Lemma 16.2, $I^k \subseteq$

    $\bigcap_\alpha (I^k V_\alpha) = \bigcap_\alpha (a^k, a \in I)V_\alpha \subseteq \bigcap_\alpha (F_v V_\alpha) = F_v$. By Proposition 11.1 of
    Chap. 1, $(I_v)^k \subseteq (I^k)_v \subseteq F_v \subseteq I_t = I$. Then $I_v \subseteq \sqrt{I} = I$, so $I_v = I$, then I
    is divisorial.
(b) It follows from (a) that $t - Spec(A) \subseteq v - Spec(A)$. The inverse inclusion is
    always true since each v-ideal is a t-ideal.
(2) Since $t - Max(A) \subseteq t - Spec(A)$, then $t - Max(A)$ is the set of the maximal
    elements of $t - Spec(A)$. Also, $v - Max(A)$ is the set of the maximal elements
    of $v - Spec(A)$. Since $t - Spec(A) = v - Spec(A)$, then $t - Max(A) =$
    $v - Max(A)$. Since $I^{-1} = A$, then $I_v = (I^{-1})^{-1} = A$. For each $M \in t -$
    $Max(A) = v - Max(A)$, $I \nsubseteq M$ since in the contrary case, $A = I_v \subseteq M_v =$
    $M_t = M$, which is absurd. Then for each $M \in t - Max(A)$, $I_t \nsubseteq M$, so
    $I_t = A$. There exists a nonzero finitely generated subideal J of I such that
    $1 \in J_v = A : J^{-1}$, then $J^{-1} \subseteq A$ so $J^{-1} = A$.

Exercise 13

(1) Let $x \in I^{-1}$. Then $xI \subseteq A$, so $xIA_\lambda \subseteq A_\lambda$, then $x \in A_\lambda : IA_\lambda = (IA_\lambda)^{-1}$. So
    $x \in \bigcap_{\lambda \in \Lambda} (IA_\lambda)^{-1}$. Conversely, let $x \in \bigcap_{\lambda \in \Lambda} (IA_\lambda)^{-1}$. For each $\lambda \in \Lambda$, $xIA_\lambda \subseteq$

A_λ, then $xI \subseteq A_\lambda$. So $xI \subseteq \bigcap_{\lambda \in \Lambda} A_\lambda = A$, then $x \in A : I = I^{-1}$. So

$$I^{-1} = \bigcap_{\lambda \in \Lambda} (IA_\lambda)^{-1}.$$

By applying the first part to I^{-1}, we find $I_v = (I^{-1})^{-1} = \bigcap_{\lambda \in \Lambda} (I^{-1}A_\lambda)^{-1}$.

But $I^{-1}A_\lambda \subseteq (IA_\lambda)^{-1}$. Indeed, let $x \in I^{-1}$. Then $xI \subseteq A$, so $xIA_\lambda \subseteq A_\lambda$, then $x \in (IA_\lambda)^{-1}$. So $(IA_\lambda)_{v_\lambda} = ((IA_\lambda)^{-1})^{-1} \subseteq (I^{-1}A_\lambda)^{-1}$. Then $\bigcap_{\lambda \in \Lambda} (IA_\lambda)_{v_\lambda} \subseteq I_v$.

(2) Let $P \in t - Spec(A)$ and $0 \neq c \in P$ a fixed element. Let $\lambda_1, \ldots, \lambda_n$ be the only indexes in Λ such that c is not invertible in $A_{\lambda_1}, \ldots, A_{\lambda_n}$. Since A_{λ_i} is a t-SFT domain, there exist nonzero finitely generated ideal $J_i \subseteq PA_{\lambda_i}$ and $k_i \in \mathbb{N}^*$ satisfying for each $a \in PA_{\lambda_i}$, $a^{k_i} \in (J_i)_{v_{\lambda_i}}$. We can suppose that $J_i = I_i A_{\lambda_i}$ with I_i a nonzero finitely generated ideal of A contained in P. Let $k = k_1 + \ldots + k_n$ and $I = cA + I_1 + \ldots + I_n$ a nonzero finitely generated ideal of A contained in P. For each $\lambda \in \Lambda \setminus \{\lambda_1, \ldots, \lambda_n\}$, $IA_\lambda = A_\lambda$. Then for each $a \in P$, $a^k \in \bigcap_{\lambda \in \Lambda} (IA_\lambda)_{v_\lambda} \subseteq I_v$.

Exercise 14 Let I be a nonzero finitely generated integral ideal of A and $J = IA[[X]] = I[[X]]$ a nonzero finitely generated integral ideal of $A[[X]]$. By hypothesis, $A[[X]] = (JJ^{-1})_v = (I[[X]](I[[X]])^{-1})_v = (I[[X]]I^{-1}[[X]])_v \subseteq ((II^{-1})[[X]])_v = (II^{-1})_v[[X]] \subseteq A[[X]]$. Then $(II^{-1})_v[[X]] = A[[X]]$, so $(II^{-1})_v = A$. Then I is v-invertible.

Exercise 15

(1) The map $s_{/A} : A \longrightarrow D$ is a homomorphism onto, with kernel M. Then $A/M \simeq D$ is an integral domain. Then M is a prime ideal of A.

(2) Since $A \subseteq T$, then $qf(A) \subseteq qf(T)$. Conversely, it suffices to show that $T \subseteq qf(A)$. Let $0 \neq a \in M$. For each $x \in T$, $x = \frac{ax}{a} \in qf(A)$ since $ax \in M \subset A$.

(3) The inclusion $A_{P \cap A} \subseteq T_P$ is clear. Let $x \in T_P$, $x = \frac{z_1}{z_2}$ with $z_1, z_2 \in T$, $z_2 \notin P$. Since M is a maximal ideal of T and $M \neq P$, then $M \nsubseteq P$. Let $a \in M \setminus P$. Then $az_1 \in M \subseteq A$ and $az_2 \in M \setminus P$. So $x = \frac{az_1}{az_2} \in A_{P \cap A}$. Then $T_P \subseteq A_{P \cap A}$.

(4) "$\implies$" Since $D \subseteq K$, then $qf(D) \subseteq K$. Conversely, let $\bar{z} \in K$, $z \in T \subseteq T_M = A_M$. Then $z = \frac{a}{b}$ with $a, b \in A$, $b \notin M$. So $bz = a$, then $\bar{b}\bar{z} = \bar{a}$ with $\bar{a}, \bar{b} \in D$ and $\bar{b} \neq 0$. Then $\bar{z} = \bar{a}/\bar{b} \in qf(D)$. So $qf(D) = K$.

  "$\impliedby$" The inclusion $A_M \subseteq T_M$ is clear. Conversely, let $x \in T_M$, $x = \frac{z_1}{z_2}$ with $z_1, z_2 \in T$ and $z_2 \notin M$. Then $\bar{z}_1, \bar{z}_2 \in T/M = K = qf(D)$ with $\bar{z}_2 \neq 0$. So $\bar{z}_1/\bar{z}_2 = \bar{d}_1/\bar{d}_2$ with $\bar{d}_1, \bar{d}_2 \in D$ and $\bar{d}_2 \neq 0$, then $d_1, d_2 \in A$ and $d_2 \notin M$. So $\bar{z}_1 \bar{d}_2 = \bar{z}_2 \bar{d}_1$, then $z_1 d_2 - z_2 d_1 = m \in M$. Dividing by $z_2 d_2$, we find $\frac{z_1}{z_2} - \frac{d_1}{d_2} = \frac{m}{z_2 d_2}$, then $x = \frac{z_1}{z_2} = \frac{d_1}{d_2} + \frac{m}{z_2 d_2}$. Since $\frac{d_1}{d_2} \in A_M$, it suffices to show that $\frac{m}{z_2 d_2} \in A_M$. But $\bar{z}_2 \bar{d}_2 \neq 0$ in the field $K = T/M$. There exists $t \in T$ such that $\bar{t}\bar{z}_2\bar{d}_2 = \bar{1}$. Then $tz_2 d_2 - 1 \in M \subset A$, so $tz_2 d_2 \in A \setminus M$. Then $\frac{m}{z_2 d_2} = \frac{tm}{tz_2 d_2} \in MA_M \subseteq A_M$. So $T_M \subseteq A_M$.

(5) T is a flat over-ring of $A \iff$ For each prime ideal P of T, $T_P = A_{P \cap A} \iff T_M = A_M \iff qf(D) = K$. We always used (3) and (4).

(6) By definition, T is a t-flat over-ring of $A \iff$ For each t-maximal ideal P of T, $T_P = A_{P \cap A}$. But by (3), this property is satisfied by each t-maximal ideal P of T other than M. Since M is a maximal ideal of T, if it is a t-ideal, then it is t-maximal. Then T is a t-flat over-ring of A if and only if in case M is a t-ideal in T, then $T_M = A_M$, which is equivalent to $qf(D) = K$. Then T is a t-flat over-ring of A that is not flat if and only if $qf(D) \neq K$ and M is not a t-ideal of T.

(7) The domain $A = D + M$ comes from the preceding construction with $T = K + M$. Let K be a field and X and Y two indeterminates over K. The domain $T = K[X, Y] = K + (X, Y)K[X, Y]$ with $M = (X, Y)K[X, Y]$ a maximal ideal of T. We showed in Example 3 of Proposition 11.10, Chap. 1, that $M_t = T$, then M is not a t-ideal of the domain T. Let D be a subring of K with $qf(D) \neq K$. By (6), T is a t-flat over-ring of $A = D + (X, Y)K[X, Y]$ but it is not flat.

Exercise 16

(1) Since $I = XB[[X]]$ is an ideal of R satisfying $\bigcap_{n:0}^{\infty} I^n = (0)$, then R is Hausdorff for its I-adic topology. Since $XB[[X]] \cap (A + XB[X]) = XB[X]$, the $XB[[X]]$-adic topology on R induces the $XB[X]$-adic topology on $A + XB[X]$. Let $f = \sum_{i:0}^{\infty} a_i X^i \in R$. For each natural integer k, let $g_k = \sum_{i:0}^{k} a_i X^i \in A + XB[X]$. Since $f - g_k = \sum_{i:k+1}^{\infty} a_i X^i \in X^{k+1} B[[X]] = I^{k+1}$, $\lim_{k \to \infty} g_k = f$. Conversely, let (g_k) be a Cauchy sequence of elements of $A + XB[X]$ and $g = g_0 + (g_1 - g_0) + (g_2 - g_1) + \ldots \in A + XB[[X]]$. Since for each $l \in \mathbb{N}$, there exists $k_0 \in \mathbb{N}$ such that for each $k \geq k_0$, $g_{k+1} - g_k \in I^l = X^l B[[X]]$, then g is well defined and $g - g_k \in I^l$. So $\lim_{k \to \infty} g_k = g$.

(2) The set $I = \{f(0); \ f \in J\}$ is an ideal of A and $J \subseteq I + XB[[X]]$. Conversely, let $a \in I$. There exists $f \in J$ such that $f(0) = a$, then $f - a \in XB[[X]] \subseteq J$ so $a = f - (f - a) \in J$. Then $I + XB[[X]] \subseteq J$.

(3) Follows from the isomorphism $R/I + XB[[X]] \simeq A/I$.

(4) "$\implies$" p is the image of P by the canonical homomorphism from $R \longrightarrow A$.
   "$\impliedby$" Put $B = Ab_1 + \ldots + Ab_s$ and $p = a_1 A + \ldots + a_n A$. We will show that the ideal $P = (a_1, \ldots, a_n, b_1 X, \ldots, b_s X)$. Let $f = \alpha_0 + \sum_{i:1}^{\infty} \alpha_i X^i \in P$.

Then $\alpha_0 \in p = (a_1, \ldots, a_n)A \subseteq (a_1, \ldots, a_n)R$ and $\alpha_1, \alpha_2, \ldots \in B$. For each $i \geq 1$, there exist $c_1^i, \ldots, c_s^i \in A$ such that $\alpha_i = \sum_{j:1}^{s} c_j^i b_j$. Then $\sum_{i:1}^{\infty} \alpha_i X^i =$

$$\sum_{i:1}^{\infty} \left(\sum_{j:1}^{s} c_j^i b_j \right) X^i = \sum_{j:1}^{s} b_j X \left(\sum_{i:1}^{\infty} c_j^i X^{i-1} \right) \in (b_1 X, \ldots, b_s X) R.$$

(5) $" \implies "$ The ring $A \simeq R/XB[[X]]$ is Noetherian. Since the ideal $XB[[X]]$ of R is finitely generated, there exist $f_1, \ldots, f_n \in B[[X]]$ such that $XB[[X]] = Xf_1 R + \ldots + Xf_n R$, then $B[[X]] = f_1 R + \ldots + f_n R$ and $B = f_1(0)A + \ldots + f_n(0)A$ is a finitely generated A-module.

$" \impliedby "$ Suppose that R is not Noetherian. The set of its ideals that are not finitely generated is nonempty. It is inductive for the inclusion. By Zorn's lemma, it has a maximal element P. By questions (2) and (4), $XB[[X]] \not\subseteq P$. Since B is a finitely generated A-module, there exist $b_1, \ldots, b_s \in B$ such that $B = b_1 A + \ldots + b_s A$, then $XB[[X]] = b_1 XA[[X]] + \ldots + b_s XA[[X]]$. Since $XB[[X]] \not\subseteq P$, there exists $i_0 \in \{1, \ldots, s\}$ such that $b_{i_0} X \notin P$. By reordering the b_i, we may suppose that $b_1 X \notin P$, then $P \subset P + b_1 XR$. By maximality of P, the ideal $P + b_1 XR$ is finitely generated. We can find a finitely generated ideal $J = (f_1, \ldots, f_n) \subseteq P$ such that $P + b_1 XR = J + b_1 XR$. We will show that $P = J + P \cap b_1 XR$. The inverse inclusion is clear. Conversely, let $f \in P$. Then $f \in J + b_1 XR$. Put $f = g + b_1 Xh$ with $g \in J$ and $h \in R$. Then $b_1 Xh = f - g \in P$ so $b_1 Xh \in P \cap b_1 XR$. Then we have the equality. We will also show that $P = J + (P : b_1 XR)b_1 XR$. Since $(P : b_1 XR)b_1 XR \subseteq P$, we have the inclusion $J + (P : b_1 XR)b_1 XR \subseteq P$. Conversely, it suffices to show that $P \cap b_1 XR \subseteq (P : b_1 XR)b_1 XR$. Let $f \in P \cap b_1 XR$. Then $f = b_1 Xg$ with $g \in R$. Since $f \in P$, then $g \in (P : b_1 XR)$, so $f \in (P : b_1 XR)b_1 XR$. So we have the equality. We will show that $P = (P : b_1 XR)$. The direct inclusion is clear. If it is strict, by maximality of P, the ideal $(P : b_1 XR)$ will be finitely generated. Then $P = J + (P : b_1 XR)b_1 XR$, is also finitely generated, which is absurd. Then we have the equality $P = (P : b_1 XR)$ so $P = J + b_1 XP$. We will show that $P = J$. Let $g \in P$. We will construct, by induction on k, a sequence $(g_k)_{k \in \mathbb{N}^*}$ of elements of J such that for each $k \in \mathbb{N}^*$, $g_k = \sum_{i:1}^{n} s_{k,i} f_i$ with for each $i \in \{1, \ldots, n\}$, $s_{k,i} \in (b_1 X)^{k-1} R$ and $g - g_1 - \ldots - g_k \in (b_1 X)^k P$. For $k = 1$, $g \in P = J + b_1 XP$, there exists $g_1 \in J$ such that $g - g_1 \in (b_1 X)P$. Put $g_1 = \sum_{i:1}^{n} s_{1,i} f_i$ with $s_{1,1}, \ldots, s_{1,n} \in R$. Suppose that $g_1, \ldots, g_k$ are constructed. Since $g - g_1 - \ldots - g_k \in (b_1 X)^k P = (b_1 X)^k (J + b_1 XP)$, there exists $g' \in J$ such that $g - g_1 - \ldots - g_k - (b_1 X)^k g' \in (b_1 X)^{k+1} P$. Put $g' = s_1 f_1 + \ldots + s_n f_n$ with $s_1, \ldots, s_n \in R$. It suffices to take $g_{k+1} = (b_1 X)^k g' = (b_1 X)^k s_1 f_1 + \ldots + (b_1 X)^k s_n f_n$, then $s_{k+1,i} = (b_1 X)^k s_i \in$

$(b_1X)^k R$ and $g - g_1 - \ldots - g_k - g_{k+1} \in (b_1X)^{k+1} P$. Since $\sum_{j:1}^{k+1} s_{j,i} - \sum_{j:1}^{k} s_{j,i} = s_{k+1,i} \in (b_1X)^k R \subseteq (XB[[X]])^k$, then for each $i \in \{1, \ldots, n\}$, the sequence $\left(\sum_{j:1}^{k} s_{j,i}\right)_k$ is Cauchy in R for the $XB[[X]]$-adic topology. Then it converges.

Put, for $1 \le i \le n$, $s_i = \lim_k \sum_{j:1}^{k} s_{j,i} \in R$ and $g' = s_1 f_1 + \ldots + s_n f_n$. We will show that $g - g' \in \bigcap_{m:0}^{\infty}(XB[[X]])^m = (0)$, then $g = g' \in J$ and $P = J$, which is absurd since P is not finitely generated. Let $m \in \mathbb{N}$ be fixed. For each integer $k \ge m$, $g - g' = (g - g_1 - \ldots - g_k) + (g_1 + \ldots + g_k - g')$ with $g - g_1 - \ldots - g_k \in (b_1X)^k P \subseteq (XB[[X]])^k \subseteq (XB[[X]])^m$ and

$$g_1 + \ldots + g_k - g' = \sum_{j:1}^{k} g_j - \sum_{j:1}^{k} s_i f_i = \sum_{j:1}^{k} \sum_{i:1}^{n} s_{j,i} f_i - \sum_{j:1}^{k} s_i f_i =$$

$\sum_{i:1}^{n} f_i(\sum_{j:1}^{k} s_{j,i} - s_i)$. Since $\lim_k \sum_{j:1}^{k} s_{ji} = s_i$ for each $i \in \{1, \ldots, n\}$, there exists an integer $k_0 \ge m$ such that for each integer $k \ge k_0$ and for each $i \in \{1, \ldots, n\}$, $\sum_{j:1}^{k} s_{j,i} - s_i \in (XB[[X]])^m$, then $g_1 + \ldots + g_k - g' = \sum_{i:1}^{n} f_i(\sum_{j:1}^{k} s_{j,i} - s_i) \in (XB[[X]])^m$, so $g - g' \in (XB[[X]])^m$.

(6) The domain $K + XL[[X]]$ is Noetherian if and only if the degree $[L : K]$ is finite. For example, $\mathbb{R} + X\mathbb{C}[[X]]$ is Noetherian but $\mathbb{Q} + X\mathbb{R}[[X]]$ is not Noetherian.

(7) (a) Suppose that the ideal $XB[[X]] = fR$ of R is principal, $f = f_1X$ with $f_1 \in B[[X]]$. Then $B[[X]] = f_1(A + XB[[X]])$, so $B = f_1(0)A$. There exists $a \in A$ such that $1 = f_1(0)a$, then $b = f_1(0)$ is invertible in B and $A = b^{-1}B = B$, which is absurd.

   (b) Let $b \in B \setminus A$, then $b = \frac{bX}{X} \in qf(R) \setminus R$. For each $n \in \mathbb{N}^*$, $Xb^n \in R$, then b is quasi-integral over R.

Exercise 17

(1) (a) Since $f(0) = a$ is invertible in $A[\frac{1}{a}]$, f is invertible in $(A[\frac{1}{a}])[[X]]$. Then $g \in A[[X]] \subseteq (A[\frac{1}{a}])[[X]] = f.(A[\frac{1}{a}])[[X]]$. There is $h \in (A[\frac{1}{a}])[[X]]$ such that $g = fh$.

   (b) The closure $\overline{fA[[X]]} = \bigcap_{n:1}^{\infty}(fA[[X]] + X^n A[[X]])$. Let $g \in \overline{fA[[X]]} \subseteq A[[X]]$. There exists $h \in (A[\frac{1}{a}])[[X]]$ such that $g = fh$. For each $n \in \mathbb{N}^*$, there exist $q_n, r_n \in A[[X]]$ such that $g = fq_n + X^n r_n$, then $f(h -$

$q_n) = X^n r_n$. Since A is an integral domain and $a = f(0) \neq 0$, the n first coefficients of $h - q_n$ are zero. Then the n first coefficients of h belong to A, so $h(X) \in A[[X]]$ and $g = fh \in f A[[X]]$. We conclude that $\overline{f A[[X]]} = f A[[X]]$.

(2) Put $f(X) = X^k f_1(X)$ with $f_1(X) \in A[[X]]$, $f_1(0) \neq 0$ and $k \geq 1$. Then

$$\overline{f A[[X]]} = \bigcap_{n:1}^{\infty} \left(f A[[X]] + X^n A[[X]] \right) = \bigcap_{n:1}^{\infty} \left(X^k f_1 A[[X]] + X^n A[[X]] \right) =$$

$$\bigcap_{n:k}^{\infty} \left(X^k f_1 A[[X]] + X^n A[[X]] \right) = X^k \bigcap_{n:k}^{\infty} \left(f_1 A[[X]] + X^{n-k} A[[X]] \right) =$$

$$X^k . \overline{f_1 A[[X]]} = X^k f_1 A[[X]] = f A[[X]].$$

Exercise 18 For each $P \in Spec(A)$, $c(fg) \subseteq P \Longleftrightarrow fg \in P[[X]] \Longleftrightarrow$ either $f \in P[[X]]$ or $g \in P[[X]] \Longleftrightarrow$ either $c(f) \subseteq P$ or $c(g) \subseteq P \Longleftrightarrow c(f)c(g) \subseteq P$.

Exercise 19

(1) In $K[X]$, $X + X^2 = X(1 + X)$ is not irreducible. In $K[[X]]$, $1 + X$ is invertible and the ideals $(X + X^2)$ and (X) are equal with $K[[X]]/(X) \simeq K$ a field. Then $(X + X^2)$ is a maximal ideal of $K[[X]]$ and $X + X^2$ is an irreducible element in $K[[X]]$.

(2) The polynomials $f = 1 + X$ and $g = X(1 + X) \in K[X]$ are not relatively prime in $K[X]$. Since f is invertible in $K[[X]]$, the common divisors of f and g are the invertible elements. These two elements are relatively prime in $K[[X]]$.

References

D.F. Anderson, A general theory of class groups. Commun. Algebra **16**(4), 805–847 (1988)

D.D. Anderson, D.F. Anderson, M. Zafrullah, Completely integrally closed Prufer v-multiplication domains. Commun. Algebra **45**(12), 5264–5282 (2017)

D.F. Anderson, A. Ryckaert, The class group of $D + M$. J. Pure Appl. Algebra **52**, 199–212 (1988)

J.T. Arnold, Krull dimension in power series rings. Trans. Am. Math. Soc. **177**, 299–304 (1973)

J.T. Arnold, J.W. Brewer, When $D[[X]]_{P[[X]]}$ is a valuation ring. Proc. Am. Math. Soc. **37**(2), 326–332 (1973)

Ch. Bakkari, Armendariz and SFT properties in subring retracts. Mediterr. J. Math. **6**, 339–345 (2009)

V. Barucci, S. Gabelli, M. Roitman, The class group of a strongly Mori domain. Commun. Algebra **22**(1), 173–211 (1994)

H. Bass, Torsion free and projective modules. Trans. Am. Math. Soc. **102**, 319–327 (1962)

A. Benhissi, A short proof for Arnold's theorem on the SFT-property. Rend. Cir. Mat. Palermo **61**(2), 199–200 (2012)

J.T. Condo, J. Coykendall, Strong convergence property of SFT rings. Commun. Algebra **27**(8), 2073–2085 (1999)

J.T. Condo, J. Coykendall, D.E. Dobbs, Formal power series rings over zero- dimensional SFT-rings. Commun. Algebra **24**(8), 2687–2698 (1996)

J. Coykendall, The SFT-property does not imply finite dimension for power series rings. J. Algebra **256**, 85–96 (2002)

J. Coykendall, Some remarks on infinite products, in *Arithmetical Properties of Commutative Rings and Monoids*. Lecture Notes in Pure and Applied Mathematics, vol. 241 (Chapman Hall/CRC Boca Raton, 2005), pp. 180–187

J. Coykendall, Progress on the dimension question for power series rings, in *Multiplicative Ideal Theory in Commutative Algebra: A Tribute to the Work of Robert Gilmer*, ed. by J.W. Brewer et al. (Springer, New York, 2006), pp. 123–135

J. Coykendall, T. Dumitrescu, An infinite product of formal power series. Bull. Math. Soc. Math. Roumanie Tome **49**(97), 31–36 (2006)

J. Coykendall, T. Dutta, Rings of very strong finite type (2013). Preprint

D.E. Dobbs, E.G. Houston, On t-spec(R[[X]]). Can. Math. Bull. **38**(2), 187–195 (1995)

T. Dutta, On a generalized notion of integral closure and VSFT domains. Dissertation, North Dakota State Faculty of Agriculture and Applied Sciences, USA, 2006

S. El Baghdadi, On a class of Prufer v-multiplication domains. Commun. Algebra **30**(2), 3723–3742 (2002)

D.E. Fields, Zero divisors and nilpotent elements in power series rings. Proc. AMS **27**(3), 427–433 (1971)

R. Gilmer, A. Grams, T. Parker, Zero-divisors in power series rings. J. Fur die Reine und Ang Mathematik Band **278/279**, 145–164 (1975)

A. Haouaoui, A. Benhissi, Zero-divisors and zero-divisor graphs of power series rings. Ricerche Mat. **65**, 1–13 (2016)

S. Hizem, A. Benhissi, When is $A + XB[[X]]$ Noetherian. C.R. Acad. Sci. Paris, Ser. I **340**, 5–7 (2005)

E. Houston, S. Malik, J. Mott, Characterizations of *-multiplication domains. Can. Math. Bull. **27**(1), 48–52 (1984)

E. Houston, M. Zafrullah, Integral domains in which each t-ideal is divisorial. Michigan Math. J. **35**, 291–299 (1984)

C.J. Hwang, Kull dimension of a completion. J. Korean Soc. Math. Educ. Ser. B: Pure Appl. Math. **11**(1), 23–27 (2004)

B.G. Kang, Prufer v-multiplication domains and the ring $R[X]_{N_v}$. J. Algebra **123**(1), 151–170 (1989)

B.G. Kang, M.H. Park, A localization of a power series ring over a valuation domain. J. Pure Appl. Algebra **140**, 107–124 (1999)

B.G. Kang, M.H. Park, A note on t-SFT rings. Commun. Algebra **34**, 3153–3165 (2006)

D.J. Kwak, Y.S. Park, On t-flat overrings. Ch. J. Math. **23**(1), 17–24 (1995)

K. Louartiti, N. Mahdou, Amalgamation algebra extensions defined by von Neumann regular and SFT conditions. Gulf J. Math. **1**(2), 105–113 (2013)

M.H. Park, Noetherian-like properties in polynomial and power series rings. J. Pure Appl. Algebra **223**, 3980–3988 (2019)

J. Querré, Sur les anneaux reflexifs. Can. J. Math **27**(6), 1222–1228 (1975)

M. Roitman, Arnold's theorem on the strongly finite type (SFT) property and the dimension of power series rings. Commun. Algebra **43**, 337–344 (2015)

A. Seidenberg, Derivations and integral closure. Pac. J. Math. **16** (1), 167–173 (1966)

P.T. Toan, B.G. Kang, Krull dimension of power series rings over non-SFT domains. J. Algebra **499**, 516–537 (2018)

P.T. Toan, B.G. Kang, Krull dimension of power series rings. J. Algebra **562**, 306–322 (2020)

Chapter 5
Nonnil-Noetherian Rings

The rings considered in this chapter are commutative with unity. The concept of Noetherian rings is one of the most important topics that is widely used in many areas including commutative algebra and algebraic geometry. The Noetherian property was originally due to the mathematician Noether who first considered a relation between the ascending chain condition on ideals and the finitely generatedness of ideals. More precisely, she showed that if A is a ring, then the ascending chain condition on ideals of A holds if and only if every ideal of A is finitely generated. The equivalence plays a significant role in simplifying the ideal structure of a ring. Due to the importance of Noetherian rings, many mathematicians have tried to use Noetherian properties in several classes of rings and attempted to generalize the notion of Noetherian rings. Nonnil-Noetherian rings and S-Noetherian rings are typical generalizations of Noetherian rings.

1 Generalities on Nonnil-Noetherian Rings

Let A be a ring. Then $Nil(A)$ is the set of nilpotent elements of A, $Spec(A)$ is its prime spectrum, and $Min(A)$ is the set of its minimal prime ideals. An ideal I of A is said to be nonnil if $I \nsubseteq Nil(A)$. We say that A is nonnil-Noetherian if each nonnil ideal of A is finitely generated. This notion is introduced and studied by Badawi (2003) under the strong hypothesis that the nilradical of the ring is a divided prime ideal. Hizem and Benhissi (2011) have extended some properties of nonnil-Noetherian rings but without any assumption on the nilradical. We should mention that nonnil-Noetherian rings have their own benefits. For example, in nonnil-Noetherian rings, we can investigate non-prime ideals because the nilradical is the intersection of prime ideals of a ring. In the first section of this chapter, we give a characterization of a nonnil-Noetherian ring A in terms of the Noetherianity of $A/Nil(A)$. Then we derive a series of corollaries extending many other properties

© The Author(s), under exclusive license to Springer Nature Switzerland AG 2022   393
A. Benhissi, *Chain Conditions in Commutative Rings*,
https://doi.org/10.1007/978-3-031-09898-7_5

of Noetherian rings. We prove the analogous of Cohen's theorem for nonnil-Noetherian rings. The first results in this section are due to Badawi (2003) and Hizem and Benhissi (2011).

Notation For a ring A, we denote by $Nil(A)$ the nilradical of A, i.e., the set of its nilpotent elements. It is also equal to the intersection of the prime ideals of A.

Example The ideal $Nil(A)$ is not always prime. For example, $Nil(\mathbb{Z}/12\mathbb{Z}) = (2\mathbb{Z}/12\mathbb{Z}) \cap (3\mathbb{Z}/12\mathbb{Z}) = 6\mathbb{Z}/12\mathbb{Z}$ is not prime.

Definition 1.1 Let A be a ring. An ideal of A is said to be nonnil if it is not contained in $Nil(A)$. We say that A is a nonnil-Noetherian ring if each nonnil ideal of A is finitely generated.

Example Each Noetherian ring is nonnil-Noetherian. We see later that the converse is false. It is clear that the two notions coincide for a reduced ring, i.e., $Nil(A) = (0)$.

The following proposition shows that nonnil-Noetherian rings can be characterized by the ascending chain condition on nonnil ideals.

Proposition 1.2 *The following assertions are equivalent for a ring A:*

(a) The ring A is nonnil-Noetherian.
(b) The ring A satisfies the acc for the nonnil ideals.
(c) Each nonempty set of nonnil ideals of A has a maximal element for the inclusion.

Proof

$"(a) \implies (b)"$ Let (I_n) be an increasing sequence of nonnil ideals of A. The ideal $I = \bigcup I_n$ is nonnil. Put $I = (x_1, \ldots, x_s)$. There exists $m \in \mathbb{N}$ such that $x_1, \ldots, x_s \in I_m$. For each $n \geq m$, $I_n = I_m$.

$"(b) \implies (c)"$ Let E be a nonempty set of nonnil ideals of A. Suppose that E does not have a maximal element. Let $I_1 \in E$. Since I_1 is not maximal in $(E, \subseteq)$, there exists $I_2 \in E$ such that $I_1 \subset I_2$. By induction, we construct a strictly increasing sequence of nonnil ideals of A, which is absurd.

$"(c) \implies (a)"$ Let I be a nonnil ideal of A and E the set of nonnil finitely generated ideals contained in I. Since $I \nsubseteq Nil(A)$, there exists $a \in I \setminus Nil(A)$, and then $(a) \in E$. By hypothesis, $(E, \subseteq)$ has a maximal element $J \subseteq I$. Suppose that $I \neq J$ and consider an element $x \in I \setminus J$. Then $J + xA \in E$ and $J \subset J + xA$, which is absurd. So $I = J$ is a finitely generated ideal.   □

The nonnil-Noetherian rings have an analogue to Cohen's theorem.

Proposition 1.3 *A ring A is nonnil-Noetherian if and only if its nonnil prime ideals are finitely generated.*

Proof Suppose that A is not nonnil-Noetherian. The set $\mathscr{F}$ of the nonnil ideals that are not finitely generated is not empty. We order $\mathscr{F}$ with the inclusion. Let $(J_\lambda)_\lambda$ be a totally ordered family of elements of $\mathscr{F}$. The ideal $J = \bigcup J_\lambda$ is nonnil. If $J = (a_1, \ldots, a_n)$, there exists λ such that $J_\lambda = (a_1, \ldots, a_n)$, which is absurd. Then $\mathscr{F}$ is inductive and so it admits a maximal element P. We will show that P is a prime ideal of A, which contradicts the hypothesis. Suppose that there exist $a, b \in A \setminus P$ such that $ab \in P$. Since $P \subset P + aA$, then $P + aA$ is nonnil, and then it is finitely generated. Put $P + aA = (\alpha_1 + ax_1, \ldots, \alpha_n + ax_n)$ with $\alpha_i \in P$ and $x_i \in A$. Consider the ideal $I = [P : a] = \{y \in A; \ ya \in P\}$. Then $P \subset I$ because of b, then I is nonnil and so finitely generated. Put $I = (\beta_1, \ldots, \beta_s)$. We will show that $P = (\alpha_1, \ldots, \alpha_n, a\beta_1, \ldots, a\beta_s)$. This is a contradiction with $P \in \mathscr{F}$. Since $\beta_i \in I$, $a\beta_i \in P$, and we always have an inclusion. Conversely, let $x \in P \subset P + aA$. There exist $u_1, \ldots, u_n \in A$ such that $x = u_1(\alpha_1 + ax_1) + \ldots + u_n(\alpha_n + ax_n) = u_1\alpha_1 + \ldots + u_n\alpha_n + (u_1x_1 + \ldots + u_nx_n)a$. It suffices to show that $(u_1x_1 + \ldots + u_nx_n)a \in (a\beta_1, \ldots, a\beta_s)$. But $(u_1x_1 + \ldots + u_nx_n)a = x - (u_1\alpha_1 + \ldots + u_n\alpha_n) \in P$, and then $u_1x_1 + \ldots + u_nx_n \in I$, by definition of I. Since $I = (\beta_1, \ldots, \beta_s)$, then $(u_1x_1 + \ldots + u_nx_n)a \in (a\beta_1, \ldots, a\beta_s)$. $\qquad\square$

Example 1 We will construct an example of a nonnil-Noetherian ring that is not Noetherian. Let $(Y_i)_{i \in \mathbb{N}}$ be a sequence of indeterminates over $\mathbb{Q}$ and $n \geq 2$ a natural integer. Consider the ideal $I = (Y_i^n; \ i \in \mathbb{N})$ of the ring $B = \mathbb{Q}[Y_i; \ i \in \mathbb{N}]$, and let $A = B/I$. The only prime ideal of A is $P = (\bar{Y}_i; \ i \in \mathbb{N})$. It is not finitely generated, and then A is not Noetherian. By the preceding proposition, since $Nil(A) = P$, then A is nonnil-Noetherian.

Now, the natural question is the relation between nonnil-Noetherian rings and S-Noetherian rings. The following two examples from Kwon and Lim (2020) show that the two concepts are not related to each other.

Example 2 Consider the product ring $A = \prod_{i:1}^{\infty} \mathbb{Z}$, its unit $\mathbb{1} = (1, 1, \ldots)$, and its element $s = (1, 0, \ldots)$. Then $S = \{\mathbb{1}, s\}$ is a multiplicative set of A, and A is S-Noetherian. Indeed, let I be an ideal of A. Then $F = sI$ is an ideal of A contained in I. Let $\pi : A \longrightarrow \mathbb{Z}$ be the first projection. Then $\pi(I)$ is an ideal of $\mathbb{Z}$, so $\pi(I) = n\mathbb{Z}$ with $n \in \mathbb{N}$. Since $sI = F = n\mathbb{Z} \times (0) \times \ldots =< (n, 0, \ldots) >$ is a principal ideal, then I is S-finite. We will now show that A is not nonnil-Noetherian. Consider the elements $e_1 = (1, 0, \ldots)$, $e_2 = (0, 1, 0, \ldots)$, $\ldots$ of A. The ideal generated by the family $\{e_i; \ i \in \mathbb{N}^*\}$ is nonnil and not finitely generated.

Example 3 Let A be any nonnil-Noetherian ring that is not Noetherian and $S = U(A)$, the group of units of A. Then A is not S-Noetherian.

Proposition 1.4 *The homomorphic image of a nonnil-Noetherian ring is nonnil-Noetherian.*

Proof Let $f : A \longrightarrow B$ be a homomorphism onto of rings with A is nonnil-Noetherian. Let J be a nonnil ideal of B. Then $J = f(f^{-1}(J))$ with $f^{-1}(J) \not\subseteq Nil(A)$. By hypothesis, $f^{-1}(J)$ is finitely generated. Then J is also finitely generated. $\square$

Example The quotient of a nonnil-Noetherian ring is nonnil-Noetherian, i.e., if A is a nonnil-Noetherian ring and I is an ideal of A, then A/I is nonnil-Noetherian.

The following theorem and its corollaries are due to Benhissi (2020).

Theorem 1.5 *Let A be a ring.*

(1) *If $Nil(A) \notin Spec(A)$, the following assertions are equivalent:*

   (i) The ring A is nonnil-Noetherian.
   (ii) The ring A is Noetherian.
   (iii) The quotient ring $A/Nil(A)$ is Noetherian, and all the minimal prime ideals of A are finitely generated.

(2) *If $Nil(A) \in Spec(A)$, then A is nonnil-Noetherian if and only if $A/Nil(A)$ is Noetherian, and all the height 1 prime ideals of A are finitely generated.*

Proof

(1) $"(i) \Rightarrow (ii)"$ For all $P \in Spec(A)$, $Nil(A) \subset P$, so P is finitely generated. Then A is Noetherian by Cohen's theorem.

$"(ii) \Rightarrow (iii)"$ Clear. $"(iii) \Rightarrow (i)"$ Let $P \in Spec(A)$ and $P_0 \in Min(A)$ with $P_0 \subseteq P$. Since $Nil(A) \subset P_0$, then $(A/Nil(A))/(P_0/Nil(A)) \simeq A/P_0$. Since $A/Nil(A)$ is Noetherian, so is A/P_0. Then its ideal P/P_0 is finitely generated. Since P_0 is finitely generated, so is P. By Cohen's theorem, A is Noetherian.

(2) $" \Rightarrow"$ A nonzero prime ideal of $A/Nil(A)$ is of the form $P/Nil(A)$ with $P \in Spec(A)$ and $Nil(A) \subset P$. By hypothesis, P is finitely generated, so is $P/Nil(A)$. By Cohen's theorem, $A/Nil(A)$ is Noetherian. Since $Nil(A) \in Spec(A)$, then $Min(A) = \{Nil(A)\}$. If P is a height 1 prime ideal of A, then $Nil(A) \subset P$ so P is finitely generated.

$" \Leftarrow"$ Let P a nonnil prime ideal of A. Then $Nil(A) \subset P$ and $\bar{P} = P/Nil(A)$ is a nonzero prime ideal of the integral domain $\bar{A} = A/Nil(A)$. Let $\bar{0} \neq \bar{x} \in \bar{P}$ and $\bar{Q} \subseteq \bar{P}$ a prime ideal of $\bar{A}$ minimal over $(\bar{x})$. Since $\bar{A}$ is Noetherian, by the principal ideal theorem, $\bar{Q}$ is of height 1. See Theorem 8.3 of Chap. 2. On the other hand, $\bar{Q} = Q/Nil(A)$ with $Nil(A) \subset Q \subseteq P$, and Q is a height 1 prime ideal of A. By the hypothesis, Q is finitely generated. By the isomorphism $(A/Nil(A))/(Q/Nil(A)) \simeq A/Q$, the domain A/Q is Noetherian. Its ideal P/Q is then finitely generated. Then P is finitely generated, and A is nonnil-Noetherian, by Proposition 1.3. $\square$

Corollary 1.6 *Let A be a ring.*

(1) *If $Nil(A) \notin Spec(A)$, then A is Noetherian if and only if A is nonnil-Noetherian.*
(2) *If $Nil(A) \in Spec(A)$, then A is Noetherian if and only if A is nonnil-Noetherian and $Nil(A)$ is finitely generated.*

Proof (2) " $\Leftarrow$ " Let $P \in Spec(A)$. If $P = Nil(A)$, it is finitely generated by hypothesis. If $P \neq Nil(A)$, then $Nil(A) \subset P$ so P is finitely generated because A is nonnil-Noetherian. We use Cohen's theorem. $\square$

Lemma 1.7 *Let $P \subset Q$ be the two prime ideals of a Noetherian ring A. If there exists a prime ideal properly between them, then there are infinitely many.*

Proof By passage to the quotient domain A/P, we may assume that $P = (0)$. Suppose that $P_1, \ldots, P_n$ are the sole prime ideals properly between (0) and Q. By the avoidance Lemma 8.4, Chap. 2, $Q \not\subseteq P_1 \cup \ldots \cup P_n$. Let $x \in Q \setminus P_1 \cup \ldots \cup P_n$. Then Q is minimal over the principal ideal (x). By the principal ideal Theorem 8.3, Chap. 2, $ht(Q) \leq 1$. But we have the chains $(0) \subset P_i \subset Q$, $1 \leq i \leq n$, of prime ideals of A. This is a contradiction. $\square$

We prove now the analogue of the preceding lemma for the nonnil-Noetherian rings.

Corollary 1.8 *Let A be a nonnil-Noetherian ring and $P \subset Q$ be the two prime ideals of A. If there exists a prime ideal of A strictly between P and Q, then there exists an infinity.*

Proof By Theorem 1.5, $P/Nil(A) \subset Q/Nil(A)$ are two prime ideals of the Noetherian ring $A/Nil(A)$. If R is a prime ideal of A such that $P \subset R \subset Q$, then $R/Nil(A)$ is a prime ideal of $A/Nil(A)$ such that $P/Nil(A) \subset R/Nil(A) \subset Q/Nil(A)$. By the preceding lemma, there exists an infinity of prime ideals of $A/Nil(A)$ between the ideals $P/Nil(A)$ and $Q/Nil(A)$. So there exist an infinity of prime ideals between P and Q. $\square$

A valuation domain is nonnil-Noetherian if and only if it is Noetherian if and only if it is a discrete valuation domain of Krull's dimension ≤ 1.

Recall that a ring is said to be chained if its lattice of ideals is totally ordered by inclusion. Note that a chained ring is local and its nilradical is a prime ideal. In fact it is the only minimal prime ideal.

Corollary 1.9 *Let A be a chained ring with maximal ideal M.*

(1) *If $Nil(A) = M$, then A is nonnil-Noetherian.*
(2) *If $Nil(A) \neq M$, then A is nonnil-Noetherian if and only if $Spec(A) = \{Nil(A), M\}$ with M a principal ideal.*

Proof Since A is chained, $Nil(A)$ is the only minimal prime ideal of A.

(1) If $Spec(A) = \{Nil(A)\}$, then by Proposition 1.3, A is nonnil-Noetherian.

(2) " $\Rightarrow$ " Since $Nil(A) \neq M$, by Theorem 1.5 (2), $A/Nil(A)$ is a Noetherian valuation domain with nonzero Krull's dimension. Then $A/Nil(A)$ is a DVR of rank 1. Thus M is of height 1 in A and $Spec(A) = \{Nil(A), M\}$. Since A is nonnil-Noetherian, M is finitely generated, so it is a principal ideal.

" $\Leftarrow$ " The only nonnil prime ideal of A is M which is principal. Then A is nonnil-Noetherian.   □

A prime ideal P of a ring A is called divided if $P \subseteq xA$ for every $x \in A \setminus P$.

Corollary 1.10 (Badawi 2003) *Let A be a ring. Suppose that $Nil(A)$ is a divided prime ideal. Then A is nonnil-Noetherian if and only if $A/Nil(A)$ is Noetherian.*

Proof By Theorem 1.5 (2), it suffices to prove that if $A/Nil(A)$ is Noetherian, then each height 1 prime ideal P of A is finitely generated. Since $A/Nil(A)$ is Noetherian, its ideal $P/Nil(A)$ is finitely generated. Put $P/Nil(A) = (\bar{x}_1, \ldots, \bar{x}_n)$ with $x_1, \ldots, x_n \in P$. If all the $x_i \in Nil(A)$, then $P/Nil(A) = (\bar{0})$ so $P = Nil(A)$ which is impossible since P is of height 1. So there is $x_{i_0} \notin Nil(A)$. Since $Nil(A)$ is divided, then $Nil(A) \subset (x_{i_0})$. Thus $P = Nil(A) + (x_1, \ldots, x_n) \subseteq (x_{i_0}) + (x_1, \ldots, x_n) = (x_1, \ldots, x_n) \subseteq P$. Then $P = (x_1, \ldots, x_n)$ which is finitely generated.   □

We will show later, in Example 2 at the beginning of Sect. 4, that the hypothesis "$Nil(A)$ is a prime ideal" is not enough in the corollary.

Krull's intersection theorem 1.11 *Let I be an ideal of a Noetherian ring A. If $b \in \bigcap_{k:1}^{\infty} I^k$, then $b \in bI$.*

Proof (Perdry 2004) Put $I = (a_1, \ldots, a_n)A$. For each $k \in \mathbb{N}^*$, $b \in I^k$, so there exists a homogeneous polynomial $P_k = P_k(X_1, \ldots, X_n) \in A[X_1, \ldots, X_n]$ of degree k such that $b = P_k(a_1, \ldots, a_n)$. For each $k \in \mathbb{N}^*$, consider the ideal $J_k = (P_1, \ldots, P_k)$ of the Noetherian ring $A[X_1, \ldots, X_n]$. Fix an integer m such that $J_{m+1} = J_m$. Then $P_{m+1} = Q_m P_1 + \ldots + Q_1 P_m$ where $Q_i = Q_i(X_1, \ldots, X_n) \in A[X_1, \ldots, X_n]$ is a homogeneous polynomial of degree i. Substituting $X_1 = a_1, \ldots, X_n = a_n$, we obtain $b = b(Q_1(a_1, \ldots, a_n) + \ldots + Q_m(a_1, \ldots, a_n))$. For $1 \leq i \leq m$, the polynomial Q_i is homogeneous of positive degree; hence $Q_i(a_1, \ldots, a_n) \in I$. From this it follows that $b \in bI$.   □

Example Let I be a proper ideal of a Noetherian integral domain. Then $\bigcap_{k:1}^{\infty} I^k = (0)$. Indeed, let $b \in \bigcap_{k:1}^{\infty} I^k$. Then $b = ba$ with $a \in I$. So $b(1-a) = 0$ with $1-a \neq 0$, then $b = 0$.

The following corollary is the analogous of Krull's intersection theorem for nonnil-Noetherian rings.

Corollary 1.12 *Let A be a nonnil-Noetherian ring and I be a proper ideal of A.*

(1) *If $Nil(A) \notin Spec(A)$, then for all $b \in \bigcap_{k=1}^{\infty} I^k$, $b \in bI$.*

(2) *If $Nil(A) \in Spec(A)$, then $\bigcap_{k=1}^{\infty} I^k \subseteq Nil(A)$.*

Proof

(1) By Theorem 1.5 (1), A is Noetherian. We use Krull's intersection theorem.
(2) Suppose now $Nil(A) \in Spec(A)$. Then $Nil(A) + I$ is a proper ideal of A (if not, then $1 = n + i$ for some $n \in Nil(A)$ and $i \in I$. Then i is the sum of a nilpotent and a unit; hence it is a unit. But this contradicts the fact that I is proper). The image $\bar{I}$ of $Nil(A) + I$ in $A/Nil(A)$ is proper. Since $A/Nil(A)$ is a Noetherian integral domain, by the Example of Krull's intersection theorem, $\bigcap_{k=1}^{\infty} \bar{I}^k = (\bar{0})$. Then $\bigcap_{k=1}^{\infty} I^k \subseteq Nil(A)$. $\square$

Proposition 1.13 *Let A be a ring. Then A is nonnil Noetherian if and only if A satisfies the ascending chain condition for finitely generated nonnil ideals.*

Proof " $\Leftarrow$ " Suppose that there is a nonnil ideal I of A not finitely generated. Let $a_1 \in I \setminus Nil(A)$. Then $(a_1) \subset I$. Let $a_2 \in I \setminus (a_1)$. Then $(a_1) \subset (a_1, a_2) \subset I$, ... Because of $a_1 \notin Nil(A)$, we have a strictly ascending chain of nonnil finitely generated ideals. $\square$

Lemma 1.14 *Let A be a Noetherian ring and B an over ring of A, i.e., $A \subseteq B \subseteq S^{-1}A$, where S is the set of the regular elements of A. Then B is Noetherian. Moreover, if $dim(A) = n < \infty$, then $dim(B) \leq n$.*

Proof

(1) Let $P \in Spec(B)$. Then $(P \cap A)B \subseteq P$. Conversely, let $p \in P \subset B \subseteq S^{-1}A$, $p = \frac{a}{s}$ with $a \in A$ and $s \in S$. Then $a = sp \in P \cap A$. So $p = \frac{a}{s} = \frac{1}{s}a \in (P \cap A)B$. Then $P = (P \cap A)B$. Since $P \cap A$ is finitely generated in A, then $P = (P \cap A)B$ is finitely generated in B by the same elements. So B is Noetherian.
(2) Let $P \in Spec(B)$. Then $ht_A(P \cap A) \leq dim A = n$. By the converse of the generalized principal ideal Theorem 8.10, Chap. 2, $P \cap A$ is minimal over an ideal $(p_1, \ldots, p_n)A$ with $p_1, \ldots, p_n \in P \cap A$. We will show that P is minimal over the ideal $(p_1, \ldots, p_n)B$. Let $Q \in Spec(B)$ be such that $(p_1, \ldots, p_n)B \subseteq Q \subseteq P$. In particular, $p_1, \ldots, p_n \in Q \cap A$. Then $(p_1, \ldots, p_n)A \subseteq Q \cap A \subseteq P \cap A$. By minimality of $P \cap A$ over $(p_1, \ldots, p_n)A$, we must have $Q \cap A = P \cap A$. By (1), $Q = (Q \cap A)B = (P \cap A)B = P$. By the generalized principal

ideal Theorem 8.7, Chap. 2, $ht_B(P) \leq n$. This is for each $P \in Spec(B)$. Then $dim(B) \leq n$. □

Example Each over ring of a PID is a PID. For example, all the sub-rings of $\mathbb{Q}$ are PID.

Indeed, let A be a PID, B an over ring of A, and I a nonzero ideal of B. Then $I \cap A = aA$. We will show that $I = aB$. Let $0 \neq x = \frac{c}{d} \in I$ with c and d are nonzero relatively prime elements in A. There exist $u, v \in A$ such that $uc + vd = 1$. Then $u\frac{c}{d} + v = \frac{1}{d}$, so $ux + v = \frac{1}{d}$. Since $c = dx \in I \cap A = aA$, there is $\alpha \in A$ such that $c = \alpha a$. Then $x = \frac{c}{d} = \frac{\alpha}{d}a = \alpha(ux + v)a \in Ba$.

Proposition 1.15 *Let A be a nonnil-Noetherian ring and B an over ring of A, i.e., $A \subseteq B \subseteq S^{-1}A$ where S is the set of regular elements of A. If $dim(A) = n$, then $dim(B) \leq n$.*

Proof Since $Nil(A) = A \cap Nil(B)$, then $\bar{A} = A/Nil(A) \subseteq B/Nil(B) = \bar{B}$. The elements of the set $\bar{S} = \{\bar{s}; s \in S\}$ are regular in $\bar{A}$. Indeed, let $s \in S$ and $a \in A$ such that $\bar{s}\bar{a} = \bar{0}$. Then $sa \in Nil(A)$. There is $n \in \mathbb{N}^*$ such that $(sa)^n = 0$. Since s^n is regular then $a^n = 0$ so $\bar{a} = \bar{0}$. We have $\bar{B} \subseteq \bar{S}^{-1}\bar{A}$. Thus $\bar{B}$ is an over ring of $\bar{A}$ with $\bar{A}$ a Noetherian ring. It is clear that $dim(\bar{A}) = dim(A) = n$ and $dim(B) = dim(\bar{B})$. By the preceding lemma, $dim(\bar{B}) \leq n$. Then $dim(B) \leq n$. □

2   The Nilradical of the Formal Power Series Ring

The results of this section are due to Hizem and Benhissi (2011).

Lemma 2.1 *Let A be a ring. Then $Nil(A[[X]]) \subseteq Nil(A)[[X]]$.*

Proof Let $f = \sum_{i:0}^{\infty} a_i X^i \in Nil(A[[X]])$. For each prime ideal P of A, $P[[X]]$ is a prime ideal of $A[[X]]$, and then $f \in P[[X]]$ so $a_0, a_1, \ldots \in P$. Then $a_0, a_1, \ldots \in \bigcap\{P; \ P \in spec(A)\} = Nil(A)$. □

The inclusion in the preceding lemma may be strict, as it is shown in the following example.

Example (Fields 1971) Let K be a field with zero characteristic, $n \geq 2$ an integer, and $A = K[Y_i, i \in \mathbb{N}]/(Y_i^n, i \in \mathbb{N})$. Let $a_i = \bar{Y}_i \in A$ for each $i \in \mathbb{N}$. It is clear that for each $i \in \mathbb{N}$, $a_i^n = 0$. Let $f(X) = \sum_{i:0}^{\infty} a_i X^i \in A[[X]]$. By Example 4 of Lemma 5.5, Chap. 4, f is not nilpotent even though all its coefficients are nilpotent with bounded index.

Proposition 2.2 *Let I be an ideal of a ring A with $I \subseteq Nil(A)$. The following assertions are equivalent:*

(a) $I[[X]] \subseteq Nil(A[[X]])$.
(b) I is an SFT-ideal of the ring A.
(c) There exists $k \in \mathbb{N}^*$ such that $x^k = 0$ for each $x \in I$.
(d) $I[[X]]$ is an SFT-ideal of the ring $A[[X]]$.

Proof The equivalence $"(b) \Longleftrightarrow (d)"$ follows from Lemma 5.9 of Chap. 4.

$"(b) \Longrightarrow (c)"$ There exist $s \in \mathbb{N}^*$ and a finitely generated ideal $F \subseteq I \subseteq Nil(A)$ such that for each $x \in I$, $x^s \in F$. Since F is finitely generated and contained in $Nil(A)$, there is $r \in \mathbb{N}^*$ such that for each $y \in F$, $y^r = 0$. Then for each $x \in I$, $x^{sr} = 0$.

$"(c) \Longrightarrow (b)"$ Take the integer k and the zero ideal.

$"(b) \Longrightarrow (a)"$ There exist $n \in \mathbb{N}^*$ and $a_1, \ldots, a_s \in I$ such that for each $x \in I$, $x^n \in (a_1, \ldots, a_s)$. Since $I \subseteq Nil(A)$, we can find $m \in \mathbb{N}^*$ such that $a_1^m = \ldots = a_s^m = 0$. For each $y \in (a_1, \ldots, a_s)$, $y^{ms} = 0$. Then for each $x \in I$, $x^{nms} = 0$. The result follows from Lemma 1.15 of Chap. 4.

$"(a) \Longrightarrow (c)"$ Suppose that there does not exist $k \in \mathbb{N}^*$ such that $x^k = 0$ for each $x \in I$. We can extract a sequence $(a_i)_{i \in \mathbb{N}^*}$ of elements of I whose indexes of nilpotence are not bounded. Let $f = \sum_{i:1}^{\infty} a_i X^{i!} \in I[[X]] \subseteq Nil(A[[X]])$. There exists an integer $k \geq 2$ such that $f^k = 0$. We will show that for each integer $n \geq k$, the coefficient of $X^{k(n!)}$ in f^k is a_n^k. Indeed, $f = g + X^{(n+1)!}h$ with $g = \sum_{i:1}^{n} a_i X^{i!}$ and $h \in A[[X]]$. The coefficient of $X^{k(n!)}$ in f^k is then equal to that of $X^{k(n!)}$ in g^k, which is a_n^k. Since $f^k = 0$, for each $n \geq k$, $a_n^k = 0$. The index of nilpotence of the sequence $(a_n)_{n \in \mathbb{N}^*}$ is then bounded, which is absurd. □

Corollary 2.3 *The following assertions are equivalent for a ring A:*

(a) $Nil(A)[[X]] = Nil(A[[X]])$.
(b) $Nil(A)$ is an SFT-ideal of the ring A.
(c) There exists $k \in \mathbb{N}^*$ such that $x^k = 0$ for each $x \in Nil(A)$.
(d) $Nil(A)[[X]]$ is an SFT-ideal of the ring $A[[X]]$.

Proof Take $I = Nil(A)$ in the preceding proposition and use Lemma 2.1. □

Corollary 2.4 *Let A be a ring and $n \in \mathbb{N}^*$. The following assertions are equivalent:*

(a) $Nil(A)[[X_1, \ldots, X_n]] = Nil(A[[X_1, \ldots, X_n]])$.
(b) $Nil(A)$ is an SFT-ideal of the ring A.
(c) $Nil(A)[[X_1, \ldots, X_n]]$ is an SFT-ideal of the ring $A[[X_1, \ldots, X_n]]$.

Proof By induction on n. The case $n = 1$ is the preceding corollary. Suppose that the equivalences are true for n. We will prove them for $n + 1$.

"$(a) \Longrightarrow (b)$" By hypothesis, $Nil(A)[[X_1, \ldots, X_{n+1}]] = Nil(A[[X_1, \ldots, X_{n+1}]])$. Then
$Nil(A)[[X_1, \ldots, X_{n+1}]] \cap A[[X_1]] = Nil(A[[X_1, \ldots, X_{n+1}]]) \cap A[[X_1]]$ so $Nil(A)[[X_1]] = Nil(A[[X_1]])$. By the preceding corollary, $Nil(A)$ is SFT.

"$(b) \Longrightarrow (c)$" Since $Nil(A)$ is SFT, by the equivalences of the induction hypothesis, the ideal $Nil(A)[[X_1, \ldots, X_n]] = Nil(A[[X_1, \ldots, X_n]])$ is SFT. We apply the preceding corollary to the ring $B = A[[X_1, \ldots, X_n]]$. Then $Nil(B)[[X_{n+1}]]$ is SFT. But the ideal $Nil(B)[[X_{n+1}]] = Nil(A[[X_1, \ldots, X_n]])[[X_{n+1}]] = Nil(A)[[X_1, \ldots, X_n]][[X_{n+1}]] = Nil(A)[[X_1, \ldots, X_n, X_{n+1}]]$. Then the ideal $Nil(A)[[X_1, \ldots, X_n, X_{n+1}]]$ is SFT.

"$(c) \Longrightarrow (a)$" By hypothesis, $Nil(A)[[X_1, \ldots, X_n, X_{n+1}]]$ is SFT. By Lemma 5.9 of Chap. 4, $Nil(A)[[X_1, \ldots, X_n]]$ is also SFT. By the equivalences of the induction hypothesis, $Nil(A)[[X_1, \ldots, X_n]] = Nil(A[[X_1, \ldots, X_n]])$ is then SFT. We apply the preceding corollary to the ring $B = A[[X_1, \ldots, X_n]]$, with $Nil(B)$ a SFT-ideal. Then $Nil(B)[[X_{n+1}]] = Nil(B[[X_{n+1}]])$. But the ideal $Nil(B)[[X_{n+1}]] = Nil(A[[X_1, \ldots, X_n]])[[X_{n+1}]]$
$= Nil(A)[[X_1, \ldots, X_n]][[X_{n+1}]] = Nil(A)[[X_1, \ldots, X_n, X_{n+1}]]$ and $Nil(B[[X_{n+1}]])$
$= Nil(A[[X_1, \ldots, X_n]][[X_{n+1}]]) = Nil(A[[X_1, \ldots, X_n, X_{n+1}]])$.
Then $Nil(A)[[X_1, \ldots, X_n, X_{n+1}]] = Nil(A[[X_1, \ldots, X_n, X_{n+1}]])$.   □

3 Krull's Dimension of the Formal Power Series Ring over a Nonnil-Noetherian Ring

The results of this section are due to Hizem and Benhissi (2011).

Theorem 3.1 *The following assertions are equivalent for any nonnil-Noetherian ring A:*

(a) For each $n \in \mathbb{N}^*$, $A[[X_1, \ldots, X_n]]$ is a SFT-ring.
(b) There exists $n \in \mathbb{N}^*$ such that $A[[X_1, \ldots, X_n]]$ is a SFT-ring.
(c) A is a SFT-ring.
(d) $Nil(A)$ is an SFT-ideal of the ring A.

Proof

"$(a) \Longrightarrow (b)$" Clear. "$(b) \Longrightarrow (c)$" $A \simeq A[X_1, \ldots, X_n]]/(X_1, \ldots, X_n)$ is a SFT-ring, by Lemma 2.1 of Chap. 4.

"$(c) \Longrightarrow (d)$" Clear. "$(d) \Longrightarrow (a)$" By Corollary 2.4, since $Nil(A)$ is a SFT-ideal of A, then $Nil(A)[[X_1, \ldots, X_n]] = Nil(A[[X_1, \ldots, X_n]])$ is a SFT-ideal of $A[[X_1, \ldots, X_n]]$.

We have $A[[X_1, \ldots, X_n]]/Nil(A[[X_1, \ldots, X_n]]) = A[[X_1, \ldots, X_n]]/Nil(A)$
$[[X_1, \ldots, X_n]] \simeq (A/Nil(A))[[X_1, \ldots, X_n]]$, which is Noetherian since the
ring $A/Nil(A)$ is Noetherian by Theorem 1.5. Let P be a prime ideal of the
ring $A[[X_1, \ldots, X_n]]$. The ideal $P/Nil(A[[X_1, \ldots, X_n]])$ is finitely generated by
$(\bar{f}_1, \ldots, \bar{f}_s)$ with $f_1, \ldots, f_s \in P$. Then $P = Nil(A[[X_1, \ldots, X_n]]) + (f_1, \ldots, f_s)$
is a SFT-ideal of $A[[X_1, \ldots, X_n]]$, by Lemma 2.2 of Chap. 4. □

Example We will construct an example of a nonnil-Noetherian ring that is not SFT.
Let $(Y_i)_{i \in \mathbb{N}}$ be a sequence of indeterminates over $\mathbb{Q}$ and $n \geq 2$ a natural integer.
Consider the ideal $I = (Y_i^n; i \in \mathbb{N})$ of the ring $B = \mathbb{Q}[Y_i^n; i \in \mathbb{N}]$. By the example
of Proposition 1.3, the ring $A = B/I$ is nonnil-Noetherian. It is not SFT. Indeed, the
unique prime ideal of A is $P = (\bar{Y}_i, i \in \mathbb{N}) = Nil(A)$. Suppose that P is SFT. By

Proposition 1.17 of Chap. 4, $P[[X]] = \sqrt{P.A[[X]]}$. Let $f = \sum_{i:0}^{\infty} \bar{Y}_i X^i \in P[[X]]$.

There exists $m \in \mathbb{N}^*$ such that $f^m \in P.A[[X]]$. Then $f^m = \bar{Y}_0 f_0 + \ldots + \bar{Y}_s f_s$ with
$f_0, \ldots, f_s \in A[[X]]$. Since $\bar{Y}_0, \ldots, \bar{Y}_s$ are nilpotent, then f is also nilpotent, which
is absurd, by the example of Lemma 2.1.

Theorem 3.2 *Let A be a nonnil-Noetherian ring.*

1. *If A is not a SFT ring, then $\dim A[[X_1, \ldots, X_n]] = \infty$.*
2. *If A is a SFT-ring, then $\dim A[[X_1, \ldots, X_n]] = \dim A + n$.*

Proof

(1) By Theorem 1.13 of Chap. 4.
(2) By Theorem 1.5, the ring $A/Nil(A)$ is Noetherian. By Corollary 2.4, since A
   is a SFT-ring, then $Nil(A[[X_1, \ldots, X_n]]) = Nil(A)[[X_1, \ldots, X_n]]$. Since all
   the prime ideals of a ring contain its nilradical, then $\dim(A/Nil(A)) = \dim A$
   and $\dim A[[X_1, \ldots, X_n]]$
   $= \dim(A[[X_1, \ldots, X_n]]/Nil(A[[X_1, \ldots, X_n]]))$
   $= \dim(A[[X_1, \ldots, X_n]]/Nil(A)[[X_1, \ldots, X_n]])$
   $= \dim(A/Nil(A))[[X_1, \ldots, X_n]]$
   $= \dim(A/Nil(A)) + n$, by Corollary 1.26 of Chap. 4.
   $= \dim(A) + n$. □

Example Let $B = \mathbb{Q}[Y_i; i \in \mathbb{N}]$, $I = (Y_i Y_j; i, j \in \mathbb{N})$; and $A = B/I$. For
each $i \in \mathbb{N}$, $\bar{Y}_i^2 = 0$, then the unique prime ideal of A is $M = (\bar{Y}_i; i \in \mathbb{N}) = Nil(A)$. By Proposition 1.3, the ring A is nonnil-Noetherian. For each $f \in Nil(A)$,
$f^2 = 0$, then $Nil(A)$ is a SFT-ideal. By Theorem 3.1, A is a SFT-ring. Since M
is not finitely generated, the ring A is not Noetherian. By the preceding theorem,
$\dim A[[X_1, \ldots, X_n]] = 0 + n = n$.

4   Transfer of the Nonnil-Noetherian Property to the Ring of Formal Power Series

The results of this section are due to Hizem and Benhissi (2011). We start by two examples.

Example 1 Let $B = \mathbb{Q}[Y_i;\ i \in \mathbb{N}^*]$, $I = (Y_i^i;\ i \in \mathbb{N}^*)$, and $A = B/I$. The unique prime ideal of A is $M = (\bar{Y}_i;\ i \in \mathbb{N}^*) = Nil(A)$. By Proposition 1.3, the ring A is nonnil-Noetherien. The ideal M is not finitely generated, and then A is not Noetherian. For each $i \in \mathbb{N}^*$, $\bar{Y}_i^i = 0$, but $\bar{Y}_i^j \neq 0$ for $j < i$. By Corollary 2.3, $Nil(A)$ is not a SFT-ideal, and then A is not a SFT-ring. The unique maximal ideal $M + XA[[X]]$ of $A[[X]]$ is not finitely generated. Indeed, in the contrary case, its image M by the canonical projection $A[[X]] \longrightarrow A$ will be finitely generated. By Lemma 2.1, $Nil(A[[X]]) \subseteq Nil(A)[[X]] = M[[X]] \subset M + XA[[X]]$. Then $A[[X]]$ is not nonnil-Noetherian.

Example 2 Let $B = \mathbb{Q}[Y_i;\ i \in \mathbb{N}]$, $I = (Y_iY_j;\ i, j \in \mathbb{N})$, and $A = B/I$. In the example of Theorem 3.2, we have always shown that A is a SFT nonnil-Noetherian ring and its unique prime ideal $M = (\bar{Y}_i;\ i \in \mathbb{N}) = Nil(A)$ is not finitely generated. The unique maximal ideal $M + XA[[X]]$ of $A[[X]]$ is not finitely generated. Indeed, in the contrary case, its image M by the canonical projection $A[[X]] \longrightarrow A$ will be finitely generated. By Lemma 2.1, $Nil(A[[X]]) \subseteq Nil(A)[[X]] = M[[X]] \subset M + XA[[X]]$. Then $A[[X]]$ is not nonnil-Noetherian.

To finish with this example, we will show that the hypothesis "The nilradical is a prime ideal" is not enough in Corollary 1.10. We will give an example of a ring A_1 such that $Nil(A_1)$ is prime and $A_1/Nil(A_1)$ is Noetherian, but A_1 is not nonnil-Noetherian. Take $A_1 = A[[X]]$, and it is not nonnil-Noetherian. By Corollary 2.3, since A is a SFT-ring, then $Nil(A[[X]]) = Nil(A)[[X]]$. By the example of Proposition 1.4, since A is nonnil-Noetherian, then $A/Nil(A)$ is Noetherian. Then $A_1/Nil(A_1) = A[[X]]/Nil(A[[X]]) = A[[X]]/Nil(A)[[X]]) \simeq (A/Nil(A))[[X]]$, which is Noetherian. Note that $Nil(A_1) = M[[X]]$ is a prime not divided since $X \notin M[[X]]$ and $M[[X]] \nsubseteq XA_1 = XA[[X]]$.

Theorem 4.1 *The following assertions are equivalent for a ring A:*

(a) *The ring A is nonnil-Noetherian and the ideal $Nil(A)$ is finitely generated.*
(b) *The ring A is Noetherian.*
(c) *The ring $A[X]$ is Noetherian.*
(d) *The ring $A[X]$ is nonnil-Noetherian.*
(e) *The ring $A[[X]]$ is Noetherian.*
(f) *The ring $A[[X]]$ is nonnil-Noetherian.*

Proof The following equivalences and implications are well known: $(b) \iff (c) \iff (e)$, $(e) \implies (f)$, and $(c) \implies (d)$. See Sects. 4 and 5 of Chap. 1.

"$(a) \Longrightarrow (b)$" Let P be a prime ideal of A, and then $Nil(A) \subseteq P$. If $P \neq Nil(A)$, then P is finitely generated since A is nonnil-Noetherian. If $P = Nil(A)$, then it is finitely generated by hypothesis. We conclude by Cohen's theorem.

"$(f) \Longrightarrow (a)$" By the example of Proposition 1.4, $A \simeq A[[X]]/(X)$ is nonnil-Noetherian. Since X is not nilpotent, $Nil(A) + XA[[X]] \not\subseteq Nil(A[[X]])$. By hypothesis, $Nil(A) + XA[[X]]$ is finitely generated. Then $Nil(A)$ is finitely generated.

"$(d) \Longrightarrow (a)$" In the same way, we prove this implication.   □

Example Let A be a non-Noetherian ring and I an ideal of A that is not finitely generated. The ideal $J = I + XA[X]$ (resp. $I + XA[[X]]$) of $A[X]$ (resp. $A[[X]]$) is nonnil since X is not nilpotent. It is not finitely generated. Indeed, if $J = (f_1(X), \ldots, f_n(X))$, then $I = (f_1(0), \ldots, f_n(0))$, which is absurd. In particular, $A[X]$ (resp. $A[[X]]$) is not nonnil-Noetherian.

Lemma 4.2 *Let A be a ring and P a prime ideal of A. The prime ideals $P[[X]] \subset P + XA[[X]]$ are consecutive in $A[[X]]$.*

Proof Let $Q \in spec(A[[X]])$ be such that $P[[X]] \subset Q \subseteq P + XA[[X]]$. There exists $f = \sum_{i\cdot 0}^{\infty} a_i X^i \in Q \setminus P[[X]]$. Let n be the smallest index such that $a_n \notin P$.

Since $Q \subseteq P + XA[[X]]$, then $a_0 \in P$ and $n \geq 1$. We have $f = \sum_{i:0}^{n-1} a_i X^i + X^n g$

with $g = a_n + a_{n+1}X + \ldots \notin P + XA[[X]]$. Then $g \notin Q$ and $\sum_{i:0}^{n-1} a_i X^i \in P[[X]]$.

So $X^n g = f - \sum_{i:0}^{n-1} a_i X^i \in Q$, then $X \in Q$. On the other hand, $P \subseteq P[[X]] \subset Q$. Then $P + XA[[X]] \subseteq Q$. So $Q = P + XA[[X]]$.   □

Example If A is an integral domain, the prime ideals $(0) \subset XA[[X]]$ are consecutive in the domain $A[[X]]$.

Proposition 4.3 *Let A be a nonnil-Noetherian ring and $P \in Spec(A[[X]])$. In the following two cases, P is finitely generated.*

1. *The ideal $p = \{f(0); \ f \in P\} \not\subseteq Nil(A)$.*
2. *A is a SFT-ring and $ht(P) \geq 2$.*

Proof

(1) Since A is nonnil-Noetherian and $p \not\subseteq Nil(A)$, then p is a finitely generated in A. By Kaplansky's theorem, Sect. 5 of Chap. 1, the ideal P is finitely generated.
(2) Suppose that P is not finitely generated. By (1), $p \subseteq Nil(A)$. Since $P \cap A \subseteq p \subseteq Nil(A) \subseteq P \cap A$, then $p = Nil(A) = P \cap A \in Spec(A)$. So

$P \subseteq p + XA[[X]] = Nil(A) + XA[[X]]$. By Corollary 2.3, since A is a SFT-ring, $Nil(A[[X]]) = Nil(A)[[X]]$ is the unique minimal prime ideal of $A[[X]]$. Then $Nil(A[[X]]) \subseteq P \subseteq Nil(A) + XA[[X]]$. But by the preceding lemma, $Nil(A)[[X]] \subset Nil(A) + XA[[X]]$ are consecutive, and then $ht(P) \leq 1$, which is absurd. $\square$

Example 1 The hypothesis $p \nsubseteq Nil(A)$ is necessary in the first point of the proposition. Let $B = \mathbb{Q}[Y_i; \ i \in \mathbb{N}^*]$, $I = (Y_i^i; \ i \in \mathbb{N}^*)$ and $A = B/I$. We see in Example 1 at the beginning of this section that A is nonnil-Noetherian, its unique prime ideal is $Nil(A) = (\bar{Y}_i; \ i \in \mathbb{N}^*)$, and the unique maximal ideal $Nil(A) + XA[[X]]$ of $A[[X]]$ is not finitely generated.

Example 2 The hypothesis $ht(P) \geq 2$ is necessary in the second point of the proposition. Let $B = \mathbb{Q}[Y_i; \ i \in \mathbb{N}]$, $I = (Y_i Y_j; \ i, j \in \mathbb{N})$ and $A = B/I$. We see in Example 2 at the beginning of this section that A is a SFT nonnil-Noetherian ring and its unique prime ideal is $M = (\bar{Y}_i; \ i \in \mathbb{N})$. Then $dim\ A = 0$. By Theorem 3.2, $dim\ A[[X]] = 1$. The height of the unique maximal ideal $M + XA[[X]]$ of $A[[X]]$ is 1, and it is not finitely generated.

5 Characterization of the Nonnil-Noetherian Property by Formal Power Series

In this section, we characterize nonnil-Noetherian rings in terms of power series. The results in this section are due to Benhissi (2020).

Lemma 5.1 *Let P be an ideal of a ring A. If P is maximal among the nonnil not finitely generated ideals of A, then P is prime.*

Proof Suppose there are $a, b \in A \setminus P$ such that $ab \in P$. Since $P \subset P + aA$, then $P + aA$ is nonnil so it is finitely generated. Put $P + aA = (\alpha_1, \ldots, \alpha_n, a)$ with $\alpha_i \in P$. Let $J = (P : a) = \{y \in A; \ ya \in P\}$. Then J is an ideal of A containing strictly P because of b, so J is nonnil and then finitely generated. Put $J = (\beta_1, \ldots, \beta_s)$. We will prove that $P = (\alpha_1, \ldots, \alpha_n, \beta_1 a, \ldots, \beta_s a)$, which is finitely generated, a contradiction. Par hypothesis, $\alpha_i \in P$ and $\beta_i \in J$, so $\beta_i a \in P$, by definition of J. This proves an inclusion. Conversely, let $z \in P \subset P + aA$. There exist $u_1, \ldots, u_n, u \in A$ such that $z = \sum_{i=1}^{n} u_i \alpha_i + ua$. It suffices to prove that $ua \in (\beta_1 a, \ldots, \beta_s a)$. But $ua = z - \sum_{i=1}^{n} u_i \alpha_i \in P$ because $z, \alpha_i \in P$. Then $u \in J$, by definition of J. Since $J = (\beta_1, \ldots, \beta_s)$, then $ua \in (\beta_1 a, \ldots, \beta_s a)$. $\square$

Lemma 5.2 *Let M be a module which is not finitely generated. Then it admits a countably generated sub-module which is not finitely generated.*

Proof Let $m_0 \in M$ then $M \neq (m_0)$. Let $m_1 \in M \setminus (m_0)$. Then $M \neq (m_0, m_1)$. Let $m_2 \in M \setminus (m_0, m_1)$. Then $M \neq (m_0, m_1, m_2)$, ... etc. Suppose that the sub-module $N = (m_0, m_1, m_2, \ldots)$ is generated by a finite subset $\{r_1, \ldots, r_s\}$. There exists $k \in \mathbb{N}$ such that $r_1, \ldots, r_s \in (m_0, m_1, \ldots, m_k)$, and then $m_{k+1} \in (m_0, m_1, \ldots, m_k)$, which is impossible. □

Lemma 5.3 *Let I be an ideal of a ring A. Then $I[[X]] = IA[[X]]$ if and only if for every countable subset S of I, there is a finitely generated ideal F of A such that $S \subseteq F \subseteq I$.*

Proof "$\Longrightarrow$" Let $(a_i)_{i \in \mathbb{N}}$ be a countable subset of elements of I. Then $f = \sum_{i=0}^{\infty} a_i X^i \in I[[X]] = IA[[X]]$. There exist $b_1, \ldots, b_n \in I$ and $g_1, \ldots, g_n \in A[[X]]$ such that $f = b_1 g_1 + \ldots + b_n g_n$. For $F = b_1 A + \ldots + b_n A$, we have $\{a_i; \ i \in \mathbb{N}\} \subseteq F \subseteq I$.

"$\Longleftarrow$" Since $I \subseteq I[[X]]$, then $IA[[X]] \subseteq I[[X]]$. Conversely, let $f = \sum_{i=0}^{\infty} a_i X^i \in I[[X]]$. There exists $F = b_1 A + \ldots + b_n A$ such that $\{a_i; \ i \subset \mathbb{N}\} \subseteq F \subseteq I$. Each $a_i = \sum_{j=1}^{n} a_{ij} b_j$ with $a_{ij} \in A$. Then $f = \sum_{j=1}^{n} b_j (\sum_{i=0}^{\infty} a_{ij} X^i) \in IA[[X]]$. □

Example 1 Let A be a nondiscrete valuation domain of Krull's dimension 1 with maximal ideal M. By the example of Lemma 1.16, Chap. 4, $M[[X]] \neq MA[[X]]$.

Example 2 A ring A is Noetherian if and only if for each ideal I of A, we have $I[[X]] = I.A[[X]]$. Indeed, if A is Noetherian, each ideal I of A in finitely generated. By Lemma 1.16, Chap. 4, we have $I[[X]] = I.A[[X]]$. Conversely, suppose that A is not Noetherian. There exists a strictly increasing sequence $(I_i)_{i \in \mathbb{N}}$ of ideals of A. For each $i \in \mathbb{N}^*$, let $a_i \in I_i \setminus I_{i-1}$. The set $I = \bigcup_{i=0}^{\infty} I_i$ is an ideal of A. By hypothesis, $I[[X]] = I.A[[X]]$. By the preceding lemma, there exists a finitely generated ideal F of A such that $\{a_i; \ i \in \mathbb{N}\} \subseteq F \subseteq I$. Since $(I_i)_{i \in \mathbb{N}}$ is increasing, there exists $k \in \mathbb{N}$ such that $F \subseteq I_k$. Then $\{a_i; \ i \in \mathbb{N}\} \subseteq I_k$, which is absurd.

Example 3 Let K be a field and $\{Y_i; \ i \in \mathbb{N}\}$ a sequence of indeterminates over K. The domain $A = K[Y_i; \ i \in \mathbb{N}]$ is not Noetherian since its ideal $I = (Y_i; \ i \in \mathbb{N})$ is not finitely generated. Suppose that $I[[X]] = I.A[[X]]$. By the preceding lemma, there exists a finitely generated ideal F of A containing the countable set $S = \{Y_i; \ i \in \mathbb{N}\}$ of I such that $F \subset I$. Then $I = F$ is finitely generated, which is absurd.

Lemma 5.4 (Arnold et al. 1977) *Let I and J be ideals of a ring A such that $J[[X]] = JA[[X]]$ and $J \subseteq \sqrt{I}$. Then there exists $n \in \mathbb{N}^*$ such that $J^n \subseteq I$.*

Proof Suppose that for each $m \in \mathbb{N}^*$, $J^m \not\subseteq I$, there exist $b_{m1}, \ldots, b_{mm} \in J$ such that the product $b_{m1} \ldots b_{mm} \notin I$. Let C be the ideal of A generated by the countable set $\{b_{mi}; \ m \in \mathbb{N}^*, 1 \leq i \leq m\}$. Then $C \subseteq J$, since J contains the generators of C, and for each $m \in \mathbb{N}^*$, $C^m \not\subseteq I$ since $b_{m1}b_{m2} \ldots b_{mm} \notin I$. By the preceding lemma, since $J[[X]] = J.A[[X]]$, there exists a finitely generated ideal F of A such that $C \subseteq F \subseteq J \subseteq \sqrt{I}$. Since F is finitely generated and $F \subseteq \sqrt{I}$, there exists $n \in \mathbb{N}^*$ such that $F^n \subseteq I$. But $C \subseteq F$, and then $C^n \subseteq I$, a contradiction.   □

We say that a ring has a Noetherian prime spectrum if it satisfies the ascending chain condition (acc) for radical ideals

Example 1 Any Noetherian ring has a Noetherian prime spectrum.

Example 2 Let A be a nondiscrete valuation domain of dimension 1 or a valuation domain of finite dimension ≥ 2. Then A is not Noetherian, but it has a finite prime spectrum. Then it has a finite number of radical ideals. So it satisfies the acc for radical ideals.

Example 3 Let A be a nondiscrete valuation domain of dimension 1. Then A has a Noetherian prime spectrum. By Example 5 of Definition 1.9, Chap. 4, A is not a SFT-ring. By Theorem 1.13, Chap. 4, $A[[X]]$ admits an infinite strictly increasing sequence of prime ideals. Then $A[[X])$ does not have a Noetherian prime spectrum.

Lemma 5.5 (Ohm and Pendleton 1968) *The following assertions are equivalent for a ring A:*

(a) The ring A has a Noetherian prime spectrum.

(b) Each radical ideal of A is the radical of a finitely generated ideal of A.

(c) Each prime ideal of A is the radical of a finitely generated ideal of A.

Proof We proceed as follow:

$"(a) \Longleftrightarrow (b) \Longleftrightarrow (c)"$.

$"(a) \Longrightarrow (b)"$ Suppose that there exists a radical ideal I of A that is not the radical of a finitely generated ideal of A. Let $a_0 \in I$. Then $\sqrt{a_0 A} \subset I$. There exists $a_1 \in I \setminus \sqrt{a_0 A}$. Then $\sqrt{a_0 A} \subset \sqrt{a_0 A + a_1 A} \subset I, \ldots$, etc. We construct a strictly increasing sequence of radical ideals of A, a contradiction.

$"(b) \Longrightarrow (a)"$ Let $(I_n)_{n \in \mathbb{N}}$ be an increasing sequence of radical ideals of A. Then $\displaystyle\bigcup_{n=0}^{\infty} I_n$ is a radical ideal of A. By hypothesis, it is the radical of a finitely generated ideal of A. Put $\displaystyle\bigcup_{n=0}^{\infty} I_n = \sqrt{a_0 A + \ldots + a_s A}$. There exists $k \in \mathbb{N}$ such that $a_0, \ldots, a_s \in I_k$. Then $\sqrt{a_0 A + \ldots + a_s A} \subseteq I_k$, so $I_n = I_k$ for each $n \geq k$.

$'(b) \implies (c)''$ Clear. $''(c) \implies (b)''$ We use the equalities $\sqrt{JK} = \sqrt{J} \cap \sqrt{K} = \sqrt{J \cap K}$, true for each couple (J, K) of ideals of A. Let $\mathscr{F}$ be the set of the radical ideals of A that are not the radicals of finitely generated ideals of A. Suppose that $\mathscr{F} \neq \emptyset$. It is easy to see that $(\mathscr{F}, \subseteq)$ is inductive. Let P be a maximal element of $\mathscr{F}$. It suffices to show that P is a prime ideal of A to obtain a contradiction. Suppose that there exist $a, b \in A \setminus P$ such that $ab \in P$. Consider the ideals $J = P + aA$ and $K = P + bA$ of A. We have $JK = (P + aA)(P + bA) = P^2 + aP + bP + abA \subseteq P$, and then $\sqrt{JK} \subseteq P$. So $\sqrt{JK} \subseteq P \subseteq J \cap K \subseteq \sqrt{JK}$, and then $P = \sqrt{JK}$. On the other hand, $P \subset J \subseteq \sqrt{J}$ and $P \subset K \subseteq \sqrt{K}$. By maximality of P, the radical ideals $\sqrt{J}$ and $\sqrt{K}$ must be the radicals of finitely generated ideals of A. Put $\sqrt{J} = \sqrt{a_1 A + \ldots + a_n A}$ and $\sqrt{K} = \sqrt{b_1 A + \ldots + b_m A}$. Then $P = \sqrt{JK} = \sqrt{J} \cap \sqrt{K} = \sqrt{a_1 A + \ldots + a_n A} \cap \sqrt{b_1 A + \ldots + b_m A} = \sqrt{(a_1 A + \ldots + a_n A)(b_1 A + \ldots + b_m A)} = \sqrt{\sum a_i b_j A}$. Then P is the radical of a finitely generated ideal, which contradicts the fact that $P \in \mathscr{F}$. We conclude that P is a prime ideal. □

Lemma 5.6 *Let A be a ring such that for every nonnil prime ideal P of A, $P[[X]] = PA[[X]]$. Then A has a Noetherian prime spectrum.*

Proof Suppose there exists a strictly increasing sequence $(I_i)_{i \in \mathbb{N}^*}$ of radical ideals of A. Then $Nil(A) = \sqrt{(0)} \subseteq I_1 \subset I_2 \subset \ldots$. Let's consider the nonnil ideal $I = \bigcup_{i=1}^{\infty} I_i$. For every $i \in \mathbb{N}^*$, choose $a_i \in I_{i+1} \setminus I_i$. Put $f = \sum_{i=1}^{\infty} a_i X^{i!} \in I[[X]]$. Let $n \geq k > 1$ be integers. The coefficient of $X^{k(n!)}$ in f^k is a_n^k. Indeed, $f = g + X^{(n+1)!} h$, with $g = \sum_{i=1}^{n} a_i X^{i!}$ and $h \in A[[X]]$. Since $k(n!) < (n+1)!$, the coefficient of $X^{k(n!)}$ in f^k is the same as the one of $X^{k(n!)}$ in g^k. But the coefficient of $X^{k(n!)}$ in the polynomial g^k is a_n^k. Since $a_n \notin I_n$ and I_n is radical, then $a_n^k \notin I_n$, so $f^k \notin I_n[[X]]$, for every $n \geq k$. Thus $f^k \notin \bigcup_{n=k}^{\infty} I_n[[X]] = \bigcup_{n=1}^{\infty} I_n[[X]]$ because the sequence $(I_n)_{n \in \mathbb{N}^*}$ is increasing. Since $IA[[X]] \subseteq \bigcup_{n=1}^{\infty} I_n[[X]]$, then for every integer $l \geq 1$, $f^l \notin IA[[X]]$, so $f \notin \sqrt{IA[[X]]}$. There exists a prime ideal $\mathscr{P}$ of $A[[X]]$ such that $IA[[X]] \subseteq \mathscr{P}$ and $f \notin \mathscr{P}$. Let $P = \mathscr{P} \cap A \in Spec(A)$. Then $PA[[X]] \subseteq \mathscr{P}$, so $f \notin PA[[X]]$. On the other hand, $I \subseteq \mathscr{P} \cap A = P$, so $I \subseteq P$, and then P is nonnil and $I[[X]] \subseteq P[[X]]$. Since $f \in I[[X]]$, then $f \in P[[X]]$. We conclude that $P[[X]] \neq PA[[X]]$: contradiction. □

Example The converse of the lemma is false. Let A be a nondiscrete valuation domain of Krull's dimension 1 and maximal ideal M. Then A has a Noetherian prime spectrum. The unique nonnil prime ideal of A is M and $M[[X]] \neq MA[[X]]$.

Theorem 5.7 *A ring A is nonnil-Noetherian if and only if for every nonnil prime ideal P of A, $P[[X]] = P.A[[X]]$.*

Proof $"\Rightarrow"$ Let P be a nonnil prime ideal of A. By hypothesis, P is finitely generated. By Lemma 1.16 of Chap. 4, $P[[X]] = P.A[[X]]$.

$"\Leftarrow"$ Suppose that A is not nonnil-Noetherian. The set $\mathscr{F}$ of nonnnil ideals of A which are not finitely generated is not empty. It is closed under unions of nonempty chains. By Zorn's lemma, $(\mathscr{F}, \subseteq)$ admits a maximal element P, and it is prime by Lemma 5.1. The quotient ring A/P is Noetherian because all the ideals of A containing strictly P are nonnil and then finitely generated by maximality of P. By the preceding lemma, the ring A has a Noetherian prime spectrum. By Lemma 5.5, there exists a finitely generated ideal I of A such that $P = \sqrt{I}$. By hypothesis, $P[[X]] = PA[[X]]$. By Lemma 5.4, there exists $n \in \mathbb{N}^*$ such that $P^n \subseteq I$. Now P/P^2 is naturally a A/P-module, since P annihilates P/P^2. We will prove that P/P^2 is a finitely generated A/P-module. In the contrary case, by Lemma 5.2, there exists a countably generated sub-module $E = (\bar{a}_i; \ i \in \mathbb{N})$ of P/P^2 which is not finitely generated, $a_i \in P$. Since $P[[X]] = PA[[X]]$, by Lemma 5.3, for the countable set $\{a_i; \ i \in \mathbb{N}\}$ of P, there exists a finitely generated ideal $F = (b_1, \ldots, b_s)$ of A such that $\{a_i; \ i \in \mathbb{N}\} \subseteq F \subseteq P$. The sub-module $M = (\bar{b}_1, \ldots, \bar{b}_s)$ of P/P^2 is finitely generated and $E \subseteq M$. Since the ring A/P is Noetherian, then M is a Noetherian A/P-module, and then E a finitely generated sub-module, which is a contradiction.

Let $\{\bar{x}_1, \ldots, \bar{x}_s\}$ be generators of the A/P-module P/P^2, where $x_i \in P$. Let's consider the ideal $H = x_1 A + \ldots + x_s A$ of A. Then $H \subseteq P$ and $P = P^2 + H$, so $P^2 = P^3 + PH$, and then $P = P^2 + H = P^3 + PH + H = P^3 + H$. Thus $P = P^3 + H$. By induction, for every $k \in \mathbb{N}^*$, $P = P^k + H$. In particular, $P = P^n + H$. Since $P^n \subseteq I \subseteq P$, then $P = P^n + H \subseteq I + H \subseteq P$. Thus $P = I + H$ is a finitely generated ideal of A, which contradicts the fact that P is not finitely generated.   □

Example Let A be a nondiscrete valuation domain of Krull's dimension 1 and maximal ideal M. Then $Nil(A) = (0)$, and M is nonnil and not finitely generated, so A is not nonnil-Noetherian and $M[[X]] \neq MA[[X]]$.

6   Nonnil-S-Noetherian Rings

The main purpose of this section is to integrate the concepts of S-Noetherian and nonnil-Noetherian rings. We start by studying the basic properties of nonnil-S-Noetherian rings. For the notions of S-finite ideals and S-Noetherian rings, the readers must refer to Chap. 1. In the sequel, a multiplicative set contains 1 and does not contain zero. The results of this section are due to Kwon and Lim (2020).

Definition 6.1 Let A be a ring and S a multiplicative set of A. We say that A is nonnil-S-Noetherian if each nonnil ideal of A is S-finite.

Example 1 If the multiplicative set S consists of units of the ring A, then the concept of S-finite ideals is the same as that of finitely generated ideals. In this case, the notion of nonnil-S-Noetherian rings coincides with that of nonnil-Noetherian rings.

Example 2 If A is reduced, then the concept of nonnil-S-Noetherian rings is precisely the same as that of S-Noetherian rings.

Example 3 Let $S_1 \subseteq S_2$ be the two multiplicative subsets of a ring A. If A is nonnil-S_1-Noetherian, then it is nonnil-S_2-Noetherian.

The reader can refer to Sect. 1 of Chap. 2 for saturated multiplicative sets.

Example 4 Let S be a multiplicative set of a ring A and S^* its saturation. Then A is nonnil-S-Noetherian if and only if it is nonnil-S^*-Noetherian. Indeed, let I be a nonnil ideal of A. There exist $r \in S^*$ and a finitely generated ideal F of A such that $rI \subseteq F \subseteq I$. By definition, $\frac{r}{1} \in U(S^{-1}A)$. There exist $a \in A$ and $s \in S$ such that $\frac{a}{s} \frac{r}{1} = \frac{1}{1}$. There is $u \in S$ such that $uar = su$. Let $s' = su \in S$. Then $s'I = uarI \subseteq uaF \subseteq F \subseteq I$. So I is S-finite.

Lemma 6.2 *Let S be a multiplicative subset of a ring A. If P is an ideal of A which is maximal among all non-S-finite nonnil ideals of A, then P is a prime ideal of A disjoint from S.*

Proof Since S is not S-finite, by Example 3 of Definition 2.1, Chap. 1, $S \cap P = \emptyset$. Suppose that P is not prime. There exist $a, b \in A \setminus P$ such that $ab \in P$. Note that $P + (a)$ is a nonnil ideal of A. By the maximality of P, $P + (a)$ is S-finite. So there exist $s \in S$, $p_1, \ldots, p_m \in P$ and $r_1, \ldots, r_m \in A$ such that $s(P + (a)) \subseteq (p_1 + ar_1, \ldots, p_m + ar_m) \subseteq P + (a)$. Note that $(P : a)$ is an ideal of A containing P and b. So $(P : a)$ is a nonnil ideal of A. By the maximality of P, $(P : a)$ is S-finite. So $t(P : a) \subseteq (q_1, \ldots, q_n) \subseteq (P : a)$ for some $t \in S$ and $q_1, \ldots, q_n \in (P : a)$.

Let $z \in P$. Then $sz = \sum_{i:1}^{m} x_i(p_i + ar_i)$ for some $x_1, \ldots, x_m \in A$. So $a(\sum_{i:1}^{m} x_i r_i) = sz - \sum_{i:1}^{m} x_i p_i \in P$. Then $\sum_{i:1}^{m} x_i r_i \in (P : a)$. Therefore we can find $c_1, \ldots, c_n \in A$ such that $t \sum_{i:1}^{m} x_i r_i = \sum_{j:1}^{n} c_j q_j$, which states that $stz = \sum_{i:1}^{m} tx_i p_i + at \sum_{i:1}^{m} x_i r_i = \sum_{i:1}^{m} tx_i p_i + \sum_{j:1}^{n} ac_j q_j$. Hence we obtain $stP \subseteq (p_1, \ldots, p_m, aq_1, \ldots, aq_n) \subseteq P$, which means that P is S-finite. However, this is a contradiction to the choice of P. Thus P is a prime ideal of A. $\square$

Now, we will give the Cohen-type theorem.

Theorem 6.3 *Let S be a multiplicative subset of a ring A. The following statements are equivalent:*

1. *The ring A is nonnil-S-Noetherian.*
2. *Every nonnil prime ideal of A (disjoint with S) is S-finite.*

Proof $"(1) \implies (2)"$ This implication is obvious. $"(2) \implies (1)"$ Let E be the set of non-S-finite nonnil ideals of A. Suppose that A is not nonnil-S-Noetherian. Then E is a nonempty partially ordered set under inclusion. Let $\{C_\alpha\}_{\alpha \in \Gamma}$ be a chain in E, and let $C = \bigcup_{\alpha \in \Gamma} C_\alpha$. Then C is a nonnil ideal of A. If C is S-finite, then we can find an element $s \in S$ and a finitely generated ideal F of A such that $sC \subseteq F \subseteq C$. Since F is finitely generated, $F \subseteq C_\beta$ for some $\beta \in \Gamma$, so $sC_\beta \subseteq F \subseteq C_\beta$. This means that C_β is S-finite, which is absurd. Therefore C is not S-finite. Clearly, C is an upper bound of $\{C_\alpha\}_{\alpha \in \Gamma}$. By Zorn's lemma, E has a maximal element P. By the preceding lemma, P is a prime ideal which is disjoint from P. However, this contradicts the hypothesis. Thus A is nonnil-S-Noetherian.   □

We next give a relation between nonnil-S-Noetherian and S-Noetherian properties in a ring.

Proposition 6.4 *Let S be a multiplicative subset of a ring A. The following statements are equivalent:*

1. *The ring A is S-Noetherian.*
2. *The ring A is nonnil-S-Noetherian and $Nil(A)$ is a S-finite ideal of A.*

Proof $"(1) \implies (2)"$ This implication is obvious. $"(2) \implies (1)"$ Let P be a prime ideal of A. If $P = Nil(A)$, then P is S-finite by the assumption. Suppose that P contains properly $Nil(A)$. Then P is a nonnil ideal of A. Since A is nonnil-S-Noetherian, then P is S-finite. By Corollary 2.2 of Chap. 1, A is S-Noetherian.   □

Since the zero ideal is S-finite for each multiplicative subset of a ring A, the preceding proposition allows us to retrieve Example 2 of Definition 6.1. That is, if a ring A is reduced, then A is S-Noetherian if and only if A is nonnil-S-Noetherian. Suppose that A is a nonnil-S-Noetherian ring that is not S-Noetherian. Then $Nil(A)$ is not a S-finite ideal. The following corollary says more.

Corollary 6.5 *Let S be a multiplicative subset of a ring A. If A is nonnil-S-Noetherian but it is not S-Noetherian, then $Nil(A)$ is the unique minimal prime ideal of A.*
Moreover, $Nil(A)$ is not a S-finite ideal.

Proof That $Nil(A)$ is not S-finite follows from the preceding proposition. Since every prime ideal of A contains $Nil(A)$, it suffices to show that $Nil(A)$ is prime. In the contrary case, each prime ideal of A is nonnil, and then it is S-finite. By Corollary 2.2 of Chap. 1, A is S-Noetherian. This is a contradiction.   □

We will give an example of a nonnil-S-Noetherian ring that is not S-Noetherian, whose unique non-S-finite prime ideal is $Nil(A)$.

Example Let $X = \{X_i;\ i \in \mathbb{N}\}$ be a set of indeterminates over an integral domain D. Consider the prime ideal $P = (X_i;\ i \in \mathbb{N})$ and the ideal $I = (X_i^2;\ i \in \mathbb{N})$ of the domain $D[X]$. Note that any prime ideal of the ring $A = D[X]/I$ contains P/I. Then the unique minimal prime ideal of A is P/I. So $Nil(A) = P/I$ and $S = A \setminus (P/I)$ is a multiplicative subset of A. Any nonnil prime ideal Q of A contains properly P/I, and then $Q \cap S \neq \emptyset$. By Example 3 of Definition 2.1, Chap. 1, the ideal Q is S-finite. By Theorem 6.3, the ring A is nonnil-S-Noetherian. Suppose that the ideal P/I is S-finite. There are $\bar{s} \in S$ and $i_1, \ldots, i_n \in \mathbb{N}$ such that $\bar{s}(P/I) \subseteq (\overline{X_{i_1}}, \ldots, \overline{X_{i_n}}) \subseteq (P/I)$. The polynomial s of $D[X]$ uses a finite number of variables $X_{j_1}, \ldots, X_{j_m}$ and its constant term $d \neq 0$. Let $k \in \mathbb{N} \setminus \{i_1, \ldots, i_n, j_1, \ldots, j_m\}$. Then $\bar{s}\overline{X_k} = \bar{f_1}\overline{X_{i_1}} + \ldots + \bar{f_n}\overline{X_{i_n}}$ with $\bar{f_1}, \ldots, \bar{f_n} \in A$. So $sX_k - f_1 X_{i_1} - \ldots - f_n X_{i_n} \in I$. Take $X_{i_1} = \ldots = X_{i_n} = X_{j_1} = \ldots = X_{j_m} = 0$, and we obtain $dX_k \in (X_i^2;\ i \in \mathbb{N} \setminus \{i_1, \ldots, i_n, j_1, \ldots, j_m\})$. This is a contradiction.

Let A be a ring and S a multiplicative subset of A. An ideal I of A is said to be *radically S-finite* if there exist an element $s \in S$ and a finitely generated subideal F of I such that $sI \subseteq \sqrt{F}$. The ring A is said to have a S-*Noetherian spectrum* if every ideal of A is radically S-finite.

Proposition 6.6 *Let S be a multiplicative subset of a ring A.*
If A is nonnil-S-Noetherian, then A has a S-Noetherian spectrum.

Proof Let I be an ideal of A. If $I \subseteq Nil(A) = \sqrt{(0)}$, then $sI \subseteq \sqrt{(0)}$ for each $s \in S$. If $I \not\subseteq Nil(A)$, there exist $s \in S$ and a finitely generated subideal F of I such that $sI \subseteq F \subseteq \sqrt{F}$. Then I is radically S-finite. So A has a S-Noetherian spectrum. $\square$

A ring A is *decomposable* if it can be written as $A = A_1 \oplus A_2$ for some nonzero rings A_1 and A_2.

Theorem 6.7 *Let A be a decomposable ring, S a multiplicative subset of A, and $\{\pi_i\}_{i \in \Lambda}$ the set of canonical epimorphisms from A to each component of decompositions of A. Then the following statements are equivalent:*

1. *The ring A is S-Noetherian.*
2. *The ring A is nonnil-S-Noetherian.*
3. *For each $i \in \Lambda$, $\pi_i(A)$ is a $\pi_i(S)$-Noetherian ring.*
4. *If $e \in A \setminus \{0, 1\}$ is an idempotent element, then every ideal of A contained in (e) is S-finite.*

Proof

$"(1) \implies (2)"$ This implication is obvious. $"(2) \implies (3)"$ Let $i \in \Lambda$. Then $A = \pi_i(A) \oplus \pi_j(A)$ for some $j \in \Lambda$. Let I be an ideal of $\pi_i(A)$. Then $I \oplus \pi_j(A)$ is a nonnil ideal of A. By hypothesis, there exist $s \in S$ and a finitely generated ideal F of A such that $s(I \oplus \pi_j(A)) \subseteq F \subseteq I \oplus \pi_j(A)$. Therefore

$\pi_i(s)I \subseteq \pi_i(F) \subseteq I$. Note that $\pi_i(F)$ is a finitely generated ideal of $\pi_i(A)$. So I is $\pi_i(S)$-finite. Thus $\pi_i(A)$ is $\pi_i(S)$-Noetherian.

"(3) $\Longrightarrow$ (4)" Let $e \in A \setminus \{0, 1\}$ be an idempotent element and I an ideal of A contained in (e). We have $A = (e) \oplus (1 - e)$. By assumption, (e) is a $\pi_i(S)$-Noetherian ring for some $i \in \Lambda$. So there exist an element $s \in S$ and a finitely generated ideal F of (e) such that $\pi_i(s)I \subseteq F \subseteq I$. Note that if J is an ideal of (e), then J can be regarded as an ideal $J \oplus (0)$ of A. So F is a finitely generated ideal of A and $sI \subseteq F \subseteq I$. Thus I is S-finite.

"(4) $\Longrightarrow$ (1)" Let I be an ideal of A. Since A is decomposable, $A = (e) \oplus (1 - e)$ for some idempotent element $e \in A \setminus \{0, 1\}$. So $I = I_e \oplus I_{1-e}$ for some ideals I_e and I_{1-e} of (e) and $(1 - e)$, respectively. Take $i, j \in \Lambda$ so that $\pi_i(A) = (e)$ and $\pi_j(A) = (1 - e)$. By the assumption, there exist $s, t \in S$ and finitely generated ideals E, F of A such that $\pi_i(s)I_e \subseteq \pi_i(E) \subseteq I_e$ and $\pi_j(t)I_{1-e} \subseteq \pi_j(F) \subseteq I_{1-e}$. Hence $stI \subseteq \pi_i(E) \oplus \pi_j(F) \subseteq I$. This implies that I is S-finite. Thus A is S-Noetherian. $\square$

Corollary 6.8 Let $A_1, \ldots, A_n$ be rings $(n \geq 2)$ and $S_1, \ldots, S_n$ multiplicative subsets of $A_1, \ldots, A_n$, respectively. Consider the product ring $A = A_1 \times \ldots \times A_n$ and its multiplicative subset $S = S_1 \times \ldots \times S_n$. The following assertions are equivalent:

1. For $i = 1, \ldots, n$, the ring A_i is S_i-Noetherian.
2. The ring A is S-Noetherian.
3. The ring A is nonnil-S-Noetherian.

Proof The equivalences are immediate consequences of the preceding theorem. $\square$

Example 1 With the notations and hypotheses of the preceding corollary, if there exists an index $k \in \{1, \ldots, n\}$ such that A_k is a nonnil-S_k-Noetherian ring that is not S_k-Noetherian, then A is never nonnil-S-Noetherian.

Indeed, suppose that for each $i \in \{1, \ldots, n\}$, the ring A_i is nonnil-S_i-Noetherian and that A_1 is not S_1-Noetherian. By Proposition 6.4, the ideal $Nil(A_1)$ is not S_1-finite in A_1. Then the ideal $Nil(A_1) \times A_2 \times \ldots \times A_n$ is nonnil in A and is not S-finite. So the ring A is not nonnil-S-Noetherian.

Example 2 The preceding corollary is not generally extended to the case of an infinite product of (nonnil-)S-Noetherian rings.

Indeed, note that $\mathbb{Z}$ is a $\{1\}$-Noetherian ring. Let $A = \prod_{i:1}^{\infty} \mathbb{Z}$ and $S = \prod_{i:1}^{\infty} \{1\}$. The ideal $I = \mathbb{Z}^{(\mathbb{N}^*)}$, of the sequences of elements of A with finite supports, is nonnil and not S-finite.

Proposition 6.9 Let $A \subseteq B$ be an extension of rings such that $IB \cap A = I$ for each ideal I of A, and let S be a multiplicative subset of A. If B is a nonnil-S-Noetherian ring, then so is A.

Proof Let I be a nonnil ideal of A. Since B is a nonnil-S-Noetherian ring and IB is a nonnil ideal of B, we can find $s \in S$ and $b_1, \ldots, b_n \in IB$ such that $s(IB) \subseteq (b_1, \ldots, b_n)B \subseteq IB$. Let F be a finitely generated subideal of I such that $(b_1, \ldots, b_n)B \subseteq FB$. Then by hypothesis, we have $sI = (sI)B \cap A \subseteq (b_1, \ldots, b_n)B \cap A \subseteq FB \cap A = F \subseteq I$. So I is S-finite. Thus A is a nonnil-S-Noetherian ring. $\square$

Proposition 6.10 *Let A be a ring and S a multiplicative subset of A.*

1. *If A is nonnil-S-Noetherian, then A_S is nonnil-Noetherian.*
2. *If A is nonnil-S-Noetherian and if S consists of regular elements of A, then for each nonnil ideal I of A, there exists $s \in S$ such that $(S^{-1}I) \cap A = (I : s)_A$.*
3. *Suppose that S consists of regular elements of A. Then A is nonnil-S-Noetherian if and only if A_S is nonnil-Noetherian and for each nonnil finitely generated ideal F of A, there exists $s \in S$ such that $(S^{-1}F) \cap A = (F : s)_A$.*

Proof

(1) An ideal of A_S is of the form $S^{-1}I$ with I an ideal of A. If $S^{-1}I$ is nonnil in A_S, then I is nonnil in A. Since A is nonnil-S-Noetherian, there exist $s \in S$ and a finitely generated ideal F of A such that $sI \subseteq F \subseteq I$. Then $S^{-1}I = S^{-1}(sI) \subseteq S^{-1}F \subseteq S^{-1}I$ so $S^{-1}I = S^{-1}F$ is a finitely generated ideal of A_S.

(2) Since S does not contain zero divisors, then $A \subseteq S^{-1}A$. Let I be any nonnil ideal of A. Since $I \subseteq (S^{-1}I) \cap A$, then $(S^{-1}I) \cap A$ is a nonnil ideal. Then $(S^{-1}I) \cap A$ is S-finite. There exist $s_1 \in S$ and a finitely generated ideal F of A such that $s_1((S^{-1}I) \cap A) \subseteq F \subseteq (S^{-1}I) \cap A$. Since F is finitely generated, there exists $s_2 \in S$ such that $s_2 F \subseteq I$. Let $s = s_1 s_2 \in S$. Then $s((S^{-1}I) \cap A) \subseteq s_2 F \subseteq I$, so $(S^{-1}I) \cap A \subseteq (I : s)_A$. Conversely, let $x \in (I : s)_A$. Then $sx \in I$, so $x = \frac{sx}{s} \in (S^{-1}I) \cap A$. Then $(S^{-1}I) \cap A = (I : s)_A$.

(3) The first inclusion follows from (1) and (2). Conversely, let I be any nonnil ideal of A. Since $I \subseteq S^{-1}I$, then $S^{-1}I$ is a nonnil ideal of $S^{-1}A$. Since $S^{-1}A$ is a nonnnil-Noetherian ring, then $S^{-1}I$ is finitely generated. Then $S^{-1}I = S^{-1}F$ with F a finitely generated ideal of A contained in I. Then $I \subseteq (S^{-1}I) \cap A = (S^{-1}F) \cap A = (F : s)_A$ for some $s \in S$. Then $sI \subseteq F \subseteq I$ and so I is S-finite. $\square$

Let A be a ring and P a prime ideal of A. Then $A \setminus P$ is a multiplicative subset of A. We end this section with a characterization of nonnil-Noetherian rings.

Proposition 6.11 *The following conditions are equivalent for a ring A.*

1. *The ring A is nonnil-Noetherian.*
2. *The ring A is nonnil-$(A \setminus P)$-Noetherian for all prime ideals P of A.*
3. *The ring A is nonnil-$(A \setminus M)$-Noetherian for all maximal ideals M of A.*

Proof The implications "$(1) \implies (2) \implies (3)$" are clear.

$(3) \implies (1)$ Let I be a nonnil ideal of A. For each maximal ideal M of A, there exist an element $s_M \in A \setminus M$ and a finitely generated ideal F_M of A such $s_M I \subseteq F_M \subseteq I$.

Consider the ideal $J = \left(s_M,\ M \in Max(A)\right)$ of A. If $J \neq A$, then there exits a maximal ideal M_0 of A such that $J \subseteq M_0$. Then $s_{M_0} \in M_0$, which is absurd. So $J = A$. We can choose a finite subset $M_1, \ldots, M_n$ of maximal ideals of A such that $(s_{M_1}, \ldots, s_{M_n}) = A$. Therefore we obtain $I = (s_{M_1}, \ldots, s_{M_n})I \subseteq F_{M_1} + \ldots + F_{M_n} \subseteq I$, which means that $I = F_{M_1} + \ldots + F_{M_n}$. Hence I is finitely generated. Thus A is a nonnil-Noetherian ring. $\square$

As an easy consequence of the preceding proposition, we obtain

Corollary 6.12 *Let (A, M) be a local ring. Then A is nonnil-Noetherian if and only if A is nonnil-$(A \setminus M)$-Noetherian.*

Problems

The first nine exercises are due to Hizem and Benhissi (2011).

Exercise 1 Let S be a multiplicative set of a nonnil-Noetherian ring A. Show that $S^{-1}A$ is nonnil-Noetherian.

Exercise 2 Show that the following assertions are equivalent for the rings A and B:

(a) The product ring $A \times B$ is nonnil-Noetherian.
(b) The product ring $A \times B$ is Noetherian.
(c) The rings A and B are Noetherian.

Exercise 3

(1) Let A be a nonnil-Noetherian ring. Show that $Spec(A)$ is Noetherian. In particular, each proper ideal of A admits a finite number minimal prime ideals.
(2) Show by examples that the converse of (1) is false.

Exercise 4 Let A be a nonnil-Noetherian ring and $P \in Spec(A)$. Show that:

(a) If P is minimal over a finitely generated ideal $I = (a_1, \ldots, a_n)$, then $ht(P) \leq n$.
(b) If $ht(P) = n \in \mathbb{N}^*$, there exist n elements $a_1, \ldots, a_n \in A$ such that P is minimal over the ideal $(a_1, \ldots, a_n)$.

Exercise 5 Show that in a nonnil-Noetherian ring A, each nonnil ideal contains a finite product of prime ideals.

Exercise 6 An ideal I of a ring A is said to be irreducible if it does not admit a decomposition of the form $I = I_1 \cap I_2$ with I_1 and I_2 two ideals of A satisfying $I \subset I_1$ and $I \subset I_2$. Show that in a nonnil-Noetherian ring A, each nonnil ideal is a finite intersection of irreducible ideals.

Exercise 7 Show that in a nonnil-Noetherian ring A, each irreducible nonnil ideal is primary.

Exercise 8 Show that in a nonnil-Noetherian ring A, each nonnil ideal I contains a power of its radical.

Exercise 9 Let A be a ring.

(a) Show that $Nil(A[X]) = Nil(A)[X]$.
(b) Show that if A is nonnil-Noetherian, then $dim\ A[X_1, \ldots, X_n] = dim\ A + n$.

Exercise 10 (Hizem and Benhissi 2005) Let $A \subset B$ be an extension of rings and $R = A + XB[[X]]$. Show that the following assertions are equivalent:

(a) The ring R is Noetherian.
(b) The ring R is nonnil-Noetherian.
(c) The ring A is Noetherian, and B is a finitely generated A-module.

Exercise 11 (Watkins 2012; Fields 1973)

(1) We consider the product ring $B = \prod_{\mathbb{N}^*} \mathbb{Q} = \{(a_i)_{i \in \mathbb{N}^*};\ a_i \in \mathbb{Q}\}$. For $a = (a_1, a_2, \ldots) \in B$, the support of a is the set $S(a) = \{i \in \mathbb{N}^*;\ a_i \neq 0\}$. Let J be the ideal of B generated by the elements $e_1 = (1, 0, 0, 0, 0, \ldots)$; $e_2 = (0, 1, 0, 0, 0, \ldots)$; $e_3 = (0, 0, 1, 0, 0, \ldots)$; $e_4 = (0, 0, 0, 1, 0, \ldots)$; $\ldots$

    (a) Show that J is formed by all the elements of B of finite supports.
    (b) Show that $J[[X]] \neq JB[[X]]$.

(2) (a) Let A be a ring and I an ideal of A such that $I[[X]] = IA[[X]]$. Show that each increasing sequence $(I_n)_{n \in \mathbb{N}}$ of ideals of A, satisfying $\bigcup_{n:0}^{\infty} I_n = I$, is stationary.
    (b) Find in the ring B, of the first question, a strictly increasing sequence $(J_n)_{n \in \mathbb{N}^*}$ of ideals such that $\bigcup_{n:1}^{\infty} J_n = J$.

(3) Let A be a valuation domain and P a nonzero prime ideal of A.

    (a) Let $(P_i)_{i \in \mathbb{N}}$ be a sequence of prime ideals of A strictly included in P with different terms such that $\bigcup_{i:0}^{\infty} P_i = P$. Show that we can extract a strictly increasing subsequence $(P_{i_k})_{k \in \mathbb{N}}$ such that $\bigcup_{k:0}^{\infty} P_{i_k} = P$.
    (b) Suppose that P is not branched. Show that $P[[X]] = PA[[X]]$ if and only if for each sequence $(P_i)_{i \in \mathbb{N}}$ of prime ideals strictly included in P, we have the strict inclusion $\bigcup_{i:0}^{\infty} P_i \subset P$.

Exercise 12 (Badawi 2003) Let R be an integral domain.

(1) Let M be a R-module. Show that $Nil(R(+)M) = (0)(+)M$. Deduce that the ideal $Nil(R(+)M)$ is finitely generated if and only if the module M is finitely generated.

(2) Suppose that R is Noetherian and its quotient field K is not a finitely generated R-module. Show that the ring $R(+)K$ is nonnil-Noetherian that is not Noetherian.

(3) Deduce that the ring $\mathbb{Z}(+)\mathbb{Q}$ is nonnil-Noetherian that is not Noetherian.

(4) Suppose that R is Noetherian with quotient field K. Let B be a ring containing K that is not a finitely generated R-module. Show that the ring $R(+)B$ is nonnil-Noetherian that is not Noetherian.

(5) Deduce that if p is a prime number, the ring $\mathbb{Z}_{(p)}(+)\mathbb{R}$ is nonnil-Noetherian that is not Noetherian.

Exercise 13 (Gilmer et al. 1999) Let A be a ring such that $A/Nil(A)$ is Noetherian. Show that the following assertions are equivalent:

(1) A is Noetherian.

(2) The minimal prime ideals of A are finitely generated.

(3) The ideal $Nil(A)$ is finitely generated.

(4) The ideal $Nil(A)$ is nilpotent, and the A-module $Nil(A)/(Nil(A))^2$ is finitely generated.

Exercise 14 Let $P \subset Q$ be the two consecutive prime ideals of a ring A. Show that the prime ideals $P + XA[[X]] \subset Q + XA[[X]]$ are consecutive in $A[[X]]$.

Exercise 15 Give $Nil(\mathbb{Z}/72\mathbb{Z})$. Is it a prime ideal?

Exercise 16 Let K be a field. Consider the domain $A = K[X, Y]$ and its ideal $I = (X^2Y, XY^2)$. Show the equality $(XY) = (X) \cap (Y)$. Deduce that $\sqrt{I} = (XY)$.

Solutions

Exercise 1 A nonnil ideal of $S^{-1}A$ is of the form $J = S^{-1}I$ with I a nonnil ideal of A. Since I is finitely generated, then so is J.

Exercise 2 By Lemma 2.2 of Chap. 2, an ideal of the product ring $A \times B$ is of the form $I = I_1 \times I_2$, where I_1 is an ideal of A and I_2 is an ideal of B. It is clear that I is finitely generated if and only if I_1 and I_2 are finitely generated. Then we have the equivalence $(b) \Longleftrightarrow (c)$. The implication $(b) \Longrightarrow (a)$ is clear. For the implication $(a) \Longrightarrow (c)$, by symmetry, it suffices to show that A is Noetherian. Let I be an ideal of A. The ideal $I \times B$ is nonnil, and then it is finitely generated in $A \times B$. So I is finitely generated in A.

Exercise 3

(1) Let P be a prime ideal of A. Then $Nil(A) \subseteq P$. If the inclusion is strict, then P is finitely generated. If $P = Nil(A) = \sqrt{0}$. We conclude by Lemma 5.5.

(2) Example 1. Let A be a nondiscrete valuation domain of dimension 1. It is not Noetherian, and then it is not nonnil-Noetherian. But A has two prime ideals. Then $Spec(A)$ is Noetherian.

Example 2. Let K be a field and $A = K[[X^{\frac{1}{\infty}}]] = \bigcup_{n=1}^{\infty} K[[X^{\frac{1}{n}}]] = \bigcup_{n=1}^{\infty} K[[X]][X^{\frac{1}{n}}]$, the Puiseux series domain with coefficients in K. The extension $K[[X]] \subset A$ is integral, so $dim(A) = dim(K[[X]]) = 1$, $Nil(A) = (0)$, the domain A is local with maximal ideal $M = (X^{\frac{1}{n}}, n \in \mathbb{N}^*)$, which is not finitely generated, so A is not nonnil-Noetherian. Since $Spec(A) = \{(0), M\}$ with $M = \sqrt{(X)}$, then A has a Noetherian spectrum.

Exercise 4

(a) Let $P_0 \subset P_1 \subset \ldots \subset P_s = P$ be any chain of prime ideals of A. In the ring $A/Nil(A)$, the prime ideal $P/Nil(A)$ is minimal over the ideal $(\bar{a}_1, \ldots, \bar{a}_n)$, and we have the chain of prime ideals $P_0/Nil(A) \subset P_1/Nil(A) \subset \ldots \subset P_s/Nil(A) = P/Nil(A)$. By Theorem 1.5, the ring $A/Nil(A)$ is Noetherian. By the generalized Krull's principal ideal Theorem 8.7 of Chap. 2, $ht(P/Nil(A)) \leq n$. Then $s \leq n$, so $ht(P) \leq n$.

(b) In the Noetherian ring $A/Nil(A)$, $ht\big(P/Nil(A)\big) = n \geq 1$. By Theorem 8.10 of Chap. 2, there exist $a_1, \ldots, a_n \in P$ such that $P/Nil(A)$ is minimal over the ideal $(\bar{a}_1, \ldots, \bar{a}_n)$. Let Q be a prime ideal of A such that $(a_1, \ldots, a_n) \subseteq Q \subseteq P$. Then $(\bar{a}_1, \ldots, \bar{a}_n) \subseteq Q/Nil(A) \subseteq P/Nil(A)$. So $Q/Nil(A) = P/Nil(A)$ and $Q = P$. Then P is minimal over the ideal $(a_1, \ldots, a_n)$.

Exercise 5 Let $\mathscr{F}$ be the set of the nonnil ideals of A that do not contain finite products of prime ideals. Suppose that $\mathscr{F} \neq \emptyset$. By Proposition 1.2, $\mathscr{F}$ admits a maximal element I. In particular, I is not prime. Let $a, b \in A \setminus I$ be such that $ab \in I$. Since $I \subset I + aA$ and $I \subset I + bA$, then $I + aA$ and $I + bA$ contain products of prime ideals. But $(I + aA)(I + bA) = I^2 + aI + bI + abA \subseteq I$. Then I contains a product of prime ideals, which is absurd.

Exercise 6 Let $\mathscr{F}$ be the set of the nonnil ideals of A that are not finite intersections of irreducible ideals. Suppose that $\mathscr{F} \neq \emptyset$. By Proposition 1.2, $\mathscr{F}$ admits a maximal element I. In particular, I is not irreducible. There exist two ideals I_1 and I_2 of A with $I \subset I_1$ and $I \subset I_2$ such that $I = I_1 \cap I_2$. The ideals I_1 and I_2 are nonnil and then are finite intersections of irreducible ideals. So is the ideal I, which is absurd.

Exercise 7 Suppose that there is an irreducible nonnil ideal I that is not primary. There exist $b, c \in A$ such that $bc \in I$ but $c \notin I$ and $b^n \notin I$ for each $n \in \mathbb{N}^*$. For each $k \in \mathbb{N}^*$, $I_k = [I : b^k] = \{y \in A; \ yb^k \in I\}$ is an ideal of A strictly containing I because of c. Then I_k is nonnil. The sequence (I_k) is increasing. By

Proposition 1.2, there exists $m \in \mathbb{N}^*$ such that $I_m = I_n$ for each integer $n \geq m$. It is clear that $I \subseteq I_m \cap (I + b^m A)$. Conversely, let $x \in I_m \cap (I + b^m A)$. There exist $a \in I$ and $r \in A$ such that $x = a + b^m r$. Since $x \in I_m$, then $x b^m \in I$ so $a b^m + r b^{2m} \in I$, and then $b^{2m} r \in I$. So $r \in I_{2m} = I_m$ and $x = a + b^m r \in I$. We conclude that $I = I_m \cap (I + b^m A)$, which contradicts the irreducibility of I.

Exercise 8 Since $I \subseteq \sqrt{I}$, then $\sqrt{I}$ is finitely generated. Put $\sqrt{I} = (x_1, \ldots, x_s)$. There exists $n \in \mathbb{N}^*$ such that $x_1^n, \ldots, x_s^n \in I$. For $m = ns$ the ideal $(\sqrt{I})^m$ is generated by the products $x_1^{r_1} \ldots x_s^{r_s}$ with $r_1 + \ldots + r_s = m$. There exists at least an index i such that $r_i \geq n$, and then $x_1^{r_1} \ldots x_s^{r_s} \in I$. So $(\sqrt{I})^m \subseteq I$.

Exercise 9

(a) Let $f = a_0 + a_1 X + \ldots + a_n X^n \in Nil(A)[X]$. Then $a_0, a_1 X, \ldots, a_n X^n$ are all nilpotent, so f is nilpotent. Conversely, we will prove by induction on $n = deg\ f$ that if $f \in Nil(A[X])$, then all its coefficients are nilpotent. If $n = 0$, then $f = a_0$ is nilpotent. Suppose that the property is true for all the polynomials of degrees $< n$. Let $f = a_0 + a_1 X + \ldots + a_n X^n \in Nil(A[X])$ and $t \in \mathbb{N}^*$ be such that $f^t = 0$. Then $a_n^t = 0$, so a_n is nilpotent, and then $f_1 = a_0 + a_1 X + \ldots + a_{n-1} X^{n-1} = f - a_n X^n$ is also nilpotent. By the induction hypothesis, $a_0, \ldots, a_{n-1}$ are nilpotent.

(b) $dim\ A[X_1, \ldots, X_n] = dim\big(A[X_1, \ldots, X_n]/Nil(A[X_1, \ldots, X_n])\big)$
$= dim\big(A[X_1, \ldots, X_n]/Nil(A)[X_1, \ldots, X_n]\big)$, by (a).
$= dim\big(A/Nil(A))[X_1, \ldots, X_n]\big)$
$= dim(A/Nil(A)) + n$, since $A/Nil(A)$ is Noetherian by Theorem 1.5.
$= dim(A) + n$.

Exercise 10 The implication $(a) \Longrightarrow (b)$ is clear, and the equivalence $(c) \Longleftrightarrow (a)$ is done in Example 2 of Corollary 4.5, Chap. 1. It remains to show the implication $(b) \Longrightarrow (c)$. Let I be an ideal of A. The ideal $I + X B[[X]]$ of R is not contained in $Nil(R)$ because of X. Then it is finitely generated. Its image I by the canonical surjection $R \longrightarrow A$ is also finitely generated. On the other hand, the ideal $X B[[X]]$ of R is not contained in $Nil(R)$ because of X. Then it is finitely generated. There exist $f_1, \ldots, f_n \in B[[X]]$ such that $X B[[X]] = X f_1 (A + X B[[X]]) + \ldots + X f_n (A + X B[[X]])$. We cancel by X, and we identify the constant terms; we obtain $B = f_1(0) A + \ldots + f_n(0) A$. Then B is a finitely generated A-module.

Exercise 11

(1) (a) An element of J is of the form $x = x_1 e_1 + \ldots + x_n e_n$ with $n \in \mathbb{N}^*$ and $x_1, \ldots, x_n \in B$. Then $S(x) \subseteq S(x_1 e_1) \cup \ldots \cup S(x_n e_n) \subseteq S(e_1) \cup \cdots \cup S(e_n) = \{1; \ldots; n\}$. Conversely, an element with finite support has the form $x = (a_1, \ldots, a_n, 0, \ldots)$
$= (a_1, 0, \ldots) e_1 + (0, a_2, 0, \ldots) e_2 + \ldots + (0, \ldots, 0, a_n, 0, \ldots) e_n \in J$.

(b) Suppose that $J[[X]] = J B[[X]]$. By Lemma 5.3, there exists a finitely generated ideal $F = (x_1, \ldots, x_n) \subseteq J$ such that $\{e_1, e_2, \ldots\} \subseteq F$. Then $J = F$. For each $i \in \mathbb{N}^*$, $S(e_i) \subseteq S(x_1) \cup \ldots \cup S(x_n)$. Then $\mathbb{N}^* \subseteq S(x_1) \cup \ldots \cup S(x_n)$, a finite set, which is absurd.

(2) (a) If $(I_n)_{n\in\mathbb{N}}$ is not stationary, we can suppose that it is strictly increasing. Let $a_i \in I_{i+1} \setminus I_i$. By Lemma 5.3, since $I[[X]] = IA[[X]]$, there exists a finitely generated ideal F of A such that $\{a_i; \ i \in \mathbb{N}\} \subseteq F \subseteq I$. There exists $k \in \mathbb{N}$ such that $F \subseteq I_k$. Then $\{a_i; \ i \in \mathbb{N}\} \subseteq I_k$, which is absurd.

(b) For each $n \in \mathbb{N}^*$, $J_n = (e_1, \ldots, e_n) = \{(a_1, a_2, \ldots, a_n, 0, \ldots); \ a_i \in \mathbb{Q}\}$.

(3) (a) We cannot have all the $P_i \subset P_0$, $i \geq 1$. Let i_1 be the first index such that $P_0 \subset P_{i_1}$. Then for each $i < i_1$, $P_i \subset P_0$. In the same way, we cannot have $P_i \subset P_{i_1}$ for each $i > i_1$. Let i_2 be the first index such that $P_{i_1} \subset P_{i_2}$. Then for each $i < i_2$, $P_i \subset P_{i_1}$. So $P_0 \subset P_{i_1} \subset P_{i_2} \subset \cdots$.

We have $\displaystyle\bigcup_{k:0}^{\infty} P_{i_k} \subseteq \bigcup_{i:0}^{\infty} P_i = P$. Since $(i_k)_{k\in\mathbb{N}}$ is strictly increasing, for each $i \in \mathbb{N}$, there exists $k \in \mathbb{N}$ such that $i < i_k$, and then $P_i \subset P_{i_k}$. So $\displaystyle\bigcup_{k:0}^{\infty} P_{i_k} = \bigcup_{i:0}^{\infty} P_i = P$.

(b) "$\Longrightarrow$" Suppose that there exists a sequence $(P_i)_{i\in\mathbb{N}}$ of prime ideals of A strictly included in P such that $\displaystyle\bigcup_{i:0}^{\infty} P_i = P$. By omitting some terms of the sequence, we may suppose that they are different. By (a), by replacing (P_i) by a subsequence, we may suppose that it is strictly increasing. Since $P[[X]] = PA[[X]]$, we have a contradiction with (2-a).

"$\Longleftarrow$" Let $(P_\lambda)_{\lambda\in\Gamma}$ be the family of the prime ideals of A strictly included in P. Since P is branched, $\displaystyle\bigcup_{\lambda\in\Lambda} P_\lambda = P$. Let $f = \displaystyle\sum_{i:0}^{\infty} a_i X^i \in P[[X]]$. For each $i \in \mathbb{N}$, there exists $\lambda_i \in \Lambda$ such that $a_i \in P_{\lambda_i}$. By hypothesis, $\displaystyle\bigcup_{i:0}^{\infty} P_{\lambda_i} \subset P$. Let $p \in P \setminus \displaystyle\bigcup_{i:0}^{\infty} P_{\lambda_i}$. For each $i \in \mathbb{N}$, $P_{\lambda_i} \subset pA$, so $f \in pA[[X]] \subseteq PA[[X]]$. Then $P[[X]] = PA[[X]]$.

Exercise 12 (1) For $k \geq 2$, $(r, m)^k = (r^k, kr^{k-1}m) = (0, 0) \iff r^k = 0$ and $kr^{k-1}m = 0 \iff r = 0$. Then $Nil(R(+)M) = (0)(+)M$.

"$\Longrightarrow$" Put $Nil(R(+)M) =< (0, b_1), \ldots, (0, b_n) >$ with $b_i \in M$. Let $b \in M$. Then $(0, b) = (a_1, c_1)(0, b_1) + \ldots + (a_n, c_n)(0, b_n) = (0, a_1 b_1) + \ldots + (0, a_n b_n)$ with $a_i \in R$ and $c_i \in M$. Then $b = a_1 b_1 + \ldots + a_n b_n$ and $M = (b_1, \ldots, b_n)$.

"$\Longleftarrow$" Put $M = (b_1, \ldots, b_n)$. For each $b \in M$, $b = a_1 b_1 + \ldots + a_n b_n$ with $a_i \in R$. Then $(0, b) = (a_1, 0)(0, b_1) + \ldots + (a_n, 0)(0, b_n)$, so $Nil(R(+)M) =< (0, b_1), \ldots, (0, b_n) >$.

(2) By (1), the ideal $Nil(R(+)K) = (0)(+)K$ is not finitely generated. Then the ring $R(+)K$ is not Noetherian. By Lemma 8.9 (2), Chap. 1, we have $Spec(R(+)K) = \{P(+)K; \ P \in spec(R)\}$. Let P be a nonzero prime ideal of R. It is generated by a finite set S. By Exercise 3, Chap. 4, $P(+)K = P(+)PK$ is generated by $S \times \{0\}$, and then it is finitely generated. Then $R(+)K$ is nonnil-Noetherian.

(3) Since $\mathbb{Q}$ is not a finitely generated $\mathbb{Z}$-module, then the ring $\mathbb{Z}(+)\mathbb{Q}$ is nonnil-Noetherian but not Noetherian.

(4) By (1), the ideal $Nil(R(+)B) = (0)(+)B$ is not finitely generated. Then the ring $R(+)B$ is not Noetherian. By Lemma 8.9 (2), Chap. 1, $Spec(R(+)B) = \{P(+)B; \ P \in Spec(R)\}$. Let P be a nonzero prime ideal of R. It is generated by a finite set S. We have $P(+)B = P(+)PB$ since the nonzero elements of P are invertible in $K \subseteq B$. By Exercise 3, Chap. 4, $P(+)PB$ is generated by $S \times \{0\}$, and then it is finitely generated. So $R(+)B$ is nonnil-Noetherian.

(5) The quotient field of $\mathbb{Z}_{(p)}$ is $\mathbb{Q} \subset \mathbb{R}$. Since $[\mathbb{Q} : \mathbb{R}] = \infty$, then $\mathbb{R}$ is not a finitely generated $\mathbb{Z}_{(p)}$-module.

Exercise 13 It is clear that (1) implies each one of the other properties.
$''(2) \implies (1)''$ Let $P \in Spec(A)$ and $P_0 \in Min(A)$ be such that $P_0 \subseteq P$. The isomorphism $\big(A/Nil(A)\big)\big/\big(P_0/Nil(A)\big) \simeq A/P_0$ shows that the ring A/P_0 is Noetherian. Its ideal P/P_0 is the finitely generated. Since P_0 is finitely generated by hypothesis, then P is also finitely generated.
$''(3) \implies (1)''$ Let $P \in Spec(A)$. Then $Nil(A) \subseteq P$. Since $A/Nil(A)$ is Noetherian, then $P/Nil(A)$ is finitely generated. But $Nil(A)$ is finitely generated by hypothesis, and then P is also finitely generated.
$''(4) \implies (1)''$ Put $N = Nil(A)$ and let $n \in \mathbb{N}^*$ be such that $N^n = (0)$. Since N/N^2 is a finitely generated A-module, there exists a finitely generated ideal $I \subseteq N$ of A such that $N = N^2 + I = (N^2 + I)^2 + I = N^4 + I = \ldots$. Since $N^n = (0)$, we obtain $N = I$ a finitely generated ideal. By the preceding implication, A is Noetherian.

Exercise 14 Let R be a prime ideal of $A[[X]]$ be such that $P + XA[[X]] \subset R \subseteq Q + XA[[X]]$. There exists $f = a + Xg \in R \setminus P + XA[[X]]$. Then $a \notin P$, but $a = f - Xg \in R \cap A$. On the other hand, $P \subseteq R \cap A \subseteq Q$, and then $R \cap A = Q$. So $Q + XA[[X]] \subseteq R$.

Exercise 15 Since $72 = 2^3.3^2$, then the prime ideals $p\mathbb{Z}$ of $\mathbb{Z}$ containing $72\mathbb{Z}$ are $2\mathbb{Z}$ and $3\mathbb{Z}$. So $Nil(\mathbb{Z}/72\mathbb{Z}) = (2\mathbb{Z}/72\mathbb{Z}) \cap (3\mathbb{Z}/72\mathbb{Z}) = (2\mathbb{Z} \cap 3\mathbb{Z})/72\mathbb{Z} = 6\mathbb{Z}/72\mathbb{Z}$ is not prime since the ideal $6\mathbb{Z}$ is not prime in $\mathbb{Z}$.

Exercise 16

(1) It is clear that $(XY) \subseteq (X) \cap (Y)$. Conversely, the ideals (X) and (Y) are prime since $K[X, Y]/(X) \simeq K[Y]$ and $K[X, Y]/(Y) \simeq K[X]$. Then the elements X and Y are irreducible and nonassociated in $K[X, Y]$ since $U(K[X, Y]) = K^*$. Let $f(X, Y) \in (X) \cap (Y)$. Then X and Y divide $f(X, Y)$. By Gauss lemma, XY divides $f(X, Y)$. So $f(X, Y) \in (XY)$. Then $(XY) = (X) \cap (Y)$.

(2) Since $(XY)^2 = (X^2Y)Y \in I$, then $XY \in \sqrt{I}$. So $(XY) \subseteq \sqrt{I}$. Conversely, $(XY) = (X) \cap (Y)$ is a radical ideal, as an intersection of two prime ideals. Since $I = (X^2Y, XY^2) \subseteq (XY)$, then $\sqrt{I} \subseteq (XY)$. So $\sqrt{I} = (XY)$.

References

J.T. Arnold, R. Gilmer, W. Heinzer, Some countability conditions in a commutative ring, Ill. J. Math. **21**, 648–665 (1977)

A. Badawi, On nonnil-Noetherian rings. Comm. Algebra **31**(4), 1669–1677 (2003)

A. Benhissi, Nonnil-Noetherian rings and formal power series. Algebra Colloq. **27**(3), 361–368 (2020)

D.E. Fields, Zero divisors and nilpotent elements in power series rings. Proc. AMS **27**(3), 427–433 (1971)

D.E. Fields, Correction to "Dimension theory in power series rings". Pacific J. Math. **49**, 616–617 (1973)

R. Gilmer, W. Heinzer, M. Roitman, Finite generation of powers of ideals. Proc. AMS **127**(11), 3141–3151 (1999)

S. Hizem, A. Benhissi, When is $A + XB[[X]]$ Noetherian. C.R. Acad. Sci. Paris Ser. I **340**, 5–7 (2005)

S. Hizem, A. Benhissi, Nonnil-Noetherian rings and the SFT-property. Rocky Mount. J. Math. **41**(5), 1483–1500 (2011)

M.J. Kwon, J.W. Lim, On nonnil-S-Noetherian rings. Mathematics **8**, 14 (2020)

J. Ohm, R.L. Pendleton, Rings with Noetherian spectrum. Duke Math. J. **35**, 631–639 (1968). Errata **35**, 875 (1968)

H. Perdry, An elementary proof of Krull's intersection theorem. Amer. Math. Monthly **111**(4), 356–357 (2004)

J.J. Watkins, Krull's dimension of polynomial and power series rings, in ed. by C. Francisco et al. *Progress in Commutative Algebra*, vol. 2 (De Gruyter, Berlin, 2012), pp. 205–219

Chapter 6
Strongly Hopfian, Endo-Noetherian, and Isonoetherian Rings

All rings considered in this chapter are commutative with identity. Let A be a ring and M a A-module. We say that M is strongly Hopfian if for each endomorphism f of M, the sequence $ker\ f \subseteq ker\ f^2 \subseteq \ldots$ is stationary. The ring A is strongly Hopfian if it is strongly Hopfian as an A-module. This is also equivalent to the fact that for each $a \in A$, the sequence $ann(a) \subseteq ann(a^2) \subseteq \ldots$ is stationary. In this chapter, we study this notion and its transfer to different extensions of a ring A.

1 Generalities on the Strongly Hopfian Rings

The results of this section are due to Hizem (2011).

Notation If a is an element of a ring A, we denote $ann(a) = \{x \in A; ax = 0\}$, an ideal of A.

Definition 1.1 We say that a ring A is strongly Hopfian if for each $a \in A$, the sequence $ann(a) \subseteq ann(a^2) \subseteq \ldots$ is stationary.

Example 1 Integral domains and Noetherian rings are strongly Hopfian.

Example 2 Let p be a prime number. The ring $\prod_{n \geq 1} \mathbb{Z}/p^n\mathbb{Z}$ is not strongly Hopfian. Indeed, let $x = (x_n)$ with $x_n = p + p^n\mathbb{Z} = \bar{p}$. Let $k \geq 1$ be a fixed integer. We will show that $ann(x^k) \subset ann(x^{k+1})$. Let $y = (y_n)$ with $y_{k+1} = 1 + p^{k+1}\mathbb{Z} = \bar{1}$ and $y_n = 0$ for $n \neq k + 1$. Then $x^{k+1} = (x_n^{k+1})$ with $x_n^{k+1} = p^{k+1} + p^n\mathbb{Z}$ for each $n \geq 1$. In particular, $x_{k+1}^{k+1} = p^{k+1} + p^{k+1}\mathbb{Z} = 0$, and then $y_{k+1}x_{k+1}^{k+1} = 0$. For $n \neq k + 1$, $y_n = 0$, then $y_n x_n^{k+1} = 0$. So $yx^{k+1} = 0$, and then $y \in ann(x^{k+1})$. On the other hand, $x^k = (x_n^k)$ with $x_n^k = p^k + p^n\mathbb{Z}$. The component number $k + 1$ in yx^k is $(1 + p^{k+1}\mathbb{Z})(p^k + p^{k+1}\mathbb{Z}) = p^k + p^{k+1}\mathbb{Z} \neq 0$. Then $y \notin ann(x^k)$.

© The Author(s), under exclusive license to Springer Nature Switzerland AG 2022
A. Benhissi, *Chain Conditions in Commutative Rings*,
https://doi.org/10.1007/978-3-031-09898-7_6

Proposition 1.2 *Let $A \subseteq B$ be an extension of rings with B strongly Hopfian. Then A is strongly Hopfian.*

Proof For each $a \in A$, $ann_A(a) = ann_B(a) \cap A$.   □

So we have the following corollary:

Corollary 1.3 *Let $A \subseteq B$ be an extension of rings with B Noetherian. Then A is strongly Hopfian.*

Proposition 1.4 (Arezzo and Robbiano 1970) *Let A be a ring and $a \in A$. The sequence $ann(a) \subseteq ann(a^2) \subseteq \ldots$ is stationary if and only if there exist two integers $m > n \geq 1$ such that $ann(a^n) = ann(a^m)$. In this case, $ann(a^n) = ann(a^{n+1}) = \ldots$.*

Proof We will show by induction that for each $k > n$, $ann(a^n) = \ldots = ann(a^k)$. The result is true for $k = n + 1$ since $ann(a^n) = ann(a^{n+1}) = \ldots = ann(a^m)$. Suppose that it is true until the order k. Let $x \in ann(a^{k+1})$. Then $xa^{k+1} = 0$, so $(xa)a^k = 0$, and then $xa \in ann(a^k) = ann(a^{k-1})$. So $0 = (xa)a^{k-1} = xa^k$, and then $x \in ann(a^k)$.   □

Lemma 1.5 *A ring A is reduced if and only if for each $a \in A$, $ann(a) = ann(a^2)$.*

Proof " $\Longrightarrow$ " Let $a \in A$ and $x \in ann(a^2)$. Then $xa^2 = 0$, so $(xa)^2 = 0$, and then $xa = 0$ and $x \in ann(a)$.

" $\Longleftarrow$ " Let $a \in Nil(A)$ and n its nilpotency index. Suppose that $n \geq 2$, then $a^{n-2}a^2 = 0$, so $a^{n-2} \in ann(a^2) = ann(a)$, and then $a^{n-1} = 0$, which is absurd. Then $n = 1$ and $a = 0$.   □

Example The equality $ann(a^2) = ann(a^3)$, for each $a \in A$, does not imply that the ring A is reduced. For example, the ring $A = \mathbb{Z}/4\mathbb{Z} = \{\bar{0}, \bar{1}, \bar{2}, \bar{3}\}$ is not reduced. It satisfies the property in an obvious way for $\bar{0}$ and $\bar{1}$ and also for $\bar{3} \in U(A)$. But $\bar{2}^2 = \bar{0} = \bar{2}^3$.

We always remarked that the integral domains are strongly Hopfian. More generally, the reduced rings have this property.

Proposition 1.6 *Each reduced ring is strongly Hopfian.*

Proof Follows from the preceding lemma.   □

Example Let B be a reduced ring. The ring $A = B[T]/(T^2)$ is strongly Hopfian and nonreduced. Indeed, let $f = a_0 + a_1 T + \ldots \in B[T]$. We will show that $ann_A(\bar{f}^3) = ann_A(\bar{f}^2)$. If $\bar{g} = b_0 + b_1 T + \ldots \in ann_A(\bar{f}^3)$, then $\overline{gf^3} = 0$, so $gf^3 \in T^2 B[T]$. But $gf^3 = (b_0 + b_1 T + \ldots)(a_0^3 + 3a_0^2 a_1 T + \ldots) = b_0 a_0^3 + (b_1 a_0^3 + 3a_0^2 a_1 b_0)T + \ldots$. Since $gf^3 \in T^2 B[T]$, then $b_0 a_0^3 = b_1 a_0^3 + 3a_0^2 a_1 b_0 = 0$. The first equality gives $(b_0 a_0)^3 = 0$ and then $b_0 a_0 = 0$. The second one becomes $b_1 a_0^3 = 0$, then $(b_1 a_0)^3 = 0$, so $b_1 a_0 = 0$. On the other hand, $gf^2 = (b_0 + b_1 T + \ldots)(a_0^2 + $

$2a_0^2 a_1 T + \ldots) = b_0 a_0^2 + (2a_0^2 a_1 b_0 + b_1 a_0^2)T + \ldots$. Its first two coefficients are zero, then $gf^2 \in T^2 B[T]$, so $\overline{gf^2} = 0$, and then $\bar{g} \in ann_A(\bar{f}^2)$.

For each ring A, the quotient ring $A/Nil(A)$ is reduced. So we have the following corollary:

Corollary 1.7 *For each ring A, the ring $A/Nil(A)$ is strongly Hopfian.*

Proposition 1.8

1. *Let A be a strongly Hopfian ring and S a multiplicative set of A. Then $S^{-1}A$ is strongly Hopfian.*

   In particular, for each $P \in spec(A)$, the ring A_P is strongly Hopfian.
2. *Conversely, if A is a quasi-local ring with $Max(A) = \{M_1, \ldots, M_n\}$ and if A_{M_i} is strongly Hopfian for each $1 \leq i \leq n$, then A is strongly Hopfian.*

Proof

(1) Let $\frac{a}{s} \in S^{-1}A$ with $a \in A$ and $s \in S$. There exists $k \in \mathbb{N}$ such that $ann(a^k) = ann(a^{k+1})$. We will show that $ann(\frac{a}{s})^k = ann(\frac{a}{s})^{k+1}$. Let $\frac{b}{t} \in ann(\frac{a}{s})^{k+1}$ with $b \in A$ and $t \in S$. Then $\frac{0}{1} = \frac{b}{t}(\frac{a}{s})^{k+1} = \frac{ba^{k+1}}{ts^{k+1}}$. There exists $u \in S$ such that $uba^{k+1} = 0$, so $ub \in ann(a^{k+1}) = ann(a^k)$. Then $uba^k = 0$, so $\frac{ba^k}{ts^k} = \frac{0}{1}$, and then $\frac{b}{t} \in ann(\frac{a}{s})^k$.

(2) Let $a \in A$. There exists $k \geq 1$ such that $ann_{A_{M_i}}((\frac{a}{1})^k) = ann_{A_{M_i}}((\frac{a}{1})^{k+1})$ for each $1 \leq i \leq n$. We will show that $ann_A(a^k) = ann_A(a^{k+1})$. Let $b \in ann_A(a^{k+1})$. Then $ba^{k+1} = 0$, so for each $1 \leq i \leq n$, $\frac{b}{1}(\frac{a}{1})^{k+1} = 0$ in A_{M_i}, and then $\frac{b}{1}(\frac{a}{1})^k = 0$. There exists $s_i \in A \backslash M_i$ such that $s_i ba^k = 0$. The ideal $(s_i; 1 \leq i \leq n) = A$ because if not, it is included in a M_{i_0}, and then $s_{i_0} \in M_{i_0}$, which is absurd. Put $1 = \alpha_1 s_1 + \ldots + \alpha_n s_n$ with $\alpha_i \in A$. Then $ba^k = \alpha_1 s_1 ba^k + \ldots + \alpha_n s_n ba^k = 0$, so $b \in ann_A(a^k)$. □

Proposition 1.9 (Yan and Liu 2010) *Let A and B be two rings. The product ring $A \times B$ is strongly Hopfian if and only if A and B are strongly Hopfian.*

Proof If $(a, b) \in A \times B, n \in \mathbb{N}^*$, then $ann_{A \times B}(a, b)^n = ann_{A \times B}(a^n, b^n) = ann_A(a^n) \times ann_B(b^n)$. So $ann_{A \times B}(a, b)^n = ann_{A \times B}(a, b)^{n+1} \iff ann_A(a^n) = ann_A(a^{n+1})$ and $ann_B(b^n) = ann_B(b^{n+1})$. □

Example (Hamed et al. 2021) Let A be a ring and I a nonzero ideal of A. The subring $A \bowtie I = \{(a, a+i); a \in A, i \in I\}$ of $A \times A$ is strongly Hopfian if and only if A is strongly Hopfian. Indeed, if A is strongly Hopfian, then so is $A \times A$. The subring $A \bowtie I$ is then strongly Hopfian. Conversely, let $a \in A$. Since $(a, a) \in A \bowtie I$, there exists $n \in \mathbb{N}$ such that $ann_{A \bowtie I}(a, a)^{n+1} = ann_{A \bowtie I}(a, a)^n$. We will show that $ann_A(a^{n+1}) = ann_A(a^n)$. Let $b \in ann_A(a^{n+1})$. Then $(b, b) \in ann_{A \bowtie I}(a, a)^{n+1}$, so $(b, b)(a, a)^n = (0, 0)$. Then $ba^n = 0$ so $b \in ann_A(a^n)$.

Corollary 1.10 (Yan and Liu 2010) *Let $A_1, \ldots, A_n$ be rings. The product ring $A_1 \times \ldots \times A_n$ is strongly Hopfian if and only if $A_1, \ldots, A_n$ are strongly Hopfian.*

Proof By induction.   □

Example The example $\prod_{n \geq 1} \mathbb{Z}/p^n \mathbb{Z}$ shows that the infinite product of strongly Hopfian rings is not necessarily strongly Hopfian.

Proposition 1.11 *Let $I_1, \ldots, I_n$ be ideals of a ring A. If $A/I_1, \ldots, A/I_n$ are strongly Hopfian, then $A/(I_1 \cap \ldots \cap I_n)$ is strongly Hopfian.*

Proof The natural homomorphism of $A/(I_1 \cap \ldots \cap I_n) \longrightarrow A/I_1 \times \ldots \times A/I_n$ is one to one. We conclude by the preceding corollary and Proposition 1.2.   □

Proposition 1.12 *Let A be a strongly Hopfian ring and I an ideal of A of the form $I = ann(S)$ with S a subset of A. Then A/I is strongly Hopfian.*

Proof Let $a \in A$. There exists $n \in \mathbb{N}^*$ such that $ann(a^n) = ann(a^{n+1})$. We show that $ann(\bar{a}^n) = ann(\bar{a}^{n+1})$. Let $\bar{b} \in ann(\bar{a}^{n+1})$. Then $\bar{b}\bar{a}^{n+1} = \bar{0}$, so $ba^{n+1} \in I = ann(S)$. Then $ba^{n+1}S = 0$, so $bS \subseteq ann(a^{n+1}) = ann(a^n)$, then $ba^n S = 0$, so $ba^n \in ann(S) = I$, and then $\bar{b}\bar{a}^n = \bar{0}$ so $\bar{b} \in ann(\bar{a}^n)$.   □

Example (Gilmer 2001) The strongly Hopfian property is not stable by passage to the quotient. Let K be a field. The integral domain $A = K[Y_i, i \in \mathbb{N}]$ is strongly Hopfian. Consider the ideal $I = (Y_0 Y_1, Y_0^2 Y_2, \ldots, Y_0^n Y_n, \ldots)$. The ring A/I is not strongly Hopfian. Indeed, let $a = \overline{Y_0}$. Then $\overline{Y_n} \in ann(a^n)\backslash ann(a^{n-1})$ for each $n \geq 2$. Since $Y_n Y_0^n \in I$, then $\overline{Y_n} a^n = 0$, but $Y_n Y_0^{n-1} \notin I$ because if not $Y_n Y_0^{n-1} = Y_0 Y_1 f_1 + \ldots + Y_0^n Y_n f_n + \ldots + Y_0^s Y_s f_s$ with $s \geq n$ and $f_1, \ldots, f_s \in A$. Taking $Y_1 = Y_2 = \ldots = Y_{n-1} = Y_{n+1} = \ldots = 0$, we obtain $Y_n Y_0^{n-1} = Y_0^n Y_n f_n (Y_0, 0, \ldots, 0, Y_n, 0, \ldots)$, and then we have $1 = Y_0 f_n (Y_0, 0, \ldots, 0, Y_n, 0, \ldots)$, which is absurd.

Proposition 1.13 *Let A be a ring such that $Nil(A) = Z(A)$, the set of the zero divisors of A. Then A is strongly Hopfian.*

Proof Let $a \in A$. Suppose that the sequence $ann(a) \subseteq ann(a^2) \subseteq \ldots$ is not stationary, and consider an integer k such that $ann(a^k) \neq ann(a^{k+1})$. There exists $x \in A$ such that $xa^k \neq 0$ and $xa^{k+1} = 0$. Then $a \in Z(A) = Nil(A)$. There exists $s \geq 1$ such that $a^s = 0$. For each $n \geq s$, $a^n = 0$. The sequence $ann(a) \subseteq ann(a^2) \subseteq \ldots$ is then stationary from the rank s, which is absurd.   □

Example Each ring A having only one prime ideal M is strongly Hopfian. Indeed, $M = Nil(A) \subseteq Z(A)$. Since A is local with maximal ideal M, then $A \setminus M = U(A)$. So $Z(A) \subseteq M = Nil(A)$, and then $Z(A) = Nil(A) = M$. By the preceding proposition, A is strongly Hopfian. For example, if K is a field, the rings $K[Y_i; i \in$

$\mathbb{N}^*]/(Y_iY_j; i, j \in \mathbb{N}^*)$, $K[X_i, i \in \mathbb{N}^*]/(X_i^i, i \in \mathbb{N}^*)$, and $K[X_i, i \in \mathbb{N}^*]/(X_i^n, i \in \mathbb{N}^*)$, with $n \geq 2$, have each one prime ideal $M = (\bar{X}_i, i \in \mathbb{N}^*)$. Then they are strongly Hopfian.

Corollary 1.14 *Let A be a ring such that for each $a \in A$, $ann(a)$ is comparable with all the ideals of A. Then A is strongly Hopfian if and only if $Nil(A) = Z(A)$.*

Proof It suffices to show that if A is strongly Hopfian, then $Z(A) \subseteq Nil(A)$. Let $a \in Z(A)$. There exists $0 \neq b \in A$ such that $ab = 0$ and then $ann(a) \neq (0)$. Since A is strongly Hopfian, the sequence $ann(a) \subseteq ann(a^2) \subseteq \ldots$ is stationary. There exists $n \geq 1$ such that $ann(a^n) = ann(a^{2n})$. We will show that $a^n A \cap ann(a^n) = (0)$. Let $x \in a^n A \cap ann(a^n)$, $x = a^n c$ with $c \in A$ and $xa^n = 0$, so $a^{2n}c = 0$, then $c \in ann(a^{2n}) = ann(a^n)$, and so $ca^n = 0$ and $x = 0$. Now, since $a^n A \cap ann(a) \subseteq a^n A \cap ann(a^n) = (0)$, then $a^n A \cap ann(a) = (0)$. By hypothesis, the ideals $a^n A$ and $ann(a)$ are comparable, and we cannot have $ann(a) \subseteq a^n A$ because $ann(a) \neq (0)$. Then $a^n A \subseteq ann(a)$, and we must have $a^n A = (0)$; then a is nilpotent. $\square$

Example Let (A, M) be a local ring with $M^2 = (0)$. Let $a \in A$. If $a = 0$, $ann(a) = A$. If $a \in U(A)$, $ann(a) = (0)$. If $a \in M \setminus (0)$, since $M^2 = (0)$, then $aM = (0)$, so $M \subseteq ann(a)$ with $ann(a) \neq A$ because $1 \notin ann(a)$, and then $ann(a) = M$. We conclude that $ann(a)$ is comparable to the ideals of A. So A is strongly Hopfian if and only if $Nil(A) = Z(A)$.

Corollary 1.15 *Let A be a ring. If the ideal (0) is primary, then $Z(A) = Nil(A)$, so A is strongly Hopfian.*

Proof Let $a \in Z(A)$. There exists $0 \neq b \in A$ such that $ab = 0$. But (0) is a primary ideal, and there exists an integer $n \geq 1$ such that $a^n = 0$, so $a \in Nil(A)$. $\square$

Proposition 1.16 *Let A be a ring. Suppose that $(0) \neq Nil(A) \subset Z(A)$ and for each $x \in Z(A) \setminus Nil(A)$, $Nil(A) \subset xA$. Then A is not strongly Hopfian.*

Proof (Anderson and Badawi 2008)

(1) We will show that $Nil(A)$ is a prime ideal of A. In the contrary case, there exist $x, y \in A \setminus Nil(A)$ such that $xy \in Nil(A)$. Let $n \geq 2$ be such that $x^n y^n = 0$. Then x and y cannot be regular elements, so $x, y \in Z(A)$. We also have $x^2 \in Z(A) \setminus Nil(A)$ and then $Nil(A) \subset x^2 A$. Put $xy = x^2 d$ with $d \in A$. Then $xd \in Nil(A)$ because $xy \in Nil(A)$, so $y - xd \notin Nil(A)$ since $y \notin Nil(A)$. We have $xy - x^2 d = 0$, then $x(y - xd) = 0$, so $y - xd \in Z(A)$. Then $y - xd \in Z(A) \setminus Nil(A)$. By hypothesis, $Nil(A) \subset (y-xd)A$, then $x Nil(A) \subset x(y - xd)A = 0A = (0)$, so $x Nil(A) = (0)$. Let $0 \neq z \in Nil(A) \subset x^2 A$. Then $z = x^2 r$ with $r \in A$. Then $xr \in Nil(A)$, so $z = x(xr) \in x Nil(A) = (0)$, and then $z = 0$, which is absurd. So $Nil(A)$ is a prime ideal.

(2) Let $x \in Z(A) \setminus Nil(A)$. Since $Nil(A)$ is prime, for each $n \geq 1$, $x^n \notin Nil(A)$,

then $Nil(A) \subset x^n A$. So $Nil(A) \subseteq \bigcap_{n:1}^{\infty} x^n A$. On the other hand, there exists

$0 \neq z \in A$ such that $xz = 0$. Since $Nil(A)$ is prime and $x \notin Nil(A)$, then

$z \in Nil(A) \subseteq \bigcap_{n:1}^{\infty} x^n A$. Put $z = x^n z_n$ with $z_n \in A$ for each $n \geq 1$. Then $z_n \in$

$ann(x^{n+1}) \setminus ann(x^n)$ because $x^{n+1} z_n = x(x^n z_n) = xz = 0$ and $x^n z_n = z \neq 0$.
$\square$

2 The Zero Dimensional Rings

Definition 2.1 Let A be a ring.

1. We say that A is π-regular if for each $x \in A$, there exist an integer $n \geq 1$ and $y \in A$ such that $x^n = x^{2n} y$.
2. We say that A is von Neumann regular if for each $x \in A$, there exists $y \in A$ such that $x = x^2 y$.

Example 1 Any product of fields $A = \prod_{i \in I} K_i$ is von Neumann regular. Indeed, let

$x = (x_i)_{i \in I} \in A$. If $x_i = 0 = 0^2.0$ and if $x_i \neq 0$, $x_i = x_i^2 \frac{1}{x_i}$. Let $y = (y_i)_{i \in I} \in A$

with $y_i = 0$ if $x_i = 0$ and $y_i = \frac{1}{x_i}$ if $x_i \neq 0$. Then $x = x^2 y$.

Example 2 A local von Neumann regular (A, M) is a field. Indeed, let $x \in M$. There exists $y \in A$ such that $x = x^2 y$ and then $x(1 - xy) = 0$. But $1 - xy$ cannot be in M, and then it is invertible. So $x = 0$ and $M = (0)$.

Example 3 Let $n \geq 2$ be an integer. The ring $\mathbb{Z}/n\mathbb{Z}$ is von Neumann regular if and only if n is square-free . Indeed, if $n = p_1 \ldots p_s$ where the p_i are distinct prime numbers, then $\mathbb{Z}/n\mathbb{Z} \simeq \mathbb{Z}/p_1\mathbb{Z} \times \ldots \times \mathbb{Z}/p_s\mathbb{Z}$, a product of fields, and then it is von Neumann regular. Suppose now that $n = p^2 m$ with $m \in \mathbb{N}^*$ and p a prime number. Suppose that $\bar{p} = \bar{p}^2 \bar{k}$ with $\bar{k} \in \mathbb{Z}/n\mathbb{Z}$. Then $\overline{pm} = \overline{nk} = \bar{0}$ with $1 \leq pm < n$, which is absurd. Then $\mathbb{Z}/n\mathbb{Z}$ is not von Neumann regular.

Example 4 Any product of von Neumann regular rings is von Neumann regular.

Proposition 2.2 *A ring A is von Neumann regular if and only if it is reduced with zero dimension.*

Proof $"\Longrightarrow"$ For each $a \in A$, there exists $x \in A$ such that $a = a^2 x$ and then $ann(a) = ann(a^2)$. By Lemma 1.5, A is reduced. We will show that each prime ideal P of A is maximal. Let $\bar{0} \neq \bar{x} \in A/P$. There exists $y \in A$ such that $x = x^2 y$

and then $\bar{x}(\bar{1} - \bar{x}\bar{y}) = \bar{0}$. Since A/P is an integral domain, $\bar{1} - \bar{x}\bar{y} = \bar{0}$, then $\bar{x}$ is invertible, and A/P is a field.

"$\Longleftarrow$" Since $dim(A) = 0$, the prime ideals of A are maximal and minimal. Let $a \in A$. Suppose that $(a) + ann(a) \neq A$ and there exists $M \in Max(A)$ such that $(a) + ann(a) \subseteq M$. By Lemma 3.11, Chap. 2, since $M \in Min(A)$ and A is reduced, there exists $s \in A \setminus M$ such that $sa = 0$ and then $s \in ann(a) \subseteq M$, which is absurd. Then $(a) + ann(a) = A$. Put $1 = \alpha a + b$ with $\alpha \in A$ and $b \in ann(a)$. By multiplying by a, we find $a = \alpha a^2 + ba = \alpha a^2$. $\qquad \square$

Proposition 2.3 *The following assertions are equivalent for a ring A:*

1. *The ring A is π-regular.*
2. *For each $a \in A$, there exist $n \in \mathbb{N}^*$ and $y \in A$ such that $a^n = a^{n+1}y$.*
3. *The quotient ring $A/Nil(A)$ is von Neumann regular.*
4. *dim $A = 0$.*

Proof "(1) $\Longrightarrow$ (2)" Let $a \in A$. There exist $n \in \mathbb{N}^*$ and $x \in A$ such that $a^n = a^{2n}x = a^{n+1}(a^{n-1}x) = a^{n+1}y$ with $y = a^{n-1}x \in A$.

"(2) $\Longrightarrow$ (1)" Let $a \in A$. There exist $n \in \mathbb{N}^*$ and $y \in A$ such that $a^n = a^{n+1}y$. Then $a^n = a^n(ay) = a^{n+1}y(ay) = a^{n+2}y^2 = \ldots = a^{2n}y^n = a^{2n}x$ with $x = y^n \in A$.

"(2) $\Longrightarrow$ (3)" Let $a \in A$. There exist $n \in \mathbb{N}^*$ and $y \in A$ such that $a^n = a^{n+1}y$. In the quotient ring $A/Nil(A)$, $\tilde{a}^n = \tilde{a}^{n+1}\tilde{y}$. If $n \geq 2$, we compute the expression $\left(\tilde{a}^n\tilde{y} - \tilde{a}^{n-1}\right)^2 = \tilde{a}^{2n}\tilde{y}^2 + \tilde{a}^{2(n-1)} - 2\tilde{a}^{2n-1}\tilde{y} = \tilde{a}^{n-1}\left(\tilde{a}^{n+1}\tilde{y}\right)\tilde{y} + \tilde{a}^{2(n-1)} - 2\tilde{a}^{2n-1}\tilde{y} = \tilde{a}^{n-1}\tilde{a}^n\tilde{y} + \tilde{a}^{2(n-1)} - 2\tilde{a}^{2n-1}\tilde{y} = \tilde{a}^{2(n-1)} - \tilde{a}^{2n-1}\tilde{y} = \tilde{a}^{2(n-1)} - \tilde{a}^{n-2}\left(\tilde{a}^{n+1}\tilde{y}\right) = \tilde{a}^{2(n-1)} - \tilde{a}^{n-2}\tilde{a}^n = 0$. Since $A/Nil(A)$ is reduced, the equality $\left(\tilde{a}^n\tilde{y} - \tilde{a}^{n-1}\right)^2 = 0$ implies $\tilde{a}^{n-1} = \tilde{a}^n\tilde{y}$. Continuing this process, we find $\tilde{a} = \tilde{a}^2\tilde{y}$, which shows that $A/Nil(A)$ is von Neumann regular.

"(3) $\Longrightarrow$ (2)" Let $a \in A$. Since $A/Nil(A)$ is von Neumann regular, there exists $x \in A$ such that $\tilde{a} = \tilde{a}^2\tilde{x}$ and then $a - a^2x \in Nil(A)$. There exists $n \in \mathbb{N}^*$ such that $(a - a^2x)^n = 0$. Then $0 = a^n(1 - ax)^n = a^n(1 - pax)$ with $p \in A$, and then $a^n = a^{n+1}px = a^{n+1}y$ with $y = px \in A$.

"(3) $\Longrightarrow$ (4)" Since $A/Nil(A)$ is von Neumann regular, by the preceding proposition, $dim(A/Nil(A)) = 0$. But $Nil(A)$ is contained in all the prime ideals of A, and then $dim(A) = dim(A/Nil(A)) = 0$.

"(4) $\Longrightarrow$ (3)" By hypothesis, $dim(A/Nil(A)) = dim(A) = 0$ with $A/Nil(A)$ reduced. By the preceding proposition, $A/Nil(A)$ is von Neumann regular. $\qquad \square$

Example 1 The rings $A[X]$ and $A[[X]]$ are never π-regular. In the contrary case, there exist $n \in \mathbb{N}^*$ and $g(X) \in A[X]$ (resp. $A[[X]]$) such that $X^n = X^{n+1}g(X)$. Then $1 = Xg(X)$, which is absurd.

Example 2 Let p be a prime number, $n \geq 2$ an integer, and $A = \mathbb{Z}/p^n\mathbb{Z}$. Then A is not von Neumann regular, but π-regular. Indeed, its unique prime ideal is $p\mathbb{Z}/p^n\mathbb{Z}$ and then $dim(A) = 0$.

Proposition 2.4 (Gilmer 2001) *A zero dimensional ring is strongly Hopfian.*

Proof Let A be a zero dimensional ring. By the preceding proposition, for each $a \in A$, there exist $n \in \mathbb{N}^*$ and $y \in A$ such that $a^n = a^{n+1}y$. Then $ann(a^n) = ann(a^{n+1})$. $\quad\square$

Examples Let K be a field. The rings $K[Y_i, i \in \mathbb{N}^*]/(Y_i^i, i \in \mathbb{N}^*)$ and $K[Y_i, i \in \mathbb{N}]/(Y_i^n, i \in \mathbb{N})$, where n is a fixed natural integer ≥ 2, have only one prime ideal, and then they are strongly Hopfian.

Corollary 2.5 (Gilmer 2001) *If a ring A has an extension with zero dimension, then A is strongly Hopfian.*

Proof We have $A \subseteq B$ with $dim\, B = 0$, and then B is strongly Hopfian. By Proposition 1.2, A is strongly Hopfian. $\quad\square$

Definition 2.6 Let A be a ring. We say that A is chained if its ideals are comparable for the inclusion.

Example 1 Valuation domains are chained.

Example 2 A homomorphic image of a valuation domain is a chained ring. Indeed, let $f : D \longrightarrow A$ be a homomorphism onto of rings with D a valuation domain. Let I and J be two ideals of A. Then $f^{-1}(I)$ and $f^{-1}(J)$ are ideals of D. For example, $f^{-1}(I) \subseteq f^{-1}(J)$. We will show that $I \subseteq J$. Let $x \in I \subseteq A$. There exists $x' \in D$ such that $f(x') = x$. Then $x' \in f^{-1}(I) \subseteq f^{-1}(J)$, so $x = f(x') \in J$.

For example, the quotient of a valuation domain is a chained ring.

Example 3 Let $n \geq 2$ be an integer. The ring $\mathbb{Z}/n\mathbb{Z}$ is chained if and only if n is the power of a prime number. Indeed, if $n = p^k$ with p a prime number and $k \geq 1$ an integer, the ideals of the ring $\mathbb{Z}/n\mathbb{Z}$ are $(\bar{0}) = p^k\mathbb{Z}/p^k\mathbb{Z} \subset p^{k-1}\mathbb{Z}/p^k\mathbb{Z} \subset \ldots \subset p\mathbb{Z}/p^k\mathbb{Z} \subset \mathbb{Z}/p^k\mathbb{Z}$. They are comparable. Conversely, if there are two different prime numbers p and q dividing n, the ideals $p\mathbb{Z}/n\mathbb{Z}$ and $q\mathbb{Z}/n\mathbb{Z}$ are not comparable in $\mathbb{Z}/n\mathbb{Z}$.

Proposition 2.7 (Khalifa 2017) *The following assertions are equivalent for a chained ring A:*

1. *The ring A is strongly Hopfian.*
2. *The ideal (0) is primary in A.*
3. *$Z(A) = Nil(A)$.*
4. *$dim(S^{-1}A) = 0$, where S is the multiplicative set of the regular elements of A.*
5. *A is imbedded in a zero dimensional ring.*

Proof $''(1) \Longrightarrow (2)''$ Let $a, b \in A$ be such that $ab = 0$ and $a \neq 0$. By hypothesis, there exists $n \geq 1$ such that $ann(b^n) = ann(b^{n+1})$. Since A is chained, the ideals (a) and (b^n) are comparable. If $(a) \subseteq (b^n)$, then $a = xb^n$ with $x \in A$. So $xb^{n+1} =$

$ab = 0$, then $x \in ann(b^{n+1}) = ann(b^n)$, so $xb^n = 0$, and then $a = 0$, which is absurd. Then $(b^n) \subseteq (a)$. Put $b^n = ac$ with $c \in A$, then $b^{n+1} = abc = 0$, so $b^{n+1} = 0$.

"(2) $\Longrightarrow$ (3)" By Corollary 1.15. "(3) $\Longrightarrow$ (4)" Let $\mathscr{P}_1 \subseteq \mathscr{P}_2$ be two prime ideals of $S^{-1}A$. Then $\mathscr{P}_1 = S^{-1}P_1$ and $\mathscr{P}_2 = S^{-1}P_2$ with $P_1 \subseteq P_2$ two prime ideals of A disjoint with S. Let $x \in \mathscr{P}_2$, $x = \frac{a}{s}$ with $a \in P_2$ and $s \in S$. If $a \in S$, then $x = \frac{a}{s} \in U(S^{-1}A)$, so $\mathscr{P}_2 = S^{-1}A$, which is absurd. Then $a \in A \setminus S = Z(A) = Nil(A) \subseteq P_1$. So $a \in P_1$ and $x = \frac{a}{s} \in S^{-1}P_1 = \mathscr{P}_1$. Then $\mathscr{P}_1 = \mathscr{P}_2$ and $dim(S^{-1}A) = 0$.

"(4) $\Longrightarrow$ (5)" The natural homomorphism $\phi : A \longrightarrow S^{-1}A$ is one to one. Indeed, let $a \in A$ be such that $\phi(a) = \frac{a}{1} = \frac{0}{1}$. There exists $s \in S$ such that $sa = 0$. But s is regular, and then $a = 0$. "(5) $\Longrightarrow$ (1)" Follows from Corollary 2.5. □

3 Auto-Injective Rings

Definition 3.1 Let A be a ring. An ideal of A is said to be essential if it has a nonzero intersection with each nonzero ideal of A.

Examples

(1) The intersection of two essential ideals is an essential ideal.
(2) An ideal containing an essential ideal is itself essential.

Lemma 3.2 *Let A be a ring. The set $S_A = \{a \in A;\ ann(a)\ is\ essential\}$ is a proper ideal of A. We say that S_A is the singular ideal of A.*

Proof The ideal $ann(0) = A$ is essential and then $0 \in S_A$. Let $a, b \in S_A$ and $c \in A$. Then $ann(a)$ and $ann(b)$ are essential, so $ann(a) \cap ann(b)$ is essential. Since $ann(a) \cap ann(b) \subseteq ann(a + b)$ and $ann(a) \subseteq ann(ca)$, then $ann(a + b)$ and $ann(ca)$ are essential. So $a + b, ca \in S_A$, and then S_A is an ideal of A. Since $ann(1) = (0)$, then $1 \notin S_A$, so $S_A \neq A$. □

Example Let p be a prime number, $n \in \mathbb{N}^*$, and $A = \mathbb{Z}/p^n\mathbb{Z}$. The nonzero ideals of A are: $p^{n-1}\mathbb{Z}/p^n\mathbb{Z} \subset p^{n-2}\mathbb{Z}/p^n\mathbb{Z} \subset \ldots \subset p\mathbb{Z}/p^n\mathbb{Z} \subset A$. The intersection of any two of them contains $p^{n-1}\mathbb{Z}/p^n\mathbb{Z} \neq (0)$. Then they are all essential. We will show that $ann(\bar{p}) = (\bar{p}^{n-1})$. Indeed, if $\bar{x} \in ann(\bar{p})$, then $\bar{x}\bar{p} = \bar{0}$, so p^n divides xp and p^{n-1} divides x in $\mathbb{Z}$. Then $\bar{p}^{n-1}$ divides $\bar{x}$ in A, so $ann(\bar{p}) \subseteq (\bar{p}^{n-1})$. The converse is clear. Then $\bar{p} \in S_A$, so $(\bar{p}) \subseteq S_A$. But $(\bar{p}) = p\mathbb{Z}/p^n\mathbb{Z}$ is the maximal ideal of A, and then $S_A = (\bar{p})$.

Lemma 3.3 *Let A be a ring:*

1. *$Nil(A) \subseteq S_A$ and $Nil(A) = (0)$ if and only if $S_A = (0)$.*
2. *If A is strongly Hopfian, then $Nil(A) = S_A$.*

Proof

(1) Let $a \in Nil(A)$. We will show that $ann(a)$ is essential. Let I be a nonzero ideal of A and $0 \neq d \in I$. There exists a smallest integer $n \geq 1$ such that $a^n d = 0$. Then $0 \neq a^{n-1} d \in I \cap ann(a)$. So $a \in S_A$.

Suppose now that $Nil(A) = (0)$. Suppose that there exists $0 \neq a \in S_A$. By definition, $I = ann(a) \cap (a) \neq (0)$. Let $0 \neq d \in I$. Then $d = ca$ with $c \in A$ and $da = 0$. Then $ca^2 = 0$, so $(ca)^2 = 0$. But $Nil(A) = (0)$, then $ca = 0$, so $d = 0$, which is absurd. Then $S_A = (0)$.

(2) Let $0 \neq a \in S_A$. Since A is strongly Hopfian, the sequence $ann(a) \subseteq ann(a^2) \subseteq \ldots$ is stationary from a certain rank n. Suppose that $a^n \neq 0$, and since S_A is an ideal of A, then $a^n \in S_A$. By definition, $I = ann(a^n) \cap (a^n) \neq (0)$. Let $0 \neq x \in I$. Then $x = da^n$ with $d \in A$ and $xa^n = 0$. So $da^{2n} = 0$, then $d \in ann(a^{2n}) = ann(a^n)$, so $da^n = 0$, and then $x = 0$, which is absurd. We conclude that $a^n = 0$ and $a \in Nil(A)$.   □

Lemma 3.4 *Let A be a ring and $x \in A$. Then $x \in Jac(A)$, the Jacobson radical of A, if and only if for each $y \in A$, $1 + xy$ is invertible.*

Proof " $\Longrightarrow$ " Suppose that there exists $y \in A$ such that $1 + xy$ is not invertible. There exists a maximal ideal M of A such that $1 + xy \in M$. By hypothesis, $x \in M$, and then $1 = (1 + xy) - xy \in M$, which is absurd.

" $\Longleftarrow$ " Suppose that $x \notin Jac(A)$. There exists a maximal ideal M of A such that $x \notin M$ and then $M \subset M + xA$. By the maximality of M, we have $M + xA = A$. Let $m \in M$ and $y \in A$ be such that $1 = m - yx$. Then $1 + yx = m \in M$. So $1 + yx$ is not invertible, which is absurd.   □

Example 1 Let $(A_i)_{i \in I}$ be a family of rings. By the lemma, $Jac(\prod_{i \in I} A_i) = \prod_{i \in I} Jac(A_i)$.

Example 2 Let A be a Noetherian ring and I an ideal of A contained in $Jac(A)$. Then $\bigcap_{k:1}^{\infty} I^k = (0)$. Indeed, let $b \in \bigcap_{k:1}^{\infty} I^k$. By Krull's intersection theorem (Chap. 5), $b \in bI$. Then $b = ba$ with $a \in I \subseteq Jac(A)$, so $(1 - a)b = 0$. By the preceding Lemma, $1 - a$ is invertible in A, and then $b = 0$.

Definition 3.5 Let A be a ring. We say that A is auto-injective if for each ideal I of A, every homomorphism $\phi : I \longrightarrow A$ of A-modules is extended to an endomorphism of A-module A.

Example The ring $\mathbb{Z}$ is not auto-injective. Indeed, the map $\phi : 2\mathbb{Z} \longrightarrow \mathbb{Z}$, defined by $\phi(2n) = n$ for each $n \in \mathbb{Z}$, is a homomorphism of $\mathbb{Z}$-modules. Suppose that it is extended to $\tilde{\phi} : \mathbb{Z} \longrightarrow \mathbb{Z}$. Then $1 = \phi(2) = \tilde{\phi}(2) = 2\tilde{\phi}(1)$ with $\tilde{\phi}(1) \in \mathbb{Z}$, which is absurd.

Proposition 3.6 *Let A be an auto-injective ring. Then $Jac(A) = S_A$.*

Proof $" \subseteq "$ Let $0 \neq d \in Jac(A)$. Suppose that there exists a nonzero ideal I of A such that $ann(d) \cap I = (0)$. The map $\phi : I \longrightarrow A$, defined by $\phi(x) = dx$ for each $x \in I$, is a homomorphism one to one of A-modules. It induces an isomorphism $\hat{\phi} : I \longrightarrow Im(\phi)$ with inverse $\hat{\phi}^{-1} : Im(\phi) \longrightarrow I \subseteq A$. Let $\psi : Im(\phi) \longrightarrow A$ be the homomorphism that satisfies $\psi(\phi(x)) = x$ for each $x \in I$. Since A is auto-injective, ψ is extended to an endomorphism of A-modules $\tilde{\psi} : A \longrightarrow A$. Put $\tilde{\psi}(1) = b$. Then for each $x \in I$, $(xd)b = \phi(x)\tilde{\psi}(1) = \tilde{\psi}(\phi(x).1) = \tilde{\psi}(\phi(x)) = \psi(\phi(x)) = x$, so $x(1 - db) = 0$. Then $I(1 - db) = 0$, but $1 - db \in U(A)$ because $d \in Jac(A)$, so $I = (0)$, which is absurd.

$" \supseteq "$ Let $d \in S_A$ and $c \in A$. We must show that $1 - cd \in U(A)$. We have $ann(d) \cap ann(1 - cd) = (0)$. Indeed, if $x \in ann(d) \cap ann(1 - cd)$, then $xd = 0$ and $x(1 - cd) = 0$, so $x = 0$. Since $ann(d)$ is essential, the ideal $ann(1 - cd) = (0)$. We can define a homomorphism of A-modules $\phi : (1 - cd)A \longrightarrow A$ by $\phi((1 - cd)x) = x$ for each $x \in A$. In particular, $\phi(1 - cd) = 1$. Since A is auto-injective, ϕ is extended to an endomorphism of A-modules $\tilde{\phi} : A \longrightarrow A$. Then $1 = \phi(1 - cd) = \tilde{\phi}(1 - cd) = (1 - cd)\tilde{\phi}(1)$, so $1 - cd \in U(A)$. ⊔

Notation Let A be a ring and I any ideal of A. The set $E = \{J$ ideal of A such that $J \cap I = (0)\}$ is inductive for the inclusion. It admits a maximal element denoted I^c and called the complement of I in A.

Suppose now that $I = ann(\alpha)$ with $\alpha \in A$. Then $I \oplus I^c$ is an essential ideal of A. Indeed, let J be a nonzero ideal of A. Suppose that $I \cap J = I^c \cap J = (0)$. Then $I^c \subset I^c \oplus J$ since $J \neq (0)$. It suffices to show that $(I^c \oplus J) \cap I = (0)$ to obtain a contradiction with the maximality of I^c in E. Let $x \in (I^c \oplus J) \cap I$, $x = a + b$ with $a \in I^c$ and $b \in J$. Then $0 = \alpha x = \alpha a + \alpha b$, so $\alpha a = -\alpha b \in I^c \cap J = (0)$, then $\alpha a = \alpha b = 0$, and so $a, b \in ann(\alpha) = I$. Then $a \in I \cap I^c = (0)$ and $b \in I \cap J = (0)$. So $x = a + b = 0$. Then $I \cap J \neq (0)$, so $I^c \cap J \neq (0)$, and then $(I \oplus I^c) \cap J \neq (0)$.

Proposition 3.7 *Let A be an auto-injective ring. Then A/S_A is a von Neumann regular ring.*

Proof Let $a \in A$. Put $I = ann(a)$ and let I^c be its complement. The map $\phi : I^c \longrightarrow A$, defined by $\phi(x) = ax$ for each $x \in I^c$, is a homomorphism one to one of A-modules since $I \cap I^c = (0)$ by definition of I^c. It induces an isomorphism $\hat{\phi} : I^c \longrightarrow Im(\phi)$, with inverse $\hat{\phi}^{-1} : Im(\phi) \longrightarrow I^c \subseteq A$. Let $\psi : Im(\phi) \longrightarrow A$ be the homomorphism that satisfies $\psi(\phi(x)) = x$ for each $x \in I^c$. Since A is auto-injective, ψ can be extended to an endomorphism of A-modules $\tilde{\psi} : A \longrightarrow A$. Put $\tilde{\psi}(1) = b$. For each $x \in I^c$, $(xa)b = \phi(x)\tilde{\psi}(1) = \tilde{\psi}(\phi(x).1) = \tilde{\psi}(\phi(x)) = \psi(\phi(x)) = x$, and then $x(a^2b - a) = 0$. So $I^c \subseteq ann(a^2b - a)$. Since $I =$

$ann(a) \subseteq ann(a^2b - a)$, then $I \oplus I^c \subseteq ann(a^2b - a)$. But $I \oplus I^c$ is an essential ideal, and then $ann(a^2b - a)$ is also an essential ideal. So $a^2b - a \in S_A$, and then $\bar{a} = \bar{a}^2\bar{b}$ in A/S_A. □

Example Let A be a reduced ring with zero dimension. Then A is not auto-injective. Indeed, by Lemma 3.3, since $Nil(A) = (0)$, then $S_A = (0)$. So $dim(A/S_A) = dim(A) \neq 0$. By Proposition 2.2, A/S_A is not von Neumann regular. By the preceding proposition, A is not auto-injective.

Theorem 3.8 (Xu 1995) *Let A be an auto-injective ring. The following assertions are equivalent:* *1. A is strongly Hopfian.* *2. $Jac(A) = Nil(A)$.* *3. $dim(A) = 0$.*

Proof $"(1) \Longrightarrow (2)"$ Let $a \in Jac(A)$. Since A is strongly Hopfian, there exists $n \geq 1$ such that $ann(a^n) = ann(a^{n+1})$. We can define a homomorphism $\phi : Aa^{n+1} \longrightarrow Aa^n$ of A-modules, by $\phi(xa^{n+1}) = xa^n$ for each $x \in A$. Since A is auto-injective, we can extend ϕ to $\tilde{\phi} : A \longrightarrow A$. In particular, $a^n = \phi(a^{n+1}) = \tilde{\phi}(a^{n+1}) = a^{n+1}\tilde{\phi}(1)$, and then $a^n(1 - a\tilde{\phi}(1)) = 0$. By Lemma 3.4, the element $1 - a\tilde{\phi}(1)$ is invertible, and then $a^n = 0$ and $a \in Nil(A)$.

$"(2) \Longrightarrow (3)"$ Since A is auto-injective, by Propositions 3.6 and 3.7, the ring $A/Jac(A) = A/S_A$ is von Neumann regular. It is zero dimensional, by Proposition 2.2. So $dim(A) = dim(A/Nil(A)) = dim(A/Jac(A)) = 0$. $"(3) \Longrightarrow (1)"$ By Proposition 2.4. □

4 Product of Zero Dimensional Rings

Theorem 4.1 (Maroscia 1974) *Let $(A_\lambda)_{\lambda \in \Lambda}$ be a family of zero dimensional rings. The following conditions are equivalent for the product ring $A = \prod_{\lambda \in \Lambda} A_\lambda$:*

1. $dim(A) = 0$.
2. $Jac(A) = Nil(A)$.
3. $Nil(A) = \prod_{\lambda \in \Lambda} Nil(A_\lambda)$.

Proof $"(1) \Longrightarrow (2)"$ Since $dim(A) = 0$, $Spec(A) = Max(A) = Min(A)$, and then $Jac(A) = Nil(A)$. $"(2) \Longrightarrow (3)"$ $Nil(A) = Jac(A) = \prod_{\lambda \in \Lambda} Jac(A_\lambda) = \prod_{\lambda \in \Lambda} Nil(A_\lambda)$ car $dim(A_\lambda) = 0$.

$"(3) \Longrightarrow (1)"$ $A/Nil(A) = \prod_{\lambda \in \Lambda} A_\lambda / \prod_{\lambda \in \Lambda} Nil(A_\lambda) \simeq \prod_{\lambda \in \Lambda} (A_\lambda/Nil(A_\lambda))$. But $A_\lambda/Nil(A_\lambda)$ is reduced with $dim(A_\lambda/Nil(A_\lambda)) = dim(A_\lambda) = 0$. By Proposition 2.2, $A_\lambda/Nil(A_\lambda)$ is von Neumann regular. Then $\prod_{\lambda \in \Lambda} (A_\lambda/Nil(A_\lambda))$ is also von Neumann regular, so zero dimensional. Then $dim(A) = dim(A/Nil(A)) = 0$. □

Example 1 A zero dimensional ring A satisfies $Jac(A) = Nil(A)$. The converse is false. For example, $Jac(\mathbb{Z}) = \bigcap_{p \text{ premier}} p\mathbb{Z} = (0) = Nil(\mathbb{Z})$. But $dim(\mathbb{Z}) = 1$.

Example 2 Let p be a prime number and $A = \prod_{n \geq 1} \mathbb{Z}/p^n\mathbb{Z}$.

Then $Jac(A) = \prod_{n \geq 1} Jac(\mathbb{Z}/p^n\mathbb{Z}) = \prod_{n \geq 1} p\mathbb{Z}/p^n\mathbb{Z}$. Let $e = (\bar{1}, \bar{1}, \ldots)$ the identity element of A. Then $pe \in Jac(A)$ and $pe \notin Nil(A)$. So $dim(A) \neq 0$.

Remark 4.2 Let $(A_\lambda)_{\lambda \in \Lambda}$ be a family of rings and $A = \prod_{\lambda \in \Lambda} A_\lambda$.

Then $Nil(A) \subseteq \prod_{\lambda \in \Lambda} Nil(A_\lambda)$. Indeed, if $x = (x_\lambda)_{\lambda \in \Lambda} \in Nil(A)$, there exists $n \in \mathbb{N}^*$ such that $x^n = 0$. For each $\lambda \in \Lambda$, $x_\lambda^n = 0$, then $x_\lambda \in Nil(A_\lambda)$. The inclusion is in general strict. In the preceding example, $pe \in \prod_{n \geq 1} Nil(\mathbb{Z}/p^n\mathbb{Z})$ but $pe \notin Nil(A)$.

If Λ is a finite set, $Nil(A) = \prod_{\lambda \in \Lambda} Nil(A_\lambda)$. Indeed, if $x = (x_\lambda)_{\lambda \in \Lambda} \in \prod_{\lambda \in \Lambda} Nil(A_\lambda)$, there exists $n \in \mathbb{N}^*$ such that for each $\lambda \in \Lambda$, $x_\lambda^n = 0$. Then $x^n = 0$. In particular, by the preceding theorem, if Λ is finite and the A_λ are zero dimensional rings, then $dim(A) = 0$.

Theorem 4.3 (Gilmer and Heinzer 1989) *Let $(A_\lambda)_{\lambda \in \Lambda}$ be a family of zero dimensional rings. The following conditions are equivalent for the product ring $A = \prod_{\lambda \in \Lambda} A_\lambda$:*

1. $dim(A) = 0$.
2. *A is imbedded in a zero dimensional ring.*
3. *The ring A is strongly Hopfian.*
4. *There exists $n \in \mathbb{N}^*$ such that the set $\Lambda_n = \{\lambda \in \Lambda; \exists x \in Nil(A_\lambda), x^n \neq 0\}$ is finite.*

Proof "(1) $\Longrightarrow$ (2)" Clear. "(2) $\Longrightarrow$ (3)" By Corollary 2.5, A is strongly Hopfian.
"(3) $\Longrightarrow$ (4)" Suppose that for each $n \in \mathbb{N}^*$, the set Λ_n is infinite. Let $n_1 \in \mathbb{N}^*$. There exist $\lambda_1 \in \Lambda$, $x_1 \in Nil(A_{\lambda_1})$, and an integer $n_2 > n_1$ such that $x_1^{n_2} = 0$ and $x_1^{n_1} \neq 0$. Also, there exist $\lambda_2 \in \Lambda \setminus \{\lambda_1\}$, $x_2 \in Nil(A_{\lambda_2})$, and an integer $n_3 > n_2 > n_1$ such that $x_2^{n_3} = 0$ and $x_2^{n_2} \neq 0$. By induction, we construct different indexes

$\lambda_1, \lambda_2, \lambda_3, \ldots \in \Lambda$ and elements $x_i \in Nil(A_{\lambda_i})$ and nonzero natural integers $n_1 <$ $n_2 < n_3 < \ldots$ such that $x_i^{n_{i+1}} = 0$ and $x_i^{n_i} \neq 0$ for each $i \in \mathbb{N}^*$. Define the element $x = (x_\lambda)_{\lambda \in \Lambda}$ of $A = \prod_{\lambda \in \Lambda} A_\lambda$ by $x_\lambda = x_i$ if $\lambda = \lambda_i$ and $x_\lambda = 0$ if not. For $i \in \mathbb{N}^*$, we consider the element e_i of A, having the coordinate of index λ_i equal to 1 and all the coordinates equal to zero. Then $e_i \in ann(x^{n_{i+1}}) \setminus ann(x^{n_i})$ for each $i \in \mathbb{N}^*$: This contradicts the fact that A is strongly Hopfian.

$"(4) \implies (1)"$ Let $n \in \mathbb{N}^*$ be such that the set $\Lambda_n = \{\lambda \in \Lambda; \exists x \in Nil(A_\lambda), x^n \neq 0\}$ is finite. Let $\Gamma = \Lambda \setminus \Lambda_n$ and $B = \prod_{\lambda \in \Gamma} A_\lambda$. Then $A =$ $B \times \prod_{\lambda \in \Lambda_n} A_\lambda$. If $x = (x_\lambda)_{\lambda \in \Gamma} \in \prod_{\lambda \in \Gamma} Nil(A_\lambda)$, then $x_\lambda^n = 0$ for each $\lambda \in \Gamma$, so $x^n = 0$. Then $\prod_{\lambda \in \Gamma} Nil(A_\lambda) \subseteq Nil(B)$. Since the inverse inclusion is true, then $Nil(B) = \prod_{\lambda \in \Gamma} Nil(A_\lambda)$. By Theorem 4.1, $dim(B) = 0$. Then A is a finite product of zero dimensional rings. So $dim(A) = 0$. $\qquad\square$

Example Let p be a prime number and $A = \prod_{n \geq 1} \mathbb{Z}/p^n\mathbb{Z}$. The ring A is not only nonzero dimensional, but it cannot be imbedded in a zero dimensional ring.

Corollary 4.4 (Maroscia 1974 and Gilmer and Heinzer 1989) *Let $(A_\lambda)_{\lambda \in \Lambda}$ be a family of zero dimensional rings. The following conditions are equivalent for the product ring $A = \prod_{\lambda \in \Lambda} A_\lambda$:*

1. $dim(A) = 0$.
2. $Jac(A) = Nil(A)$.
3. $Nil(A) = \prod_{\lambda \in \Lambda} Nil(A_\lambda)$.
4. *A is imbedded in a zero dimensional ring.*
5. *The ring A is strongly Hopfian.*
6. *There exists $n \in \mathbb{N}^*$ such that the set $\Lambda_n = \{\lambda \in \Lambda; \exists x \in Nil(A_\lambda), x^n \neq 0\}$ is finite.*

5 Injection in a Zero Dimensional Ring

Proposition 5.1 *The following assertions are equivalent for a ring A:*

1. *The ring A is π-regular.*
2. *For each $a \in A$, there exists an idempotent $e \in A$ such that $a + (1 - e)$ is invertible in A and $a(1 - e)$ is nilpotent.*

3. *For each $a \in A$, there exists $b \in A$ such that $a + b$ is invertible and ab is nilpotent.*
4. *For each $a \in A$, there exist $n \in \mathbb{N}^*$, a unity u of A, and an idempotent e of A such that $a^n = ue$.*

Proof "(1) $\Longrightarrow$ (2)" Let $a \in A$. There exist $n \in \mathbb{N}^*$ and $x \in A$ such that $a^{2n}x = a^n$. The element $e = a^n x$ is idempotent since $e^2 = a^{2n}x^2 = (a^{2n}x)x = a^n x = e$, so is $1-e$. Then $(a(1-e))^n = a^n(1-e)^n = a^n(1-e) = a^n(1-a^nx) = a^n - a^{2n}x = 0$, so $a(1-e)$ is nilpotent. Suppose that $a+(1-e)$ is not invertible and then it is contained in a maximal ideal M of A. The nilpotent element $a(1 - e) \in M$, then either $a \in M$ or $1 - e \in M$ so $a, 1 - e \in M$. Then $1 = (1 - a^nx) + a^nx = (1 - e) + a^nx \in M$, which is absurd.

"(2) $\Longrightarrow$ (3)" Clear. "(3) $\Longrightarrow$ (4)" Let $a \in A$. There exist $b \in A$ and $n \in \mathbb{N}^*$ such that $a + b \in U(A)$ and $(ab)^n = 0$. Suppose that $u = a^n + b^n \notin U(A)$, there exists $M \in Max(A)$ such that $a^n + b^n \in M$, and since $a^n b^n = (ab)^n = 0 \in M$, then either $a^n \in M$ or $b^n \in M$. So $a^n, b^n \in M$, then $a, b \in M$, and so $a + b \in M$, which is absurd. Since $a^n u = a^n(a^n + b^n) = a^{2n} + a^n b^n = a^{2n}$, then $u^{-1}a^n = u^{-2}a^{2n}$, and the element $e = u^{-1}a^n \in A$ is then idempotent. Finally, $a^n = ue$.

"(4) $\Longrightarrow$ (1)" By Proposition 2.3, it suffices to show that $dim(A) = 0$. Let $P_1 \subseteq P_2$ be two prime ideals of A. For each $a \in P_2$, there exist $n \in \mathbb{N}^*$, $u \in U(A)$, and an idempotent element e of A such that $a^n = ue$. Since $e = u^{-1}a^n \in P_2$, then $1 - e \notin P_2$, but $e(1 - e) = 0 \in P_1$, then $e \in P_1$ and $a^n = ue \in P_1$, so $a \in P_1$. Then $P_1 = P_2$. $\qquad \square$

Lemma 5.2 (Arapovic 1983) *Let A be a zero dimensional ring and $\{M_\alpha\}_\alpha$ the maximal spectrum of A. For each α, let $\psi_\alpha : A \longrightarrow A_{M_\alpha}$ be the natural homomorphism, defined by $\psi_\alpha(x) = \frac{x}{1}$, for each $x \in A$:*

1. *For each α, $ker(\psi_\alpha)$ is a M_α-primary ideal.*
2. *$\bigcap_\alpha ker(\psi_\alpha) = (0)$.*

Proof

(1) Let $a, b \in A$ be such that $ab \in ker(\psi_\alpha)$. Then $\psi_\alpha(ab) = \frac{ab}{1} = \frac{0}{1}$. There exists $s \in A \setminus M_\alpha$ such that $sab = 0 \in M_\alpha$. For example, $a \in M_\alpha \in Min(A)$. There exist $t \in A \setminus M_\alpha$ and $n \geq 1$ such that $ta^n = 0$ and then $\frac{a^n}{1} = \frac{0}{1}$. So $a^n \in ker(\psi_\alpha)$, and then $ker(\psi_\alpha)$ is primary. Let $a \in ker(\psi_\alpha)$. Then $\psi_\alpha(a) = \frac{a}{1} = \frac{0}{1}$. There exists $s \in A \setminus M_\alpha$ such that $sa = 0 \in M_\alpha$, then $a \in M_\alpha$. So $ker(\psi_\alpha) \subseteq M_\alpha$, then $\sqrt{ker(\psi_\alpha)} \subseteq M_\alpha$. Conversely, let $a \in M_\alpha \in Min(A)$. By Lemma 3.11, Chap. 2, there exist $s \in A \setminus M_\alpha$ and $n \geq 1$ such that $sa^n = 0$, then $\frac{a^n}{1} = \frac{0}{1}$, so $a^n \in ker(\psi_\alpha)$ and then $a \in \sqrt{ker(\psi_\alpha)}$. So $\sqrt{ker(\psi_\alpha)} = M_\alpha$.

(2) Let $x \in \bigcap_\alpha ker(\psi_\alpha) \subseteq \bigcap_\alpha M_\alpha = Nil(A)$ since $dim(A) = 0$. Let $I = xA + ann(x)$. If $I \neq A$, there exists α_0 such that $I \subseteq M_{\alpha_0}$ and then $ann(x) \subseteq M_{\alpha_0}$. But $\psi_{\alpha_0}(x) = \frac{x}{1} = \frac{0}{1}$ in $A_{M_{\alpha_0}}$. There exists $t_{\alpha_0} \notin M_{\alpha_0}$ such that $t_{\alpha_0}x = 0$, then $t_{\alpha_0} \in ann(x) \subseteq M_{\alpha_0}$, which is absurd. Then $I = A$ and $1 = a_1x + b$

with $a_1 \in A$ and $b \in ann(x)$. Multiplying by x, we find $x = a_1 x^2$. Then $x = a_1 x^2 = a_2 x^4 = \ldots = a_k x^{2^k} = \ldots$ with $a_k \in A$. Since x is nilpotent, then it is zero. $\qquad \square$

Theorem 5.3 (Arapovic 1983) *A ring A is imbedded in a zero dimensional ring if and only if there exists a family $\{Q_\lambda\}_{\lambda \in \Lambda}$ of primary ideals of A satisfying*

1. $\bigcap_{\lambda \in \Lambda} Q_\lambda = (0).$

2. For each $a \in A$, there exists an integer $n_a \geq 1$ such that $a^{n_a} \notin \bigcup_{\lambda \in \Lambda} (P_\lambda \setminus Q_\lambda)$,

with $P_\lambda = \sqrt{Q_\lambda}$.

Proof $" \Longrightarrow "$ (1) Let B be a zero dimensional ring admitting A as a sub-ring and $\{M_\lambda\}_{\lambda \in \Lambda}$ the maximal spectrum of B. For $\lambda \in \Lambda$, let $\psi_\lambda : B \longrightarrow B_{M_\lambda}$ be the natural homomorphism, defined by $\psi_\lambda(x) = \frac{x}{1}$ for each $x \in B$. By the preceding lemma, $ker(\psi_\lambda)$ is an ideal M_λ-primary of B and $\bigcap_{\lambda \in \Lambda} ker(\psi_\lambda) = (0)$. Then $Q_\lambda = A \cap ker(\psi_\lambda)$ is an ideal $P_\lambda = A \cap M_\lambda$-primary of A and $\bigcap_{\lambda \in \Lambda} Q_\lambda = A \cap \bigcap_{\lambda \in \Lambda} ker(\psi_\lambda) = (0)$.

(2) Let $a \in A$ and $\{M_\lambda\}_{\lambda \in \Gamma}$ the subfamily of $Max(B)$ such that $a \in M_\lambda$ for each $\lambda \in \Gamma$. It suffices to show that there exists an integer $n_a \geq 1$ such that $a^{n_a} \notin \bigcup_{\lambda \in \Gamma}(P_\lambda \setminus Q_\lambda)$. By Proposition 5.1, there exists an idempotent $e \in B$ such that $a + (1 - e) \in U(B)$ and $a(1 - e) \in Nil(B)$. For each $\lambda \in \Gamma$, $1 - e \notin M_\lambda$. Consider the multiplicative set $S = B \setminus \bigcup_{\lambda \in \Gamma} M_\lambda$ of B. Then $1 - e \in S$, so $\frac{1-e}{1} \in U(S^{-1}B)$, and then $\frac{a}{1} \in Nil(S^{-1}B)$. There exists an integer $n_a \geq 1$ such that $\frac{a^{n_a}}{1} = \frac{0}{1}$ in $S^{-1}B$. There exists $s \in S$ such that $sa^{n_a} = 0$ in B. For each $\lambda \in \Gamma$, $\frac{a^{n_a}}{1} = \frac{0}{1}$ in B_{M_λ}, then $a^{n_a} \in A \cap ker(\psi_\lambda) = Q_\lambda$, so $a^{n_a} \notin P_\lambda \setminus Q_\lambda$. Then $a^{n_a} \notin \bigcup_{\lambda \in \Gamma}(P_\lambda \setminus Q_\lambda)$.

$" \Longleftarrow "$ Let $(Q_\lambda)_{\lambda \in \Lambda}$ be a family of primary ideals of A satisfying the conditions (1) and (2). Let $\phi : A \longrightarrow \bar{A} = \prod_{\lambda \in \Lambda}(A/Q_\lambda)_{P_\lambda/Q_\lambda}$, the map defined by $\phi(a) = \left(\frac{a+Q_\lambda}{1+Q_\lambda}\right)_{\lambda \in \Lambda}$ for each $a \in A$. Then ϕ is a homomorphism one to one. Indeed, let $a \in ker(\phi)$. For each $\lambda \in \Lambda$, $\frac{a+Q_\lambda}{1+Q_\lambda} = \frac{0+Q_\lambda}{1+Q_\lambda}$. There exists $s_\lambda \notin P_\lambda$ such that $(s_\lambda + Q_\lambda)(a + Q_\lambda) = 0 + Q_\lambda$. Then $as_\lambda \in Q_\lambda$ so $a \in Q_\lambda$. Then $a \in \bigcap_{\lambda \in \Lambda} Q_\lambda = (0)$.

We identify A to a sub-ring of $\bar{A}$. For each $a \in A$, we define the idempotent e_a of $\bar{A}$ whose component of index λ is 1 if $a \notin P_\lambda$ and 0 if not. Then $a + (1 - e_a)$ is a regular element of $\bar{A}$. Indeed, let $\left(\frac{x_\lambda + Q_\lambda}{s_\lambda + Q_\lambda}\right)_{\lambda \in \Lambda} \in \bar{A}$ such that $\left(\left(\frac{a+Q_\lambda}{1+Q_\lambda}\right)_{\lambda \in \Lambda} + (1 - e_a)\right)\left(\frac{x_\lambda + Q_\lambda}{s_\lambda + Q_\lambda}\right)_{\lambda \in \Lambda} = 0$ with $x_\lambda \in A$ and $s_\lambda \notin P_\lambda$. If $a \notin P_\lambda$, we find $\frac{ax_\lambda + Q_\lambda}{s_\lambda + Q_\lambda} = \frac{0+Q_\lambda}{1+Q_\lambda}$.

There exists $s'_\lambda \notin P_\lambda$ such that $(ax_\lambda + Q_\lambda)(s'_\lambda + Q_\lambda) = 0 + Q_\lambda$. Then $(s'_\lambda a)x_\lambda \in Q_\lambda$ with $s'_\lambda a \notin P_\lambda$, then $x_\lambda \in Q_\lambda$, so $\frac{x_\lambda + Q_\lambda}{s_\lambda + Q_\lambda} = \frac{0 + Q_\lambda}{1 + Q_\lambda}$. If $a \in P_\lambda$, we find $\frac{(a+1)x_\lambda + Q_\lambda}{s_\lambda + Q_\lambda} = \frac{0 + Q_\lambda}{1 + Q_\lambda}$. There exists $s'_\lambda \notin P_\lambda$ such that $((a + 1)x_\lambda + Q_\lambda)(s'_\lambda + Q_\lambda) = 0 + Q_\lambda$. Then $s'_\lambda (a + 1)x_\lambda \in Q_\lambda$ with $s'_\lambda (a + 1) \notin P_\lambda$, so $x_\lambda \in Q_\lambda$, then $\frac{x_\lambda + Q_\lambda}{s_\lambda + Q_\lambda} = \frac{0 + Q_\lambda}{1 + Q_\lambda}$. We will show that $\left(a(1 - e_a)\right)^{n_a} = 0$. If $a \notin P_\lambda$, the coordinate of index λ of e_a is 1, and then that of $a(1 - e_a)$ is 0. If $a \in P_\lambda$, the coordinate of index λ of e_a is 0, then that of $1 - e_a$ is 1. But by (2), $a^{n_a} \notin P_\lambda \setminus Q_\lambda$, then $a^{n_a} \in Q_\lambda$. So $\left(\frac{a+Q_\lambda}{1+Q_\lambda}\right)^{n_a} = \frac{a^{n_a}+Q_\lambda}{1+Q_\lambda} = \frac{0+Q_\lambda}{1+Q_\lambda}$. Let $A_1 = A[e_a; a \in A] \subseteq \tilde{A}$. Then $A \subseteq A_1$ is an integral extension since $e_a^2 - e_a = 0$. Let S be the set of regular elements of A_1 and $\tilde{A} = S^{-1}A_1$ its total ring of fractions. Then $A \subseteq A_1 \subseteq \tilde{A}$. We will show that $dim(\tilde{A}) = 0$. Suppose that there exist two prime ideals $P_1 \subset P_2$ in $\tilde{A}$ and then $P_1 \cap A_1 \subset P_2 \cap A_1$. Since $A \subseteq A_1$ is an integral extension, $P_1 \cap A \subset P_2 \cap A$. Let $a \in P_2 \cap A \setminus P_1 \cap A$ and e_a be the idempotent element of A_1 such that $a(1 - e_a)$ is nilpotent and $a + (1 - e_a)$ is regular. Then $a(1 - e_a) \in P_1$, so $1 - e_a \in P_1$, and then $a + (1 - e_a) \in P_2$, which is absurd since $a + (1 - e_a) \in U(\tilde{A})$. $\qquad\square$

Example 1 An integral domain is imbedded in its quotient field.

Example 2 Let P be the set of the prime numbers. The ideals $(p^2\mathbb{Z})_{p \in P}$ of the domain $\mathbb{Z}$ satisfy the conditions of the theorem. Indeed, for each $p \in P$, $p^2\mathbb{Z}$ is an ideal $p\mathbb{Z}$-primary and $\bigcap_{p \in P} p^2\mathbb{Z} = (0)$. Then we have the condition (1).

(2) For each $p \in P$, $\pm 1 \notin p\mathbb{Z} \setminus p^2\mathbb{Z}$. Let $a \in \mathbb{Z} \setminus \{\pm 1\}$, $a = \pm p_1^{\alpha_1} \ldots p_n^{\alpha_n}$ with $\alpha_1 \geq 1$ and the $p_i \in P$ being different. For $1 \leq i \leq n$, $a^2 \in p_i^2\mathbb{Z}$, then $a^2 \notin p_i\mathbb{Z} \setminus p_i^2\mathbb{Z}$. For each $p \in P \setminus \{p_1, \ldots, p_n\}$, $a^2 \notin p\mathbb{Z}$, then $a^2 \notin p\mathbb{Z} \setminus p^2\mathbb{Z}$. So $a^2 \notin \bigcup_{p \in P} (p\mathbb{Z} \setminus p^2\mathbb{Z})$.

Example 3 Let (V, M) be a valuation domain with dimension ≥ 2. Let P be a nonzero prime ideal of V different from M and $0 \neq a \in P$. We will show that the chained ring $A = V/aM$ cannot be imbedded in a zero dimensional ring. Note first that $\bar{a} \neq \bar{0}$ because if $a = am$ with $m \in M$, then $1 = m \in M$, which is absurd. The ideal $(\bar{0})$ is not primary. Indeed, in the contrary case, for $x \in M \setminus P$, $\bar{a}\bar{x} = \overline{ax} = \bar{0}$ with $\bar{a} \neq \bar{0}$. There exists $k \geq 1$ such that $\bar{x}^k = \bar{0}$, then $x^k \in aM \subseteq P$, so $x \in P$, which is absurd. Let J be an ideal of V containing strictly aM. There exists $b \in J \setminus aM$, and then $\frac{b}{a} \notin M$. Since V is a valuation domain, $\frac{a}{b} \in V$. Then $a = b\frac{a}{b} \in bV \subseteq J$. So $\bar{a}$ belongs to all the nonzero ideals of A. Let $\{Q_\lambda\}_{\lambda \in \Lambda}$ be a family of primary ideals of A. For each $\lambda \in \Lambda$, $Q_\lambda \neq (\bar{0})$, then $\bar{a} \in Q_\lambda$. So $\bigcap_{\lambda \in \Lambda} Q_\lambda \neq (0)$. The ring A does not satisfy the condition (1) of the theorem. By Proposition 2.7, the ring A is not strongly Hopfian. We can also note that since (0) is not a primary ideal, by Proposition 2.7, the chained ring A is not strongly Hopfian.

Example 4 The condition (1) of the theorem is not sufficient for A to be imbedded in a zero dimensional ring. Let p be a prime number and $A = \prod_{n \geq 1} \mathbb{Z}/p^n\mathbb{Z}$. By the example of Theorem 4.3, A is not imbedded in a zero dimensional ring. For $n \in \mathbb{N}^*$, let e_n be the element of A with the component number n being zero and all the others being equal to 1. The ideal $e_n A$ is formed by the elements of A whose component numbers n are zero. It is primary. Indeed, let $x = (\bar{x}_n)_{n \in \mathbb{N}^*}$ and $y = (\bar{y}_n)_{n \in \mathbb{N}^*} \in A$ be such that $xy \in e_n A$ and $x \notin e_n A$. Then p^n divides $x_n y_n$ and does not divide x_n, and then p divides y_n. So p^n divides y_n^n, then $\bar{y}_n^n = 0$ and $y^n \in e_n A$. The ideal $\bigcap_{n \geq 1} e_n A = (0)$ since an element of the intersection must have all its components zero.

6 Polynomials Over a Strongly Hopfian Ring

The results of this section are due to Hmaimo et al. (2007).

Lemma 6.1 *Let A be a ring and $P(X) = a_0 + a_1 X + \ldots + a_r X^r \in A[X]$ a polynomial of degree r. Suppose that there exists an integer $n \geq 1$ such that $ann(a_i^n) = ann(a_i^{n+1})$, for each $i \in \{0, 1, \ldots, r\}$. Then $ann(P(X)^{(r+1)n+1}) = ann(P(X)^{(r+1)n})$.*

Proof By induction on r.

(1) For $r = 0$, $P(X) = a_0$ with $ann_A(a_0^{n+1}) = ann_A(a_0^n)$. Let $H(X) = b_0 + b_1 X + \ldots + b_s X^s \in ann_{A[X]}(a_0^{n+1})$. Then $a_0^{n+1}b_0 = \ldots = a_0^{n+1}b_s = 0$, so $b_0, \ldots, b_s \in ann_A(a_0^{n+1}) = ann_A(a_0^n)$. Then $a_0^n H(X) = a_0^n b_0 + a_0^n b_1 X + \ldots + a_0^n b_s X^s = 0$, so $H(X) \in ann_{A[X]}(a_0^n)$.

Suppose in the sequel that the property is true until the order $r - 1$, and put $P(X) = a_0 + a_1 X + \ldots + a_{r-1} X^{r-1} + a_r X^r$, then $P_{r-1}(X) = a_0 + a_1 X + \ldots + a_{r-1} X^{r-1}$.

Let $H(X) = b_0 + b_1 X + \ldots + b_s X^s \in ann(P(X)^{(r+1)n+1})$.

(2) We will show that $b_s, b_{s-1}, \ldots, b_0 \in ann(a_r^n)$. Since $H(X)P(X)^{(r+1)n+1} = 0$, then $b_s a_r^{(r+1)n+1} = 0$, so $b_s \in ann(a_r^{(r+1)n+1}) = ann(a_r^n)$. Then $a_r^n H(X)P(X)^{(r+1)n+1} = 0$, and since $a_r^n b_s = 0$, then $a_r^n b_{s-1} a_r^{(r+1)n+1} = 0$, so $b_{s-1} \in ann(a_r^{(r+2)n+1}) = ann(a_r^n)$. By this method, we show that the coefficients $b_s, b_{s-1}, \ldots, b_0 \in ann(a_r^n)$. Then $a_r^n H(X) = a_r^n (b_0 + b_1 X + \ldots + b_s X^s) = 0$. We deduce that $a_r^n H(X) = 0$.

(3) We have $a_r^{n-1} H(X) P_{r-1}(X)^{rn} = 0$. Indeed, $P(X)^{(r+1)n+1} = (a_r X^r + P_{r-1}(X))^{(r+1)n+1} = \sum_{k:0}^{(r+1)n+1} C_{(r+1)n+1}^k a_r^k X^{kr} (P_{r-1}(X))^{(r+1)n+1-k}$. By

hypothesis, $0 = H(X)P(X)^{(r+1)n+1} = \sum_{k:0}^{n-1} C_{(r+1)n+1}^k H(X)a_r^k X^{kr}$

$\left(P_{r-1}(X)\right)^{(r+1)n+1-k}$. We multiply by a_r^{n-1}, we find the equality $H(X)a_r^{n-1}$ $\left(P_{r-1}(X)\right)^{(r+1)n+1} = 0$, and then $a_r^{n-1}H(X) \in ann\left(P_{r-1}(X)^{(r+1)n+1}\right)$.

Let $deg\ P_{r-1}(X) = d \le r - 1$. By the induction hypothesis, $ann\left(P_{r-1}(X)^{(d+1)n+1}\right) = ann\left(P_{r-1}(X)^{(d+1)n}\right)$. But $(r + 1)n + 1 > rn \ge (d + 1)n$, and then $ann\left(P_{r-1}(X)^{(r+1)n+1}\right) = ann\left(P_{r-1}(X)^{rn}\right)$. So $a_r^{n-1}H(X)P_{r-1}(X)^{rn} = 0$. We have always observed before that $\sum_{k:0}^{n-1} C_{(r+1)n+1}^k H(X)a_r^k X^{kr}\left(P_{r-1}(X)\right)^{(r+1)n+1-k} = 0$. Multiplying by a_r^{n-2} and noting that $a_r^{n-1}H(X)P_{r-1}(X)^{rn} = 0$, we obtain the equality $H(X)a_r^{n-2}\left(P_{r-1}(X)\right)^{(r+1)n+1} = 0$. Then $a_r^{n-2}H(X) \in ann\left(P_{r-1}(X)^{(r+1)n+1}\right) = ann\left(P_{r-1}(X)^{rn}\right)$. So $a_r^{n-2}H(X)P_{r-1}(X)^{rn} = 0$.

By induction, we show that $H(X)P_{r-1}(X)^{rn} = 0$.

(4) Now, $H(X)\left(P(X)\right)^{(r+1)n} = H(X)\left(a_r X^r + P_{r-1}(X)\right)^{(r+1)n}$

$= H(X)\sum_{k:0}^{(r+1)n} C_{(r+1)n}^k a_r^k X^{kr} P_{r-1}(X)^{(r+1)n-k}$

$= \sum_{k:0}^{n-1} C_{(r+1)n}^k a_r^k X^{kr} H(X) P_{r-1}(X)^{(r+1)n-k} + \sum_{k:n}^{(r+1)n} C_{(r+1)n}^k a_r^k H(X) X^{kr}$

$P_{r-1}(X)^{(r+1)n-k} = 0$. The first expression is zero since $H(X)P_{r-1}(X)^{rn} = 0$, and the second is zero since $a_r^n H(X) = 0$. Then $H(X) \in ann\left(P(X)^{(r+1)n}\right)$. □

Theorem 6.2 *The ring $A[X]$ is strongly Hopfian if and only if A is strongly Hopfian.*

Proof "$\Longrightarrow$" By Proposition 1.2. "$\Longleftarrow$" By the preceding lemma.   □

Corollary 6.3 *Let A be a strongly Hopfian ring. Then so is $A[X_1, \ldots, X_n]$.*

Proof By induction.   □

Corollary 6.4 (Hizem 2011) *Let A be a strongly Hopfian ring and I any set of indexes. Then the ring $R = A[X_i; i \in I]$ is strongly Hopfian.*

Proof Let $f \in R$. There exists a finite number of variables $X_{i_1}, \ldots, X_{i_n}$ such that $f \in R_n = A[X_{i_1}, \ldots, X_{i_n}]$. By the preceding corollary, the sequence $ann_{R_n}(f) \subseteq ann_{R_n}(f^2) \subseteq \cdots$ is stationary. There exists $k \in \mathbb{N}^*$ such that $ann_{R_n}(f^k) = ann_{R_n}(f^{k+1})$. We will show that $ann_R(f^k) = ann_R(f^{k+1})$. Let $g \in ann_R(f^{k+1})$. If $g \in R_n$, then $g \in ann_R(f^k)$. If $g \in R_{n+1}$, we have $g = g_r X_{i_{n+1}}^r + g_{r-1} X_{i_{n+1}}^{r-1} + \ldots + g_1 X_{i_{n+1}} + g_0$ with $g_i \in R_n$ and $r \in \mathbb{N}^*$. Since $0 = gf^{k+1} = f^{k+1}g_r X_{i_{n+1}}^r + f^{k+1}g_{r-1} X_{i_{n+1}}^{r-1} + \ldots + f^{k+1}g_1 X_{i_{n+1}} + f^{k+1}g_0$, then

$f^{k+1}g_i = 0$ for $0 \leq i \leq r$, so $g_i \in ann_{R_n}(f^{k+1}) = ann_{R_n}(f^k)$, and then $gf^k = 0$. Since the number of variables involved in the expression of g is finite, after a finite number of steps, we obtain $gf^k = 0$.   □

Example Let $A \subseteq B$ be an extension of rings. Then $R = A + (X_i, i \in I)B[X_i, i \in I]$ is strongly Hopfian if and only if B is strongly Hopfian.

   " $\Longleftarrow$" By the preceding corollary, $B[X_i, \ i \in I]$ is strongly Hopfian. Since $R \subseteq B[X_i, \ i \in I]$, then R is strongly Hopfian.

   " $\Longrightarrow$" Let $b \in B$. Fix $j \in I$ and $k \in \mathbb{N}^*$. Then $ann_R(X_jb)^k = ann_A(b^k) + (X_i, i \in I)ann_B(b^k)[X_i, \ i \in I]$. Indeed, let $g \in R$, $g = a + f$ with $a \in A$ and $f \in (X_i, i \in I)B[X_i, \ i \in I]$. So $g \in ann_R(bX_j)^k \Longleftrightarrow b^kX_j^k(a + f) = 0 \Longleftrightarrow b^k(a + f) = 0 \Longleftrightarrow b^ka = 0$ and $b^kf = 0$ since the constant term of f is zero. But $b^kf = 0 \Longleftrightarrow$, and the coefficients of f belong to $ann_B(b^k)$. Then we have the equality. Now, since the sequence $\left(ann_R(X_jb)^k\right)_{k\in\mathbb{N}^*}$ is stationary, so is the sequence $\left(ann_B(b^k)\right)_{k\in\mathbb{N}^*}$.

Corollary 6.5 *If A is a strongly Hopfian ring, then so is the Laurent polynomial ring $A[X, X^{-1}]$.*

Proof Let $f(X) = a_{-r}X^{-r} + \ldots + a_0 + \ldots + a_sX^s \in A[X, X^{-1}]$. Since A is strongly Hopfian, there exists $n \in \mathbb{N}^*$ such that $ann(a_i^n) = ann(a_i^{n+1})$ for $-r \leq i \leq s$. We will show that $ann\left(f(X)^{(r+s+1)n}\right) = ann\left(f(X)^{(r+s+1)n+1}\right)$. Let $g(X) = b_{-k}X^{-k} + \ldots + b_0 + \ldots + b_lX^l \in A[X, X^{-1}]$ be such that $g(X)f(X)^{(r+s+1)n+1} = 0$. The products $X^kg(X) = g_1(X) \in A[X]$ and $X^rf(X) = f_1(X) \in A[X]$ are polynomials of degree $r + s$. We have $g_1(X)f_1(X)^{(r+s+1)n+1} = 0$. By Lemma 6.1, $g_1(X)f_1(X)^{(r+s+1)n} = 0$.

   Then $g(X)f(X)^{(r+s+1)n} = 0$.   □

Notation Let $\mathscr{A} = (A_n)_{n\in\mathbb{N}}$ be an increasing sequence of rings and $A = \bigcup_{n\in\mathbb{N}} A_n$ their union. Let $\mathscr{A}[X]$ be the set of polynomials whose coefficient of X^i belongs to A_i for each $i \in \mathbb{N}$. Then $\mathscr{A}[X]$ is a sub-ring of $A[X]$.

Corollary 6.6 (Hamed et al. 2021) *Let $\mathscr{A} = (A_n)_{n\in\mathbb{N}}$ be an increasing sequence of rings and $A = \bigcup_{n\in\mathbb{N}} A_n$ their union. Then A is strongly Hopfian if and only if $\mathscr{A}[X]$ is strongly Hopfian.*

Proof " $\Longrightarrow$" By Theorem 6.2, $A[X]$ is strongly Hopfian. By Proposition 1.2, since $\mathscr{A}[X] \subseteq A[X]$, then $\mathscr{A}[X]$ is strongly Hopfian.

   " $\Longleftarrow$" Let $a \in A$. There exists $p \in \mathbb{N}$ such that $a \in A_p$. Since $aX^p \in \mathscr{A}[X]$, there exists $n \in \mathbb{N}$ such that $ann_{\mathscr{A}[X]}(aX^p)^{n+1} = ann_{\mathscr{A}[X]}(aX^p)^n$. We will show that $ann_A(a^{n+1}) = ann_A(a^n)$. Let $b \in A$ be such that $ba^{n+1} = 0$. There exists $m \in \mathbb{N}$ such that $b \in A_m$, then $bX^m \in \mathscr{A}[X]$ and $bX^m(aX^p)^{n+1} = 0$. So $bX^m(aX^p)^n = 0$ and then $ba^n = 0$.   □

7 Formal Power Series Over a Strongly Hopfian Bounded Ring

The results of this section are due to Khalifa (2017).

Proposition 7.1 *Let A be a zero dimensional SFT-ring. The total ring of fractions $Tot(A[[X]])$ is zero dimensional, so $A[[X]]$ is strongly Hopfian.*

Proof By Proposition 1.20 of Chap. 4, $Min(A[[X]]) = \{P[[X]], P \in Spec(A)\}$, and by Corollary 1.22 of the same chapter, the other prime ideals of $A[[X]]$ contain the regular element X. By Lemma 3.11 (2) of Chap. 2, the only prime ideals of $A[[X]]$ that disjoint with the set S of its regular elements are the minimal prime ideals. Then the Krull dimension of the ring $Tot(A[[X]]) = S^{-1}A[[X]]$ is zero. By Corollary 2.5, $A[[X]]$ is strongly Hopfian. $\square$

Example 1 Let $A = \mathbb{F}_2[Y_i; i \in \mathbb{N}]/(Y_i^2; i \in \mathbb{N})$. The only prime ideal of A is $M = (\bar{Y}_i; i \in \mathbb{N})$. It is not finitely generated. Then A is not Noetherian, is not reduced, and is zero dimensional. By the example of Proposition 1.19, Chap. 4, A is a SFT-ring. Then $A[[X]]$ is strongly Hopfian.

Example 2 The SFT property is not necessary in the proposition for $A[[X]]$ to be strongly Hopfian. Let $(K_i)_{i\in\mathbb{N}}$ be a sequence of fields. By Example 1 of Definition 2.1, the product ring $A = \prod_{i:0}^{\infty} K_i$ is von Neumann regular. By Proposition 2.2, A is reduced and zero dimensional. The formal power series ring $A[[X]]$ is also reduced, and then it is strongly Hopfian. By Example 1 of Definition 1.9, Chap. 4, A is not a SFT-ring.

Lemma 7.2 (Fields 1971) *Let A be a ring.*

1. *Let $f = \sum_{i:0}^{\infty} a_i X^i \in A[[X]]$. Suppose that there exists $t \in \mathbb{N}$ such that a_t is regular in A and a_i is nilpotent for $0 \le i < t$. Then f is regular in $A[[X]]$.*
2. *Suppose that in A, Let A be a strongly Hopfian ring. Then so Then a zero divisor in $A[[X]]$ has all its coefficients zero divisors in A; i.e., if $Z(A) = Nil(A)$, then $Z(A[[X]]) \subseteq Nil(A)[[X]]$.*

Proof

1. Let $g = \sum_{i:0}^{t-1} a_i X^i$ and $h = \sum_{i:t}^{\infty} a_i X^i$ with $g = 0$ if $t = 0$. Then $f = g + h$ with g nilpotent. Let $T = Tot(A)$ be the total ring of fractions of A and $S = \{X^i; i \in \mathbb{N}\}$: a multiplicative set of $T[[X]]$ formed by regular elements. If $B = S^{-1}T[[X]]$, then $A[[X]] \subseteq T[[X]] \subseteq B$. We have $h = X^t h'$ with

$h' = \sum_{i:t}^{\infty} a_i X^{i-t} \in A[[X]]$, and then h and h' are associated with B. The constant term of h' is a_t, a regular element in A, so invertible in T. Then h' is invertible in $T[[X]]$ and then in B. So h is invertible in B, and then $f = g + h$ is invertible in B, so regular in $A[[X]]$.

2. Let $f = \sum_{i:0}^{\infty} a_i X^i$ be a zero divisor in $A[[X]]$, and suppose that one of its coefficients is regular. Let $t = inf\{i \in \mathbb{N}; \ a_i \text{ is regular in } A\}$. For each $i < t$, a_i is a zero divisor in A and then nilpotent. We conclude by (1). $\square$

Proposition 7.3 (Hizem 2011) *Let A be a ring. Suppose that $Z(A) = Nil(A)$ is an SFT-ideal of A. Then $A[[X]]$ is strongly Hopfian.*

Proof By the preceding lemma and Corollary 2.4 of Chap. 5, we have $Z(A[[X]]) \subseteq Nil(A)[[X]] = Nil(A[[X]]) \subseteq Z(A[[X]])$. Then $Z(A[[X]]) = Nil(A[[X]])$. By Proposition 1.13, the ring $A[[X]]$ is strongly Hopfian. $\square$

Example Let $A = K[Y_i; i \in \mathbb{N}]/(Y_i Y_j; i, j \in \mathbb{N})$ with K a field. Its prime spectrum is reduced to $P = (\bar{Y}_i, i \in \mathbb{N})$. Then $Z(A) = Nil(A) = P$. For each $f \in Nil(A)$, $f^2 = 0$, then $Nil(A)$ is a SFT-ideal. By the preceding proposition, $A[[X]]$ is strongly Hopfian.

Notations Let A be a ring.

(1) Let $a \in A$. We denote by $\mu(a)$ the smallest integer $n \geq 1$ such that $ann(a^{n+1}) = ann(a^n)$ if it exists. In the contrary case, $\mu(a) = \infty$.

Example Let a be a regular element. Then $ann(a) = ann(a^2) = \ldots = (0)$, so $\mu(a) = 1$.

Let $\mu(A) = sup\{\mu(a); a \in A\} = sup\{\mu(a); a \in Z(A)\}$, since $\mu(a) = 1$ if a is a regular element. We say that A is strongly Hopfian bounded if $\mu(A) < \infty$.

Example Let $A \subseteq B$ be an extension of rings. Then $\mu(A) \leq \mu(B)$. In particular, a sub-ring of a strongly Hopfian bounded ring is strongly Hopfian bounded.

(2) Let $a \in Nil(A)$. We denote by $\eta(a)$ the index of nilpotence of a; i.e., the smallest integer $n \geq 1$ such that $a^n = 0$. Let $\eta(A) = sup\{\eta(a); a \in Nil(A)\}$.

Example Let $A \subseteq B$ be an extension of rings. Then $Nil(A) \subseteq Nil(B)$, so $\eta(A) \leq \eta(B)$.

Example By Lemma 1.5, a reduced ring is strongly Hopfian bounded. If A is a reduced ring, then $A[[X_1, \ldots, X_n]]$ is also reduced and then strongly Hopfian bounded.

Let Λ be any set of indexes. The ring $A[[X_\alpha, \alpha \in \Lambda]]_1 = \bigcup_F A[[X_\alpha, \alpha \in F]]$, where F ranges among the finite subsets of Λ. If A is reduced, $A[[X_\alpha, \alpha \in \Lambda]]_1$ is also reduced and then strongly Hopfian bounded.

Lemma 7.4 *Let A be a ring.*

1. *For each $a \in Nil(A)$, we have $\eta(a) = \mu(a)$.*
2. *We have $\eta(A) \leq \mu(A)$, and if $Nil(A) = Z(A)$, then $\eta(A) = \mu(A)$.*
3. *If S is a multiplicative set of A, then $\mu(S^{-1}A) \leq \mu(A)$.*
   In particular, if A is strongly Hopfian bounded, so is $S^{-1}A$.
4. *$\mu(Tot(A)) = \mu(A)$ and $\eta(Tot(A)) = \eta(A)$.*

Proof

(1) Since $a^{\eta(a)} = 0$, then $ann(a^{\eta(a)}) = ann(a^{\eta(a)+1}) = A$, so $\mu(a) \leq \eta(a)$. Suppose that $\mu(a) < \eta(a)$. Since $a^{\mu(a)+1+\eta(a)-(\mu(a)+1)} = a^{\eta(a)} = 0$ with $\eta(a) - (\mu(a) + 1) \geq 0$, then $a^{\eta(a)-(\mu(a)+1)} \in ann(a^{\mu(a)+1}) = ann(a^{\mu(a)})$. So $a^{\mu(a)+\eta(a)-(\mu(a)+1)} = 0$, and then $a^{\eta(a)-1} = 0$, which contradicts the minimality of $\eta(a)$.

(2) y (1), we have $\eta(A) = sup\{\eta(a); a \in Nil(A)\} = sup\{\mu(a); a \in Nil(A)\} \leq sup\{\mu(a); a \in Z(A)\} = \mu(A)$.

(3) Suppose that $\mu(A) = n < \infty$. Let $\frac{a}{s} \in S^{-1}A$ and $\frac{b}{t} \in ann(\frac{a}{s})^{n+1}$ with $a, b \in A$ and $s, t \in S$. Then $\frac{ba^{n+1}}{ts^{n+1}} = \frac{0}{1}$. There exists $u \in S$ such that $uba^{n+1} = 0$, then $ub \in ann(a^{n+1}) = ann(a^n)$. So $uba^n = 0$, then $\frac{b}{t}(\frac{a}{s})^n = \frac{0}{1}$ so $\frac{b}{t} \in ann(\frac{a}{s})^n$.

(4) Since $A \subseteq Tot(A) = S^{-1}A$ where S is the set of the regular elements of A, then $\mu(A) \leq \mu(Tot(A))$. By (3), $\mu(Tot(A)) \leq \mu(A)$. Then we have the equality.

Also, since $A \subseteq Tot(A)$, then $\eta(A) \leq \eta(Tot(A))$. If $\eta(A) = \infty$, we have the result. In the finite case, put $\eta(A) = n$. Let $\frac{a}{s} \in Nil(Tot(A))$ with $a \in A$ and $s \in S$. There exists $k \in \mathbb{N}^*$ such that $(\frac{a}{s})^k = \frac{0}{1}$. There exists $t \in S$ such that $ta^k = 0$ and then $a^k = 0$. So $a \in Nil(A)$, then $a^n = 0$, and so $(\frac{a}{s})^n = \frac{0}{1}$. Then $\eta(Tot(A)) \leq n$. □

Example A ring with only one prime ideal is zero dimensional; then it is strongly Hopfian. But it is not necessarily strongly Hopfian bounded. Indeed, if K is a field, the only prime ideal of the ring $A = K[X_i, i \in \mathbb{N}^*]/(X_i^i, i \in \mathbb{N}^*)$ is $P = (\bar{X}_i, i \in \mathbb{N}^*) = Nil(A)$. Since $\eta(A) \leq \mu(A)$ and $\eta(A) = \infty$, then $\mu(A) = \infty$.

Notation By Lemma 1.15 of Chap. 4, if I is an ideal of a ring A and if $k \in \mathbb{N}^*$ satisfy for each $x \in I$, $x^k = 0$, then there exists an integer $n_k \geq 1$ such that for each $f \in I[[X]]$, $f^{n_k} = 0$. Now, if $\eta(A) < \infty$, for each $x \in Nil(A)$, $x^{\eta(A)} = 0$; then for each $f \in Nil(A)[[X]]$, $f^{n_{\eta(A)}} = 0$.

Proposition 7.5 *Let A be a ring with only one prime ideal. The following assertions are equivalent:*

1. *The ring $A[[X]]$ is strongly Hopfian bounded.*
2. *The ring A is strongly Hopfian bounded.*
3. *A is a SFT-ring.*
4. *The ring $Tot(A[[X]])$ is zero dimensional strongly Hopfian bounded.*
5. *$A[[X]]$ is imbedded in a zero dimensional strongly Hopfian bounded ring.*

In this case, $\mu(A[[X]]) \leq n_{\mu(A)}$.

Proof $"(1) \Longrightarrow (2)"$ A sub-ring of a strongly Hopfian bounded ring is strongly Hopfian bounded.

$"(2) \Longrightarrow (3)"$ Put $\eta(A) \leq \mu(A) = n < \infty$. The unique prime ideal of A coincides with $Nil(A)$. For each $x \in Nil(A)$, $x^n = 0$. Then $Nil(A)$ is a SFT-ideal. By Proposition 1.19 of Chap. 4, A is a SFT-ring.

$"(3) \Longrightarrow (4)"$ By Proposition 7.1, $Tot(A[[X]])$ is zero dimensional. Since A is local with maximal ideal $Nil(A)$, then $U(A) = A \setminus Nil(A)$, so $Z(A) = Nil(A)$. By Lemma 7.2 (2) and Corollary 2.4 of Chap. 5, we have $Z(A[[X]]) \subseteq Nil(A)[[X]] = Nil(A[[X]]) \subseteq Z(A[[X]])$. Then $Nil(A)[[X]] = Nil(A[[X]]) = Z(A[[X]])$. By the preceding lemma, $\mu(A[[X]]) = \eta(A[[X]])$. By Corollary 2.4 of Chap. 5, there exists $n \in \mathbb{N}^*$ such that for each $x \in Nil(A)$, $x^n = 0$. By Lemma 1.15 of Chap. 4, $\eta(A[[X]]) < \infty$. By the preceding lemma, $\mu(Tot(A[[X]]) = \mu(A[[X]]) < \infty$, and $Tot(A[[X]])$ is strongly Hopfian bounded.

$"(4) \Longrightarrow (5)"$ We have $A[[X]] \subseteq Tot(A[[X]])$ with $Tot(A[[X]])$ zero dimensional strongly Hopfian bounded.

$"(5) \Longrightarrow (1)"$ A sub-ring of a strongly Hopfian bounded ring is strongly Hopfian bounded.

Since $Nil(A)[[X]] = Nil(A[[X]]) = Z(A[[X]])$, then $\mu(A[[X]]) = \eta(A[[X]]) \leq n_{\mu(A)}$. □

Lemma 7.6 *Let $(A_\alpha)_{\alpha \in \Lambda}$ be a family of strongly Hopfian bounded rings. If the set $\{\mu(A_\alpha), \alpha \in \Lambda\}$ is bounded, the product ring $A = \prod_{\alpha \in \Lambda} A_\alpha$ is strongly Hopfian bounded.*

Proof Put $n = sup\{\mu(A_\alpha), \alpha \in \Lambda\} \in \mathbb{N}^*$. Let $a = (a_\alpha)_{\alpha \in \Lambda} \in A$ and $b = (b_\alpha)_{\alpha \in \Lambda} \in ann_A(a^{n+1})$. For each $\alpha \in \Lambda$, $b_\alpha \in ann_{A_\alpha}(a_\alpha^{n+1}) = ann_{A_\alpha}(a_\alpha^n)$. Then $b \in ann_A(a^n)$. □

Example The condition $\{\mu(A_\alpha), \alpha \in \Lambda\}$ is bounded and is necessary in the lemma.

Indeed, if p is a prime number and $n \in \mathbb{N}^*$, the ring $\mathbb{Z}/p^n\mathbb{Z}$, is finite, then it is strongly Hopfian bounded. For $1 \leq k \leq n$, $ann(\bar{p}^k) = (\bar{p}^{n-k})$, then $ann(\bar{p}) \subset ann(\bar{p}^2) \subset \ldots \subset ann(\bar{p}^n) = A$, so $\mu(\mathbb{Z}/p^n\mathbb{Z}) > n$. By Example 1 of Definition 1.1, the ring $\prod_{n \geq 1} \mathbb{Z}/p^n\mathbb{Z}$ is not even strongly Hopfian.

Theorem 7.7 *The following assertions are equivalent for a zero dimensional ring A:*

1. *The ring $A[[X]]$ is strongly Hopfian bounded.*
2. *The ring A is strongly Hopfian bounded.*
3. *$A[[X]]$ is imbedded in a zero dimensional strongly Hopfian bounded ring.*

Proof A sub-ring of a strongly Hopfian bounded ring is strongly Hopfian bounded. Then we have the implications $''(1) \Longrightarrow (2)''$ and $''(3) \Longrightarrow (1)''$.

$''(2) \Longrightarrow (3)''$ Let $\{M_\alpha\}_{\alpha \in \Lambda}$ be the maximal spectrum of the ring A. For each $\alpha \in \Lambda$, let $s_\alpha : A \longrightarrow A_{M_\alpha}$ be the canonical homomorphism defined by $s_\alpha(a) = \frac{a}{1}$ for each $a \in A$. By Lemma 5.2, $\bigcap\limits_{\alpha \in \Lambda} ker \ s_\alpha = (0)$. We define the homomorphism $\psi_\alpha : A[[X]] \longrightarrow A_{M_\alpha}[[X]]$ as follows: if $f = \sum\limits_{i:0}^{\infty} a_i X^i \in A[[X]]$, $\psi_\alpha(f) = \sum\limits_{i:0}^{\infty} s_\alpha(a_i) X^i$. Then $ker \ \psi_\alpha = (ker \ s_\alpha)[[X]]$. We define, also, the homomorphism $\psi : A[[X]] \longrightarrow \prod\limits_{\alpha \in \Lambda} A_{M_\alpha}[[X]]$ by $\psi(f) = (\psi_\alpha(f))_{\alpha \in \Lambda}$ for each $f \in A[[X]]$. Then $ker \ \psi = \bigcap\limits_{\alpha \in \Lambda} ker \ \psi_\alpha = (\bigcap\limits_{\alpha \in \Lambda} ker \ s_\lambda)[[X]] - (0)$. So ψ is one to one. If we denote T_α the total ring of fractions of $A_{M_\alpha}[[X]]$, then $A[[X]]$ is imbedded in the product ring $T = \prod\limits_{\alpha \in \Lambda} T_\alpha$. By Lemma 7.4 (3), for each $\alpha \in \Lambda$, $\mu(A_{M_\alpha}) \leq \mu(A)$, and then A_{M_α} is strongly Hopfian bounded with a unique prime ideal since M_α is a minimal prime of A. By Proposition 7.5, T_α is a zero dimensional strongly Hopfian bounded ring. By Lemma 7.4 (2), $\eta(T_\alpha) \leq \mu(T_\alpha) \leq n_{\mu(A)}$. By Lemma 7.6, T is strongly Hopfian bounded. The set $\Lambda_{n_{\mu(A)}} = \{\lambda \in \Lambda; \ \exists x \in Nil(T_\lambda), x^{n_{\mu(A)}} \neq 0\}$ is nonempty. By Corollary 4.4, $dim(T) = 0$. $\square$

Example (Hizem 2011) In the theorem, we cannot replace "strongly Hopfian bounded" by "strongly Hopfian." Indeed, let K be a field with zero characteristic and I the ideal of $K[X_i, Y_i, i \geq 1]$ generated by the X_i^2, Y_i^2 for $i \geq 1$ and $Y_n \sum\limits_{i_1+...+i_n=k} X_{i_1} \ldots X_{i_n}$ for $n \geq 1$ and $k \geq 1$. The prime spectrum of the ring $A = K[X_i, Y_i, i \geq 1]/I$ is reduced to the ideal $(\bar{X}_i, \bar{Y}_i, i \geq 1)$. Then A is strongly Hopfian since it is zero dimensional. We will show that $A[[T]]$ is not strongly Hopfian. Let $f = \sum\limits_{i \geq 1} \overline{X_i} T^i \in A[[T]]$. By construction, for each $n \geq 1$, $\overline{Y_n} f^n = 0$. We will show that $\overline{Y_n} f^{n-1} \neq 0$. The first nonzero coefficient of f^{n-1} is $(n-1)! \overline{X_1 \ldots X_{n-1}}$. If $\overline{Y_n (n-1)! X_1 \ldots X_{n-1}} = 0$, since $(n-1)!$ is invertible in K, taking $X_i = 0$ for $i \geq n$ and $Y_i = 0$ for $i \neq n$, we obtain $Y_n X_1 \ldots X_{n-1} \in (X_1^2, \ldots, X_{n-1}^2, Y_n^2) K[X_1, \ldots, X_{n-1}, Y_n]$. This is absurd. $\square$

Corollary 7.8 *The following assertions are equivalent for a ring A:*

1. $A[[X]]$ is imbedded in a zero dimensional strongly Hopfian bounded ring.
2. A is imbedded in a zero dimensional strongly Hopfian bounded ring.

Proof "(1) $\implies$ (2)" Let B be a zero dimensional strongly Hopfian bounded ring containing $A[[X]]$. Then $A \subset A[[X]] \subseteq B$.

"(2) $\implies$ (1)" Let B be a zero dimensional strongly Hopfian bounded ring containing A. By the preceding theorem, $B[[X]]$ is imbedded in a zero dimensional strongly Hopfian bounded ring T. Then $A[[X]] \subseteq B[[X]] \subseteq T$. $\qquad\square$

8 Formal Power Series Over a Chained Ring

Lemma 8.1 *The maximal ideal of a chained ring is either principal or idempotent.*

Proof Let A be a chained ring with maximal ideal M. Suppose that M is not principal. For each $a \in M, aA \subset M$. Let $b \in M \setminus aA$. Then $aA \subset bA$, so $a = bc$ with $c \in A \setminus U(A) = M$. Then $a \in M^2$ and $M = M^2$. $\qquad\square$

Notation Let $f \in A[X]$ (resp. $A[[X]]$). We denote $c(f)$ the ideal de A generated by all the coefficients of f.

Lemma 8.2 (Gilmer et al. 1975) *Let A be a chained ring and f and $g \in A[X]$. Then $c(fg) = c(f)c(g)$.*

Proof Since A is chained, $c(f)$ (resp. $c(g)$) is generated by a coefficient a of f (resp. a coefficient b of g). Then $f = af_1$ and $g = bg_1$ with one of the coefficients of f_1 (resp. g_1) are equal to 1. Then $c(f_1) = c(g_1) = A, c(f) = c(af_1) = ac(f_1) = aA, c(g) = bA$, and $c(fg) = c(abf_1g_1) = abc(f_1g_1)$. Let M be the maximal ideal of A. Then $f_1, g_1 \notin M[X]$, so $f_1g_1 \notin M[X]$. One of the coefficients of f_1g_1 is invertible in A, and then $A_{f_1g_1} = A$ and $c(fg) = abA = c(f)c(g)$. $\qquad\square$

Lemma 8.3 *Let (A, M) be a zero dimensional chained ring with $M^2 = M$. Let $f \in A[[X]]$ be a zero divisor that is not nilpotent. Then $c(f) = M$ and for each $g \in A[[X]]$ such that $fg = 0$, and for each coefficient α of g, we have $\alpha f = 0$.*

Proof Since M is the unique prime ideal of A, $Nil(A) = M = Z(A)$. By Lemma 7.2, $Z(A[[X]]) \subseteq Nil(A)[[X]] = M[[X]]$. Since f is a zero divisor, then $f \in M[[X]]$, so $c(f) \subseteq M$. If $c(f) \subset M$, an element $b \in M \setminus c(f)$ is nilpotent, and if $c(f) \subset bA$, then $f = bh$ with $h \in A[[X]]$. Then f is nilpotent, which is absurd. So $c(f) = M$. Let $0 \neq g \in A[[X]]$ be such that $fg = 0$ and α a nonzero coefficient of g. We will show that $\alpha c(f) = \alpha M = (0)$. The element α is the coefficient of a monomial X^e in g. Let $m \in M$. Put $f = \displaystyle\sum_{i:0}^{\infty} a_i X^i$. Since A is chained, $M = c(f) = \bigcup_{i:0}^{\infty}(a_i)$, and then $M = M^2 = \bigcup_{i:0}^{\infty}(a_i^2)$. There exists an index l such that $m \in (a_l^2)$. Put $k = e + l$. We have $m \in (a_l^2) \subseteq a_l M = a_l M^k = a_l(\bigcup_{i:0}^{\infty}(a_i^k)) = \bigcup_{i:0}^{\infty}(a_l a_i^k)$. There exists $s \in \{a_i, i \in \mathbb{N}\}$ such that $m \in (a_l s^k)$.

Since s is nilpotent, we define a ring homomorphism $\phi : A[[X]] \longrightarrow A[X]$, by $\phi(h(X)) = h(sX)$. Since $fg = 0$, then $\phi(fg) = 0$. By the preceding lemma, $c(\phi(f))c(\phi(g)) = c(\phi(f)\phi(g)) = c(\phi(fg)) = 0$. Since αX^e is a monomial of g, and then $\alpha s^e X^e$ is a monomial of $\phi(g)$. Also, $a_l X^l$ is a monomial of f, then $a_l s^l X^l$ is a monomial of $\phi(f)$. Since $c(\phi(f))c(\phi(g)) = 0$, then $\alpha s^e a_l s^l = 0$, so $\alpha a_l s^k = \alpha a_l s^{e+l} = 0$. But $m \in (a_l s^k)$, and then $\alpha m = 0$. □

Theorem 8.4 (Khalifa 2017) *Let A be a chained ring. Then $A[[X]]$ is strongly Hopfian if and only if A is strongly Hopfian.*

Proof "$\Longrightarrow$" A sub-ring of a strongly Hopfian ring is strongly Hopfian.

"$\Longleftarrow$" By Proposition 2.7, $Tot(A)$ is a zero dimensional chained ring and then strongly Hopfian. Replacing A by $Tot(A)$, we may suppose that A is zero dimensional. Let M be the unique prime ideal of A. By Lemma 8.1, M is either principal or idempotent.

First Case: M is principal. By Cohen theorem, A is Noetherian. Then $A[[X]]$ is also Noetherian and then strongly Hopfian.

Second Case: $M^2 = M$. Since M is the unique prime ideal of A, then $Nil(A) = M = Z(A)$. By Lemma 7.2, $Z(A[[X]]) \subseteq Nil(A)[[X]] = M[[X]]$. Let $f \in A[[X]]$. If f is regular, then $ann(f) = ann(f^2) = \ldots = (0)$. If f is nilpotent of index n, then $ann(f^n) = ann(f^{n+1}) = \ldots = A[[X]]$. Suppose that f is a zero divisor that is not nilpotent. By the preceding lemma, $c(f) = c(f^2) = M$, and there exists $0 \neq a \in A$ such that $af = 0$. Then $aM = (0)$, so $M \subseteq ann(a)$. Conversely, let $b \in ann(a)$. Then $ab = 0$, so $b \in Z(A) = M$. Then $ann(a) = M$. We will show that $ann(f) = ann(f^2)$. Let $g = \sum_{i:0}^{\infty} b_i X^i \in ann(f^2)$. By the preceding lemma, for each $i \in \mathbb{N}$, $b_i f^2 = 0$, then $b_i M = (0)$, so $M \subseteq ann(b_i)$. Suppose that there exists an index i_0 such that $aA \subset b_{i_0} A$ and then $a = b_{i_0} m$ with $m \in A \setminus U(A) = M \subseteq ann(b_{i_0})$. Then $a = b_{i_0} m = 0$, which is absurd. Since A is chained, for each $i \in \mathbb{N}$, $b_i A \subset aA$. But $af = 0$, then $b_i f = 0$, for each $i \in \mathbb{N}$, so $gf = 0$ and $g \in ann(f)$. □

Proposition 8.5 (Khalifa 2017) *The following assertions are equivalent for a chained ring A:*

1. *The ring $A[[X]]$ is strongly Hopfian bounded.*
2. *The ring A is strongly Hopfian bounded.*
3. *The ring A is strongly Hopfian and $Nil(A)$ is an SFT ideal.*

Proof *"(1)* $\implies$ *(2)"* A sub-ring of a strongly Hopfian bounded ring is strongly Hopfian bounded.

"(2) $\implies$ *(3)"* By Proposition 2.7, $Nil(A) = Z(A)$, and by Lemma 7.4 (2), $\eta(A) = \mu(A) < \infty$. Then for each $x \in Nil(A)$, $x^{\mu(A)} = 0$. So $Nil(A)$ is a SFT ideal.

"(3) $\implies$ *(1)"* By Proposition 2.7, $Tot(A)$ is a zero dimensional chained ring, then it is strongly Hopfian, and $Nil(Tot(A))$ is a SFT ideal. Replacing A by $Tot(A)$, we may suppose that A is zero dimensional. We conclude by Proposition 7.5. □

Proposition 8.6 (Hamed et al. 2021) *Let $A \subseteq B$ be an extension of rings. Then the ring $A + XB[[X]]$ is strongly Hopfian if and only if $B[[X]]$ is strongly Hopfian.*

Proof *"* $\Longleftarrow$ *"* $A + XB[[X]]$ is a sub-ring of $B[[X]]$. We conclude by Proposition 1.2.

" $\implies$ *"* Let $f \in B[[X]]$. Then $Xf \in A + XB[[X]]$. By hypothesis, there exists $n \in \mathbb{N}^*$ such that $ann_{A+XB[[X]]}(Xf)^{n+1} = ann_{A+XB[[X]]}(Xf)^n$. We will show that $ann_{B[[X]]}(f)^{n+1} = ann_{B[[X]]}(f)^n$. Let $g \in B[[X]]$ be such that $gf^{n+1} = 0$. Since $Xg \in A + XB[[X]]$ and $Xg(Xf)^{n+1} = 0$, then $Xg(Xf)^n = 0$, so $gf^n = 0$. □

Corollary 8.7 (Hamed et al. 2021) *Let $A \subseteq B$ be an extension of rings with B chained. Then $A + XB[[X]]$ is strongly Hopfian if and only if B is strongly Hopfian.*

Proof By Theorem 8.4, B is strongly Hopfian if and only if $B[[X]]$ is strongly Hopfian. By the preceding proposition, $A + XB[[X]]$ is strongly Hopfian if and only if B is strongly Hopfian. □

9 Formal Power Series Over a Ring with Nonzero Characteristic

The results of this section are due to Hizem (2011).

Lemma 9.1 *Let A be a ring whose characteristic is a prime number p, and let $f = \sum_{i \geq 0} a_i X^i \in A[[X]]$. Suppose that there exists $k \in \mathbb{N}^*$ such that for each integer $i \geq 0$, $ann(a_i^k) = ann(a_i^{k+1})$. Then the chain $ann(f) \subseteq ann(f^2) \subseteq \ldots$ is stationary.*

Proof Let $s \in \mathbb{N}$ be such that $p^s \geq k$. We will show that $ann(f^{p^{s+1}}) = ann(f^{p^s})$. By the example of Lemma 1.14 of Chap. 4, $f^{p^{s+1}} = \sum_{i \geq 0} a_i^{p^{s+1}} X^{ip^{s+1}}$. Let $g = \sum_{i \geq 0} b_i X^i \in ann(f^{p^{s+1}})$. Then $b_0 a_0^{p^{s+1}} = 0$, so $b_0 \in ann(a_0^{p^{s+1}}) = ann(a_0^k) = $

$ann(a_0^{p^s})$. The relation $a_0^k g f^{p^{s+1}} = 0$ can be written $\left(\sum_{i\geq 0} a_0^k b_i X^i\right)\left(\sum_{i\geq 0} a_i^{p^{s+1}} X^{ip^{s+1}}\right)$

$= 0$. Since $a_0^k b_0 = 0$, cancel by X, we obtain $\left(\sum_{i\geq 1} a_0^k b_i X^{i-1}\right)\left(\sum_{i\geq 0} a_i^{p^{s+1}} X^{ip^{s+1}}\right) =$

0. Then $b_1 a_0^k a_0^{p^{s+1}} = 0$, so $b_1 \in ann(a_0^{p^s})$. By induction, for each $n \in \mathbb{N}$, $b_n \in$ $ann(a_0^{p^s})$. Then $g a_0^{p^s} = 0$. We show that for each $i \geq 0$, $g a_i^{p^s} = 0$ by noting that $0 = g f^{p^{s+1}} = g(a_0^{p^{s+1}} + a_1^{p^{s+1}} X^{p^{s+1}} + \ldots) = X^{p^{s+1}} g(a_1^{p^{s+1}} + \ldots)$ and then $g(a_1^{p^{s+1}} + \ldots) = 0$. $\qquad\square$

Example Let $A = \mathbb{F}_2[X_i; i \geq 1]/(X_1^2)$. The characteristic of A is 2. The ring is neither Noetherian nor reduced. We will show that for each $\overline{f} \in A$, $ann(\overline{f}^4) = ann(\overline{f}^2)$, and then $A[[X]]$ is strongly Hopfian. Indeed, there exists $n \in \mathbb{N}^*$ such that $f \in \mathbb{F}_2[X_1, \ldots, X_n]$. Then $f = f_0(X_2, \ldots, X_n) + f_1(X_2, \ldots, X_n)X_1 + \ldots + f_s(X_2, \ldots, X_n)X_1^s$. Since $\overline{X_1}^2 = 0$, then $\overline{f} = \overline{f_0(X_2, \ldots, X_n)} + \overline{f_1(X_2, \ldots, X_n)X_1}$. Since the characteristic of A is 2, we have $\overline{f}^2 = \overline{f_0(X_2, \ldots, X_n)}^2$ and $\overline{f}^4 = \overline{f_0(X_2, \ldots, X_n)}^4$. Let $\overline{g} \in ann(\overline{f}^4)$. Then $\overline{g} = \overline{g_0(X_2, \ldots, X_r)} + \overline{g_1(X_2, \ldots, X_r)X_1}$ with $r \geq n$ (by completing with lost variables). Then $0 = \overline{g}\overline{f}^4 = \overline{f_0(X_2, \ldots, X_n)^4 g_0(X_2, \ldots, X_r)} + \overline{f_0(X_2, \ldots, X_n)^4 g_1(X_2, \ldots, X_r)X_1}$. We have $f_0(X_2, \ldots, X_n)^4 g_0(X_2, \ldots, X_r) + f_0(X_2, \ldots, X_n)^4 g_1(X_2, \ldots, X_r)X_1 \in X_1^2 \mathbb{F}_2[X_i; i \geq 1]$. Then $f_0(X_2, \ldots, X_n)^4 g_i(X_2, \ldots, X_r) = 0$ for $i = 1, 2$. Since $\mathbb{F}_2[X_i; i \geq 1]$ is an integral domain, then either $f_0 = 0$ and, in this case, $ann(\overline{f}^2) = ann(\overline{f}^4) = A$, or $f_0 \neq 0$ and, in this case, $g_0 = g_1 = 0$, so $g = 0$ and $ann(\overline{f}^2) = ann(\overline{f}^4) = 0$.

Proposition 9.2 *Let A be a ring with characteristic $n = p_1 \ldots p_t$, where the p_i are different prime numbers, and let $f = \sum_{i\geq 0} a_i X^i \in A[[X]]$. Suppose that there exists $k \in \mathbb{N}^*$ such that for each $i \geq 0$, $ann(a_i^k) = ann(a_i^{k+1})$. Then the chain $ann(f) \subseteq ann(f^2) \subseteq \ldots$ is stationary.*

Proof For $1 \leq j \leq t$, the characteristic of the ring $A/p_j A$ is p_j. Let $s_j : A \longrightarrow A/p_j A$ be the canonical homomorphism and $\varphi_j : A[[X]] \longrightarrow (A/p_j A)[[X]]$ the homomorphism, defined for each $g = \sum_{i\geq 0} b_i X^i \in A[[X]]$, by $\varphi_j(g) = \sum_{i\geq 0} s_j(b_i)X^i$. Then $ker\varphi_j = (p_j A)[[X]] = p_j A[[X]]$. For $1 \leq j \leq t$, $ann(s_j(a_i)^k) = ann(s_j(a_i)^{k+1})$ in $A/p_j A$. Indeed, let $s_j(b) \in ann(s_j(a_i)^{k+1})$. Then $ba_i^{k+1} \in p_j A$, so $\widehat{p_j} ba_i^{k+1} \in nA = 0$, where $\widehat{p_j} = \prod_{l\neq j} p_l$. But $ann(a_i^k) = ann(a_i^{k+1})$, and then $\widehat{p_j} ba_i^k = 0$. In the ring $A/p_j A$, for each $l \neq j$, the element $s_j(p_l)$ is invertible. Indeed, since $p \gcd(p_j, p_l) = 1$, there exist $u, v \in \mathbb{Z}$ such that $up_j + vp_l = 1$ and then $s_j(p_l)s_j(v) = 1$. We deduce that $s_j(b)s_j(a_i)^k = 0$ and $s_j(b) \in ann(s_j(a_i)^k)$. By the preceding lemma, there exists $r \in \mathbb{N}^*$ such

that $ann(\varphi_j(f)^r) = ann(\varphi_j(f)^{r+1})$ for $j \in \{1, 2, \ldots, t\}$. We will show the equality $ann(f^r) = ann(f^{r+1})$. Let $g \in ann(f^{r+1})$. Then $gf^{r+1} = 0$. For $1 \leq j \leq t$, $\varphi_j(g)\varphi_j(f)^{r+1} = 0$, then $\varphi_j(g)\varphi_j(f)^r = 0$, so $\varphi_j(gf^r) = 0$. Then $gf^r \in p_j A[[X]]$. Put $gf^r = \sum_{i \geq 0} b_i X^i$ with $b_i \in p_1 A \cap \ldots \cap p_t A$ for each $i \geq 0$.
But $p_1 A \cap \ldots \cap p_t A = p_1 \ldots p_t A = nA = (0)$. Indeed, let $x \in p_1 A \cap \ldots \cap p_t A$, $x = p_1 c_1$ with $c_1 \in A$. In $A/p_2 A$, $0 = s_2(x) = s_2(p_1)s_2(c_1)$. Since $s_2(p_1)$ is invertible, then $s_2(c_1) = 0$, so $c_1 \in p_2 A$, then $c_1 = p_2 c_2$ with $c_2 \in A$, and so $x = p_1 p_2 c_2$. By induction, $x = p_1 p_2 \ldots p_t c_t = nc_t = 0$ with $c_t \in A$. Then $gf^r = 0$. □

Corollary 9.3 *Let A be ring with characteristic $n = p_1 \ldots p_t$, where the p_i are different prime numbers. If A is strongly Hopfian bounded, then $A[[X]]$ is strongly Hopfian bounded.*

Corollary 9.4 *Let A be ring with characteristic $n = p_1 \ldots p_t$, where the p_i are different prime numbers. If A is strongly Hopfian bounded, then $A[[X_1, \ldots, X_n]]$ is strongly Hopfian bounded.*

Corollary 9.5 *Let A be ring with characteristic $n = p_1 \ldots p_t$, where the p_i are different prime numbers. If A is strongly Hopfian bounded, then $A[[X_\alpha; \alpha \in \Lambda]]_1$ is strongly Hopfian bounded.*

Proof Let $f \in R = A[[X_\alpha; \alpha \in \Lambda]]_1$ and $n \in \mathbb{N}^*$ be such that $f \in R_n = A[[X_1, \ldots, X_n]]$. Since R_n is strongly Hopfian, there exists $k \in \mathbb{N}^*$ such that $ann_{R_n}(f^k) = ann_{R_n}(f^{k+1})$. We will show that $ann_R(f^k) = ann_R(f^{k+1})$. Indeed, let $g \in ann_R(f^{k+1})$. If $g \in R_n$, then $g \in ann_{R_n}(f^k)$. If $g \in R_{n+1}$, we can write $g = \sum_{r:0}^{\infty} g_r X_{n+1}^r$ with $g_r \in R_n$. Since $0 = gf^{k+1} = \sum_{r:0}^{\infty} g_r f^{k+1} X_{n+1}^r$, then $g_r f^{k+1} = 0$ for each $r \in \mathbb{N}$, so $g_r \in ann_R(f^{k+1}) = ann_{R_n}(f^k)$. Then $gf^k = 0$. We continue by induction on the number of variables in g to get $gf^k = 0$. □

10 Study of an Example

The results of this section are due to Kerr (1983) and Hizem (2011).

We construct an example of a ring A whose characteristic is either a prime number or zero, which is neither reduced nor Noetherian such that $A[[X]]$ is strongly Hopfian bounded. Let K be an integral domain, $\{X_{ij}; \ i, j \in \mathbb{N}^*, j \leq i\}$ a family of indeterminates, I the homogeneous ideal of $K[\{X_{ij}\}]$ generated by the $\{X_{ij}X_{kl}X_{nm}\}$, and $\{X_{ij}X_{ik}, j \neq k\}$ and $A = K[\{X_{ij}\}]/I$. Then A is neither reduced nor Noetherian with the same characteristic as K. We represent the family of the $x_{ij} = \overline{X_{ij}}$ by the matrix

$$x_{11}$$
$$x_{21} \ x_{22}$$
$$x_{31} \ x_{32} \ x_{33}$$
$$\vdots \qquad \ddots$$

The product of any three elements is zero, and the product of two distinct elements on the same line is zero. The ring A is graduated by $\deg x_{ij} = 1$ and $\deg \alpha = 0$ if $\alpha \in K$. We have $A = A_0 \oplus A_1 \oplus A_2$ where A_i is formed by the homogeneous elements of A of degree i. The family $\{x_{ij}\}$ is a basis of A_1 over K, and the family $\{x_{ij}x_{kl}; (i \neq k) \text{ or } (i = k \text{ and } j = l)\}$ is a basis of A_2 over K. Note that $(A_1 A)^3 = 0$. We will show that the ideal $A_1 A$ contains all the zero divisors of A.

Lemma 10.1 *Let S be a nonempty subset of A with $S \nsubseteq A_1 A$. Then $ann_A(S) = (0)$.*

In particular, the zero divisors of A are contained in $A_1 A$.

Proof Let $p, q \in A$ be such that $pq = 0$. We write $p = p_0 + p_1 + p_2$ and $q = q_0 + q_1 + q_2$, where $p_i, q_i \in A_i$. Then $p_0 q_0 = p_0 q_1 + p_1 q_0 = p_0 q_2 + p_1 q_1 + p_2 q_0 = 0$. If $p \notin A_1 A$, then $p_0 \neq 0$, so $q_0 = q_1 = q_2 = 0$, and then $q = 0$. □

Remarks 10.2

(1) By construction, $(A_1 A)^3 = 0$. If $p \in A_1 A$, then $p^3 = 0$, so $ann_A(p^3) = ann_A(p^4)$.
(2) If $p \in A \setminus A_1 A$, then for each $n \in \mathbb{N}^*$, $p^n \notin A_1 A$ and $ann_A(p^n) = (0)$.

  Indeed, if $p^n \in A_1 A$, then $p^{3n} = 0$, so p is nilpotent, and then it is a zero divisor. So $p \in A_1 A$, which is absurd.

Proposition 10.2 *The ring $A[[X]]$ is strongly Hopfian bounded.*

Proof Let $f \in A[[X]]$. If $f \in (A_1 A)[[X]]$, then $f^3 \in (A_1 A)^3[[X]] = (0)$, so $ann(f^3) = ann(f^4) = A[[X]]$. If $f = \sum_{i \geq 0} p_i X^i \notin (A_1 A)[[X]]$, there exists $i \in \mathbb{N}$ such that $p_i \notin A_1 A$. Let i be the smallest index with this property. For each $k < i$, $p_k \in A_1 A$, then $p_k^3 = 0$. By Lemma 10.1, p_i is regular and a_k is nilpotent for each $k < i$. By Lemma 7.3 (1), f is a regular element of $A[[X]]$ and then $ann(f^n) = (0)$ for each $n \in \mathbb{N}^*$. □

11 Other Criteria for the Strong Hopfianity

Lemma 11.1 (O'Malley 1970) *Let A be a ring and a a regular element of A. Then,*

$$a\left(\bigcap_{n:0}^{\infty} a^n A\right) = \bigcap_{n:0}^{\infty} a^n A.$$

Proof Let $x \in \bigcap_{n:0}^{\infty} a^n A$. For each $n \in \mathbb{N}$, there exists $d_n \in A$ such that $x = a^n d_n$. In particular, $x = a d_1 = a^n d_n$ for each $n \geq 1$, then $a(d_1 - a^{n-1} d_n) = 0$, and, since a is regular, $d_1 = a^{n-1} d_n$. So, $d_1 \in \bigcap_{k:0}^{\infty} a^k A$ and then $x = a d_1 \in a(\bigcap_{k:0}^{\infty} a^k A)$. $\qquad \square$

Lemma 11.2 (O'Malley 1970) *Let A be a ring, $f \in A[[X]]$ a formal power series with constant term b_0, and I an ideal of A. If $b_0 I = I$, then $I \subseteq \bigcap_{k:0}^{\infty} f^k A[[X]]$.*

Proof It suffices to show that $I \subseteq f A[[X]]$. Indeed, since $b_0 I = I$, by induction, $b_0^k I = I$ for each $k \in \mathbb{N}^*$. By the particular case, if g is a formal power series of $A[[X]]$ with constant term b_0^k, then $I \subseteq g A[[X]]$. For $g = f^k$, we have $I \subseteq f^k A[[X]]$. Put $f = \sum_{i:0}^{\infty} b_i X^i \in A[[X]]$. Let $t \in I$. We must find a formal power series $g = \sum_{i:0}^{\infty} c_i X^i \in A[[X]]$ such that $t = fg$, i.e., (*) $c_0 b_0 = t$, $c_0 b_1 + c_1 b_0 = 0$, $c_0 b_2 + c_1 b_1 + c_2 b_0 = 0, \ldots, c_0 b_k + c_1 b_{k-1} + \ldots + c_k b_0 = 0, \ldots$

We will define the sequence (c_i) by induction. Since $t \in I = b_0 I$, there exists $c_0 \in I$ such that $t = b_0 c_0$. Also, $c_0 \in I = b_0 I$, and there exists $d_0 \in I$ such that $c_0 = b_0 d_0$. Take $c_1 = -d_0 b_1 \in I$, then $c_0 b_1 + c_1 b_0 = c_0 b_1 - d_0 b_1 b_0 = b_1(c_0 - d_0 b_0) = 0$. Suppose defined the elements $c_0, c_1, \ldots, c_{k-1} \in I$ such that the first k equations of (*) are satisfied. Since $I = b_0 I$ for $0 \leq i \leq k - 1$, there exists $s_i \in I$ such that $c_i = b_0 s_i$. Let $c_k = -(s_0 b_k + s_1 b_{k-1} + \ldots + s_{k-1} b_1) \in I$. We multiply the two sides by b_0, and we find $c_0 b_k + c_1 b_{k-1} + \ldots + c_{k-1} b_1 + c_k b_0 = 0$. $\qquad \square$

Lemma 11.3 (Khalifa 2017) *Let A be a ring and a a regular element of A.*
If $\bigcap_{n:0}^{\infty} a^n A = (0)$, then $\bigcap_{n:0}^{\infty} (X - a^n) A[[X]] = (0)$.

Proof Let $f = \sum_{i:0}^{\infty} b_i X^i \in \bigcap_{n:0}^{\infty} (X - a^n) A[[X]]$. Then $b_0 \in \bigcap_{n:0}^{\infty} a^n A = (0)$, so $b_0 = 0$. On the other hand, for each $n \in \mathbb{N}$, there exists $f_n \in A[[X]]$ such that $f = (X - a^n) f_n$ and then $-a^n f_n(0) = f(0) = b_0 = 0$. Since a is regular, $f_n(0) = 0$ so $f_n = X f_n'$ with $f_n' \in A[[X]]$. Since $f = (X - a^n) f_n$, then $\sum_{i:1}^{\infty} b_i X^{i-1} = (X - a^n) f_n'$, so $\sum_{i:1}^{\infty} b_i X^{i-1} \in \bigcap_{n:0}^{\infty} (X - a^n) A[[X]]$. By the first step, $b_1 = 0$. By induction, we show that all the b_i are zero and then $f = 0$. $\qquad \square$

Theorem 11.4 (Eakin and Sathaye 1976) *Let A be a ring. Consider the subsets:*

$I_c(A) = \{a \in A;$ *there exists an A-endomorphism of $A[[X]]$ satisfying $\sigma(X) = a + X\}$,*

$I_1(A) = \{a \in A;$ *there exists a A-homomorphism $\sigma : A[[X]] \longrightarrow A$ satisfying $\sigma(X) = a\}$,*

$I_2(A) = \{a \in A;$ *there exist $m, n, i \in \mathbb{N}^*$, $1 \leq i \leq m$ and a A-homomorphism $\sigma : A[[X_1, \ldots, X_m]] \longrightarrow A[[Y_1, \ldots, Y_n]]$ satisfying $\sigma(X_i) = a + f$ with $f \in (Y_1, \ldots, Y_n)\}$.*

Then $I_c(A) = I_1(A) = I_2(A)$ is an ideal of A contained in the Jacobson radical of A and containing the nilradical of A.

Proof

(1) We will show the inclusions $I_c(A) \subseteq I_1(A) \subseteq I_2(A) \subseteq I_c(A)$. Let $a \in I_c(A)$. There exists a A-endomorphism σ de $A[[X]]$ such that $\sigma(X) = a + X$. Let $\tau : A[[X]] \longrightarrow A$ be the A-homomorphism, defined by $\tau(f) = f(0)$ for each $f \in A[[X]]$. Then $\tau \circ \sigma : A[[X]] \longrightarrow A$ is a A-homomorphism such that $\tau \circ \sigma(X) = \tau(a + X) = a$. Then $a \in I_1(A)$ so $I_c(A) \subseteq I_1(A)$. The inclusion $I_1(A) \subseteq I_2(A)$ is clear since a A-homomorphism of $A[[X]] \longrightarrow A$ can be seen as a A-homomorphism of $A[[X]] \longrightarrow A[[Y]]$. Let $a \in I_2(A)$ and $\sigma : A[[X_1, \ldots, X_m]] \longrightarrow A[[Y_1, \ldots, Y_n]]$ be a A-homomorphism such that $\sigma(X_i) = a + f$ with $1 \leq i \leq m$ and $f \in (Y_1, \ldots, Y_n)$. The restriction of σ to $A[[X_i]]$ is a A-homomorphism $\sigma_1 : A[[X_i]] \longrightarrow A[[Y_1, \ldots, Y_n]]$ such that $\sigma_1(X_i) = a + f$. By [Benhissi 2003; Chap. 9, Theorem 2.5], there exists an A-automorphism τ of $A[[X_i]]$ such that $\tau(X_i) = a + X_i$. Then $a \in I_c(A)$ so $I_2(A) \subseteq I_c(A)$.

(2) Let $a, b \in I_c(A)$. There exist A-homomorphisms $\sigma : A[[X]] \longrightarrow A$ and $\tau : A[[Y]] \longrightarrow A$ such that $\sigma(X) = a$ and $\tau(Y) = b$. We can extend simultaneously σ and τ to A-homomorphism $\psi : A[[X, Y]] = (A[[X]])[[Y]] \longrightarrow A$. Indeed, an element f of $A[[X, Y]]$ can be written $f = \sum_{i:0}^{\infty} f_i Y^i$ with $f_i \in A[[X]]$. Put $\psi(f) = \tau(\sum_{i:0}^{\infty} \sigma(f_i) Y^i)$. Let r and s be any two elements of A and Z an indeterminate. Since $rX + sY$ is a series of $A[[X, Y]]$ with zero constant term, there exists a A-homomorphism $\rho : A[[Z]] \longrightarrow A[[X, Y]]$ such that $\rho(Z) = rX + sY$. Indeed, for $f = \sum_{i:0}^{\infty} a_i Z^i \in A[[Z]]$, let $\rho(f) = \sum_{i:0}^{\infty} a_i (rX + sY)^i$. The A-homomorphism $\psi \circ \rho : A[[Z]] \longrightarrow A$ satisfies $\psi \circ \rho(Z) = \psi(rX + sY) = r\psi(X) + s\psi(Y) = r\sigma(X) + s\tau(Y) = ra + sb$, then $ra + sb \in I_c(A)$, so $I_c(A)$ is an ideal of A.

(3) Let $a \in I_c(A)$. There exists a A-endomorphism σ of $A[[X]]$ such that $\sigma(X) = a + X$. By [12; Chap. 9, Theorem 2.4], σ is an A-automorphism of $A[[X]]$. Since $X \in Jac(A[[X]])$, the Jacobson radical of $A[[X]]$, then $a + X = \sigma(X) \in Jac(A[[X]]) = Jac(A) + XA[[X]]$, so $a \in Jac(A)$. Then $I_c(A) \subseteq Jac(A)$. $\square$

Other Proof (O'Malley 1970) For each $b \in A$, $1 + bX \in U(A[[X]])$. Since σ is an A-automorphism, then $\sigma(1 + bX) = 1 + b(a + X) = 1 + ab + bX \in U(A[[X]]) = U(A) + XA[[X]]$, so $1 + ab \in U(A)$, and then $a \in Jac(A)$.

Let $a \in Nil(A)$, the nilradical of A. There exists $n \in \mathbb{N}$ such that $a^{n+1} = 0$. The map $\tau : A[[X]] \longrightarrow A$, defined for $f = \sum_{i:0}^{\infty} b_i X^i$ by $\tau(f) = \sum_{i:0}^{n} b_i a^i$, is a A-homomorphism and $\tau(X) = a$. Then $a \in I_c(A)$ so $Nil(A) \subseteq I_c(A)$. $\square$

Example 1 We have $I_c(\mathbb{Z}) = Jac(\mathbb{Z}) = \cap\{p\mathbb{Z};\ p \text{ prime number}\} = (0)$. There is no $\mathbb{Z}$-endomorphism σ of $\mathbb{Z}[[X]]$ such that $\sigma(X) = 1 + X$.

Example 2 (Eakin et al. 1987) Let A be any ring and Y an indeterminate over A. Then $I_c(A[Y]) = Nil(A[Y]) = Nil(A)[Y] = Jac(A[Y])$. In particular, $I_c(A[Y]) = (0)$ if and only if A is reduced. Indeed, $Nil(A[Y]) \subseteq I_c(A[Y]) \subseteq Jac(A[Y])$. Let $f \in Jac(A[Y])$ and then $1 + Yf \in U(A[Y])$. So all the coefficients of f are nilpotent in A, and then $f \in Nil(A[Y])$. For example, $I_c((\mathbb{Z}/4\mathbb{Z})[Y]) = (\bar{2})$.

Lemma 11.5 *Let A be a ring. If $a \in I_c(A)$ is a regular element of A, then*
$$\bigcap_{n:0}^{\infty} a^n A = (0).$$

Proof Since $a \in I_c(A)$, there exists an A-endomorphism σ of $A[[X]]$ such that $\sigma(X) = a + X$. By [12; Chap. 9, Theorem 2.4], σ is an automorphism. Then $\bigcap_{n:0}^{\infty} (a + X)^n A[[X]] = \bigcap_{n:0}^{\infty} \sigma(X)^n A[[X]] = \sigma(\bigcap_{n:0}^{\infty} X^n A[[X]]) = \sigma(0) = 0$. By Lemma 11.1, if $I = \bigcap_{n:0}^{\infty} a^n A$, then $aI = I$. By Lemma 11.2, $I \subseteq \bigcap_{n:0}^{\infty} (a + X)^n A[[X]] = (0)$, then $I = (0)$. $\square$

Lemma 11.6 (Khalifa 2017) *Let A be a ring, $a \in I_c(A)$ a regular element, and $\sigma : A[[X]] \longrightarrow A$ be a A-homomorphism such that $\sigma(X) = a$. Then $\ker \sigma = (X - a)A[[X]]$.*

Proof Since $\sigma(X - a) = 0$, then $(X - a)A[[X]] \subseteq \ker\sigma$. Conversely, let $f = \sum_{i:0}^{\infty} b_i X^i \in \ker\sigma$. For each integer $j \geq 0$, let $f = \sum_{i:0}^{j} b_i X^i +$

$X^{j+1}(\sum_{i:j+1}^{\infty} b_i X^{i-j-1})$. Denoting $\alpha_j = \sigma(\sum_{i:j+1}^{\infty} b_i X^{i-j-1}) \in A$, then $\sum_{i:0}^{j} b_i a^i +$

$a^{j+1}\alpha_j = 0$ (1). So $b_0 + a\alpha_0 = 0$, and for $j \geq 1$, $\sum_{i:0}^{j-1} b_i a^i + a^j \alpha_{j-1} = 0$ (2). By

subtraction between (1) and (2), we find for $j \geq 1$, $b_j a^j + a^{j+1}\alpha_j - a^j \alpha_{j-1} = 0$.

Since a is regular, $b_j = \alpha_{j-1} - a\alpha_j$. Then $f = \sum_{i:0}^{\infty} b_i X^i = -a\alpha_0 + \sum_{j:1}^{\infty}(\alpha_{j-1} -$

$a\alpha_j)X^j = X\sum_{j:0}^{\infty} \alpha_j X^j - a\sum_{j:0}^{\infty} \alpha_j X^j = (X-a)\sum_{j:0}^{\infty} \alpha_j X^j \in (X-a)A[[X]]$. $\quad\square$

Theorem 11.7 (Khalifa 2017) *Let A be a strongly Hopfian bounded ring. Suppose that $I_c(A)$ contains a regular element of A. Then $A[[X]]$ is strongly Hopfian bounded and $\mu(A[[X]]) = \mu(A)$.*

Proof Put $\mu(A) = k$ and let $a \in I_c(A)$ be a regular element of A. By Lemma 11.5, $\bigcap_{n:0}^{\infty} a^n A = (0)$. By Theorem 11.4, for each $n \in \mathbb{N}^*$, there exists a A-homomorphism $\sigma_n : A[[X]] \longrightarrow A$ such that $\sigma_n(X) = a^n$. By Lemma 11.6, $ker\, \sigma_n = (X - a^n)A[[X]]$. Let $f \in A[[X]]$ and $g \in ann_{A[[X]]}(f^{k+1})$. Then $gf^{k+1} = 0$, so $\sigma_n(g)\sigma_n(f)^{k+1} = 0$, and then $\sigma_n(g) \in ann_A(\sigma_n(f)^{k+1}) = ann_A(\sigma_n(f)^k)$. So $\sigma_n(gf)^k = 0$, then $gf^k \in ker\, \sigma_n = (X - a^n)A[[X]]$. So $gf^k \in \bigcap_{n:0}^{\infty}(X - a^n)A[[X]] = (0)$, by Lemma 11.3. $\quad\square$

Corollary 11.8 (Khalifa 2017) *The following assertions are equivalent for any ring A:*

1. *The ring $A[[X]]$ is strongly Hopfian bounded.*
2. *A is a sub-ring of a strongly Hopfian bounded ring B such that $I_c(B)$ contains a regular element of B.*

Proof "(1) $\Longrightarrow$ (2)" We have $A \subset A[[X]]$, X is a regular element of $A[[X]]$, and $X \in I_c(A[[X]])$ since the map $\sigma : A[[X, Y]] \longrightarrow A[[X]]$, defined by $\sigma(f(X, Y)) = f(X, X)$, is a $A[[X]]$-homomorphism and $\sigma(Y) = X$.

"(2) $\Longrightarrow$ (1)" By the preceding theorem, $B[[X]]$ is strongly Hopfian bounded. So is the sub-ring $A[[X]]$. $\quad\square$

Corollary 11.9 (Khalifa 2017) *If the ring $A[[X]]$ is strongly Hopfian bounded, so is $A[[X, Y]]$ and $\mu(A[[X, Y]]) = \mu(A[[X]])$.*

Proof Follows from the preceding theorem, since $X \in I_c(A[[X]])$ is a regular element of $A[[X]]$. $\quad\square$

Corollary 11.10 (Khalifa 2017) *If the ring $A[[X]]$ is strongly Hopfian bounded, so is $A[[X_1, \ldots, X_n]]$ and $\mu(A[[X_1, \ldots, X_n]]) = \mu(A[[X]])$.*

Proof By induction. $\qquad\qquad\qquad\qquad\qquad\qquad\qquad\qquad\qquad\qquad\qquad\qquad$ $\square$

Corollary 11.11 (Khalifa 2017) *If the ring $A[[X]]$ is strongly Hopfian bounded, so is $A[[(X_i)_{i \in I}]]_1 = \bigcup_F A[[(X_i)_{i \in F}]]$, where F ranges over the finite subsets of I, and $\mu(A[[(X_i)_{i \in I}]]_1) = \mu(A[[X]])$.*

Proof Put $\mu(A[[X]]) = k$ and let $f, g \in A[[(X_i)_{i \in I}]]_1$ be such that $gf^{k+1} = 0$. There exist $i_1, \ldots, i_n \in I$ such that $f, g \in A[[X_{i_1}, \ldots, X_{i_n}]]$. Since $A[[X_{i_1}, \ldots, X_{i_n}]]$ is strongly Hopfian bounded and $\mu(A[[X_{i_1}, \ldots, X_{i_n}]]) = \mu(A[[X]]) = k$, then $gf^k = 0$. $\qquad\qquad\qquad\qquad\qquad\qquad$ $\square$

12 Power Series Armendariz Rings

Let A be an integral domain. Given two formal power series $f = \sum_{i=0}^{\infty} a_i X^i$ and $g = \sum_{i=0}^{\infty} b_i X^i \in A[[X]]$ satisfying $fg = 0$. Since $A[[X]]$ is an integral domain, either $f = 0$ or $g = 0$, then $a_i b_j = 0$ for each $i, j \in \mathbb{N}$. Rings with this property are called power series Armendariz. The class of such rings contains reduced rings. The Nagata idealization is a very useful tool to construct rings in this class. These rings will be used in the following sections.

Definition 12.1 Let A be a ring.

1. We say that A an Armendariz (for the polynomials) ring if given two polynomials $f = \sum_{i=0}^{m} a_i X^i$ and $g = \sum_{i=0}^{n} b_i X^i \in A[X]$ satisfying $fg = 0$ and then $a_i b_j = 0$ for each i and j.

2. We say that A is a power series Armendariz ring if given two formal power series $f = \sum_{i=0}^{\infty} a_i X^i$ and $g = \sum_{i=0}^{\infty} b_i X^i \in A[[X]]$ satisfying $fg = 0$ and then $a_i b_j = 0$ for each $i, j \in \mathbb{N}$.

The definition for the polynomial case is given by Rege and Chhawchharia (1997). They use the name of Armendariz since he demonstrates in 1974 that the reduced rings satisfy this property. The formal power series case is a natural generalization of this notion. In this section, we are interested rather to the ring of formal power series. We start by some examples.

Example 1 A power series Armendariz ring is an Armendariz (for polynomials) ring.

Example 2 A sub-ring of an Armendariz (resp. power series Armendariz) ring is Armendariz (resp. power series Armendariz).

Example 3 (Anderson and Camillo 1998) Let Y and T be two indeterminates over a field K and $n \geq 2$ an integer. The ring $A = K[Y, T]/(Y^n, T^n)$ is not Armendariz. Indeed, let $f(X) = \bar{Y} - \bar{T}X \in A[X]$ and $g(X) = \bar{Y}^{n-1} + \bar{Y}^{n-2}\bar{T}X + \ldots + \bar{T}^{n-1}X^{n-1} \in A[X]$. Then $f(X)g(X) = \bar{Y}^n - \bar{T}^nX^n = 0$ with $\bar{T}\bar{Y}^{n-1} \neq 0$. This example and the following one show also that the quotient of a power series Armendariz (integral domain) is not necessarily Armendariz.

Example 4 (Lee and Wong 2003) Let s and t be two indeterminates over the field $\mathbb{F}_3$. The ring $\mathbb{F}_3[s, t]/(s^3, s^2t^2, t^3)$ is not Armendariz. Indeed, $0 = (\bar{s} + \bar{t}X)^3 = (\bar{s} + \bar{t}X)(\bar{s}^2 + 2\bar{s}\bar{t}X + \bar{t}^2X^2)$. But $\bar{s}^2\bar{t} \neq 0$.

Example 5 (Rege and Chhawchharia 1997) The ring $A = (\mathbb{Z}/8\mathbb{Z})(+)(\mathbb{Z}/8\mathbb{Z})$ is not Armendariz. Indeed, let $f(X) = (\bar{4}, \bar{0}) + (\bar{4}, \bar{1})X \in A[X]$. Then $\left(f(X)\right)^2 = 0$ but $(\bar{4}, \bar{0})(\bar{4}, \bar{1}) - (\bar{0}, \bar{4}) \neq 0$.

The following example is demonstrated in a different way for the polynomial case by E. Armendariz at 1974. After this example, M.B. Rege and S. Chhawchharia attribute in 1997 the name of Armendariz to the class of rings having this property.

Example 6 A reduced ring A is power series Armendariz. Indeed, let $f, g \in A[[X]]$ be such that $fg = 0$. If $P \in Spec(A)$, then $P[[X]] \in Spec(A[[X]])$ and $fg = 0 \in P[[X]]$, and then either f or $g \in P[[X]]$. If a is a coefficient of f and b is a coefficient of g, then $ab \in P$. This is for each $P \in Spec(A)$, and then $ab \in Nil(A) = (0)$.

Example 7 Let A be a reduced ring and X a finite set of indeterminates over A. The rings $A[X] \subset A[[X]]$ are also reduced and then power series Armendariz. Indeed, let $f \in A[[X]]$ and $n \in \mathbb{N}^*$ be such that $f^n = 0$. For each $P \in Spec(A)$, $P[[X]] \in Spec(A[[X]])$ and $f^n = 0 \in P[[X]]$, and then $f \in P[[X]]$. So $f \in \bigcap \{P[[X]]; \; P \in Spec(A)\} = Nil(A)[[X]] = (0)$. Then $f = 0$.

Other Proof We will give another proof for the fact that a reduced ring A is Armendariz. Let $f, g \in A[X]$ be such that $fg = 0$. By the content formula, there exists $n \in \mathbb{N}^*$ such that $c(f)^{n+1}c(g) = c(fg)c(f)^n = (0)$. Then $\left(c(f)c(g)\right)^{n+1} \subseteq c(f)^{n+1}c(g) = (0)$. Since A is reduced, $c(f)c(g) = (0)$. Then for any coefficients a of f and b of g, $ab = 0$. □

Example 8 (Anderson and Camillo 1998) Let (A, M) be a local ring with $M^2 = (0)$. Then A is Armendariz. Indeed, let $f(X), g(X) \in A[X]$ be such that

$f(X)g(X) = 0$. By the content formula, there exist $m, n \in \mathbb{N}^*$ such that $c(f)^{n+1}c(g) = c(fg)c(f)^n = 0$ and $c(f)c(g)^{m+1} = c(fg)c(g)^m = 0$. If $f \notin M[X]$, one of its coefficients is invertible, then $c(f) = A$ and $c(g) = 0$, and the result is clear. If $g \notin M[X]$, one of its coefficients is invertible, then $c(g) = A$ and $c(f) = 0$, and the result is also clear. If f and $g \in M[X]$, for any coefficients a of f and b of g, we have $ab \in M^2 = (0)$.

The result is not true for a local ring (A, M) with $M^3 = (0)$. Indeed, let Y and T be two indeterminates over a field K. By Example 3, the ring $A = K[Y, T]/(Y^2, T^2)$ is not Armendariz. It is local with maximal ideal $M = (\bar{Y}, \bar{T})$. We have $M^2 = (\bar{Y}^2, \bar{Y}\bar{T}, \bar{T}^2) = (\bar{Y}\bar{T})$ and $M^3 = (\bar{Y}^2\bar{T}, \bar{Y}\bar{T}^2) = (0)$.

A small modification of a result of Anderson and Camillo (1998), we construct in the following example a nonreduced power series Armendariz ring.

Example 9 (Hizem 2011) Let B be any ring. Then the nonreduced ring $A = B[T]/(T^2)$ is power series Armendariz if and only if B is reduced. Indeed, the natural homomorphism of $B \longrightarrow A = B[T]/(T^2)$ is one to one. Then B is a sub-ring of A. Suppose that A is (power series) Armendariz. Let $b \in B$ be such that $b^2 = 0$. Then $0 = b^2 - \bar{T}^2 X^2 = (b - \bar{T}X)(b + \bar{T}X)$. Since A is Armendariz, $b\bar{T} = 0$, then $bT \in T^2 B[T]$, so $b = 0$. Then B is reduced. Conversely, A is a free B-module of basis $\{1, u\}$ with $u = \bar{T}$. Then $A[[X]]$ is a free $B[[X]]$-module with basis $\{1, u\}$. Let $f(X)$ and $g(X) \in A[[X]]$ be such that $f(X)g(X) = 0$. Put $f(X) = f_0(X) + uf_1(X)$ and $g(X) = g_0(X) + ug_1(X)$ with $f_i(X)$ and $g_i(X) \in B[[X]]$, $0 \le i \le 1$. Since $u^2 = 0$, then $0 = f(X)g(X) = f_0(X)g_0(X) + (f_0(X)g_1(X) + f_1(X)g_0(X))u$, so $f_0(X)g_0(X) = 0$ and $f_0(X)g_1(X) + f_1(X)g_0(X) = 0$. We multiply the second equality by $f_0(X)$ and by $g_0(X)$, and we find $f_0(X)^2g_1(X) = 0$ and $f_1(X)g_0(X)^2 = 0$. Then $(f_0(X)g_1(X))^2 = (f_1(X)g_0(X))^2 = 0$. Since $B[[X]]$ is reduced, $f_0(X)g_0(X) = f_0(X)g_1(X) = f_1(X)g_0(X) = 0$. But B is power series Armendariz, and then each coefficient of $f_i(X)$ annihilates each coefficient of $g_j(X)$, $0 \le i, j \le 1$, except perhaps the coefficients of $f_1(X)$ and $g_1(X)$. Since $u^2 = 0$, each coefficient of $f(X) = f_0(X) + uf_1(X)$ annihilates each coefficient of $g(X) = g_0(X) + ug_1(X)$. Then A is power series Armendariz.

The transfer of the power series Armendariz property to Nagata idealization requires some supplementary hypotheses.

Proposition 12.2 (El Ouarrachi and Mahdou 2018) *Let (A, M) be a local ring and E a A-module satisfying $ME = (0)$. Then $A(+)E$ is power series Armendariz if and only if A is power series Armendariz.*

Proof " $\Longrightarrow$ " A is a sub-ring of $A(+)E$ via the homomorphism $\phi : A \longrightarrow A(+)E$, defined by $\phi(a) = (a, 0)$ for each $a \in A$.

" $\Longleftarrow$ " Let $f = \sum_{i:0}^{\infty}(a_i, e_i)X^i$ and $g = \sum_{j:0}^{\infty}(b_j, f_j)X^j \in (A(+)E)[[X]]$ be such that $fg = 0$. For each $k \in \mathbb{N}$, $(0, 0) = \sum_{i+j=k}(a_i, e_i)(b_j, f_j) = \sum_{i+j=k}(a_ib_j, a_if_j +$

$$b_j e_i) = \left(\sum_{i+j=k} a_i b_j, \sum_{i+j=k} a_i f_j + b_j e_i \right).$$ Then for each $k \in \mathbb{N}$, $\sum_{i+j=k} a_i b_j = 0$ and

$$\sum_{i+j=k} a_i f_j + b_j e_i = 0.$$

We want to show that for each $i, j \in \mathbb{N}$, $(0,0) = (a_i, e_i)(b_j, f_j) = (a_i b_j, a_i f_j + b_j e_i)$. That is for each $i, j \in \mathbb{N}$, $a_i b_j = 0$, and $a_i f_j + b_j e_i = 0$.

Let $f_A = \sum_{i:0}^{\infty} a_i X^i$ and $g_A = \sum_{j:0}^{\infty} b_j X^j \in A[[X]]$. Then $f_A g_A = \sum_{k:0}^{\infty} c_k X^k$

with $c_k = \sum_{i+j=k} a_i b_j = 0$, by hypothesis. Then $f_A g_A = 0$. Since A is power series Armendariz, for each $i, j \in \mathbb{N}$, $a_i b_j = 0$. This is the first condition that we want to prove. It remains to show that for each $i, j \in \mathbb{N}$, $a_i f_j + b_j e_i = 0$.

First Case: For each $i, j \in \mathbb{N}$, a_i and $b_j \in M$. Since $ME = (0)$, we have the result.

Second Case: One of the a_i or the b_j does not belong to M. Without a loss of generality, we may suppose that one of the a_i does not belong to M. Let i_0 be the smallest natural integer such that $a_{i_0} \notin M$. Since (A, M) is local, a_{i_0} is invertible in A, and, for each $i < i_0$, $a_i \in M$, then $a_i E = (0)$. Since for each $j \in \mathbb{N}$, $a_{i_0} b_j = 0$, then $b_j = 0$. It suffices then to show that for each $i, j \in \mathbb{N}$, $a_i f_j = 0$. We will show that for each $j \in \mathbb{N}$, $f_j = 0$. The hypothesis becomes for each $k \in \mathbb{N}$, $\sum_{i+j=k} a_i f_j = 0$.

Suppose that the f_j, $j \in \mathbb{N}$, are not all zero and consider the smallest index j_0 such that $f_{j_0} \neq 0$. By hypothesis, $\sum_{i+j=i_0+j_0} a_i f_j = 0$. If $i < i_0$, then $a_i \in M$, so $a_i f_j = 0$. If $i > i_0$, then $j < j_0$, so $f_j = 0$ and $a_i f_j = 0$. Then $a_{i_0} f_{j_0} = 0$. But a_{i_0} is invertible in A, and then $f_{j_0} = 0$, which is absurd.   □

Example 1 Let K be a field and E a nonzero K-vector space. Since K is local with maximal ideal zero and $(0)E = (0)$, the ring $K(+)E$ is power series Armendariz. It is not reduced since for each $0 \neq x \in E$, $(0, x)(0, x) = (0, 0)$.

Example 2 Let K be a field. The ring $K[[X]]$ is local with maximal ideal (X) and power series Armendariz since it is an integral domain. The additive group K is endowed with the structure of a $K[[X]]$-module for the multiplication $f(X) * k = f(0)k$ for each $f(X) \in K[[X]]$ and $k \in K$ and $(X) * K = (0)$. By the preceding proposition, $K[[X]](+)K$ is power series Armendariz. It is never reduced since $(0, 1)(0, 1) = (0, 0)$.

Example 3 Let (A, M) be a local ring. By the preceding proposition, A is power series Armendariz if and only if $A(+)(A/M)$ is power series Armendariz. We consider the following particular cases:

(1) For each local integral domain (A, M), the ring $A(+)(A/M)$ is power series Armendariz. It is never reduced since $(0, \bar{1})(0, \bar{1}) = (0, \bar{0})$.

(2) Let K be a field and $n \geq 2$ an integer. The ring $A = K[Y, T]/(Y^n, T^n)$ is local with maximal ideal $M = (Y, T)/(Y^n, T^n)$. Since A is not Armendariz, the ring $A(+)(A/M)$ is not power series Armendariz.

Example 4 (Bakkari 2009) Let A be a power series Armendariz and E a A-module. The ring $A(+)E$ is not necessarily power series Armendariz. Indeed, let K be a field. By the preceding proposition, the ring $A = K(+)K$ is power series Armendariz. We will show that the ring $A(+)A$ is not even Armendariz. Let $f(X) = ((0, 1), (0, 0)) + ((0, 0), (1, 0))X \in A[X]$ and $g(X) = ((0, 1), (0, 0)) + ((0, 0), (-1, 0))X \in A[X]$. Then $fg = 0$ but $((0, 0), (1, 0))((0, 1), (0, 0)) = ((0, 0), (0, 1)) \neq 0$.

Note that by Lemma 8.9 of Chap. 1, the ring $A = K(+)K$ is local with maximal ideal $M = (0)(+)K \neq (0)$. But $MA = M \neq (0)$. Then the hypothesis in the preceding proposition is not satisfied.

Let A be a ring and M a A-module. Recall from Sect. 9 of Chap. 1 that $M[[X]]$ is a $A[[X]]$-module for the natural operations. We start by a very useful lemma.

Lemma 12.3 *Let A be a ring and M a A-module. We have the canonical isomorphism of rings $(A(+)M)[[X]] \simeq A[[X]](+)M[[X]]$.*

Proof The map $\phi : (A(+)M)[[X]] \longrightarrow A[[X]](+)M[[X]]$, defined by

$$\phi\left(\sum_{i:0}^{\infty}(a_i, m_i)X^i\right) = \left(\sum_{i:0}^{\infty} a_i X^i, \sum_{i:0}^{\infty} m_i X^i\right),$$ is an additive bijection with $\phi(1, 0) =$

$(1, 0)$. Let $f(X) = \sum_{i:0}^{\infty}(a_i, m_i)X^i \in (A(+)M)[[X]]$ and $g(X) = \sum_{j:0}^{\infty}(b_j, n_j)X^j \in$

$(A(+)M)[[X]]$. Then $f(X)g(X) = \sum_{n:0}^{\infty} c_n X^n$ with $c_n = \sum_{i+j=n} (a_i, m_i)(b_j, n_j) =$

$\sum_{i+j=n} (a_i b_j, a_i n_j + b_j m_i)$

$= \left(\sum_{i+j=n} a_i b_j, \sum_{i+j=n} a_i n_j + b_j m_i\right).$

Then $\phi(f(X)g(X)) = \left(\sum_{n:0}^{\infty}(\sum_{i+j=n} a_i b_j)X^n, \sum_{n:0}^{\infty}(\sum_{i+j=n} a_i n_j + b_j m_i)X^n\right)$

$= \left(\sum_{n:0}^{\infty}(\sum_{i+j=n} a_i b_j)X^n, \sum_{n:0}^{\infty}(\sum_{i+j=n} a_i n_j)X^n + \sum_{n:0}^{\infty}(\sum_{i+j=n} b_j m_i)X^n\right)$

$= \left((\sum_{i:0}^{\infty} a_i X^i)(\sum_{j:0}^{\infty} b_j X^j), (\sum_{i:0}^{\infty} a_i X^i)(\sum_{j:0}^{\infty} n_j X^j) + (\sum_{j:0}^{\infty} b_j X^j)(\sum_{i:0}^{\infty} m_i X^i)\right)$

$= \left(\sum_{i:0}^{\infty} a_i X^i, \sum_{i:0}^{\infty} m_i X^i\right)\left(\sum_{j:0}^{\infty} b_j X^j, \sum_{j:0}^{\infty} n_j X^j\right) = \phi(f(X))\phi(g(X)).$ □

Proposition 12.4 (Rege and Chhawchharia 1997) *Let A be an integral domain and I an ideal of A. Suppose that the quotient ring A/I is power series Armendariz. Then the ring $A(+)(A/I)$ is power series Armendariz.*

Proof Let $f(X)$ and $g(X) \in (A(+)(A/I))[[X]] \simeq A[[X]](+)(A/I)[[X]]$ be such that $fg = 0$. By the preceding lemma, $f(X) = \sum_{i:0}^{\infty}(a_i, \bar{u}_i)X^i = (f_0(X), \overline{f_1(X)})$ with $f_0(X) = \sum_{i:0}^{\infty} a_i X^i \in A[[X]]$ and $\overline{f_1(X)} = \sum_{i:0}^{\infty} \bar{u}_i X^i \in (A/I)[[X]] \simeq A[[X]]/I[[X]]$. Also, $g(X) = \sum_{j:0}^{\infty}(b_j, \bar{v}_j)X^j = (g_0(X), \overline{g_1(X)})$.

Then $f_0(X)g_0(X) = 0$ and $\overline{f_0(X)g_1(X) + f_1(X)g_0(X)} = 0$. Since $A[[X]]$ is an integral domain, either $f_0(X) = 0$ or $g_0(X) = 0$. By symmetry, we may suppose that $f_0(X) = 0$ and then $a_i = 0$ for each $i \in \mathbb{N}$. The second equality becomes $\overline{f_1(X)g_0(X)} = 0$. But by hypothesis, the ring A/I is power series Armendariz, and then $\bar{u}_i \bar{b}_j = 0$ for each $i, j \in \mathbb{N}$. So $(a_i, \bar{u}_i)(b_j, \bar{v}_j) = (a_ib_j, \overline{a_iv_j + u_ib_j}) = 0$ for each $i, j \in \mathbb{N}$. □

We can generalize the preceding proposition in the following way:

Proposition 12.5 (Rege and Chhawchharia 1997) *Let A be a reduced ring and I an ideal of A such that the quotient ring A/I is also reduced. Then the ring $A(+)(A/I)$ is power series Armendariz.*

Proof Let $f(X)$ and $g(X) \in (A(+)(A/I))[[X]] \simeq A[[X]](+)(A/I)[[X]]$ be such that $fg = 0$. By Lemma 12.3, $f(X) = \sum_{i:0}^{\infty}(a_i, \bar{u}_i)X^i = (f_0(X), \overline{f_1(X)})$ and $g(X) = \sum_{i:0}^{\infty}(b_i, \bar{v}_i)X^i = (g_0(X), \overline{g_1(X)})$. Then $f_0(X)g_0(X) = 0$ and $\overline{f_0(X)g_1(X) + f_1(X)g_0(X)} = 0$. Since A is reduced, then it is power series Armendariz, so $a_ib_j = 0$ for each $i, j \in \mathbb{N}$. We multiply the second equality by $\overline{g_0(X)}$, and we find $\overline{f_1(X)g_0(X)^2} = 0$. Since the ring $(A/I)[[X]]$ is reduced, $\overline{f_1(X)g_0(X)} = 0$. Then $\bar{u}_i\bar{b}_j = 0$ for each $i, j \in \mathbb{N}$. Also, multiplying the second equality by $\overline{f_0(X)}$, we find $\overline{g_1(X)f_0(X)} = 0$, and then $\bar{a}_i\bar{v}_j = 0$ for each $i, j \in \mathbb{N}$. So $(a_i, \bar{u}_i)(b_j, \bar{v}_j) = (a_ib_j, \overline{a_iv_j + u_ib_j}) = 0$ for each $i, j \in \mathbb{N}$. □

Example 1 By the preceding proposition, if A is a reduced ring, then $A(+)A$ is power series Armendariz. It suffices to take $I = (0)$ in the proposition. The result is not true for any ring A. Just reconsider the examples $A = \mathbb{Z}/8\mathbb{Z}$ or $A = K(+)K$ with K a field.

Example 2 Let A be a von Neumann regular ring (i.e., for each $a \in A$, there exists $b \in A$ such that $a = a^2b$) and I an ideal of A. It is clear that A/I is also a von

Neumann regular ring. By Proposition 2.2, the rings A and A/I are reduced. By the preceding proposition, $A(+)(A/I)$ is power series Armendariz.

Definition 12.6 Let A be a ring and M a A-module. We say that M is a power series Armendariz module if given two formal power series $f(X) = \sum_{i=0}^{\infty} a_i X^i \in A[[X]]$ and $g(X) = \sum_{j=0}^{\infty} m_j X^j \in M[[X]]$ satisfying $f(X)g(X) = 0$ and then $a_i m_j = 0$ for each $i, j \in \mathbb{N}$.

Let A be a ring and M a A-module. Recall the following definitions:

1. An element x of M is said to be torsion if there exists $0 \neq a \in A$ such that $ax = 0$.
2. The module M is said to be torsion if all its elements are torsion.
3. The module M is said to be torsion-free if 0 is the unique torsion element.

Examples

(1) An element x of M is torsion-free if and only if it is free over A.
(2) A free module M over an integral domain A is torsion-free. Indeed, let (e_i) be a basis of M and $x \in M$ a torsion element. Put $x = \sum \alpha_i e_i$ where the $\alpha_i \in A$ are almost zero. There exists $0 \neq a \in A$ such that $ax = 0$. Then $ax = \sum a\alpha_i e_i = 0$, so $a\alpha_i = 0$, and then $\alpha_i = 0$ for each i and $x = 0$.
(3) Each sub-module of a torsion-free module is torsion-free.
(4) Each sub-module of a torsion module is torsion.

Let $T(M)$ be the set of torsion elements of M. Then
(1) M is torsion-free $\iff T(M) = (0)$. (2) M is torsion $\iff T(M) = M$. (3) In general, $(0) \subseteq T(M) \subseteq M$. If A is an integral domain, then $T(M)$ is a torsion sub-module of M.

Examples Take $A = \mathbb{Z}$.

(1) For $M = \mathbb{Z}$, $T(M) = (0)$.
(2) For $M = \mathbb{Z}/n\mathbb{Z}$ with $n \geq 2$, $T(M) = M$ since for each $\bar{x} \in M$, $n\bar{x} = 0$.
(3) For $M = \mathbb{Z} \times \mathbb{Z}/3\mathbb{Z}$, $T(M) = (0) \times \mathbb{Z}/3\mathbb{Z}$ since if $0 \neq a \in \mathbb{Z}$ and $(x, \bar{y}) \in \mathbb{Z} \times \mathbb{Z}/3\mathbb{Z}$ are such that $a(x, \bar{y}) = (0, \bar{0})$, then $ax = 0$ so $x = 0$. Conversely, $3(0, \bar{y}) = (0, \bar{0})$.

Proposition 12.7 *Let A be a reduced ring and M a torsion-free A-module. Then M is power series Armendariz module .*

Proof Let $f(X) = \sum_{i=0}^{\infty} a_i X^i \in A[[X]]$ and $g(X) = \sum_{j=0}^{\infty} m_j X^j \in M[[X]]$ be such that $f(X)g(X) = 0$. Then $a_0 m_0 = 0$. Suppose by induction that $a_0 m_0 =$

$\ldots = a_0 m_k = 0$. The coefficient of X^{k+1} in $f(X)g(X)$ is $\sum\limits_{i:0}^{k+1} a_i m_{k+1-i} = 0$.
We multiply this equality by a_0, and we find $a_0^2 m_{k+1} = 0$. If $m_{k+1} = 0$, then $a_0 m_{k+1} = 0$. If not, since M is torsion-free, then $a_0^2 = 0$. But A is a reduced ring, then $a_0 = 0$ and $a_0 m_{k+1} = 0$. Then we have showed that $a_0 g(X) = 0$. Suppose by induction that $a_0 g(X) = \ldots = a_n g(X) = 0$. Then $\left(f(X) - \sum\limits_{i:0}^{n} a_i X^i \right) g(X) =$
$f(X)g(X) - \sum\limits_{i:0}^{n} X^i a_i g(X) = 0$. Then $X^{n+1} \left(\sum\limits_{i:n+1}^{\infty} a_i X^{i-n-1} \right) g(X) = 0$, so
$\left(\sum\limits_{i:n+1}^{\infty} a_i X^{i-n-1} \right) g(X) = 0$. By the preceding step, $a_{n+1} g(X) = 0$.   □

Example A vector space is power series Armendariz.
   We give now another generalization of Proposition 12.4.

Proposition 12.8 (Anderson and Camillo 1998) *Let A be an integral domain and M a A-module. Then $A(+)M$ is a power series Armendariz ring if and only if the module M is power series Armendariz.*

Proof " $\Longrightarrow$ " Let $f(X) = \sum\limits_{i=0}^{\infty} a_i X^i \in A[[X]]$ and $g(X) = \sum\limits_{j=0}^{\infty} m_j X^j \in M[[X]]$
be such that $f(X)g(X) = 0$. By Lemma 12.3, $\left(A(+)M \right)[[X]] \simeq A[[X]](+)M[[X]]$
and the product $\left(\sum\limits_{i=0}^{\infty} (a_i, 0) X^i \right) \left(\sum\limits_{j=0}^{\infty} (0, m_j) X^j \right) = \left(\sum\limits_{i=0}^{\infty} a_i X^i, 0 \right) \left(0, \sum\limits_{j=0}^{\infty} m_j X^j \right) =$
$\left(f(X), 0 \right) \left(0, g(X) \right) = \left(0, f(X)g(X) \right) = (0, 0)$. Since $A(+)M$ is a power series Armendariz ring, for each $i, j \in \mathbb{N}$, $(a_i, 0)(0, m_j) = (0, 0)$, and then $(0, a_i m_j) = (0, 0)$. So for each $i, j \in \mathbb{N}$, $a_i m_j = 0$. Then M is power series Armendariz.

   " $\Longleftarrow$ " Let $f(X) = \sum\limits_{i=0}^{\infty} (a_i, m_i) X^i$ and $g(X) = \sum\limits_{i=0}^{\infty} (b_j, n_j) X^j \in$
$(A(+)M)[[X]]$ be such that $f(X)g(X) = 0$. By Lemma 12.3, $f(X) = \left(f_1(X), f_2(X) \right)$ with $f_1 = \sum\limits_{i=0}^{\infty} a_i X^i \in A[[X]]$ and $f_2 = \sum\limits_{i=0}^{\infty} m_i X^i \in M[[X]]$.
Also, $g(X) = \left(g_1(X), g_2(X) \right)$ with $g_1 = \sum\limits_{j=0}^{\infty} b_j X^j \in A[[X]]$ and $g_2 =$
$\sum\limits_{j=0}^{\infty} n_j X^j \in M[[X]]$. In the ring $A[[X]](+)M[[X]]$, we have $(0, 0) = f(X)g(X) =$
$\left(f_1(X), f_2(X) \right) \left(g_1(X), g_2(X) \right) = \left(f_1(X)g_1(X), f_1(X)g_2(X) + g_1(X)f_2(X) \right)$.
Then $f_1 g_1 = 0$ and $f_1 g_2 + g_1 f_2 = 0$. Since $A[[X]]$ is an integral domain, for

example, $f_1 = 0$. Then for each $i \in \mathbb{N}$, $a_i = 0$. The second equality becomes $g_1 f_2 = 0$. Since M is power series Armendariz, for each $i, j \in \mathbb{N}$, $b_i m_j = 0$ so $(a_i, m_i)(b_j, n_j) = (a_i b_j, a_i n_j + b_j m_i) = (0, 0)$. Then the ring $A(+)M$ is power series Armendariz. $\qquad\square$

Corollary 12.9 (Anderson and Camillo 1998) *Let A be an integral domain and M a torsion-free A-module. Then $A(+)M$ is a power series Armendariz ring.*

Proof Follows from the two preceding propositions. $\qquad\square$

Example 1 We retrieve the following example. Let K be a field and M a nonzero K-vector space. The ring $K(+)M$ is power series Armendariz. It is not reduced since for each $0 \neq x \in M$, $(0, x)(0, x) = (0, 0)$.

Example 2 Let $A \subseteq B$ be an extension of integral domains. Then $A(+)B$ is a power series Armendariz ring.

We study now the transfer of the power series Armendariz property to the direct product of rings.

Proposition 12.10 (El Ouarrachi and Mahdou 2018) *Let $A_1, \ldots, A_n$ be a finite number of rings. The product $A = A_1 \times \ldots \times A_n$ is power series Armendariz if and only if the rings $A_1, \ldots, A_n$ are power series Armendariz.*

Proof By induction, it suffices to show the equivalence for $n = 2$.

"$\Longrightarrow$" By symmetry, it suffices to show that A_1 is power series Armendariz. Let

$$f = \sum_{i:0}^{\infty} a_i X^i \text{ and } g = \sum_{j:0}^{\infty} b_j X^j \in A_1[[X]] \text{ be such that } fg = 0. \text{ For each } k \in \mathbb{N},$$

$$\sum_{i+j=k} a_i b_j = 0. \text{ Let } f_1 = \sum_{i:0}^{\infty} (a_i, 0)X^i \text{ and } g_1 = \sum_{j:0}^{\infty} (b_j, 0)X^j \in (A_1 \times A_2)[[X]].$$

Then $f_1 g_1 = \sum_{k:0}^{\infty} c_k X^k$ with $c_k = \sum_{i+j=k} (a_i, 0)(b_j, 0) = \left(\sum_{i+j=k} a_i b_j, 0\right) = (0, 0)$.

So $f_1 g_1 = 0$. Since $A_1 \times A_2$ is power series Armendariz, for each $i, j \in \mathbb{N}$, $(a_i, 0)(b_j, 0) = (0, 0)$, then $a_i b_j = 0$.

"$\Longleftarrow$" Let $f(X) = \sum_{i:0}^{\infty} (a_i, e_i)X^i$ and $g(X) = \sum_{j:0}^{\infty} (b_j, f_j)X^j \in A[[X]]$ be

such that $fg = 0$. Let $f_1(X) = \sum_{i:0}^{\infty} a_i X^i \in A_1[[X]]$; $f_2(X) = \sum_{i:0}^{\infty} e_i X^i \in$

$A_2[[X]]$; $g_1(X) = \sum_{j:0}^{\infty} b_j X^j \in A_1[[X]]$ and $g_2(X) = \sum_{j:0}^{\infty} f_j X^j \in A_2[[X]]$. Put

$$f(X)g(X) = \sum_{k:0}^{\infty} c_k X^k. \text{ For each } k \in \mathbb{N}, (0,0) = c_k = \sum_{i+j=k} (a_i, e_i)(b_j, f_j) =$$

$$\left(\sum_{i+j=k} a_i b_j, \sum_{i+j=k} e_i f_j \right). \text{ For each } k \in \mathbb{N}, \sum_{i+j=k} a_i b_j = 0 \text{ and } \sum_{i+j=k} e_i f_j = 0. \text{ So}$$

$f_1(X)g_1(X) = 0$ and $f_2(X)g_2(X) = 0$. But the rings A_1 and A_2 are power series Armendariz. Then for each $i, j \in \mathbb{N}$, $a_i b_j = 0$ and $e_i f_j = 0$ so $(a_i, e_i)(b_j, f_j) = (a_i b_j, e_i, f_j) = (0, 0)$. Then $A_1 \times A_2$ is power series Armendariz. $\qquad \square$

We always have showed by some examples that power series Armendariz property does not pass to the quotient. But for PID, the result is true, as can be seen in the following example:

Example 1 (Rege and Chhawchharia 1997) In many steps

(1) Let p be a prime number and $k \in \mathbb{N}^*$. The ring $\mathbb{Z}/p^k\mathbb{Z}$ is power series Armendariz. Indeed, let $f(X)$ and $g(X) \in \mathbb{Z}[[X]]$ be such that the images in $(\mathbb{Z}/p^k\mathbb{Z})[[X]]$ satisfy $\overline{f(X)}.\overline{g(X)} = 0$. Then p^k divides $f(X)g(X)$ in $\mathbb{Z}[[X]]$. On the other hand, $f(X) = p^r f'(X)$ and $g(X) = p^s g'(X)$ with $r, s \in \mathbb{N}$ and $f'(X)$ and $g'(X) \in \mathbb{Z}[[X]]$ where the GCD of the coefficients of $f'(X)$ (resp. $g'(X)$) is not divisible by p. Then p^k divides $f(X)g(X) = p^{r+s} f'(X)g'(X)$ in $\mathbb{Z}[[X]]$. Suppose that $r + s < k$ and then p divides the product $f'(X)g'(X)$ in $\mathbb{Z}[[X]]$. Put $f'(X) = \sum_{i:0}^{\infty} a_i X^i$ and $g'(X) = \sum_{i:0}^{\infty} b_i X^i$. Let a_{i_0} and b_{j_0} be the first coefficients of $f'(X)$ and of $g'(X)$ not divisible by p. The coefficient of $X^{i_0+j_0}$ in $f'(X)g'(X)$ is $\sum_{i+j=i_0+j_0} a_i b_j$. Since p divides a_i for $0 \leq i < i_0$ and divides b_j for $0 \leq j < j_0$, then p divides $a_{i_0} b_{i_0}$, which is absurd. Then $r + s \geq k$. For each coefficient a of $f(X)$ and each coefficient b of $g(X)$, p^k divides ab, and then $\bar{a}\bar{b} = 0$ in $\mathbb{Z}/p^k\mathbb{Z}$.

(2) Let $n \geq 2$ be an integer. Then $n = p_1^{k_1} \ldots p_s^{k_s}$, where the p_i are different primes and the $k_i \geq 1$. By the Chinese remainder theorem, $\mathbb{Z}/n\mathbb{Z} \simeq \mathbb{Z}/p_1^{k_1}\mathbb{Z} \times \ldots \times \mathbb{Z}/p_s^{k_s}\mathbb{Z}$. By the preceding proposition, the ring $\mathbb{Z}/n\mathbb{Z}$ is power series Armendariz. Moreover, if the integer n has a square factor, the ring $\mathbb{Z}/n\mathbb{Z}$ is not reduced. Indeed, $n = q^2 n'$ with $q \geq 2$ and $n' \in \mathbb{N}^*$. Then $\overline{qn'} \neq 0$ and $(\overline{qn'})^2 = 0$.

(3) In the same way, we show that if A is a PID and I is an ideal of A, the quotient ring A/I is power series Armendariz. The result is not true for a UFD. It suffices to see the example $K[Y, T]/(Y^n, T^n)$ where Y and T are two indeterminates over a field K and $n \geq 2$ is an integer.

Example 2 (Rege and Chhawchharia 1997) Another example shows that the power series Armendariz property does not pass to the quotient. By the preceding example, the ring $\mathbb{Z}/8\mathbb{Z}$ is power series Armendariz. By Proposition 12.4, the ring $\mathbb{Z}(+)(\mathbb{Z}/8\mathbb{Z})$ is power series Armendariz. Let $\phi : \mathbb{Z}(+)(\mathbb{Z}/8\mathbb{Z}) \longrightarrow$

$(\mathbb{Z}/8\mathbb{Z})(+)(\mathbb{Z}/8\mathbb{Z})$ be the map defined by $\phi(a, \bar{x}) = (\bar{a}, \bar{x})$ for each $(a, \bar{x}) \in \mathbb{Z}(+)(\mathbb{Z}/8\mathbb{Z})$. Then ϕ is a homomorphism onto. Then $\mathbb{Z}(+)(\mathbb{Z}/8\mathbb{Z})/\ker \phi \simeq (\mathbb{Z}/8\mathbb{Z})(+)(\mathbb{Z}/8\mathbb{Z})$. But we have always shown that the ring $(\mathbb{Z}/8\mathbb{Z})(+)(\mathbb{Z}/8\mathbb{Z})$ is not even Armendariz.

Example 3 (Bakkari 2009) Another example showing that if A is a power series Armendariz local ring and E is a A-module, the ring $A(+)E$ is not necessarily power series Armendariz. Let $n \geq 2$ be an integer. The ring $A = \mathbb{Z}/2^n\mathbb{Z}$ is local with maximal ideal $M = (\bar{2})$. By Example 1, the ring A is power series Armendariz. But the ring $A(+)A$ is not even Armendariz. Indeed, let $f(X) = \left(\overline{2^{n-1}}, \bar{0} \right) + \left(\overline{2^{n-1}}, \bar{1} \right)X \in (A(+)A)[X]$. Then $\left(f(X) \right)^2 = 0$ with $\left(\overline{2^{n-1}}, \bar{0} \right)\left(\overline{2^{n-1}}, \bar{1} \right) = \left(\bar{0}, \overline{2^{n-1}} \right) \neq 0$. Note that $MA = M \neq (0)$. Then the hypothesis in Proposition 12.2 is not satisfied.

Corollary 12.11 (El Ouarrachi and Mahdou 2018) *Let A be a ring and $n \in \mathbb{N}^*$. Then A^n is power series Armendariz if and only if A is power series Armendariz.*

Definition 12.12 An ideal I of a ring A is said to be reduced if for all $x \in I$ and $n \in \mathbb{N}^*$, when $x^n = 0$, then $x = 0$.

Example Let $n \geq 2$ be an integer. The product ring $A = \mathbb{Z}/2^n\mathbb{Z} \times \mathbb{Z}/3\mathbb{Z}$ is not reduced since $(\bar{2}, \bar{0})^n = (\bar{0}, \bar{0})$. But its ideal $I = (0) \times \mathbb{Z}/3\mathbb{Z}$ is reduced. Indeed, if $(\bar{0}, \bar{x})^k = (\bar{0}, \bar{0})$ with $x \in \mathbb{Z}$ and $k \in \mathbb{N}^*$, then 3 divides x^k, so 3 divides x, then $(\bar{0}, \bar{x}) = (\bar{0}, \bar{0})$.

Proposition 12.13 (El Ouarrachi and Mahdou 2018) *Let A be a ring and I a reduced ideal of A. If the quotient ring A/I is power series Armendariz, so is the ring A.*

Proof Let $f = \sum_{i:0}^{\infty} a_i X^i$ and $g = \sum_{j:0}^{\infty} b_j X^j \in A[[X]]$ be such that $fg = 0$.

For each $k \in \mathbb{N}$, $\sum_{i+j=k} a_i b_j = 0$ (*). Put $\bar{f} = \sum_{i:0}^{\infty} \bar{a}_i X^i$ and $\bar{g} = \sum_{j:0}^{\infty} \bar{b}_j X^j \in (A/I)[[X]]$. Then $\bar{f}\bar{g} = 0$. Since A/I is power series Armendariz, for each $i, j \in \mathbb{N}$, $\bar{a}_i \bar{b}_j = 0$, then $a_i b_j \in I$ (**). We will show by induction on $i + j$ that $a_i b_j = 0$. By (*), if $i + j = 0$, $a_0 b_0 = 0$. Suppose that for $i + j < k$, $a_i b_j = 0$. By (*),

$$\sum_{i+j=k}^{k} a_i b_j = 0 \text{ so } \sum_{i:0} a_i b_{k-i} = 0.$$ We multiply this equality by a_0. By the induction

hypothesis, $a_0 b_{k-i} = 0$ for $1 \leq i \leq k$, then $a_0^2 b_k = 0$, and so $(a_0 b_k)^2 = 0$. By (**),

$a_0 b_k \in I$, then $a_0 b_k = 0$ since I is reduced. Then $\sum_{i:1}^{k} a_i b_{k-i} = 0$. We multiply this equality by a_1. By the induction hypothesis, $a_1 b_{k-i} = 0$ for $2 \leq i \leq k$, then $a_1^2 b_{k-1} = 0$, and so $(a_1 b_{k-1})^2 = 0$ with $a_1 b_{k-1} \in I$ by (**). Since I is reduced, $a_1 b_{k-1} = 0$. By induction, we show that $a_0 b_k = a_1 b_{k-1} = \ldots = a_k b_0 = 0$. □

Example To see the necessity of the hypothesis on the ideal I, we consider a ring A that is not power series Armendariz and a prime ideal P of A. The quotient ring A/P is an integral domain and then power series Armendariz. The ring A is necessarily not reduced. Then $(0) \neq Nil(A) \subseteq P$. So P is not reduced.

Notation Let A be a ring and e an idempotent element of A. Then $1 - e$ is idempotent and eA is a commutative ring with unity e.

Example Let e and f be two idempotent elements of a ring A. Then $eA = fA$ if and only if $e = f$. Indeed, $e = fa$ with $a \in A$. Then $ef = f^2 a = fa = e$. So $ef = e$. Also, $ef = f$. Then $e = f$.

Proposition 12.14 (El Ouarrachi and Mahdou 2018) *Let A be a ring and e an idempotent element of A. Then A is power series Armendariz if and only if the rings eA and $(1 - e)A$ are power series Armendariz.*

Proof " $\Longrightarrow$ " Since $eA \subseteq A$ and $(1 - e)A \subseteq A$, the rings eA and $(1 - e)A$ are power series Armendariz.

" $\Longleftarrow$ " Let $f = \sum_{i:0}^{\infty} a_i X^i$ and $g = \sum_{j:0}^{\infty} b_j X^j \in A[[X]]$ be such that $fg = 0$.

Put $f_1 = ef = \sum_{i:0}^{\infty} ea_i X^i \in eA[[X]]$, $g_1 = eg = \sum_{j:0}^{\infty} eb_j X^j \in eA[[X]]$, $f_2 = (1 - e)f = \sum_{i:0}^{\infty} (1 - e)a_i X^i \in (1 - e)A[[X]]$, and $g_2 = (1 - e)g = \sum_{j:0}^{\infty} (1 - e)b_j X^j \in (1 - e)A[[X]]$. Then $f_1 g_1 = efg = 0$ and $f_2 g_2 = (1 - e)fg = 0$. By hypothesis, for each $i, j \in \mathbb{N}$, $ea_i b_j = 0$, and $(1 - e)a_i b_j = 0$, and then $a_i b_j = ea_i b_j + (1 - e)a_i b_j = 0$. Then the ring A is power series Armendariz. □

Corollary 12.15 (El Ouarrachi and Mahdou 2018) *The following assertions are equivalent for a ring A:*

1. *The ring A is power series Armendariz.*
2. *There exists an idempotent element e of A such that the rings eA and $(1 - e)A$ are power series Armendariz.*
3. *For each idempotent element e of A, the rings eA and $(1 - e)A$ are power series Armendariz.*

Proof By the preceding proposition, the condition (1) is equivalent to each one of the two others. □

13 Endo-Noetherian Rings

A module M is said to be endo-Noetherian if it satisfies the ascending chain condition for the kernels of the endomorphisms; i.e., each increasing sequence of the form $Ker(f_1) \subseteq Ker(f_2) \subseteq \cdots$ is stationary, where $(f_k)_{k \in \mathbb{N}^*}$ is a sequence of endomorphisms of M. A ring A is said to be endo-Noetherian if it is endo-Noetherian as a A-module; i.e., each chain of annihilators $ann(a_1) \subseteq ann(a_2) \subseteq \cdots$ is stationary, for each sequence $(a_k)_{k \in \mathbb{N}^*}$ of elements of A. The class of endo-Noetherian rings contains the Noetherian rings. In this section, we study different properties of an endo-Noetherian ring. We show that if A is a ring and S a multiplicative set of A formed by regular elements, then A is endo-Noetherian if and only if $S^{-1}A$ is endo-Noetherian. We construct an example of a ring A with a multiplicative set S such that $S^{-1}A$ is Noetherian (then endo-Noetherian), but A is not endo-Noetherian. We study the endo-Noetherian property in the case of direct product of rings. We show that if $(A_\lambda)_{\lambda \in \Lambda}$ is a family of rings and $A = \prod_{\lambda \in \Lambda} A_\lambda$ their direct product, then A is endo-Noetherian if and only if Λ is a finite set, and for each $\lambda \in \Lambda$, the ring A_λ is endo-Noetherian. Denote by $(*)$ the following condition: Each increasing sequence of the form $(I : a_1) \subseteq (I : a_2) \subseteq \cdots$ is stationary, where I is an ideal of A and $(a_k)_k$ is a sequence of elements of A. We show that A satisfies the condition $(*)$ if and only if A/I is endo-Noetherian for each ideal I of A. We also study this property on the amalgamation $A \bowtie I = \{(a, a+i); \ a \in A, i \in I\}$, called the duplication of the ring A through the ideal I, which is a sub-ring of $A \times A$. This construction was studied, in the general case, and via different points of view, by M. D'Anna and M. Fontana. Recall that an ideal I is said to be regular if it contains a regular element of A. We show that if I is a regular ideal of A, then A is endo-Noetherian if and only if $A \bowtie I$ is endo-Noetherian. The results of this section are due to Gouaid et al. (2020).

Recall that if a is an element of a ring A, then $ann_A(a) = \{x \in A; \ ax = 0\}$ is an ideal of A.

Definition 13.1 We say that a ring A is endo-Noetherian if each increasing sequence of the form: $ann(a_1) \subseteq ann(a_2) \subseteq \cdots$ is stationary, where $(a_k)_k$ is a sequence of elements of A.

Example 1 An integral domain is endo-Noetherian.

Example 2 It is clear that each Noetherian ring is endo-Noetherian. The converse is false. Let K be a field and V a K-vector space of infinite dimension. By Corollary 8.11 of Chap. 1, the ring $K(+)V$ is not Noetherian. But it is endo-Noetherian. Note that given an element $(k, m) \in K(+)V \setminus (0, 0)$, then

$$ann_{K(+)V}(k, m) = \begin{cases} (0, 0) & \text{if } k \neq 0 \\ 0(+)V & \text{if } k = 0 \end{cases}$$

Indeed, let $(l, n) \in ann_{K(+)V}(k, m)$ and then $(k, m)(l, n) = (kl, kn + lm) = (0, 0)$. If $k \neq 0$, then $l = 0$ and $n = 0$. If $k \neq 0$, then $m \neq 0$ and $lm = 0$, so $l = 0$ and n arbitrary.

Example 3 Each endo-Noetherian ring is strongly Hopfian. We show later that the converse is false.

Let S be a multiplicative set of a ring A. In the following proposition, we are interested to the transfer of the endo-Noetherian property between A and $S^{-1}A$.

Proposition 13.2 *Let A be a ring and S a multiplicative set of A formed by regular elements. Then A is endo-Noetherian if and only if $S^{-1}A$ is endo-Noetherian.*

Proof "$\implies$" Let $(\frac{a_k}{s_k})_k$ be a sequence of elements of $S^{-1}A$ satisfying $ann_{S^{-1}A}(\frac{a_1}{s_1}) \subseteq ann_{S^{-1}A}(\frac{a_2}{s_2}) \subseteq \cdots$. We will show that $ann_A(a_k) \subseteq ann_A(a_{k+1})$ for each $k \in \mathbb{N}$. Let $b \in ann_A(a_k)$. Since $\frac{b}{1} \in ann_{S^{-1}A}(\frac{a_k}{s_k}) \subseteq ann_{S^{-1}A}(\frac{a_{k+1}}{s_{k+1}})$, then $b \in ann_A(a_{k+1})$ because S is formed by regular elements. Since A is endo-Noetherian, there exists a natural integer n such that for each integer $k \geq n$, $ann(a_k) = ann(a_n)$. Let $k \geq n$ and $\frac{b}{t} \in ann(\frac{a_k}{s_k})$. Then $ba_k = 0$, so $ba_n = 0$, and then $\frac{b}{t} \in ann(\frac{a_n}{s_n})$. Then the sequence $ann_{S^{-1}A}(\frac{a_1}{s_1}) \subseteq ann_{S^{-1}A}(\frac{a_2}{s_2}) \subseteq \cdots$ is stationary. So the ring $S^{-1}A$ is endo-Noetherian.

"$\Longleftarrow$" Let $(a_k)_k$ be a sequence of elements of A satisfying $ann(a_1) \subseteq ann(a_2) \subseteq \cdots$. The elements of S are regular, and then for each $k \in \mathbb{N}^*$, $ann_{S^{-1}A}(\frac{a_k}{1}) \subseteq ann_{S^{-1}A}(\frac{a_{k+1}}{1})$. By hypothesis, there exists a natural integer n satisfying for each $k \geq n$, $ann_{S^{-1}A}(\frac{a_n}{1}) = ann_{S^{-1}A}(\frac{a_k}{1})$. Let now $k \geq n$ and $\alpha \in ann(a_k)$. Then $\frac{\alpha}{1} \in ann_{S^{-1}A}(\frac{a_k}{1}) = ann_{S^{-1}A}(\frac{a_n}{1})$. Then $\alpha a_n = 0$ since S is formed by regular elements. Then $ann(a_n) = ann(a_k)$ for each $k \geq n$. So A is endo-Noetherian. $\square$

We show later that the condition on the multiplicative set S of A is necessary in the preceding proposition. The following result furnishes a sufficient condition on the localizations of a ring A in order that A itself is endo-Noetherian:

Proposition 13.3 *Let A be a ring with finite maximal spectrum $M_1, \ldots, M_n$. If all the rings A_{M_i}, $1 \leq i \leq n$, are endo-Noetherian, then A is endo-Noetherian.*

Proof Let $(a_k)_k$ be a sequence of elements of A such that $ann_A(a_1) \subseteq ann_A(a_2) \subseteq \cdots$. It is easy to see that for $1 \leq i \leq n$, $ann_{A_{M_i}}(\frac{a_1}{1}) \subseteq ann_{A_{M_i}}(\frac{a_2}{1}) \subseteq \cdots$. There exists a natural integer m such that for each $1 \leq i \leq n$ and for each $k \geq m$, $ann_{A_{M_i}}(\frac{a_k}{1}) = ann_{A_{M_i}}(\frac{a_m}{1})$. We will prove that for each integer $k \geq m$, $ann_A(a_k) = ann_A(a_m)$. Let $k \geq m$ and $b \in ann_A(a_k)$. Then $ba_k = 0$. For each $1 \leq i \leq n$, $\frac{b}{1}\frac{a_k}{1} = 0$ in A_{M_i}, then $\frac{b}{1}\frac{a_m}{1} = 0$, so there exists $s_i \in A \setminus M_i$ satisfying $s_i ba_m = 0$. The ideal $(s_1, \ldots, s_n)A$ is not included in any maximal ideal M_j, $1 \leq j \leq n$, and then $(s_1, \ldots, s_n)A = A$. Put $1 = c_1 s_1 + \ldots + c_n s_n$ with $c_1, \ldots, c_n \in A$. Then $ba_m = c_1 s_1 ba_m + \ldots + c_n s_n ba_m = 0$, so $b \in ann_A(a_m)$. We conclude that A is endo-Noetherian. $\square$

We show later that the finiteness condition of the maximal spectrum of the ring A is necessary in the preceding proposition.

Proposition 13.4 *Let* $(A_\lambda)_{\lambda \in \Lambda}$ *be a family of rings. The product ring* $A = \prod_{\lambda \in \Lambda} A_\lambda$ *is endo-Noetherian if and only if the set* Λ *of indexes is finite and, for each* $\lambda \in \Lambda$, *the ring* A_λ *is endo-Noetherian.*

Proof "$\Longrightarrow$"

(1) Suppose that Λ is infinite. It contains an infinite countable set $\Lambda' = \{\lambda_1, \lambda_2, \dots\}$. Consider the sequence $(a_j)_{j \in \mathbb{N}^*}$ of elements of A, where $a_j = (a_{j,\lambda})_{\lambda \in \Lambda}$ is defined by $a_{j,\lambda} = 0$ for each $\lambda \in \Lambda \setminus \Lambda'$, $a_{j,\lambda_k} = 1$ for each $k \geq j + 1$, and $a_{j,\lambda_k} = 0$ for $1 \leq k \leq j$. We will show that the sequence $(ann(a_j))_{j \in \mathbb{N}^*}$ is increasing. Fix $j \in \mathbb{N}^*$. Let $x = (x_\lambda)_{\lambda \in \Lambda} \in ann(a_j)$. Then $x_{\lambda_k} = 0$ for each $k \geq j + 1$, so $x_{\lambda_k} a_{j+1,\lambda_k} = 0$ for each $k \geq j + 2$ so $x \in ann(a_{j+1})$. Then $ann(a_j) \subseteq ann(a_{j+1})$. This inclusion is strict. Indeed, let $b = (b_\lambda)_{\lambda \in \Lambda}$ be the element of A defined by $b_\lambda = 0$ for each $\lambda \in \Lambda \setminus \{\lambda_{j+1}\}$ and $b_{\lambda_{j+1}} = 1$. Then $b_{\lambda_{j+1}} a_{j+1,\lambda_{j+1}} = 1.0 = 0$, so $b \in ann(a_{j+1})$. But $b_{\lambda_{j+1}} a_{j,\lambda_{j+1}} = 1.1 = 1$, then $b \notin ann(a_j)$. The ring A is not endo-Noetherian, which is absurd.

(2) We will show that if the ring $A = \prod_{i=1}^{n} A_i$ is endo-Noetherian, then $A_1, \dots, A_n$ are endo-Noetherian. By induction, we may suppose that $n = 2$. It suffices to show that if $A \times B$ is endo-Noetherian, then so is A. Consider a sequence $ann(a_1) \subseteq ann(a_2) \dots (1)$ in A. Then $ann(a_1) \times B \subseteq ann(a_2) \times B \dots$ in $A \times B$. So $ann(a_1, 0) \subseteq ann(a_2, 0) \dots (2)$ in $A \times B$. By hypothesis, (2) is stationary, and then (1) is stationary.

"$\Longleftarrow$" By induction, it suffices to show that the product $R = A \times B$ of two endo-Noetherian rings is endo-Noetherian. Let $(a_1, b_1), \dots$ be a sequence of elements of R satisfying $ann_R((a_1, b_1)) \subseteq \cdots$. It is easy to see that $ann_A(a_1) \subseteq ann_A(a_2) \subseteq \cdots$ and $ann_B(b_1) \subseteq ann_B(b_2) \cdots$. Since A and B are endo-Noetherian, there exists $n \in \mathbb{N}^*$ such that for each $k \geq n$, $ann_A(a_k) = ann_A(a_n)$ and $ann_B(b_k) = ann_B(b_n)$. For each $k \geq n$, $ann_R((a_k, b_k)) = ann_R((a_n, b_n))$, so $R = A \times B$ is endo-Noetherian. $\square$

Example 1 Let $(K_i)_{i \in \mathbb{N}}$ be an infinite sequence of fields and $A = \prod_{i \in \mathbb{N}} K_i$. By Proposition 2.2, the ring A is zero dimensional. By Proposition 2.4, it is strongly Hopfian. But by the preceding proposition, A is not endo-Noetherian.

Example 2 We construct an example of a ring A and a multiplicative set S of A such that A_S is Noetherian (then endo-Noetherian), but A is not endo-Noetherian. Then the condition S formed by regular elements in Proposition 13.2 is necessary. Let B be a Noetherian ring and C a ring that is not endo-Noetherian (e.g., an infinite product of fields). By the preceding proposition, $A = B \times C$ is not endo-Noetherian.

Consider the multiplicative set $S = \{(1,1); (1,0)\}$ of A. For each ideal $I \times J$ of $A = B \times C$, the ideal $I \times 0$ is finitely generated because B is Noetherian. Since $(1.0)(I \times J) = I \times 0 \subseteq I \times J$, then $I \times J$ is S-finite, and then A is S-Noetherian. By Lemma 5.4 of Chap. 1, A_S is Noetherian. Note that the element $(1,0)$ of S is a zero divisor since $(1,0)(0,1) = (0,0)$.

Example 3 By the preceding proposition, the ring $A = \prod_{i:1}^{\infty} \mathbb{F}_2$ is not endo-Noetherian. By Example 1 of Lemma 5.4, Chap. 1, for each prime ideal P of A, the ring A_P is a field and then Noetherian. This example shows the necessity of the finiteness condition of maximal spectrum of the ring A in Proposition 13.3.

Example 4 The quotient of an endo-Noetherian ring is not in general endo-Noetherian. Indeed, let K be a field and $A = K[X_0, X_1, \ldots]$, the ring of polynomials in an infinitely many variables. Since A is an integral domain, then it is endo-Noetherian. Let I be the ideal of A generated by $X_0 X_1, X_0^2 X_2, \cdots$. By the example of Proposition 1.12, the ring A/I is not strongly Hopfian. Then it is not endo-Noetherian.

Definition 13.5 A ring A satisfies the *accr* if for each ideal N of A and for each finitely generated ideal I of A, the increasing sequence of ideals $(N :_A I^k)_{k \in \mathbb{N}^*}$ is stationary.

Theorem 13.6 (Radu 1981) *Let A be a ring. The formal power series ring $A[[X]]$ (resp. polynomials ring $A[X]$) satisfies the accr if and only if A is Noetherian.*

Proof Suppose that A is not Noetherian, and consider a strictly increasing sequence $I_0 \subset I_1 \subset \ldots$ of ideals in A. Let $J = \left\{ f = \sum_{i:0}^{\infty} a_i X^i \in A[[X]]; \ a_i \in I_i \right\}$. Then J is an ideal of $A[[X]]$. Indeed, the stability by addition is clear. Let $f = \sum_{i:0}^{\infty} a_i X^i \in J$ and $g = \sum_{i:0}^{\infty} b_i X^i \in A[[X]]$. Then $fg = \sum_{n:0}^{\infty} c_n X^n$ with $c_n = \sum_{i+j=n} a_i b_j$. Since $a_i \in I_i \subseteq I_n$, then $a_i b_j \in I_n$ and $c_n \in I_n$. So $fg \in J$. The sequence $(J : X) \subseteq (J : X^2) \subseteq \ldots$ of ideals in $A[[X]]$ is strictly increasing. Indeed, if $a \in I_{i+1} \setminus I_i$, then $aX^{i+1} \in J$, but $aX^i \notin J$, so $a \in (J : X^{i+1}) \setminus (J : X^i)$. Then A is Noetherian.

For the ring $A[X]$, we consider $J = \left\{ f = \sum_{i:0}^{n} a_i X^i \in A[X]; \ n \in \mathbb{N}, a_i \in I_i, 0 \leq i \leq n \right\}$. $\qquad\square$

Lemma 13.7 (Lu 1988) *Let $I = (b_1, \ldots, b_t)$ be a finitely generated ideal of a ring A. If k and $n \in \mathbb{N}^*$ are such that $n > kt$, then $I^n = (b_1^k, \ldots, b_t^k) I^{n-k}$.*

Proof Put $J = (b_1^k, \ldots, b_t^k)I^{n-k}$. Then $J \subseteq I^k I^{n-k} = I^n$. Conversely, the ideal I^n is generated by the set $T = \{b_1^{\lambda_1} b_2^{\lambda_2} \ldots b_t^{\lambda_t}; \; \lambda_i \geq 0, \lambda_1 + \ldots + \lambda_t = n\}$. For each element $x = b_1^{\lambda_1} b_2^{\lambda_2} \ldots b_t^{\lambda_t} \in T$, there exists at least one index i, $1 \leq i \leq t$, such that $\lambda_i > k$. Indeed, in the contrary case, $n = \lambda_1 + \ldots + \lambda_t \leq k + \ldots + k = kt < n$, which is absurd. Then $x = b_i^k(b_1^{\lambda_1} b_2^{\lambda_2} \ldots b_i^{\lambda_i - k} \ldots b_t^{\lambda_t}) \in J$ since $\lambda_1 + \lambda_2 + \ldots + (\lambda_i - k) + \ldots + \lambda_t = n - k$. We conclude that $I^n = J$.   □

In the following proposition, we show that in the definition of *accr* rings, we can replace the finitely generated ideal by a principal ideal. To prove this proposition, note that for each ideals I, J, and N of a ring A: (1) if $I \subseteq J$, then $N : J \subseteq N : I$.

(2) $(N : I + J) = (N : I) \cap (N : J)$, and (3) $(N : IJ) = (N : I) : J$. Indeed, let $x \in A$. Then $x \in (N : IJ) \Longleftrightarrow x(IJ) \subseteq N \Longleftrightarrow (xJ)I \subseteq N \Longleftrightarrow xJ \subseteq N : I \Longleftrightarrow x \in (N : I) : J$.

Proposition 13.8 (Lu 1988) *A ring A satisfies accr if and only if for each ideal N of A and each element $b \in A$, the increasing sequence $(N :_A b^k)_{k \in \mathbb{N}^*}$ is stationary.*

Proof Let N be any ideal of A and $I = (b_1, \ldots, b_t)$ a finitely generated ideal of A. There exists $k \in \mathbb{N}^*$ such that for each i, $1 \leq i \leq t$, $(N : b_i^k) = (N : b_i^{k+j})$ for each integer $j \geq 1$. Then $N : (b_1^k, \ldots, b_t^k) = \bigcap_{i:1}^{t} (N : b_i^k) = \bigcap_{i:1}^{t} (N : b_i^{k+j}) = N :$ $(b_1^{k+j}, \ldots, b_t^{k+j})$ for each integer $j \geq 1$. Let $n > kt$ be an integer. We will show that the sequence $(N :_A I^i)_{i \in \mathbb{N}^*}$ is stationary at the rank n. By the preceding lemma, $I^n = (b_1^k, \ldots, b_t^k)I^{n-k}$. Then for each $j \geq 1$, $N : I^n = \left(N : (b_1^k, \ldots, b_t^k)\right) : I^{n-k} = \left(N : (b_1^{k+j}, \ldots, b_t^{k+j})\right) : I^{n-k} = N : (b_1^{k+j}, \ldots, b_t^{k+j})I^{n-k} \supseteq N : I^{n+j} \supseteq N : I^n$ since $(b_1^{k+j}, \ldots, b_t^{k+j})I^{n-k} \subseteq I^{k+j}I^{n-k} = I^{n+j} \subseteq I^n$. Then $N : I^n = N : I^{n+j}$, for each $j \geq 1$.   □

Notation A ring A satisfies the property $(*)$ if each increasing sequence of ideals of A of the form $(I : a_1) \subseteq (I : a_2) \subseteq \cdots$ is stationary, where I is an ideal of A and $(a_k)_{k \in \mathbb{N}^*}$ is a sequence of elements of A.

Noetherian rings satisfy the property $(*)$. We show later that the converse is false. It is also clear that if a ring A satisfies the property $(*)$, then A is endo-Noetherian (Take $I = (0)$) and satisfies the condition *accr*, by the preceding proposition. The following example shows that an endo-Noetherian ring does not satisfy necessarily the condition $(*)$.

Example Let K be a field and $A = K[Y_i; i \in \mathbb{N}]$, the polynomial ring in an infinitely countable indeterminates over K. Let X be an indeterminate over A. By Theorem 13.6, since A is not Noetherian, the ring $A[\![X]\!]$ is not *accr*. In particular, $A[\![X]\!]$ does not satisfy the condition $(*)$. Note that the ring $A[\![X]\!]$ is endo-Noetherian since it is an integral domain.

Corollary 13.9 *The formal power series ring $A[[X]]$ (resp. polynomials $A[X]$) satisfies the condition $(*)$ if and only if the ring A is Noetherian.*

Proof Since the ring $A[[X]]$ (resp. $A[X]$) satisfies the condition $(*)$, it satisfies *accr*. By Theorem 13.6, the ring A is Noetherian. □

The following theorem relies the endo-Noetherian rings with the condition $(*)$:

Theorem 13.10 *The following assertions are equivalent for a ring A:*

1. *The quotient ring A/I is endo-Noetherian for each ideal I of A.*
2. *The ring A satisfies the condition $(*)$.*

Proof $"(1) \implies (2)"$ Let I be an ideal of A and $a_1, a_2, \ldots \in A$ such that $(I : a_1) \subseteq (I : a_2) \subseteq \cdots$. Then $ann_{A/I}(\overline{a_1}) \subseteq ann_{A/I}(\overline{a_2}) \subseteq \cdots$. Indeed, let $\overline{x} \in ann_{A/I}(\overline{a_k})$. Then $xa_k \in I$, so $x \in (I : a_k) \subseteq (I : a_{k+1})$. Then $xa_{k+1} \in I$ and $\overline{x} \in ann_{A/I}(\overline{a_{k+1}})$. Now, since A/I is endo-Noetherian, there exists a natural integer n satisfying $ann_{A/I}(\overline{a_k}) = ann_{A/I}(\overline{a_n})$ for each integer $k \geq n$. We will show that $(I : a_k) = (I : a_n)$ for each $k \geq n$. Let $b \in (I : a_k)$. Then $\overline{b}\overline{a_k} = \overline{0}$ in A/I. Then $\overline{b} \in ann_{A/I}(\overline{a_k}) = ann_{A/I}(\overline{a_n})$, so $ba_n \in I$. Then $b \in (I : a_n)$.

$"(2) \implies (1)"$ Let I be an ideal of A and $a_1, a_2, \ldots$ a sequence of elements of A satisfying $ann_{A/I}(\overline{a_1}) \subseteq ann_{A/I}(\overline{a_2}) \subseteq \cdots$. We will show that $(I : a_1) \subseteq (I : a_2) \subseteq \cdots$. Let $b \in (I : a_k)$. Then $\overline{b} \in ann_{A/I}(\overline{a_k}) \subseteq ann_{A/I}(\overline{a_{k+1}})$, so $b \in (I : a_{k+1})$. By hypothesis, there exists a natural integer n satisfying $(I : a_k) = (I : a_n)$ for each $k \geq n$. Then $ann_{A/I}(\overline{a_k}) = ann_{A/I}(\overline{a_n})$ for each integer $k \geq n$. □

The preceding theorem allows us to construct an example of a non-Noetherian ring that satisfies the condition $(*)$.

Example Let K be a field and M a K-vector space of infinite dimension. By Corollary 8.11 of Chap. 1, the ring $A = K(+)M$ is not Noetherian. By the example of Lemma 6.2, Chap. 2, a proper ideal of A is of the form $I = 0(+)N$ with N a sub-space of M, then $A/I = (K(+)M)/(0(+)N) \simeq K(+)(M/N)$. By Example 1 of Definition 13.1, for each ideal I of A, the ring A/I is endo-Noetherian. By the preceding theorem, the ring A satisfies $(*)$.

Proposition 13.11 *A homomorphic image of a ring satisfying the condition $(*)$ satisfies the condition $(*)$.*

Proof Let $f : A \to B$ be a homomorphism onto of rings with A satisfying the condition $(*)$. Let J be an ideal of B and $b_1, b_2, \ldots \in B$ satisfying $(J :_B b_1) \subseteq (J :_B b_2) \subseteq \cdots$. There exist $a_1, a_2, \ldots \in A$ such that $f(a_i) = b_i$ for each $i \in \mathbb{N}^*$. We will show that $(f^{-1}(J) :_A a_1) \subseteq (f^{-1}(J) :_A a_2) \subseteq \cdots$. Let $x \in (f^{-1}(J) :_A a_k)$. Then $xa_k \in f^{-1}(J)$, so $f(x) \in (J :_B f(a_k)) = (J :_B b_k) \subseteq (J :_B b_{k+1})$. Then $f(xa_{k+1}) = f(x)b_{k+1} \in J$, so $xa_{k+1} \in f^{-1}(J)$, and then $x \in (f^{-1}(J) :_A a_{k+1})$. Since A satisfies $(*)$, there exists a natural integer n satisfying $(f^{-1}(J) :_A a_k) = (f^{-1}(J) :_A a_n)$ for each integer $k \geq n$. We will show that $(J :_B b_k) =$

$(J :_B b_n)$ for each $k \geq n$. Let $y \in (J :_B b_k)$. Then $yb_k \in J$. Since f is onto, there exists $x \in A$ such that $y = f(x)$. Then $f(xa_k) \in J$, so $xa_k \in f^{-1}(J)$, and then $x \in (f^{-1}(J) :_A a_k) = (f^{-1}(J) :_A a_n)$. So $xa_n \in f^{-1}(J)$, then $yb_n \in J$ and $y \in (J :_B b_n)$.   □

Example The quotient of a ring satisfying the condition (∗) satisfies (∗).

Notation (D'Anna and Fontana 2007) Let A be a ring and I an ideal of A. The sub-ring of the product ring $A \times A$, denoted by $A \bowtie I = \{(a, a+i); \ a \in A, i \in I\}$, is called the duplication of A through I. It is a particular case of the amalgamation, introduced in Sect. 7 of Chap. 1 with $f = id_A : A \longrightarrow B = A$ and $J = I$.

We end this section by the following theorem:

Theorem 13.12 *Let A be a ring and I a regular ideal of A. The following assertions are equivalent:*

1. *The ring A is endo-Noetherian.*
2. *The product ring $A \times A$ is endo-Noetherian.*
3. *The duplication $A \bowtie I$ is an endo-Noetherian ring.*

Proof "(1) $\Longrightarrow$ (2)" Follows from Proposition 13.4.

"(2) $\Longrightarrow$ (3)" Let $(a_1, a_1 + i_1), (a_2, a_2 + i_2), \ldots \in A \bowtie I$ satisfying $ann_{A \bowtie I}(a_1, a_1 + i_1) \subseteq ann_{A \bowtie I}(a_2, a_2 + i_2) \subseteq \cdots$. We will show that $ann_{A \times A}(a_k, a_k + i_k) \subseteq ann_{A \times A}(a_{k+1}, a_{k+1} + i_{k+1})$ for each $k \geq 1$. Let $(a, b) \in ann_{A \times A}(a_k, a_k + i_k)$ and $i \in I$ be a regular element of A. Since $(i, i)(a, b) = (ia, ia + i(b - a)) \in A \bowtie I$, then $(i, i)(a, b) \in ann_{A \bowtie I}(a_{k+1}, a_{k+1} + i_{k+1})$. But (i, i) is regular in $A \times A$, and then $(a, b) \in ann_{A \times A}(a_{k+1}, a_{k+1} + i_{k+1})$. By hypothesis, there exists $n \in \mathbb{N}^*$ such that for each $k \geq n$, $ann_{A \times A}(a_k, a_k + i_k) = ann_{A \times A}(a_n, a_n + i_n)$ and then $(A \bowtie I) \cap ann_{A \times A}(a_k, a_k + i_k) = (A \bowtie I) \cap ann_{A \times A}(a_n, a_n + i_n)$. So $ann_{A \bowtie I}(a_k, a_k + i_k) = ann_{A \bowtie I}(a_n, a_n + i_n)$, for each integer $k \geq n$.

"(3) $\Longrightarrow$ (1)" This implication is true even if I is not a regular ideal. Let $a_1, a_2, \ldots \in A$ be elements such that $ann_A(a_1) \subseteq ann_A(a_2) \subseteq \cdots$. We will show that $ann_{A \bowtie I}(a_k, a_k) \subseteq ann_{A \bowtie I}(a_{k+1}, a_{k+1})$ for each k. Let $(x, x + i) \in ann_{A \bowtie I}(a_k, a_k)$ with $x \in A$ and $i \in I$. Then $(x, x + i)(a_k, a_k) = (0, 0)$, so the elements $x, x + i \in ann_A(a_k) \subseteq ann_A(a_{k+1})$. Then $(x, x + i) \in ann_{A \bowtie I}(a_{k+1}, a_{k+1})$. Since the ring $A \bowtie I$ is endo-Noetherian, there exists $n \in \mathbb{N}^*$ such that for each integer $k \geq n$, $ann_{A \bowtie I}(a_k, a_k) = ann_{A \bowtie I}(a_n, a_n)$. We will show that $ann_A(a_k) = ann_A(a_n)$ for each $k \geq n$. Let $b \in ann_A(a_k)$. Since $(b, b) \in ann_{A \bowtie I}(a_k, a_k)$, then $(b, b)(a_n, a_n) = (0, 0)$. So $ba_n = 0$, and then $b \in ann_A(a_n)$.   □

14   Polynomials and Formal Power Series Over an Endo-Noetherian Ring

The goal of this section is to find necessary and sufficient conditions for formal power series (resp. polynomials) rings to be endo-Noetherian. The reader can refer to Sect. 12 for the notion of Armendariz rings. The results are due to Gouaid et al. (2020).

Let S be a nonempty set of a ring A. The annihilator of S is the ideal of A denoted $ann_A(S) = \{a \in A;\ as = 0\}$.

Lemma 14.1 *Let A be an Armendariz ring, $f(X) \in A[X]$, and S the set of the coefficients of $f(X)$. Then $\big(ann_A(S)\big)[X] = ann_{A[X]}(f)$.*

Proof Let $g(X) = \sum_{i:0}^{n} b_i X^i \in A[X]$. Then $g(X) \in \big(ann_A(S)\big)[X] \Longleftrightarrow b_0, \dots, b_n \in ann_A(S) \Longleftrightarrow \forall a \in S, \forall i \in \{0 \dots, n\}, ab_i = 0 \Longleftrightarrow g(X)f(X) = 0 \Longleftrightarrow g(X) \in ann_{A[X]}(f)$. $\square$

In the following theorem, we give necessary and sufficient condition for $A[X]$ to be endo-Noetherian when A is an Armendariz ring.

Theorem 14.2 *The following assertions are equivalent for an Armendariz ring A:*

1. *The ring $A[X]$ is endo-Noetherian.*
2. *The ring A satisfies the acc property for the annihilators of finite sets.*
3. *The ring A satisfies the acc property for the annihilators of finitely generated ideals.*

Proof $"(1) \Longrightarrow (2)"$ Let $(S_i)_{i \in \mathbb{N}^*}$ be a sequence of nonempty finite subsets of A satisfying $ann_A(S_1) \subseteq ann_A(S_2) \subseteq \cdots$. Then $\big(ann_A(S_1)\big)[X] \subseteq \big(ann_A(S_2)\big)[X] \subseteq \cdots$. For each $i \in \mathbb{N}^*$, we consider a polynomial $f_i(X) \in A[X]$ whose S_i is the set of coefficients. By the preceding lemma, $ann_{A[X]}(f_1) \subseteq ann_{A[X]}(f_2) \subseteq \cdots$. Since $A[X]$ is endo-Noetherian, there exists $n \in \mathbb{N}^*$ such that $ann_{A[X]}(f_k) = ann_{A[X]}(f_n)$ for each integer $k \geq n$. Then $ann_A(S_k) = ann_A(S_n)$ for each $k \geq n$.

$"(2) \Longrightarrow (1)"$ Let $(f_i)_{i \in \mathbb{N}^*}$ be a sequence of elements of $A[X]$ satisfying $ann_{A[X]}(f_1) \subseteq ann_{A[X]}(f_2) \subseteq \cdots$. For each $i \in \mathbb{N}^*$, let S_i be the set of the coefficients of f_i. By the preceding lemma, $\big(ann_A(S_1)\big)[X] \subseteq \big(ann_A(S_2)\big)[X] \subseteq \cdots$. Then $ann_A(S_1) \subseteq ann_A(S_2) \subseteq \cdots$. By (2), there exists $n \in \mathbb{N}^*$ such that $ann_A(S_k) = ann_A(S_n)$ for each $k \geq n$. Then $ann_{A[X]}(f_n) = ann_{A[X]}(f_k)$ for each integer $k \geq n$.

$"(2) \Longleftrightarrow (3)"$ Follows from the fact that if I is an ideal of A generated by a set S, then $ann_A(I) = ann_A(S)$. $\square$

Lemma 14.3 *Let A be a power series Armendariz ring, $f(X) \in A[[X]]$, and S the set of the coefficients of $f(X)$. Then $\big(ann_A(S)\big)[[X]] = ann_{A[[X]]}(f)$.*

Proof Let $g(X) = \sum\limits_{i:0}^{\infty} b_i X^i \in A[[X]]$. Then $g(X) \in (ann_A(S))[X] \Longleftrightarrow \forall i \in$ $\mathbb{N}, b_i \in ann_A (S) \Longleftrightarrow \forall a \in S, \forall i \in \mathbb{N}, ab_i = 0 \Longleftrightarrow g(X)f(X) = 0 \Longleftrightarrow$ $g(X) \in ann_{A[[X]]} f.$ □

An ideal generated by a countably set is said to be countably generated.

Theorem 14.4 *The following assertions are equivalent for a power series Armendariz ring A:*

1. *The ring $A[\![X]\!]$ is endo-Noetherian.*
2. *The ring A satisfies acc for the annihilators of the countably sets.*
3. *The ring A satisfies acc for the annihilators of the countably generated ideals.*
4. *For each sequence $(f_k)_{k \in \mathbb{N}^*}$ of elements of $A[\![X]\!]$, $f_k = \sum\limits_{j \geq 0} a_{k,j} X^j$, satisfying*

   $ann_{A[\![X]\!]}(f_1) \subseteq ann_{A[\![X]\!]}(f_2) \subseteq \cdots$, *there exists $n \in \mathbb{N}$ such that for each natural integer $k \geq n$, $\bigcap\limits_{j \geq 0} ann_A(a_{k,j}) = \bigcap\limits_{j \geq 0} ann_A(a_{n,j})$.*

Proof $''(1) \Longrightarrow (2)''$ Let $(S_i)_{i \in \mathbb{N}^*}$ be a sequence of countably subsets of A satisfying $ann_A(S_1) \subseteq ann_A(S_2) \subseteq \cdots$. Then $(ann_A(S_1))[[X]] \subseteq (ann_A(S_2))[[X]] \subseteq \cdots$. For each $i \in \mathbb{N}^*$, we consider a series $f_i(X) \in A[[X]]$ whose set of coefficients is equal to S_i. By the preceding lemma, $ann_{A[[X]]}(f_1) \subseteq ann_{A[[X]]}(f_2) \subseteq \cdots$. Since the ring $A[[X]]$ is endo-Noetherian, there exists $n \in \mathbb{N}^*$ such that $ann_{A[X]}(f_k) = ann_{A[X]}(f_n)$ for each integer $k \geq n$. Then $ann_A(S_k) = ann_A(S_n)$ for each $k \geq n$.

$''(2) \Longrightarrow (1)''$ Let $(f_i)_{i \in \mathbb{N}^*}$ be a sequence of elements of $A[[X]]$ satisfying $ann_{A[[X]]}(f_1) \subseteq ann_{A[[X]]}(f_2) \subseteq \cdots$. For each $i \in \mathbb{N}^*$, let S_i be the set of the coefficients of the series f_i. By the preceding lemma, $(ann_A(S_1))[[X]] \subseteq (ann_A(S_2))[[X]] \subseteq \cdots$. Then $ann_A(S_1) \subseteq ann_A(S_2) \subseteq \cdots$. By (2), there exists $n \in \mathbb{N}^*$ such that $ann_A(S_k) = ann_A(S_n)$ for each $k \geq n$. Then $ann_{A[[X]]}(f_n) = ann_{A[[X]]}(f_k)$ for each integer $k \geq n$.

$''(2) \Longleftrightarrow (3)''$ Clear. $''(1) \Longleftrightarrow (4)''$ By the preceding lemma, for each $f = \sum\limits_{j \geq 0} a_j X^j \in A[\![X]\!]$, $ann_{A[\![X]\!]}(f) = ann_A(a_i, i \in \mathbb{N})[\![X]\!] = (\bigcap\limits_{j \geq 0} ann_A(a_j))[\![X]\!]$. Then $A[\![X]\!]$ is endo-Noetherian if and only if for each sequence $(f_k)_{k \in \mathbb{N}^*}$ of elements of $A[\![X]\!]$, with $f_k = \sum\limits_{i \geq 0} a_{k,j} X^j$, such that $ann_{A[\![X]\!]}(f_1) \subseteq ann_{A[\![X]\!]}(f_2) \subseteq \cdots$, there exists $n \in \mathbb{N}$ such that for each integer $k \geq n$, $\bigcap\limits_{j \geq 0} ann_A(a_{k,j}) = \bigcap\limits_{j \geq 0} ann_A(a_{n,j})$. □

15 Polynomials and Formal Power Series Over a PF-Ring

Definition 15.1 An ideal I of a ring A is said to be pure if for each $x \in I$, there exists $y \in I$ such that $xy = x$.

Proposition 15.2 (Al-Ezeh 1987) *The intersection of a finite number of pure ideals is a pure ideal.*

Proof Let $I_1, \ldots, I_n$ be pure ideals of a ring A and $x \in J = I_1 \cap \ldots \cap I_n$. There exist $y_1 \in I_1, \ldots, y_n \in I_n$ such that $xy_i = x$, $1 \leq i \leq n$. Then $y = y_1 \ldots y_n \in J$ and $xy = x$. $\square$

Definition 15.3 A ring A is said to be PF if for each $a \in A$, the annihilator $(0 : a)_A$ of the element a is a pure ideal; i.e., for each $x \in (0 : a)$, there exists $y \in (0 : a)$ such that x=xy.

Proposition 15.4 (Al-Ezeh 1987) *A PF-ring A is reduced.*

Proof Suppose that there exists $0 \neq a \in Nil(A)$. Let $n \geq 1$ be the smallest natural integer such that $a^n = 0$. Then $n \geq 2$ and $a \in (0 : a^{n-1})_A$. Since $(0 : a^{n-1})_A$ is a pure ideal of A, there exists $b \in (0 : a^{n-1})_A$ such that $ab = a$. Then $0 = ba^{n-1} = (ba)a^{n-2} = aa^{n-2} = a^{n-1}$, which contradicts the minimality of n. $\square$

Lemma 15.5 *Let A be a reduced ring and $f = \displaystyle\sum_{i:0}^{\infty} f_i X^i \in A[[X]]$. Then f is a regular element in $A[[X]]$ if and only if $\displaystyle\bigcap_{i:0}^{\infty}(0 : f_i) = (0)$.*

Proof " $\Longrightarrow$ " Let $a \in \displaystyle\bigcap_{i:0}^{\infty}(0 : f_i)$. Then $af = \displaystyle\sum_{i:0}^{\infty} af_i X^i = 0$. Since f is regular, $a = 0$.

" $\Longleftarrow$ " Let $g = \displaystyle\sum_{i:0}^{\infty} g_i X^i \in A[[X]]$ be such that $fg = 0$. By Example 6 of Definition 12.1, A is power series Armendariz. For each $j \in \mathbb{N}$, $g_j \in \displaystyle\bigcap_{i:0}^{\infty}(0 : f_i) = (0)$, and then $g = 0$. $\square$

Theorem 15.6 (Al-Ezeh 1987) *Let A be a ring. Then $A[X]$ is a PF-ring if and only if A is a PF-ring.*

Proof " $\Longrightarrow$ " Let $a \in A$ and $b \in (0 : a)_A \subseteq (0 : a)_{A[X]}$. Since $A[X]$ is a PF-ring,

there exists $g(X) = \sum_{i:0}^{n} c_i X^i \in (0 : a)_{A[X]}$ such that $bg(X) = b$. Then $bc_0 = b$ and

$c_0 a = 0$.

" $\Longleftarrow$ " Let $f = \sum_{i:0}^{m} a_i X^i \in A[X]$ and $h = \sum_{i:0}^{n} h_i X^i \in (0 : f)$. Then

$fh = 0$. By Proposition 15.4, A is reduced, and then $a_i h_j = 0$ for each $i, j \in \mathbb{N}$.

Put $J = \bigcap_{j:0}^{n}(0 : h_j)_A$. Since A is a PF-ring, the ideals $(0 : h_j)$ are pure,

and by Proposition 15.2, J is also pure in A. Since $a_0, a_1, \ldots, a_m \in J$, there
exist $b_0, b_1, \ldots, b_m \in J$ such that $a_0 b_0 = a_0, a_1 b_1 = a_1, \ldots, a_m b_m = a_m$.
We want to find an element $c \in J$ such that $c(a_0 + a_1 X + \ldots + a_m X^m) = a_0 + a_1 X + \ldots + a_m X^m$. First, $a_0 b_0 = a_0$; put $c_0 = b_0$. On the other hand,
$(a_0 + a_1 X)(b_0 + b_1 - b_1 b_0) = a_0 b_0 + a_0 b_1 - a_0 b_1 b_0 + a_1 b_0 X + a_1 b_1 X - a_1 b_1 b_0 X = a_0 + a_0 b_1 - a_0 b_1 + a_1 b_0 X + a_1 X - a_1 b_0 X = a_0 + a_1 X$. Put $c_1 = b_0 + b_1 - b_1 b_0$.
We also have $(a_0 + a_1 X + a_2 X^2)(c_1 + b_2 - b_2 c_1) = a_0 c_1 + a_0 b_2 - a_0 b_2 c_1 + a_1 c_1 X + a_1 b_2 X - a_1 b_2 c_1 X + a_2 c_1 X^2 + a_2 b_2 X^2 - a_2 b_2 c_1 X^2 = a_0 + a_0 b_2 - a_0 b_2 + a_1 X + a_1 b_2 X - a_1 b_2 X + a_2 c_1 X^2 + a_2 X^2 - a_2 c_1 X^2 = a_0 + a_1 X + a_2 X^2$. Put
$c_2 = c_1 + b_2 - b_2 c_1, \ldots, c_m = c_{m-1} + b_m - c_{m-1} b_m$. Note that $c_0, c_1, \ldots, c_m \in J$.
Suppose by induction that $(a_0 + a_1 X + \ldots + a_k X^k)c_k = a_0 + a_1 X + \ldots + a_k X^k$. Then
$(a_0 + a_1 X + \ldots + a_k X^k + a_{k+1} X^{k+1})c_{k+1} = (a_0 + a_1 X + \ldots + a_k X^k + a_{k+1} X^{k+1})(c_k + b_{k+1} - c_k b_{k+1}) = (a_0 + a_1 X + \ldots + a_k X^k)c_k + (a_0 + a_1 X + \ldots + a_k X^k)(b_{k+1} - c_k b_{k+1}) + a_{k+1} X^{k+1}(c_k + b_{k+1} - c_k b_{k+1}) = a_0 + a_1 X + \ldots + a_k X^k + a_{k+1} c_k X^{k+1} + a_{k+1} b_{k+1} X^{k+1} - c_k a_{k+1} b_{k+1} X^{k+1} = a_0 + a_1 X + \ldots + a_k X^k a_{k+1} X^{k+1}$. Then for
$t = 0, 1, \ldots, m, (a_0 + a_1 X + \ldots + a_t X^t)c_t = a_0 + a_1 X + \ldots + a_t X^t$. Le $c = c_m$,

then $cf = f$ and $c \in J = \bigcap_{j:0}^{n}(0 : h_j)_A$, and so $c \in (0 : h)_{A[X]}$. $\qquad\square$

Corollary 15.7 (Kim 1988) *Let A be a Noetherian ring. Then $A[[X]]$ is a PF-ring if and only if A is a PF-ring.*

Proof " $\Longrightarrow$ " (True even if A is not Noetherian). Let $a \in A$ and $b \in (0 : a)_A \subseteq$

$(0 : a)_{A[[X]]}$. There exist $g = \sum_{i:0}^{\infty} c_i X^i \in (0 : a)_{A[[X]]}$ such that $bg = b$. Then

$bc_0 = b$ and $ag = 0$, so $ac_0 = 0$, and then $c_0 \in (0 : a)_A$.

" $\Longleftarrow$ " Let $h = \sum_{i:0}^{\infty} h_i X^i \in A[[X]]$ and $f = \sum_{i:0}^{\infty} a_i X^i \in (0 : h)_{A[[X]]}$. Then

$fh = 0$. Since A is reduced, $a_i h_j = 0$ for each $i, j \in \mathbb{N}$. Since A is Noetherian,
the contents $c(f)$ and $c(h)$ of the series f and h are finitely generated ideals. Put

$c(f) = (a_0, \ldots, a_m)$ and $c(h) = (h_0, \ldots, h_n)$. Let $J = \bigcap_{j:0}^{n}(0 : h_j)_A$. Since A

is a PF-ring, the $(0 : h_j)$ are pure ideals, and by Proposition 15.2, J is also pure in A. Since $a_0, a_1, \ldots, a_m \in J$, there exist $b_0, b_1, \ldots, b_m \in J$ such that $a_0 b_0 = a_0, a_1 b_1 = a_1, \ldots, a_m b_m = a_m$. As in the proof of the preceding theorem, we can find an element $c \in J$ such that $c(a_0 + a_1 X + \ldots + a_m X^m) = a_0 + a_1 X + \ldots + a_m X^m$. Since $c(h) = (h_0, h_1, \ldots, h_n)$, then $c \in (0 : h)_{A[[X]]}$. On the other hand, $ca_0 = a_0, ca_1 = a_1, \ldots, ca_m = a_m$. Since $c(f) = (a_0, a_1, \ldots, a_m)$, then $ca = a$ for each $a \in c(f)$, so $cf = f$. $\qquad\square$

Theorem 15.8 (Al-Ezeh 1988) *Let A be a ring. Then $A[[X]]$ is a PF-ring if and only if for any countably subsets $S = \{b_0, b_1, \ldots\}$ and $T = \{a_0, a_1, \ldots\}$ of A such that $S \subseteq (0 : T)$, there exists $c \in (0 : T)$ such that $b_i c = b_i$ for each $i \in \mathbb{N}$.*

Proof "$\Longrightarrow$" Since $A[[X]]$ is a PF-ring, then $A[[X]]$ and A are reduced. Let $S = \{b_0, b_1, \ldots\}$ and $T = \{a_0, a_1, \ldots\}$ be two countably subsets of A such that $S \subseteq (0 : T)$. Put $f = \sum_{i:0}^{\infty} a_i X^i$ and $g = \sum_{i:0}^{\infty} b_i X^i \in A[[X]]$, then $a_i b_j = 0$ for each $i, j \in \mathbb{N}$, so $fg = 0$ and $g \in (0 : f)$. Since $A[[X]]$ is a PF-ring, there exists $h = \sum_{i:0}^{\infty} c_i X^i \in (0 : f)$ such that $g = hg$ and then $g(h - 1) = 0$ and $fh = 0$. So, $a_i c_j = 0$ for each $i, j \in \mathbb{N}$ and in particular $c_0 \in (0 : T)$. Also, $b_i(c_0 - 1) = 0$ for each $i \in \mathbb{N}$, then $b_i c_0 = b_i$.

"$\Longleftarrow$" By the proof of the first implication of the preceding corollary, A is a PF-ring, and then it is reduced. Let $f = \sum_{i:0}^{\infty} a_i X^i$ and $g = \sum_{i:0}^{\infty} b_i X^i \in A[[X]]$ be such that $g \in (0 : f)$. Then $fg = 0$. So $a_i b_j = 0$ for each $i, j \in \mathbb{N}$. Let $S = \{b_0, b_1, \ldots\}$ and $T = \{a_0, a_1, \ldots\}$. Then $S \subseteq (0 : T)$. By hypothesis, there exists $c \in (0 : T)$ such that $b_i c = b_i$ for each $i \in \mathbb{N}$. Then $c \in (0 : f)$ and $cg = g$. $\qquad\square$

16 Isonoetherian Valuation Domains

Definition 16.1 (Facchini and Nazemian 2016) Let A be a ring and M a A-module.

1. We say that M is isonoetherian if for each increasing sequence $M_1 \subseteq M_2 \subseteq \ldots$ of sub-A-modules of M, there exists a rank n such that $M_n \simeq_A M_i$ (an isomorphism of A-modules) for each integer $i \geq n$.
2. The ring A is said to be isonoetherian if the A-module A is isonoetherian; i.e., for each increasing sequence $I_1 \subseteq I_2 \subseteq \ldots$ of ideals of A, there exists a rank n such that $I_i \simeq_A I_n$ for each integer $i \geq n$.

Example 1 Each Noetherian module is isonoetherian. In particular, each Noetherian ring is isonoetherian.

Example 2 (Khalifa 2018) Let E be vector space. Then (i) E is isonoetherian $\Longleftrightarrow$ (ii) E is Noetherian $\Longleftrightarrow$ (iii) $dim(E) < \infty$. Indeed, the equivalence (ii) $\Longleftrightarrow$ (iii) is proved in Example 1 of Definition 3.1, Chap. 2. The implication (ii) $\Longrightarrow$ (i) is clear. To prove (i) $\Longrightarrow$ (iii), suppose that $dim(E) = \infty$, and let $(e_i)_{i \in \mathbb{N}^*}$ be a free family of elements of E. Put $E_i = (e_1, \ldots, e_i)$ for each $i \in \mathbb{N}^*$. Then $E_1 \subset E_2 \subset \ldots$ and $E_i \not\simeq E_{i+1}$ for each $i \in \mathbb{N}^*$. So E is not isonoetherian, which is absurd.

Lemma 16.2 *Let A be a ring, $a, b \in A \setminus (0)$ and I an ideal of A.*

1. *If $ann_A(a) = ann_A(b)$, then $aA \simeq_A bA$.*
2. *If A is an integral domain, then $aA \simeq_A bA$.*
3. *If $I \simeq_A cA$ for some $c \in A$, then I is a principal ideal of A.*

Proof

(1) Consider the map $\varphi : aA \to bA$

$$x = ar \mapsto \varphi(x) = \varphi(ar) = br$$

If $r, t \in A$ are such that $ar = at$, then $a(r - t) = 0$. So $r - t \in ann_A(a) = ann_A(b)$. Then $(r - t)b = 0$ so $rb = tb$. Then φ is well defined. Moreover, φ is a homomorphism onto of A-modules. Let $x = ar \in Ker(\varphi) \Leftrightarrow \varphi(x) = 0 \Leftrightarrow br = 0 \Leftrightarrow r \in ann_A(b) = ann_A(a) \Leftrightarrow ar = 0 \Leftrightarrow x = 0$. Then φ is one to one, so φ is an isomorphism of A-modules and $aA \simeq_A bA$.

(2) Since A is an integral domain, then $ann_A(a) = 0 = ann_A(b)$. By (1), $aA \simeq_A bA$.

(3) Let $c \in A$ and $\varphi : cA \longrightarrow I$ an isomorphism of A-modules. Then $I = \varphi(cA) = \varphi(c)A$, which is a principal ideal of A. $\qquad\square$

Lemma 16.3 (Facchini and Nazemian 2016) *A valuation domain is isonoetherian if and only if it satisfies the acc for the ideals that are not finitely generated.*

Proof $'' \Longrightarrow ''$ Suppose that an isonoetherian valuation domain A admits a strictly increasing sequence $I_1 \subset I_2 \subset \ldots$ of ideals that are not finitely generated. For each $j \geq 1$, let $a_j \in I_{j+1} \setminus I_j$. Then $I_j \subset a_j A \subseteq I_{j+1}$. So $I_1 \subset a_1 A \subseteq I_2 \subset a_2 A \subseteq I_3 \subset \ldots$. Since A is isonoetherian, there exist integers $k, n \geq 1$ such that $I_n \cong_A a_k A$. By the preceding lemma, the ideal I_n is principal, which is absurd.

$'' \Longleftarrow ''$ Let A be a valuation domain satisfying the acc for nonfinitely generated ideals and $(0) \neq I_1 \subseteq I_2 \subseteq \ldots$ an increasing sequence of ideals of A.

First case: There exists $n \geq 1$ such that for each $i \geq n$, I_i is finitely generated. Since A is a valuation domain, I_i is a principal ideal for each $i \geq n$. By the preceding lemma, $I_i \cong_A I_n$ for each $i \geq n$.

Second case: For each $n \geq 1$, there exists $i > n$ such that I_i is not principal. We can extract from the sequence $(0) \neq I_1 \subseteq I_2 \subseteq \ldots$ an increasing subsequence of ideals that are not principal. After a new numerating, we may suppose that it is the sequence $(0) \neq I_1 \subseteq I_2 \subseteq \ldots$ itself. Since A is a valuation domain, the ideals I_i are not finitely generated. By hypothesis, there exists $n \geq 1$ such that for each $i \geq n$, $I_i = I_n$. $\qquad \square$

Lemma 16.4 (M. Khalifa) *Let P and Q be two prime ideals of a valuation domain A. If $P \simeq_A Q$, then $P = Q$.*

Proof Suppose that $P \neq Q$. For example, $Q \nsubseteq P$. Let $a \in Q \setminus P$. By hypothesis, there exists an isomorphism of A-modules $f : Q \longrightarrow P$. For each $r \in Q$, $af(r) = f(ar) = rf(a)$, then $f(r) = r\frac{f(a)}{a} = r\alpha$ where $0 \neq \alpha = \frac{f(a)}{a} \in K$, the quotient field of A. In particular, $P = f(Q) = Q\alpha$. Since A is a valuation domain, either $\alpha \in A$ or $\alpha^{-1} \in A$. If $\alpha^{-1} \in A$, then $Q = \alpha^{-1}P \subseteq AP = P$, a contradiction. If $\alpha \in A$, since $a \in Q$, then $f(a) = \alpha a \in P$. But P is a prime ideal and $a \notin P$, and then $\alpha \in P = Q\alpha$. So $1 \in Q$, which is absurd. Then $P = Q$. $\qquad \square$

Lemma 16.5 *Let A be a valuation domain and I a proper ideal of A. The ideal $P = \bigcap_{n \geq 1} I^n$ of A is prime.*

Proof Since $P \subseteq I$, then $P \neq A$. Let $a, b \in A \setminus P$. There exist $n, m \geq 1$ such that $a \notin I^n$ and $b \notin I^m$. Then $I^n \subset (a)$, $I^m \subset (b)$ and $I^n b \subseteq (a)(b) = (ab)$. Suppose that $I^n b = (ab)$. There exists $x \in I^n$ such that $ab = xb$. Then $a = x \in I^n$, which is absurd. So $bI^n \subset (ab)$, which implies that $I^{n+m} \subseteq bI^n \subset (ab)$ so $ab \notin I^{n+m}$. Then $ab \notin \bigcap_{k \geq 1} I^k = P$. $\qquad \square$

Definition 16.6 Let A be a valuation domain.

1. We say that a prime ideal P of A is branched if there exists a P-primary ideal Q of A such that $Q \neq P$.
2. We say that the domain A is discrete if for each primary ideal Q of A, there exists $n \in \mathbb{N}^*$ such that $Q = (\sqrt{Q})^n$.

Lemma 16.7 *Let A be a valuation domain. Then A is discrete if and only if for each branched prime ideal P of A, $P^2 \neq P$.*

Proof "$\Longrightarrow$" Let P be a branched prime ideal of A. By definition, there exists a P-primary ideal Q of A such that $Q \neq P$. Since A is discrete, there exists $n \in \mathbb{N}^*$ such that $Q = (\sqrt{Q})^n = P^n$. Suppose that $P^2 = P$, then $P^n = P$, so $Q = P$, which is absurd. Then $P^2 \neq P$.

"$\Longleftarrow$" Let Q be a primary ideal of A. Then $P = \sqrt{Q}$ is a prime ideal of A. Suppose that $Q \neq P^n$ for each $n \geq 1$. In particular, $Q \neq P$, and then P is branched.

By hypothesis, $P \neq P^2$. If $Q \subseteq P^n$ for each $n \geq 1$, then $Q \subseteq I := \bigcap_{n \geq 1} P^n$, which is

a prime ideal of A, by the preceding lemma. Then $P = \sqrt{Q} \subseteq \sqrt{I} = I \subseteq P^2 \subseteq P$
so $P = P^2$, a contradiction. There exists $n \geq 1$ such that $Q \not\subseteq P^n$. But A is
a valuation domain, and then $P^n \subset Q$. Let $k \geq 1$ be the smallest integer such
that $P^k \subset Q$. Then $P^k \subset Q \subset P^{k-1}$. Let $a \in P^{k-1} \setminus Q$ and $b \in Q \setminus P^k$.
Since A is a valuation domain, $P^k \subset bA \subseteq Q \subset aA \subseteq P^{k-1}$. Put $b = ac$ with
$c \in A$. Since $ac = b \in Q$ and $a \notin Q$ with Q a P-primary ideal, then $c \in P$. So
$b = ac \in P^{k-1}P = P^k$, a contradiction. Then $Q = P^n$ for some integer $n \geq 1$. $\quad\square$

Lemma 16.8 *Let A be a finite dimensional valuation domain. Then A is discrete if
and only if for each nonzero prime ideal P of A, $P \neq P^2$.*

Proof $"\Longleftarrow"$ Follows from the preceding lemma.

$\quad"\Longrightarrow"$ Let P be a nonzero prime ideal of A. Since the dimension of A is finite, P
has a predecessor Q. By the preceding lemma, it suffices to show that P is branched.

Particular Case: P is the maximal ideal of A. Let $x \in P \setminus Q$. Then $x^2 \notin Q$,
so $Q \subset x^2A \subset P$. The second inclusion is strict because of x. Since A is a
valuation domain, the ideal $\sqrt{x^2A}$ is prime and $Q \subset x^2A \subseteq \sqrt{x^2A} \subseteq P$. But
$Q \subset P$ are consecutive, and then $\sqrt{x^2A} = P$. We will show that x^2A is primary.
Let $y, z \in A$ be such that $yz \in x^2A$ and $y \notin x^2A$. Suppose that $z \notin \sqrt{x^2A} = P$,
which is a maximal ideal of A, then $P + zA = A$. Put $1 = p + za$ with $p \in P$ and
$a \in A$. There exists $n \in \mathbb{N}^*$ such that $p^n \in x^2A$. Then $1 = (p + za)^n = p^n + zb$
with $b \in A$ and $y = yp^n + yzb \in x^2A$, which is absurd.

General Case: Since $Q \subset P$ are consecutive in A, $QA_P \subset PA_P$ are also
consecutive in A_P. Since A_P is a valuation domain with maximal ideal PA_P, by
the particular case, there exists an ideal PA_P-primary R of A_P with $R \subset PA_P$.
We will show that $R \cap A \subset P$, $\sqrt{R \cap A} = P$, and $R \cap A$ is primary in A.
Indeed, $R \cap A \subseteq PA_P \cap A = P$. If $R \cap A = P$, then $P \subseteq R$ so $PA_P \subseteq R$,
which is absurd. Then $R \cap A \subset P$. We also have $\sqrt{R \cap A} \subseteq P$. Conversely, let
$x \in P \subseteq PA_P = \sqrt{R}$. There exists $n \in \mathbb{N}^*$ such that $x^n \in R$, then $x^n \in R \cap A$,
so $x \in \sqrt{R \cap A}$. Then $\sqrt{R \cap A} = P$. Let $y, z \in A$ be such that $yz \in R \cap A$
and $y \notin R \cap A$. Then $yz \in R$ and $y \notin R$, so $z \in \sqrt{R} = PA_P$, and then
$z \in PA_P \cap A = P$. $\quad\square$

Lemma 16.9 (Facchini and Nazemian 2016) *An isonoetherian valuation domain
is of rank ≤ 2.*

Proof Suppose that there exists an isonoetherian valuation domain A of rank ≥ 3.
There exists a chain $0 \subset P \subset Q \subset M$ of prime ideals of A. Let $0 \neq c \in P$,
$b \in Q \setminus P$, and $a \in M \setminus Q$. Then $c \in P \subset Q \subseteq \bigcap_{n \geq 0} a^nA$ so $ca^{-n} \in A$ for

each $n \geq 0$. The ideal $I = (ca^{-n}; n \geq 0)$ is not finitely generated. Indeed, in the
contrary case, I is principal and there exists $k \geq 0$ such that $I = ca^{-k}A$. Since

$ca^{-k-1} \in ca^{-k}A$, then $a^{-1} \in A$, so $1 = a.a^{-1} \in MA = M$, which is absurd. Then I is not finitely generated. For each $n \geq 0$, $b^{-n}cA \subseteq b^{-n}I \subseteq b^{-n-1}cA$. Indeed, $c = c.1 = c.a^{-0} \in I$, and then $b^{-n}c \in b^{-n}I$. We obtain the inclusion $b^{-n}cA \subseteq b^{-n}I$. Since $b \in Q \subseteq \bigcap_{n \geq 0} a^n A$, then $ba^{-n} \in A$ and $bca^{-n} \in cA$ for each $n \geq 0$. Then $bI = b(ca^{-n}; n \geq 0) \subseteq cA$ so $I \subseteq b^{-1}cA$. We have the second inclusion $b^{-n}I \subseteq b^{-n-1}cA$ for each $n \geq 0$. Since $b^n \notin P$ and $c \in P$, then $cA \subset b^n A$. So $b^{-n}cA \subset A$, and then $b^{-n}cA$ is an integral ideal of A for each $n \geq 0$. Then $cA \subseteq I \subseteq b^{-1}cA \subseteq b^{-1}I \subseteq b^{-2}cA \subseteq \ldots$ is an increasing sequence of integral ideals of A. Since A is isonoetherian, there exists $n \geq 0$ such that $b^{-n}I \simeq_A cb^{-n}A$. By Lemma 16.2 (3), the ideal $b^{-n}I$ is principal. Then I itself is a principal ideal of A, a contradiction. Then the rank of A is ≤ 2. $\square$

Lemma 16.10 (Facchini and Nazemian 2016) *Every isonoetherian valuation domain is discrete.*

Proof Let A be an isonoetherian valuation domain. By the preceding lemma, the rank of A is ≤ 2. We may suppose that A is not a field. Let $(0) \subset P \subseteq M$ be the prime spectrum of A. By Lemma 16.8, it suffices to show that $P^2 \neq P$ and $M^2 \neq M$. Suppose that either $P^2 = P$ or $M^2 = M$. Then A is not Noetherian, so it does not satisfy the acc on the finitely generated (principal) ideals. There exists $(0) \neq a_1 A \subset a_2 A \subset \ldots$ a strictly increasing sequence of principal ideals of A. For each $n \geq 1$, let $b_n \in M$ be such that $a_n = a_{n+1}b_n$. Suppose that $M^2 = M$. Then M is not principal, and then $a_n M$ is not principal. So $a_1 M \subseteq a_2 M \subseteq \ldots$ is an increasing sequence of ideals of A that are not finitely generated. By Lemma 16.3, since A is isonoetherian, there exists $n \geq 1$ such that $a_n M = a_{n+1}M$. Then $a_{n+1}b_n M = a_{n+1}M$, which implies that $b_n M = M$. Since $b_n \in M$, then $1 \in M$, a contradiction. Then $M^2 \neq M$ so $P^2 = P$. Since $(PA_P)^2 = P^2 A_P = PA_P$, the valuation domain A_P is not Noetherian. Let $(r_n)_{n \geq 1}$ be a sequence of nonzero elements of $PA_P = P$ such that $r_n A_P \subset r_{n+1}A_P$ for each $n \geq 1$. Then $r_n P \subseteq r_{n+1}P$. Since $P^2 = P$, P is not principal, and then $r_n P$ is not principal. By Lemma 16.3, there exists $k \geq 1$ such that $r_k P = r_{k+1}P$. The strict inclusion $r_n A_P \subset r_{n+1}A_P$ implies that there exists $s \in A \setminus P$ and $c \in P$ such that $sr_k = r_{k+1}c$. Then $sr_{k+1}P = sr_k P = r_{k+1}cP$, so $sP = cP \subseteq cA \subseteq P = P^2 \subseteq sP$, and then $P = cA$ is principal, a contradiction. So A is discrete. $\square$

Theorem 16.11 (Facchini and Nazemian 2016) *A valuation domain is isonoetherian if and only if it is discrete of rank ≤ 2.*

Proof " $\Longrightarrow$ " By the preceding lemma, an isonoetherian valuation domain is discrete, and by Lemma 16.9, its rank ≤ 2.

" $\Longleftarrow$ " Let A be a discrete valuation domain of rank ≤ 2. If the rank of A is ≤ 1, then A is Noetherian, so isonoetherian. Suppose that the rank of A is 2. Let $(0) \subset P \subset M$ be the prime spectrum of A. The valuation domains A_P and A/P are discrete of rank 1 and then Noetherian. By Lemma 16.3, it suffices to show that each ideal of A that is not finitely generated is an ideal of A_P. Let I be such an ideal of A. Suppose that $P \subset I$, then I/P is a nonzero ideal of A/P, so principal. There exists $a \in I \setminus P$ such that $I/P = \overline{a}(A/P)$ so $I = aA + P = aA$, a contradiction. Then $I \subseteq P$. By Lemma 16.8, since A is discrete, $M^2 \neq M$. By Lemma 8.1, M is principal since it is the maximal ideal of a valuation domain. Put $M = aA$ with $a \in M$. Let $x \in I$. By hypothesis, $I \neq xA$. Let $z \in I \setminus xA$. Then $xA \subset zA \subset I$. So $x = zy$ with $y \in M$. Then $I = IM = Ia$. By induction, $I = Ia^n$ for each $n \geq 1$. By Lemma 16.5, the ideal $\bigcap_{n \geq 1} a^n A$ of A is prime with $(0) \neq I \subseteq \bigcap_{n \geq 1} a^n A \subset M$. Then $\bigcap_{n \geq 1} a^n A = P$. Let $s \in A \setminus P$. There exists an integer $n_s \geq 1$ such that $s \notin a^{n_s} A$, so $a^{n_s} A \subset sA$, and since $a^{n_s} I = I$, then $I = sI$. So $s^{-1} I = I$ for each $s \in A \setminus P$. Then $I = I A_P$ is an ideal of the domain A_P. $\qquad\qquad\square$

17 Isonoetherian Modules

We prove the analogue of Lemma 16.2 for the modules.

Lemma 17.1 *Let A be a ring, M a A-module and $x, y \in M \setminus (0)$. If $ann_A(x) = ann_A(y)$, then $Ax \simeq_A Ay$.*

Proof Let the map $\varphi : Ax \longrightarrow Ay$
$$z = ax \longmapsto \varphi(z) = \varphi(ax) = ay.$$

Let $a, a' \in A$. Then $ax = a'x \Longleftrightarrow (a - a')x = 0 \Longleftrightarrow a - a' \in ann_A(x) = ann_A(y) \Longleftrightarrow ay = a'y$. So φ is well defined and one to one. It is also clear that φ is a homomorphism onto of A-modules. Then φ is an isomorphism. $\qquad\square$

Proposition 17.2 (Facchini and Nazemian 2016) *Let A be a rank one discrete valuation domain with quotient field K.*

1. *The proper sub-modules of the A-module K are cyclic.*
2. *The A-module K is isonoetherian.*

Proof

(1) Let $M = aA$ be the unique maximal ideal of A. For each $0 \neq x \in K$, there exists a unique element $u_x \in U(A)$ and a unique integer $n_x \in \mathbb{Z}$ such that $x = u_x a^{n_x}$. Let N be a nonzero proper sub-A-module of K. The set $E = \{n_x; \, 0 \neq x \in N\}$ is a nonempty subset of $\mathbb{Z}$. Suppose that E is not lower bounded. Let $k \in \mathbb{Z}$. There exists $0 \neq x \in N$ such that $n_x < k$, $x = ua^{n_x}$ with $u \in U(V)$. Then $a^k = a^{n_x} a^{k - n_x} = xu^{-1} a^{k - n_x} \in N$. For each $k \in \mathbb{Z}$ and $u \in U(V)$, $ua^k \in N$. Then $N = K$, which is absurd. Then E is necessarily lower bounded. Let n be its smallest element and $0 \neq b \in N$ be such that $n = n_b \in E$. Then $b^n \in N$ so $Ab^n \subseteq N$. For each $0 \neq x \in N$, $x = ua^{n_x}$ with

$u \in U(A)$ and $n_x \geq n$. Then $x = ua^n a^{n_x - n} \in Aa^n$. So $N = Aa^n$ is a cyclic sub-module.

(2) Let $(0) \neq N_1 \subseteq N_2 \subseteq \ldots$ be an increasing sequence of proper sub-modules of the A-module K. By (1), all the N_i are cyclic. By the preceding lemma, $N_i \simeq_A N_1$ for each $i \geq 1$. Then the A-module K is isonoetherian. $\square$

Proposition 17.3 (Khalifa 2018) *Let A be a ring, M an isonoetherian A-module, and $z \in M$ be such that $ann_A(z) = (0)$. Then the ring A is isonoetherian.*

Proof Let $I_1 \subseteq I_2 \subseteq \ldots$ be an increasing sequence of ideals of A. Then $I_1 z \subseteq I_2 z \subseteq \ldots$ is an increasing sequence of sub-A-modules of M. By hypothesis, there exists $n \in \mathbb{N}^*$ such that for each integer $i \geq n$, $I_i z \simeq_A I_n z$. Fix an integer $i \geq n$ and an isomorphism $\varphi : I_i z \longrightarrow I_n z$ of A-modules. We will construct an isomorphism $\psi : I_i \longrightarrow I_n$ of A-modules. If $r \in I_i$, then $\varphi(rz) \in I_n z$, so $\varphi(rz) = \lambda_r z$ with $\lambda_r \in I_n$. Put $\psi(r) = \lambda_r$. The map ψ is well defined: if $a, b \in I_n$ are such that $\varphi(rz) = az$ and $\varphi(rz) = bz$, then $az = bz$. So $a - b \in ann_A(z) = (0)$, and then $a = b$. We will show that ψ is linear. Let $r, a \in I_i$, $\psi(r + a)z = \varphi((r + a)z) = \varphi(rz) + \varphi(az) = \psi(r)z + \psi(a)z = (\psi(r) + \psi(a))z$. Then $\psi(r+a) - \psi(r) - \psi(a) \in ann_A(z) = (0)$. So $\psi(r + a) = \psi(r) + \psi(a)$. Let $b \in A$, $\psi(rb)z = \varphi(rbz) = b\varphi(rz) = b\psi(r)z$. Then $\psi(rb) - b\psi(r) \in ann_A(z) = (0)$. So $\psi(rb) = b\psi(r)$. Then ψ is A-linear.

Injectivity of ψ: Let $r \in I_i$ be such that $\psi(r) = 0$. Then $\varphi(rz) = \psi(r)z = 0$. So $rz = 0$, and then $r \in ann_A(z) = (0)$. So $r = 0$, then ψ is one to one.

Surjectivity of ψ: Let $y \in I_n$. Then $yz \in I_n z$. There exists $r \in I_i$ such that $yz = \varphi(rz) = \psi(r)z$. Then $\psi(r) - y \in ann_A(z) = (0)$ and $y = \psi(r)$. So ψ is an isomorphism. $\square$

18 u-Isonoetherian Extensions of Rings

Definition 18.1 (Khalifa 2018) Let $A \subseteq B$ be an extension of rings. We say that $A \subseteq B$ is u-isonoetherian if for each increasing sequence $W_1 \subseteq W_2 \subseteq \ldots$ of sub-A-modules of B, there exists $n \in \mathbb{N}^*$ such that for each integer $i \geq n$, there exists $u_i \in U(B)$ satisfying $W_i = u_i W_n$.

Examples Let $A \subseteq B$ be an extension of rings.

(1) The extension $A \subseteq A$ is u-isonoetherian if and only if A is Noetherian.
(2) If the A-module B is Noetherian, the extension $A \subseteq B$ is u-isonoetherian.
(3) If the extension $A \subseteq B$ is u-isonoetherian, the A-module B is isonoetherian.

Indeed, if $W \subseteq \Omega$ are two sub-A-modules of B and $u \in U(S)$ is such that $\Omega = uW$, then $\Omega \simeq_A W$. In fact, the map $W \longrightarrow \Omega$, which associates with each $x \in W$, the element ux, is an isomorphism of A-modules.

Proposition 18.2 (Khalifa 2018) *Let A be a rank one discrete valuation domain with quotient field K. The extension $A \subset K$ is u-isonoetherian.*

Proof Let $(0) \neq N_1 \subseteq N_2 \subseteq \ldots$ be an increasing sequence of sub-A-modules of K. If there exists $i \in \mathbb{N}^*$ such that $N_i = K$, for each integer $j \geq i$, $N_j = K = 1.N_i$ with $1 \in U(K)$. If not, by Proposition 17.2, for each $i \in \mathbb{N}^*$, $N_i = A\alpha_i$ is cyclic. Since $N_1 = \alpha_1 A \subseteq N_i = \alpha_i A$, then $\alpha_1 = \beta_i \alpha_i$ with $0 \neq \beta_i \in A$. Then $N_1 = A\alpha_1 = A\beta_i\alpha_i = \beta_i N_i$ and $\beta_i \in U(K)$ for each $i \in \mathbb{N}^*$.   □

Example For each prime number p, $\mathbb{Z}_{(p)}$ is a rank one discrete valuation domain with quotient field $\mathbb{Q}$. Then the extension $\mathbb{Z}_{(p)} \subset \mathbb{Q}$ is u-isonoetherian.

Lemma 18.3 (Khalifa 2021) *Let A be an integral domain with quotient field K and H a A-sub-module of K. For each $M \in Max(A)$, the set denoted $H_M = \{\frac{x}{s}; x \in H, s \in A \setminus M\}$ is a sub-A_M-module of K containing H and $H = \bigcap\{H_M; M \in Max(A)\}$.*

Proof It is clear that for each $M \in Max(A)$, H_M is a sub-A_M-module of K. For each $x \in H$, $x = \frac{x}{1} \in H_M$, then $H \subseteq H_M$. Conversely, let x be an element in the intersection. For each $M \in Max(A)$, there exists $s_M \in A \setminus M$ such that $s_M x \in H$. Let J be the ideal of A generated by the family $\{s_M, M \in Max(A)\}$. Then $Jx \subseteq H$. Suppose that $J \neq A$. There exists $M \in Max(A)$ such that $J \subseteq M$. Then $s_M \in M$, which is absurd. Then $J = A$ and $x = 1.x \in Jx \subseteq H$. So we have the equality.   □

Notations

(1) For each integral domain A with quotient field K, $a_1, \ldots, a_t, b_1, \ldots, b_l \in A - (0)$ and $m_1, \ldots, m_t \in \mathbb{Z} - 0$, we denote:

$$A \prod_j a_j^{m_j} \prod_i b_i^{\mathbb{Z}} := \{aa_1^{n_1} \ldots a_t^{n_t} b_1^{k_1} \ldots b_l^{k_l}, a \in A, n_j \geq m_j, k_i \in \mathbb{Z}\},$$

which is clearly a sub-module of the A-module K
and

$$Ab_1^{\mathbb{Z}} \ldots b_l^{\mathbb{Z}} := \{ab_1^{k_1} \ldots b_l^{k_l}, a \in A, k_i \in \mathbb{Z}\}$$

(2) The symbol $\cong_A$ means an isomorphism of A-modules.

Lemma 18.4 (Khalifa 2021) *Let A be a semi-local PID with quotient field K, $M_1, \ldots, M_s$ the maximal ideals of A and a_j be a generator of each M_j. For each nonzero proper sub-module W of the A-module K, there exists a unique nonempty subset H of $\{1, \ldots, s\}$ and integers $(m_j)_{j \in H}$ such that*

$$W = A \prod_{j \in H} a_j^{m_j} \prod_{j \notin H} a_j^{\mathbb{Z}}.$$

Proof For each $1 \leq j \leq s$, denote $V_j = A_{M_j}$ which is a rank one discrete valuation domain with quotient field K. So W_{M_j} is a sub-module of the V_j-module K and $W = \cap_j W_{M_j}$ by Lemma 18.3. Denote $H = \{1 \leq j \leq s \text{ such that } W_{M_j} \neq K\}$. Since $W = \cap_j W_{M_j}$ and $W \neq K$, H is nonempty. For each $j \in H$, there exists a unique integer m_j such that $W_{M_j} = V_j a_j^{m_j}$ (see the proof of Proposition 17.2). Denote $\Omega = A \prod_{j \in H} a_j^{m_j} \prod_{j \notin H} a_j^{\mathbb{Z}}$. It is clear that Ω is a sub-module of the A-module K and $\Omega_{M_j} = K = W_{M_j}$ for each $j \notin H$. Moreover, $\Omega_{M_j} = A_{M_j} a_j^{m_j} = V_j a_j^{m_j} = W_{M_j}$ for each $j \in H$. Hence, $W = \cap_j W_{M_j} = \cap_j \Omega_{M_j} = \Omega$. $\qquad \square$

Theorem 18.5 (Khalifa 2021) *Let A be an integral domain with quotient field K. The following statements are equivalent:*

1. *The A-module K is isonoetherian.*
2. *The ring extension $A \subseteq K$ is u-isonoetherian.*
3. *The ring A is a semi-local PID.*

Proof $"(1) \implies (2)"$ Let $W_1 \subseteq W_2 \subseteq \ldots$ be an ascending chain of nonzero proper sub-modules of the A-module K. Since $W_1 \neq 0$, let $0 \neq z \in W_1$. So $A \subseteq W_1 z^{-1} \subseteq W_2 z^{-1} \subseteq \ldots$ is an ascending chain of sub-modules of the A-module K. So there exists an index n such that for all $i > n$, $W_i z^{-1} \cong_A W_n z^{-1}$. Let so $\varphi_i \cdot W_i z^{-1} \longrightarrow W_n z^{-1}$ be an isomorphism of A-modules, $0 \neq x \in W_i z^{-1}$, and $a, b \in A$ such that $x = a/b$. Thus, $b\varphi_i(x) = \varphi_i(bx) = \varphi_i(a) = au_i$ where $u_i = \varphi_i(1) \in K - 0$ and so $\varphi_i(x) = u_i x$. Then $W_i z^{-1} = u_i W_n z^{-1}$ and hence $W_i = u_i W_n$.

$"(2) \implies (3)"$ Let I be a nonzero ideal of A and $0 \neq a \in I$. Then $aA \subseteq I \subseteq A \subseteq Ia^{-1} \subseteq Aa^{-1} \subseteq Ia^{-2} \subseteq Aa^{-2} \subseteq \ldots$ is an ascending chain of sub-modules of the A-module K. Thus, $Ia^{-n} = uAa^{-n}$ for some positive integer n and some $0 \neq u \in K$. So $I = uA$ is a principal ideal of A. Then A is a PID. Assume that A has infinitely many maximal ideals $M_1, M_2, \ldots$, and let a_i be a generator of M_i. Clearly, the set denoted $Aa_1^{\mathbb{Z}} \ldots a_n^{\mathbb{Z}} := \{aa_1^{m_1} \ldots a_n^{m_n}, a \in A \text{ and } m_i \in \mathbb{Z}\}$ is a sub-module of the A-module K. Thus, $Aa_1^{\mathbb{Z}} \subseteq Aa_1^{\mathbb{Z}} a_2^{\mathbb{Z}} \subseteq \ldots$ is an ascending chain of sub-modules of the A-module K. So there exists an index n and $0 \neq u \in K$ such that $Aa_1^{\mathbb{Z}} \ldots a_n^{\mathbb{Z}} = uAa_1^{\mathbb{Z}} \ldots a_n^{\mathbb{Z}} a_{n+1}^{\mathbb{Z}}$. Then $cAa_1^{\mathbb{Z}} \ldots a_n^{\mathbb{Z}} = dAa_1^{\mathbb{Z}} \ldots a_n^{\mathbb{Z}} a_{n+1}^{\mathbb{Z}}$ for some $c, d \in A - 0$. For each positive integer k, there exist $d_k \in A$ and $p_{k,1}, \ldots, p_{k,n} \in \mathbb{Z}$ such that $da_{n+1}^{-k} = cd_k a_1^{p_{k,1}} \ldots a_n^{p_{k,n}}$. So $d \in a_{n+1}^k A$ (because for each nonempty subset Δ of $\{1, .., n\}$ and each $\{q_i\}_{i \in \Delta}$ positive integers, $A = Aa_{n+1}^k + A \prod_{i \in \Delta} a_i^{q_i}$), and thus $d \in \cap_k a_{n+1}^k A = 0$, a contradiction. Then A is semi-local.

$"(3) \implies (1)"$ Let $\{M_1, \ldots, M_s\}$ be the set of maximal ideals of D and $W_1 \subseteq W_2 \subseteq \ldots$ be an ascending chain of nonzero proper sub-modules of the A-module K. By Lemma 18.4, for each $i \geq 1$:

$$W_i = A \prod_{j \in H_i} a_j^{m_{i,j}} \prod_{j \notin H_i} a_j^{\mathbb{Z}}$$

for some integers $(m_{i,j})_{j \in H_i}$ where $H_i = \{1 \leq j \leq s | (W_i)_{M_j} \neq K\}$ (nonempty set), and each a_j is a generator of M_j. It is clear that $\ldots \subseteq H_{i+1} \subseteq H_i \subseteq \ldots \subseteq H_1 \subseteq \{1, \ldots, s\}$ is a descending chain of subsets of $\{1, \ldots, s\}$. Thus, there exists an index n such that $H_i = H_n$ for all $i > n$. Let $d_i = \prod_{j \in H_n} a_j^{m_{i,j}-m_{i+1,j}} \in K - 0$. Hence

$$W_{i+1} \cong_A d_i W_{i+1} = \prod_{j \in H_n} a_j^{m_{i,j}-m_{i+1,j}} A \prod_{j \in H_n} a_j^{m_{i+1,j}} \prod_{j \notin H_n} a_j^{\mathbb{Z}}$$

$$= A \prod_{j \in H_n} a_j^{m_{i,j}} \prod_{j \notin H_n} a_j^{\mathbb{Z}} = W_i.$$

$\square$

Remarks 18.6 (Khalifa 2021)

(1) In the proof of "(3) $\implies$ (1)" of Theorem 18.5, we can show that $d_i \in A$. Indeed, let $j \in H_n$ and $i \geq n$. We have $(W_i)_{M_j} = a_j^{m_{i,j}} V_j \subseteq (W_{i+1})_{M_j} = a_j^{m_{i+1,j}} V_j$, and so $a_j^{m_{i,j}-m_{i+1,j}} \in V_j = A_{M_j}$. Thus, $s_j a_j^{m_{i,j}-m_{i+1,j}} \in A$ for some $s_j \in A - M_j$. Then $m_{i,j} - m_{i+1,j} \geq 0$ because $s_j \notin M_j$. It follows that $a_j^{m_{i,j}-m_{i+1,j}} \in A$ and hence $d_i \in A$.

(2) We deduce that if A is an integral domain with quotient field K, then the ring extension $A \subseteq K$ is u-isonoetherian if and only if for each ascending chain $W_1 \subseteq W_2 \subseteq \ldots$ of sub-modules of the A-module K, there exists an index n such that for all $i \geq n$, $W_i = u_i W_{i+1}$ for some nonzero element u_i of A.

19   Isonoetherian Rings of the Form $A + XB[X]/A + XB[[X]]$

Facchini and Nazemian (2016), proved that a valuation domain is isonoetherian if and only if it is discrete with Krull dimension ≤ 2 (Theorem 16.11). The local case of discrete valuation domain corresponds in the global case to the so-called generalized Dedekind domain. Unfortunately, they proved that the two-dimensional generalized Dedekind domain $\mathbb{Z} + X\mathbb{Q}[[X]]$ is not isonoetherian.

Example (Facchini and Nazemian 2016) The ring $\mathbb{Z}+X\mathbb{Q}[[X]]$ is not isonoetherian. Indeed, assume that it is isonoetherian. Let $\mathbb{P} = \{p_1, p_2, \ldots\}$ be the set of prime numbers. For each positive integer n, let $S_n = < \{p_i, i \leq n\} >$ be the multiplicative subset of $\mathbb{Z}$ generated by $(p_i)_{i \leq n}$, $\mathbb{Z}_{S_n}$ the localization of $\mathbb{Z}$ by the multiplicative subset S_n of $\mathbb{Z}$ and $I_n = X\mathbb{Z}_{S_n} + X^2\mathbb{Q}[[X]]$. Since $S_1 \subset S_2 \subset \ldots$, $I_1 \subseteq I_2 \subseteq \ldots$ is an ascending chain of ideals of the ring $R := \mathbb{Z}+X\mathbb{Q}[[X]]$. So there exists an index n such that $I_n \cong_R I_{n+1}$. Let $\varphi : I_{n+1} \longrightarrow I_n$ be an isomorphism of R-modules and $c_k = p_{n+1}^k$ for each positive integer k. Let $a \in \mathbb{Z}, b \in S_n$, and $f \in \mathbb{Q}[[X]]$

and $\varphi(X) = X\frac{a}{b} + X^2 f$ (if $a \neq 0$, we can assume that a and b are coprime, i.e., $\gcd(a, b) = 1$). We have $\varphi(\frac{1}{c_k}X^2) = \frac{1}{c_k}X\varphi(X)$ and $\varphi(\frac{1}{c_k}X^2) = X\varphi(\frac{1}{c_k}X)$. Thus, $\varphi(\frac{1}{c_k}X) = \frac{1}{c_k}\varphi(X) = X\frac{a}{c_k b} + X^2\frac{1}{c_k}f$ and so $\frac{a}{c_k b} \in \mathbb{Z}_{S_n}$. Then c_k divides da in $\mathbb{Z}$ for some $d \in S_n$. Since c_k is coprime with d, c_k divides a in $\mathbb{Z}$. It follows that $a \in \cap_k c_k \mathbb{Z} = \cap_k p_{n+1}^k \mathbb{Z} = 0$. Thus, $a = 0$. So $\varphi(X) = X^2 f$. For all $\alpha \in \mathbb{Z}_{S_{n+1}}$ and $\xi \in \mathbb{Q}[[X]]$:

$$X\varphi(X\alpha + X^2\xi) = \varphi(X^2\alpha + X^3\xi) = (X\alpha + X^2\xi)\varphi(X) = (X\alpha + X^2\xi)X^2 f$$

because $X\alpha + X^2\xi \in X\mathbb{Q}[[X]] \subseteq R$.

Then $\varphi(g) = gXf$ for all $g \in I_{n+1}$. It follows that $I_n = I_{n+1}Xf$ and so $X\mathbb{Z}_{S_n} + X^2\mathbb{Q}[[X]] \subseteq Xf(X\mathbb{Z}_{S_{n+1}} + X^2\mathbb{Q}[[X]]) \subseteq X^2\mathbb{Q}[[X]]$. Hence, $\mathbb{Z}_{S_n} = 0$, a contradiction.

Motivated by Facchini and Nazemian's work, Khalifa studied the "isonoetherian" and the "ACC_d on ideals" properties in rings of the form $A + XB[X] / A + XB[[X]]$. A new algebraic concept appears in 2017 that is related (in some way) to the isonoetherian concept. Following Dastanpour and Ghorbani (2018), we say that a ring A (not necessarily commutative) satisfies divisibility on ascending chain of ideals (for short, satisfies ACC_d on ideals) if, for every ascending chain $I_1 \subseteq I_2 \subseteq \ldots$ of ideals of A, there exists an index n such that for all $i \geq n$, $I_i = u_i I_{i+1}$ for some $a_i \in A$.

Proposition 19.2 *An integral domain that satisfies ACC_d on ideals is isonoetherian.*

Proof Let A be an integral domain satisfying ACC_d on ideals and $I_1 \subseteq I_2 \subseteq \ldots$ be an ascending chain of nonzero ideals of A. By hypothesis, there exists an index n such that for all $i \geq n$, $I_i = a_i I_{i+1}$ for some $a_i \in A$. Since $I_i \neq 0$, $a_i \neq 0$. Clearly, the map $\varphi_i : I_{i+1} \longrightarrow I_i$ defined by $\varphi_i(x) = a_i x$ is an isomorphism of A-modules. Then $I_i \cong_A I_{i+1}$ for all $i \geq n$. Hence, A is an isonoetherian ring. $\square$

Bastida and Gilmer (1973), give a description of ideals of the ring $D + M$, where V is a valuation domain of the form $K + M$ (M is the maximal ideal of V and K a field) and D is a proper sub-ring of K. Inspired from Bastia and Gilmer's work, Khalifa (2021) finds the same description for the more general following situation:

Lemma 19.3 (Khalifa 2021) *Let S be a PID of the form $K + M$ where M is a nonzero maximal ideal of S and K is a field. Let D be an integral domain with quotient field K and $R = D + M$. Then every ideal I of R has the form $I = Wa + Ma$ for any generator a of the ideal IS.*

Proof Let $a \in S$ such that $IS = aS$ and $W = \{u \in K \text{ such that } ua \in I\}$. Clearly, W is a sub-module of the D-module K and $Wa \subseteq I$. Since M is a common ideal of R and S, $Ma \subseteq (IS)M = IM \subseteq I$, and so $Wa + Ma \subseteq I$. Conversely, let $z \in I$. Since $z \in IS = aS$, $z = a(u + y)$ for some $u \in K$ and $y \in M$. Thus, $ua = z - ay \in I + Ma \subseteq I$ and so $u \in W$. Then $z = ua + ya \in Wa + Ma$. Hence, $I = Wa + Ma$. $\square$

Lemma 19.4 (Khalifa 2021) *Let S be a PID of the form $K + M$ where M is a nonzero maximal ideal of S and K is a field. Let D be an integral domain with quotient field K and $R = D + M$. Let α be a generator of M in S and $W \subseteq \Omega$ be two nonzero sub-modules of the D-module K. If $W\alpha + M\alpha \cong_R \Omega\alpha + M\alpha$, then $Wu = \Omega$ for some $u \in K$.*

Proof Let $\varphi : W\alpha + M\alpha \longrightarrow \Omega\alpha + M\alpha$ be an isomorphism of R-modules. For any $z \in W\alpha + M\alpha$, $\alpha^2 \varphi(z) = \varphi(\alpha^2 z) = z\varphi(\alpha^2)$, and so $\varphi(z) = z\frac{\varphi(\alpha^2)}{\alpha^2}$. Let $v \in \Omega$ and $c \in S$ such that $\varphi(\alpha^2) = v\alpha + c\alpha^2$. For all $x \in W$, $\varphi(x\alpha) = x\alpha\frac{\varphi(\alpha^2)}{\alpha^2} = xv + xc\alpha \in \Omega\alpha + M\alpha \subseteq M$, and so $xv \in K \cap M = 0$. Thus, $v = 0$ and $\varphi(z) = zc$ for all $z \in W\alpha + M\alpha$. Then $\Omega\alpha + M\alpha = \varphi(W\alpha + M\alpha) = (W\alpha + M\alpha)c$. It follows that $\Omega + M = (W + M)c$. Let $u \in K$ and $d \in S$ such that $c = u + d\alpha$. Thus, $\Omega + M \subseteq Wu + Wd\alpha + Mc \subseteq Wu + M$ and so $\Omega \subseteq Wu$ because $K \cap M = 0$. Conversely, if $y \in W$, then $yu + yd\alpha = yc \in \Omega + M$, and so $yu \in \Omega$ again because $K \cap M = 0$. Then $Wu \subseteq \Omega$ and hence $\Omega = Wu$. $\square$

Domains of the form $D + M$ were studied by several authors because they are a big source of construction of examples and counterexamples. M. Khalifa studied the isonoetherian and "ACC_d on ideals" properties in rings of construction $D + M$.

Theorem 19.5 (Khalifa 2021) *Let S be a PID of the form $K + M$ where M is a nonzero maximal ideal of S and K is a field. Let D be an integral domain with quotient field K and $R = D + M$. The following statements are equivalent:*

1. *The ring R is isonoetherian.*
2. *The ring R satisfies ACC_d on ideals.*
3. *$D \subseteq K$ is u-isonoetherian.*

Proof $"(3) \Rightarrow (2)"$ Let $I_1 \subseteq I_2 \subseteq \ldots$ be an ascending chain of ideals of R. Since the ring S is Noetherian, there exists an index n such that $I_i S = I_n S$ for all $i \geq n$. Let a be a generator of the ideal $I_n S$ (and so of the ideal $I_i S$ when $i > n$) of S. By Lemma 19.3, $I_i = W_i a + Ma$ for each $i \geq n$ where $W_i = \{u \in K$ such that $ua \in I_i\}$. Since $W_1 \subseteq W_2 \subseteq \ldots$ is an ascending chain of sub-modules of the D-module K, there exists an index $m \geq n$ such that for all $i \geq m$, $W_i = d_i W_{i+1}$ for some $0 \neq d_i \in D$ (so $d_i \in R$) by Remark 18.6-(2). Note that $K - 0 \subset U(S)$ and so $M = uM$ for all $0 \neq u \in K$. Then $I_i = W_i a + Ma = d_i W_{i+1} a + d_i Ma = d_i(W_i a + Ma) = d_i I_{i+1}$ for all $i \geq m$.

$"(2) \Rightarrow (1)"$ By Proposition 19.2.

$"(1) \Rightarrow (3)"$ Let α be a generator of M in S. If W is a sub-module of the D-module K, then $W\alpha + M\alpha$ is an ideal of the ring R. Let $0 \neq W_1 \subseteq W_2 \subseteq \ldots$ be an ascending chain of sub-modules of the D-module K. Then $W_1\alpha + M\alpha \subseteq W_2\alpha + M\alpha \subseteq \ldots$ is an ascending chain of ideals of R. So there exists a positive integer n such that $W_i\alpha + M\alpha \cong_R W_n\alpha + M\alpha$ for all $i \geq n$. By Lemma 19.4, $W_i = u_i W_n$ for some $0 \neq u_i \in K$. $\square$

Immediately, we deduce and summarize.

Corollary 19.6 (Khalifa 2021) *Let A be an integral domain with quotient field $K \neq A$. The following statements are equivalent:*

1. *The ring $A + XK[X]$ is isonoetherian.*
2. *The ring $A + XK[X]$ satisfies ACC_d on ideals.*
3. *The ring $A + XK[[X]]$ is isonoetherian.*
4. *The ring $A + XK[[X]]$ satisfies ACC_d on ideals.*
5. *$A \subset K$ is u-isonoetherian.*
6. *A is a semi-local principal ideal domain.*
7. *The A-module K is isonoetherian.*

Proof For the equivalences "(1) $\Leftrightarrow$ (2) $\Leftrightarrow$ (5)," apply Theorem 19.5 for the PID $S = K[X]$, $M = XK[X]$, and $D = A$ (so $R = D + M = A + M = A + XK[X]$.

For the equivalences "(3) $\Leftrightarrow$ (4) $\Leftrightarrow$ (5)", apply Theorem 19.5 for the PID $S = K[[X]]$, $M = XK[[X]]$, and $D = A$ (so $R = D + M = A + M = A + XK[[X]]$.

By Theorem 18.5, "(5) $\Leftrightarrow$ (6) $\Leftrightarrow$ (7)". □

For each ring A and each nonempty subset F of A, the annihilator of F in A is $ann_A(F) = \{a \in A, aF = 0\}$. Note for all nonempty $F \subseteq A$, $ann_A(F)$ is an ideal of A and $ann_A(F) = ann_A(< F >)$ where $< F >$ is the ideal of A generated by F.

Lemma 19.7 *Let F and H two nonempty subsets of a ring A and I and J two ideals of A:*

1. *If $F \subseteq H$, then $ann_A(H) \subseteq ann_A(F)$.*
2. *If $I \cong_A J$, then $ann_A(I) = ann_A(J)$.*
3. *$I \subseteq ann_A(ann_A(I))$.*
4. *$ann_A(I) = ann_A(ann_A(ann_A(I)))$.*

Proof

(1) Let $a \in ann_A(H)$. Thus, $aH = 0$. Since $F \subseteq H$, $aF \subseteq aH = 0$ and so $aF = 0$. Then $a \in ann_A(F)$.

(2) Let $\varphi : I \to J$ be an isomorphism of A-modules. The result follows from the fact that $aJ = a\varphi(I) = \varphi(aI)$ for all $a \in A$.

(3) Since $I.ann_A(I) = 0$, then $I \subseteq ann_A(ann_A(I))$.

(4) By (3), $ann_A(I) \subseteq ann_A(ann_A(ann_A(I)))$. Since $I \subseteq ann_A(ann_A(I))$, then we have $ann_A(ann_A(ann_A(I))) \subseteq ann_A(I)$ by (1). □

We say a ring A satisfies DCC (i.e., descending chain conditions) on annihilators if for each descending chain $I_1 \supseteq I_2 \supseteq \ldots$ of ideals that are annihilators of subsets of A there an index n such that $I_i = I_n$ for all $i \geq n$.

Lemma 19.8 (Facchini and Nazemian 2016) *If A is an isonoetherian ring, then A satisfies DCC on annihilators.*

Proof Let $I_1 \supseteq I_2 \supseteq \ldots$ be a descending chain of ideals that are annihilators of subsets of A. By Lemma 19.7-(4), $I_i = ann_A(ann_A(I_i))$ for all $i \geq 1$. Since

$ann_A(I_1) \subseteq ann_A(I_2) \subseteq \ldots$ is an ascending chain of ideals of A, there exists an index n such that for all $i > n$, $ann_A(I_i) \cong_A ann_A(I_n)$. By Lemma 19.7-(2), we have $ann_A(ann_A(I_i)) = ann_A(ann_A(I_n))$ for all $i > n$. Hence, $I_i = I_n$ for all $i > n$.   □

Let $A \subseteq B$ an extension of rings. Note that if W is a sub-module of the A-module B, then $XW + X^2B[[X]]$ is an ideal of the ring $A + XB[[X]]$. To study the isonoetherian property in the ring $A + XB[[X]]$, we will test chain of ideals of the form $XW + X^2B[[X]]$.

Lemma 19.9 (Khalifa 2018 and Khalifa 2021) *If* $W \subseteq \Omega$ *are two nonzero sub-modules of the* A*-module* B *such that* $ann_B W = ann_B \Omega$ *and* $XW + X^2B[[X]] \cong_{A+XB[[X]]} X\Omega + X^2B[[X]]$*, then* $W = u\Omega$ *for some unit* u *of* B*.*

Proof Let $\varphi : XW + X^2B[[X]] \longrightarrow X\Omega + X^2B[[X]]$ be an isomorphism of modules over $A + XB[[X]]$. Set:

$$\varphi(X^2) = tX + X^2\rho \quad and \quad \varphi^{-1}(X^2) = wX + X^2\theta$$

where $t \in \Omega$, $w \in W$, and $\rho, \theta \in B[[X]]$. For all $z \in W$:

$$X^2\varphi(zX) = \varphi(zX^3) = zX\varphi(X^2) = zX(tX + X^2\rho)$$

and so $\varphi(zX) = zt + zX\rho$. Thus, $zt = 0$ because $\varphi(zX) \in XB[[X]]$ and $zt \in B$. Then $Wt = \Omega t = 0$ and $\varphi(zX) = zX\rho$ for all $z \in W$. It follows that $W\rho_0 \subseteq \Omega$ where $\rho = \rho_0 + \rho_1 X + \ldots (\rho_i \in B)$. With a similar argument, $Ww = \Omega w = 0$ and $\varphi^{-1}(\alpha X) = \alpha X\theta$ for all $\alpha \in \Omega$. So $\Omega\theta_0 \subseteq W$ where $\theta = \theta_0 + \theta_1 X + \ldots (\theta_i \in B)$.

$$X^3 = X\varphi(\varphi^{-1}(X^2)) = X\varphi(wX + X^2\theta) = X\varphi(wX) + X\theta\varphi(X^2) = XwX\rho + X\theta(tX + X^2\rho).$$

So $X = w\rho + t\theta + X\theta\rho$. Thus, $1 = w\rho_1 + t\theta_1 + \rho_0\theta_0$. Since $t^2 = tw = w^2 = 0$, $t = t\rho_0\theta_0$ and $w = w\rho_0\theta_0$. Then $1 = \rho_0\theta_0(1 + t\theta_1 + w\rho_1)$ and so ρ_0, θ_0 are units of B. For all $z \in W$, $zX^2 = X\varphi^{-1}(\varphi(zX)) = X\varphi^{-1}(zX\rho) = \varphi^{-1}(zX^2\rho) = X\rho\varphi^{-1}(zX) = X\rho zX\theta$ and so $z = z\rho\theta$. Then $W = W\rho_0\theta_0 \subseteq \Omega\theta_0 \subseteq W$ and hence $W = \Omega\theta_0$.   □

Theorem 19.10 (Khalifa 2018) *If the ring* $A + XB[[X]]$ *is isonoetherian, then* $A \subseteq B$ *is u-isonoetherian.*

Proof Let $W_1 \subseteq W_2 \subseteq \ldots$ be an ascending chain of nonzero sub-modules of the A-module B. Since $XW_i[[X]] \subseteq XW_{i+1}[[X]]$, then we have $ann_{A+XB[[X]]}(XW_i[[X]]) \supseteq ann_{A+XB[[X]]}(XW_{i+1}[[X]])$ a descending chain of annihilators of $A + XB[[X]]$. By Lemma 19.8, $A + XB[[X]]$ satisfies DCC on annihilators, and so there exists an index n such that $ann_{A+XB[[X]]}(XW_n[[X]]) = ann_{A+XB[[X]]}(XW_i[[X]])$ for all $i \geq n$. Then $ann_B(W_n) = ann_B(W_i)$ for all $i > n$. Since $XW_1 + X^2B[[X]] \subseteq XW_2 + X^2B[[X]] \subseteq \ldots$ is an ascending chain of ideals of $A + XB[[X]]$, there exists an index $m \geq n$ such that

$XW_i + X^2 B[[X]] \cong_{A+XB[[X]]} XW_m + X^2 B[[X]]$ (isomorphism of $(A+XB[[X]])$-modules) for all $i \geq m$. By Lemma 19.9, $W_i = u_i W_m$ for some unit u_i of B. Hence, $A \subseteq B$ is u-isonoetherian. □

Theorem 19.11 (Khalifa 2021) *If the ring $A + XB[X]$ is isonoetherian, then $A \subseteq B$ is u-isonoetherian.*

Proof The same proofs of Lemma 19.9 and Theorem 19.10 work also for the ideals of the form $XW + X^2 B[X]$ of the ring $A + XB[X]$. □

Clearly, if A is a ring, then the ring extension $A \subseteq A$ is u-isonoetherian if and only if A is Noetherian. So the converse of Theorem 19.10 is true in case $A = B$.

Corollary 19.12 (Khalifa 2018 and Khalifa 2021) *The following statements are equivalent for a ring A:*

1. *The ring $A[[X]]$ is isonoetherian.*
2. *The ring $A[X]$ is isonoetherian.*
3. *The ring A is Noetherian.*

Corollary 19.13 (Khalifa 2021) *Let A be an integral domain and $n \geq 2$. The following statements are equivalent:*

1. *The ring $A[[X_1, \ldots, X_n]]$ is isonoetherian.*
2. *The ring $A[[X_1, \ldots, X_n]]$ satisfies ACC_d on ideals.*
3. *The ring $A[[X]]$ is isonoetherian.*
4. *The ring $A[[X]]$ satisfies ACC_d on ideals.*
5. *The ring $A[X_1, \ldots, X_n]$ is isonoetherian.*
6. *The ring $A[X_1, \ldots, X_n]$ satisfies ACC_d on ideals.*
7. *The ring $A[X]$ is isonoetherian.*
8. *The ring $A[X]$ satisfies ACC_d on ideals.*
9. *The ring A is Noetherian.*

Proof By Hilbert basis's theorem, $"(9) \Rightarrow (i)"$ for each $i = 1, \ldots, 8$ because a Noetherian domain satisfies ACC_d on ideals (so is isonoetherian).

$"(3) \Rightarrow (9)"$ By Corollary 19.12.

$"(4) \Rightarrow (9)"$ Let $I_1 \subseteq I_2 \subseteq \ldots$ be an ascending chain of nonzero ideals of A. Then $I_1 + XA[[X]] \subseteq I_2 + XA[[X]] \subseteq \ldots$ is an ascending chain of ideals of $A[[X]]$. So there exists an index n such that for all $i \geq n$, $I_i + XA[[X]] = f_i(I_{i+1} + XA[[X]])$ for some $f_i = a_i + b_i X + \ldots \in A[[X]]$ (where $a_i, b_i \in A$). Thus, $I_i = a_i I_{i+1}$ and $X = f_i(r_i + Xg_i)$ (for some $r_i \in I_{i+1}$ and $g_i \in A[[X]]$). It follows that $r_i a_i = 0$ and $1 = r_i b_i + a_i c_i$ (where c_i is the constant term of g_i). Then $a_i = a_i^2 c_i$ and so $a_i = 0$ or a_i is a unit of A. Assume that $a_i = 0$, then $f = X h_i$ and so $I_i + XA[[X]] = X h_i(I_{i+1} + XA[[X]]) \subseteq XA[[X]]$. Thus, $I_i = 0$, a contradiction. Then a_i is a unit of A and so f_i is a unit of $A[[X]]$. We have so $I_i + XA[[X]] = I_{i+1} + XA[[X]]$. Thus, $I_i = I_{i+1}$. The $I_i = I_n$ for all $i \geq n$.

$"(2) \Rightarrow (1)"$ By Proposition 19.2.

$"(1) \Rightarrow (9)"$ If $A[[X_1, \ldots, X_n]]$ is isonoetherian, then $A[[X_1, \ldots, X_{n-1}]]$ is Noetherian by the implication $"(3) \Rightarrow (9)"$ and so A is Noetherian.

Hence, $"(1) \Leftrightarrow (2) \Leftrightarrow (3) \Leftrightarrow (4) \Leftrightarrow (9)"$.

With a similar argument, we show that $"(5) \Leftrightarrow (6) \Leftrightarrow (7) \Leftrightarrow (8) \Leftrightarrow (9)"$.   □

When B is the quotient field of A, the converse of Theorem 19.10 is true by Corollary 19.6. In general, the converse of Theorem 19.10 is false. The example needs the following proposition:

Proposition 19.14 *Let S be a rank one valuation domain of the form $K + M$ where M is the nonzero maximal ideal of S and K is a field. Let D be an integral domain with quotient field K and $R = D + M$. If D is a k-rank valuation domain, then R is a $(k + 1)$-rank valuation domain.*

Proof Since M is a nonzero common ideal of R and S, R and S have the same quotient field F. Let $0 \neq x \in F$. Thus, x or $x^{-1} \in S$. Assume that $x \in S$. If $x \in M$, then $x \in R$. If $x \notin M$, then x is a unit of S and so $x^{-1} \in S$. Let $u, v \in K$ and $y, z \in M$ such that $x = u + y$ and $x^{-1} = v + z$. Since $1 = xx^{-1} = uv + uz + yx^{-1}$ and $K \cap M = 0$, $1 = uv$. Since D is a valuation domain with quotient field K, u or $v \in D$. Then x or $x^{-1} \in R$. With a similar argument, we prove that x or $x^{-1} \in R$ when $x^{-1} \in S$. Then R is a valuation domain. Since M is an ideal of R and $R/M \cong D$, M is a prime ideal of R. Since M is comparable to each ideal of R, $\dim R = \operatorname{ht}_R(M) + \dim(R/M) = \operatorname{ht}_R(M) + \dim(D)$. It suffices to show that $\operatorname{ht}_R(M) = 1$. It suffices to show that 0 is the only prime ideal of R strictly contained in M. Let P be a prime ideal of R strictly contained in M and $x \in M - P$. Thus, $x^n \in M - P$ for all positive integers n. So $P \subset x^n R$ for all n. Thus, $P \subseteq \cap_n x^n S = 0$ because a rank one discrete valuation domain is Archimedean. Hence, $P = 0$.   □

In the next, we give an example of u-isonoetherian ring extension $A \subset B$ such that the ring $A + XB[[X]]$ (respectively, $A + XB[X]$) is not isonoetherian.

Example (Khalifa 2021) Let F be a field, Y and Z are algebraically independent indeterminates over F, and $V = F + YF[[Y]] + ZF((Y))[[Z]]$. Let $K = F((Y))$ which is a field, $S = K[[Z]]$ which is a rank one discrete valuation domain (so a PID) with maximal ideal $M = ZK[[Z]]$. Let $D = F[[Y]]$. Since D is a rank one discrete valuation domain with quotient field K, the ring extension $D \subset K$ is u-isonoetherian by Proposition 18.2. Since $V = D + M$, the ring V satisfies ACC_d on ideals by Lemma 19.5. By Proposition 19.14, V is a rank 2 discrete valuation domain. Let $0 \subset P \subset N$ be the prime spectrum of V.

Claim 1:   Each sub-module of the V-module V_P is comparable to P.

Indeed, let W be a sub-module of the V-module V_P such that $W \nsubseteq P$. Let so $b \in W - P$. The unique maximal ideal of V_P is $PV_P = P$. Thus, b is a unit of V_P and for all $x \in P$, $x = xb^{-1}b \in PW \subseteq VW \subseteq W$. Then $P \subset W$.

Claim 2:   The ring extension $V \subset V_P$ is u-isonoetherian.

Indeed, let $0 \neq W_1 \subseteq W_2 \subseteq \ldots$ be an ascending chain of sub-modules of the V-module V_P.

Case 1:   Assume that $P \subset W_n$ for some index n. Then $W_n/P \subseteq W_{n+1}/P \subseteq \ldots$ is an ascending chain of sub-modules of the V/P-module $(V_P)/P$ (which is a field isomorphic to the quotient field of V/P). By Proposition 18.2, $V/P \subset (V_P)/P$ is u-isonoetherian. So there exists an index $m \geq n$ such that for all $i \geq m$, $W_i/P = \overline{u_i} W_m/P$ ($\overline{u_i}$ is the class of u_i modulo P) for some $u_i \in V - P$ by Remark 18.6-(2). If $y \in W_m$, then $u_i y - x \in P$ for some $x \in W_i$. Since $P \subset W_i$, $u_i y \in W_i$ and so $u_i W_m \subseteq W_i$. Conversely, if $x \in W_i$, then $x - u_i y \in P$ for some $y \in W_m$. Since $u_i W_m$ is a sub-module of the V-module V_P, $u_i W_m$ is comparable to P by Claim 1. Since $u_i \notin P$ and $W_m \nsubseteq P$, $P \subseteq u_i W_m$ and so $x \in u_i W_m$. Then $W_i = u_i W_m$ for all $i > m$. Since $u_i \in V - P$, u_i is a unit of V_P.

Case 2:   Assume that all $W_i \subseteq P$. Then all W_i are ideals of V. Since V satisfies ACC_d on ideals, there exists an index n such that $W_i = r_i W_{i+1}$ for all $i \geq n$ and for some $r_i \in V$. Assume that there exist integers $n \leq i_1 < i_2 \ldots$ such that r_{i_k} is not a unit of V_P (so $r_{i_k} \in PV_P = P$). Then $W_{i_1} = r_{i_1} W_{i_1+1} = r_{i_1} r_{i_1+1} \ldots r_{i_2} W_{i_2+1} = r_{i_1} r_{i_1+1} \ldots r_{i_k} W_{i_k+1} \subseteq P^k V_P = P^k$ for all $k \geq 1$ and so $W_{i_1} = 0$, a contradiction. Hence, there exists $m \geq n$ such that r_i is a unit of V_P for all $i \geq m$.

Claim 3:   The ring $V + XV_P[[X]]$ is not isonoetherian.

Indeed, let $a \in P - 0$ and $b \in N - P$. Thus, $aV \subseteq P \subset J \subseteq N$ where $J = bV$. So there exist $u_1, a_2, \ldots \subset P$ such that $a = a_1 b = a_2 b^2 = \ldots = a_n b^n = \ldots$. Set $I_i = a_i V$ for all $i \geq 1$. We have a strictly ascending chain $I_1 \subset I_2 \subset \ldots$ of ideals of V. Since $I_i V_P \subseteq P \subset J$, $I_1 + XJ + X^2 V_P[[X]] \subset I_2 + XJ + X^2 V_P[[X]] \subset \ldots$ is an ascending chain of ideals of $V + XV_P[[X]]$. Suppose that there exists an index n such that $I_n + XJ + X^2 V_P[[X]] \cong_{V+XV_P[[X]]} I_{n+1} + XJ + X^2 V_P[[X]]$.

Let $\varphi : I_n + XJ + X^2 V_P[[X]] \longrightarrow I_{n+1} + XJ + X^2 V_P[[X]]$ be a such isomorphism. There exist $r \in I_{n+1}$, $c \in I_n$, $\alpha, d \in J$ and $\rho, \theta \in V_P[[X]]$ such that:
$$\varphi(X^2) = r + \alpha X + X^2 \rho \quad \varphi^{-1}(X^2) = c + dX + X^2 \theta$$
For all $z \in I_n$, $X^2 \varphi(z) = \varphi(X^2 z) = z\varphi(X^2) = zr + z\alpha X + z\rho X^2$ and so $r = \alpha = 0$. Thus, $\varphi(X^2) = X^2 \rho$ and $\varphi(z) = z\rho$ for all $z \in I_n$. For all $z \in J$, $X\varphi(zX) = \varphi(zX^2) = z\varphi(X^2) = zX^2 \rho$ and so $\varphi(zX) = zX\rho$. Then $\varphi(f) = f\rho$ for all $f \in I_n + XJ + X^2 V_P[[X]]$. With similar argument, $c = d = 0$ and $\varphi^{-1}(g) = g\theta$ for all $g \in I_{n+1} + XJ + X^2 V_P[[X]]$. Thus, $1 = \rho\theta$ because $X^2 = \varphi(\varphi^{-1}(X^2)) = \varphi(X^2 \theta) = X^2 \theta\rho$. Then $(I_n + XJ + X^2 V_P[[X]])\rho = I_{n+1} + XJ + X^2 V_P[[X]]$, and so $J\rho_0 \subseteq J \subseteq \rho_0 J + I_n V_P \subseteq \rho_0 J + P = \rho_0 J$ because $\rho_0 J$ is an ideal of V and necessarily contains P since ρ_0 is a unit of V_P, where ρ_0 is the constant term of ρ. Then $J = \rho_0 J$ and so $bV = b\rho_0 V$. Thus, $V = \rho_0 V$. It follows that ρ_0 is a unit of V and ρ is a unit of $V + XV_P[[X]]$. Thus, $I_n + XJ + X^2 V_P[[X]] = I_{n+1} + XJ + X^2 V_P[[X]]$ and then $I_n = I_{n+1}$, a contradiction. Then $I_i + XJ + X^2 V_P[[X]] \ncong_{V+XV_P[[X]]} I_{i+1} + XJ + X^2 V_P[[X]]$ for all $i \geq 1$.

The same proof of Claim 3 works also for the ring $V + XV_P[X]$.

Lemma 19.15 *Let A be an integral domain and $I \subset J$ two ideals of A such that $I \cong_A J$ and $0 \neq a \in J$. Then there exists $0 \neq b \in I$ such that $aI = bJ$.*

Proof Let $\varphi : J \longrightarrow I$ be an isomorphism of A-modules and set $0 \neq b = \varphi(a) \in I$. For all $x \in J$, $a\varphi(x) = \varphi(ax) = x\varphi(a) = xb$ and so $aI = bJ$. □

Theorem 19.16 (Dastanpour and Ghorbani 2018) *A valuation domain A satisfies ACC_d on ideals if and only if A is discrete with rank ≤ 2.*

Proof "$\Longrightarrow$" Combine Proposition 19.2 and Theorem 16.11.

"$\Longleftarrow$" Let A be a discrete valuation domain with rank ≤ 2. Let $I_1 \subseteq I_2 \subseteq \ldots$ be an ascending chain of nonzero ideals of A. By Theorem 16.11, A is isonoetherian, and so there exists an index n such that $I_i \cong_A I_{i+1}$ for all $i > n$. Fix $i \geq n$ and search $r_i \in A$ such that $I_i = r_i I_{i+1}$. By Lemma 19.15, $I_i = cI_{i+1}$ for some $0 \neq c$ in the quotient field of A. If $c \in A$, then $I_i = cI_{i+1}$ (we take $r_i = c$). If $c^{-1} \in A$, then $I_{i+1} = c^{-1}I_i \subseteq I_i$ and so $I_i = I_{i+1}$ (we take $r_i = 1$). □

Proposition 19.17 (Khalifa 2021) *If the ring $A[[X]]$ (respectively $A[X]$) satisfies ACC_d on ideals, then so is A.*

Proof Let $I_1 \subseteq I_2 \subseteq \ldots$ be an ascending chain of nonzero ideals of A. Thus, $I_1[[X]] \subseteq I_2[[X]] \subseteq \ldots$ is an ascending chain of ideals of $A[[X]]$. So there exists an index n such that for all $i \geq n$, $I_i[[X]] = f_i I_{i+1}[[X]]$ for some $f_i \in A[[X]]$. Then $I_i = r_i I_{i+1}$ for all $i \geq n$ where r_i is the constant term of f_i. □

The converse is false: an example of a ring A that satisfies ACC_d on ideals but $A[[X]]$ and $A[X]$ do not satisfy ACC_d on ideals.

Example (Khalifa 2021) Let A be a discrete valuation domain with rank 2. By Proposition 19.16, A satisfies ACC_d on ideals. Since A is not Noetherian, $A[[X]]$ and $A[X]$ do not satisfy ACC_d on ideals by Corollary 19.13.

Now, we study the isonoetherian property in polynomial and power series ring with an infinite countably set of indeterminates. Let $\mathscr{X} = (X_n)_{n \geq 1}$ be an infinite countably set of algebraically independent indeterminates over A. Denote $A[[\mathscr{X}]] = \cup_{n \geq 1} A[[X_1, \ldots, X_n]]$ and $A[\mathscr{X}] = \cup_{n \geq 1} A[X_1, \ldots, X_n]$.

Theorem 19.18 (Khalifa 2021) *If A is an integral domain, then each of the rings $A[[\mathscr{X}]]$ and $A[\mathscr{X}]$ is never isonoetherian.*

Proof The ring $A[[\mathscr{X}]]$ is not Noetherian because $(X_1) \subset (X_1, X_2) \subset \ldots$ is a strictly ascending chain of ideals of $A[[\mathscr{X}]]$. We claim that $(X_1, \ldots, X_n, X_{n+1}) \not\cong_{A[[\mathscr{X}]]} (X_1, \ldots, X_n)$ for all $n \geq 1$. Assume that $(X_1, \ldots, X_n, X_{n+1}) \cong_{R[[\mathscr{X}]]} (X_1, \ldots, X_n)$ for some index n. By Lemma 19.15, $X_{n+1}(X_1, \ldots, X_n) = (X_1, \ldots, X_n, X_{n+1})f$ for some $0 \neq f \in (X_1, \ldots, X_n)$. There exist $h_1, \ldots, h_n \in (X_1, \ldots, X_n)$ and $g_1, \ldots, g_n \in (X_1, \ldots, X_n, X_{n+1})$ such that $X_{n+1}X_i = fg_i$ and $fX_i = X_{n+1}h_i$ for each $1 \leq i \leq n$. Thus, $fX_i^2 = fg_ih_i$ and so $X_i^2 = g_ih_i$. Note that there exists a positive integer $s \geq n+1$ such that $f, g_1, \ldots, g_n, h_1, \ldots, h_n \in B := A[[X_1, \ldots, X_s]]$. We can read elements of B as elements of $A[[(X_j)_{1 \leq j \neq i \leq s}]][[X_i]]$. Since $X_i^2 = g_ih_i$ and X_i is a prime element of B, g_i is a unit of B or h_i is a unit of B, or $g_i, h_i \in X_iB - X_i^2B$. If g_i is a unit of B (and so a unit of $A[[\mathscr{X}]]$), then $(X_1, \ldots, X_n) = (X_1, \ldots, X_n, X_{n+1})X_i$ and so $X_1 \in (X_1 X_i, \ldots, X_{n+1} X_i)$ which implies that $1 = 0$, a contradiction. If h_i is a unit, then $X_i(X_1, \ldots, X_n) = (X_1, \ldots, X_n, X_{n+1})$, and so $1 = 0$, a contradiction. If g_i and h_i are not units, then $X_{n+1} = f\rho$ (for some unit ρ of $A[[\mathscr{X}]]$), and so $(X_1, \ldots, X_n) = (X_1, \ldots, X_n, X_{n+1})$. Hence, $1 = 0$, a contradiction. The same proof works also for the ring $A[\mathscr{X}]$. $\qquad\square$

Corollary 19.19 (Khalifa 2021) *If A is an integral domain, then each of the rings $A[[\mathscr{X}]]$ and $A[\mathscr{X}]$ does not satisfy ACC_d on ideals.*

Proof Combine 19.18 and Proposition 19.2. $\qquad\square$

It follows that a directed union (respectively an ascending chain) of Noetherian domains need not to be isonoetherian.

Example (Khalifa 2021) For any field K and any infinite countably set of algebraically independent indeterminates $\mathscr{X} = (X_n)_{n \geq 1}$ over K, the ring $K[[\mathscr{X}]]$ (also the ring $K[\mathscr{X}]$) is not isonoetherian.

20 Isonoetherian Nagata's Idealization Ring

Let A be a ring and M a A-module. It was shown, in Corollary 8.11 of Chap. 1, that the ring $A(+)M$ is Noetherian if and only if A is Noetherian and M is finitely generated. Khalifa (2021), studied the isonoetherian property in the ring $A(+)M$, and he obtained some partial results on this subject. Recall that a module M is said to be divisible if for each regular element of a of A and each element $x \in M$, there exists $y \in M$ such that $x = ay$.

Lemma 20.1 (Khalifa 2021) *Let A be an integral domain and M a divisible A-module. If the ring $A(+)M$ is isonoetherian, then A is an isonoetherian ring, and M is an isonoetherian module.*

Proof Let $I_1 \subseteq I_2 \subseteq \ldots$ be an ascending chain of nonzero ideals of A. Thus, $I_1(+)M \subseteq I_2(+)M \subseteq \ldots$ is an ascending chain of nonzero ideals of $A(+)M$. Then there exists an index n such that for all $i \geq n$, $I_i(+)M \cong_{A(+)M} I_n(+)M$.

Claim: If $I \subseteq J$ are two proper ideals of A such that $I(+)M \cong_{A(+)M} J(+)M$ and then $I \cong_A J$.

Indeed, let $\varphi : I(+)M \longrightarrow J(+)M$ be an isomorphism of $(A(+)M)$-modules. We define the map $\psi : I \longrightarrow J$ as follows: for each $a \in I$, $\psi(a)$ is the first component of $\varphi(a, 0)$. Thus, ψ is a homomorphism of A-modules. One to one: let $a \in I$ be such that $\psi(a) = 0$. Then $\varphi(a, 0) = (0, x)$, for some $x \in M$, and so $\varphi((a, 0)\varphi(a, 0)) = (\varphi(a, 0))^2 = (0, x)^2 = (0, 0)$. Thus, $\varphi(a^2, 0) = (a, 0)\varphi(a, 0) = (0, 0)$ and so $a^2 = 0$. Then $a = 0$. Then ψ is one to one. Onto: let $0 \neq b \in J$. Since φ is onto, $(b, 0) = \varphi(a, y)$ for some $a \in I$ and $y \in M$. Since M is divisible, $y = b^2 z$ for some $z \in M$. If $a = 0$, then

$(b, 0) = \varphi(0, b^2 z) = \varphi((b^2, 0)(0, z)) = (b^2, 0)\varphi(0, z)$, and so $b = b^2 c$ for some $c \in J$. Thus, b is invertible in A and so $J = A$, a contradiction. Then $a \neq 0$ and let so $\alpha \in M$ be such that $y = a\alpha$. Then $(b, 0) = \varphi(a, a\alpha) = \varphi((a, 0)(1, \alpha)) = (1, \alpha)\varphi(a, 0) = (1, \alpha)(\psi(a), .)$. So $b = \psi(a)$ and hence ψ is an isomorphism. Then $I \cong_A J$.

By the claim, $I_i \cong_A I_n$ for all $i \geq n$ and so A is isonoetherian. Let $N_1 \subseteq N_2 \subseteq \ldots$ be an ascending chain of nonzero sub-modules of M. Thus, $0(+)N_1 \subseteq 0(+)N_2 \subseteq \ldots$ is an ascending chain of nonzero ideals of $A(+)M$. Then there exists an index n such that for all $i \geq n$, $0(+)N_i \cong_{A(+)M} 0(+)N_n$. Easily, one can show that if $N \subseteq W$ are two sub-modules of M such that $0(+)N \cong_{A(+)M} 0(+)W$, then $N \cong_A W$. Thus, $N_i \cong_A N_n$ for all $i \geq n$. Then M is an isonoetherian module. $\quad\square$

Lemma 20.2 *Let A be an integral domain and M be an A-module. The following statements are equivalent:*

1. *Every ideal of $A(+)M$ has the form $I(+)M$ for some ideal I of A or $0(+)N$ for some sub-module N of M.*
2. *The module M is divisible.*

Proof $"(1) \Rightarrow (2)"$ Let $0 \neq a \in A$ and $x \in M$. Let $\mathscr{J}$ be the principal ideal of $A(+)M$ generated by (a, x). Since $(a, x) \in \mathscr{J}$ and $a \neq 0$, there exists an ideal I of A such that $\mathscr{J} = I(+)M$ and so $a \in I$. Thus, $(a, 0) \in \mathscr{J}$ and so $(a, 0) = (a, x)(b, y)$ for some $(b, y) \in A(+)M$. Then $a = ab$ and $0 = ay + bx$. It follows that $b = 1$ and $x = -ay$.

$"(2) \Rightarrow (1)"$ By Lemma 6.2 of Chap. 2, it suffices to show that each ideal of $A(+)M$ is comparable to $0(+)M$. Let $\mathscr{J}$ be a nonzero ideal of $A(+)M$ and assume that $\mathscr{J} \nsubseteq 0(+)M$. Let so $(a, z) \in \mathscr{J}$ such that $a \neq 0$. Let $x \in M$. Since M is divisible, there exists $y \in M$ such that $x = ay$. So $(0, x) = (0, ay) = (a, z)(0, y) \in \mathscr{J}$. Thus $0(+)M \subseteq \mathscr{J}$. $\quad\square$

Theorem 20.3 (Khalifa 2021) *Let A be a Noetherian domain and M a divisible A-module. The ring $A(+)M$ is isonoetherian if and only if M is isonoetherian.*

Proof $"\Rightarrow"$ By Lemma 20.1. $"\Leftarrow"$ By Lemma 20.2, ideals of $A(+)M$ have the form $I(+)M$ or $0(+)N$ for some ideal I of A or for some sub-module N of M. Let $\mathscr{J}_1 \subseteq \mathscr{J}_2 \subseteq \ldots$ be an ascending chain of ideals of $A(+)M$.

Case 1: Assume that $\mathscr{J}_n$ has the form $I(+)M$ for some index n and some nonzero ideal I of A. For all $i \geq n$, $\mathscr{J}_i = I_i(+)M$ for some ideal I_i of A. Thus, $I_1 \subseteq I_2 \subseteq \ldots$ is an ascending chain of ideals of A. Since A is Noetherian, there exists $m \geq n$ such that $I_i = I_m$ for all $i \geq m$. Then $\mathscr{J}_i = \mathscr{J}_m \cong_{A(+)M} \mathscr{J}_m$ for all $i \geq m$.

Case 2: Assume that for all $i \geq 1$, $\mathscr{J}_i = 0(+)N_i$ for some sub-module N_i of M. Since M is isonoetherian and $N_1 \subseteq N_2 \subseteq \ldots$, there exists an index n such that $N_i \cong_A N_n$ for all $i \geq n$. Hence, $\mathscr{J}_i = 0(+)N_i \cong_{A(+)M} 0(+)N_n = \mathscr{J}_n$ for each $i \geq n$. $\quad\square$

An example of an isonoetherian ring of the form $A(+)M$ that is not Noetherian:

Example (Khalifa 2021) Let A be a rank one discrete valuation domain with quotient field K. The ring A is Noetherian and the A-module K is divisible. Also, the A-module K is isonoetherian because all its proper sub-modules are cyclic. Hence the ring $A(+)K$ is isonoetherian by Theorem 20.3 but not Noetherian by Corollary 8.11 of Chap. 2.

Definition 20.4 (Khalifa 2021) We say that a A-module M satisfies strongly divisibility on ascending chain of sub-modules (for short, strongly ACC_d on sub-modules) if for every ascending chain $M_1 \subseteq M_2 \subseteq \ldots$ of sub-modules of M, there exists an index n such that for all $i \geq n$, $M_i = a_i M_{i+1}$ for some $a_i \in A$.

Example Let A be a rank one discrete valuation domain with quotient field K. By Proposition 17.2, proper sub-modules of the A-module K are cyclic. Hence, the A-module K satisfies strongly ACC_d on sub-modules.

Theorem 20.5 (Khalifa 2021) *Let A be an integral domain and M be a divisible module over A. The following statements are equivalent:*

1. *The ring $A(+)M$ satisfies ACC_d on ideals.*
2. *The ring A satisfies ACC_d on ideals and M satisfies strongly ACC_d on sub-modules.*

Proof "(1) $\Rightarrow$ (2)" Let $I_1 \subseteq I_2 \subseteq \ldots$ be an ascending chain of ideals of A. Thus, $I_1(+)M \subseteq I_2(+)M \subseteq \ldots$ is an ascending chain of ideals of $A(+)M$. Thus, there exists an index n such that for all $i \geq n$, $I_i(+)M = (a_i, x_i)(I_{i+1}(+)M)$ for some $(a_i, x_i) \in A(+)M$. Note that if $I \subseteq J$ are two ideals of A and $(a, x) \in A(+)M$ such that $I(+)M = (a, x)(J(+)M)$, then $I = aJ$. So $I_i = a_i I_{i+1}$ for all $i \geq n$. Thus, the ring A satisfies ACC_d on ideals. Let $N_1 \subseteq N_2 \subseteq \ldots$ be an ascending chain of sub-modules of M. Thus, $0(+)N_1 \subseteq 0(+)N_2 \subseteq \ldots$ is an ascending chain of ideals of $A(+)M$. Thus, there exists an index m such that for all $i \geq m$, $0(+)N_i = (b_i, y_i)(0(+)N_{i+1})$ for some $(b_i, y_i) \in A(+)M$. Note that if $N \subseteq W$ are two sub-modules of M and $(b, y) \in A(+)M$ such that $0(+)N = (b, y)(0(+)W)$, then $N = bW$. So $N_i = b_i N_{i+1}$ for all $i \geq m$. Thus, the module M satisfies strongly ACC_d on sub-modules.

"(2) $\Rightarrow$ (1)" Let $\mathscr{I}_1 \subseteq \mathscr{I}_2 \subseteq \ldots$ be an ascending chain of ideals of $A(+)M$. By Lemma 20.2, ideals of $A(+)M$ have the form $I(+)M$ or $0(+)N$ for some ideal I of A or for some sub-module N of M:

Case 1: Assume that for some index k, $\mathscr{I}_k$ has the form $I(+)M$ for some nonzero ideal I of A. Thus, $\mathscr{I}_i = I_i(+)M$ for all $i \geq k$ and for some nonzero ideal I_i of A. Since $I_k \subseteq I_{k+1} \subseteq \ldots$, there exists an index $m \geq k$ such that $I_i = a_i I_{i+1}$ for all $i \geq m$ and for some $a_i \in A$. Thus $(a_i, 0)\mathscr{I}_{i+1} = (a_i, 0)(I_{i+1}(+)M) \subseteq a_i I_{i+1}(+)M = I_i(+)M = \mathscr{I}_i$. Conversely, let $b \in I_i$ and $x \in M$. So $b = a_i a$ and $x = a_i y$ for some $a \in I_{i+1}$ and some $y \in M$. Thus, $(b, x) = (a_i a, a_i y) = (a_i, 0)(a, y) \in (a_i, 0)(I_{i+1}(+)M) = (a_i, 0)\mathscr{I}_{i+1}$. Then $\mathscr{I}_i = (a_i, 0)\mathscr{I}_{i+1}$ for all $i \geq m$.

Case 2:   Assume that for all i, $\mathscr{J}_i$ has not the form $I(+)M$ for some nonzero ideal I of A. Thus, $\mathscr{J}_i = 0(+)N_i$ for all $i \geq 1$ and for some sub-module N_i of M. Since $N_1 \subseteq N_2 \subseteq \ldots$, there exists an index n such that $N_i = a_i N_{i+1}$ for all $i \geq n$ and for some $a_i \in A$. Then $\mathscr{J}_i = 0(+)N_i = (a_i, 0)(0(+)N_{i+1}) = (a_i, 0)\mathscr{J}_{i+1}$ for all $i \geq n$.   □

An example of a ring A that satisfies ACC_d on ideals and an A-module M that satisfies strongly ACC_d on sub-modules but the ring $A(+)M$ does not satisfy ACC_d on ideals:

Example (Khalifa 2021) Let A be a discrete valuation domain with rank 2 and a prime spectrum $0 \subset P \subset N$. By Theorem 19.17, A satisfies ACC_d on ideals, and so the A-module A satisfies strongly ACC_d on sub-modules. Let $0 \neq a \in P$, $b \in N - P$. Set $I_1 = aA \subset P \subset bA$ and so $I_1 = bI_2, I_2 = bI_3, \ldots$ for some ideals $I_2 \subset I_3 \subset \ldots \subset P$ of A. Suppose that $I_n(+)A = (c, x)(I_{n+1}(+)A)$ for some index n and some $c, x \in A$. Thus, $I_n = cI_{n+1}$ and $A = cA + I_{n+1}x = cA + P = cA + N$. So c is a unit of A and then $I_n = I_{n+1}$, a contradiction. Hence, the ring $A(+)A$ does not satisfy ACC_d on ideals. It follows that the hypothesis "M is a divisible module" is necessary in Theorem 20.5.

Proposition 20.6 (Khalifa 2021) *Let A be a quasi-local ring and M be a nonzero finitely generated module over A. The following statements are equivalent:*

1. *The ring $A(+)M$ satisfies ACC_d on ideals.*
2. *The ring A is Noetherian.*
3. *The ring $A(+)M$ is Noetherian.*

Proof "(1) $\Rightarrow$ (2)" Let $I_1 \subseteq I_2 \subseteq \ldots$ be an ascending chain of proper ideals of A. Thus, $I_1(+)M \subseteq I_2(+)M \subseteq \ldots$ is an ascending chain of ideals of $A(+)M$. So there exists an index n such that for all $i \geq n$, $I_i(+)M = (a_i, x_i)(I_{i+1}(+)M)$ for some $(a_i, x_i) \in A(+)M$. So $I_i = a_i I_{i+1}$ and $M = a_i M + I_{i+1}x_i = J_i M$ (where $J_i = a_i A + I_{i+1}$) for all $i \geq n$. By Nakayama's Lemma 8.1 of Chap. 2, $J_i = A$ and so a_i is a unit of A. Thus, $I_i = I_{i+1}$ for all $i \geq n$. Then A is Noetherian.
    "(2) $\Rightarrow$ (3)" By Corollary 8.11 of Chap. 1. "(3) $\Rightarrow$ (1)" Trivial.   □

Problems

Exercise 1 (Hizem 2011) Let A be a strongly Hopfian ring. Show that for each $a \in A$ and each ideal I of A, the sequence $(ann(a^n I))_n$ is stationary.

Exercise 2 (Hizem 2011)

(1) Let R be a ring and M a R-module. Show that the ring $R(+)M$ is strongly Hopfian if and only if the ring R is strongly Hopfian and, for each $r \in R$, the sequence $ann_M(r) \subseteq ann_M(r^2) \subseteq \ldots$ is stationary.

(2) Let K be a field, $R = K[[X]]$ and $M = R[Y_i, i \in \mathbb{N}^*]/(Y_i X^i, i \in \mathbb{N}^*)$. Show that the ring $R(+)M$ is not strongly Hopfian.

Exercise 3 Let A be a ring.

(1) Let $(I_\alpha)_{\alpha \in \Lambda}$ be a family of ideals of A and $I = \bigcap_{\alpha \in \Lambda} I_\alpha$. Show that the quotient ring A/I is imbedded in the product $\prod_{\alpha \in \Lambda} A/I_\alpha$.

(2) Show that a reduced ring is imbedded in a zero dimensional ring.
(3) Let Q be a P-primary ideal of A. Show that $dim(A/Q)_{P/Q} = 0$ and the ring A/Q is imbedded in $(A/Q)_{P/Q}$.
(4) Deduce that if (0) is a finite intersection of primary ideals of A, then A is imbedded in a zero dimensional ring, so A is strongly Hopfian.
(5) Show by an example that the finiteness of the number of primary ideals in (4) is necessary.

Exercise 4 (Fields 1971) Let A be a ring.

(1) Let $x, y \in A$. Show that

   a) x is a zero divisor if and only if there exists $n \in \mathbb{N}^*$ such that x^n is a zero divisor.

   b) If y is nilpotent, then $x - y$ is a zero divisor if and only if x is a zero divisor.

(2) Let Q be a strongly P-primary ideal of A (i.e., Q contains a power of P) and then $Q[[X]]$ is $P[[X]]$-primary.
(3) Deduce that if (0) is a finite intersection of strongly primary ideals in A, then $A[[X]]$ is strongly Hopfian.

Exercise 5 Let A be a ring and $(Q_\lambda)_{\lambda \in \Lambda}$ a family of ideals of A.

(1) Show that this family satisfies condition (2) of Theorem 5.3 if and only if for each subset $\Gamma \subseteq \Lambda$, $\sqrt{\bigcap_{\lambda \in \Gamma} Q_\lambda} = \bigcap_{\lambda \in \Gamma} \sqrt{Q_\lambda}$.

(2) Deduce that if the equality is not satisfied for each subset $\Gamma \subseteq \Lambda$, then it is not satisfied for each countably subset of Λ.

Exercise 6 (Hamed et al. 2021) Let A and B be two rings, J an ideal of B, and $f : A \longrightarrow B$ a homomorphism. Recall from Sect. 7 of Chap. 1 that the amalgamation of A with B along J with respect to f is the sub-ring of $A \times B$ denoted by $A \bowtie^f J = \{(a, f(a) + j); a \in A, j \in J\}$:

(1) a) Show that A is imbedded in $A \bowtie^f J$ and if $J = (0)$, then $A \bowtie^f J \simeq A$.
   b) Show that $A \bowtie^f J / f^{-1}(J) \times (0) \simeq f(A) + J$.
(2) Show that if A and $f(A) + J$ are strongly Hopfian, then so is $A \bowtie^f J$.

(3) Suppose that there exists an ideal I of A such that $f^{-1}(J) = ann(I)$. Show that $A \bowtie^f J$ is strongly Hopfian if and only if A and $f(A) + J$ are strongly Hopfian.

(4) Show by an example that if $A \bowtie^f J$ is strongly Hopfian, $f(A) + J$ is not necessarily strongly Hopfian.

(5) Show by an example that if A is strongly Hopfian, $A \bowtie^f J$ is not necessarily strongly Hopfian.

Exercise 7 (Hamed et al. 2021) Let A be a ring and S a multiplicative set of A. We say that A is S-strongly Hopfian if for each $a \in A$, there exists $s \in S$ and $n \in \mathbb{N}$ such that for each integer $k \geq n$, we have $s.ann(a^k) \subseteq ann(a^n)$.

(1) Show that

    a) If A is strongly Hopfian, then it is S-strongly Hopfian.

    b) If S is formed by units of A, then A is S-strongly Hopfian if and only if it is strongly Hopfian.

(2) Suppose that S is formed by regular elements of A. Show that A is S-strongly Hopfian if and only if it is strongly Hopfian.

(3) Let $A \subseteq B$ be an extension of rings and S a multiplicative set of A. Show that if B is strongly Hopfian, then so is A.

Exercise 8 (Hamed et al. 2021) Let A be a ring and S a multiplicative set of A. We say that A satisfies the condition S-accr if for each ideal I of A and each $a \in A$, there exist $s \in S$ and $n \in \mathbb{N}$ such that $s(I : a^k) \subseteq (I : a^n)$ for each integer $k \geq n$.

(1) Show that A satisfies the condition S-accr if and only if for each ideal I of A such that $S \cap I = \emptyset$, the ring A/I is $\overline{S}$-strongly Hopfian, where $\overline{S} = \{\overline{s} \in A/I, s \in S\}$.

(2) Deduce that the two following conditions are equivalent:

    (i) The ring A satisfies the condition accr.

    (ii) For each ideal I of A, the ring A/I is strongly Hopfian.

(3) Let J be a subset of A and $I = ann(J)$. Suppose that A is S-strongly Hopfian and $S \cap I = \emptyset$. Show that A/I is $\overline{S}$-strongly Hopfian, where $\overline{S} = \{\overline{s} \in A/I; s \in S\}$.

Exercise 9 (Hamed et al. 2021) Let A be a ring and S a multiplicative subset of A. Show that if A is S-strongly Hopfian, $S^{-1}R$ is strongly Hopfian.

Exercise 10 (Hizem 2011) Let A be a Noetherian ring and Λ a set of indexes. Show that the ring $R = A[[X_\alpha; \alpha \in \Lambda]]_1$ is strongly Hopfian.

Exercise 11 (Hizem 2011) Let A be a ring.

(1) Show that for each $a \in A$, $ann_{A[[X]]}(a) = ann_A(a)[[X]]$.
(2) Suppose that A is strongly Hopfian.

(a) Show that for each $a \in A$, the sequence $ann_{A[[X]]}(a) \subseteq ann_{A[[X]]}(a^2) \subseteq \ldots$ is stationary.
(b) Let $f \in A[[X]]$ and $n \in \mathbb{N}^*$ be such that $f^n \in A$. Show that the sequence $ann_{A[[X]]}(f) \subseteq ann_{A[[X]]}(f^2) \subseteq \ldots$ is stationary.

Exercise 12 (Gilmer et al. 1975) Let A be a zero dimensional chained ring. Show that if f is a zero divisor in the ring $A[[X]]$, there exists $0 \neq a \in A$ such that $af = 0$.

Exercise 13 (Hizem 2011) Let A be a ring. We define the map $\Psi : A[[X]] \to A$, by $\Psi(f) = f(0)$ for each $f \in A[[X]]$.

(1) Show that if $f \in A[[X]]$ and $n \in \mathbb{N}^*$ are such that $\Psi(ann(f^n)) = \Psi(ann(f^{n+1}))$, then $ann(f^n) = ann(f^{n+1})$.
(2) Deduce that if $f \in A[[X]]$, the sequence $(\Psi(ann(f^n)))_n$ is stationary if and only if the sequence $(ann(f^n))_n$.

Exercise 14 (Hizem 2011) Let A be a ring and $f = \sum\limits_{i=0}^{\infty} a_i X^i \in A[[X]]$. Suppose that there exists $k \in \mathbb{N}^*$ satisfying $ann(a_i^k) = ann(a_i^{k+1})$ for each $i \in \mathbb{N}$. Show that $ann(f^{k+1}) \subseteq ann_A\{a_i^k a_{i-1}^{k-1} a_{i-2}^{k-1} \ldots a_0^{k-1}; i \in \mathbb{N}\}[[X]]$.

Exercise 15 (Kim et al. 2006) Let A be a ring and $ann_A(2^A) = \{ann_A(U); U \subseteq A\}$.

(1) Show that for each $V \subset A[[X]]$, $ann_{A[[X]]}(V) \cap A = ann_A(V) = ann_A(A_V)$, where $A_V = \bigcup\limits_{f \in V} A_f$.
(2) Show that we can define a map $\Psi : ann_{A[[X]]}(2^{A[[X]]}) \to ann_A(2^A)$, by $\Psi(I) = I \cap A$ for each $I \in ann_{A[[X]]}(2^{A[[X]]})$, and that is onto.
(3) Let $U \subseteq A$. Show that $ann_A(U)[[X]] = ann_{A[[X]]}(U)$.
(4) Show that the map $\Phi : ann_A(2^A) \to ann_{A[[X]]}(2^{A[[X]]})$, defined by $\Phi(J) = J[[X]]$ for each $J \in ann_A(2^A)$, is one to one.

We recall that A is power series Armendariz if for each $f = \sum\limits_{i \geq 0} a_i X^i$ and $g = \sum\limits_{i \geq 0} b_i X^i \in A[[X]]$ such that $fg = 0$, we have $a_i b_j = 0$ for each i et j.

(5) Give an example.

(6) Show that the following assertions are equivalent:

(a) The ring A is power series Armendariz.
(b) For each integer $n \geq 2$ and for each $f_1, \ldots, f_n \in A[[X]]$ such that $f_1 \ldots f_n = 0$, we have $a_1 \ldots a_n = 0$ where a_i is any coefficient of f_i.
(c) The map Φ is bijective.
(d) The map Ψ is bijective.

And in this case, the maps Φ and Ψ are inverse to each other.

(7) Suppose that A is power series Armendariz. Show that $A[[X]]$ satisfies acc on the annihilators if and only if A has the same property.

Exercise 16 (Hizem 2011)

(1) Let A be a power series Armendariz ring. Show that the following assertions are equivalent:

(a) The ring $A[[X]]$ is strongly Hopfian.
(b) For each $f = \sum_{i \geq 0} a_i X^i \in A[[X]]$, there exists $n \in \mathbb{N}^*$ such that we have the equality $\bigcap_{(i_1, \ldots, i_n) \in \mathbb{N}^n} ann(a_{i_1} \ldots a_{i_n}) = \bigcap_{(i_1, \ldots, i_{n+1}) \in \mathbb{N}^{n+1}} ann(a_{i_1} \ldots a_{i_{n+1}})$.

(c) For each countably generated ideal I of A, the sequence $(ann(I^n))_n$ is stationary.

(2) Application: Let B be a reduced ring and $A = B[T]/(T^2)$. We recall from Example 1 of Definition 12.1 that A is a nonreduced power series Armendariz ring. Deduce that the ring $A[[X]]$ is strongly Hopfian.

Exercise 17 Let A be a ring. Show that $Bool(A[[X]]) = Bool(A)$.

Exercise 18 (Anderson and Camillo 2002) Let A be a ring. We say that A is decomposable (resp. uniquely decomposable) if each element $x \in A$ can be written (in a unique way) in the form $x = u + e$ with $u \in U(A)$ and $e \in Bool(A)$.

(1) Show that if A is decomposable and if I is an ideal of A, then A/I is decomposable.
(2) Show that $A[[X]]$ is decomposable if and only if A is decomposable.
(3) Show that $A[[X]]$ is uniquely decomposable if and only if A is uniquely decomposable.

Exercise 19 (Anderson and Camillo 1998) Show that a ring A is Armendariz if and only if the polynomials ring $A[X]$ is Armendariz.

Exercise 20 (Anderson and Camillo 1998) Suppose that (0) is P-primary in a ring A with $P^2 = (0)$. Show that A is power series Armendariz. Hint: Use Exercise 4.

Exercise 21 Let A be a ring.

(1) Show that the set $Z(A)$ of the zero divisors of A contains a prime ideal of A.
(2) Give an example of a ring A where $Z(A)$ is not an ideal of A.

Solutions

Exercise 1 There exists $k \geq 1$ such that for each $n \geq k$, $ann(a^n) = ann(a^k)$. Let $x \in ann(a^n I)$. Then $xa^n I = (0)$, so $xI \subseteq ann(a^n) = ann(a^k)$, then $xIa^k = (0)$, and so $x \in ann(a^k I)$.

Exercise 2

(1) Recall that in $R(+)M = R \times M$, the operations are $(r, m) + (r', m') = (r + r', m + m')$ and $(r, m)(r', m') = (rr', rm' + r'm)$. In particular, $(r, m)^k = (r^k, kr^{k-1}m)$ for each $k \in \mathbb{N}$. The zero element is $(0, 0)$ and the identity is $(1, 0)$.

  "$\Longrightarrow$" Let $r \in R$. Then $ann(r, 0)^k = ann(r^k, 0) = \{(r', m') \in R(+)M; (r', m')(r^k, 0) = (0, 0)\} = \{(r', m') \in R(+)M; (r'r^k, r^k m') = (0, 0)\} = \{(r', m') \in R(+)M; r' \in ann_R(r^k), m' \in ann_M(r^k)\} = ann_R(r^k) \times ann_M(r^k)$. The sequences $(ann_R(r^k))_k$ and $(ann_M(r^k))_k$ are stationary. In particular, the ring R is strongly Hopfian.

  "$\Longleftarrow$" Let $(r, m) \in R(+)M$. Let $k \geq 1$ be such that the sequences $(ann_R(r^n))_n$ and $(ann_M(r^n))_n$ are stationary at the rank k. We will show the equality $ann(r, m)^{k+1} = ann(r, m)^{k+2}$, i.e., $ann(r^{k+1}, (k + 1)r^k m) = ann(r^{k+2}, (k + 2)r^{k+1}m)$.

  Let $(r', m') \in ann(r^{k+2}, (k + 2)r^{k+1}m)$. Then $(r', m')(r^{k+2}, (k + 2)r^{k+1}m) = (0, 0) \Longleftrightarrow (r'r^{k+2}, (k + 2)r'r^{k+1}m + r^{k+2}m') = (0, 0) \Longleftrightarrow r'r^{k+2} = 0$, and $(k + 2)r'r^{k+1}m + r^{k+2}m' = 0$. Then $r' \in ann_R(r^{k+2}) = ann_R(r^k)$, then $r^{k+2}m' = 0$, and then $m' \in ann_M(r^{k+2}) = ann_M(r^{k+1})$. Compute $(r', m')(r^{k+1}, (k + 1)r^k m) = (r'r^{k+1}, r^{k+1}m' + (k + 1)r'r^k m) = (0, 0)$. Then we have the result.

(2) The ring $R = K[[X]]$ is Noetherian and then strongly Hopfian. The ideal $I = (Y_i X^i, i \in \mathbb{N}^*)$ of the ring $R[Y_i, i \in \mathbb{N}^*]$ satisfies $I \cap R = (0)$ and then $R \subset M$. We will show that the sequence $(ann_M(X^n))_n$ is strictly increasing and then the ring $R(+)M$ is not strongly Hopfian. Indeed, $\bar{Y}_n \in ann(X^n) \setminus ann(X^{n-1})$ for each $n \geq 1$. Note that $Y_n X^n \in I$ and then $\bar{Y}_n X^n = \bar{0}$, so $\bar{Y}_n \in ann_M(X^n)$. Suppose that $\bar{Y}_n \in ann(X^{n-1})$, then $\bar{Y}_n X^{n-1} = \bar{0}$, so $Y_n X^{n-1} \in I$. Put $Y_n X^{n-1} = Y_1 X f_1 + \ldots + Y_{n-1} X^{n-1} f_{n-1} + Y_n X^n f_n + Y_{n+1} X^{n+1} f_{n+1} + \ldots + Y_s X^s f_s$ with $s \geq n$ and $f_i \in R[Y_i, i \in \mathbb{N}^*]$. Put $Y_1 = \ldots = Y_{n-1} = Y_{n+1} = \ldots = 0$, we obtain $Y_n X^{n-1} = Y_n X^n f_n(0, \ldots, 0, Y_n, 0, \ldots)$, and then $1 = X f_n(0, \ldots, 0, Y_n, 0, \ldots)$ with $f_n(0, \ldots, 0, Y_n, 0, \ldots) \in K[[X]][Y_n] \subset K[[X, Y_n]]$, which is absurd.

Exercise 3

(1) The map $\psi : A/I \longrightarrow \prod_{\alpha \in \Lambda} A/I_\alpha$, defined by $\psi(x + I) = (x + I_\alpha)_{\alpha \in \Lambda}$, is well

defined and injective. Indeed, $x + I = y + I \Longleftrightarrow x - y \in I = \bigcap_{\alpha \in \Lambda} I_\alpha \Longleftrightarrow$

$\forall \alpha \in \Lambda, x - y \in I_\alpha \Longleftrightarrow \forall \alpha \in \Lambda, x + I_\alpha = y + I_\alpha.$

(2) Since A is reduced, $(0) = Nil(A) = \cap\{P; \ P \in Spec(A)\}$. By (1), A is

imbedded in $\prod_{P \in Spec(A)} A/P$. But the integral domain A/P is imbedded in its

quotient field K_P. Since $\prod_{P \in Spec(A)} K_P$ is von Neumann regular, then it is zero

dimensional.

(3) A prime ideal of A/Q contained in P/Q is of the form R/Q with $R \in Spec(A)$
and $Q \subseteq R \subseteq P$, then $P = \sqrt{Q} \subseteq R \subseteq P$, so $R = P$. Then the ring
$(A/Q)_{P/Q}$ has only one prime ideal; then it is zero dimensional. The map $\psi :$
$A/Q \longrightarrow (A/Q)_{P/Q}$, defined by $\psi(a + Q) = \frac{a+Q}{1+Q}$, is a homomorphism one
to one. Indeed, if $\frac{a+Q}{1+Q} = \frac{0+Q}{1+Q}$, there exists $s \notin P$ such that $(s + Q)(a + Q) =$
$0 + Q$, then $sa \in Q$, so $a \in Q$ and $a + Q = 0 + Q$.

(4) Put $(0) = Q_1 \cap \ldots \cap Q_n$ with Q_i a P_i-primary ideal of A. By (1), the ring A is
imbedded in $(A/Q_1) \times \ldots \times (A/Q_n)$, which is also imbedded in $(A/Q_1)_{P_1/Q_1} \times$
$\ldots \times (A/Q_n)_{P_n/Q_n}$, a finite product of zero dimensional ring; then it is zero
dimensional.

A sub-ring of a zero dimensional ring is strongly Hopfian.

(5) Let p be prime number and $A = \prod_{n \geq 1} \mathbb{Z}/p^n \mathbb{Z}$. By Example of Theorem 4.3, the

ring A is not imbedded in a zero dimensional ring. For $n \in \mathbb{N}^*$, let e_n be the
element of A whose n-th component is zero and the other components are equal
to 1. By Example 1 of Theorem 5.3, the ideal $e_n A$ is primary and $\bigcap_{n \geq 1} e_n A = (0)$.

Exercise 4

(1) (a) $" \Longleftarrow "$ There exists $0 \neq t \in A$ such that $x^n t = 0$. Let $k > 0$ be the
smallest integer such that $x^k t = 0$. Then $1 \leq k \leq n$ and $x^{k-1} t \neq 0$, so
$x(x^{k-1}t) = x^k t = 0$.
  (b) Let $n \in \mathbb{N}^*$ be such that $y^n = 0$. We have $x^n - y^n = (x-y)(x^{n-1}+x^{n-2}y+$
$\ldots + xy^{n-2} + y^{n-1})$. Put $z = x^{n-1} + x^{n-2}y + \ldots + xy^{n-2} + y^{n-1} \in A$,
then $x^n = (x - y)z$.
  $" \Longrightarrow "$ Let $t \neq 0$ be such that $(x - y)t = 0$. Then $x^n t = 0$. By (a), x is a zero
divisor.
  $" \Longleftarrow "$ There exists $t \neq 0$ such that $xt = 0$ and then $(x - y)zt = x^n t = 0$.
If $zt \neq 0$, then $x - y$ is a zero divisor. If not, $0 = zt = (x^{n-1} + x^{n-2}y + \ldots +$
$xy^{n-2}+y^{n-1})t = y^{n-1}t$. Let k be the smallest natural integer such that $y^k t = 0$.

Then $1 \leq k \leq n-1$ and $y^{k-1}t \neq 0$. So $(x-y)y^{k-1}t = xy^{k-1}t + y^k t = 0$ with $y^{k-1}t \neq 0$, and then $x-y$ is a zero divisor.

(2) We know that $P[[X]]$ is a prime ideal. Since $Q \subseteq P$, then $Q[[X]] \subseteq P[[X]]$, so we have $\sqrt{Q[[X]]} \subseteq P[[X]]$. Conversely, let $k \in \mathbb{N}^*$ be such that $P^k \subseteq Q$. Then $(P[[X]])^k \subseteq P^k[[X]] \subseteq Q[[X]]$, so $P[[X]] \subseteq \sqrt{Q[[X]]}$. So we have the equality $\sqrt{Q[[X]]} = P[[X]]$. It remains to show that $Q[[X]]$ is primary. We have $A[[X]]/Q[[X]] \simeq (A/Q)[[X]]$. Suppose that there exists $f = a_0 + a_1 X + \ldots \in A[[X]]$ such that $\bar{f}$ is a zero divisor that is not nilpotent in $(A/Q)[[X]]$. For each $n \in \mathbb{N}^*$, $f^n \notin Q[[X]]$, and then $f \notin P[[X]]$. Let $m = \inf\{i \in \mathbb{N}; a_i \notin P\}$. Then $a_0, \ldots, a_{m-1} \in P = \sqrt{Q}$ and $a_m \notin P$, so $\bar{a}_0, \ldots, \bar{a}_{m-1}$ are nilpotent in A/Q, but $\bar{a}_m$ is not. Let $\bar{h} = \bar{a}_0 + \bar{a}_1 X + \ldots + \bar{a}_{m-1}X^{m-1}$. Then $\bar{h}$ is nilpotent in $(A/Q)[[X]]$. By the preceding question, $\bar{f} - \bar{h}$ is a zero divisor in $(A/Q)[[X]]$, then its initial term $\bar{a}_m$ is a zero divisor in A/Q so it is nilpotent since Q is primary. We obtain a contradiction.

(3) Let $(0) = Q_1 \cap \ldots \cap Q_n$ be a finite decomposition in strongly primary ideals of A. By (2), $(0) = Q_1[[X]] \cap \ldots \cap Q_n[[X]]$ is a finite decomposition in strongly primary ideals of $A[[X]]$. By the preceding exercise, $A[[X]]$ is strongly Hopfian.

Exercise 5

(1) " $\Longrightarrow$ " The first inclusion is always true. Conversely, let $x \in \bigcap_{\lambda \in \Gamma} \sqrt{Q_\lambda}$. By (2), there exists an integer $n \geq 1$ such that $x^n \notin \bigcup_{\lambda \in \Lambda} (\sqrt{Q_\lambda} \setminus Q_\lambda)$. Then for each $\lambda \in \Gamma$, $x^n \in Q_\lambda$. So $x^n \in \bigcap_{\lambda \in \Gamma} Q_\lambda$, and then $x \in \sqrt{\bigcap_{\lambda \in \Gamma} Q_\lambda}$.

" $\Longleftarrow$ " Let $x \in A$ and $\Gamma = \{\lambda \in \Lambda; x \in \sqrt{Q_\lambda}\}$. It suffices to find an integer $n \geq 1$ such that $x^n \notin \bigcup_{\lambda \in \Gamma} (\sqrt{Q_\lambda} \setminus Q_\lambda)$. By hypothesis, $x \in \bigcap_{\lambda \in \Gamma} \sqrt{Q_\lambda} = \sqrt{\bigcap_{\lambda \in \Gamma} Q_\lambda}$. There exists an integer $n \geq 1$ such that $x^n \in \bigcap_{\lambda \in \Gamma} Q_\lambda$. Then $x^n \notin \bigcup_{\lambda \in \Gamma} (\sqrt{Q_\lambda} \setminus Q_\lambda)$.

(2) There exists $x \in \bigcap_{\lambda \in \Gamma} \sqrt{Q_\lambda} \setminus \sqrt{\bigcap_{\lambda \in \Gamma} Q_\lambda}$. For each integer $n \geq 1$, there exists $\lambda_n \in \Gamma$ such that $x^n \notin Q_{\lambda_n}$. Then $x \in \bigcap_{n:1}^{\infty} \sqrt{Q_{\lambda_n}} \setminus \sqrt{\bigcap_{n:1}^{\infty} Q_{\lambda_n}}$.

Exercise 6

(1) (a) The map $i : A \longrightarrow A \bowtie^f J$, defined by $i(a) = (a, f(a))$, is a homomorphism one to one. If $J = (0)$, then i is an isomorphism.

(b) The projection from $A \bowtie^f J \subseteq A \times B$ to B is a homomorphism with image $f(A) + J$ and kernel $f^{-1}(J) \times (0)$. Then $A \bowtie^f J/f^{-1}(J) \times (0) \simeq f(A) + J$.

(2) Let $(a, f(a) + j) \in A \bowtie^f J$. There exists $n \in \mathbb{N}$ such that $ann_A(a^{n+1}) = ann_A(a^n)$ and $ann_{f(A)+J}(f(a) + j)^{n+1} = ann_{f(A)+J}(f(a) + j)^n$. Since $ann_{A \bowtie^f J}(a, f(a) + j)^n = ann_{A \bowtie^f J}(a^n, (f(a) + j)^n) = ann_A(a^n) \times ann_{f(A)+J}(f(a) + j)^n$, we have the result.

(3) " $\Longrightarrow$ " $A \subseteq A \bowtie^f J$. We conclude by Proposition 1.2. By (1), $A \bowtie^f J/f^{-1}(J) \times (0) \simeq f(A) + J$ with $f^{-1}(J) \times (0) = ann(I) \times (0) = ann(I \times B)$. By Proposition 1.12, $f(A) + J$ is strongly Hopfian.

(4) Let K be a field, $A = K[X_i; i \in \mathbb{N}]$, $I = (X_0 X_1, X_0^2 X_2, \ldots X_0^n X_n, \ldots)$, $B = A/I$, $f : A \to B$ the canonical surjection, and $J = 0$. By (1), $A \bowtie^f J \simeq A$ is an integral domain, so strongly Hopfian. But $f(A) + J = B$ is not strongly Hopfian, by the example of Proposition 1.12.

(5) Let $A \subseteq B$ be an extension of rings with A strongly Hopfian and B not strongly Hopfian. Take $J = XB[X]$ and $f : A \longrightarrow B[X]$ the natural injection. By (1), $A \bowtie^f J \simeq f(A) + J = A + XB[X]$. By the example of Corollary 6.4, $A + XB[X]$ is not strongly Hopfian.

Exercise 7

(1) Clear. (2) Let $a \in R$. There exist $s \in S$ and $n \in \mathbb{N}$ such that $s.ann(a^k) \subseteq ann(a^n)$ for each $k \geq n$. We will show that $ann(a^{n+1}) = ann(a^n)$. Let $b \in ann(a^{n+1})$. Then $sba^n = 0$. Since s is regular, $ba^n = 0$, and then $b \in ann(a^n)$.

(3) Let $a \in A$. There exist $s \in S$ and $n \in \mathbb{N}$ such that for each $k \geq n$, $s.ann_B(a^k) \subseteq ann_B(a^n)$, then $s(ann_B(a^k) \cap A) \subseteq ann_B(a^n) \cap A$, so $s.ann_A(a^k) \subseteq ann_A(a^n)$.

Exercise 8

(1) " $\Longrightarrow$ " Let I be an ideal of A such that $S \cap I = \emptyset$ and $\overline{a} \in A/I$. By hypothesis, there exist $s \in S$ and $n \in \mathbb{N}$ such that for each integer $k \geq n$, $s(I :_R a^k) \subseteq (I :_R a^n)$. We will show that $\overline{s}.ann(\overline{a}^k) \subseteq ann(\overline{a}^n)$. Let $\overline{\alpha} \in ann_{R/I}\overline{a}^k$. Then $\overline{\alpha a^k} = \overline{0}$. So $s\alpha \in (I : a^n)$, and then $\overline{s\alpha a^n} = \overline{0}$ in A/I. So $\overline{s\alpha} \in ann_{A/I}(\overline{a}^n)$.

" $\Longleftarrow$ " Let I be an ideal of R and $a \in A$.

First Case: $S \cap I = \emptyset$. By hypothesis, there exist $s \in S$ and $n \in \mathbb{N}$ such that for each integer $k \geq n$, $\overline{s}.ann_{A/I}(\overline{a}^k) \subseteq ann(\overline{a}^n)$. Let $\alpha \in (I :_A a^k)$. Then $\overline{\alpha a^k} \in I$, so $\overline{s\alpha a^n} = \overline{0}$ in A/I. Then $s\alpha a^n \in I$, so $s\alpha \in (I :_A a^n)$.

Second Case: $S \cap I \neq \emptyset$. Let $s \in S \cap I$. Then $s(I : a^{k+1}) \subseteq (I : a^k)$ for each $k \in \mathbb{N}$, and then A satisfies the condition S-accr.

(2) Take $S \subseteq U(A)$.

(3) Let $\overline{a} \in A/I$. There exist $s \in S$ and $n \in \mathbb{N}$ such that for each $k \geq n$, $s.ann(a^k) \subseteq ann(a^n)$. We will show that $\overline{s}.ann_{A/I}(\overline{a}^k) \subseteq ann_{A/I}(\overline{a}^n)$. Let

$\overline{b} \in ann_{A/I}(\overline{a}^k)$. Then $ba^k \in I$, so $ba^k J = 0$, and then $bJ \subseteq ann(a^k)$. So $sba^n J = 0$, then $sba^n \in I$ and $\overline{s}\overline{b}\overline{a}^n = \overline{0}$, and so $\overline{s}\overline{b} \in ann_{A/I}(\overline{a}^n)$.

Exercise 9 Let $\frac{a}{s} \in S^{-1}R$ with $a \in A$ and $s \in S$. There exist $s' \in S$ and $n \in \mathbb{N}$ such that for each $k \geq n$, $s'.ann(a^k) \subseteq ann(a^n)$. We will show that $ann(\frac{a}{s})^{n+1} = ann(\frac{a}{s})^n$. Let $\frac{b}{t} \in ann(\frac{a}{s})^{n+1}$. Then $\frac{ba^{n+1}}{ts^{n+1}} = 0$. There exists $r \in S$ such that $rba^{n+1} = 0$ and then $rb \in ann(a^{n+1})$. So $s'rb \in ann(a^n)$ and then $(s'r)ba^n = 0$ with $s'r \in S$. So $\frac{b}{t}(\frac{a}{s})^n = 0$.

Exercise 10 Let $f \in R$. There exists $n \in \mathbb{N}^*$ such that $f \in R_n = A[[X_1, \ldots, X_n]]$. Since R_n is Noetherian, there exists $k \in \mathbb{N}^*$ such that $ann_{R_n}(f^k) = ann_{R_n}(f^{k+1})$. We will show that $ann_R(f^k) = ann_R(f^{k+1})$. Let $g \in ann_R(f^{k+1})$. If $g \in R_n$, then $g \in ann_{R_n}(f^k)$. If $g \in R_{n+1}$, then $g = \sum_{r:0}^{\infty} g_r X_{n+1}^r$ with $g_r \in R_n$.

Since $0 = gf^{k+1} = \sum_{r:0}^{\infty} g_r f^{k+1} X_{n+1}^r$, then $g_r f^{k+1} = 0$ for each $r \in \mathbb{N}$, so $g_r \in ann_R(f^{k+1}) = ann_{R_n}(f^k)$. Then $gf^k = 0$. We continue by induction on the number of variables in g to obtain $gf^k = 0$.

Exercise 11

(1) Let $f = \sum_{i:0}^{\infty} b_i X^i \in A[[X]]$. Then $f \in ann_{A[[X]]}(a) \iff af = \sum_{i:0}^{\infty} ab_i X^i = 0 \iff \forall i \in \mathbb{N}, ab_i = 0 \iff f \in ann_A(a)[[X]]$.

(2) (a) Let $n \in \mathbb{N}^*$ be such that $ann_A(a^n) = ann_A(a^{n+1})$.
Then $ann_A(a^n)[[X]] = ann_A(a^{n+1})[[X]]$, so $ann_{A[[X]]}(a^n) = ann_{A[[X]]}(a^{n+1})$.

(b) Since $f^n \in A$ and A is strongly Hopfian, there exists $k \in \mathbb{N}^*$ such that $ann_A(f^{nk}) = ann_A(f^{n(k+1)})$. Then $ann_{A[[X]]}(f^{nk}) = ann_{A[[X]]}(f^{n(k+1)})$.

Exercise 12 Let M be the unique prime ideal of A. Then $Nil(A) = M = Z(A)$. By Lemma 7.2 (2), $Z(A[[X]]) \subseteq Nil(A)[[X]] = M[[X]]$. If $f = 0$, take $a = 1$.

If $f = \sum_{i:0}^{\infty} a_i X^i \neq 0$, the ideal $A_f = (a_0, a_1, \ldots)$ of A is contained in M. If A_f is nilpotent of index k, then $k \geq 2$. Let a be a nonzero element of A_f^{k-1}. Then $af = 0$. Suppose that A_f is not nilpotent. We will show that $A_f = M$ and $M^2 = M$. Indeed, if $A_f \subset M$, an element $m \in M \setminus A_f$ is nilpotent, and if $A_f \subset mA$, then A_f is nilpotent, which is absurd. If $M^2 \subset M$, we consider an element $x \in M \setminus M^2$, and then $M^2 \subset xA$. So $A_f^2 = M^2 \subset xA$, and then A_f is nilpotent, which is also absurd. Since f is a zero divisor, there exists $0 \neq g \in A[[X]]$ such that $fg = 0$. Let c be a nonzero coefficient of g. We will show that $cA_f = cM = (0)$. The element

c is the coefficient of a monomial X^e in g. Let $m \in M$. Since A is chained, $M = A_f = \bigcup_{i:0}^{\infty}(a_i)$, then $M = M^2 = \bigcup_{i:0}^{\infty}(a_i^2)$. There exists an index l such that $m \in (a_l^2)$.

Put $k = e + l$. We have $m \in (a_l^2) \subseteq a_l M = a_l M^k = a_l(\bigcup_{i:0}^{\infty}(a_i^k)) = \bigcup_{i:0}^{\infty}(a_l a_i^k)$.

There exists $s \in \{a_i, i \in \mathbb{N}\}$ such that $m \in (a_l s^k)$. Since s is nilpotent, we have a homomorphism of rings $\phi : A[[X]] \longrightarrow A[X]$, defined by $\phi(h(X)) = h(sX)$. Since $fg = 0$, then $\phi(fg) = 0$. By Lemma 8.2, $A_{\phi(f)} A_{\phi(g)} = A_{\phi(f)\phi(g)} = A_{\phi(fg)} = 0$. Since cX^e is a monomial of g, then $cs^e X^e$ is a monomial of $\phi(g)$. Also, $a_l X^l$ is a monomial of f, and then $a_l s^l X^l$ is a monomial of $\phi(f)$. Since $A_{\phi(f)} A_{\phi(g)} = 0$, then $cs^e a_l s^l = 0$, so $ca_l s^k = ca_l s^{e+l} = 0$. But $m \in (a_l s^k)$, then $cm = 0$.

Exercise 13

(1) Let $g_0 = \sum_{i \geq 0} b_{i,0} X^i \in ann(f^{n+1})$. We will show that $g_0 f^n = 0$. Since $b_{0,0} \in \Psi(ann(f^{n+1})) = \Psi(ann(f^n))$, there exists $g_1 = \sum_{i \geq 0} b_{i,1} X^i \in ann(f^n)$ such that $b_{0,1} = b_{0,0}$. Then $g_0 f^n = f^n(g_0 - g_1) = f^n \cdot \sum_{i \geq 1}(b_{i,0} - b_{i,1})X^i$. So $ord(g_0 f^n) \geq 1$. Since $(g_0 - g_1)f^{n+1} = 0$, then $f^{n+1} \cdot \sum_{i \geq 1}(b_{i,0} - b_{i,1})X^{i-1} = 0$, so $b_{1,0} - b_{1,1} \in \Psi(ann(f^{n+1})) = \Psi(ann(f^n))$. There exists $g_2 = \sum_{i \geq 0} b_{i,2} X^i \in ann(f^n)$ such that $b_{0,2} = b_{1,0} - b_{1,1}$. Also, $g_0 f^n = f^n(g_0 - g_1 - Xg_2) = f^n(\sum_{i \geq 1}(b_{i,0} - b_{i,1})X^i - \sum_{i \geq 0} b_{i,2} X^{i+1}) = f^n \cdot \sum_{i \geq 2}(b_{i,0} - b_{i,1} - b_{i-1,2})X^i$. Then $ord(g_0 f^n) \geq 2$. But $(g_0 - g_1 - Xg_2)f^{n+1} = 0$, then $f^{n+1} \cdot \sum_{i \geq 2}(b_{i,0} - b_{i,1} - b_{i-1,2})X^i = 0$ so $b_{2,0} - b_{2,1} - b_{1,2} \in \Psi(ann(f^{n+1})) = \Psi(ann(f^n))$. There exists $g_3 = \sum_{i \geq 0} b_{i,3} X^i \in ann(f^n)$ such that $b_{0,3} = b_{2,0} - b_{2,1} - b_{1,2}$. Also,

$$g_0 f^n = f^n(g_0 - g_1 - Xg_2 - X^2 g_3) = f^n(\sum_{i \geq 2}(b_{i,0} - b_{i,1} - b_{i-1,2})X^i -$$

$$\sum_{i \geq 0} b_{i,3} X^{i+2}) = f^n \sum_{i \geq 3}(b_{i,0} - b_{i,1} - b_{i-1,2} - b_{i-2,3})X^i. \text{ Then } ord(g_0 f^n) \geq 3.$$

By induction, we show that $ord(g_0 f^n) \geq k$ for each $k \in \mathbb{N}$. Then $g_0 f^n = 0$.

(2) " $\Longrightarrow$ " Let $n \in \mathbb{N}^*$ be such that $\Psi(ann(f^n)) = \Psi(ann(f^{n+1}))$. By (1), $ann(f^n) = ann(f^{n+1})$.

" $\Longleftarrow$ " Let $n \in \mathbb{N}^*$ be such that for each $k \geq n$, $ann(f^k) = ann(f^n)$. Then $\Psi(ann(f^k)) = \Psi(ann(f^n))$.

Exercise 14

Let $g = \sum_{i=0}^{\infty} b_i X^i \in ann(f^{k+1})$. Then $gf^{k+1} = 0$, so $b_0 a_0^{k+1} = 0$, and then $b_0 \in ann(a_0^{k+1}) = ann(a_0^k)$. Multiplying the equality $gf^{k+1} = 0$ by

a_0^k and simplifying by X, we find $a_0^k g_1 f^{k+1} = 0$ with $g_1 = b_1 + b_2 X + \dots$.
Then $b_1 a_0^{2k+1} = 0$, so $b_1 \in ann(a_0^{2k+1}) = ann(a_0^k)$, and then $b_1 a_0^k = 0$.
By induction, we prove that $a_0^k b_i = 0$ for each $i \in \mathbb{N}$, and then $a_0^k g = 0$. Put
$f = a_0 + f_1$ with $f_1 = a_1 X + a_2 X^2 + \dots$, then $0 = gf^{k+1} = g(a_0 + f_1)^{k+1} =$
$\sum_{i=0}^{k+1} C_{k+1}^i a_0^i g f_1^{k+1-i}$. Since $a_0^k g = 0$, multiplying by a_0^{k-1}, we find $a_0^{k-1} g f_1^{k+1} = 0$.

Let $\tilde{f}_1 = \frac{f_1}{X}$. Then $(a_0^{k-1} g) \tilde{f}_1^{k+1} = 0$. The coefficients of $\tilde{f}_1$ are coefficients of
f. By the first step, $a_1^k(a_0^{k-1} g) = 0$. Again, $\tilde{f}_1 = a_1 + a_2 X + \dots = a_1 + f_2$
with $f_2 = a_2 X + \dots$. Then $0 = a_0^{k-1} g f \tilde{f}_1^{k+1} = a_0^{k-1} g(a_1 + f_2)^{k+1} =$
$a_0^{k-1} g \sum_{i:0}^{k+1} C_{k+1}^i a_1^i f_2^{k+1-i}$. Since $a_1^k a_0^{k-1} g = 0$, multiplying by a_1^{k-1}, we find
$a_0^{k-1} g a_1^{k-1} f_2^{k+1} = 0$. Let $\tilde{f}_2 = \frac{f_2}{X}$. Then $(a_1^{k-1} a_0^{k-1} g) \tilde{f}_2^{k+1} = 0$. By the first
step, $a_2^k(a_1^{k-1} a_0^{k-1} g) = 0$. By the same way, we obtain $a_3^k a_2^k a_1^{k-1} a_0^{k-1} g = 0$,
and by induction, we have $a_i^k a_{i-1}^{k-1} a_{i-2}^{k-1} \dots a_0^{k-1} g = 0$ for each $i \in \mathbb{N}$. Then $g \in$
$\bigcap_{i:0}^{\infty} ann_A \{a_i^k a_{i-1}^{k-1} a_{i-2}^{k-1} \dots a_0^{k-1}\}[[X]] = ann_A \{a_i^k a_{i-1}^{k-1} a_{i-2}^{k-1} \dots a_0^{k-1}; i \in \mathbb{N}\}[[X]]$.

Exercise 15

(1) $ann_{A[[X]]}(V) \cap A = \{b \in A; bg = 0, \forall g \in V\} = ann_A(V)$. Also, $ann_A(A_V)$
$= \{b \in A; ba = 0, \forall a \in A_V\} = \{b \in A; bf = 0, \forall f \in V\} = ann_A(V)$. Then
we have the result.

(2) By (1), Ψ takes its values in $ann_A(2^A)$. Let $J \in ann_A(2^A)$. There exists $V \subseteq A$
such that $J = ann_A(V)$. By (1), $\Psi(ann_{A[[X]]}(V)) = ann_{A[[X]]}(V) \cap A = ann_A(V) = J$. Then Ψ is onto.

(3) Let $f = \sum_{i \geq 0} a_i X^i \in A[[X]]$. Then $f \in ann_{A[[X]]}(U) \iff fU = 0 \iff$
$a_i U = 0$ for each $i \in \mathbb{N} \iff a_i \in ann_A(U)$ for each $i \in \mathbb{N} \iff f \in$
$ann_A(U)[[X]]$. Then we have the equality.

(4) By (3), the map Φ takes its values in $ann_{A[[X]]}(2^{A[[X]]})$. It is clear that it is one
to one.

(5) Reduced rings are power series Armendariz.

(6) "$(b) \implies (a)$" Clear. "$(a) \implies (b)$" Let $f_1, \dots, f_n \in A[[X]]$ be such that
$f_1 \dots f_n = 0$. We denote a_i any coefficient of f_i. We have $f_1(f_2 \dots f_n) = 0$,
and then $a_1 \alpha = 0$ for each coefficient α of $f_2 \dots f_n$. This implies that
$a_1 f_2 \dots f_n = 0$, so $(a_1 f_2)(f_3 \dots f_n) = 0$. Then $a_1 a_2 \beta = 0$ for each coefficient
β of $f_3 \dots f_n$. By induction, we obtain the result.
    "$(a) \implies (c)$" Let $V \subseteq A[[X]]$ and $f = \sum_{i \geq 0} a_i X^i \in A[[X]]$. Then $f \in$
$ann_{A[[X]]}(V) \iff$ for each $g = \sum_{i \geq 0} b_j X^j \in V$, $fg = 0 \iff$ for each $i, j \in \mathbb{N}$,
$b_j a_i = 0 \iff$ for each $i \in \mathbb{N}$, and $a_i \in ann_A(A_V) \iff f \in ann_A(A_V)[[X]]$.
Then $ann_{A[[X]]}(V) = ann_A(A_V)[[X]] = \Phi(ann_A(A_V))$. So Φ is onto.

"$(c) \implies (a)$" Let $f = \sum_{i \geq 0} a_i X^i$ and $g = \sum_{i \geq 0} b_j X^j \in A[[X]]$ be such that $fg = 0$. Since $ann_{A[[X]]}(f) \in ann_{A[[X]]}(2^{A[[X]]})$, there exists $F \subseteq A$ such that we have $ann_{A[[X]]}(f) = \Phi(ann_A(F)) = ann_A(F)[[X]]$. Then $g \in ann_A(F)[[X]]$, so $b_j \in ann_A(F) \subset ann_{A[[X]]}(f)$ for each $j \in \mathbb{N}$. Then $b_j f = 0$ for each $j \in \mathbb{N}$ so $a_i b_j = 0$ for each $i, j \in \mathbb{N}$.

"$(c) \implies (d)$" Let I and $J \in ann_{A[[X]]}(2^{A[[X]]})$ be such that $\Psi(I) = \Psi(J)$. Then $I \cap A = J \cap A$. Since Φ is onto, there exist I_1 and $J_1 \in ann_A(2^A)$ such that $I = \Phi(I_1) = I_1[[X]]$ and $J = \Phi(J_1) = J_1[[X]]$. Then $I_1 = J_1$ so $I = J$.

"$(d) \implies (c)$" Let $J = ann_{A[[X]]}(V) \in ann_{A[[X]]}(2^{A[[X]]})$ with $V \subseteq A[[X]]$. By (1), $J \cap A = ann_{A[[X]]}(V) \cap A = ann_A(A_V)$. By (3), we have $ann_A(A_V)[[X]] = ann_{A[[X]]}(A_V) \in ann_{A[[X]]}(2^{A[[X]]})$ and $ann_A(A_V)[[X]] \cap A = ann_A(A_V)$. Then we have the equality $\Psi(J) = \Psi(ann_A(A_V)[[X]])$. Since Ψ is one to one, $J = ann_A(A_V)[[X]] = \Phi(ann_A(A_V))$. Then Φ is onto.

Let $J \in ann_A(2^A)$. Then $\Psi \circ \Phi(J) = \Psi(J[[X]]) = J[[X]] \cap A = J$, so $\Psi \circ \Phi = id$.

Let $I \in ann_{A[[X]]}(2^{A[[X]]})$. Then $I = ann_{A[[X]]}(V)$ with $V \subseteq A[[X]]$. So $\Phi \circ \Psi(I) = \Phi(I \cap A) = (I \cap A)[[X]]$. By (1), $I \cap A = ann_{A[[X]]}(V) \cap A = ann_A(A_V)$, and then $(I \cap A)[[X]] = ann_A(A_V)[[X]] = ann_{A[[X]]}(V)$ since A is power series Armendariz. Then $\Phi \circ \Psi(I) = ann_{A[[X]]}(V) = I$ and $\Phi \circ \Psi = id$.

(7) The maps Φ and Ψ are increasing bijections between the sets $ann_{A[[X]]}(2^{A[[X]]})$ and $ann_A(2^A)$.

Exercise 16

(1) Since the ring A is power series Armendariz, the maps Φ and Ψ, defined in the preceding exercise, are inverse to each other. For $g \in A[[X]]$, $ann_{A[[X]]}(g) = \Phi \circ \Psi(ann_{A[[X]]}(g)) = \Phi(ann_{A[[X]]}(g) \cap A) = \Phi(ann_A(g)) = ann_A(g)[[X]]$. Then we have $ann_{A[[X]]}(g) = ann_A(g)[[X]]$.

Let $g = \sum_{i \geq 0} b_i X^i \in A[[X]]$ and $k \in \mathbb{N}^*$. Then $ann_A(g^k) = \bigcap_{(i_1,...,i_k) \in \mathbb{N}^k} ann_A(b_{i_1} \ldots b_{i_k})$.

Indeed, let $a \in A$. Since A is power series Armendariz, $a \in ann_A(g^k) \iff ag^k = 0 \iff \forall (i_1, \ldots, i_k) \in \mathbb{N}^k, ab_{i_1} \ldots b_{i_k} = 0 \iff a \in \bigcap_{(i_1,...,i_k) \in \mathbb{N}^k} ann_A(b_{i_1} \ldots b_{i_k})$.

1. "$(a) \implies (b)$" Let $f \in A[[X]]$. There exists $n \in \mathbb{N}^*$ such that we have $ann_{A[[X]]}(f^n) = ann_{A[[X]]}(f^{n+1})$, then $ann_A(f^n)[[X]] = ann_A(f^{n+1})[[X]]$, so $ann_A(f^n) = ann_A(f^{n+1})$. Put $f = \sum_{i \geq 0} a_i X^i$, then

$$\bigcap_{(i_1,...,i_n) \in \mathbb{N}^n} ann(a_{i_1} \ldots a_{i_n}) = \bigcap_{(i_1,...,i_{n+1}) \in \mathbb{N}^{n+1}} ann(a_{i_1} \ldots a_{i_{n+1}}).$$

2. "$(b) \implies (c)$" Let $I = (a_0, a_1, \ldots)$ be a countably generated ideal of A. For each $k \in \mathbb{N}^*$, $I^k = (a_{i_1} \ldots a_{i_k}; (i_1, \ldots, i_k) \in \mathbb{N}^k)$. Then $ann_A(I^k) =$

$\bigcap_{(i_1,\ldots,i_k)\in\mathbb{N}^k} ann_A(a_{i_1}\ldots a_{i_k})$. Let $f = \sum_{i\geq 0} a_i X^i \in A[[X]]$. By hypothesis, there

exists $n \in \mathbb{N}^*$ such that we have the equality $\bigcap_{(i_1,\ldots,i_n)\in\mathbb{N}^n} ann(a_{i_1}\ldots a_{i_n}) =$

$\bigcap_{(i_1,\ldots,i_{n+1})\in\mathbb{N}^{n+1}} ann(a_{i_1}\ldots a_{i_{n+1}})$. Then $ann(I^n) = ann(I^{n+1})$. The sequence

becomes stationary at the rank n.

3. $"(c) \implies (a)"$ Let $f = \sum_{i\geq 0} a_i X^i \in A[[X]]$, and consider the

countably generated ideal $I = (a_0, a_1, \ldots)$ of A. There exists $n \in \mathbb{N}^*$ such that $ann_A(I^n) = ann_A(I^{n+1})$. Then $\bigcap_{(i_1,\ldots,i_n)\in\mathbb{N}^n} ann(a_{i_1}\ldots a_{i_n})$

$= \bigcap_{(i_1,\ldots,i_{n+1})\in\mathbb{N}^{n+1}} ann(a_{i_1}\ldots a_{i_{n+1}})$. So we have $ann_A(f^n) = ann_A(f^{n+1})$.

We obtain the equalities $ann_{A[[X]]}(f^n) = ann_A(f^n)[[X]] = ann_A(f^{n+1})$ $[[X]] = ann_{A[[X]]}(f^{n+1})$. So the ring $A[[X]]$ is strongly Hopfian.

(2) The natural homomorphism $B \longrightarrow A = B[T]/(T^2)$ is one to one, and then B is a sub-ring of A. Moreover, A is a free B-module with basis $\{1, u\}$ where $u = \bar{T}$. Let $I = (f_i; i \in \mathbb{N})$ be a countably generated ideal of A. Put $f_i = a_i + ub_i$ with a_i and $b_i \in B$. It suffices to show that the sequence $(ann_A(I^n))_n$ is constant. Let $n \geq 2$ be an integer, $g \in ann_A(I^n)$, and $g = c + du$ with c and $d \in B$. For each $(i_1, \ldots, i_n) \in \mathbb{N}^n$, $gf_{i_1}\ldots f_{i_n} = 0$, then $ca_{i_1}\ldots a_{i_n} = 0$, and $c\sum a_{j_1}\ldots a_{j_{n-1}}b_{j_n} + da_{i_1}\ldots a_{i_n} = 0$. In particular, $ca_i^n = 0$ for each $i \in \mathbb{N}$, but B is reduced, and then $ca_i = 0$. So $da_i^n = 0$, and then $da_i = 0$ for each $i \in \mathbb{N}$. So $gf_i = 0$ for each $i \in \mathbb{N}$, then $g \in ann_A(I)$.

Exercise 17 Let $f = \sum_{i:0}^{\infty} a_i X^i \in A[[X]]$ be such that $f^2 = f$. Then (0) $a_0^2 = a_0$;
(1) $a_0a_1 + a_1a_0 = a_1$; (2) $a_0a_2 + a_1a_1 + a_2a_0 = a_2$; ...; (n) $a_0a_n + a_1a_{n-1} + \ldots + a_na_0 = a_n$; ... The equation (0) shows that $a_0 \in Bool(A)$. Multiplying (1) by a_0, we find $a_0a_1 + a_1a_0 = a_1a_0$, then $a_0a_1 = 0$, and, by (1), $a_1 = 0$. Multiply (2) by a_0, we find $a_0a_2 + a_2a_0 = a_2a_0$, then $a_0a_2 = 0$, and, by (2), $a_2 = 0$. Suppose that $a_1 = \ldots = a_{n-1} = 0$. Multiplying the equation (n) by a_0, we find $a_0a_n + a_0a_n = a_0a_n$, then $a_0a_n = 0$, and, by (n), $a_n = 0$. Then $f = a_0 \in Bool(A)$.

Exercise 18

(1) Let $x \in A$, $x = u + e$ with $u \in U(A)$ and $e^2 = e$. Then $\bar{x} = \bar{u} + \bar{e}$ with $\bar{u} \in U(A/I)$ and $\bar{e}^2 = \bar{e}$.
(2) $" \implies "$ $A \simeq A[[X]]/(X)$ is decomposable by (1).

$" \Longleftarrow "$ Let $f = \displaystyle\sum_{i:0}^{\infty} a_i X^i \in A[[X]]$. Then $a_0 = u + e$ with $u \in U(A)$

and $e \in Bool(A) = Bool(A[[X]])$. Then $f = e + (u + \displaystyle\sum_{i:1}^{\infty} a_i X^i)$ with $u +$

$\displaystyle\sum_{i:1}^{\infty} a_i X^i \in U(A[[X]])$.

(3) It suffices to show the uniqueness of the decompositions.

$" \Longrightarrow "$ Clear because $A \subseteq A[[X]]$. $" \Longleftarrow "$ Suppose that $u_1 + e_1 = u_2 + e_2$ with $u_1, u_2 \in U(A[[X]])$ and $e_1, e_2 \in Bool(A[[X]]) = Bool(A)$. Then $u_1(0) + e_1 = u_2(0) + e_2$ with $u_1(0)$ and $u_2(0) \in U(A)$. By the uniqueness in A, $u_1(0) = u_2(0)$ and $e_1 = e_2$. Then $u_1 = u_2$.

Exercise 19 $" \Longleftarrow "$ A sub-ring of an Armendariz ring is Armendariz.

$" \Longrightarrow "$ Let $f(T)$ and $g(T) \in A[X][T]$ be such that $f(T)g(T) = 0$. Put $f(T) = f_0(X) + f_1(X)T + \ldots + f_n(X)T^n$ and $g(T) = g_0(X) + g_1(X)T + \ldots + g_m(X)T^m$ with $f_i(X), g_i(X) \in A[X]$. Let $k = 1 + deg\, f_0(X) + \ldots + deg\, f_n(X) + deg\, g_0(X) + \ldots + deg\, g_m(X)$, where the degrees are taken with respect to X and the degree of the zero polynomial is zero. Then $f(X^k) = f_0(X) + f_1(X)X^k + \ldots + f_n(X)X^{kn}$. The degree of a monomial in $f_i(X)X^{ki}$ is $< k + ki = k(i+1) \leq$ to the degree of each monomial in $f_{i+1}(X)X^{k(i+1)}$. Then the set of the coefficients of $f(X^k)$ is equal to the set of the coefficients of $f_0(X), \ldots, f_n(X)$. Since $f(T)g(T) = 0$, then $f(X^k)g(X^k) = 0$. But the ring A is Armendariz, and then each coefficient of $f_i(X)$ annihilates each coefficient of $g_j(X)$, so $f_i(X)g_j(X) = 0$.

Exercise 20 Let $f(X) = \displaystyle\sum_{i:0}^{\infty} a_i X^i$ and $g(X) = \displaystyle\sum_{j:0}^{\infty} b_j X^j \in A[[X]]$ be such that $f(X)g(X) = 0$. If either $f(X) = 0$ or $g(X) = 0$, then $a_i b_j = 0$ for each $i, j \in \mathbb{N}$. Suppose that $f(X) \neq 0$ and $g(X) \neq 0$. Then $f(X)$ and $g(X)$ are zero divisors. By Exercise 4, (0) is $P[[X]]$-primary in the ring $A[[X]]$. Then $f(X)$ and $g(X) \in P[[X]]$, so $a_i b_j \in P^2 = (0)$ for each $i, j \in \mathbb{N}$.

Exercise 21

(1) The set $S = A \setminus Z(A)$ of the regular elements of A is multiplicative. There exists a prime ideal P of A disjoint with S. Then $P \subseteq Z(A)$.
(2) Let $A = \mathbb{Z}/6\mathbb{Z} = \{\bar{0}, \bar{1}, \bar{2}, \bar{3}, \bar{4}, \bar{5}\}$. Then $Z(A) = \{\bar{0}, \bar{2}, \bar{3}, \bar{4}\}$ and $U(A) = \{\bar{1}, \bar{5}\}$. Since $\bar{2} + \bar{3} = \bar{5} \notin Z(A)$, then $Z(A)$ is not an ideal of A.

References

H. Al-Ezeh, On some properties of polynomial rings. Int. J. Math. Math. Sci. **10**(2), 311–314 (1987)

H. Al-Ezeh, Two properties of the power series ring. Int. J. Math. Math. Sci. **11**(1), 9–14 (1988)

D.F. Anderson, A. Badawi, On the zero-divisor graph of a ring. Commun. Algebra **36**, 3073–3092 (2008)

D.D. Anderson, V.P. Camillo, Armendariz rings and Gaussian rings. Commun. Algebra **26**(7), 2265–2272 (1998)

D.D. Anderson, V.P. Camillo, Commutative rings whose elements are a sum of a unit and idempotent. Commun. Algebra **30**(7), 3327–3336 (2002)

M. Arapovic, On the embedding of a commutative ring into a 0-dimensional commutative ring. Glasnik Mat. **18**, 53–59 (1983)

D. Arezzo, L. Robbiano, Sul completato di un anello rispetto ad un ideale di tipo finito. Rendiconti del Seminario Matematico della Università di Padova **44**, 133–154 (1970)

E. Armendariz, A note on extension of Baer and P.P. rings. J. Aust. Math. Soc. **18**, 470–473 (1974)

Ch. Bakkari, Armendariz and SFT properties in subring retracts. Mediterr. J. Math. **6**, 339–345 (2009)

E. Bastida, R. Gilmer, Overrings and divisorial ideals of rings of the form $D + M$. Michigan Math. J. **20**, 79–95 (1973)

A. Benhissi, *Les anneaux de séries formelles*. Queen's Papers in Pure and Applied Mathematics, vol. 124 (Kingston, Ontario, 2003)

M. D'Anna, M. Fontana, An amalgamated duplication of a ring along an ideal: the basic properties. J. Algebra Appl. **6**, 443–459 (2007)

R. Dastanpour, A. Ghorbani, Divisibility on chain of submodules. Commun. Algebra **46**(6), 2305–2318 (2018)

P. Eakin, A. Sathaye, R-endomorphisms of $R[[X]]$ are essentially continuous. Pac. J. Math. **66**(1), 83–87 (1976)

P. Eakin, A. Sathaye, R. Pervine, Analytic extensions of commutative ring. J. Algebra **110**, 431–448 (1987)

M. El Ouarrachi, N. Mahdou, On power series Armendariz rings. Palestine J. Math. **7**(Special Issue I), 79–87 (2018)

A. Facchini, Z. Nazemian, Modules with chain conditions up to isomorphism. J. Algebra **453**, 578–601 (2016)

D. Fields, Zero divisors and nilpotent elements in power series rings. Proc. AMS **27**, 427–433 (1971)

R. Gilmer, A new criterion for embeddability in a zero-dimensional commutative ring, in *Ideal Theoritic Methods in Commutative Algebra*. Lecture Notes in Pure and Applied Math., ed. by J.A. Huckaba, D.D. Anderson, Ira J. Papick, vol. 220 (Marcel Dekker, 2001)

R. Gilmer, A. Grams, T. Parker, Zero divisors in power series rings. J. Reine Angew. Math. **278/279**, 145–164 (1975)

R. Gilmer, W. Heinzer, On the imbedding of a direct product into a zero-dimensional commutative ring. Proc. AMS **106**(3), 631–636 (1989)

B. Gouaid, A. Hamed, A. Benhissi, Endo-Noetherian rings. Annali di Matematica Pura ed Applicata **199**, 563–572 (2020)

A. Hamed, B. Gouaid, A. Benhissi, Rings satisfying the strongly Hopfian and S-strongly Hopfian properties. Math. Rep. **23**(4), 383–395 (2021)

S. Hizem, Formal power series over strongly Hopfian rings. Commun. Algebra **39**, 279–291 (2011)

A. Hmaimo, A. Kaidi, E. Sanchez Campos, Generalized fitting modules and rings. J. Algebra **308**, 199–214 (2007)

M. Khalifa, Power series over strongly Hopfian bounded rings. Commun. Algebra **45**(8), 3587–3593 (2017)

M. Khalifa, Isonoetherian power series rings. Commun. Algebra **46**(6), 2451–2458 (2018)

M. Khalifa, Isonoetherian power series rings II. Commun. Algebra **49**(10), 4447–4468 (2021)

J.H. Kim, A note on the quotient ring R((X)) of the power series ring R[[X]]. J. Korean Math. Soc. **25**(2), 265–271 (1988)

N.K. Kim, K.H. Lee, Y. Lee, Power series rings satisfying a zero divisor property. Commun. Algebra **34**, 2205–2218 (2006)

J.W. Kerr, Very long chains of annihilator ideals. Israel J. Math. **46**(3), 197–204 (1983)

T.K. Lee, T.L. Wong, On Armendariz rings. Houst. J. Math. **29**(3), 583–593 (2003)

C.-P. Lu, Modules satisfying accr on a certain type of colons. Pac. J. Math. **131**(2), 303–318 (1988)

P. Maroscia, Sur les anneaux de dimension zero. Atti. Accad. Naz. Lincei. Rend. Cl. Sci. Fis. Mat. Natur. **56**, 451–459 (1974)

M.J. O'Malley, R-automorphisms of $R[[X]]$. Proc. Lond. Math. Soc. (3) **20**, 60–78 (1970)

N. Radu, Descompunerea primara in inele comutative. Lectii de Algebra III (Universitatea din Bucuresti, 1981)

M.B. Rege, S. Chhawchharia, Armendariz rings. Proc. Jpn. Acad. **73**(Ser. A), 14–17 (1997)

J. Xu, When is a self-injective ring zero-dimensional, in *Zero-Dimensional Rings*. Lecture Notes Pure Appl. Math., ed. by D.F. Anderson, D.E. Dobbs, vol. 171 (Marcel Dekker, 1995)

X.F. Yan, Z.K. Liu, Extensions of generalized fitting modules. J. Math. Res. Exposition **30**(3), 407–414 (2010)

Index

© The Author(s), under exclusive license to Springer Nature Switzerland AG 2022
A. Benhissi, *Chain Conditions in Commutative Rings*,
https://doi.org/10.1007/978-3-031-09898-7

Printed in the United States
by Baker & Taylor Publisher Services

Printed in the United States
by Baker & Taylor Publisher Services